Kinematics and Dynamics of Machines

Ashraf Omran

Kinematics and Dynamics of Machines

Modeling and Design with MATLAB®, Simulink® and Simscape™ Multibody™

Springer

Ashraf Omran
Facilities Engineering Department
Massachusetts Maritime Academy
Buzzards Bay, MA, USA

Department of Mechanical Engineering
Johns Hopkins University
Baltimore, MD, USA

ISBN 978-3-031-72700-9 ISBN 978-3-031-72701-6 (eBook)
https://doi.org/10.1007/978-3-031-72701-6

This Springer imprint is published by the registered company Springer Nature Switzerland AG
The registered company address is: Gewerbestrasse 11, 6330 Cham, Switzerland

Preface

The *Theory of Machines*, which consists of the kinematics and dynamics of individual mechanisms, is a crucial course in mechanical engineering and closely related topics. The principles in this textbook are tools and building blocks for basic machine design, which is crucial to a mechanical engineer's understanding. They are also the fundamental elements that are necessary when using common industry-specific software such as MATLAB®, Simulink®, and Simscape™ as these are engineering-specific programs that work based on inputs in terms of kinematics and dynamics.

The degree to which a student understands this material will directly affect their future understanding of more complicated subject matter and their ability to use and manipulate specific software that is increasingly used in engineering industries.

During my teaching of the kinematics and dynamics of machines, I have noticed that students need a new textbook that is not just a mix of theory and mathematical expressions that can be difficult to understand using stereotypical engineering examples—tractors, bulldozers, transmissions, and internal combustion engines. The principles can be difficult to explain if the student is unfamiliar with these machines.

Another issue relates to many students' lack of ability to use programs that are essential to visualizing, designing, simulating, and analyzing mechanical systems such as MATLAB®, Simulink®, and Simscape™, which are very common in the industry. Often, the students will have familiarity with the programs' existence and even how to perform certain functions with them, but little understanding of how to use the programs to their fullest potential. This is a disservice to students because these programs can be used to increase technical skills in modeling and reinforce concepts learned in the Kinematics and Dynamics of Machines course.

After considering the above problems, I developed a solution in the form of a new textbook. This textbook comprises the same subject matter as previous textbooks but is a version with less emphasis on mathematical explanations and more on graphical representations. Furthermore, more relevant examples, such as household items and gymnastic equipment, are used along with the more traditional examples mentioned previously.

The most pronounced difference in the textbook is the addition of basic tutorials, guides, explanations, and exercises in some commonly used engineering programs such as MATLAB®, Simulink®, and Simscape™. The objective of this supplemental information is to give the students familiar with these programs a more practical way to use them for basic machine design and to introduce them to those without experience. This not only builds the students' technical skills (an important quality sought out by prospective employers) but also reinforces material specific to this textbook.

The text teaches students how to create a system so that they can manipulate the model to better understand what the mechanism is experiencing. Whether this means getting a three-dimensional perspective of the mechanism, simulating motion, or even analyzing the system, the exercises better inform the students of what it is they are truly looking at in engineering terms. Combining this new perspective on how to properly convey conceptual information through the use of better examples, more visuals, and the integration of relevant software makes this textbook a valuable tool in the understanding of machine design.

Concerning this product's marketability, there are few, if any, alternatives to this one. Current textbooks do not make the connection between the material covered and how it is implemented in practice. By using this textbook, students will no longer need to attend a separate class to learn the applicable software. They will learn more of the skills that industry leaders look for. By adding these teachings about relevant software to the textbook itself, a student will not just have a better understanding conceptually but will also be able to create these systems on a professional level sought out by employers. This will make the institution and its graduates more desirable to the engineering world. I hope it will lead to a trend for other courses to follow suit, resulting in the publication of more textbooks about a new style of learning.

For the reasons given above, this text fills a gap in the market at the mechanical engineering level in Introduction to Kinematics and Dynamics of Machines. Filling this gap has necessitated the difficult task of creating a textbook that blends both theory and practice–instilling the viewpoints of both academic engineering and business practitioners.

This textbook is organized into 12 chapters. It starts in Chap. 1 with an introduction to machines' dynamics and kinematics and explains some of the common examples of machines to be used later in the text. In Chap. 2, I define kinematic machines' terminology and the machines' degrees of freedom. Chapter 3 introduces kinematic diagrams. Chapter 4 presents Simscape™ Multibody™ as a tool to simulate machines. Chapters 5–7 present graphical, analytical, and model-based techniques to calculate the position, velocity, acceleration, and jerk for any node on the machine. Chapter 8 introduces trajectory design. Chapters 9 and 10 present the kinematic analysis of cams and gears. Finally, Chaps. 11 and 12 introduce the dynamical model of any machine. Students who are going to use this textbook should have previously studied college algebra, vector analysis, trigonometry courses, statics, and dynamics plus have a good background in MATLAB® scripting.

I have chosen the textbook structure above based on my teaching of the Kinematics and Dynamics of Machines course at Texas A&M University and teaching Mechanics of Solids at Illinois Institute of Technology and Massachusetts Maritime Academy. I have designed the book with the intention of its being used in the easiest possible way by students who come from any background. Hopefully, this is a foundational textbook that will serve the mechanical engineering student well in their journey to success.

Solutions to the end-of-chapter problems are provided for instructors who adopt this book for use in teaching their courses and can be found at https://sites.google.com/springernature.com/extramaterial/lecturer-material.

Buzzards Bay/Baltimore, USA Ashraf Omran

Acknowledgements

I would like to thank the people below for helping edit the textbook:

James Colley
Aaron Neuwirth
Lynette Stewart
Malak Saad
Lesly Carreon
Jessica Lybarger
Liam Nielsen
David Balderas
Jeremey Broadcap

I would also like to thank those named below for developing the 3D figures, animation videos, and sketches:

Pranav Dhola
Kenneth Manalansan
Andrew Godfrey

I would like to thank the reviewers for their help and comments that improved the quality of this textbook.

Contents

1 Machines Are Everywhere 1
- 1.1 Introduction 1
- 1.2 Purpose 1
- 1.3 Kinematics and Dynamics of Machines 2
- 1.4 Applications of Machines 4
  - 1.4.1 Household Machines 4
  - 1.4.2 Kitchen Appliances 12
  - 1.4.3 Gymnasium Equipment 18
  - 1.4.4 Bicycle, Car, and Aircraft 24
  - 1.4.5 Other Machines 37
- 1.5 Book Summary and Expected Outcome 41
- 1.6 Summary of the Book's Expected Outcomes 43
- References 43

2 Mechanism and Kinematics Fundamentals 45
- 2.1 Introduction 45
- 2.2 Linkage, Mechanism, and Machine 45
  - 2.2.1 Rigid Body 48
  - 2.2.2 Point Mass 49
  - 2.2.3 Link 49
  - 2.2.4 Ground Link or Fixed Link 50
  - 2.2.5 Joint 51
  - 2.2.6 Kinematic Pair and Kinematic Chain 51
- 2.3 Kinematic Motion 62
  - 2.3.1 Coordinate Systems 62
  - 2.3.2 Types of 2D Motion 65
    - 2.3.2.1 Point Mass Motion 65
    - 2.3.2.2 Link Motion 66
  - 2.3.3 Point Mass Degrees of Freedom 67
  - 2.3.4 Link Degrees of Freedom (DoF) 68
  - 2.3.5 Joint Degrees of Freedom (DoF) 70
    - 2.3.5.1 Revolute Joint 71
    - 2.3.5.2 Prismatic Joint 72
    - 2.3.5.3 Screw Joint 75
    - 2.3.5.4 Pin-in-a-Slot Joint 75

2.3.5.5 Cam Joint 76
2.3.5.6 Gear Joint 76
2.3.6 Linkage Degrees of Freedom 76
2.4 Chapter Summary 84
Problems 84
References 95

3 Mechanisms and Machines Fundamentals 97
3.1 Introduction 97
3.2 Kinematic Diagram 98
3.3 Four-Bar Mechanism 106
3.3.1 Grashof's Law 107
3.4 Slider–Crank Four-Bar Mechanism 112
3.5 Inverted Slider–Crank Four-Bar Mechanism ... 113
3.6 Five-Bar Mechanism 116
3.7 Six-Bar Mechanism 116
3.8 Special Types of Mechanisms 119
3.8.1 Straight-Line Mechanism 120
3.8.2 Parallelogram Mechanism 120
3.8.3 Quick-Return Mechanism 121
3.8.4 Scotch Yoke Mechanism 122
3.8.5 Intermittent Mechanisms 123
3.8.5.1 Geneva Mechanism 123
3.8.5.2 Ratchet–Pawl Mechanism 125
3.9 Actuators 127
3.9.1 Electric Motor System 128
3.9.2 Hydraulic System 129
3.10 Chapter Summary 131
Problems 131
References 137

4 Kinematics of Machines Using Simulink® and Simscape™ Multibody™ 139
4.1 Introduction 139
4.2 MATLAB® 140
4.3 Simulink® 140
4.4 Simulink® Sources Library 145
4.4.1 Constant Block 145
4.4.2 Signal Editor Block 146
4.4.3 Clock Block 148
4.5 Simulink® Sinks Library 149
4.5.1 Scope Block 149
4.5.2 Terminator Block 152
4.6 Simulink® Math Operations Library 153
4.7 Simscape™ Multibody™ 158
4.7.1 Simscape™ Multibody™ Joints 161
4.7.2 Simscape™ Multibody™ Rigid Body Links or Solids 163

4.7.3 Simscape™ Multibody™ Frames and Transformation 167
4.7.3.1 Local Frames on Joints and Links 174
4.7.4 Simscape™ Multibody™ Machine Assembly . . . 182
4.7.5 Simulink® Subsystems and Masks 189
4.8 Chapter Summary 196
Problems 196
References 200

5 Position and Displacement Analysis 201
5.1 Introduction 201
5.2 Position 202
5.3 Displacements 208
5.3.1 Linear Displacement 208
5.3.2 Angular Displacement 209
5.4 Simulink® Simscape™ Multibody™ Motion Sensor and Actuator 213
5.5 Displacement Graphical Analysis 221
5.6 Displacement Analytical Analysis 225
5.6.1 Four-Bar Mechanism Analytical Analysis 226
5.6.2 Slider–Crank Mechanism Analytical Analysis . . . 228
5.6.3 Inverted Slider–Crank Mechanism Analytical Analysis 232
5.6.4 Five and Beyond Bar Mechanisms Analytical Analysis 239
5.7 Node of Interest Position 241
5.8 Limiting Position Analysis 245
5.9 Coupler Curve 246
5.10 Chapter Summary 248
Problems 248
References 253

6 Velocity Analysis 255
6.1 Introduction 255
6.2 Translational Velocity and Angular Velocity 256
6.2.1 Translational Velocity 256
6.2.2 Angular Velocity 261
6.2.3 Sliding Velocity 263
6.2.4 Rolling Velocity 264
6.3 Single Link Connected to Ground 266
6.3.1 Hinged to Ground Link 266
6.3.2 Pin-in-a-Slot to Ground 269
6.4 Absolute and Relative Velocities 270
6.5 Simscape™ Multibody™ Velocity Sensor and Actuator Blocks 272
6.6 Graphical Velocity Analysis 277
6.6.1 Rigid Body Instantaneous Center of Zero Velocity 277

6.6.2 Kinematic Pair Instantaneous Centers of Velocity . . . 278
6.6.3 Kennedy's Theorem . . . 280
6.6.4 Four-Bar Mechanism Graphical Velocity Analysis . . . 282
6.6.5 Slider–Crank Mechanism Graphical Velocity Analysis . . . 286
6.6.6 Inverted Slider–Crank Mechanism Graphical Velocity Analysis . . . 290
6.6.7 General Mechanism Graphical Velocity Analysis . . . 295
6.7 Analytical Velocity Analysis . . . 295
6.7.1 Four-Bar Mechanism Analytical Velocity Analysis . . . 295
6.7.2 Slider–Crank Mechanism Analytical Velocity Analysis . . . 298
6.7.3 Inverted Slider–Crank Mechanism Analytical Velocity Analysis . . . 299
6.8 Chapter Summary . . . 302
Problems . . . 302
References . . . 307

7 Acceleration and Jerk Analysis . . . 309
7.1 Introduction . . . 309
7.2 Translational Acceleration and Angular Acceleration . . . 310
7.2.1 Translational Acceleration . . . 310
7.2.2 Angular Acceleration . . . 315
7.2.3 Absolute Polar Acceleration Components . . . 319
7.3 Absolute and Relative Acceleration . . . 323
7.3.1 Single Link Hinged to Ground . . . 330
7.3.2 Single Link Pin-in-a-Slot to Ground . . . 337
7.4 Simscape™ Multibody™ Acceleration Sensor and Actuator Blocks . . . 339
7.5 Graphical Acceleration Analysis . . . 345
7.5.1 Four-Bar Mechanism Graphical Acceleration Analysis . . . 346
7.5.2 Slider–Crank Mechanism Graphical Velocity Analysis . . . 352
7.5.3 Inverted Slider–Crank Mechanism Graphical Velocity Analysis . . . 361
7.5.4 General Mechanism Graphical Acceleration Analysis . . . 367
7.6 Analytical Acceleration Analysis . . . 368
7.6.1 Four-Bar Mechanism Analytical Acceleration Analysis . . . 368
7.6.2 Slider–Crank Mechanism Analytical Velocity Analysis . . . 373
7.6.3 Inverted Slider–Crank Mechanism Analytical Velocity Analysis . . . 377

7.7 Jerk 382
7.8 Chapter Summary 389
Problems 389
References 395

8 Motion Trajectory Design 397
8.1 Introduction 397
8.2 Input–Output Block Diagram of SVAJ 398
8.3 Hydraulic Actuators and Corresponding Motion Limits . . . 400
8.3.1 Hydraulic Circuit 400
8.3.2 Power Losses 405
8.3.2.1 Pump Losses 405
8.3.2.2 Hose and Valve Losses 407
8.3.2.3 Cylinder Losses 415
8.3.3 Hydraulic Subsystem SVAJ Constraints 417
8.4 Limitations of Electrical Motors 420
8.5 Trajectory Generation 423
8.6 Recall Simulink® 432
8.7 Animation Using Simulink® SimScape™ Multibody™ . . . 442
8.8 Chapter Summary 448
Problems 449
References 452

9 Cam Analysis 453
9.1 Introduction 453
9.2 Different Types of Cams and Cam Followers 453
9.2.1 Cam Mechanism Types Based on Cam Link 457
9.2.2 Cam Mechanism Types Based on Follower Link 458
9.2.3 Cam Mechanism Types Based on Cam Joint 459
9.3 Follower Motion Design 460
9.3.1 Follower Motion Segments 460
9.3.2 Polynomial CEP Motion 472
9.3.2.1 Step CEP Motion or Constant Displacement CEP Motion 473
9.3.2.2 Ramp CEP Motion or Constant Velocity CEP Motion 473
9.3.2.3 Higher-Order Polynomial Motion 475
9.3.3 Harmonic CEP Motion 480
9.3.4 Cycloidal CEP Motion 481
9.3.5 Constant Velocity CPM Motion 481
9.3.6 Constant Acceleration CPM Motion 482
9.4 Cam Sizing 482
9.4.1 Cam Geometric Parameters 483
9.4.2 Cam Profile Design 488
9.4.2.1 Knife-Edge Follower Using Graphical Method 488
9.4.2.2 Knife-Edge Follower Using Analytical Method 491

9.4.2.3 Cam Mechanism Design and Limitations 494
9.5 Implementing Cams Using Simscape™ Multibody™ 496
9.6 Examples of Machines Utilizing Cam Mechanism 506
9.7 Chapter Summary 509
Problems 510
References 514

10 Gear Analysis 515
10.1 Introduction 515
10.2 Gear Types 515
10.2.1 Spur Type 516
10.2.2 Rack Type 516
10.2.3 Internal Type 518
10.2.4 Helical Type 519
10.2.5 Worm Type 520
10.2.6 Bevel Type 520
10.3 Spur Gear Design and Analysis 521
10.3.1 Configurations and Parameters of Spur Gear Teeth 521
10.3.2 Spur Gear Tooth Profile 523
10.3.3 Spur Gear Design 530
10.3.4 Spur Gear Kinematics 531
10.4 Simple Gear Trains 541
10.5 Compound Gear Trains 546
10.6 Planetary Gear Trains 547
10.7 Differential 552
10.8 Chapter Summary 554
Problems 555
References 561

11 Statics of Machines 563
11.1 Introduction 563
11.2 External Forces 563
11.3 External Moment 565
11.4 Contact Forces and Moments 568
11.5 Free Body Diagram 571
11.6 Static Equilibrium 573
11.7 Calculating Equilibrium Forces and Moments Using Simscape™ Multibody™ 581
11.8 Chapter Summary 588
Problems 588
References 594

12 Dynamics of Machines 595
12.1 Introduction 595
12.2 Mass 596
12.3 Center of Gravity 599

12.4 Mass Moment of Inertia 604
12.5 Dynamic Model 609
12.5.1 Newton's Second Law 609
12.5.2 Four-Bar Mechanism Dynamic Model. 615
12.5.3 Slider–Crank Mechanism Dynamic Model 620
12.6 Calculating Dynamic Forces and Dynamic Moments using Simscape™ Multibody™ 624
12.7 Chapter Summary 628
Problems 631
References 634

Appendix A: Metal Material Density 635

Appendix B: Geometric Properties 639

Index 643

1 Machines Are Everywhere

1.1 Introduction

In this chapter, we will first speak about the purpose of this textbook. After that, we will introduce the definition of machines, kinematics, and dynamics along with each one's importance in the modeling, simulation, and design of machines. We are going to introduce and explain some machines in terms of understanding their assembly, functionality, and articulation through written explanation, graphic representation (sketches and pictures), and animated videos. These explained machines will be used throughout this textbook as applied examples for our kinematic and dynamic analysis to help in the modeling, simulation, and design phases. At the end of the chapter, we will outline the flow of this textbook as well as the outcomes that we expect to reach throughout its twelve chapters.

1.2 Purpose

This textbook presents a new approach to teaching and studying the kinematics and dynamics of machines by being a theory-based version with less emphasis on mathematical explanations. This is accomplished using graphics and animation via contemporary software, specifically, MATLAB®, and Simulink®, and Simscape™ Multibody™.

This textbook's objective is to give students hands-on experience with relevant engineering software skills in the forms of modeling, analysis, simulation, and animation at the same time as they are learning the kinematics and dynamics of machines. Studying these relevant software skills concurrently to traditional mechanics material will facilitate new ways of learning, as with these tools students will be able to create their own systems to better understand and optimize machine design.

Another objective of this textbook is to take advantage of the fact that machines are not simply restricted to generic factory examples such as the packing machine or agricultural-construction machines like tractors. The machines discussed in this textbook are much more relevant and accessible to students and could exist in the household, kitchen, or gym. This textbook aims to bridge the gap between an academic degree and rare marketable skills.

Supplementary Information The online version contains supplementary material available at https://doi.org/10.1007/978-3-031-72701-6_1.

A. Omran, *Kinematics and Dynamics of Machines*, https://doi.org/10.1007/978-3-031-72701-6_1

1.3 Kinematics and Dynamics of Machines

A machine is defined as any device that transfers energy from a source to a sink to accomplish a specific goal. The machine's energy source could be manual from humans or animals. The source could also be electrical, wind, chemical, or any other type of energy. The machine's energy sink is some kind of work to be done such as lifting, cutting, or completing a specific job. Humans use machines to achieve better performance or accomplish something that they cannot perform manually. For example, early humans invented simple devices such as the hand axe to butcher and skin game, dig in the soil, and cut wood or other plant materials (see Fig. 1.1). Another example of a classic simple device is the inclined plane (ramp), which can help to move heavy loads such as those that the ancient Egyptians used in building and construction.

Later, humans started to come up with more sophisticated multibody assemblies such as the noria (waterwheel) that were first used by the ancient Egyptians and Sumerians. A noria is a hydro-powered scoop wheel that lifts water into a small aqueduct to carry it to villages. For thousands of years, the waterwheel has been used as a machine that transfers energy from moving water into a wheel's mechanical rotational energy. There are many more examples of early invented machines such as the pulley, wheel and axle, screw, and shadoof water lifting device.

Ideas and inventions of new machines started to grow rapidly in every application, especially with the start of the Industrial Revolution, whereby new manufacturing processes came into existence (1760–1840) in Europe and the United States. During this time, steam and water power were used as energy sources to run machines that replaced many manual jobs. With these revolutionary machines, manufacturers could achieve higher performance as well as save time and money. Four particular technologies became very important during the Industrial Revolution: textiles, steam power, iron making, and the invention of machine tools.

The kinematics and dynamics of machines is a science that helps us to analyze and design machines to have better performance and more applications for new ideas. During the history of the kinematics and dynamics of machines science, important engineers, and scientists have come to the

Fig. 1.1 Early human using axe. Designed by the author and implemented by Kenneth Manalansan. Courtesy of Kenneth Manalansan

forefront such as Archimedes (287–212 BCE), Leonardo da Vinci (1452–1519), James Watt (1736–1819), James J. Sylvester (1814–1897), Alfred B. Kempe (1849–1922), Franz Grashof (1826–1893), Franz Reuleaux (1829–1905), Pafnuty L. Chebyshev (1821–1894), Samuel Roberts (1827–1913), Arthur Cayley (1821–1895), Ferdinand Freudenstein (1926–2006), George N. Sandor (1929–1996), Jacques Denavit (1930–2013), Richard S. Hartenberg (1907–1997), and more to be counted and more to come. Each one of these people significantly contributed to the kinematics and dynamics of machines science, either by invention or with a groundbreaking theory. More details about the contribution of each are available in Refs. [1–3]. We recommend the student reads the contribution/contributions of these visionaries and how they came up with their theories and/or inventions.

Today, there are many annually published kinematics and dynamics books, inventions, journals, and conferences [4–8, 10]. *Kinematics and Dynamics of Machines* is a course that most mechanical engineering schools teach to their students. There is a huge market for the design and optimization of machines. Additionally, machine graphic design interests many non-engineering individuals. If you look anywhere online, you can see that many graphic designers like showing their work as pieces of art.

Now, we are going to speak more about what exactly the kinematics, statics, and dynamics of machines are. How can we use these branches of mechanics to design and analyze machines?

A machine can be a single-body or a multibody assembly. The body could be flexible or rigid. This textbook addresses machines that consist only of rigid bodies. A rigid body means that an acting force will affect neither the body's form nor its volume. Later, in Chap. 2, we will explain the difference between a rigid body and a flexible body in more detail.

Machine design has phases and processes that may take a lot of time. The design steps could be summarized as follows:

- Identifying the need for the desired machine.
- Selecting the possible mechanisms to implement into the desired machine.
- Analyzing the kinematics and dynamics of the machine.
- Selecting materials.
- Designing the elements' shapes and dimensions.
- Selecting the actuators' types and sizes.
- Modifying and optimizing.
- Making detailed drawings.
- Prototyping a model.
- Testing the prototype model.
- Validating and verifying the model.
- Modifying based on the test results and analysis.
- Production.

In this textbook, we will focus on the kinematic and dynamic analysis of machines as a very important part of the design process. Figure 1.2 shows the study branches for the rigid body mechanics that we use in the machine design process. Rigid body mechanics is divided into statics and dynamics. Statics studies rigid bodies at rest or in constant motion. Dynamics studies rigid bodies with acceleration. Dynamics is divided into kinematics and kinetics. Kinematics describes the motion of bodies, while kinetics studies the forces and torques that cause these motion changes.

The first step in design is to consider the kinematics of machines, which is a study of motion without including any forces or moments. Studying the motion of a machine means studying the position, velocity, acceleration, and perhaps the jerk of each element. The designer starts from the desired motion to be accomplished, such as lifting a load from one position to another at a specific time. The first step in design is to determine the kinematics of machine analysis to calculate each individual element's motion.

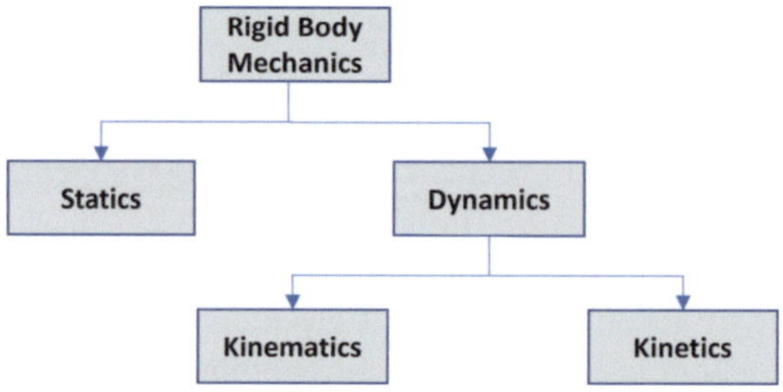

Fig. 1.2 Rigid body mechanics branches of study

We will introduce some terminologies when we speak about the kinematics of machines such as links, joints, linkages, cams, and gears, which will all be defined and used across this textbook in motion analysis.

After that, we introduce the statics of machines, which studies forces and moments when the machine is at a stationary position. The statics of machines is very important for stress analyses because it tells us how to design each element's dimensions appropriately.

Finally, we look at the dynamics of machines to study the required forces and/or torques to generate the desired motion. In other words, the dynamics of machines is the study of motion concerning forces and moments.

1.4 Applications of Machines

In this section, we speak about different types of machines and mechanisms as multibody assemblies with the help of sketches and animation videos. For each machine, we will explain its major elements, articulation, and functionality. These machines and mechanisms will be used all over this textbook in our kinematic and dynamic analysis in the form of solved examples or problems. The reader is recommended to watch the animation video of each machine while reading the machine description below. Also, the reader can return to the relevant chapter section to read about any specific machine when the machine is shown in one of the solved examples or in one of the problems.

Now, we are going to show how you can break down a machine or mechanism as a multibody that consists of single bodies and mechanical assemblies. The single body could be a standalone body or could be a set of bodies welded together. A mechanical assembly is a set of bodies that allows a specific relative motion between at least two single bodies. This explanation will be clearer when we go over the upcoming examples.

1.4.1 Household Machines

Nowadays, people use machines to make their lives easier and to save time. Each household (Fig. 1.3) and office contains many machines. For example, look around your house at machines such as ironing boards, chairs, doors, windows, or even the flushing toilet in your bathroom. We are going to discuss seven machines in the household: door closer, folding chair, casement window, toilet flush valve, ironing board, hydraulic door closer, and umbrella. There are many more machines in the average household, and the reader is encouraged to think of additional examples of machines beyond the ones mentioned in this chapter.

Fig. 1.3 Household area. Designed by the author and implemented by Kenneth Manalansan. Courtesy of Kenneth Manalansan

Door Closer Mechanism: Figure 1.4 shows a door closer mechanism as a multibody assembly with the listed elements in Table 1.1.

The door closer provides an additional mechanism for the door so that it can close softly even if someone decides to slam it. The mechanism enforces a speed limit for how fast the door can close. Sometimes, an oil-filled hydraulic damper can also be added in the forearm element to increase the slowing effectiveness.

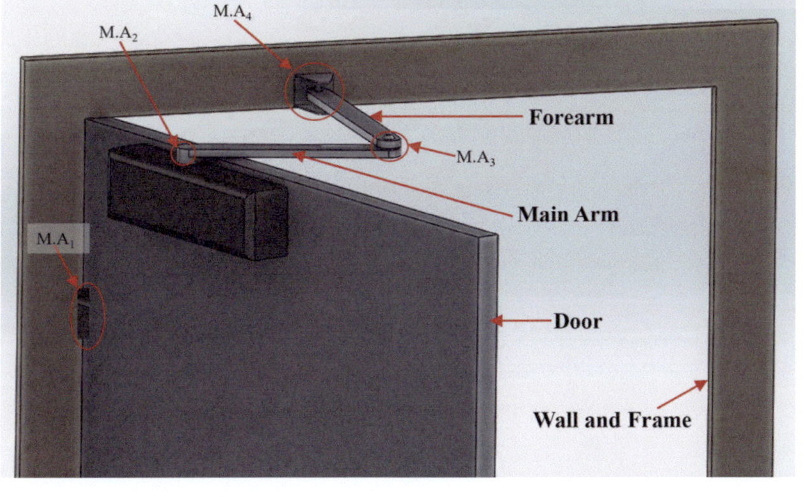

Fig. 1.4 Door closer mechanism. Figure and associated Door_4Bar_Mechanism.mp4 video designed by the author and implemented by both the author and Pranav Dhola. Both courtesy of Pranav Dhola

Table 1.1 Multibody elements of door closer machine

Rigid bodies	Mechanical assemblies (M.A)
• Wall frame	• $M.A_1$: Between the wall frame and the door allowing relative rotational motion only
• Door	• $M.A_2$: Between the door and the main arm allowing relative rotational motion only
• Main arm	• $M.A_3$: Between the main arm and forearm allowing relative rotational motion only
• Forearm	• $M.A_4$: Between the forearm and the wall frame allowing relative rotational motion only

When someone pushes the door to open it (rotate the door clockwise), the forearm rotates clockwise as well because of the coupler (main arm) and the three mechanical assemblies $M.A_2$, $M.A_3$, and $M.A_4$. The door closer mechanism reduces the motion by creating a resistance force between the door and the main arm. It is a simple mechanism as an assembly, but there are more aspect ratios and parameters that need to be designed using kinematic and dynamic analysis as we will discuss later in this textbook.

Folding Chair Mechanism: Figure 1.5 shows a picture of a folding chair mechanism as a symmetric multibody consisting of the listed elements in Table 1.2.

When the different parts of a folding chair mechanism move, the mechanism changes its shape. Sometimes this change of shape serves to transfer force from one place to another. Other times, the change of shape serves to change the direction of a force. For machines like this folding chair, the change in shape is simply the utility and practicality of the mechanism itself. Chairs take up a lot of space when they're not being used. This mechanism folds the chair into a shape that is more easily stored.

The seat is attached to the front leg and back leg on single pins ($M.A_3$ and $M.A_4$) that allow for rotation. As the front legs are rotated toward the back legs, the seat rotates with both legs. Rotation

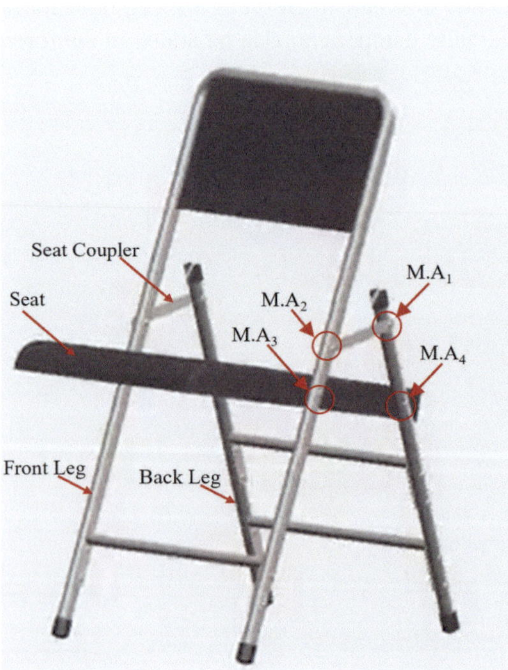

Fig. 1.5 Folding chair mechanism . Figure and associated Folding_Chair_Mechanism.mp4 video designed by the author and implemented by both the author and Pranav Dhola. Both courtesy of Pranav Dhola

Table 1.2 Multibody elements of folding chair machine

Rigid bodies	Mechanical assemblies (M.A)
• Back leg	• $M.A_1$: Between the back leg and the seat coupler allowing rotational motion
• Seat coupler	• $M.A_2$: Between the seat coupler and the front leg allowing rotational motion
• Front leg	• $M.A_3$: Between the front leg and seat allowing rotational motion
• Seat	• $M.A_4$: Between the seat and the back leg allowing rotational motion

stops at a position where the links are all roughly parallel to one another, resulting in a nearly flat arrangement for storage. Two other pieces (seat couplers) are connected to the front leg and the back leg with single rotating pins ($M.A_1$ and $M.A_2$). They help keep everything located while this rotation is taking place. If you look for it in the video, you will notice that the two top pieces stay fairly parallel (symmetry) to the seat. They help keep everything in place through the entire range of motion. Finally, the chair's backrest, which is part of the front leg, comes to rest firmly on the back legs when the mechanism is fully extended and ready to be used as a chair.

Casement Window Machine: The machine shown in Fig. 1.6, often referred to as a casement window, is a multibody assembly that consists of elements that are listed in Table 1.3.

The top of the yellow window frame shown below holds a piece of glass and is attached to the gray casement with two pins ($M.A_6$), allowing it to rotate or "hinge". The way this particular mechanism is arranged, the yellow frame can only rotate outward. When used together with a set of gears and some sliding joints, these pins enable the window frame and glass to open from the bottom

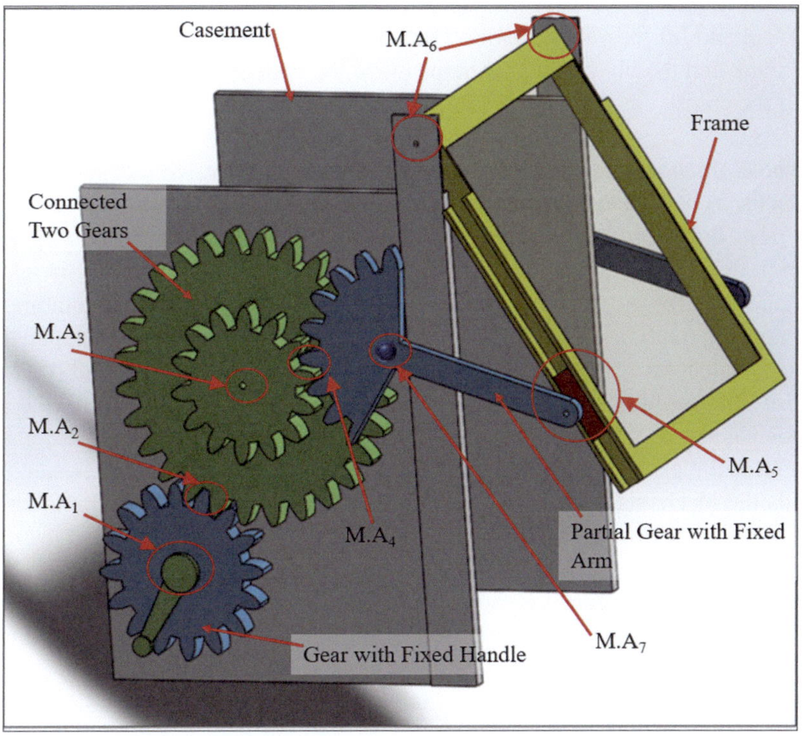

Fig. 1.6 Casement window machine. Figure and associated Casement_Window_Mechanism.mp4 video designed by the author and implemented by both the author and Pranav Dhola. Both courtesy of Pranav Dhola

Table 1.3 Multibody elements of casement window machine

Rigid bodies	Mechanical assemblies (M.A)
• Frame	• M.A$_1$: Between the gear with a fixed handle and the casement allowing rotational motion
• Casement	• M.A$_2$: Between the gear with a fixed handle and the bigger gear from the connected two gears allowing the gear motion
• Gear with fixed handle	• M.A$_3$: Between the two connected gears and the casement allowing rotational motion
• Two connected gears through a shaft	• M.A$_4$: Between the partial gear with a fixed arm and the smaller gear from the connected two gears allowing the gear motion
• Partial gear with fixed arm	• M.A$_5$: Between the partial gear with a fixed arm and the frame allowing both transitional motion and sliding motion
• Slider	• M.A$_6$: Between the frame and the casement allowing rotational motion
	• M.A$_7$: between the partial gear with a fixed arm and the casement allowing rotational motion

of the casement until the slider runs out of space in the channel on the sides of the yellow frame. It would take a very different mechanism, but the yellow window frame could be set up to rotate inward too.

To activate the casement window machine, there is a green handle fixed to a smaller blue gear. Turning this handle will rotate the small blue gear in either direction, causing it to rotate the larger green gear in the opposite direction along with the smaller green gear that is fixed to its center. This smaller gear rotates the blue arm which pushes the red slider through the channel in the side of the yellow window frame. Notice that the pin at the end of the blue arm connecting it to the red slider allows the arm to rotate relative to the yellow frame. This rotation allows the mechanism to transform the rotational motion of the gear into the sliding motion needed to move the red piece through the yellow channel and lift the yellow frame. The difference between the sizes of the small blue gear with the hand crank and the big green gear helps move the window frame with less effort from whoever is working the crank.

Toilet Flush Valve Mechanism: Figure 1.7 shows the toilet flush valve mechanism as a multibody assembly that consists of elements that are listed in Table 1.4.

When you push down on the handle, the toilet will be flushed and refilled by the water held above the bowl in the tank. The tank then fills and will be ready to be flushed again when needed.

The flush handle on the outside of the tank is pushed down, subsequently rotating a handle arm upward. This upward rotation raises the lift rod that pulls the flush valve flapper up. The entire motion happens through four rotating joints (mechanical assemblies 1, 2, 3, and 4). The toilet flush valve mechanism allows the water stored in the tank to be quickly emptied into the toilet bowl. A siphoning action caused by the sudden change in pressure and the properly shaped and sized piping empties the toilet and replaces the water just released from the tank with fresh water. The weight of the flapper pulls the mechanism back down, sealing the flush valve between the toilet and the tank, enabling the tank to fill again with fresh water.

Ironing Board Mechanism: Figure 1.8 shows an ironing board machine. This is another example of a mechanism whose purpose is to change its shape through a multibody assembly. It consists of the elements in Table 1.5.

The ironing table machine has two legs made of round pipe that are attached (roughly in the middle) by a single pin which allows for rotation. Another round pipe is fixed on the bottom ends of each leg to provide a larger footprint for the ironing board to stand on without rocking too much. Rods are attached to the upper ends of the legs under the board. These rods are located by channels

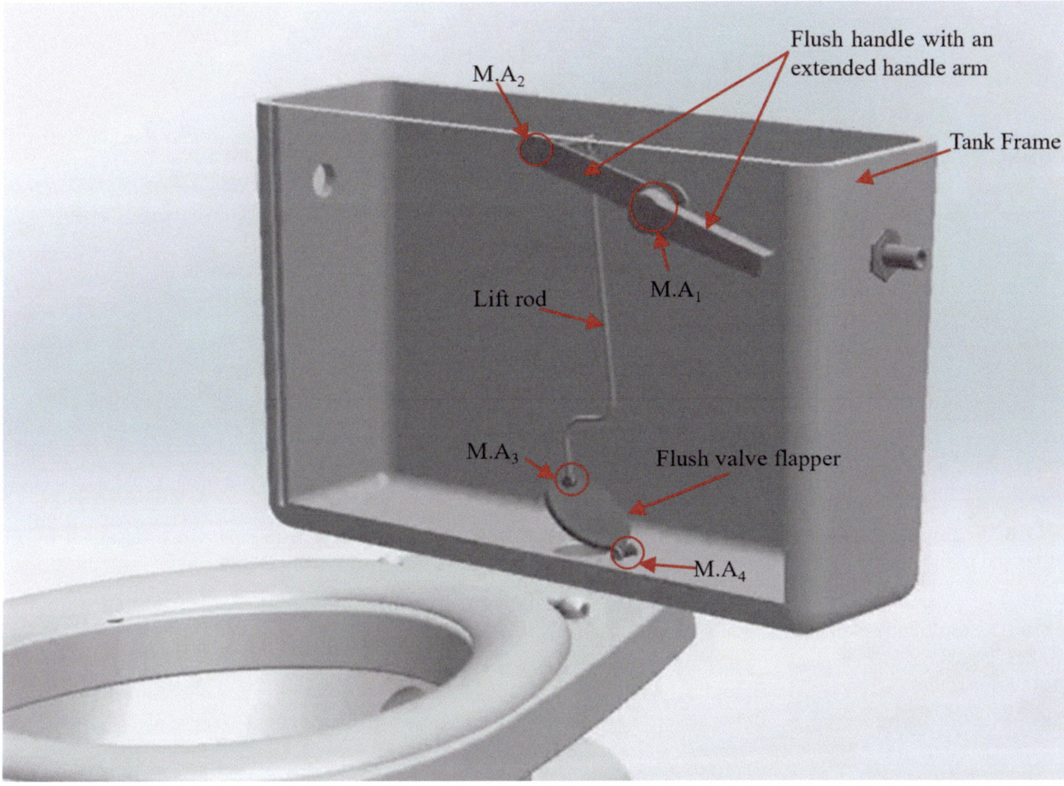

Fig. 1.7 Toilet flush valve mechanism. Figure and associated Flush_Toilet_Mechanism.mp4 video designed by the author and implemented by both the author and Pranav Dhola. Both courtesy of Pranav Dhola

Table 1.4 Multibody elements of toilet flush valve machine

Rigid bodies	Mechanical assemblies (M.A)
• Stationary tank frame	• M.A$_1$: Between the flush handle and the tank frame allowing rotational motion
• Flush handle with an extended handle arm	• M.A$_2$: Between the handle arm with the lift rod allowing rotational motion
• Lift rod	• M.A$_3$: Between the lift rod and flush valve flapper allowing rotational motion
• Flush valve flapper	• M.A$_4$: Between the flush valve flapper and the tank frame allowing rotational motion

attached to the board which only allow sliding movement in a horizontal direction. As the "feet" are rotated inward toward each other, the rods in the channel slide toward each other, raising the height of the board. When the desired height is reached, the rods can be locked into place in the channels and the board is now ready for use. When it is no longer needed, the rods are released from their position in the channels and the legs are rotated away from each other until the mechanism's profile is flat, minimizing the storage space required.

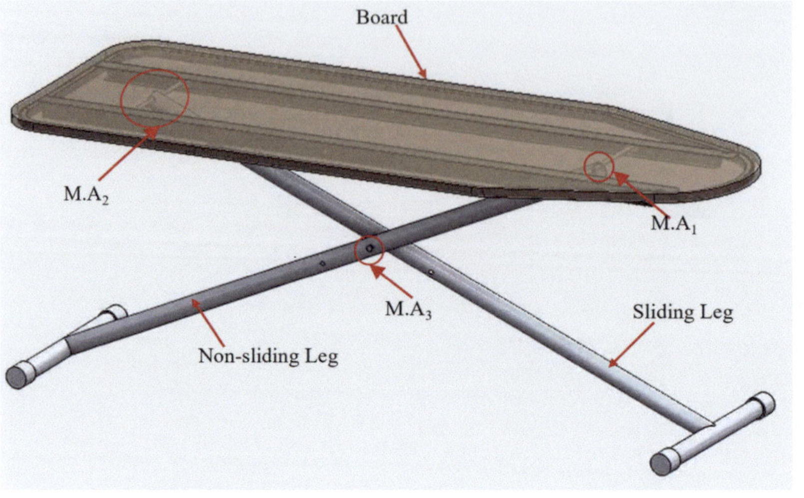

Fig. 1.8 Ironing board mechanism. Figure and associated Ironing_Table_Mechanism.mp4 video designed by the author and implemented by both the author and Pranav Dhola. Both courtesy of Pranav Dhola

Table 1.5 Multibody elements of ironing board machine

Rigid bodies	Mechanical assemblies (M.A)
1. Board	1. M.A$_1$: Between the board and the non-sliding leg allowing rotational motion
2. Non-sliding leg	2. M.A$_2$: Between the board and the sliding leg allowing rotational motion plus sliding motion
3. Sliding leg	3. M.A$_3$: Between the non-sliding leg and the sliding leg allowing rotational motion

Hydraulic Gate Closer Mechanism: A gate is shown in Fig. 1.9, probably a familiar sight at the entrance of some houses as a multibody assembly that consists of the elements in Table 1.6.

The gate itself is fairly straightforward. It is attached to a stationary post by two hinges (M.A$_1$), allowing it to swing back and forth in opening and closing motions. The bit we are concerned with here is the mechanism that pulls it back to the closed position after it has been swung open. A piston–cylinder assembly connected between the bottom of the frame's gate and the stationary post provides the closing motion. The piston–cylinder assembly consists of a piston element, a cylinder element, and the connection to make the piston slide inside the cylinder (M.A$_3$).

First, a simple, rigid structure is fixed to the stationary post. This one has two horizontal elements that hold a pin. The pin (M.A$_4$) connects to the cylinder end of the piston–cylinder assembly. This pin allows the component to pivot. The piston end of the piston–cylinder assembly is then attached to the bottom of the gate's frame through another pin (M.A$_2$) allowing some rotation on this end as well. Inside the cylinder is a spring and hydraulic oil. When the gate is opened, it swings on the hinges away from the rigid structure and pulls the piston out of the cylinder. As the piston is pulled out of the cylinder, it compresses the spring. When the gate is released, the spring pushes the piston back inside the cylinder toward the stationary post. A spring stiff enough to pull the heavy gate closed will cause the gate to move too fast and slam shut. The oil in the cylinder slows the movement down just enough such that the gate still closes completely, but without damaging anything or being too noisy. Another scenario to open and close this gate is to connect the piston–cylinder assembly to a hydraulic actuator that can be remotely controlled.

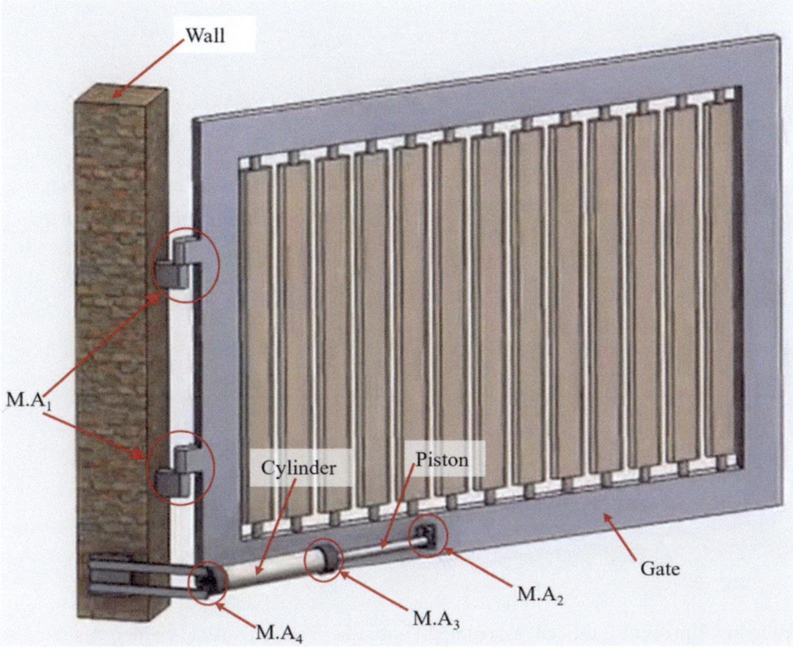

Fig. 1.9 Hydraulic gate closer mechanism. Figure and associated Hydraulic_Door_Opener_Mechanism.mp4 video designed by the author and implemented by both the author and Pranav Dhola. Both courtesy of Pranav Dhola

Table 1.6 Multibody elements of hydraulic gate mechanism

Rigid bodies	Mechanical assemblies (M.A)
1. Wall	1. $M.A_1$: Between the wall and the gate allowing rotational motion
2. Gate	2. $M.A_2$: Between the gate and the piston allowing rotational motion
3. Cylinder	3. $M.A_3$: Between the cylinder and the piston allowing the piston to slide inside the cylinder
4. Piston	4. $M.A_4$: Between the cylinder and the wall allowing rotational motion

Umbrella Mechanism: Figure 1.10 shows an umbrella mechanism as a multibody assembly that consists of the elements in Table 1.7.

The umbrella mechanism is another example of a mechanism whose purpose is to change its shape. In other words, the purpose of the umbrella mechanism is to fold and unfold the umbrella to provide shade or shelter. In the event of a rainstorm, it is difficult to imagine a more practical and useful device than an umbrella, but imagine the inconvenience of one that didn't fold up into a neat, easily carried bundle! Even a stationary umbrella for outdoor tables and the like would be virtually unusable if it could not be collapsed and secured in the event of strong wind.

To fold the umbrella, the hand is held on the runner element and slid down the central shaft. This is possible thanks to the assembly $M.A_3$. During the sliding and closing process, the ribs will rotate toward the central shaft because of the stretchers (couplers) thanks to $M.A_1$, $M.A_2$, and $M.A_4$. To unfold the umbrella, the runner is slid up so that the ribs will rotate away from the central shaft. It is a simple mechanism that needs a careful design regarding the sizes, dimensions, and materials. The kinematics and dynamics of machines turns out to be very critical design phases to achieve better performance. Manufacturers with high standards will use engineers to make their produced umbrellas high quality and with optimal performance.

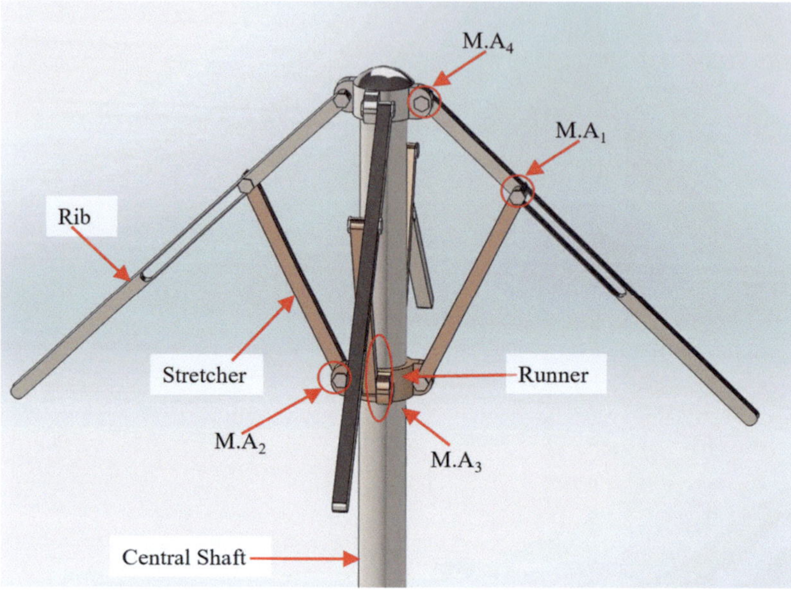

Fig. 1.10 Umbrella mechanism. Figure and associated Umbrella_Mechanism.mp4 video designed by the author and implemented by both the author and Pranav Dhola. Both courtesy of Pranav Dhola

Table 1.7 Multibody elements of Umbrella mechanism

Rigid bodies	Mechanical assemblies (M.A)
1. Central shaft	1. Four $M.A_1$: Between each rib and each stretcher allowing only relative rotational motion
2. Four ribs	2. Four $M.A_2$: Between each stretcher and the runner allowing only relative rotational motion
3. Four stretchers	3. One $M.A_3$: Between the runner and the central shaft allowing only sliding motion of the runner on the central shaft
4. Runner	4. Four $M.A_4$: Between each rib and the central shaft allowing only relative rotational motion

1.4.2 Kitchen Appliances

The kitchen, as shown in Fig. 1.11, is a room that we stop inside every day to at least make coffee or a sandwich. There are many machines in the kitchen, ranging from the mixer for making our smoothies to the toaster for making our toast.

We are going to talk about four machines that can be found in most kitchens: the potato peeler, potato chopper, planetary mixer, and corkscrew. There are many more machines in the kitchen. The reader is encouraged to think of additional examples of kitchen machines.

Potato Peeler: Figure 1.12 shows the potato peeler machine as a multibody assembly that consists of the elements listed in Table 1.8.

The potato peeler truly fits the description of a "kitchen gadget". A potato, or other similarly sized vegetable or fruit, is held firmly in place and peeled as the name of the machine implies. After the peel has been removed, the process can be repeated resulting in ribbons of potato, zucchini, apple, etc.

Fig. 1.11 Machines inside the kitchen. Designed by the author and implemented by Kenneth Manalansan. Courtesy of Kenneth Manalansan

First, the potato is pushed down onto an array of spikes fixing it to the yellow disk on the base of the machine. Then, the single spike of the brown holding arm is pushed into the top of the potato by sliding the brown arm down the stationary post fixed to the base. The blue peeling arm has a replaceable blade attached to its end; this is what peels the potato. It is moved up the stationary post and located at the top of the potato to begin the process.

An electric motor located in the base of the peeler powers two parts of the machine. First, it turns the yellow disk installed in the base. Since there are several spikes at this end, the potato is fixed to the disk and it turns at the same speed as the yellow disk. The top brown arm is attached to the potato with a single spike, which helps hold the potato upright but allows it to spin freely. While this potato spinning is taking place, the motor is also turning a screw inside the machine which pulls the blue arm down the fixed post. This in turn pulls a knife attached to the end of the blue arm through the potato's skin. As the potato spins, the knife spirals around the outside removing the skin in the process. When the blue arm reaches the bottom of the stationary post, it contacts a switch that turns off the electric motor.

Potato Chopper: If you happen to be making French fries, Fig. 1.13 shows an irreplaceable machine as a symmetric multibody consisting of the elements listed in Table 1.9.

This particular type of chopper uses what is often called a blade and a plate. The blade is the rectangular grid-like arrangement of sharp metal pieces fixed to the machine's frame that does the

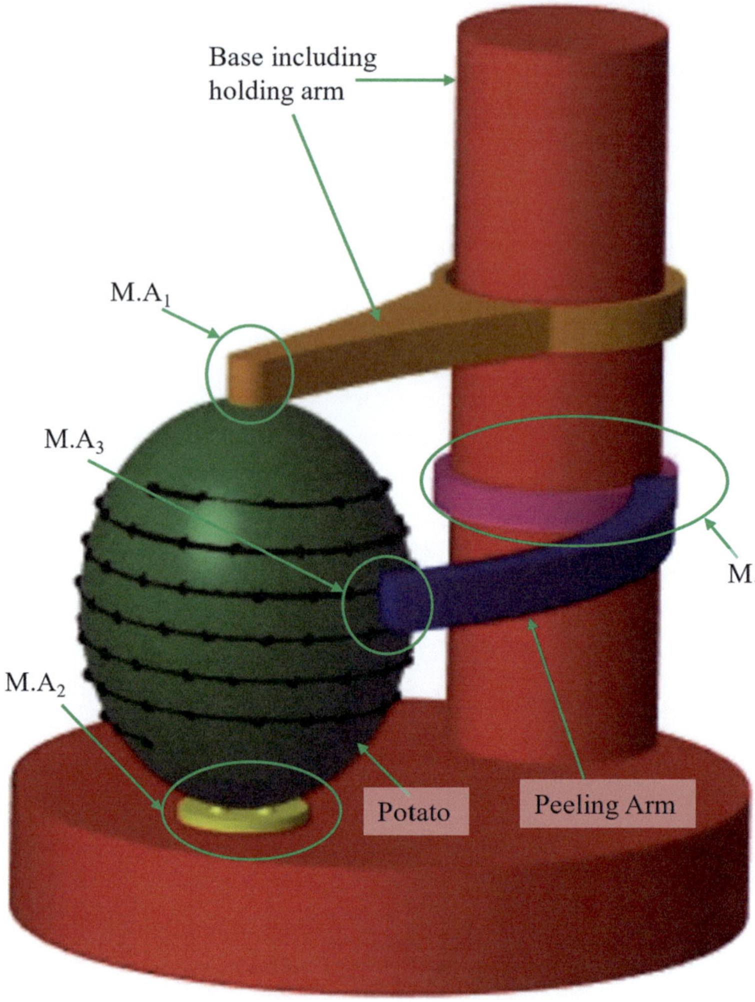

Fig. 1.12 Potato peeler machine. Figure and associated Potato_Peeling_Mechanism.mp4 video designed by the author and implemented by both the author and Pranav Dhola. Both courtesy of Pranav Dhola

Table 1.8 Multibody elements of potato peeler machine

Rigid bodies	Mechanical assemblies (M.A)
1. Base including the holding arm	1. M.A$_1$: Between the holding arm and the potato allowing rotational motion
2. Peeling arm	2. M.A$_2$: Between the base and the potato allowing rotational motion
3. Potato or any other similarly sized vegetable or fruit	3. M.A$_3$: Between the peeling arm and the potato allowing the cutting edge to slide on the potato's surface
	4. M.A$_4$: Between the base and the peeling arm allowing the rotational motion and the sliding motion of the peeling arm on the base column

cutting. The plate is a solid, heavy piece that is specifically shaped to fit neatly into the grid of the blade, allowing it to forcefully drive the potato through and complete the cut.

We now look more closely at how the force needed to drive the plate (and the potato) is applied. The handle is pulled up, opening a space between the blade and the plate where a whole potato is placed. Notice that a handle is fixed between two long curved elements and that these elements are

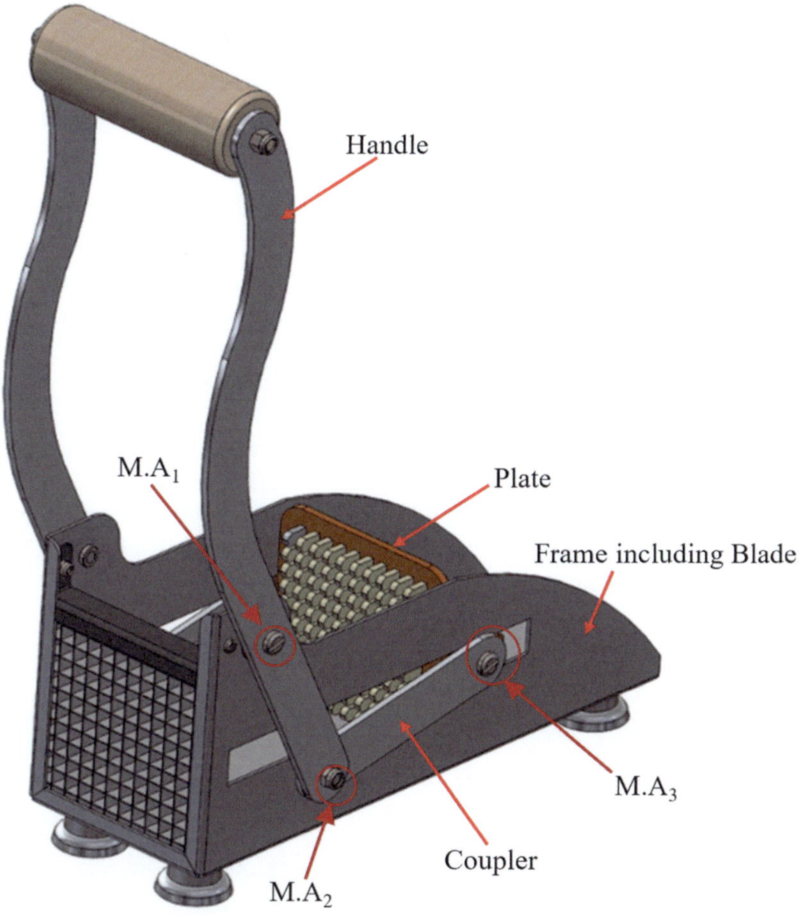

Fig. 1.13 Potato chopper machine. Figure and associated Potato_Chopper_Mechanism.avi video designed by the author and implemented by both the author and Pranav Dhola. Both courtesy of Pranav Dhola

Table 1.9 Multibody elements of potato chopper machine

Rigid bodies	Mechanical assemblies (M.A)
1. Frame including the blade	1. $M.A_1$: Between the frame and the handle allowing the rotational motion
2. Handle	2. $M.A_2$: Between the handle and the coupler allowing the rotational motion
3. Coupler	3. $M.A_3$: Between the coupler, the frame, and the plate allowing the sliding motion and the rotational motion
4. Plate	

attached to the stationary frame by pins ($M.A_1$) that allow rotation. The opposite ends of the long curved elements are attached to shorter straight elements (couplers) with one pin each ($M.A_2$) and then attached to the plate with one pin each ($M.A_3$) as well. Slots in the frame allow the pins attached to the plate a clear path to move through ($M.A_3$). When the handle is pushed down, it pivots on the pins in the frame, rotating the ends of the long curved elements toward the blade. This rotation pulls the smaller straight element which in turn pulls on the plate, driving the potato through the blade.

It is interesting to notice a few details. The distance from the handle where the force is applied to the pin attaching it to the frame is much greater than the distance from the pin in the frame to the smaller element. Also, the long elements are not only curved for looks. This curve locates the handle in a different place than it would be if the long elements were straight. These are both examples of "mechanical advantage". Lastly, notice that there are elements on both sides of the handle. If there were only elements on one side of the handle, it would cause the pins that connect the elements to wobble, and they would not be able to transfer the force as they should. Having elements on both sides of the handle ensures that the force pushing down on it is distributed evenly and is directly applied to drive the potato through the blade.

Planetary Mixer Machine: Figure 1.14 shows the assembly of the planetary mixer machine as a symmetric multibody consisting of the listed elements in Table 1.10.

The planetary mixer is a more recent kitchen device that is used in baking when there is a need to mix flour with eggs, butter, and maybe sugar. The planetary mixer has a motor that rotates the paddles inside the bowl. Now, we are going to understand how the planetary mixer machine works by looking at the attached video.

The motor rotates both the attached flywheel and the blue drive shaft that has an extended blue arm (carrier) to orbit both the paddle shafts (planets) around the axes of the two base gears (sun). In other words, the base gears, the carrier, and the paddle gears are like the planets rotating around the sun. The sun-planets mechanism is called a planetary gear, and we will discuss it further in Chap. 10. As seen in Fig. 1.14, the handle consists of the blue arm, purple coupler, blue part, and magenta

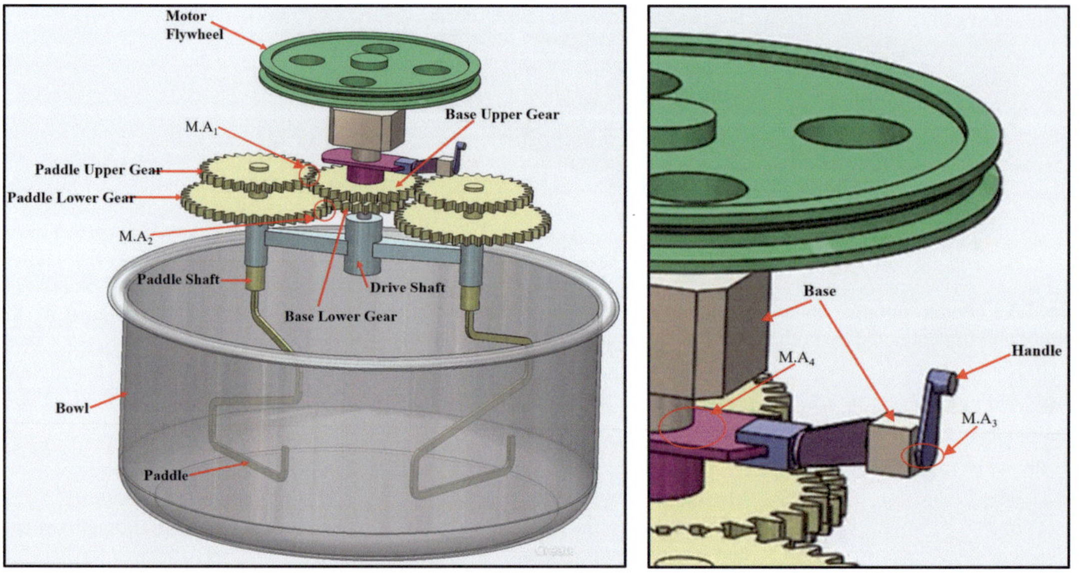

Fig. 1.14 Planetary mixer machine. Figure and associated Planetary_Mixer_Mechanism.mp4 video designed by the author and implemented by both the author and Pranav Dhola. Both courtesy of Pranav Dhola

Table 1.10 Multibody elements of planetary mixer machine

Rigid bodies	Mechanical assemblies (M.A)
1. Base and the bowl as a single stationary element	1. $M.A_1$: Between the paddle upper gear and the base upper gear allowing gear motion
2. Handle attached to the base lower gear and the base upper gear	2. $M.A_2$: Between the paddle lower gear and the base lower gear allowing gear motion
3. Flywheel attached to drive shaft	3. $M.A_3$: Between the handle and base allowing rotational motion
4. Paddle shaft attached to the paddle upper gear and the paddle lower gear plus the paddle	4. $M.A_4$: Between the handle and the base allowing sliding motion

part that is attached to both the base lower gear and base upper gear. The mechanical assembly M.A$_3$ allows the rotation between the handle and the base. When the handle is down, the upper gears (base and paddle) are engaged while the lower gears are disengaged. When the handle is up, the lower gears (base and paddle) are engaged while the upper gears (base and paddle) are disengaged. In either case, whether the handle is up or down, the paddle gears spin around the paddle shaft as well as rotate around the drive shaft. The spinning rotational speed relative to the motor rotational speed depends on the ratio of the base gear size and the paddle gear size. In the attached video, you can see that when the upper gears are engaged, there is no spinning since the upper paddle and base gears have the same size and dimensions. As an engineering design, we should run a kinematic and dynamic analysis to design the proper size and configuration for the elements, as well as to select an adequate motor for this machine.

Corkscrew Machine: Figure 1.15 shows the assembly of the corkscrew machine as a multibody assembly that consists of the listed elements in Table 1.11.

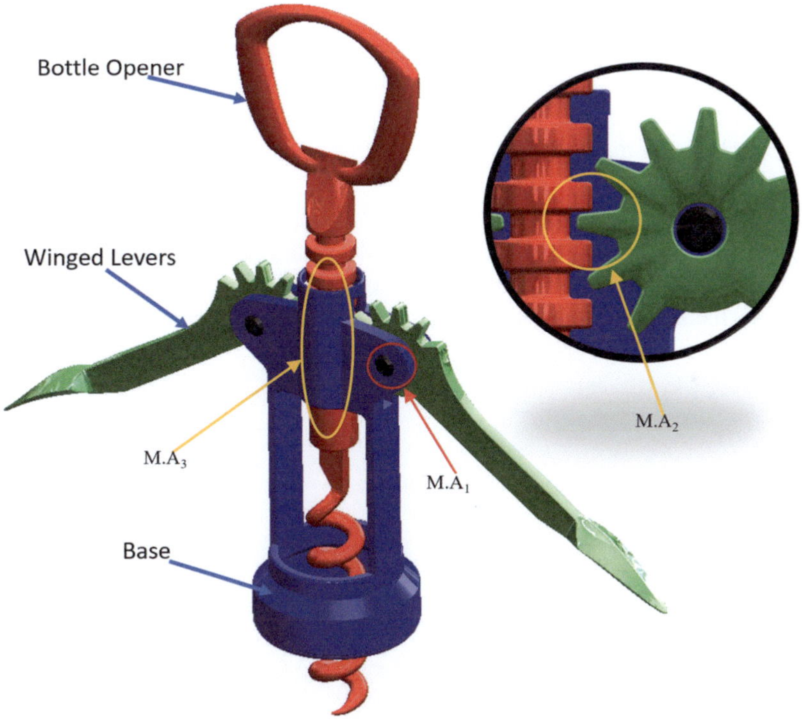

Fig. 1.15 Corkscrew machine. Figure and associated Corkscrew_Mechanism.mp4 video designed by the author and implemented by both the author and Pranav Dhola. Both courtesy of Pranav Dhola

Table 1.11 Multibody elements of corkscrew machine

Rigid bodies	Mechanical assemblies (M.A)
1. Base to hold the bottle cork and as a self-centering device	1. M.A$_1$: Between the winged lever and the base allowing rotational motion
2. Two winged levers as thin arms with partial gears	2. M.A$_2$: Between the winged lever and the bottle opener allowing worm–gear motion
3. Bottle opener with a top as a handle, a middle as a worm–gear, and an end as a non-sliding worm	3. M.A$_3$: Between the bottle opener and base allowing sliding motion

Some bottle manufacturers seal drinking bottles with a cork. A corkscrew is a machine that facilitates opening the bottle by drawing the cork out. The process starts by holding the corkscrew on top of the bottle whereby the sharp tip of the non-sliding worm is pointed toward the middle of the cork. Hold the bottle with one hand. Use the second hand to rotate the handle clockwise to insert the worm into the cork. You will notice that the bottle opener slides down in $M.A_3$ and the worm gets deeper inside the cork. Continue turning the handle until the worm is all way inside the cork. You will notice that the winged levers rotate counterclockwise (all the way up) because of $M.A_2$ and $M.A_1$. Now, take your two hands away. Grip the two winged levers and rotate them clockwise (all the way down). The bottle opener will slide through $M.A_3$ all the way up because of $M.A_2$. Finally, the cork comes away with the worm. It is an easy process that is safe from too much hassle, compared to attempting to open the bottle using a regular knife. Now, enjoy whatever is inside the bottle! As a design engineer, you will have more parameters and shapes to select from. The kinematics and dynamics of machines helps to design every machine, as we will discuss in upcoming chapters.

1.4.3 Gymnasium Equipment

Another place that is full of machines is the gym, in which every device used for working out is a machine; see Fig. 1.16. We are going to explore four examples of machines inside the gym. There are many more machines in the gym. The reader is encouraged to think of other examples of additional machines here.

When a person uses a gym device, both the person's anatomy and the gym machine become an integrated mechanism. The human body is considered a multibody assembly for the purpose of kinematic and dynamic analysis. The multibody human anatomy has single rigid bodies (such as the right upper arm segment) as well as mechanical assemblies (such as the knee joint) that allow for

Fig. 1.16 Gymnasium machines. Designed by the author and implemented by Kenneth Manalansan. Courtesy of Kenneth Manalansan

specific motion between two rigid bodies. Figure 1.17 shows how we break down the human anatomy into a multibody that consists of elements as listed in Table 1.12.

There are many more elements to consider when treating the human anatomy as a multibody. For our current purposes, we will only consider the elements that a human uses during fitness routines.

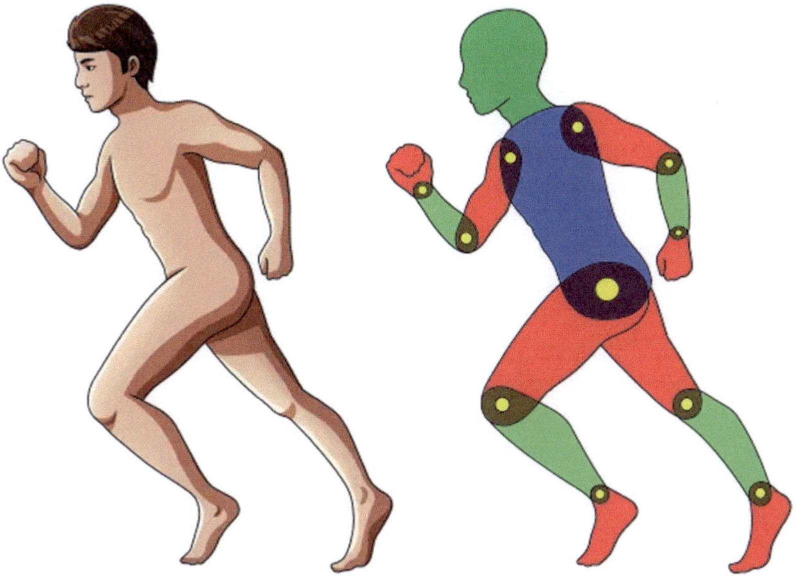

Fig. 1.17 Human anatomy as multibody for machine analysis. Designed by the author and implemented by Kenneth Manalansan. Courtesy of Kenneth Manalansan

Table 1.12 Multibody elements of human anatomy

Rigid bodies	Mechanical assemblies (M.A)
1. Torso	1. $M.A_1$ or left glenohumeral joint: to allow the rotational motion between the left upper arm segment and the torso
2. Right upper arm segment	2. $M.A_2$ or right glenohumeral joint: to allow the rotational motion between the right upper arm segment and the torso
3. Left upper arm segment	3. $M.A_3$ or left elbow joint: to allow the rotational motion between the left upper arm segment and the left lower arm segment
4. Right lower arm segment	4. $M.A_4$ or right elbow: to allow the rotational motion between the right upper arm segment and the right lower arm segment
5. Left lower arm segment	5. $M.A_5$ or left wrist: to allow the rotational motion between the left-hand segment and the left lower arm segment
6. Right upper leg segment	6. $M.A_6$ or right wrist: to allow the rotational motion between the right-hand segment and the right lower arm segment
7. Left upper leg segment	7. $M.A_7$ or left hip: to allow the rotational motion between the left upper leg segment and the torso
8. Right lower leg segment	8. $M.A_8$ or right hip: to allow the rotational motion between the right upper leg segment and the torso
9. Left lower leg segment	9. $M.A_9$ or left knee: to allow the rotational motion between the left upper leg segment and the left lower leg segment
10. Left foot	10. $M.A_{10}$ or right knee: to allow the rotational motion between the right upper leg segment and the right lower leg segment
11. Right foot	11. $M.A_{11}$ or left ankle: to allow the rotational motion between the left foot segment and the left lower leg segment
12. Left hand	12. $M.A_{12}$ or right ankle: to allow the rotational motion between the right foot segment and the right lower leg segment
13. Right hand	

Now, we will go over four examples of gym machines that integrate with human anatomy to create mechanisms. There are animation videos as well as an explanation for each mechanism.

Hammer Strength Machine: Figure 1.18 shows a person sitting on a hammer strength gym machine. The assembly of the person along with the hammer machine constitutes a mechanism. When a person sits on this machine, there is symmetry between the right and left sections of the mechanism's multi-body assembly. Any of these two sections consists of the elements listed in Table 1.13.

The hammer strength machine incorporates weight plates, rather than stacked weights, as a means for the lifter to add weight to their workout. It uses leverage and resistance similar to a free-weight exercise. For this reason, the hammer strength machine is often called a hybridizing machine. The primary target of the hammer machine is to isolate the chest, arm, and shoulder muscles for an effective workout.

When a person sits on the hammer machine, his/her torso should be upright, such that it is fixed with the back part of the machine's frame. The left hand and right hand should hold the left and right movement arm, respectively. The connection between the human hands and the machine is assumed to be a mechanical assembly, $M.A_3$, that allows for rotational motion. The person, along with the machine, then becomes an integrated assembly or a mechanism.

The person-hammer mechanism starts by lifting the movement arm all the way up by applying an upward pushing force from the arms, which leads to a rotation of the movement arm about the pivot ($M.A_4$), causing the green weight to go down. After that, the person pulls the movement arm back down, causing the green weight to move upwards. During upward and downward movement,

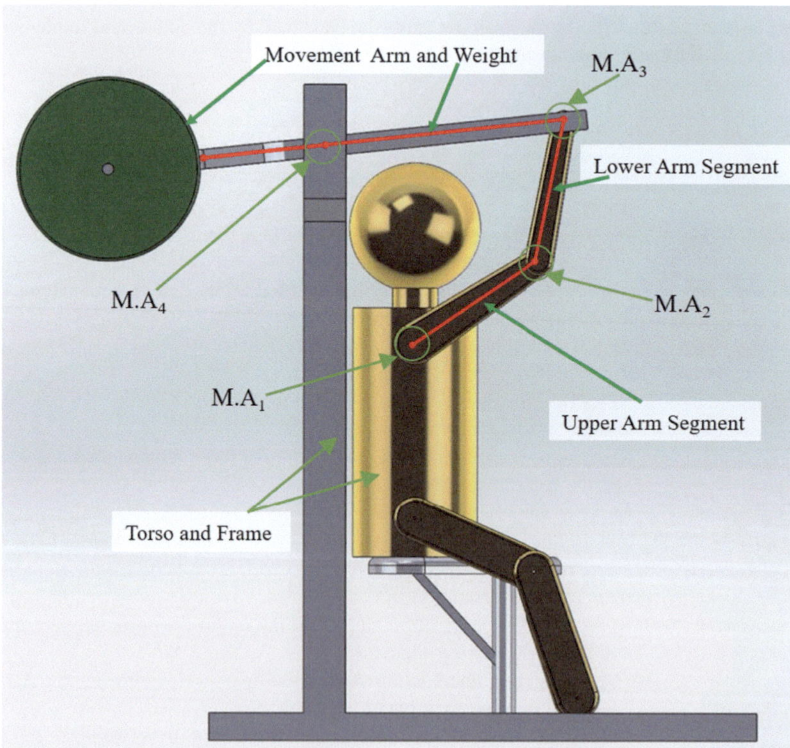

Fig. 1.18 Hammer strength gym machine. Figure and associated Hammer_Strength_Machine.mp4 video designed by the author and implemented by both the author and Pranav Dhola. Both courtesy of Pranav Dhola

Table 1.13 Multibody elements of hammer strength gym machine

Rigid bodies	Mechanical assemblies (M.A)
1. Torso-frame as a single stationary element	1. $M.A_1$: Between the torso-frame and the upper arm segment that allows the rotational motion
2. Upper arm segment	2. $M.A_2$: Between the lower arm segment and the upper arm segment that allows the rotational motion
3. Lower arm segment	3. $M.A_3$: Between the arm-weight and the lower arm segment that allows the rotational motion
4. Arm-weight; movement arm and the weight	4. $M.A_4$: Between the arm-weight and the torso-frame that allows the rotational motion

the upper arm segment moves up and down, respectively, while the lower arm segment is considered a coupler. The machine designer should consider the lengths and dimensions of the frame and the movement arm. Sometimes, the machine can be adjusted to fit the person's dimensions to achieve better performance.

Leg Press Machine: Figure 1.19 shows a person performing sets on a leg press machine. The assembly of the person along with the leg press machine constitutes a mechanism. When you perform sets on this machine, there is symmetry between the right and left sections of the mechanism's multibody assembly. Any of these two sections consists of the elements listed in Table 1.14.

The leg press machine is developed to isolate and exercise the hamstrings, quads, and even calf muscles. It's a great option for a leg day exercise. To use the leg press machine, a person should sit with their back and head flat against the seat and with their feet positioned shoulder-width apart on the middle of the platform. The feet should be kept neutral, meaning your toes should be directly above your heels. The exercise starts by rotating the upper leg segment to slide the footplate-weight on the torso-frame using the lower leg segment as a coupler. There are four mechanical assemblies,

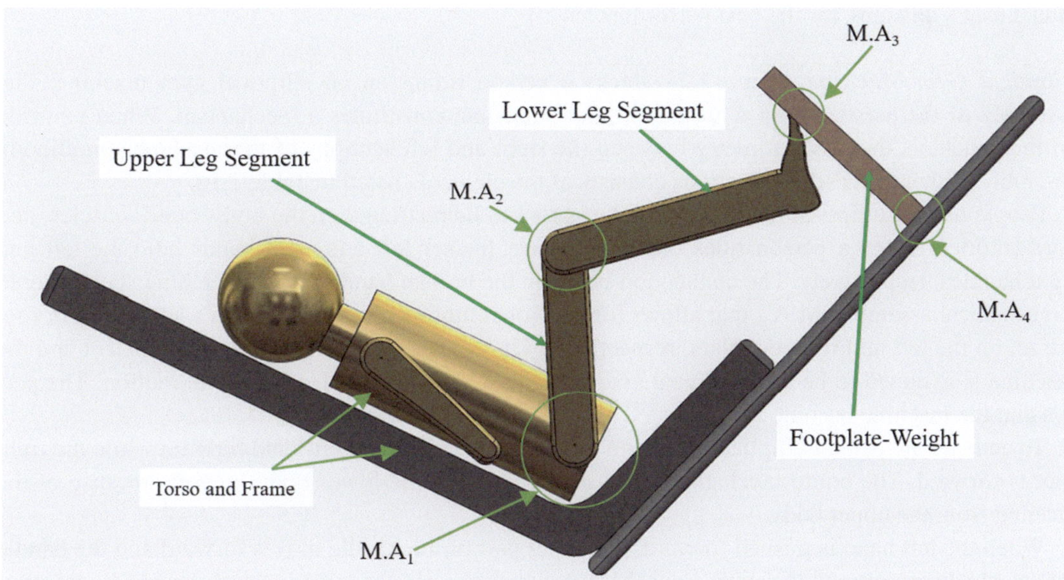

Fig. 1.19 Leg press machine. Figure and associated Leg Press_Machine.mp4 video designed by the author and implemented by both the author and Pranav Dhola. Both courtesy of Pranav Dhola

Table 1.14 Multibody elements of leg press machine

Rigid bodies	Mechanical assemblies (M.A)
1. Torso-frame as a single stationary element	1. M.A_1: Between the torso-frame and the upper leg segment that allows the rotational motion
2. Upper leg segment	2. M.A_2: Between the lower leg segment and the upper leg segment that allows the rotational motion
3. Lower leg segment	3. M.A_3: Between the footplate-weight and the lower leg segment that allows the rotational motion
4. Footplate-weight	4. M.A_4: Between the footplate-weight and the torso-frame that allows the sliding motion

M.A_1, M.A_2, M.A_3, and M.A_4 that all help to keep the assembly together with relative motions. A rotational motion is allowed in the case of M.A_1, M.A_2, and M.A_3. A sliding motion is allowed in M.A_4. The size and dimensions of the machine elements can be adjusted to fit the user's body.

Stationary Bike Machine: Figure 1.20 shows a person riding on a stationary bike machine. The assembly of the person along with the stationary bike machine constitutes a mechanism. When you ride on this machine, there is symmetry between the right and left sections of the mechanism's multibody assembly. Any of these two sections consists of the elements listed in Table 1.15.

The stationary bike is a simulation of the bicycle. It is well known that exercising on a stationary bike for half an hour every day builds cardiovascular and muscular endurance and improves aerobic capacity. The exercise starts with a person sitting on the machine as shown in Fig. 1.20. The person sits on the bike's seat with their feet resting on the pedals. The upper leg segment is connected to the bike's base through M.A_4. The feet are attached to the pedals, constituting the mechanical assembly M.A_2 which allows for rotational motion between the crank and the lower leg segment. The person partially rotates the upper leg segment around M.A_4, which causes a full crank rotation around M.A_1, while the lower leg segment is working as a coupler between M.A_2 and M.A_3. The stationary bike includes the option to make adjustments, ensuring that dimensions and aspect ratios fit the individual user's anatomy for the best performance.

Elliptical Gym Machine: Figure 1.21 shows a person riding on an elliptical gym machine. The assembly of the person along with the elliptical machine constitutes a mechanism. When you ride on this machine, there is symmetry between the right and left sections of the machine's multibody assembly. Either of these two sections consists of the elements listed in Table 1.16.

People use the elliptical machine as a fitness tool to help strengthen the body, build muscles, and burn calories. When a person rides on the machine, his/her left and right hands hold the left and right handles, respectively. The connection between the human hands and the machine is considered a mechanical assembly, M.A_2, that allows for rotational motion. Also, the person's left and right foot are set on the left and right couplers, respectively. The connection between the person's feet and the machine is assumed to be a mechanical assembly, M.A_7, that allows for rotational motion. The person and the machine become an integrated assembly or a mechanism.

To activate the person-elliptical mechanism, the rider pushes the left-hand forward while the right foot is forward. The entire mechanism then moves, forcing the lower body to work because of the pushing from the upper body.

When the left hand is pushed forward, the upper part of the handle moves forward and the handle rotates clockwise around the pivot joint (M.A_1), which pushes the lower part of the handle to move backward. The coupler is moved backward because of the pivot joint (M.A_8). The crank rotates

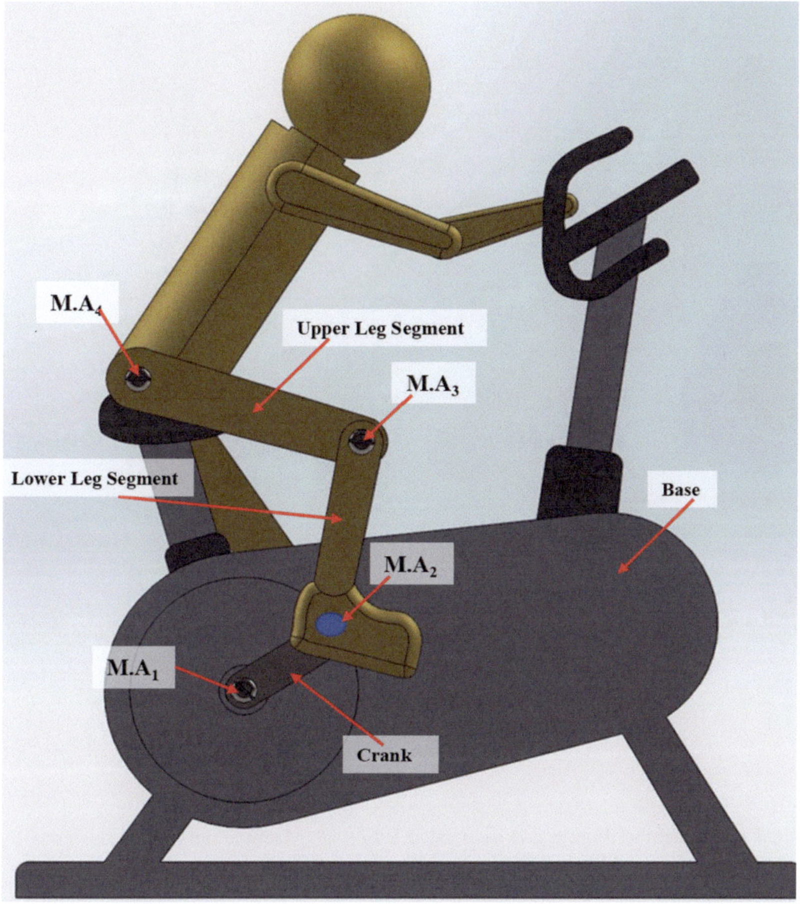

Fig. 1.20 Stationary bike machine. Figure and associated Stationary_Bicycle_Machine.mp4 video designed by the author and implemented by both the author and Pranav Dhola. Both courtesy of Pranav Dhola

Table 1.15 Multibody elements of stationary bike machine

Rigid bodies	Mechanical assemblies (M.A)
1. Base	1. M.A$_1$: Between the base and the crank allowing the rotational motion
2. Upper leg segment	2. M.A$_2$: Between the crank and the lower leg segment allowing the rotational motion
3. Lower leg segment	3. M.A$_3$: Between the lower leg segment and the upper leg segment allowing the rotational motion
4. Crank	4. M.A$_4$: Between the upper leg segment and the base allowing the rotational motion

counterclockwise because of the two pivots (M.A$_9$ and M.A$_{10}$). The pivot M.A$_7$ forces the body's lower part to move because of the coupler movement.

The entire machine design has special aspect ratios and dimensions between the crank, coupler, handler, and base. The design engineer should select these aspect ratios and dimensions using the kinematics and dynamics of the machines.

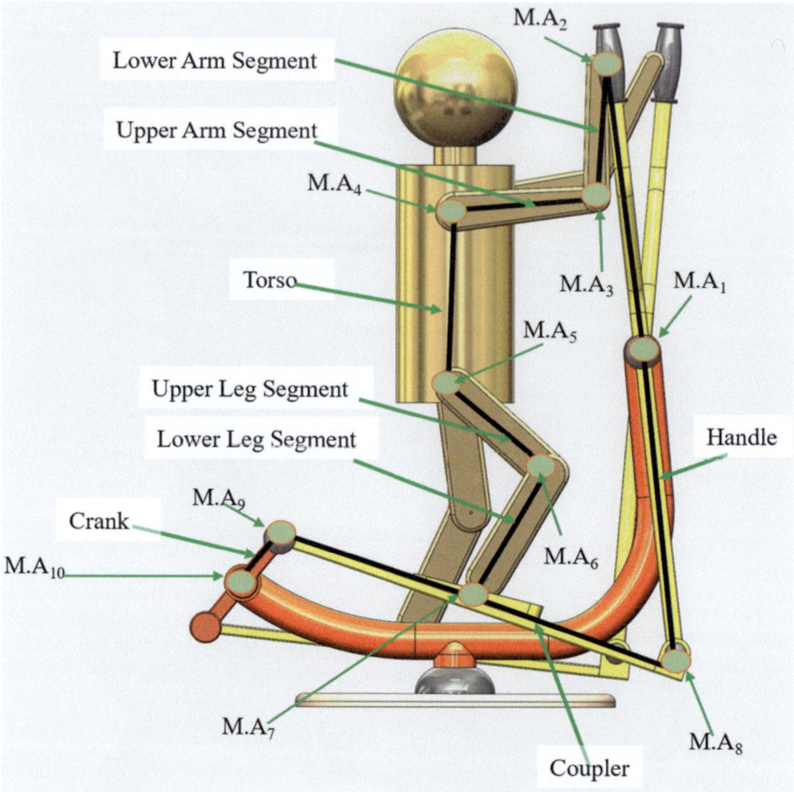

Fig. 1.21 Elliptical gym machine. Figure and associated Elliptical_Machine.mp4 video designed by the author and implemented by both the author and Pranav Dhola. Both courtesy of Pranav Dhola

Table 1.16 Multibody elements of elliptical gym machine

Rigid bodies	Mechanical assemblies (M.A)
1. Base	1. $M.A_1$: Between the handle and the base allowing rotational motion
2. Handle or handlebar	2. $M.A_2$: Between the handle and the lower arm segment allowing rotational motion
3. Lower arm segment	3. $M.A_3$: Between the lower arm segment and the upper arm segment allowing rotational motion
4. Upper arm segment	4. $M.A_4$: Between the upper arm segment and the torso allowing rotational motion
5. Torso	5. $M.A_5$: Between the torso and the upper leg segment allowing rotational motion
6. Lower arm segment	6. $M.A_6$: Between the lower leg segment and the upper leg segment allowing rotational motion
7. Upper arm segment	7. $M.A_7$: Between the lower leg segment and the coupler allowing rotational motion
8. Crank	8. $M.A_8$: Between the handle and the coupler allowing rotational motion
9. Coupler	9. $M.A_9$: Between the crank and the coupler allowing rotational motion
	10. $M.A_{10}$: Between the crank and the base allowing rotational motion

1.4.4 Bicycle, Car, and Aircraft

When you walk in the street, ride in a car, or go to the airport, your eyes would catch the bicycles, cars, and aircraft that make modern transportation possible. As an example, Fig. 1.22 shows some mechanisms inside the car and shows some of the mechanisms inside the aircraft. Upon seeing these

Fig. 1.22 Mechanisms inside a car. Designed by the author and implemented by Kenneth Manalansan. Courtesy of Kenneth Manalansan

machines, you may be interested to consider the complex mechanisms that are functioning to make our lives easier in many situations. We are going to list some of these mechanisms. The reader is encouraged to think of additional examples of machines in bicycles, cars, and/or aircraft (Fig. 1.23).

Bicycle Rider: Figure 1.24 shows a person riding, or pedaling, a bicycle. The obvious goal of maintaining balance while riding a bicycle is aided by the propulsion provided by pedaling. The assembly of the human anatomy with the bicycle provides a mechanism as a multibody that consists of the listed elements in Table 1.17.

The act of getting the bicycle's rear wheel to turn is fairly straightforward. Pushing down on the pedals turns a toothed disk that pulls the links of a chain. The chain transfers this rotary motion, or torque, to another (in this case smaller) toothed disk which is rigidly fixed to the rear wheel. This causes the rear wheel to spin forward, providing propulsion to the bike and rider. The difference in the size of the larger disk at the crank and the smaller disk fixed to the rear wheel is important in influencing how the bicycle performs. These ratios exploit what is referred to as "mechanical advantage", and form an interesting topic to be discussed later.

Fig. 1.23 Mechanisms inside an aircraft. Designed by the author and implemented by Kenneth Manalansan. Courtesy of Kenneth Manalansan

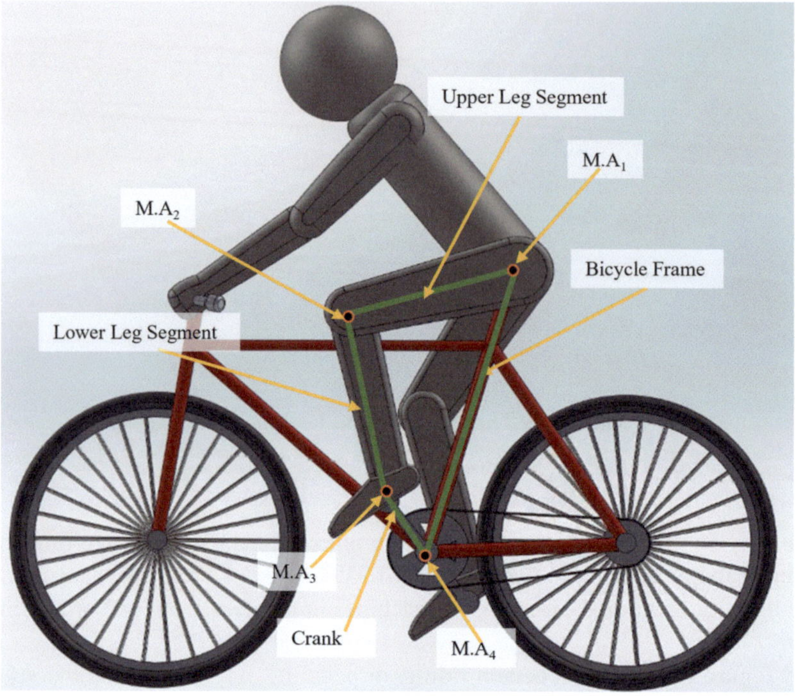

Fig. 1.24 Bicycle rider mechanism. Figure and associated Bicycle_Rider_Mechanism.mp4 video designed by the author and implemented by both the author and Pranav Dhola. Both courtesy of Pranav Dhola

Table 1.17 Multibody elements of bicycle rider mechanism

Rigid bodies	Mechanical assemblies (M.A)
1. Bicycle frame	1. $M.A_1$: Between the bicycle frame and the upper leg segment allowing rotational motion
2. Upper leg segment	2. $M.A_2$: Between the upper leg segment and the lower leg segment allowing rotational motion
3. Lower leg segment	3. $M.A_3$: Between the lower leg segment and the crank allowing rotational motion
4. Crank	4. $M.A_4$: Between the bicycle frame and the crank that allowing rotational motion

Note that in this machine, we use the mechanical assembly $M.A_1$ to describe the connection between the human anatomy and the bicycle as an assumption for our analysis. A similar assumption is used in the case of the mechanical assemblies $M.A_2$ and $M.A_3$. Also, we consider the human leg segments as rigid bodies (see Fig. 1.17)

Car Wiper Mechanism: Figure 1.25 shows a car wiper mechanism as a multibody that consists of the listed elements in Table 1.18.

Unless there is wonderful weather, you cannot drive without a wiper mechanism in your car. This wiper mechanism makes the life of the driver easier by removing rain, snow, dirt, pollen, frost, and other debris on the windscreen quickly and smoothly at the push of a button. The more expensive the car, the more sophisticated and luxurious the wiper mechanism is installed. Designers need to use kinematic and dynamic analysis to maximize the quality of the wiper mechanism for a better selling point.

With just a push of a button, the car's wiper blades start rocking and cleaning the front car window. In some cars, there is a rear car wiper mechanism as well. This wiper blades' rocking motion is generated from a mechanism such as the one shown in Fig. 1.25. A motor fixed to the car frame generates a complete rotational motion that drives the crank. Both the connecting rod and the left

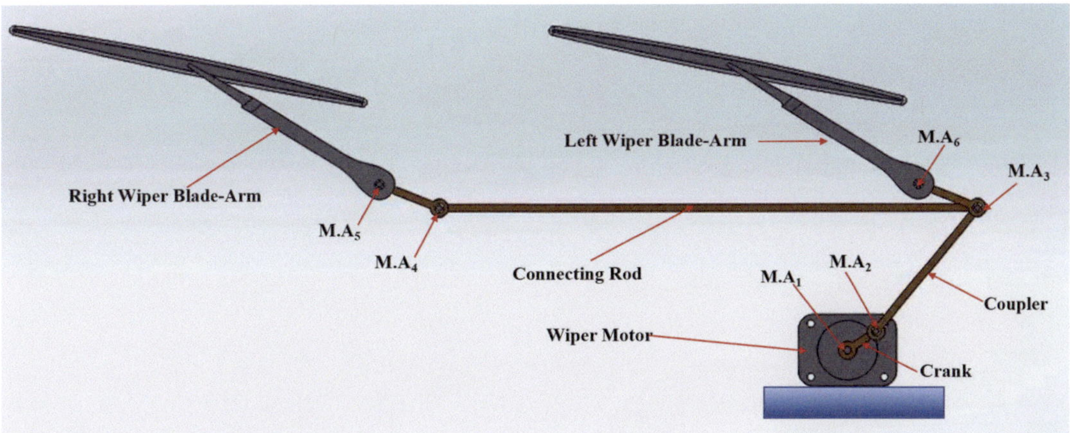

Fig. 1.25 Car wiper mechanism. Figure and associated Car_Wiper_Mechanism.mp4 video designed by the author and implemented by both the author and Pranav Dhola. Both courtesy of Pranav Dhola

Table 1.18 Multibody elements of car wiper mechanism

Rigid bodies	Mechanical assemblies (M.A)
1. Car frame	1. M.A$_1$: Between the wiper motor (fixed to the frame) and the crank allowing relative rotational motion only
2. Crank	2. M.A$_2$: Between the crank and the coupler allowing relative rotational motion only
3. Coupler	3. M.A$_3$: Between the coupler and both the left wiper blade-arm plus the connecting rod allowing relative rotational motion only
4. Right wiper blade-arm	4. M.A$_4$: Between the connecting rod and the right wiper blade-arm allowing relative rotational motion only
5. Left wiper blade-arm	5. M.A$_5$: Between the right wiper blade-arm and the car frame allowing relative rotational motion only
6. Connecting rod	6. M.A$_6$: Between the left wiper blade-arm and the car frame allowing relative rotational motion only

wiper blade-arm start a rocking motion through the coupler thanks to M.A$_2$ and M.A$_3$. The right wiper blade-arm starts rocking from the connecting rod thanks to M.A$_4$. Both the left and right wiper blade-arms are connected to the car frame through M.A$_6$ and M.A$_5$, respectively. The main purpose of the wiper mechanism is to convert the complete rotational motion coming from the motor into a blade rocking motion. There are many design parameters and dimensions that require kinematic and dynamic analysis to determine; car manufacturers ask design engineers to perform this analysis with the ultimate goal in mind being performance optimization.

Car Hood Mechanism: Figure 1.26 shows a car hood mechanism as a multibody that consists of the listed elements in Table 1.19.

A car hood is a hinged door that rests over the engine to provide easy access to the engine bay for purposes such as light repair, maintenance, or oil changes. A car hood mechanism serves the purpose of keeping the hood in an open position during regular engine checking. For security, there is a latch under the hood that keeps the hood flush against the car's body in case of heavy wind or any car accident. After releasing the latch, we can check the engine or change the oil by lifting the hood. The hood mechanism now serves its purpose. This car hood mechanism must be designed using kinematic and dynamic analyses with more optimization design targets as listed in [9]. Note, for better

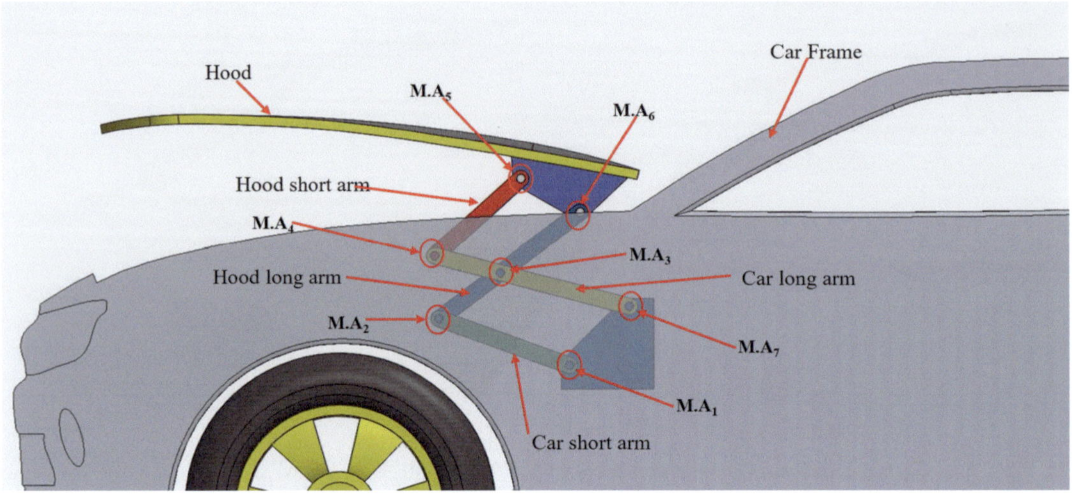

Fig. 1.26 Car hood mechanism. Figure and associated Car_Hood_Mechanism.mp4 video designed by the author and implemented by both the author and Pranav Dhola. Both courtesy of Pranav Dhola

Table 1.19 Multibody elements of car hood mechanism

Rigid bodies	Mechanical assemblies (M.A)
1. Car frame	1. M.A$_1$: Between the car frame and the car short arm allowing relative rotational motion only
2. Car short arm	2. M.A$_2$: Between the car short arm and the hood long arm allowing relative rotational motion only
3. Car long arm	3. M.A$_3$: Between the car long arm and the hood long arm allowing relative rotational motion only
4. Hood	4. M.A$_4$: Between the car long arm and the hood short arm allowing relative rotational motion only
5. Hood short arm	5. M.A$_5$: Between the hood and the hood short arm allowing relative rotational motion only
6. Hood long arm	6. M.A$_2$: Between the car long arm and the hood allowing relative rotational motion only
	7. M.A$_1$: Between the car frame and the car long arm allowing relative rotational motion only

visualization, the aspect ratio between the car hood mechanism dimension and the car dimension in Fig. 1.26 is much larger than the actual aspect ratio. In the actual design, there is a spring that helps to smooth the opening and closing of the hood. This spring is attached from the car frame to the end of the car's short arm. This spring stores potential energy during the closing of the hood and releases it when the hood is opened. Thus, the spring helps to both open the hood with less effort and to close it more gently. We did not count this spring as part of the mechanism since it is not a rigid body; it is out of the scope of this textbook.

Car Window Mechanism: Figure 1.27 shows a car window mechanism as a multibody that consists of the listed elements in Table 1.20.

The window mechanism, which we may call the scissor lift mechanism, allows the window to move up and down. When you rotate the window handle, the window moves up. Once you stop rotating the handle, the window stays where it is; that is, it holds its position. You can fully close or fully open the window with this mechanism, as well as partially open or close the window.

The window mechanism can be activated manually by a handle or automatically by a motor. When the handle is rotated (thanks to M.A$_1$), the partial gear rotates with the attached arm thanks

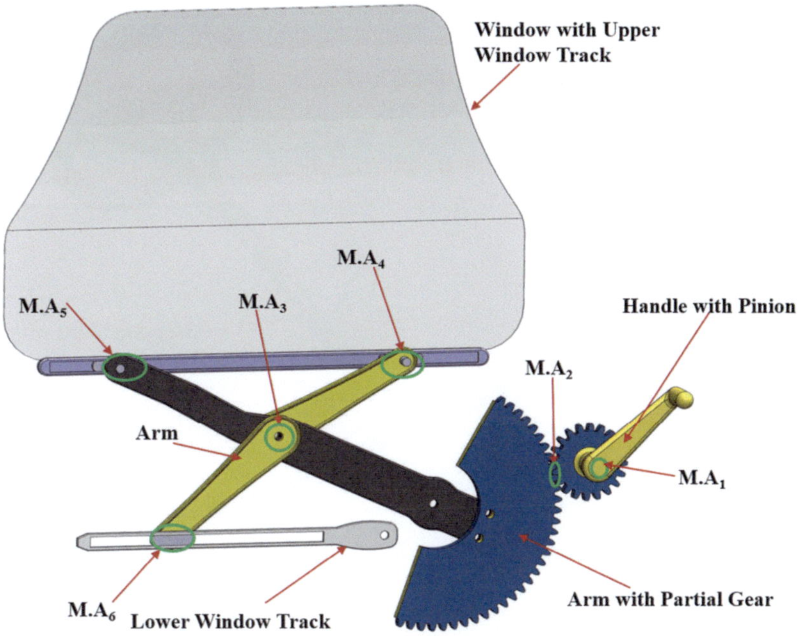

Fig. 1.27 Car window mechanism. Figure and associated Car Window Mechanism.mp4 video designed by the author and implemented by both the author and Pranav Dhola. Both courtesy of Pranav Dhola

Table 1.20 Multibody elements of car window mechanism

Rigid bodies	Mechanical assemblies (M.A)
1. Lower window track attached to the car door	1. M.A$_1$: Between the car door and the handle with the pinion allowing relative rotational motion only
2. Window with upper window track	2. M.A$_3$: Between the pinion and the partial gear allowing relative gear motion only
3. Arm	3. M.A$_3$: Between the arm and the arm with partial gear allowing relative rotational motion only
4. Arm with partial gear	4. M.A$_4$: Between the arm and the upper window track allowing relative rotational and sliding motion only
5. Handle with pinion	5. M.A$_5$: Between the arm with partial gear and the upper window track allowing relative rotational and sliding motion only
	6. M.A$_6$: Between the arm and the lower window track allowing relative rotational motion only

to M.A$_2$. The window moves up or down depending on the direction of the handle's rotation clockwise or counterclockwise. The other arm is simply for balance. The window mechanism requires the designer to come up with a kinematic and dynamic model to select suitable elements in terms of materials, shapes, and dimensions. Even optimizing an existing design requires a kinematic and dynamic model.

Car Steering Mechanism: Figure 1.28 shows a car steering mechanism as a multibody that consists of the listed elements in Table 1.21.

When you drive on a road with a curve and must turn left or right with your car, the steering mechanism is responsible for accomplishing this goal. As you drive the car, you simply turn the steering wheel left to turn the car toward the left or right to turn the car toward the right. In other

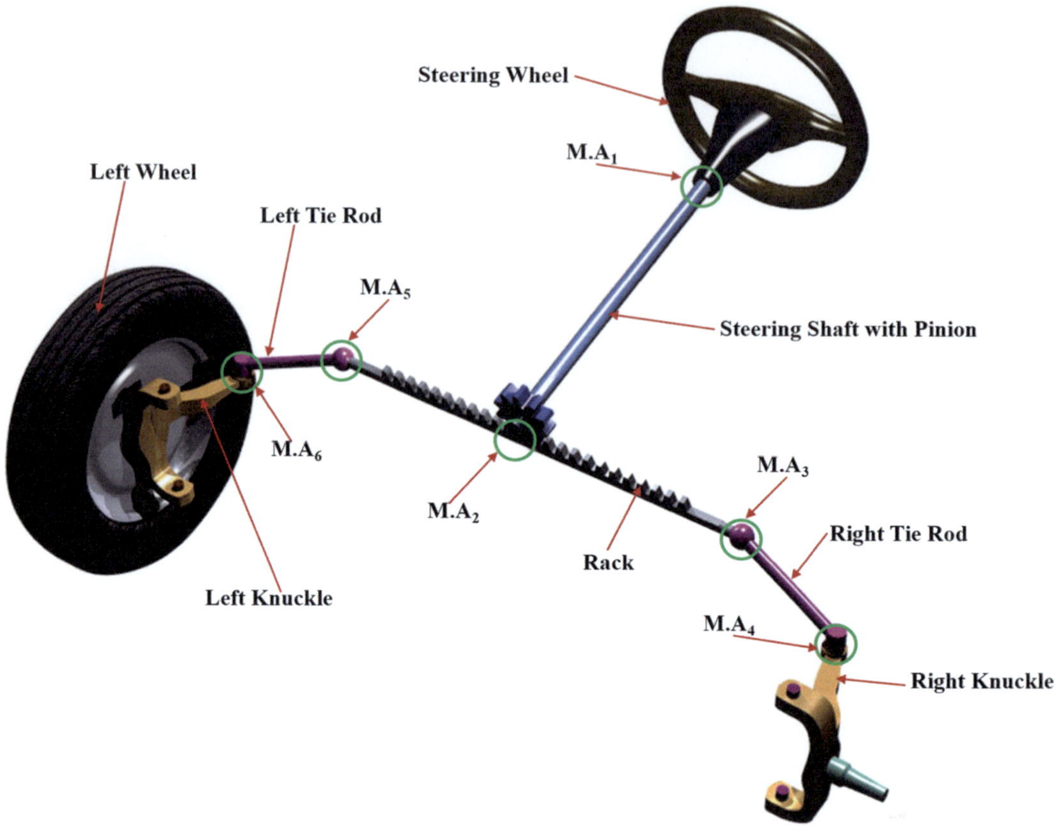

Fig. 1.28 Car steering mechanism. Figure and associated Car_Steering_Mechanism.mp4 video designed by the author and implemented by both the author and Pranav Dhola. Both courtesy of Pranav Dhola

Table 1.21 Multibody elements of car steering mechanism

Rigid bodies	Mechanical assemblies (M.A)
1. Steering wheel	1. $M.A_1$: Between the steering wheel and the steering shaft
2. Steering shaft with pinion	2. $M.A_2$: Between the pinion and the rack
3. Rack	3. $M.A_3$: Between the rack and right tie rod
4. Right tie rod	4. $M.A_4$: Between the right tie rod and the right knuckle
5. Left tie rod	5. $M.A_5$: Between the rack and left tie rod
6. Left knuckle attached to left wheel	6. $M.A_6$: Between the left tie rod and the left knuckle
7. Right knuckle attached to right wheel	

words, the steering mechanism provides vehicle turning per the will of the driver by converting the steering wheel's rotary movement into an angular turn of the front wheels.

The mechanism starts to move when the driver rotates the steering wheel. The steering shaft rotates along with the pinion that is on the top of the rack thanks to $M.A_1$ and $M.A_2$. The rotation of the pinion causes a linear transitional motion in the rack that leads to a rotation of both the left and right knuckles, through the force provided by the tie rods (as couplers). The steering mechanism should be designed to be easy to handle (driver's side) and to be stable (car's side). Thus, the steering wheel should not be tiresome to turn, and it should provide directional stability when the car makes a turn. The kinematics and dynamics of machines comes as a very important phase to create the design or to improve an existing design.

Car Suspension Mechanism: Figure 1.29 shows a car suspension mechanism as a multibody that consists of the listed elements in Table 1.22. When a luxurious car that has great suspension is driven over a road with irregularities such as bumps or dips, the riders feel almost nothing from the inside. On the other hand, drivers of cheaper cars would probably feel considerable bumping and shaking. The suspension system is responsible for ensuring that whoever is inside the car has a smooth ride by absorbing the energy from road bumps or by supporting the car from road dips. Various mechanisms are proposed to accomplish this shock observation rule; Fig. 1.29 is one such example. When the car encounters a road bump, the tire recoils upwards due to spring compression. Both the lower control arm and upper control arm are jacked up and rotated; thanks to $M.A_3$ and $M.A_4$. This leads to a compression of the spring and a pushing of the piston inside the cylinder thanks to $M.A_6$. The spring absorbs impact forces from bumps, while dampers control the spring oscillations. A similar scenario occurs in the case of dips, except that the tire would now recoil downward due to spring extension to fill the dip and provide support for the vehicle. A car's suspension mechanism requires a detailed design for the best riding quality. The kinematics and dynamics of machines comes to be a very important phase in such design.

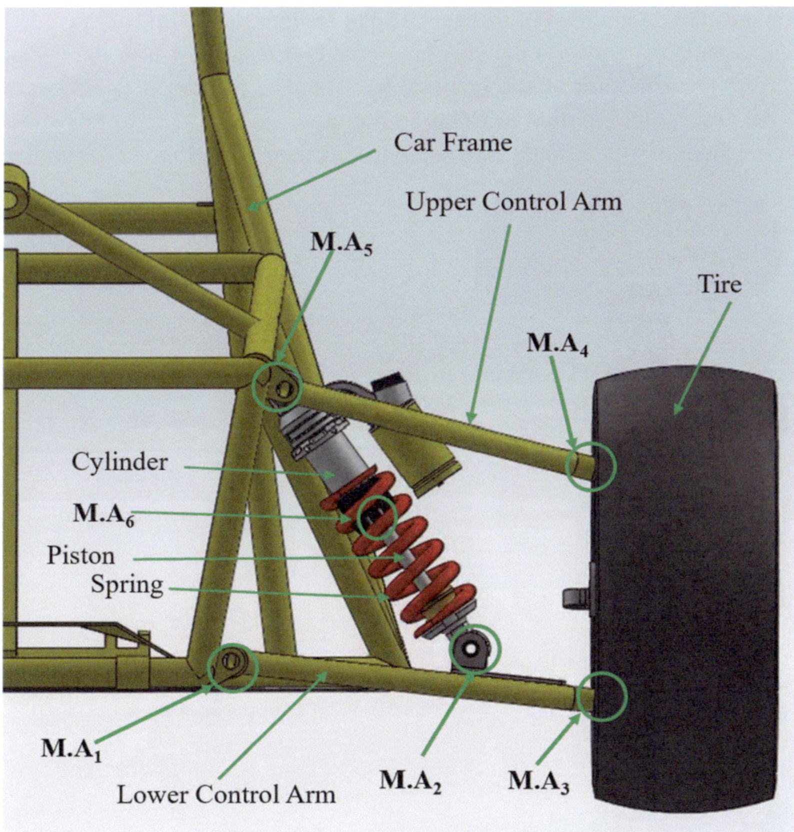

Fig. 1.29 Car suspension mechanism. Figure and associated Car Suspension Mechanism.mp4 video designed by the author and implemented by both the author and Pranav Dhola. Both courtesy of Pranav Dhola

Table 1.22 Multibody elements of car suspension mechanism

Rigid bodies	Mechanical assemblies (M.A)
1. Car frame	1. $M.A_1$: Between the car frame and lower control arm
2. Upper control arm	2. $M.A_2$: Between the lower control arm and piston
3. Tire	3. $M.A_3$: Between the lower control arm and tire
4. Lower control arm	4. $M.A_4$: Between the upper control arm and tire
5. Cylinder	5. $M.A_5$: Between the upper control arm, cylinder, and car frame
6. Piston	6. $M.A_6$: Between the cylinder and the piston

Four-Stroke Engine Mechanism: One of the most interesting and relevant topics in mechanical engineering is the car engine, specifically the piston engine. We are going to introduce some of the basic mechanisms of the four-stroke engine as shown in Fig. 1.30. Even if you have not seen the four-stroke engine before, you probably have some experience with it. There are four connecting rods as well as four individual pistons that are assembled onto a single crankshaft through pivots. The four pistons are also assembled to slide inside four cylinders. There are two valves in each cylinder: the intake valve and the exhaust valve. The camshaft is driven from the crankshaft using a timing belt, camshaft sprocket, and crankshaft sprocket. There are eight cams (four exhaust cams and four intake cams) placed along the camshaft. The shapes of these cams serve to open and close each cylinder's valves within a specific time. There is a very particular sequence of when the pistons will slide up or down inside each cylinder, and when the valves will open or close. We recommend the reader watch the animation video of this four-stroke engine.

The engine as a machine converts the chemical potential energy or heat energy of the fuel into kinetic energy (rotational motion of the crankshaft). There is a more detailed explanation of the four-stroke engine process and motion in Chap. 9. For now, we are going to focus more on a single piston–cylinder mechanism and a single cam valve mechanism as multibody assemblies.

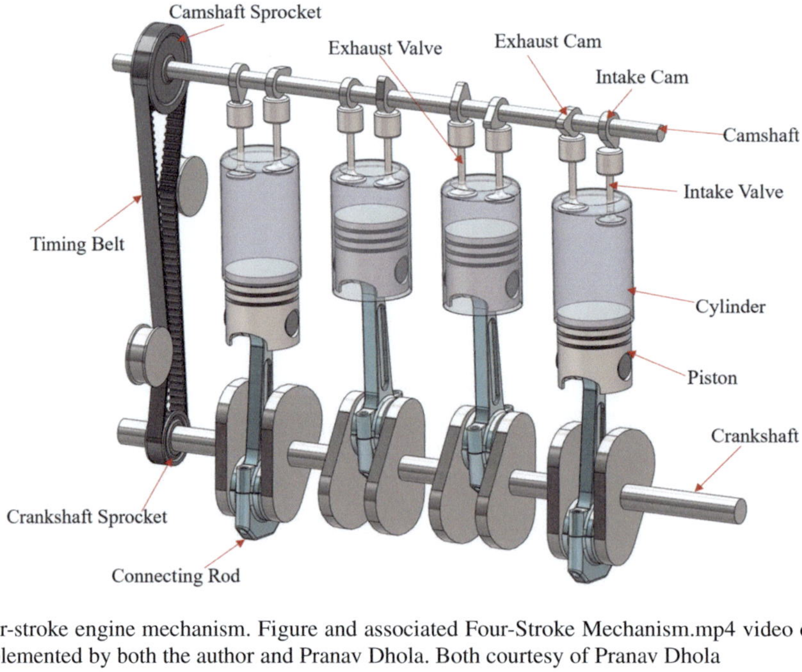

Fig. 1.30 Four-stroke engine mechanism. Figure and associated Four-Stroke Mechanism.mp4 video designed by the author and implemented by both the author and Pranav Dhola. Both courtesy of Pranav Dhola

Single Piston–Cylinder Engine Mechanism: Figure 1.31 is a single piston–cylinder engine machine that is the basis for internal combustion engines, which power the majority of motor vehicles in use today. The specific purpose that this machine serves is to convert high pressure in the cylinder into reciprocal (up and down) motion. This reciprocal motion is then converted into rotary motion. The single-cylinder engine mechanism is a multibody assembly that consists of the listed elements in Table 1.23.

To begin, a high-pressure event takes place inside the cylinder and pushes down on the piston. This high-pressure event can be from an explosion of gasoline, steam pressure from boiling water, compressed air, or other sources. The pressure event occurs just after the piston moves away from the top of the cylinder. Then, the connecting rod (coupler) is used to transmit the force of the pressure event to the crankshaft, resulting in rotational torque. This torque is then used to turn the car's wheels, generate electricity, and power other components and systems of the car. The reader is recommended to watch the attached animation video of the mechanism.

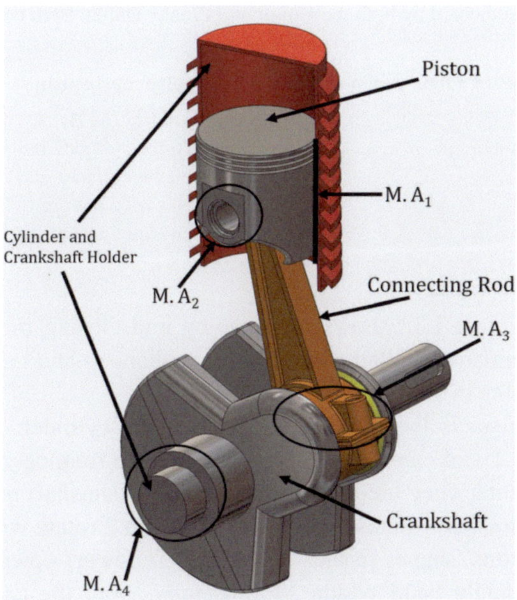

Fig. 1.31 Single piston–cylinder engine mechanism. Figure and associated Single Piston-Cylinder Mechanism.mp4 video designed by the author and implemented by both the author and Pranav Dhola. Both courtesy of Pranav Dhola

Table 1.23 Multibody elements of single piston–cylinder engine mechanism

Rigid bodies	Mechanical assemblies (M.A)
1. Cylinder and crankshaft holder	1. M.A$_1$: Between the cylinder and the piston that allows only the relative sliding motion of the piston inside the cylinder
2. Piston	2. M.A$_2$: Between the piston and the connecting rod that allows the relative rotational motion only
3. Connecting rod	3. M.A$_3$: Between the connecting rod and the crankshaft that allows the relative rotational motion only
4. Crankshaft	4. M.A$_4$: Between the crankshaft and the crankshaft holder that allows the relative rotational motion only

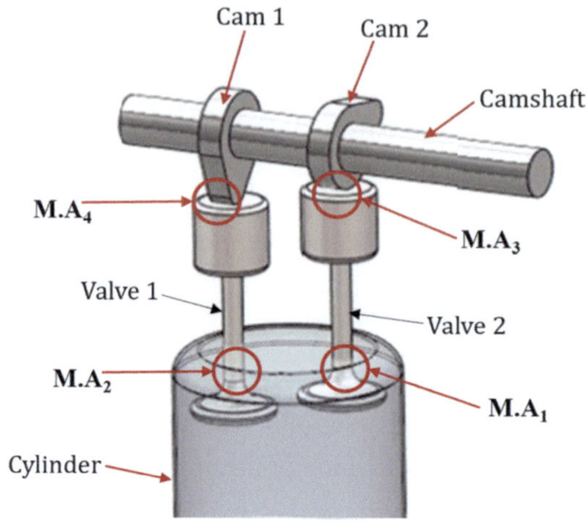

Fig. 1.32 Single piston–cylinder engine valve mechanism. Figure and associated Four-Stroke Mechanism.mp4 video designed by the author and implemented by both the author and Pranav Dhola. Both courtesy of Pranav Dhola

Table 1.24 Multibody elements of single piston-cylinder engine valve mechanism

Rigid bodies	Mechanical assemblies (M.A)
1. Camshaft with cam 1 and cam 2	1. $M.A_1$: Between the cylinder and valve 2
2. Valve 1	2. $M.A_2$: Between the cylinder and valve 1
3. Valve 2	3. $M.A_3$: Between cam 2 and valve 2
4. Cylinder	4. $M.A_4$: Between cam 1 and valve 1

Cam Valve Mechanism: Figure 1.32 shows the opening and closing mechanisms of the inlet and exhaust valves. This mechanism is called a cam valve mechanism and is a multibody assembly that consists of the listed elements in Table 1.24.

The cam valve mechanism is located near the end of the cylinder. There are two uncommon shapes which we call cam 1 and cam 2. These two cams have rotating irregular cylindrical shapes that have different alignments. They are used in transforming camshaft rotary motion into the valve linear motion. When the camshaft rotates, both cam 1 and cam 2 rotate with the same angular speed. The irregularity in these cams' shapes pushes the valves (follower) down for a specific time; more accurately, for a specific angle band width. More details about the cam motion and profile are explained in Chap. 9.

Steam Engine Straight-Line Mechanism: Figure 1.33 shows a steam engine straight-line mechanism. The input is the piston's straight-line motion, and the output is the wheel's rotational motion. The input starts by heating water until it converts to steam. The increase in the volume from water to steam creates high pressure that pushes the pistons to the drive wheels that operate the locomotive. There are twelve elements connected with eleven mechanical assemblies. We are not going to point to each element or each mechanical assembly in this mechanism.

Aircraft Foldable Stair Mechanism: Figure 1.34 shows an example of an aircraft foldable stair mechanism consisting of a set of scissor mechanisms assembled as a multibody assembly that consists of the listed elements in Table 1.25.

An airstair or an aircraft foldable stair mechanism is a set of steps built into a small-sized aircraft so that the passengers can get in and out of the plane as shown in Fig. 1.34. The folding mechanism

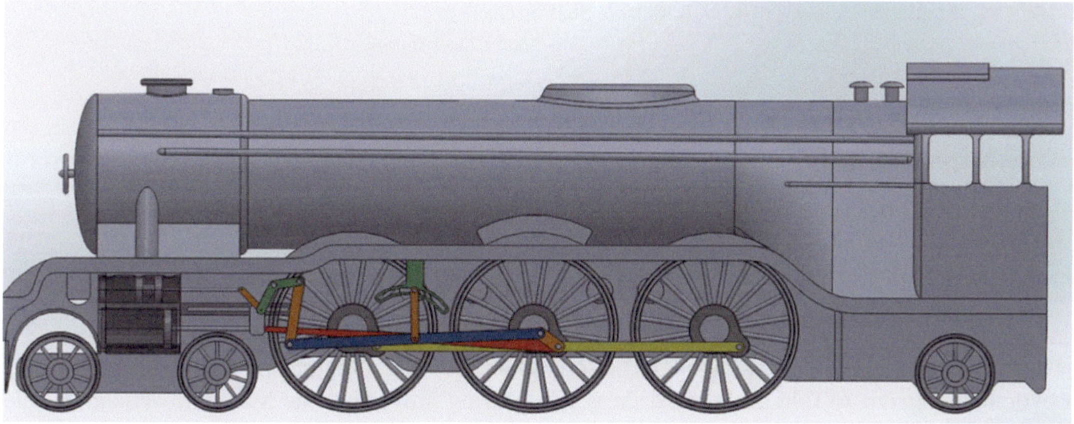

Fig. 1.33 Steam engine straight-line mechanism. Figure and associated Steam Engine Train Mechanism.mp4 video designed by the author and implemented by both the author and Pranav Dhola. Both courtesy of Pranav Dhola

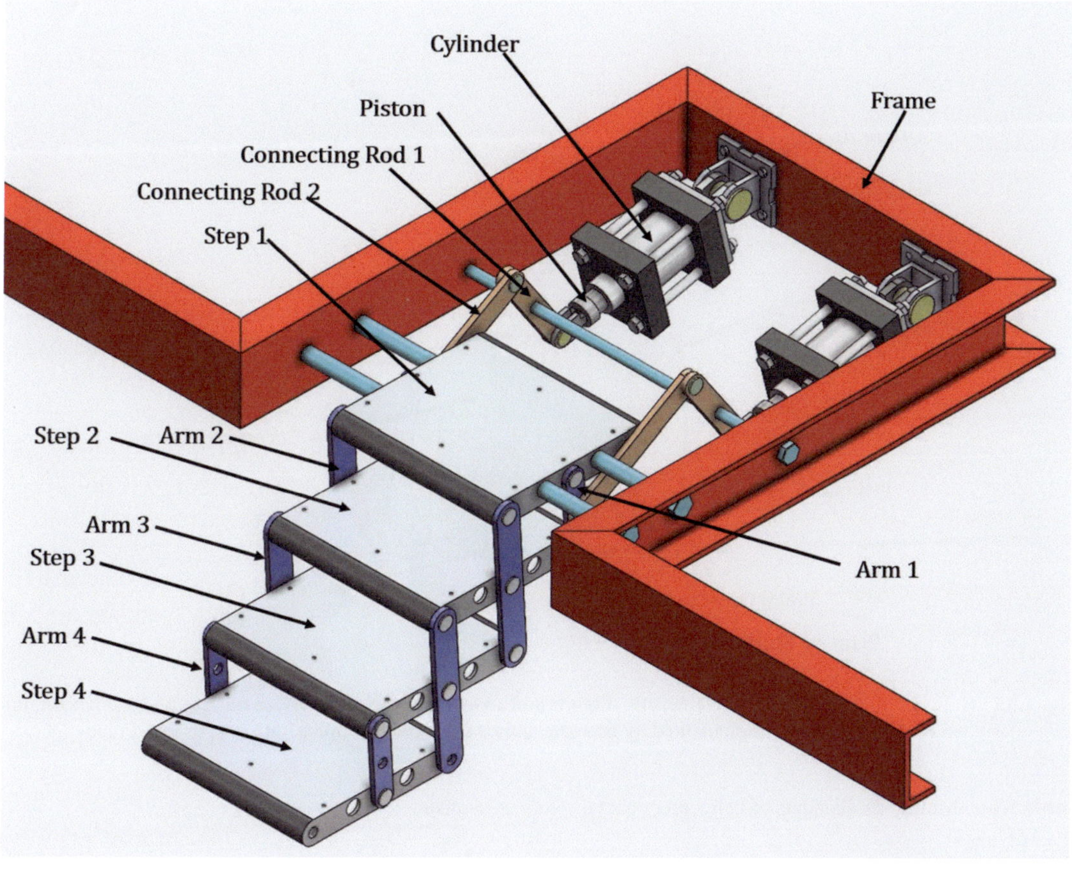

Fig. 1.34 Aircraft foldable stair mechanism. Figure and associated Folding Staircase Mechanism.mp4 video designed by the author and implemented by both the author and Pranav Dhola. Both courtesy of Pranav Dhola

Table 1.25 Multibody elements of aircraft foldable Stair mechanism

Rigid bodies	Mechanical assemblies (M.A)
• Aircraft frame	• $M.A_1$: Between the aircraft frame and the cylinder
• Cylinder	• $M.A_2$: Between the cylinder and the piston
• Piston	• $M.A_3$: Between the piston and the connecting rod 1
• Connecting rods 1	• $M.A_4$: Between the aircraft frame and the connecting rod 1
• Connecting rod 2	• $M.A_5$: Between connecting rod 1 and connecting rod 2
• Four right arms 1, 2, 3, and 4	• From $M.A_6$ to $M.A_{15}$ between the right arms and the steps
• Four left arms 1, 2, 3, and 4	• From $M.A_{16}$ to $M.A_{25}$ between the left arms and the steps
• Four steps 1, 2, 3, and 4	• $M.A_{26}$ between connecting rod 2 and step 1 along with arm 1

saves the space that these steps take up. It also reduces the drag when the aircraft is flying. There is a hydraulic actuator to fold and unfold the stairs. During unfolding, the hydraulic actuator pushes the piston out of the cylinder to unfold the steps through the rotational motions of connecting rods and arms and vice versa. Each arm with a connected step is called a scissor mechanism. In Fig. 1.34, there are four scissor mechanisms (4 steps with four arms on each side).

Helicopter Coaxial Rotors Mechanism: Figure 1.35 shows an example of a helicopter coaxial rotors mechanism as a multibody assembly that consists of the listed elements in Table 1.26.

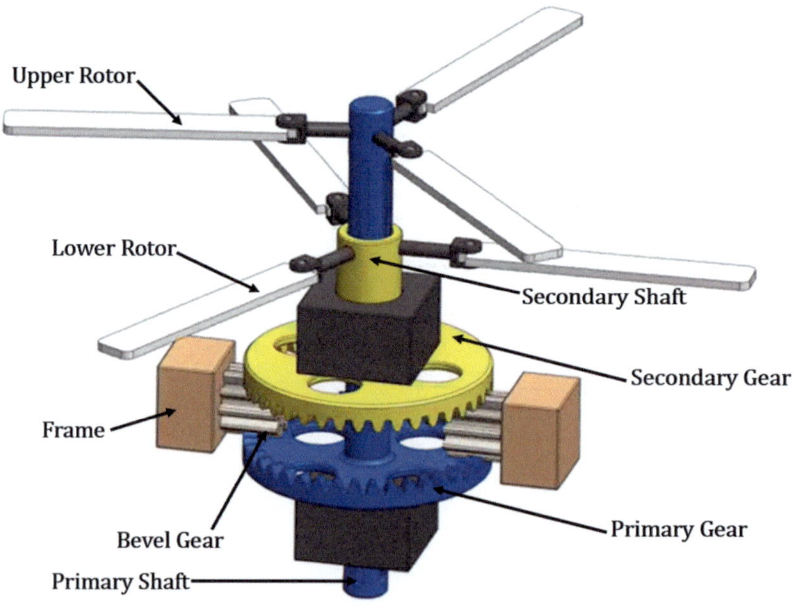

Fig. 1.35 Helicopter coaxial rotors mechanism. Figure and associated Helicopter Coaxial Rotor Mechanism.mp4 video designed by the author and implemented by both the author and Pranav Dhola. Both courtesy of Pranav Dhola

Table 1.26 Multibody elements of helicopter coaxial rotors mechanism

Rigid bodies	Mechanical assemblies (M.A)
• Helicopter frame • Primary shaft holding primary gear and lower rotor • Secondary shaft holding the secondary gear and upper rotor • Two bevel gears	All the mechanical assemblies are gear ones that will be explained in Chap. 10

Some helicopters have coaxial rotors, which are pairs of rotors mounted one above the other on two concentric shafts. The two rotors are contra-rotating: they rotate in opposite directions. The coaxial rotors are designed for more stability and maneuverability, and to produce less noise than the single rotor. Also, a helicopter with coaxial rotors does not need a tail rotor. The lower rotor is set on the primary shaft that is driven from the engine. The upper rotator is set on a secondary shaft. This secondary shaft is driven from the primary shaft through the primary gear, secondary gear, and two bevel gears. There are more aspects to designing coaxial rotors mechanism that requires the kinematics and dynamics of machines.

1.4.5 Other Machines

Skid Steer Bucket Lift Machine

Figure 1.36 shows the skid steer bucket lift machine as a multibody assembly that consists of the listed elements in Table 1.27.

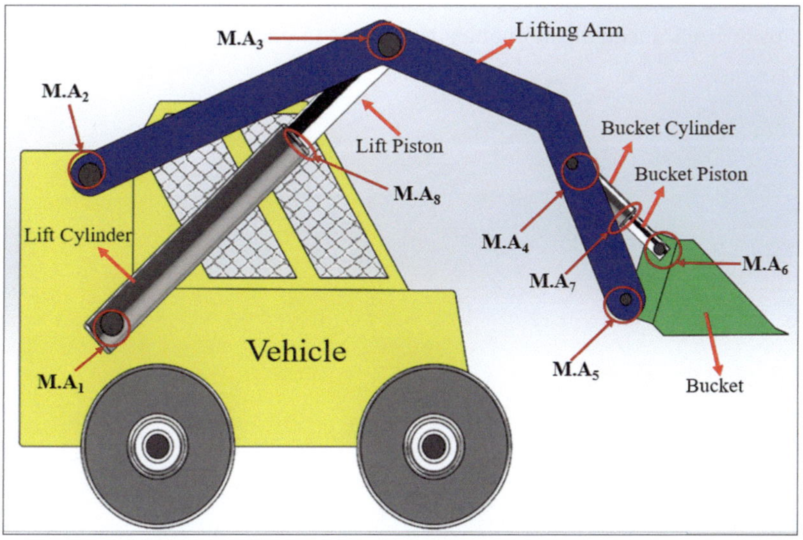

Fig. 1.36 Skid steer bucket lift machine. Figure and associated Skid Steer Articulated Bucket Lift Mechanism.mp4 video designed by the author and implemented by both the author and Pranav Dhola. Both courtesy of Pranav Dhola

Table 1.27 Multibody elements of skid steer bucket lift machine

Rigid bodies	Mechanical assemblies (M.A)
• Vehicle frame	• $M.A_1$: Between the vehicle frame and the lift cylinder
• Lifting arm	• $M.A_2$: Between the vehicle frame and the lifting arm
• Lifting cylinder	• $M.A_3$: Between the lifting arm and the lift piston
• Lifting piston	• $M.A_4$: Between the lifting arm and the bucket cylinder
• Bucket cylinder	• $M.A_5$: Between the lifting arm and the bucket
• Bucket piston	• $M.A_6$: Between the bucket and the bucket piston
• Bucket	• $M.A_7$: Between the bucket piston and the bucket cylinder
	• $M.A_8$: Between the lift piston and the lift cylinder

To lift the bucket to a specific position, there are two hydraulic subsystems in the skid steer bucket lift machine: the lift hydraulic subsystem and the bucket hydraulic subsystem. When the lift hydraulic subsystem pushes the lift piston to slide out from the lift cylinder thanks to $M.A_3$, the lifting arm rotates counterclockwise because of the three mechanical assemblies $M.A_1$, $M.A_2$, and $M.A_3$. The same scenario happens clockwise when the lift hydraulic subsystem pulls the lift piston to slide into the lift cylinder. Rotating the lifting arm counterclockwise or clockwise results in bringing the bucket all the way up or all the way down, respectively. The bucket hydraulic subsystem uses $M.A_4$, $M.A_5$, $M.A_6$, and $M.A_7$ to rotate the bucket counterclockwise or clockwise around the mechanical assembly $M.A_5$. The two rotational motions of the lifting arm and bucket give the driver the ability to orient the bucket in any way to perform the required job. The machine's accessibility depends on the shape and size of each element, and on each mechanical assembly in the skid steer bucket lift mechanism. Note that these two hydraulic subsystems are controlled by switches and joysticks inside the vehicle which are designed to be accessible for the operator's ease of use.

Dumpster Truck

Figure 1.37 shows the dumpster truck mechanism as a multibody assembly that consists of the listed elements in Table 1.28.

Dumpster trucks are commonly used for transporting large volumes of materials inside their beds. The dumpster truck mechanism enables the driver to empty the material out of the bed via a

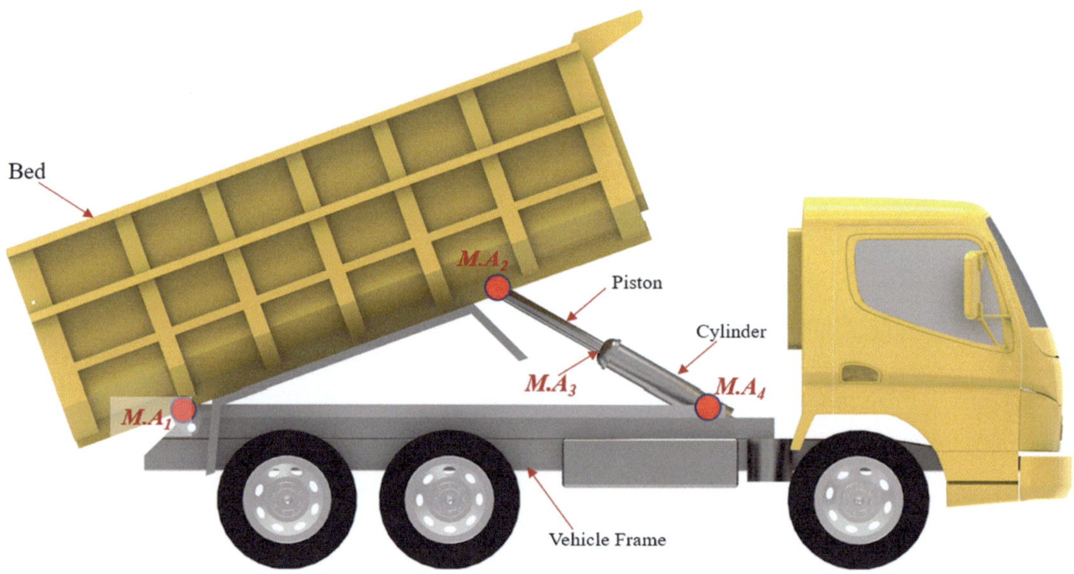

Fig. 1.37 Dumpster truck mechanism. Figure and associated Dumpster Truck Mechanism.mp4 video designed by the author and implemented by both the author and Pranav Dhola. Both courtesy of Pranav Dhola

Table 1.28 Multibody elements of dumpster truck mechanism

Rigid bodies	Mechanical assemblies (M.A)
• Vehicle frame	• $M.A_1$: Between the vehicle frame and the bed
• Bed	• $M.A_2$: Between the bed and the piston
• Cylinder	• $M.A_3$: Between the piston and the cylinder
• Piston	• $M.A_4$: Between the cylinder and the vehicle frame

hydraulic subsystem. The driver has a switch or joystick to control the hydraulic subsystem. When the hydraulic subsystem is asked to push the piston and slide it out from the cylinder, the bed rotates counterclockwise thanks to $M.A_1$, $M.A_2$, $M.A_3$, and $M.A_4$. Then, gravity pushes the material (such as gravel or asphalt) out of the bed, dropping it to the ground. After emptying the bed, the driver commands the hydraulic subsystem to slide the piston inside the cylinder, causing the bed to rotate clockwise and return to its initial position.

Pitcher Water Pump

Figure 1.38 shows the pitcher water pump mechanism as a multibody assembly that consists of the listed elements in Table 1.29.

The pitcher water pump is used to bring water up from the ground. A person lifts the tip of the handle all the way up, in turn rotating it counterclockwise around $M.A_4$. This leads to the piston-valve being pushed all the way down due to the connecting rod acting as a coupler between the handle and the piston-valve thanks to the $M.A_1$, $M.A_2$, and $M.A_3$. The process runs by creating a vacuum that pulls all the water up. The pitcher water pump is one of the basic machines that farmers and other rural people use. Many companies work on improving the pitcher water pump based on kinematic and dynamic models for better water lifting capability.

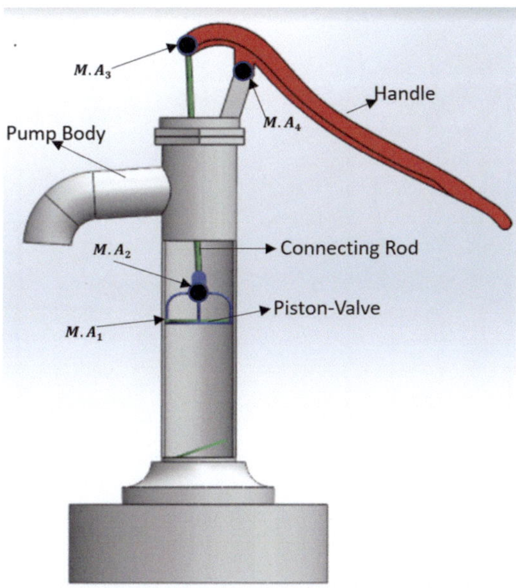

Fig. 1.38 Pitcher water pump mechanism. Figure and associated Hand Pump Mechanism.mp4 video designed by the author and implemented by both the author and Pranav Dhola. Both courtesy of Pranav Dhola

Table 1.29 Multibody elements of pitcher water pump mechanism

Rigid bodies	Mechanical assemblies (M.A)
• Pump body	• $M.A_1$: Between the pump body and the piston-valve
• Piston-valve	• $M.A_2$: Between the piston-valve and the connecting rod
• Connecting rod	• $M.A_3$: Between the connecting rod and the handle
• Handle	• $M.A_4$: Between the handle and the pump body

Sewing Mechanism

Figure 1.39 shows the sewing machine as a multibody assembly that consists of the listed elements in Table 1.30.

The shaft rotation as a part of $M.A_1$ is powered either electrically by a motor or manually. This causes the roll to rotate, moving the needle bar periodically up and down. The needle bar slides through $M.A_3$, which keeps it in rectilinear motion. The coupler connects the roll with the needle bar through $M.A_2$ and $M.A_4$. The sewing mechanism is a neat design to convert rotational motion into a sliding motion that allows a needle to push a thread through a piece of cloth or leather many times in quick succession.

Throughout this textbook, we will return to the above examples as well as explain any newly introduced mechanisms or machines. For example, in Chap. 9, we will introduce a few machines that use the cam mechanism. In Chap. 10, we will introduce some machines that use gears, and so on.

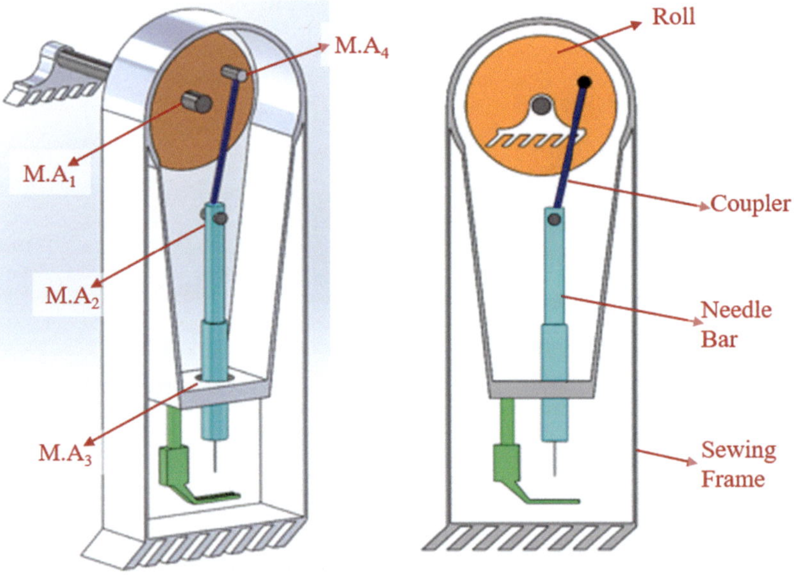

Fig. 1.39 Sewing mechanism. Designed by the author and implemented by both the author and Pranav Dhola. Courtesy of Pranav Dhola

Table 1.30 Multibody elements of sewing machine

Rigid bodies	Mechanical assemblies (M.A)
• Sewing frame	• $M.A_1$: Between the sewing frame and the roll
• Roll	• $M.A_2$: Between the connecting rod and the needle bar
• Coupler	• $M.A_3$: Between the needle bar and the sewing frame
• Needle bar	• $M.A_4$: Between the roll and the connecting rod

1.5 Book Summary and Expected Outcome

This textbook starts by introducing the student to various machines and to the kinematic and dynamic terminologies that engineers use to design and model such machines. This textbook shows different kinematic and dynamic analysis techniques for machine design and modeling, including graphical technique, analytical technique, and model-based simulation techniques. The graphic technique is based on traditional drawing techniques using triangles, parallel straight edges, compasses, and protractors. The drawing can be done at a 1:1 scale but is more often done with a lower scale to fit the drawing page. The graphic method is quite time-consuming. Often, the graphic method's results are not accurate enough to justify its use, since the drawing and measurements rely heavily on hand and eye. However, some educators and students prefer this method since it provides good visibility and more insight into a problem. Recently, the graphical method has been applied through CAD software.

The second method is the analytical method, whereby a mathematical closed-form solution is developed based on physical laws and assumptions. The analytical method is precise, but it becomes more complicated for high-level problems. We introduce MATLAB® scripting to build a program that substitutes the machine design parameters and the actuators' inputs into the analytical solution and plots the results.

Model-based simulation is a technique that's offered by advanced software. We are going to focus on using Simscape™ and Simscape™ Multibody™ in MATLAB® and Simulink®. The model-based simulation carries both the animation component of the graphical technique and the solution component (as values and plots) of the analytical technique. This book consists of twelve chapters.

This chapter introduces kinematics and dynamics. It explains why kinematic and dynamic modeling, analysis, and simulation are important for machine design. We identify and explain the functionality of some common machines found in everyday life, ranging from familiar household appliances to cars. Finally, the chapter summarizes the book's entire outline.

Chapter 2: This chapter gives an understanding of what we mean by linkages, machines, and mechanisms. We look at how to break down the linkage model into basic elements such as point masses, links, joints, and nodes. It provides an understanding of how to identify the ground link in a machine or mechanism. There is an introduction to 2D coordinate systems and different types of motion. The chapter gives an understanding of the degrees of freedom for individual elements. Lastly, there is an introduction to Gruebler's equation and mechanism degrees of freedom.

Chapter 3: This chapter introduces the kinematic diagram of machines, including the basic symbols of various links and joints. It shows that four-bar mechanisms, slider–crank mechanisms, and inverted slider–crank mechanisms are major, closed-chain mechanisms. There is a presentation of Grashof's law to identify if each link in a four-bar mechanism is a crank and rocker. There is also an introduction to five-bar and six-bar mechanisms. The chapter presents special types of mechanisms such as the quick return mechanism. There's an introduction to the systematic breakdown of complex machines into kinematic diagrams, as well as general ways to sketch free body diagrams for common machines in varying applications such as gymnasium equipment and household appliances.

Chapter 4: This chapter serves as an introduction to scripting and functions in MATLAB®. It provides an understanding of how to build a Simulink® model to simulate a mathematical script. We look at understanding the source, sink, and math libraries in Simulink®, as well as how to use Simscape™ Multibody™ with joints, solids, and frames. Lastly, we look at how to model and simulate a machine in Simscape™ Multibody™.

Chapter 5: This chapter explains the position and displacement of mechanisms and machines in both linear and angular terms. The chapter gives an introduction to the process of developing mathematical models of displacement using graphical, analytical, and block diagram techniques. There's also a discussion of the limitation of a machine's movement and the expected coupler curve.

Chapter 6: This chapter explains the absolute and relative velocities of mechanisms and machines in both linear and angular terms. It gives an introduction to the process of developing mathematical models of velocity using graphical, analytical, and block diagrams. It also discusses using instantaneous center techniques to determine the velocity of a node.

Chapter 7: This chapter explains the absolute and relative accelerations of mechanisms and machines in both linear and angular terms. It describes the sliding, normal, tangential, and Coriolis components of acceleration. It also gives an introduction to the process of developing mathematical models of acceleration using graphical, analytical, and block diagram techniques. Finally, jerk rate analysis is covered.

Chapter 8: This chapter gives an introduction to graphically describing input and output models for displacement, velocity, acceleration, and jerk rate (SVAJ) using the block diagram technique. It explains the kinematic study of different kinds of actuators found in common machines and their physical limitations. Finally, mathematical techniques to avoid discontinuities in design trajectory and lessons on generating trajectories for models in Simscape™ Multibody™ with animation are examined.

Chapter 9: This chapter shows different types of cams and cam followers. It defines the configuration parameters and performance parameters of a cam mechanism. We introduce methods for plotting the displacement, velocity, acceleration, and jerk of a cam (SVAJ cam diagram) and how to design a cam mechanism using a specified SVAJ diagram. We provide examples of cam mechanism usage in machines and a method to implement cams in Simscape™ Multibody™.

Chapter 10: This chapter serves as an introduction to various gears used in common machines and the performance parameters of different teeth and mesh configurations. It includes explanations for the kinematic models of simple, compound, and planetary gear trains, as well as an introduction to transmissions and differentials.

Chapter 11: This chapter introduces the equilibrium conditions in terms of forces and moments for the purpose of calculating how to support a machine's weight and inertia load. There is also an explanation of the free body diagram and internal acting forces and moments. We show how to measure equilibrium forces and moments using Simscape™ Multibody™.

Chapter 12: This chapter gives an introduction to the mass, center of gravity, and mass moment of inertia of a single link. We explain the concept of inertial load and how to develop a dynamic model of a machine using analytical equations and Simscape™ Multibody™.

1.6 Summary of the Book's Expected Outcomes

- Understand different types of machines and how they articulate using written explanations, sketches, pictures, and animated videos.
- Learn the kinematic and dynamic terminologies that engineers use to model and design machines.
- Learn graphical, analytical, and model-based techniques to estimate the position, velocity, acceleration, and jerk of any node on a specific machine.
- Learn how to calculate the mass, center of gravity, and mass moment of inertia using hand calculation.
- Learn how to develop static and dynamic models of machines from Newton's laws and using Simscape™ Multibody™ software.
- Learn how to use MATLAB®, Simulink® and Simscape™ Multibody™ for the simulation and animation of machines for a better understanding and characterization of the design concepts of a machine's motion.
- Practice with solved example problems that apply analysis techniques on some of the presented example machines and view the attached animation videos.
- A set of end-of-chapter problems is listed. In these problems, many applications of machines are introduced. The problems are focused on physics, mathematics, and computer-based analyses.

References

1. "A Brief Illustrated History of Machines and Mechanisms (History of Mechanism and Machine Science)," Emilio B. Paz, Marco Ceccarelli, Javier E. Otero, and José L. Muñoz Sanz, Springer, 10th Edition, 2010.
2. "The History Of The Machine," Sigvard Strandh, 1989, 1st Edition.
3. "A Bulleted/Pictorial History of Mechanisms and Machines", R.L. Williams II, Internet Publication, December 2020.
4. www.ohio.edu/mechanical-faculty/williams/html/PDF/HistoryOfMechanisms.pdf
5. "Design of Machinery," Robert L. Norton, McGraw-Hill; 5th Edition, 2011.
6. "Theory of Machines and Mechanisms," John Uicker, Gordon Pennock and Joseph Shigley, Oxford University Press; 5th Edition, 2016.
7. "Machines & Mechanisms: Applied Kinematic Analysis", David Myszka, Pearson; 4 edition, 2011.
8. "Robot Dynamics Algorithms," Roy Featherstone, Springer, 1st Edition, 1987.
9. "Physics for Engineers and Scientists," Hans Ohanian and John Markert, W. W. Norton & Company; 3rd Edition, 2006.
10. "Linkage Mechanism Optimization and Sensitivity Analysis of an Automotive Engine Hood," Onur Denizhan and Meng-Sang Chew, International Journal Of Automotive Science And Technology, 2018, VOL. 2, NO:1, 7–16

2 Mechanism and Kinematics Fundamentals

Chapter Objectives

1. Understanding the differences between linkages, machines, and mechanisms.
2. Understanding how to break down linkages into nodes, links, and joints as basic elements.
3. Understanding how to identify the ground link in a machine or mechanism.
4. Introduction to 2D coordinate systems and different types of motion.
5. Understanding the degrees of freedom for individual elements.
6. Introduction to Grübler's equation and linkage degrees of freedom.

2.1 Introduction

This chapter covers the terminology used in the analysis of the kinematics of machines. Topics include an introduction to the linkage model and the differences between a machine and a mechanism, followed by a discussion of the kinematic elements of any linkage, such as *point mass*, *link*, and *joint*. Finally, a discussion about 2D motion and degrees of freedom is introduced.

2.2 Linkage, Mechanism, and Machine

Definitions

Linkage:

A mechanical linkage is an assembly where bodies are connected to manage forces and motions [1–3].

Supplementary Information The online version contains supplementary material available at https://doi.org/10.1007/978-3-031-72701-6_2.

A. Omran, *Kinematics and Dynamics of Machines*, https://doi.org/10.1007/978-3-031-72701-6_2

Mechanism:

A mechanism is a linkage that has at least one stationary body or is attached to a moveable or stationary frame of reference. The mechanism transforms energy provided by a power source into mechanical work. In this book, we shall study mechanisms that only contain bodies with no deformation, otherwise called rigid links [1–3].

Machine:

A machine is a mechanism or collection of mechanisms that transmit a force across a distance to perform mechanical work [1–3].

Engineering in modern practice is often very complicated and involves many moving or stationary bodies that must work together to perform specific tasks. Take the internal combustion engine in Fig. 2.1 for example. These modern engines are composed of multiple pistons working together to convert combustion energy (burning fuel) into mechanical energy (rotating shaft). At first glance, it may seem to be an intimidating task as everything must work not only as it should but also in reference to other components. The engine shown in Fig. 2.1 is an inline four-cylinder engine that has two pairs of pistons traveling in opposite directions to each other along parallel axes to rotate the crankshaft.

The first step in analyzing such a system is to simplify things by just considering a one-piston–cylinder mechanism (slider–crank mechanism) with a simplified graphical representation as shown in Fig. 2.2. The mechanism in Fig. 2.2 can be broken down into a set of major elements as follows:

- L_1: Cylinder plus crankshaft holder, as a single, stationary rigid body.
- L_2: Piston, as a rigid body, to move up and down inside the cylinder.

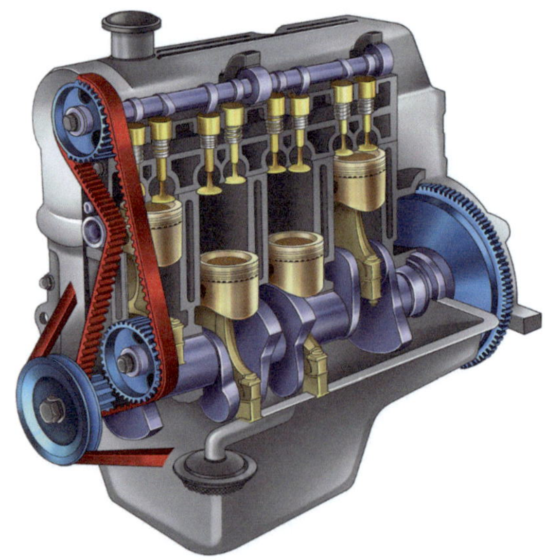

Fig. 2.1 Internal combustion engine as a machine. Designed by the author and implemented by Kenneth Manalansan. Courtesy of Kenneth Manalansan

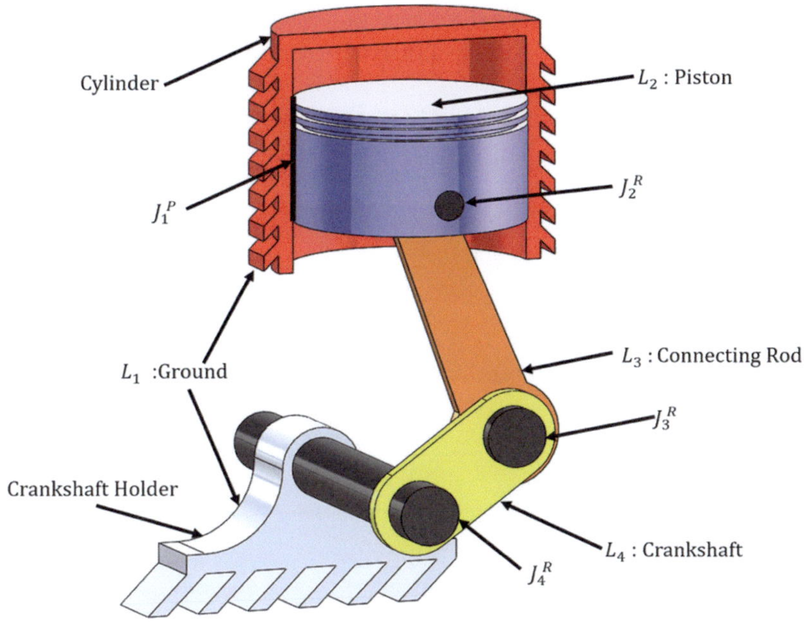

Fig. 2.2 Single Piston–cylinder mechanism. Designed by the author and implemented by both the author and Pranav Dhola. Courtesy of Pranav Dhola

- L_3: Connecting rod, as a rigid body, connecting the piston to the crankshaft.
- L_4: Crankshaft, as a rigid body, to deliver the required mechanical motion.

The above four elements are assumed to be rigid and thus can be called *links*. The symbol L_i refers to link number i in the mechanism. The cylinder is for the fuel combustion process. The crankshaft holder represents a fixed point of rotation for the crankshaft. Neither the cylinder nor the crankshaft holder moves with the local motion of the piston–cylinder mechanism. The two of them are stationary with respect to the local motion of the piston–cylinder mechanism. For that reason, we consider these two elements as a single rigid body L_1 and we may label them as the *engine block*. Additionally, a set of elements that connect to the above four links are called *joints*.

Engineering Notes

Prismatic Joint—A mechanical connection between two links that allows transitional motion along a single axis only. More details are provided in Sects. 2.2.5 and 2.3.5.

Revolute Joint—A mechanical connection between two links that only allows rotational motion about a single axis (a hinge is a classic example). More details are provided in Sects. 2.2.5 and 2.3.5.

In this mechanism, there are:

- J_1^P : Prismatic joint 1, interface between the piston (L_2) and the engine block (cylinder) (L_1).
- J_2^R : Revolute joint 2, interface between the piston (L_2) and the connecting rod (L_3).

- J_3^R : Revolute joint 3, interface between the connecting rod (L_3) and the crankshaft (L_4).
- J_4^R : Revolute joint 4, interface to hold the crankshaft (L_4) to the engine block (crankshaft holder) (L_1).

Where J_i^P refers to the prismatic joint number i in the mechanism and J_i^R refers to the revolute joint number i in the mechanism. We will return to this example later in this book but first, we need to understand the key aspects of how to analyze and design this system or any other machine or mechanism. Now, we are going to study the major mechanism's kinematic elements which are broken down into two categories known as *links* and *joints*.

2.2.1 Rigid Body

Definition

Rigid Body:

Physically speaking, a rigid body is an idealization of a solid body in which deformation is neglected. Mathematically speaking, the distance between any two points on a rigid body remains constant regardless of any force being applied [4].

The definition of a rigid body is an assumption for an ideal case. Engineers use rigid bodies to simplify physical and mathematical analyses. In some applications, the deformation of a body can be neglected as it will not impact the design goal. In other cases, when the deformation cannot be ignored, the system will be considered a deformable body.

Figure 2.3 shows an example of two similar mechanisms made from two different materials. The gray mechanism has no noticeable deformation after applying a force and/or a torque while the red one has a noticeable change in the shape of each link. The gray mechanism is considered a rigid body mechanical assembly and the red mechanism is considered a deformable one.

In general, a rigid body can be defined as a collection of points whose relative distance from one another does not change with any external forces or movements. In this book, we are only going to cover rigid body elements (links and joints). When considering a rigid body, we need to identify the points of interest that can connect a joint/link or an end-effector (input node or output node). We call these points of interest *nodes*. If there is just one point of interest, we consider the rigid body to be a ***point mass***. When the rigid body has at least two nodes, we consider it a *link*.

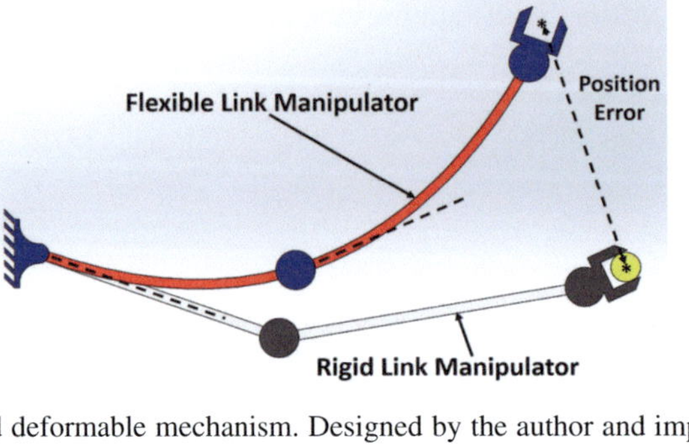

Fig. 2.3 Rigid mechanism and deformable mechanism. Designed by the author and implemented by both the author and Pranav Dhola. Courtesy of Pranav Dhola

2.2.2 Point Mass

Definition

Point Mass:

A *point mass* is an abstract *point* in space where the entire *mass* of the object has been concentrated. The moment of inertia or the mass distribution is not considered in the case of point mass [5].

Figure 2.4 provides an example of the revolution of the earth around the sun. The earth–sun model has one major gravitational force emitting from the sun's core. The earth moves as one body around the sun in an elliptical or semi-circular orbit. Since the earth can be considered one semi-spherical body, we can reduce this system to a single point which is the center of the earth's core. The system can thus be considered as two point masses: one for the earth and another for the sun, each found at their respective centers of gravity. We can apply the same technique when we model the satellite maneuver around the earth. Many other examples can also be considered as a point mass such as magnetic ball levitation.

2.2.3 Link

Definition

Link:

Each rigid body in a mechanism that moves relative to another rigid body is called a kinematic link. A rigid body does not undergo deformation while transmitting force [1–3].

When a rigid body has more than one node, we call the rigid body a *link*. A link is made up of at least two nodes, one being the center of gravity of the rigid body and any additional point attached

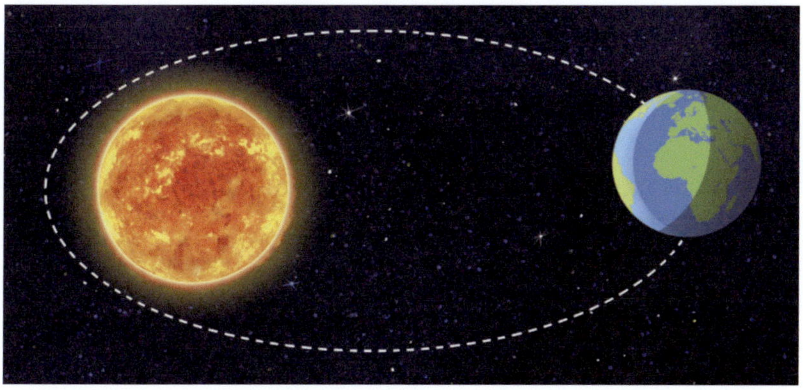

Fig. 2.4 Earth–sun motion. Designed by the author and implemented by Kenneth Manalansan. Courtesy of Kenneth Manalansan

to another link or any other point of interest on the rigid body such as geometry. For example, in Fig. 2.2, we can see the connecting rod as a rigid body that moves relative to both the piston and the crankshaft. The connecting rod has two points of interest located at revolute joint 2 (J_2^R) and revolute joint 3 (J_3^R). This connecting rod does not undergo deformation while the piston is pushing the rod up and down. We consider the connecting rod to be a *link* because it satisfies all the conditions for a link. Similar examples can be applied to the piston, crankshaft, and cylinder.

2.2.4 Ground Link or Fixed Link

Definition

Ground Link or Fixed Link:

A ground link is a set of nodes that are connected to the mechanism but do not move with the mechanism's motion. These nodes are considered as a fixed connection to the mechanism.

For motion to occur, there needs to be a link that remains stationary relative to the other links. Mechanisms often have a ground link that is not necessarily stationary with respect to a global coordinate system; however, the ground link can be regarded as a fixed local coordinate system to facilitate analysis of the mechanism. You can think of these links as the ground that we walk on every day. Imagine a playground with a swing set and a child swinging as shown in Fig. 2.5. You will notice that the frame of the swing set does not move with the child. The swing seat and the child move together in relation to the frame by means of the joint that connects the two. Because the frame is attached to the ground with fixed or welded joints, we can say that the ground and the swing set frame are the same link because no motion is conveyed.

Fig. 2.5 Swing set. Designed by the author and implemented by Kenneth Manalansan. Courtesy of Kenneth Manalansan

This *ground* idea is key to making any mechanism work and there will always be at least one connection to the ground in every mechanism. For example, the piston–cylinder mechanism in Fig. 2.2 has two relatively stationary rigid bodies (the cylinder and the crankshaft holder). These two rigid bodies do not participate in the mechanism's local motion. For that reason, we merge the cylinder and the crankshaft holder to be Link 1 (L_1) which is the fixed or ground link of the piston–cylinder mechanism. We will talk more about grounds later when we discuss kinematic diagrams, but it is important to understand how motion in a mechanism can be achieved.

2.2.5 Joint

Definition

Joint:

A joint is a connection between two links (at a common node) that allows relative motion between them [1–3].

A mechanism performs its desired task through the relative motion of different links. The interface that connects two links is called a *joint*. To clarify our understanding of the joint, let's look at the medical diagram of a human arm, as shown in Fig. 2.6. The arm can be thought of as a simple mechanism despite being a complex anatomical system.

We can identify two of the major links—humerus and ulna—connected by a joint, the elbow. For the case in Fig. 2.6, the elbow not only connects the humerus and ulna but also gives some freedom for the ulna to move with respect to the humerus. The job of this joint (elbow) is to translate motion and force from one link to another in a base–follower array. For our analysis of machines, we assume the connection interface is one element that we will call a joint. Additionally, we will not be concerned with force until Chaps. 11 and 12.

2.2.6 Kinematic Pair and Kinematic Chain

Definitions

Kinematic Pair:

A kinematic pair consists of two rigid links connected by an ideal single joint [1–3].

Kinematic Chain:

A kinematic chain is a linkage modeled as a network of rigid links and ideal joints [1–3].

A mechanical linkage is an assembly of links connected through joints to manage forces and movement. The linkage needs at least two links and one joint to function. Each set of two links connected with a joint is called a *kinematic pair*. When the linkage can be modeled as an array of kinematic pairs, it is called a *kinematic chain*.

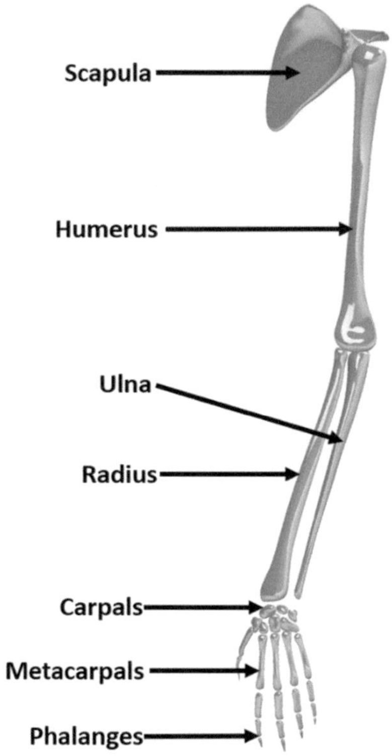

Fig. 2.6 Human arm machine. Designed by the author and implemented by both the author and Pranav Dhola. Courtesy of Pranav Dhola

Kinematic Pair

Figure 2.2 shows a kinematic chain that can be broken down into kinematic pairs as shown in Fig. 2.7. These kinematic pairs are:

1. ***Kinematic Pair 1***
   The crankshaft Link4 L_4, the ground Link1 L_1 (crankshaft holder), and the revolute joint J_4^R form the first kinematic pair.
2. ***Kinematic Pair 2***
   The piston Link2 L_2, ground Link1 L_1 (cylinder), and prismatic joint J_1^P form the second kinematic pair.
3. ***Kinematic Pair 3***
   The connecting rod Link3 L_3, the crankshaft Link4 L_4 and the revolute joint J_3^R form the third kinematic pair.
4. ***Kinematic Pair 4***
   The piston Link2 L_2, the connecting rod Link3 L_3, and the revolute joint J_2^R form the fourth kinematic pair.

Closed Chain and Open Chain

There are two categories of kinematic chains—open kinematic chains and closed kinematic chains. Figure 2.8 shows an example of closed kinematic chains that consist of

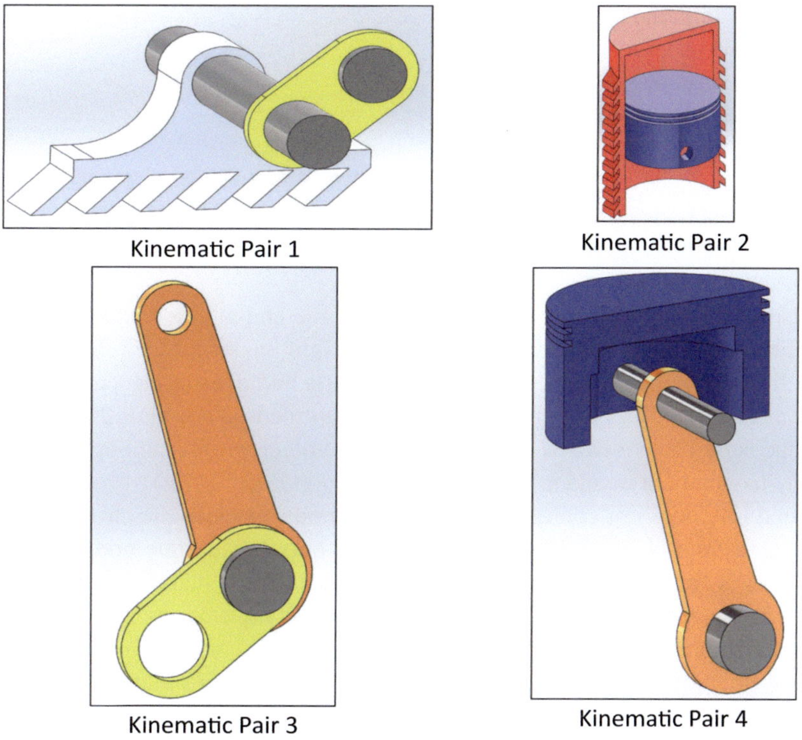

Fig. 2.7 Kinematic pairs of the piston–cylinder mechanism. Designed by the author and implemented by both the author and Pranav Dhola. Courtesy of Pranav Dhola

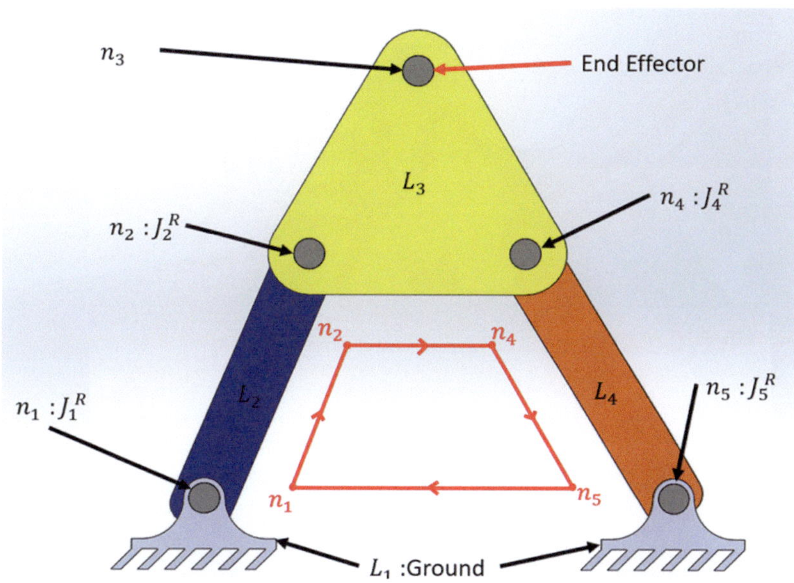

Fig. 2.8 Example of a closed kinematic chain. Designed by the author and implemented by both the author and Pranav Dhola. Courtesy of Pranav Dhola

1. Ground Link1 L_1 with the two nodes n_1 and n_5.
2. Link2 L_2 with two nodes n_1 and n_2.
3. Link3 L_3 with three nodes, n_2, n_3, and n_4.
4. Link4 L_4 with two links n_4 and n_5.
5. Revolute joint J_1^R at node n_1 between Link1 L_1 and Link2 L_2.
6. Revolute joint J_2^R at node n_2 between Link2 L_2 and Link3 L_3.
7. Revolute joint J_4^R at node n_4 between Link3 L_3 and Link4 L_4.
8. Revolute joint J_5^R at node n_5 between Link4 L_4 and Link1 L_1.

This mechanism's purpose is to manage the motion of the end-effector node n_3 that is located on the end-link Link L_3. There is a closed loop formed by rigid links L_1, L_2, L_3, and L_4, which are connected by the links n_1, n_2, n_4, and n_5. Since the loop starts and ends at the same link (ground link), we call this a *closed loop*. The end link has two nodes located on the closed loop. For that reason, we call this kinematic chain a *closed kinematic chain*. In other words, although n_3 is free, this node is controlled by having both n_2 and n_4 inside the closed loop.

When the end link does not connect by at least two nodes back in the closed loop, it is referred to as an *open kinematic chain*. Figure 2.9 shows L_5 as only having one node n_3 connected to the chain and node n_5 is free from any connection. It should be noted that n_5 is still important because it describes the link's geometry. In this case, L_5 is not part of any closed kinematic chain and is considered an open kinematic chain.

In the real world, most robotic machines are open kinematic chains along with other systems like cranes, while most manufacturing machinery is usually closed kinematic chains. The main concern about the open kinematic chain is the flexibility of the end link which may lead to inaccurate motion control.

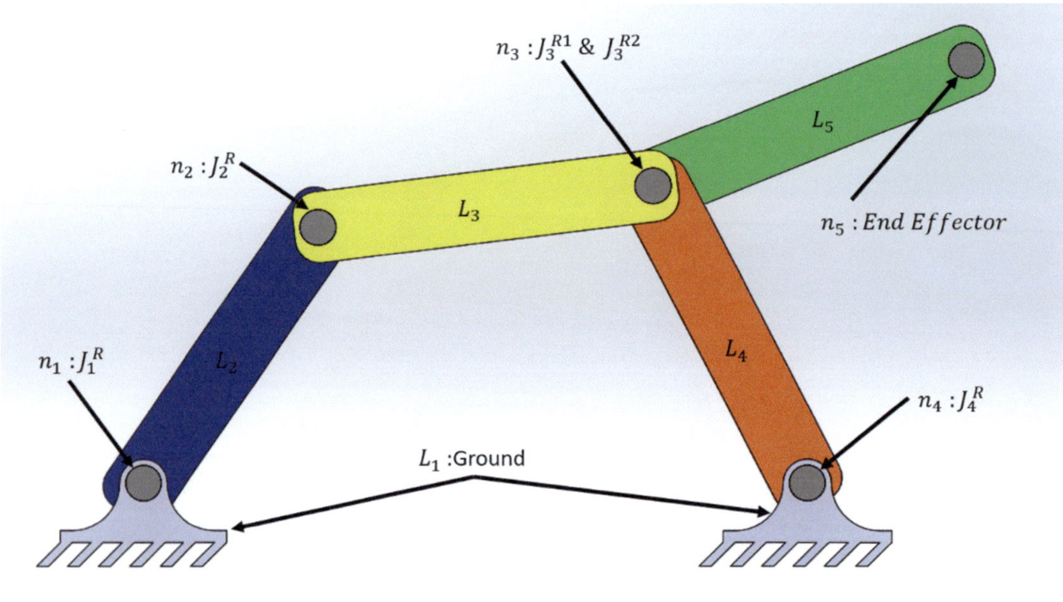

Fig. 2.9 Example of an open kinematic chain. Designed by the author and implemented by both the author and Pranav Dhola. Courtesy of Pranav Dhola

Engineering Tips–Coincident Joints

In Fig. 2.9, joint J_3^{R1} and joint J_3^{R2} are coincident in the same location. We consider them as two separate joints even though they are at the same location or the same node. The same case can be repeated in some other machines or mechanisms. Again, the definition of a joint is an interface between two links only.

Example 2.1

For the **door holder linkage** shown in Fig. 2.10, specify:

1. Links and specify the ground link.
2. Joints.
3. Nodes.
4. Kinematic pairs.
5. Closed or open chain.

Fig. 2.10 Door holder linkage. Designed by the author and implemented by both the author and Pranav Dhola. Courtesy of Pranav Dhola

Solution

As shown in Fig. 2.11

1. Four links:
   - Link1 L_1, the door frame that consists of the wall plus the blue arm. This link is considered the ground link.
   - Link2 L_2, the brown door body.
   - Link3 L_3, the green arm.
   - Link4 L_4, the yellow arm.

2. Four nodes between every two links:
   - Node n_1 between L_1 and L_2.
   - Node n_2 between L_2 and L_3.
   - Node n_3 between L_3 and L_4.
   - Node n_4 between L_4 and L_1.
   - Node n_5 as the end-effector of L_2.
3. Four revolute joints
   - Revolute joint J_1^R between L_1 and L_2.
   - Revolute joint J_2^R between L_2 and L_3.
   - Revolute joint J_3^R between L_3 and L_4.
   - Revolute joint J_4^R between L_4 and L_1.
4. Four kinematic pairs
   - Pair 1, L_1 connected to L_2 through J_1^R.
   - Pair 2, L_2 connected to L_3 through J_2^R.
   - Pair 3, L_3 connected to L_4 through J_3^R.
   - Pair 4, L_4 connected to L_1 through J_4^R.
5. The mechanism is a closed chain because the end link L_2 has two nodes n_1 and n_2 inside the closed loop passing by n_1, n_2, n_3, and n_4.

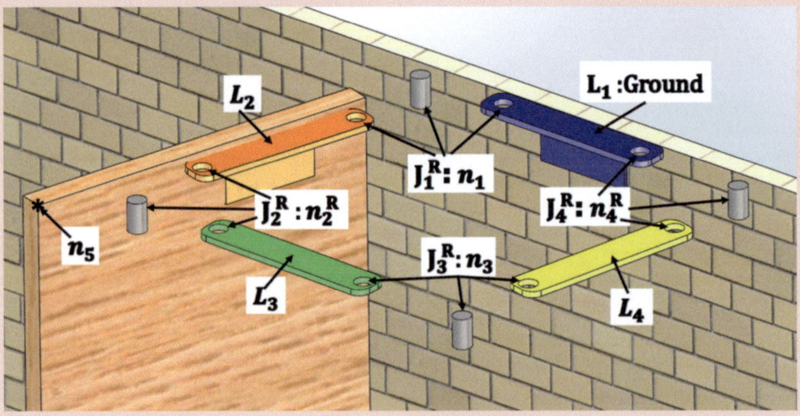

Fig. 2.11 Free body diagram of a door holder linkage. Designed by the author and implemented by both the author and Pranav Dhola. Courtesy of Pranav Dhola

Example 2.2

For the **car window lift machine** shown in Figs. 2.12 and 2.13, specify the nodes, links, and joints. Additionally, identify the ground link and the ground nodes.

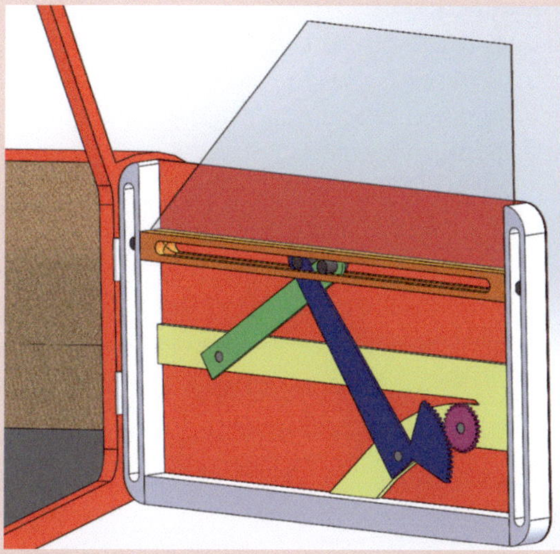

Fig. 2.12 Car window lift machine (3D). Figure and associated Car Window Mechanism.mp4 video designed by the author and implemented by both the author and Pranav Dhola. Both courtesy of Pranav Dhola

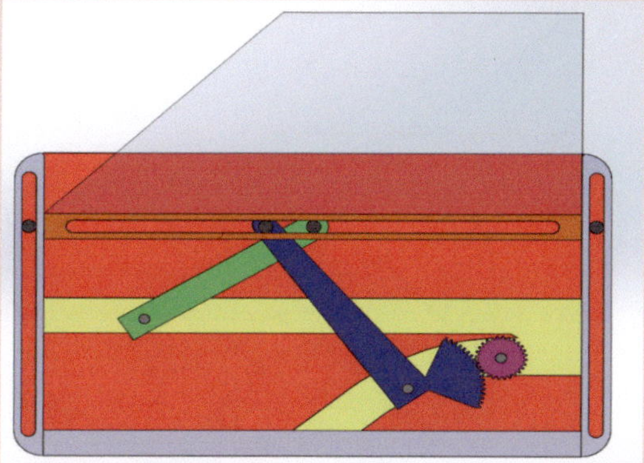

Fig. 2.13 Car window lift machine (2D). Figure and associated Car Window Mechanism.mp4 video designed by the author and implemented by both the author and Pranav Dhola. Both courtesy of Pranav Dhola

Solution

As shown in Fig. 2.14, the mechanism is broken down into

1. Five links:
   - Link1 L_1, the door red frame that consists of the wall plus the gray arm and the yellow arms. Link1 L_1 is the ground link.
   - Link2 L_2, the blue gear rod.
   - Link3 L_3, the green connecting rod.
   - Link4 L_4, the purple gear.
   - Link5 L_5, the glass window plus the orange arm.

2. Eight joints:
   - Revolute joint J_1^R between L_1 and L_2.
   - Revolute joint J_2^R between L_1 and L_3.
   - Revolute joint J_3^R between L_1 and L_4.
   - Gear joint J_4^G between L_2 and L_4.
   - Pin-in-a-slot joint J_5^S between L_2 and L_5.
   - Pin-in-a-slot joint J_6^S between L_3 and L_5.
   - Pin-in-a-slot joint J_7^S between L_1 and L_5.
   - Pin-in-a-slot joint J_8^S between L_1 and L_5.

Note that the two pin-in-a-slot joints J_7^S and J_8^S can be replaced by one prismatic joint.

3. Ten nodes:
   - Node n_1 between L_1 and L_2.
   - Node n_2 between L_1 and L_3.
   - Node n_3 between L_1 and L_4.
   - Node n_4 between L_2 and L_4.
   - Node n_5 between L_2 and L_5.
   - Node n_6 between L_3 and L_5.
   - Node n_7 between L_1 and L_5.
   - Node n_8 between L_1 and L_5.
   - Node n_9 end-effector on L_5.
   - Node n_{10} end-effector on L_5.

   All located nodes (n_1, n_2, n_3, n_7, and n_8) on L_1 are ground nodes.

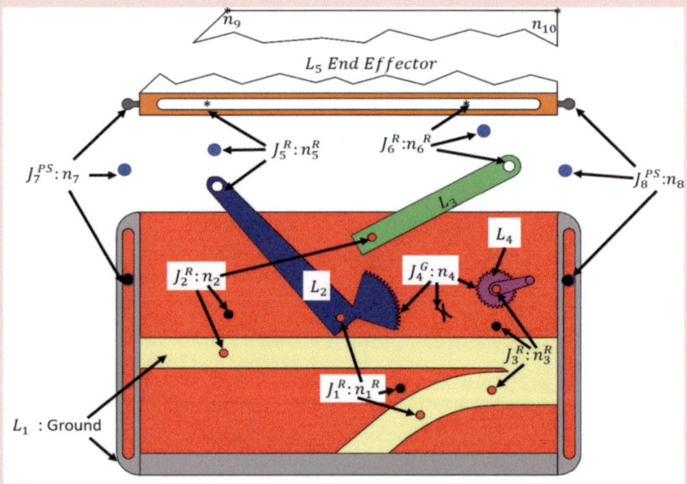

Fig. 2.14 Free body diagram of a car window lift machine. Figure and associated Car Window Mechanism.mp4 video designed by the author and implemented by both the author and Pranav Dhola. Both courtesy of Pranav Dhola

Example 2.3

For the shown **excavator lift machine** in Fig. 2.15, specify the linkage nodes, links, and joints. Additionally, identify the ground link and the ground nodes.

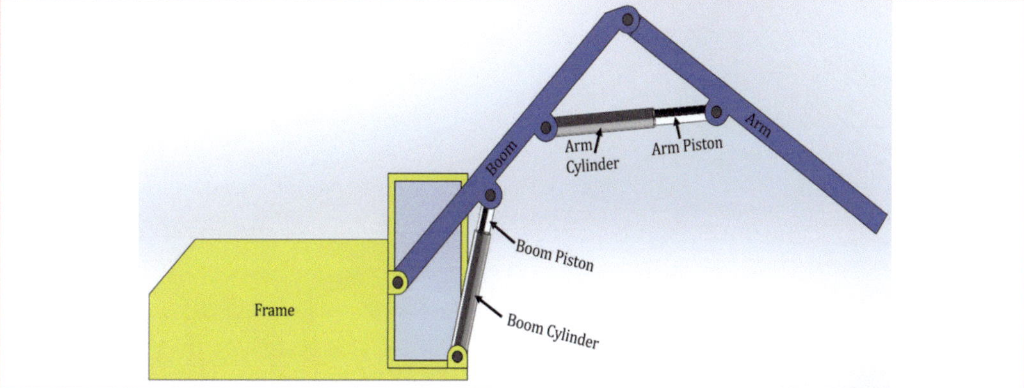

Fig. 2.15 Excavator lift machine. Designed by the author and implemented by both the author and Pranav Dhola. Courtesy of Pranav Dhola

Solution

As shown in Fig. 2.16, the mechanism can be broken down into

1. Seven links:
   - Link1 L_1, the yellow excavator frame that is considered as the ground link.
   - Link2 L_2, the blue boom.
   - Link3 L_3, the blue arm.
   - Link4 L_4, the black boom cylinder.
   - Link5 L_5, the gray boom piston.
   - Link6 L_6, the black arm cylinder.
   - Link7 L_7, the gray arm piston.

2. Eight joints:
   - Revolute joint J_1^R between L_1 and L_4.
   - Revolute joint J_2^R between L_1 and L_2.
   - Revolute joint J_3^R between L_2 and L_5.
   - Revolute joint J_4^R between L_2 and L_6.
   - Revolute joint J_5^R between L_2 and L_3.
   - Revolute joint J_6^R between L_3 and L_7.
   - Prismatic joint J_7^P between L_4 and L_5.
   - Prismatic joint J_8^P between L_6 and L_7.

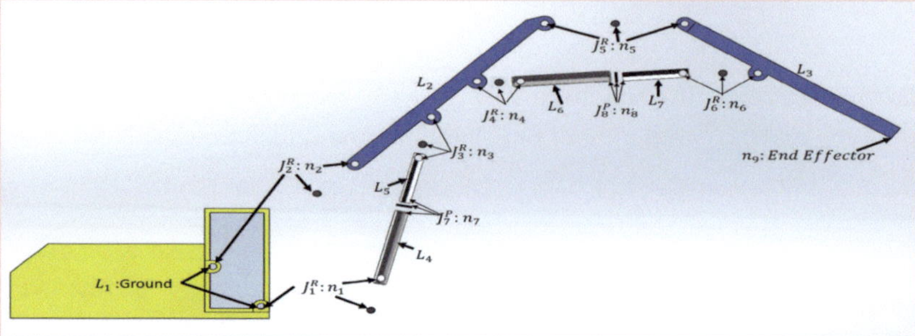

Fig. 2.16 Free body diagram of an excavator lift machine. Designed by the author and implemented by both the author and Pranav Dhola. Courtesy of Pranav Dhola

3. Nine nodes:
   - Node n_1 between L_1 and L_4.
   - Node n_2 between L_1 and L_2.
   - Node n_3 between L_2 and L_5.
   - Node n_4 between L_2 and L_6.
   - Node n_5 between L_2 and L_3.
   - Node n_6 between L_3 and L_7.
   - Node n_7 between L_4 and L_5.
   - Node n_8 between L_6 and L_7.
   - Node n_9 between end-effector on L_3.

   All located nodes (n_1 and n_2) on L_1 are ground nodes.

Example 2.4

Specify whether the **three-arm robot** linkage in Fig. 2.17 is a closed kinematic chain or an open kinematic chain. Explain the reasoning behind your decision.

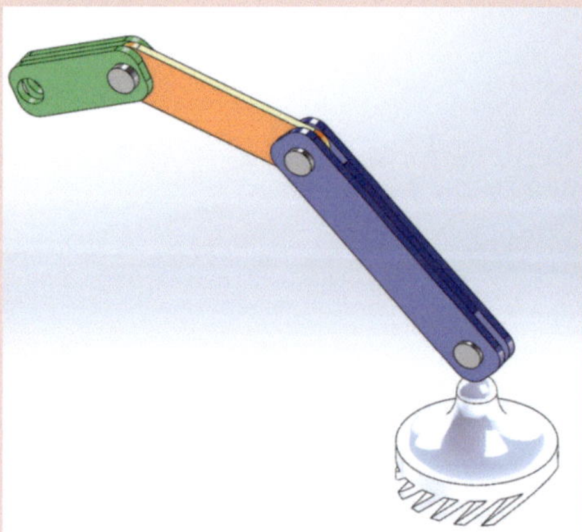

Fig. 2.17 Three-arm robot. Designed by the author and implemented by both the author and Pranav Dhola. Courtesy of Pranav Dhola

Solution

As shown in Fig. 2.18, the mechanism has four links: L_1 (ground), L_2, L_3, and L_4. There are three revolute joints: J_1^R, J_2^R, and J_3^R. There are four nodes: n_1, n_2, n_3, and n_4. The mechanism is an open chain since the end-effector link L_4 is connected to the rest of the mechanism only by node n_3.

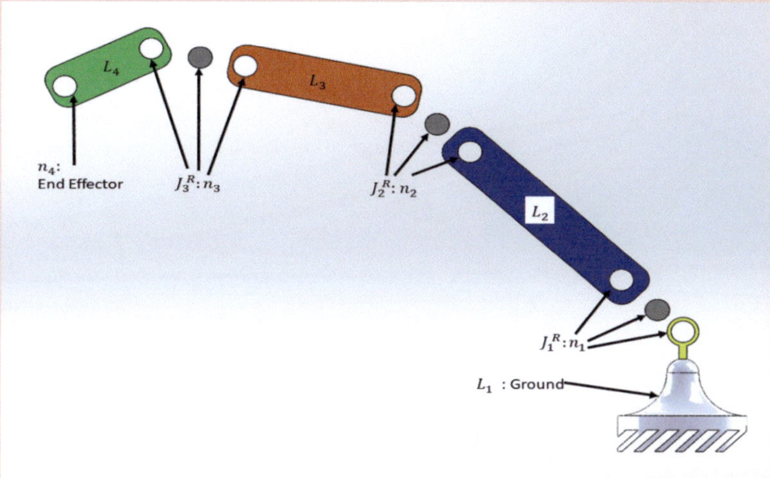

Fig. 2.18 Free body diagram of three-arm robot. Designed by the author and implemented by both the author and Pranav Dhola. Courtesy of Pranav Dhola

Fig. 2.19 Skid steer bucket lift machine. Figure and associated Skid Steer Articulated Bucket Lift Mechanism.mp4 designed by the author and implemented by both the author and Pranav Dhola. Both courtesy of Pranav Dhola

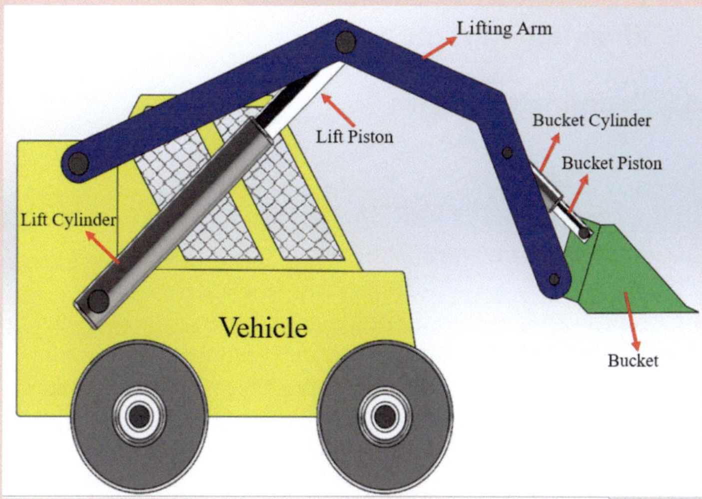

Example 2.5

Specify whether the **skid steer bucket lift machine**, shown in Fig. 2.19, is a closed kinematic chain or an open kinematic chain. Explain the reasoning behind your decision.

Solution

As shown in Fig. 2.20, the mechanism has seven links: L_1 (ground), L_2, L_3, L_4, L_5, $L_{6,}$ and L_7. There are eight joints: J_1^R, J_2^R, J_3^R, J_4^R, J_5^R, J_6^R, J_7^P, and J_8^P. There are nine nodes: n_1, n_2, n_3, n_4, n_5, n_6, n_7, $n_{8,}$ and n_9. The mechanism is a closed chain since the end-effector link L_7 is connected to the rest of the mechanism by two nodes n_5 and n_6.

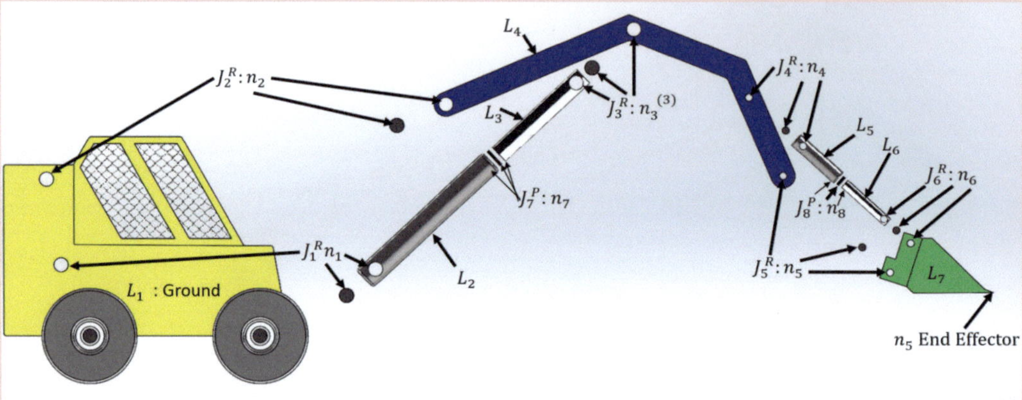

Fig. 2.20 Free body diagram of a skid steer bucket lift machine. Figure and associated Skid Steer Articulated Bucket Mechanism.mp4 video designed by the author and implemented by both the author and Pranav Dhola. Both courtesy of Pranav Dhola

2.3 Kinematic Motion

2.3.1 Coordinate Systems

Definition

Coordinate System:

A coordinate system is an arbitrary set of generally orthogonal fixed axes, each corresponding to a single variable that uses one or more numbers, or coordinates, to uniquely determine the position of the points or other geometrics elements. This system can be set up as either single or multi-dimensional with two and three-dimensional systems being the most common [1–3].

A coordinate system uniquely and numerically determines the position of a point. For kinematic analysis, we usually use planar or flat surface (two-dimensional or 2D) coordinate systems, or spatial (three-dimensional or 3D) coordinate systems. Figure 2.21 shows the difference between a 2D domain and a 3D domain. Cartesian and polar coordinate systems are the most common reference coordinate systems used in 2D kinematics analysis.

In 2D, the Cartesian coordinate system consists of a pair of perpendicular lines (x-axis and y-axis) on a plane. The intersection point of the x-axis and y-axis is called the ***origin***. The axes are usually drawn to determine the location of any point as shown in Fig. 2.22. A point in the plane with coordinates **(*a*, *b*)** is ***a*** units to the right of the y-axis and ***b*** units up from the x-axis if ***a*** and ***b*** are positive

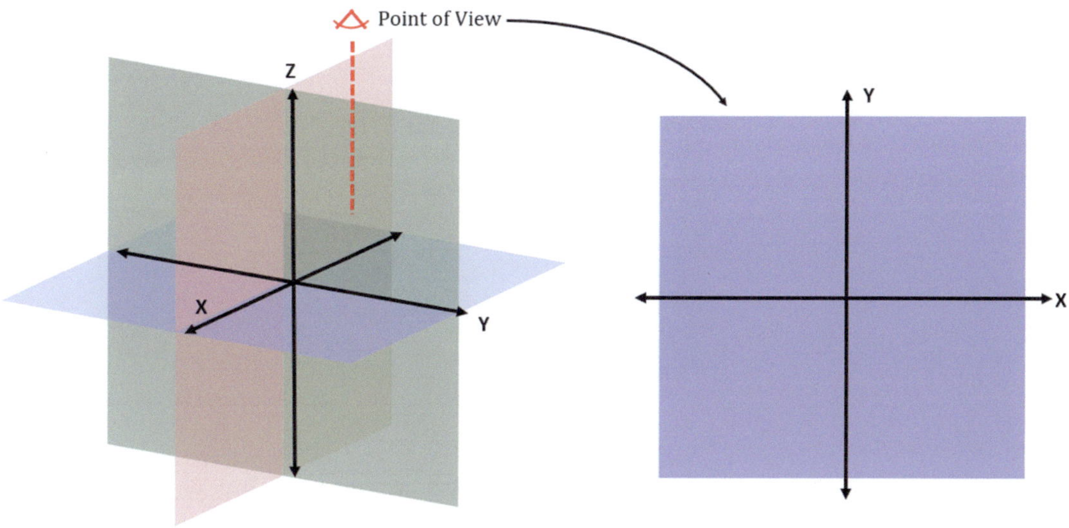

Fig. 2.21 2D and 3D spaces. Designed by the author and implemented by both the author and Pranav Dhola. Courtesy of Pranav Dhola

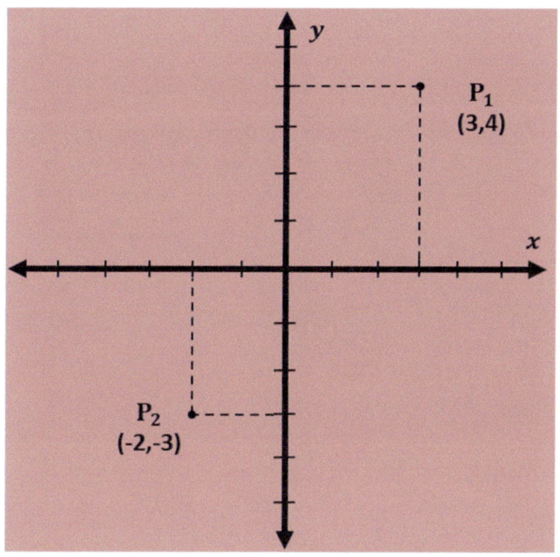

Fig. 2.22 Cartesian coordinate in 2D

numbers. As shown in Fig. 2.22, P_1 has coordinates (3, 4) and P_2 has coordinates (−1, −3). In 3D, the z-axis is added so that there are three axes all perpendicular to one another as shown in Fig. 2.23.

For a point on the plane, we can also use the polar coordinate system where the r-coordinate is the distance (always positive) from the point to the pole, and the θ-coordinate is the counterclockwise angle (from 0° to 360°) between the x-axis and a line joining the origin to the point as shown in Fig. 2.24. The cylindrical and spherical coordinate systems are the 3D extension to the polar coordinate systems. This book will focus on 2D analysis—we will not cover spherical or cylindrical coordinate systems.

Equation 2.1 represents the mathematical transformation from Cartesian to polar:

$$\begin{aligned} x &= r\cos(\theta) \quad y = r\sin(\theta) \\ r &= \sqrt{x^2 + y^2} \; \theta = \tan\left(\tfrac{y}{x}\right) \end{aligned} \tag{2.1}$$

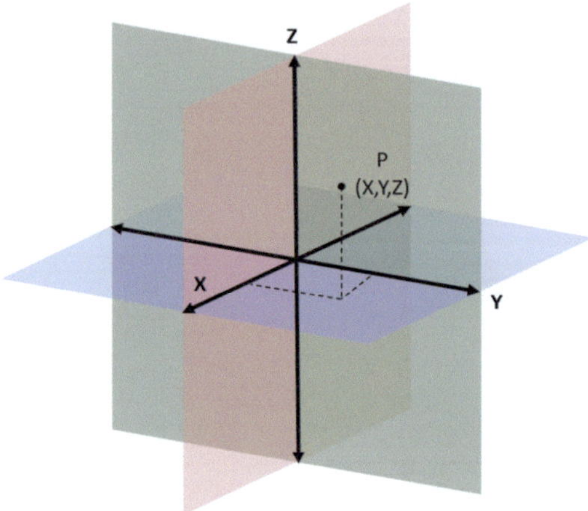

Fig. 2.23 Cartesian coordinate in 3D. Designed by the author and implemented by both the author and Pranav Dhola. Courtesy of Pranav Dhola

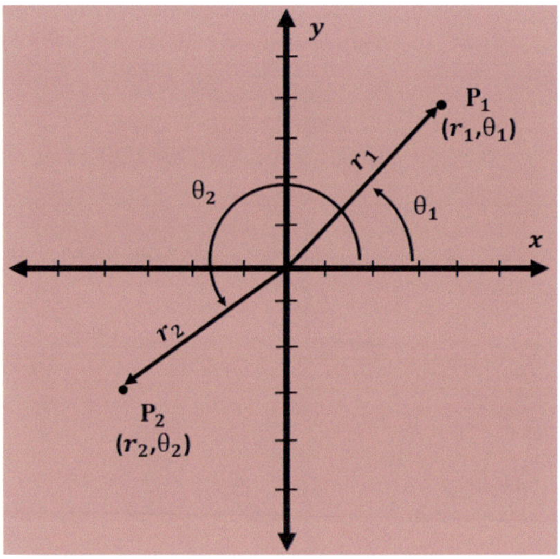

Fig. 2.24 Polar coordinate system

Engineering Tips—A Helping Hand

A word concerning 3D domains:

It is accepted practice in the industry to use coordinate systems that follow a specific ordination of the axes (*X*, *Y*, *Z*). Technically, you can define a coordinate system any way you like, but this becomes complicated when working with other designers or engineers who use a different system. Hence, a standard model (3D space shown previously) is used.

The right-hand rule, known as John Ambrose Fleming's right-hand rule, can be used as a reminder of how the axes should look as shown in Fig. 2.25. Take your index finger (on your right hand) and point straight ahead. Next, take your middle finger and point it to the left so it is perpendicular to your index finger. Stick your thumb straight up as if you were giving a thumbs up to a fellow engineer. If your index finger is the *x*-axis and your middle finger is the *y*-axis, then your thumb is in the direction of the *z*-axis. Silly, but a very useful trick.

Fig. 2.25 Fleming's right-hand rule. Designed by the author and implemented by both the author and Pranav Dhola. Courtesy of Pranav Dhola

2.3.2 Types of 2D Motion

Definition

Motion:

Motion is the change in a node's position versus time and is often abbreviated to $x(t)$ where x is the variable in question and t is time. Velocity is the rate of change of the position with respect to time. Acceleration is the rate of change of velocity with respect to time [1–3].

In 2D, motion can be itemized as

1. Translational Motion: Represented by variation in $x(t)$ and $y(t)$.
2. Rotational Motion: Represented by variation in $\theta(t)$.
3. Complex Motion (Translational + Rotational): Represented by variation in $x(t)$, $y(t)$ and $\theta(t)$.

2.3.2.1 Point Mass Motion

For the point mass, the motion is represented as the variation of this point's position or location with time. This variation is called *trajectory* or *position trajectory*. In the case of a 2D Cartesian system, the variation is expressed as $x_p(t)$ and $y_p(t)$, where x_p is the instantaneous x-coordinate of the point mass, y_p is the instantaneous y-coordinate of the point mass, and t is instantaneous time.

Velocity is the rate of change of position with respect to time. In 2D Cartesian, the point's velocity is defined by $v_{xp}(t)$ and $v_{yp}(t)$, where v_{xp} is the instantaneous velocity in the x-coordinate direction of the point mass, v_{yp} is the instantaneous velocity in the y-coordinate direction of the point mass and t is instantaneous time.

Acceleration is the rate of change of velocity with respect to time. In 2D Cartesian, the point's acceleration is defined by $a_{xp}(t)$ and $a_{yp}(t)$, where a_{xp} is the instantaneous acceleration in the x-coordinate direction of the point mass, a_{yp} is the instantaneous velocity in the y-coordinate direction of the point mass and t is instantaneous time. The velocity and acceleration can be expressed as [1–3]:

$$\begin{aligned} v_{xp}(t) &= \frac{\mathrm{d}x_p(t)}{\mathrm{d}t} && \text{and } v_{yp}(t) = \frac{\mathrm{d}y_p(t)}{\mathrm{d}t} \\ a_{xp}(t) &= \frac{\mathrm{d}v_{xp}(t)}{\mathrm{d}t} = \frac{\mathrm{d}^2x_p(t)}{\mathrm{d}t^2} && \text{and } a_{yp}(t) = \frac{\mathrm{d}v_{yp}(t)}{\mathrm{d}t} = \frac{\mathrm{d}^2y_p(t)}{\mathrm{d}t^2} \end{aligned} \tag{2.2}$$

Figure 2.26 shows how a point mass moves over a flat plane from point $P_i(1, 5)$ at time zero seconds ($t = 0$) to $P_f(5, 1)$ 10 s later ($t = 10$) with a position trajectory. The velocity and acceleration components are computed from the first and second derivatives of the position components as listed in Eq. 2.2. The time variation of the motion or the trajectory is given in Fig. 2.27 and Fig. 2.28 in terms of the position's components, the velocity's components, and the acceleration's components.

Note, the example in Fig. 2.26 was not included as an example-solution because later we will introduce the method to calculate the derivatives of the position to obtain the velocity and acceleration plus the way to plot their time histories using Simulink® in Chap. 4 and beyond.

2.3.2.2 Link Motion

For a link, the motion is represented as the variation of the link's configuration with time. Again, a link has at least ***two nodes***. This variation of the configuration is represented by:

1. The location changes of a specific node i with respect to time (translation motion).
2. The change in the angle (change in rotation) between line connected node i and node j where $j \neq i$ with respect to time.

Node i may be the link's center of gravity or any arbitrary node that belongs to the link. Figure 2.29 shows the variation of a link from the initial configuration until the final configuration. The variation is mathematically described as the variation of $x_1(t)$ and $y_1(t)$ as a function of time as well as the variation of the angle $\theta(t)$ with respect to time. The first node n_1 changes the location from $n_1(x_{n1}(t=0), y_{n1}(t=0)) = n_1(1,5)$ at $t = 0$ s to $n_1(x_{n1}(t=10), y_{n1}(t=10)) = n_1(5,1)$ at $t = 10$ s . The angle of the line connecting n_1 by n_2 changes from 30° at $t = 0$ s to 120° at $t = 10$ s as shown in Figs. 2.29 and 2.30.

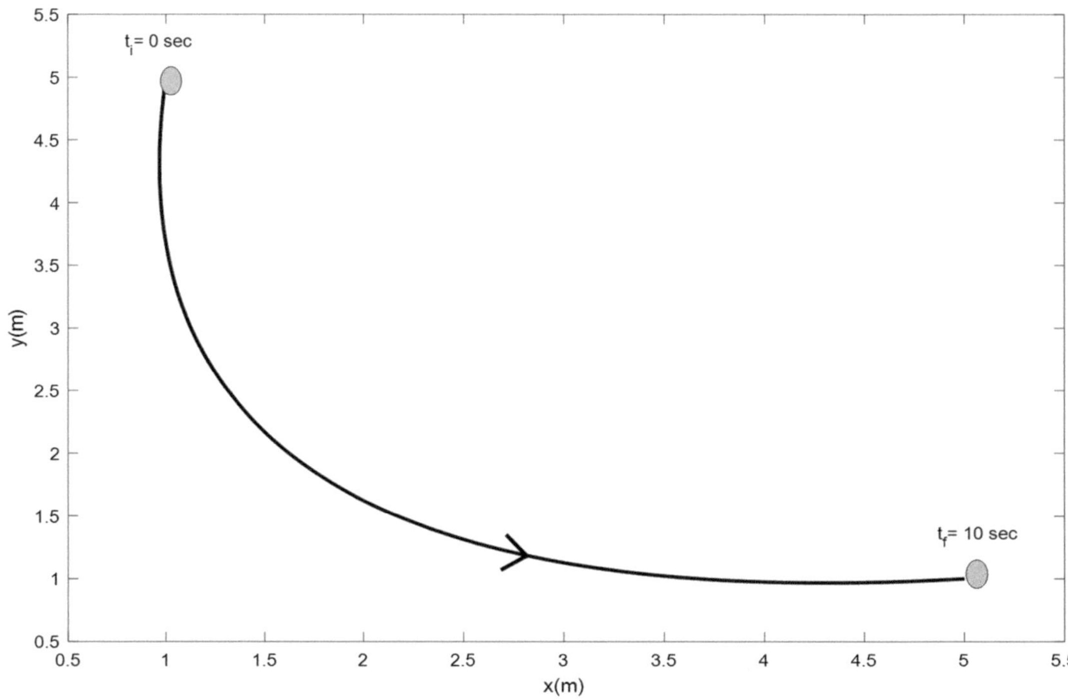

Fig. 2.26 Position trajectory

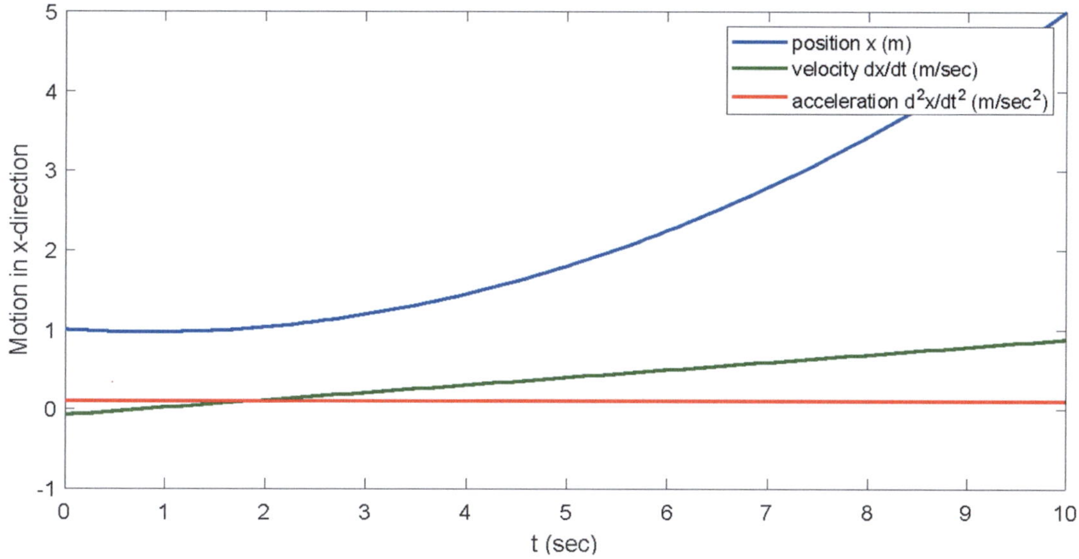

Fig. 2.27 Motion in *x*-coordinate with time

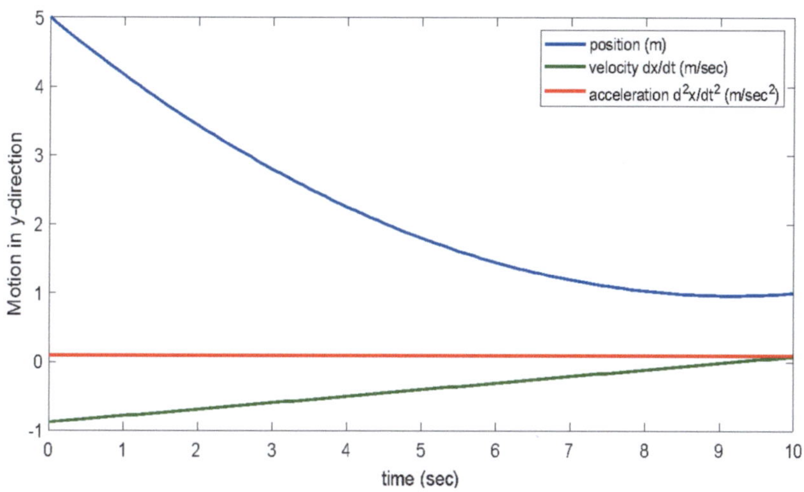

Fig. 2.28 Motion in *y*-direction with time

2.3.3 Point Mass Degrees of Freedom

Definition

Point Mass DoF:

The point mass has 2-DoF in 2D and 3-DoF in 3D. The point mass has only translational motion.

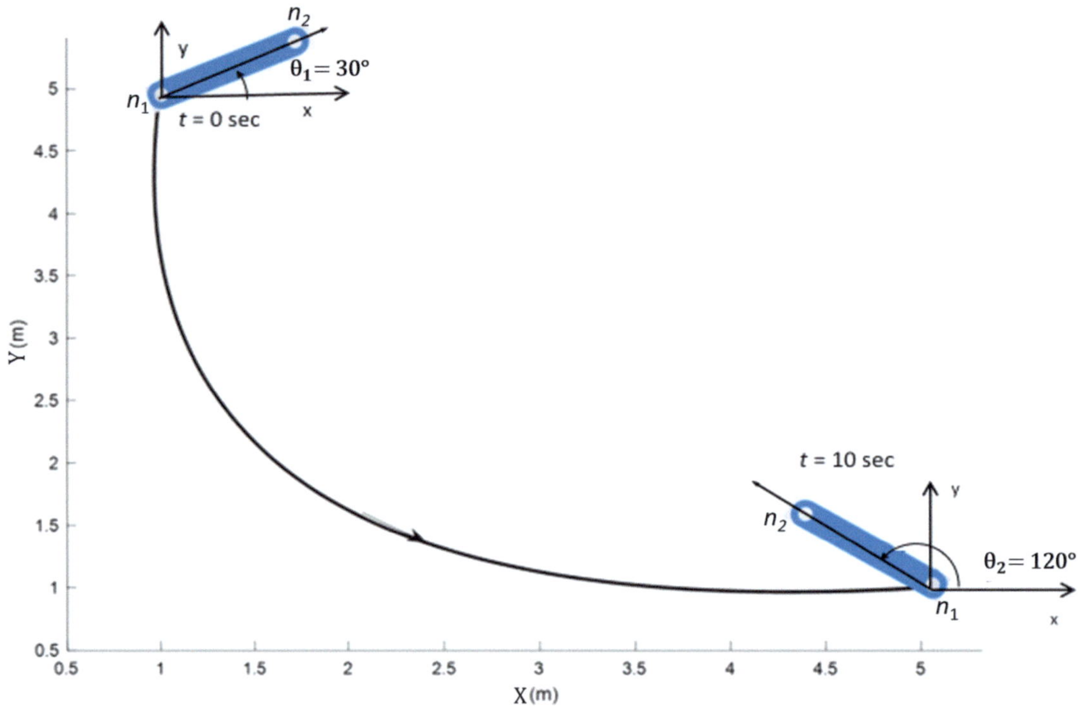

Fig. 2.29 Motion of a link with two nodes

Assuming a single point mass, the degree of freedom of this point mass can be viewed as the minimum number of coordinates required to specify the point mass configuration. In 2D, if anybody with a single point mass is free to move everywhere on a flat surface, we need two coordinates to specify this motion as up-down $y(t)$ and left–right $x(t)$ as shown in Fig. 2.31. In this case, we say that this point mass has 2DoF. In 3D, it will be three coordinates x-axis, y-axis, and z-axis or 3DoF.

2.3.4 Link Degrees of Freedom (DoF)

Definition

Link Degrees of Freedom (DoF):

The link has 3-DoF in the 2D; two translational and one rotational. The link has 6-DoF in the 3D, three translational, and three rotational.

Let's assume that we have a tablet that is set on a tabletop as shown in Fig. 2.32. The tabletop is considered as the 2D plane with an x-axis and y-axis. The tablet is considered as a single link with 4 nodes (n_1, n_2, n_3, and n_4). The degree of freedom of this link can be viewed as the minimum number of coordinates/parameters (translational and rotational) required to specify this link's configuration on the 2D plane with respect to time t. The link is free to move and rotate. The movement is called

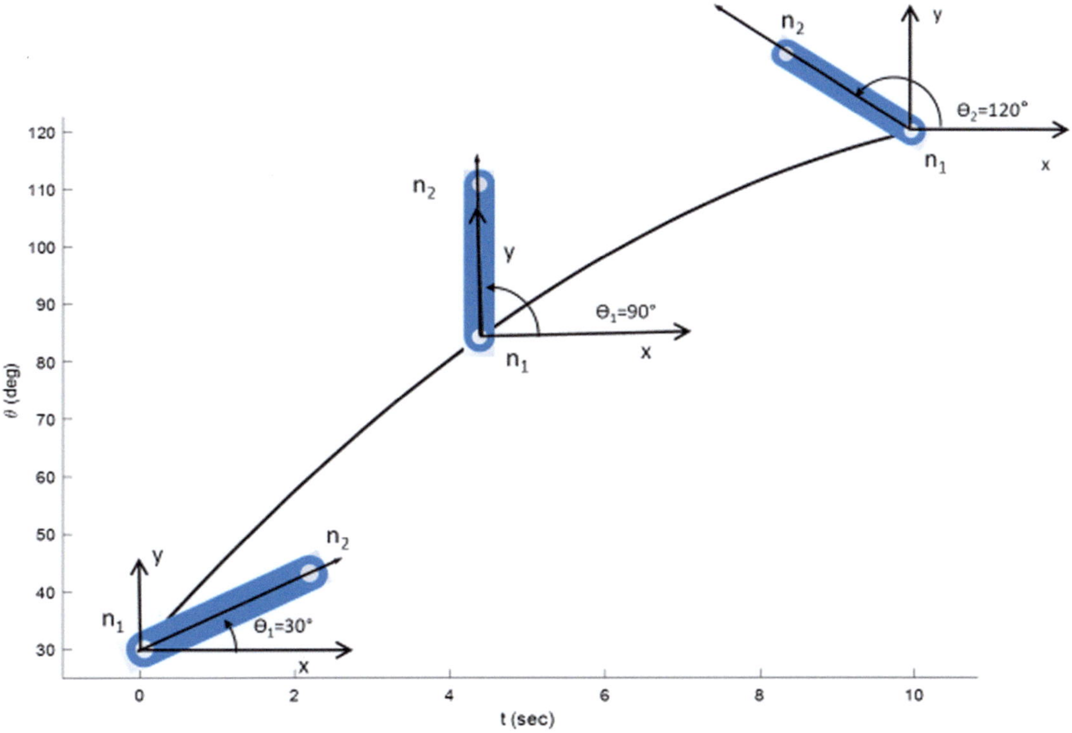

Fig. 2.30 Variation of rotation w.r.t. time

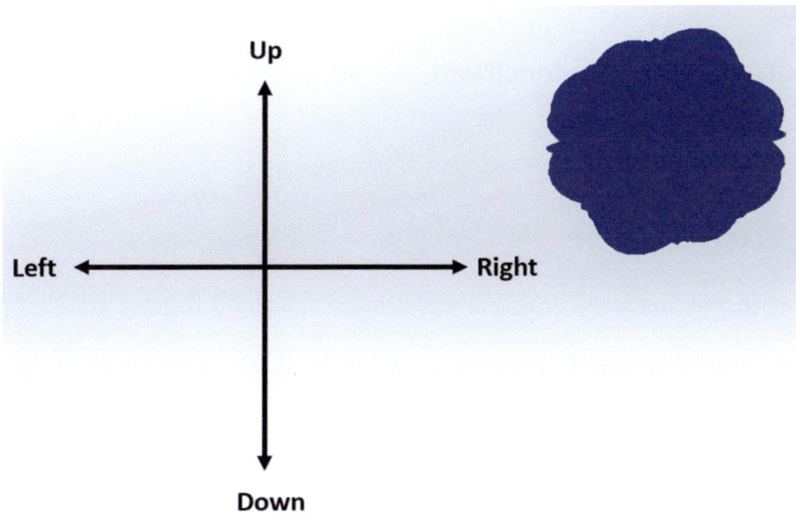

Fig. 2.31 Ball DoF in 2D

Fig. 2.32 Link DoF in 2D

the translational motion that requires two parameters or two coordinates $x(t)$ and $y(t)$ of any arbitrary node on the link. Node n_1 is selected to have the translational coordinates $x_{n1}(t)$ and $y_{n1}(t)$. The rotational motion requires one parameter, that is the rotational angle between node n_l and any of the other nodes (node n_2, node $n_{3,}$ or node n_4). Node n_2 and node n_l are selected to measure the rotational motion $\theta(t)$. As shown in Fig. 2.33, $x_{n1} = 5$ m and $y_{n1} = 1$ m at $t = 0$ s and $x_{n1} = 1$ m and $y_{n1} = 5$ m at $t = 10$ s while the rotation angle changes from $\theta = 30°$ at $t = 0$ s to $\theta = 60°$ at $t = 10$ s.

2.3.5 Joint Degrees of Freedom (DoF)

Definition

Joint Degrees of Freedom (DoF):

A joint connects two links; the *base* and the *follower*. The joint restricts some of the follower's DoF. In other words, the joint has a negative DoF. The joint can be active if there is an externally automated actuator or manual actuator to provide the motion. Otherwise, the joint is considered a passive joint.

Figure 2.34 shows two separate links in 2D. Link1 is the base and Link2 is the follower. If these two links are not connected by any joint, each link has 3 DoF. We cannot consider the two separate links as a linkage because they are not connected. Figure 2.35 shows how the joint is used to connect two links in 2D to provide a kinematic pair.

Let's assume that we are going to use a revolute joint to connect these two links. The revolute joint allows rotation around the z-axis by angle (θ) and it prevents any relative translation motion

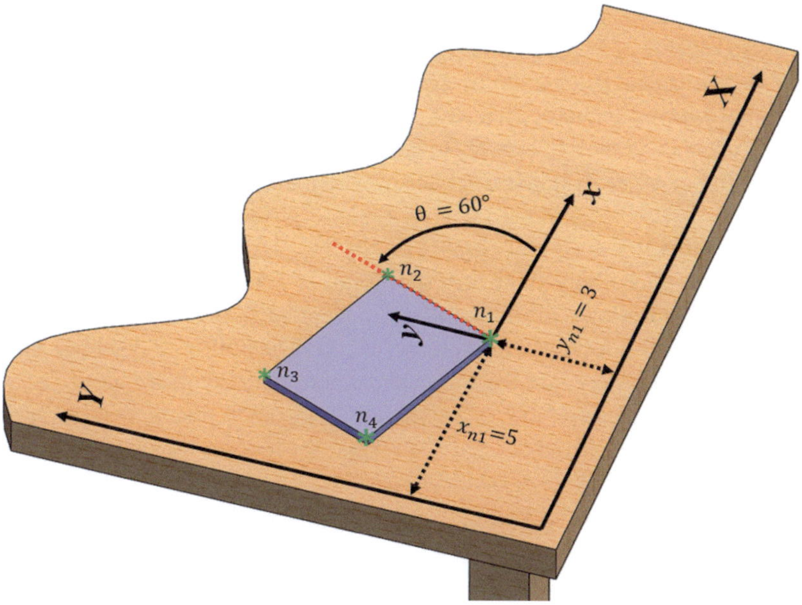

Fig. 2.33 Free link motion in 2D. Designed by the author and implemented by both the author and Pranav Dhola. Courtesy of Pranav Dhola

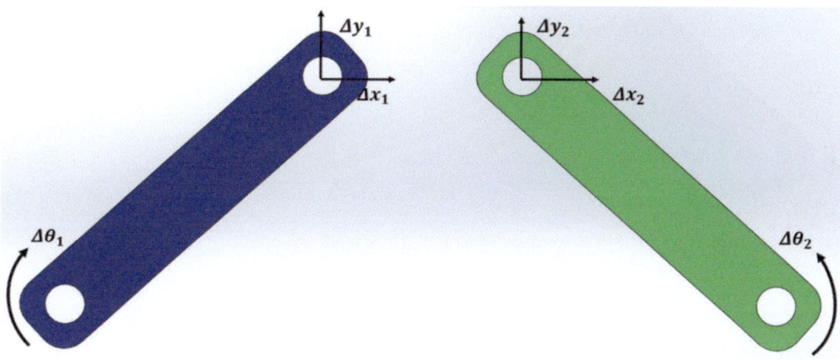

Fig. 2.34 Two separate links DoF

between the two links ($x_{12}-x_{22}=0$ and $y_{12}-y_{22}=0$). Another way to formulate this is that the joint keeps the same DoF of the base and the joint takes some DoF from the follower. For example, the revolute joint in Fig. 2.35 reduces the DoF of the follower so the entire linkage model will have 3 (Base DoF) + 3 (Follower DoF) – 2 (joint DoF) = 4 DoF.

From the previous example, we can conclude that the joint adds a negative DoF to the total linkage model. Table. 2.1 shows different types of common joints with their equivalent negative DoF. All the listed joints are more suited to 2D kinematics analysis.

2.3.5.1 Revolute Joint

A revolute joint (also called pin joint or hinge joint) is a one-degree-of-freedom kinematic pair used in mechanisms. The revolute joint allows only the relative rotational motion between two links or it restricts two degrees of freedom of the follower in the case of 2D. The follower can't relatively translate away from the base. The revolute joint has many applications such as the door hinges in Fig. 2.10 and the excavator in Fig. 2.15. The revolute joint can be actively powered by an actuator (such as a hand-driven (manual) or a motor-driven (automated)) directly on the joint. The revolute joint can be passive with no actuator attached. Figure 2.36 shows the shape of a revolute joint. In this book, we will use the symbol J_i^R to refer to a revolute joint with an index i, where $i = 1, 2, \ldots, N$ and N is the total number of joints in the mechanism where the joint belongs.

2.3.5.2 Prismatic Joint

A prismatic joint (also called slider joint or piston joint) is a 1 DoF kinematic pair used in mechanisms. The prismatic joint allows a linear sliding movement between two links and restricts two degrees of freedom from the follower in the case of 2D. The follower does not have any relative rotation with respect to the base. Also, the follower cannot relatively move perpendicular to the sliding motion with respect to the base. The prismatic joint has many practical applications such as the

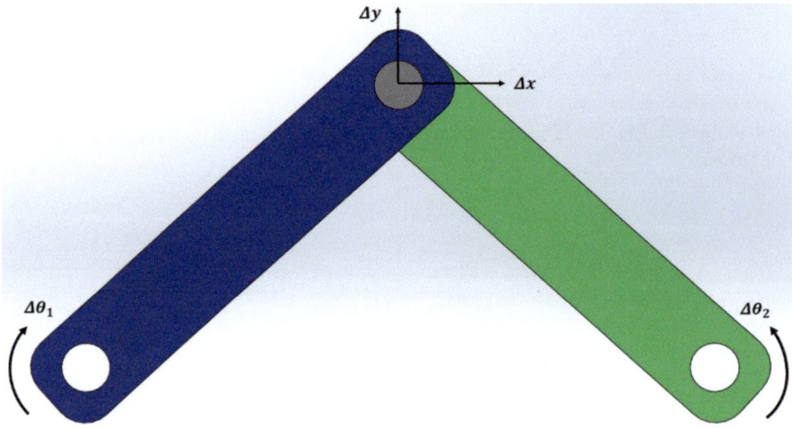

Fig. 2.35 Revolute kinematic pair DoF. Designed by the author and implemented by both the author and Pranav Dhola. Courtesy of Pranav Dhola

Table 2.1 DoF of a joint in 2D. Figures are designed by the author and implemented by both the author and Pranav Dhola. Courtesy of Pranav Dhola

Joint Type	DoF	Shape
Revolute, Pin, or Hinge joint	−2	**Fig. 2.36** Revolute joint 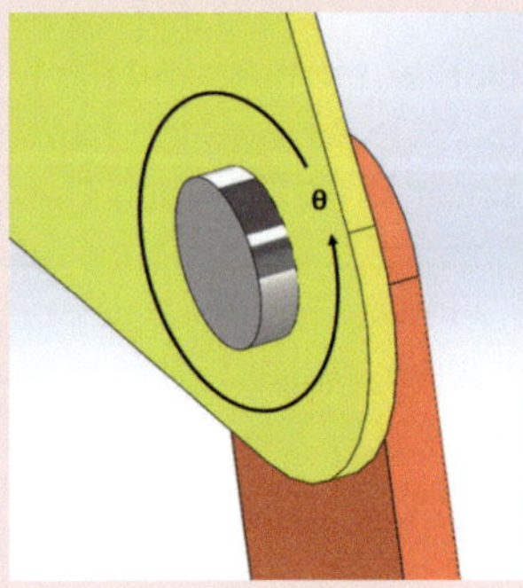

(continued)

Table 2.1 (continued)

Joint Type	DoF	Shape
Prismatic, Sliding, or Piston joint	−2	**Fig. 2.37** Prismatic joint 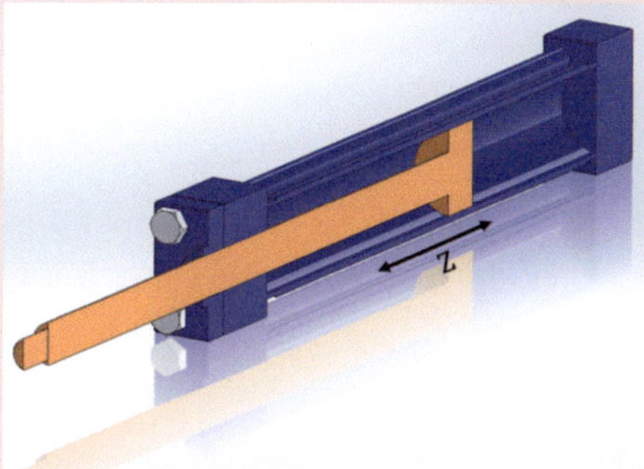 **Fig. 2.38** Sliding joint 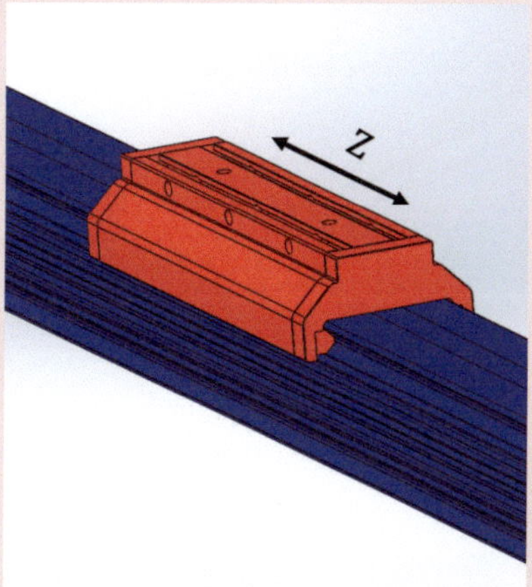
Threaded or Screw joint	−2	**Fig. 2.39** Screw joint 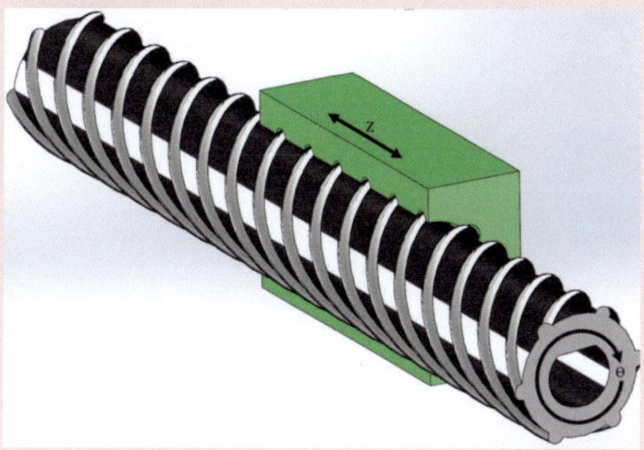

(continued)

Table 2.1 (continued)

Joint Type	DoF	Shape
Pin-in-Slot joint	−1	**Fig. 2.40** Pin-in-a-slot joint 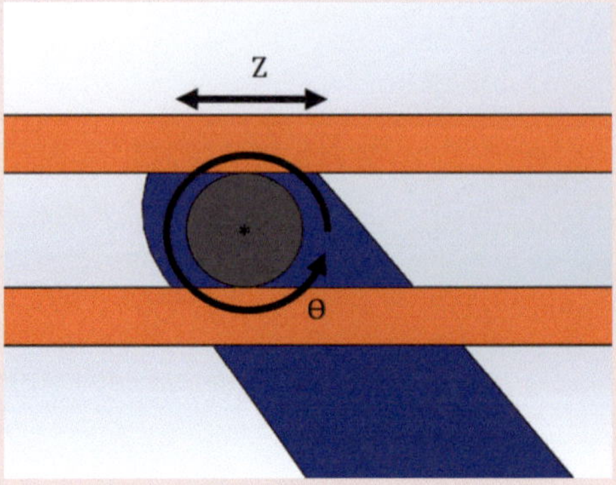
Cam joint	−1	**Fig. 2.41** Cam joint 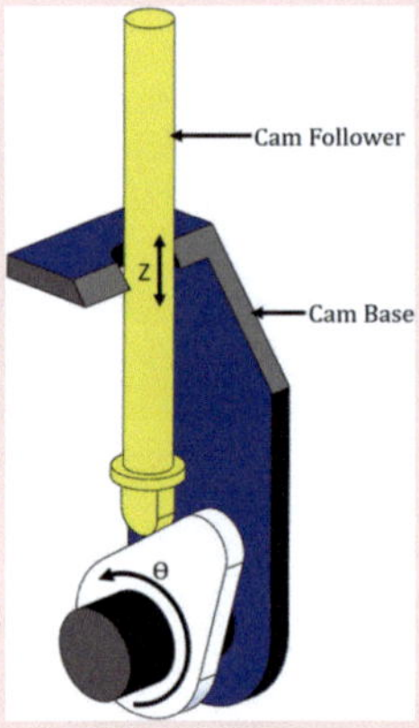
Gear joint	−1	**Fig. 2.42** Gear joint 

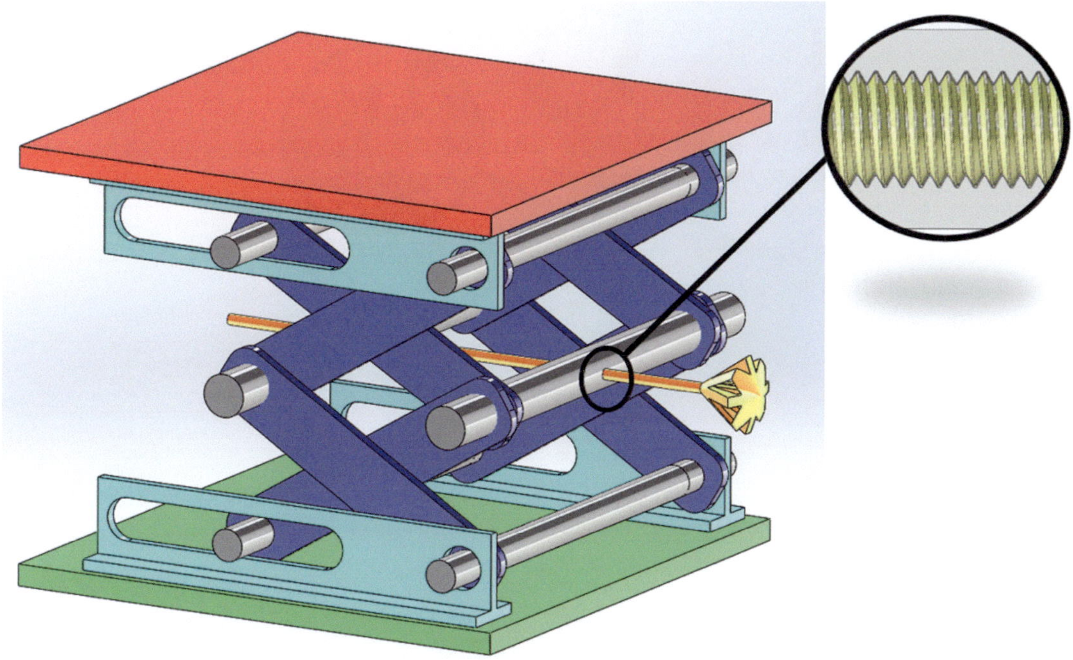

Fig. 2.43 Lift table machine. Designed by the author and implemented by both the author and Pranav Dhola. Courtesy of Pranav Dhola

hydraulic system in the skid steer in Fig. 2.19 and the prismatic joint in the piston–cylinder mechanism in Fig. 2.2.

The prismatic joint can be active by attaching an actuator such as a hydraulic, pneumatic, or hand-driven power directly to the joint. The prismatic joint can be passive with no actuator attached as well. Figures 2.37 and 2.38 show the shape of a prismatic joint. In this book, we will use the symbol J_i^P to refer to a prismatic joint with an index i, where $i = 1, 2, \ldots, N$ and N is the total number of joints in the mechanism where the joint belongs.

2.3.5.3 Screw Joint

A screw joint (also called a thread joint) is a 1 DoF kinematic pair used in mechanisms. The screw joint allows only a single-axis translation by utilizing the threads of the threaded rod and restricts two degrees of freedom from the follower in the case of 2D. The screw joint has many practical applications such as the lift table in Fig. 2.43 and the wing corkscrew in Fig. 2.44. The screw joint can be active or passive. Figure 2.39 shows the shape of a screw joint. In this book, we will use the symbol J_i^S to refer to a screw or thread joint with an index i, where $i = 1, 2, \ldots, N$ and N is the total number of joints in the mechanism where the joint belongs.

2.3.5.4 Pin-in-a-Slot Joint

A pin-in-a-slot joint has one translational and one rotational degree of freedom. One prismatic motion provides the translational degree of freedom. One revolute motion provides the rotational degree of freedom. Prismatic and revolute axes are mutually orthogonal as shown in Fig. 2.40. One common application for the pin-in-a-slot joint is the car window mechanism shown in Fig. 2.12. In this book, we will use the symbol J_i^{PS} to refer to a pin-in-a-slot joint with an index i, where $i = 1, 2, \ldots, N$ and N is the total number of joints in the mechanism where the joint belongs.

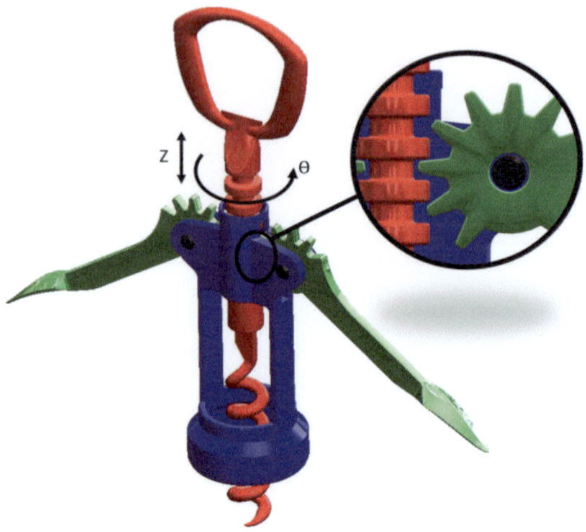

Fig. 2.44 Wing corkscrew machine. Figure and associated Corkscrew_Mechanism.mp4 video designed by the author and implemented by both the author and Pranav Dhola. Both courtesy of Pranav Dhola

2.3.5.5 Cam Joint

A cam joint is a rotating and sliding piece used especially for transforming rotary motion into linear motion. As shown in Fig. 2.41, the cam has two parts: cam base and cam follower. When the cam base rotates, the follower slides up and down. The cam joint allows only rolling and sliding and restricts 1 DoF from the follower in the case of 2D. One common application for the cam is the camshaft of an automobile, which transfers the engine's rotary motion into the reciprocating motion of the valves as shown in Fig. 2.45. We will cover more details about different types of cams and the different cam configurations in Chap. 9. In this book, we will use the symbol J_i^C to refer to a cam joint with an index i, where $i = 1, 2, \ldots, N$ and N is the total number of joints in the mechanism where the joint belongs.

2.3.5.6 Gear Joint

Figure 2.42 shows a gear joint. Gears are mechanisms that mesh together via teeth and are used to transmit rotary motion from one shaft to another. A gear joint allows for simultaneous rotation and sliding and allows only for rolling and sliding as it takes away 1 DoF from the follower in the case of 2D. The gear joint has many applications such as the transmission and the car window example in Fig. 2.12. We will cover more details about different types of gears and the different configurations in Chap. 10. In this book, we will use the symbol J_i^G to refer to a gear joint with an index i, where $i = 1, 2, \ldots, N$ and N is the total number of joints in the mechanism where the joint belongs.

2.3.6 Linkage Degrees of Freedom

The DoF for a linkage reflects the number of position/orientation variables required to determine the position and orientation of links in a mechanism. Also, the degrees of freedom reflect the number of velocity variables required to determine the velocity of the links in the mechanism. Additionally, the DoF of a linkage is the number of independent inputs required to precisely position all links of the mechanism with respect to the frame as functions of time. The number of degrees of freedom of a mechanism is also called *mobility*, and it is given the symbol (M).

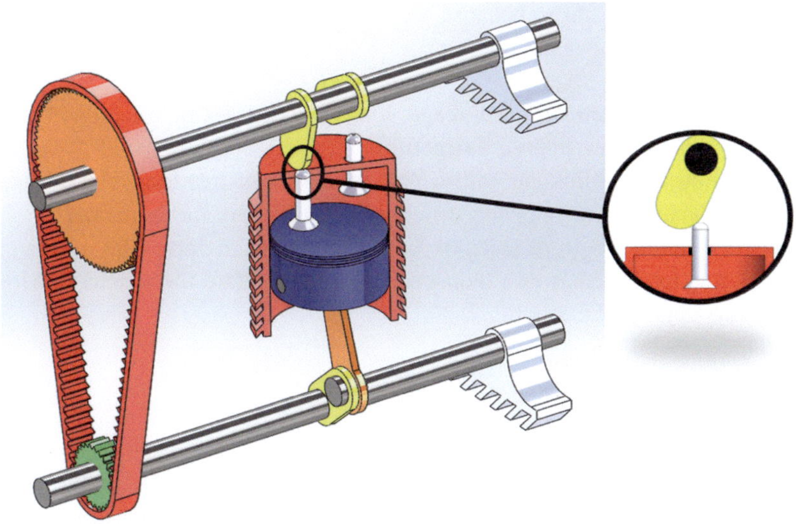

Fig. 2.45 Cam piston engine mechanism. Designed by the author and implemented by both the author and Pranav Dhola. Courtesy of Pranav Dhola

1 DoF mechanisms are very common in engineering practice. In a 1 DoF mechanism, one link position must be known, along with the link lengths, to determine the positions of all of the other links. Similarly, one link velocity, along with all of the link positions, must be known to determine the velocities of all of the other links. Finally, one link acceleration, along with all of the link velocities, must be known to determine all of the other link accelerations.

In 2D systems, each link has 3 DoF and each joint has -n DoF, where $n=1, 2$, or 3 depending on the type of joint. The joint with −2 DoF is called a primary joint J_p. The joint with −1 DoF is called a higher-order joint and is also called a half joint J_h.

In this book, we are not going to rely on these terminologies. We are going to assign each joint with a generic mechanical name: revolute joint J_i^R, prismatic joint J_i^P, pin-in-a-slot joint J_i^{PS}, screw joint J_i^S, gear joint J_i^G, and cam joint J_i^C each with an index i, where $i=1, 2, \ldots, N$ and N is the total number of joints in the mechanism where the joint belongs.

The mathematical formula to find a DoF for any linkage is given by:

1. Count 3 DoF for each link (except the ground link).
2. Count 0 DoF for the ground link.
3. Count −2 DoF for revolute joint J_i^R, screw joint J_i^S, and prismatic joint J_i^P.
4. Count −1 DoF for pin-in-a-slot joint J_i^{PS}, gear joint J_i^G, and cam joint J_i^C.
5. Add all the DoF to get the total linkage's DoF.

Grübler's equation above is used to calculate the degrees of freedom of any planar linkage as [1–3]:

$$M = DoF = 3(n_l - 1) - 2n_P - n_h \tag{2.3}$$

where

n_l = total number of links in the mechanism

n_P = total number of primary joints: revolute joints J_i^R, screw joints J_i^S, and prismatic joints J_i^P.

n_h = total number of high-order joints: pin-in-a-slot joints J_i^{PS}, gear joints J_i^G, and cam joints J_i^C.

In Eq. 2.3, we have $(n_l - 1)$ not n_l because we delete the ground link from the DoF calculation. From Grübler's equation, we can state:

1. A **truss** or stationary mechanism is a linkage with zero degrees of freedom. The mechanism is unable to move and is thus considered a structure.
2. A **pre-loaded structure** is a linkage with negative degrees of freedom. The pre-loaded structure as a mechanism is unable to move or the structure is a stationary mechanism.
3. A **mechanism** is a linkage with degrees of freedom more than zero. The mechanism with multiple degrees of freedom needs more than one driver to operate the machine. Most mechanical machines have one degree of freedom.

Example 2.6

For the shown linkage model in Fig. 2.46, specify whether the linkage is a structure, a pre-loaded structure, or a machine. If it is a machine, specify the DoF.

Solution

Figure 2.47 shows the free body diagram of the four-bar linkage. The number of links is $n_l=4$. Note the two ground nodes are considered to be a single link, which we call the ground. The ground link is stationary with 0 DoF. The number of joints is 4 and all of them are revolute joints so $n_p=4$ and $n_h=0$. Substituting directly into Grübler's equation:

$$M = 3(n_l - 1) - 2n_p - n_h = 3x(4 - 1)-2x4-0 = 1.$$

Example 2.7

For the shown linkage model in Fig. 2.48, specify if the linkage is a structure, pre-loaded structure, or a machine. If it is a machine, specify the DoF.

Solution

Figure 2.49 shows the free body diagram of the linkage model in Fig. 2.48. The number of links is $n_l = 3$. The number of joints is 3 and all of them are revolute joints so $n_p = 3$ and $n_h = 0$. Substituting directly into Grübler's equation:

$$M = 3(n_l - 1) - 2n_p - n_h = 3x(3 - 1)-2x3-0 = 0$$

when $M = 0$, the linkage model in Fig. 2.41 does not move and we call the linkage model a structure model.

Example 2.8

For the shown linkage model in Fig. 2.50, specify whether the linkage is a structure, pre-loaded structure, or a machine. If it is a machine, specify the DoF.

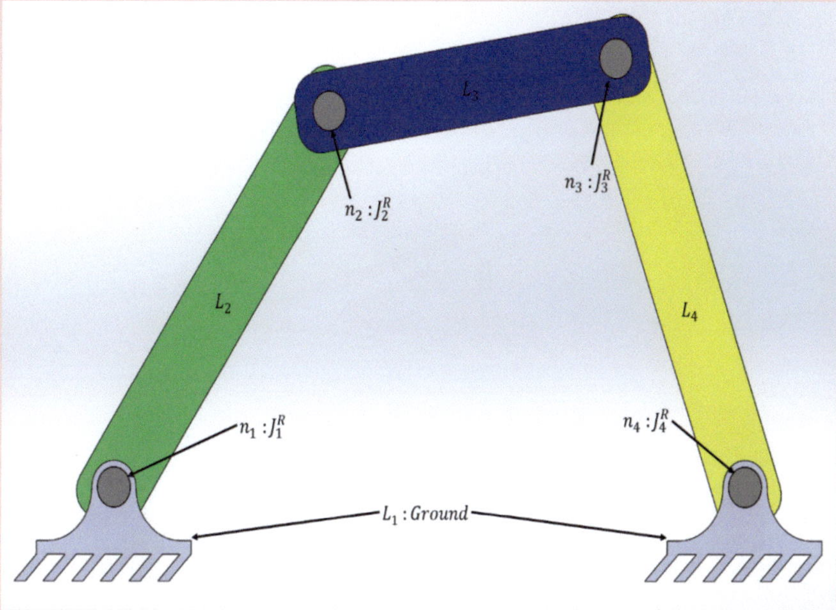

Fig. 2.46 Four-bar linkage. Designed by the author and implemented by both the author and Pranav Dhola. Courtesy of Pranav Dhola

Fig. 2.47 Free body diagram of a four-bar linkage. Designed by the author and implemented by both the author and Pranav Dhola. Courtesy of Pranav Dhola

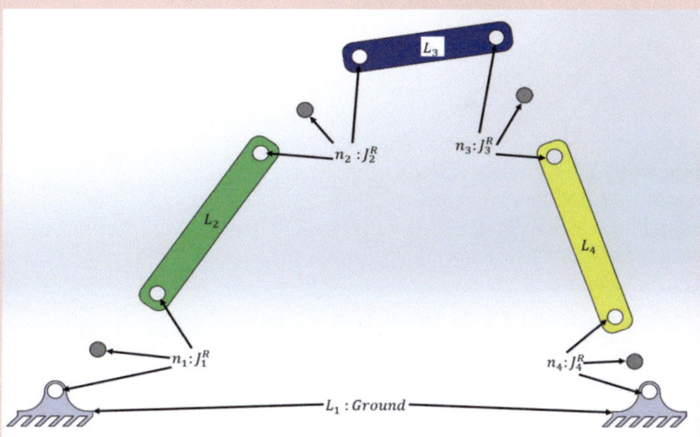

Solution

Figure 2.51 shows the free body diagram of the linkage model in Fig. 2.50. The number of links is $n_l=2$. The number of joints is 2, and all of them are revolute joints so $n_p=2$ and $n_h=0$. Substituting directly into Grübler's equation:

$$M = 3(n_l - 1) - 2n_p - n_h = 3x(2 - 1) - 2x2 - 0 = -1$$

when M is negative, the linkage model in Fig. 2.50 does not move and we call the linkage model a pre-loaded structure model.

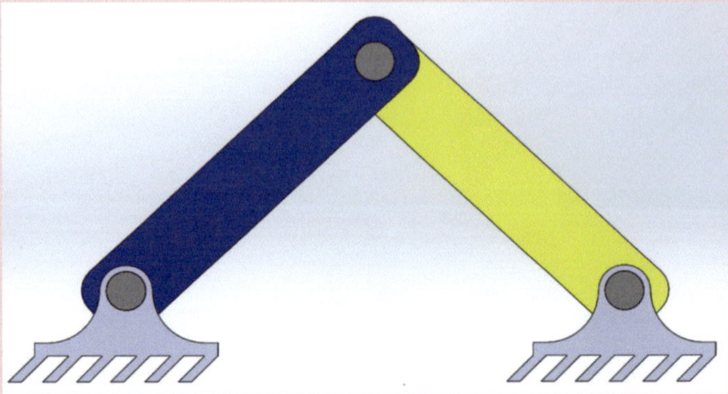

Fig. 2.48 Three-bar linkage. Designed by the author and implemented by both the author and Pranav Dhola. Courtesy of Pranav Dhola

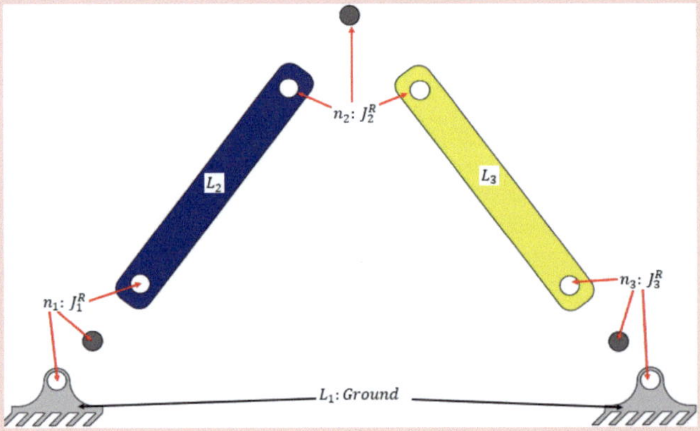

Fig. 2.49 Free body diagram of a closed chain three-bar linkage. Designed by the author and implemented by both the author and Pranav Dhola. Courtesy of Pranav Dhola

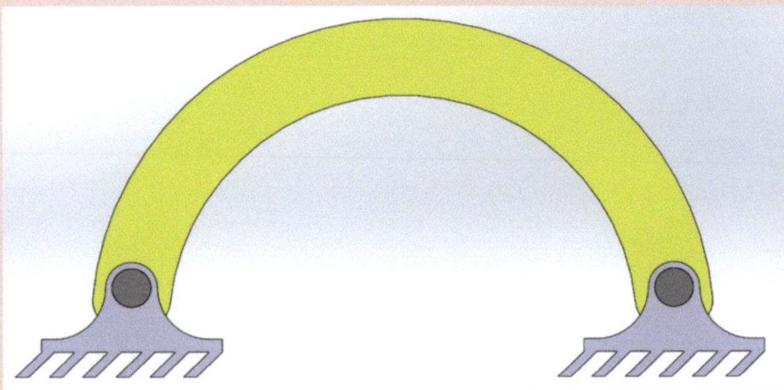

Fig. 2.50 Closed chain two-bar linkage. Designed by the author and implemented by both the author and Pranav Dhola. Courtesy of Pranav Dhola

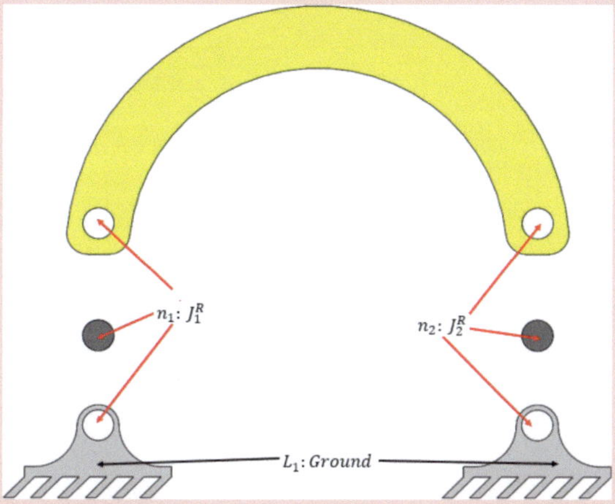

Fig. 2.51 Free body diagram of closed chain two-bar linkage. Designed by the author and implemented by both the author and Pranav Dhola. Courtesy of Pranav Dhola

Example 2.9

For the shown linkage model in Fig. 2.52, specify whether the linkage is a structure, a pre-loaded structure, or a machine. If it is a machine, specify the DoF.

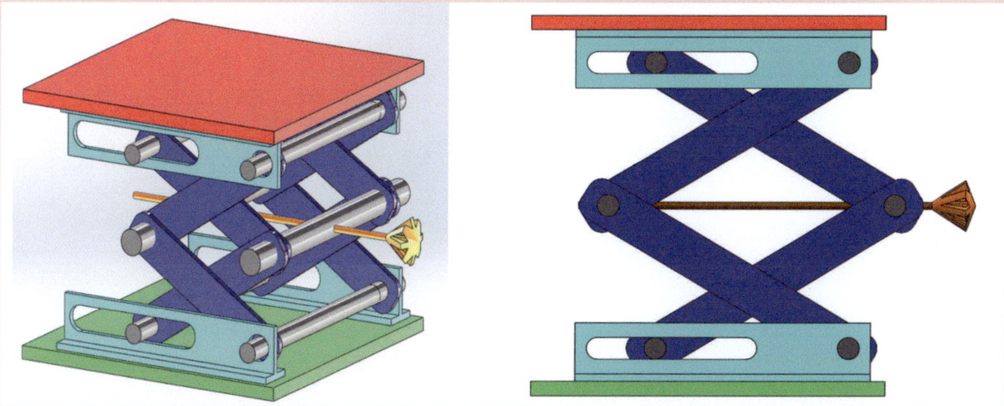

Fig. 2.52 Scissor lift table. Designed by the author and implemented by both the author and Pranav Dhola. Courtesy of Pranav Dhola

Solution

Figure 2.53 shows the free body diagram of the scissor lift mechanism. The number of links is $n_l=6$. The number of joints is seven: five revolute joints ($J_1^R, J_2^{R1}, J_2^{R2}, J_3^R and J_4^R$), one screw joint ($J_4^S$), and two pin-in-a-slot joints (J_4^{PS}). That leads to $n_p=6$ (five revolute joints plus one screw joint) and $n_h=2$ (two pin-in-a-slot). Substituting directly into Grübler's equation:

$$M = 3(n_l - 1) - 2n_p - n_h = 3x(6 - 1) - 2x6 - 1x2 = 1$$

when $M=1$, then the mechanism has 1 DOF.

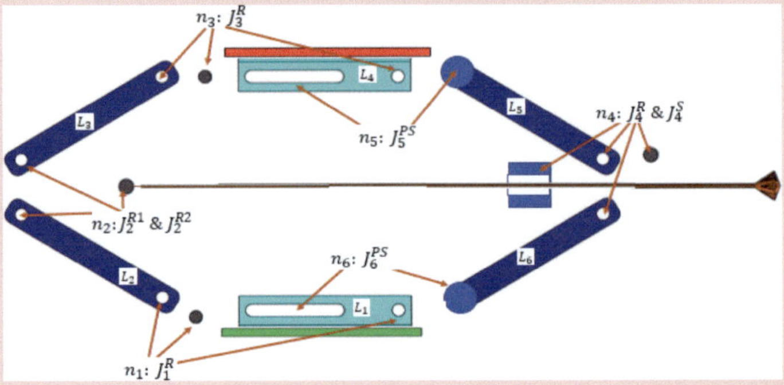

Fig. 2.53 Free body diagram of a scissor lift. Designed by the author and implemented by both the author and Pranav Dhola. Courtesy of Pranav Dhola

Example 2.10

For the shown linkage model in Fig. 2.54, specify whether the linkage is a structure, pre-loaded structure, or machine. If it is a machine, specify the DoF.

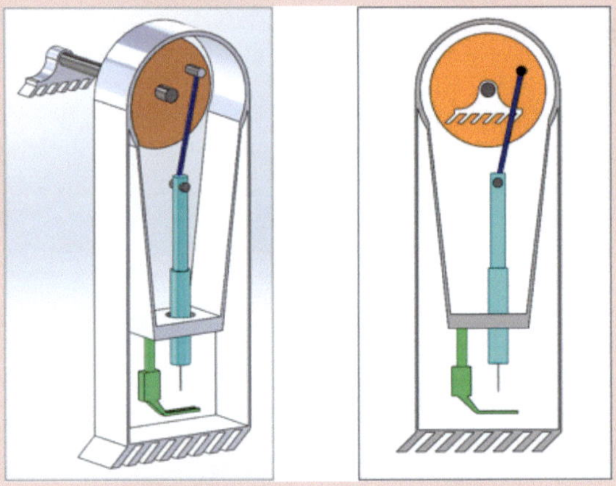

Fig. 2.54 Sewing machine. Designed by the author and implemented by both the author and Pranav Dhola. Courtesy of Pranav Dhola

Solution

Figure 2.55 shows the free body diagram of the sewing mechanism. The number of links is four or $n_l=4$. The number of joints is four: three revolute joints and one prismatic joint or $n_p=4$ and $n_h=0$. Substituting directly into Grübler's equation:

$$M = 3(n_l - 1) - 2n_p - n_h = 3x(4 - 1){-}2x4{-}0 = 1$$

When $M=1$, the mechanism has one degree of freedom. When the crankshaft rotates, the mechanism will lead to pushing the pin arm up and down.

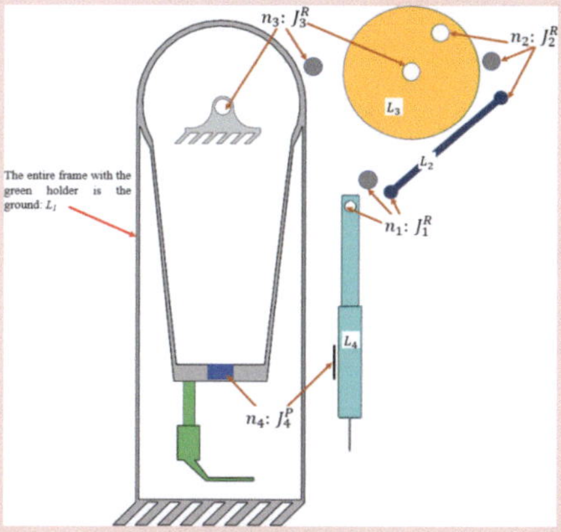

Fig. 2.55 Free body diagram of a sewing machine. Designed by the author and implemented by both the author and Pranav Dhola. Courtesy of Pranav Dhola

Example 2.11

For the shown linkage model in Fig. 2.56, specify whether the linkage is a structure, preloaded structure, or machine. If it is a machine, specify the DoF.

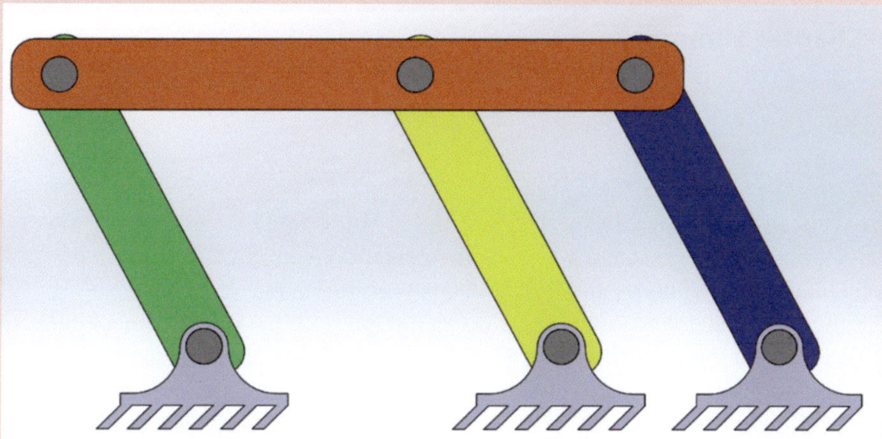

Fig. 2.56 Closed chain five-bar mechanism. Designed by the author and implemented by both the author and Pranav Dhola. Courtesy of Pranav Dhola

Solution

Figure 2.57 shows the free body diagram of the linkage model in Fig. 2.56. The number of links is $n_l=5$. The number of joints is 6 and all of them are revolute joints so $n_p=6$ and $n_h=0$. Substituting directly into Grübler's equation:

$$M = 3(n_l - 1) - 2n_p - n_h = 3x(5 - 1) - 2x6 - 0 = 0$$

Grübler's equation states that the mechanism has no degrees of freedom. From our understanding of the mechanism, there is redundancy, and the mechanism has one degree of freedom.

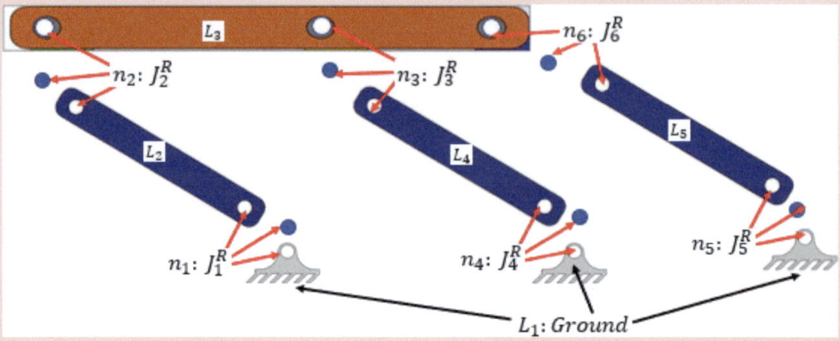

Fig. 2.57 Free body diagram of a closed chain five-bar mechanism. Designed by the author and implemented by both the author and Pranav Dhola. Courtesy of Pranav Dhola

From the previous examples, we can come to the following conclusion:

1. A four-bar linkage has a minimum number of 4 links needed to build a mechanism. Two- and three-bar linkages cannot build a mechanism as listed in Examples 2.6, 2.7 and 2.8.
2. Grübler's equation sometimes fails to provide the actual DoF for redundant mechanisms or machines as listed in Example 2.11.

2.4 Chapter Summary

Outcome

1. Definitions of nodes, joints, and links.
2. Breaking down any machine or mechanism into a free body diagram.
3. Point mass has one node and 2 DoF in any planar domain.
4. A link has at least 2 nodes and 3 DoF in any planar domain.
5. Revolute, prismatic, and screw joints restrict 2DoF from the follower in a kinematic pair.
6. Cam, gear, and slot joints restrict 1 DoF from the follower in a kinematic pair.
7. Use Grübler's equation to easily estimate a machine's or mechanism's DoF.
8. The minimum number of joints for any mechanism is 4 links or bars.

Problems

For each problem, on the picture of each linkage, specify:

- links.
- ground link.
- joints.
- nodes.
- if the linkage is closed or open chain.
- if the linkage is either a structure, pre-loaded structure or a machine.
- if the linkage is a machine or mechanism, specify the DoF.

Note, to understand the mechanism, it is recommended to watch the provided animation videos with the textbook. You can use 2D projection in your solution.

Problem 1: Flush Toilet Mechanism (Fig. 2.58).

Problem 2: Folding Chair (Fig. 2.59).

Problem 3: Ironing Table (Fig. 2.60).

Problem 4: Potato Chopper (Fig. 2.61).

Problem 5: Corkscrew (Fig. 2.62).

Problem 6: Stationary Bike (Fig. 2.63).

Problem 7: Car Hood (Fig. 2.64).

Problem 8: Dumpster Truck (Fig. 2.65).

Problem 9: Pitcher Water Pump (Fig. 2.66).

Problem 10: Landing Gear (Fig. 2.67).

Problem 11: Casement Window (Fig. 2.68).

Problem 12: Single piston–cylinder (Fig. 2.69).

Problem 13: Pumpjack (Fig. 2.70).

Problem 14: Man Toy (Fig. 2.71).

Problem 15: Sewing Machine (Fig. 2.72).

Problem 16: Car Wiper (Fig. 2.73).

Problem 17: Hydraulic Door (Fig. 2.74).

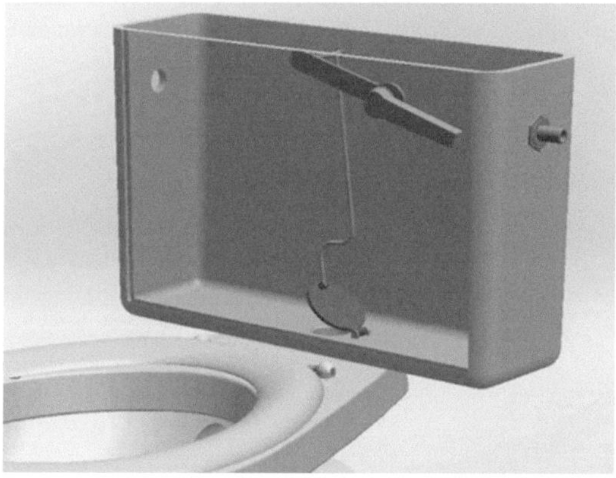

Fig. 2.58 Toilet flush valve mechanism. Figure and associated Flush_Toilet_Mechanism.mp4 video designed by the author and implemented by both the author and Pranav Dhola. Both courtesy of Pranav Dhola

Fig. 2.59 Folding chair mechanism. Figure and associated Folding_Chair_Mechanism.mp4 video designed by the author and implemented by both the author and Pranav Dhola. Both courtesy of Pranav Dhola

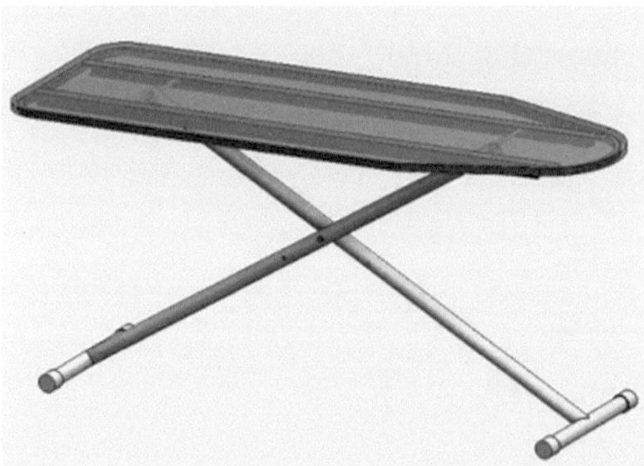

Fig. 2.60 Ironing board mechanism. Figure and associated Ironing_Table_Mechanism.mp4 video designed by the author and implemented by both the author and Pranav Dhola. Both courtesy of Pranav Dhola

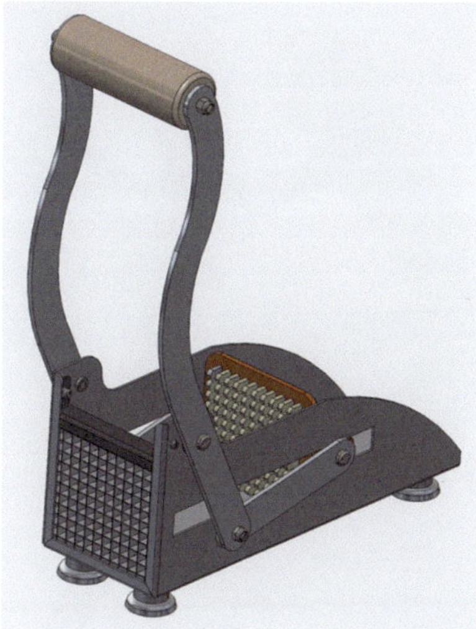

Fig. 2.61 Potato chopper machine. Figure and associated Potato_Chopper_Mechanism.avi video designed by the author and implemented by both the author and Pranav Dhola. Both courtesy of Pranav Dhola

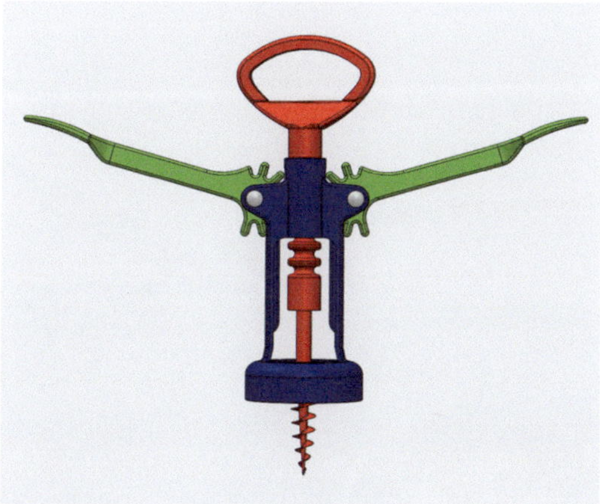

Fig. 2.62 Corkscrew machine. Figure and associated Corkscrew Mechansim.mp4 video designed by the author and implemented by both the author and Pranav Dhola. Both courtesy of Pranav Dhola

Fig. 2.63 Stationary bike machine. Figure and associated Stationary_Bicycle_Machine.mp4 video designed by the author and implemented by both the author and Pranav Dhola. Both courtesy of Pranav Dhola

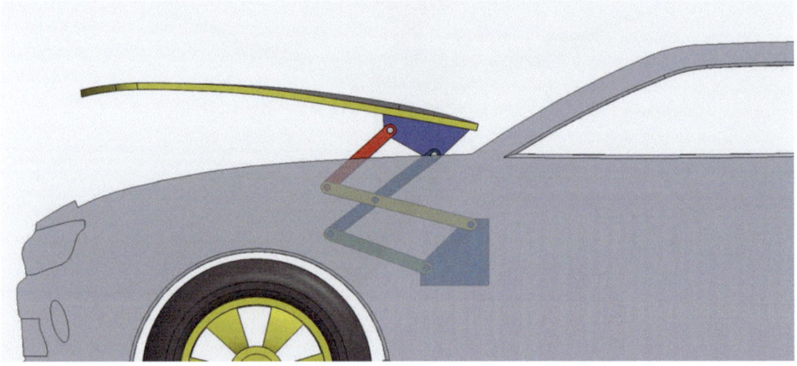

Fig. 2.64 Car hood mechanism. Figure and associated Car_Hood_Mechanism.mp4 video designed by the author and implemented by both the author and Pranav Dhola. Both courtesy of Pranav Dhola

Fig. 2.65 Dumpster truck mechanism. Figure and associated Dumpster Truck Mechanism.mp4 video designed by the author and implemented by both the author and Pranav Dhola. Both courtesy of Pranav Dhola

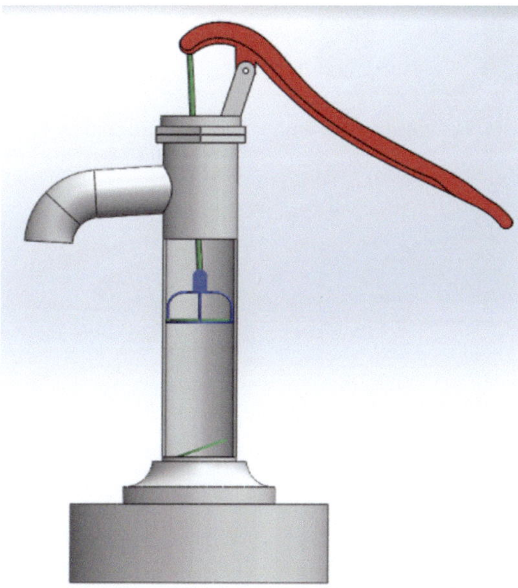

Fig. 2.66 Pitcher water pump mechanism. Designed by the author and implemented by both the author and Pranav Dhola. Courtesy of Pranav Dhola

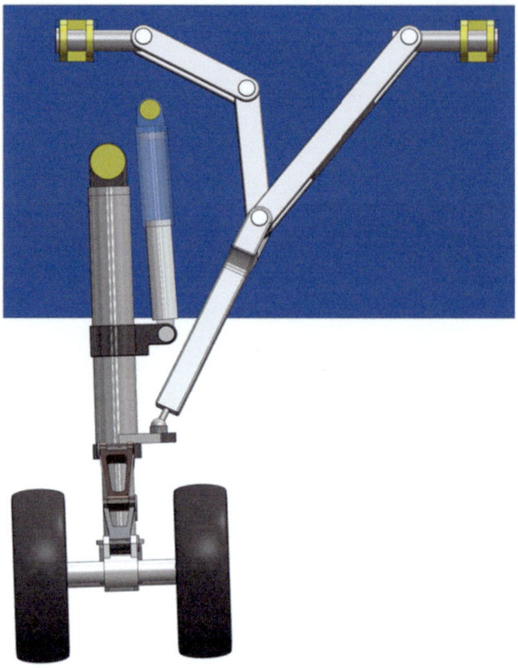

Fig. 2.67 Landing gear mechanism. Designed by the author and implemented by both the author and Pranav Dhola. Courtesy of Pranav Dhola

Fig. 2.68 Car window mechanism. Figure and associated Casement_Window_Mechanism.mp4 video designed by the author and implemented by both the author and Pranav Dhola. Both courtesy of Pranav Dhola

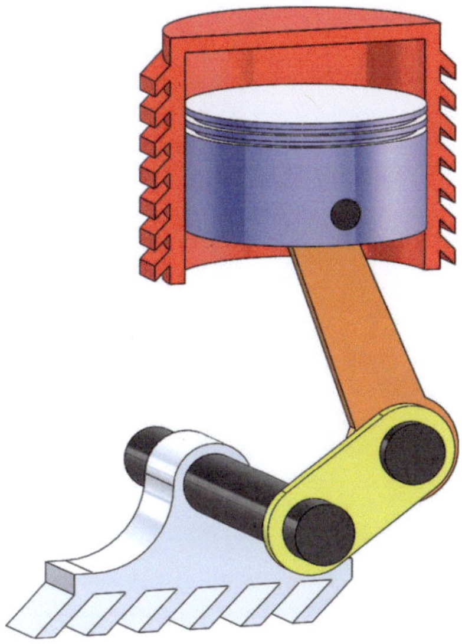

Fig. 2.69 Single piston–cylinder engine mechanism. Designed by the author and implemented by both the author and Pranav Dhola. Courtesy of Pranav Dhola

Fig. 2.70 Pumpjack mechanism. Designed by the author and implemented by both the author and Pranav Dhola. Courtesy of Pranav Dhola

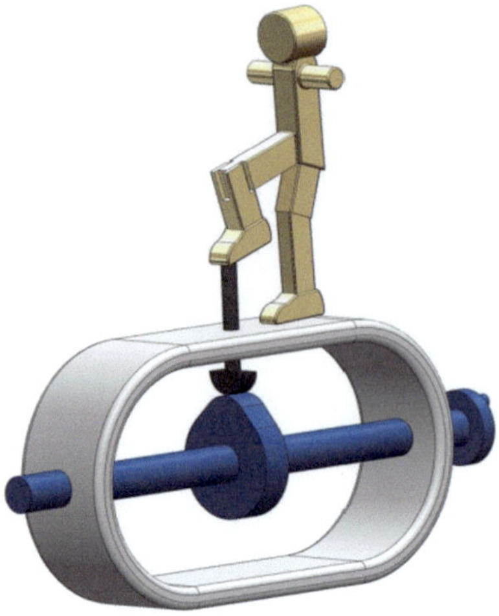

Fig. 2.71 Cam mechanism in Man-Toy. Designed by the author and implemented by both the author and Pranav Dhola. Courtesy of Pranav Dhola

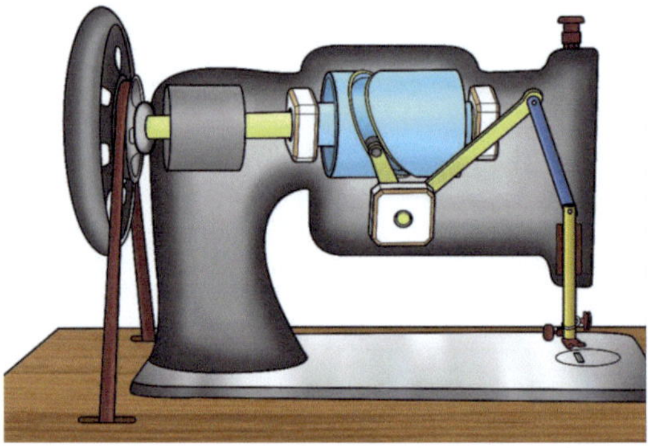

Fig. 2.72 Sewing machine. Figure and associated Sewing Machine.mp4 video designed by the author and implemented by both the author and Pranav Dhola. Both courtesy of Pranav Dhola

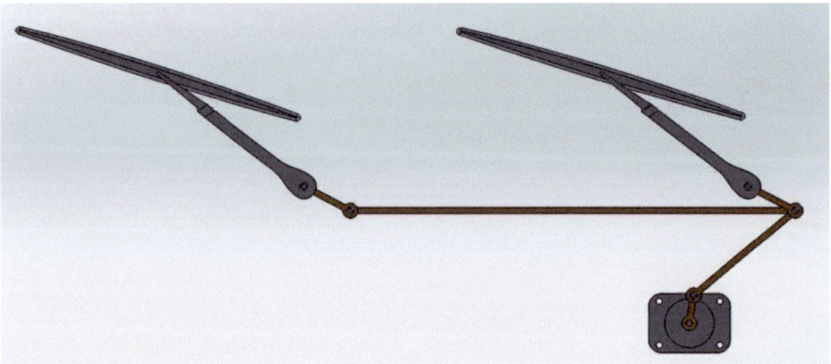

Fig. 2.73 Car wiper mechanism. Figure and associated Car_Wiper_Mechanism.mp4 video designed by the author and implemented by both the author and Pranav Dhola. Both courtesy of Pranav Dhola

Fig. 2.74 Hydraulic gate closer mechanism. Designed by the author and implemented by both the author and Pranav Dhola. Courtesy of Pranav Dhola

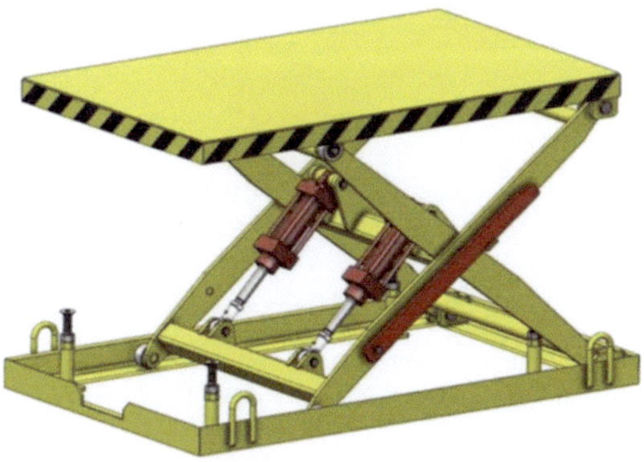

Fig. 2.75 Scissor lift table mechanism. Figure and associated Scissor Lift Table Mechanism.mp4 video designed by the author and implemented by both the author and Pranav Dhola. Both courtesy of Pranav Dhola

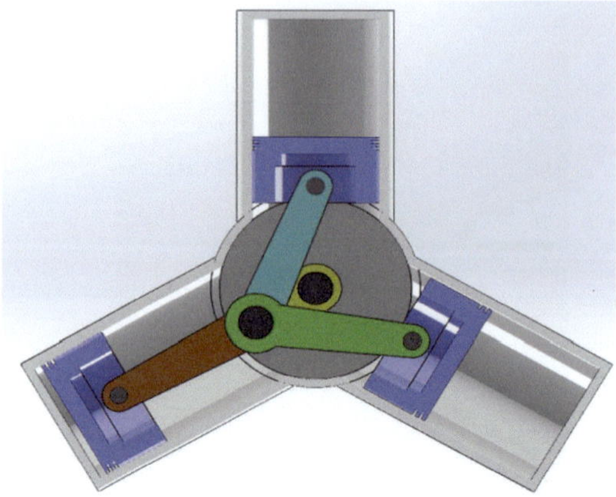

Fig. 2.76 Radial engine mechanism. Designed by the author and implemented by both the author and Pranav Dhola. Courtesy of Pranav Dhola

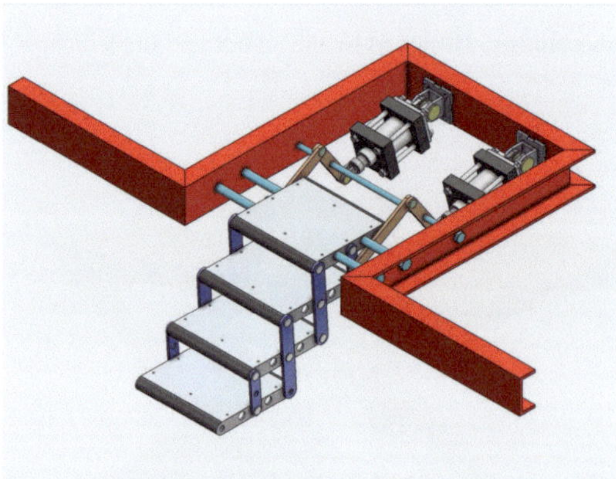

Fig. 2.77 Aircraft foldable stair mechanism. Figure and associated Folding Staircase Mechanism.mp4 video designed by the author and implemented by both the author and Pranav Dhola. Both courtesy of Pranav Dhola

Problem 18: Lift Table (Fig. 2.75).

Problem 19: Radial Engine (Fig. 2.76).

Problem 20: Aircraft Folding Stair (Fig. 2.77).

References

1. *Design of Machinery*, Robert L. Norton, McGraw-Hill; 5th Edition, 2011.
2. *Theory of Machines and Mechanisms*, John Uicker, Gordon Pennock and Joseph Shigley, Oxford University Press; 5th Edition, 2016.
3. *Machines & Mechanisms: Applied Kinematic Analysis*, David Myszka, Pearson; 4th Edition, 2011.
4. *Robot Dynamics Algorithms*, Roy Featherstone, Springer, 1st Edition,1987.
5. *Physics for Engineers and Scientists*, Hans Ohanian and John Markert, W. W. Norton & Company; 3rd Edition, 2006.

3 Mechanisms and Machines Fundamentals

Chapter Objectives

1. Understanding and sketching kinematic diagrams for machines and mechanisms.
2. Understanding and distinguishing between four-bar mechanisms, slider–crank mechanisms, and inverted slider–crank mechanisms in real-world applications.
3. Understanding and distinguishing the difference between crank, rocker, and coupler using Grashof's law.
4. Determining and analyzing high-order and special types of closed-chain mechanisms.
5. Understanding and analyzing motor actuators and hydraulic actuators.
6. Understanding and analyzing the energy flow to a certain mechanism from sources such as electricity or fuel consumption.

3.1 Introduction

In this chapter, we introduce the kinematic diagrams of machines including the basic symbols of various nodes, links, and joints. After that, we show that four-bar mechanisms, slider–crank mechanisms, and inverted slider–crank mechanisms are major closed-chain mechanisms. Grashof's law is introduced. This law presents the necessary condition for a mechanism to be a Grashof linkage. That is, one link in the linkage must be capable of 360° rotation. An introduction to five-bar mechanisms, six-bar mechanisms, and special types of mechanisms such as the quick-return mechanism is also presented. Lastly, we present an introduction to machine actuators for electric motors and hydraulic piston cylinders. We will make use of the nomenclature L_1, L_2, L_3, etc. to identify the links in a mechanism.

Supplementary Information The online version contains supplementary material available at https://doi.org/10.1007/978-3-031-72701-6_3.

A. Omran, *Kinematics and Dynamics of Machines*, https://doi.org/10.1007/978-3-031-72701-6_3

3.2 Kinematic Diagram

Definition

Kinematic Diagram:

A kinematic diagram, or kinematic scheme, is a simplified and handy visualization method to illustrate the connectivity of links and joints of a mechanism or machine rather than the fully detailed shape of each part. Links are often presented as geometric objects, such as lines, triangles, or squares, that support schematic versions of the joints of a mechanism or machine. A kinematic diagram is sometimes called a joint map or a skeleton diagram [1–3].

Nowadays, we use engineering graphic software tools such as SolidWorks and AutoCAD to design and analyze machines. These software sets provide a high-quality visualization that can be very close to the real machines. The computational time and hourly effort of these sets are expensive and not efficient in the early design phases. For early analysis, we need a handy or simple sketch for the basic elements such as nodes, joints, and links. This kind of sketch or diagram should avoid detailed graphics. We call such diagrams ***kinematic diagrams***.

A kinematic diagram has three major elements:

1. ***Joints***: Table 3.1 lists the equivalent kinematic diagram shape of each joint. As you can see, there is a difference when some joints are connected between two non-ground links or between a ground link and a non-ground link. There are two views (front and side) of the gear joint. The side view is very important when we sketch the kinematic or schematic diagram of a transmission later in Chap. 10.
2. ***Nodes***: There are two types of nodes: joint nodes and end-effector nodes. Each joint is considered a node. The end-effector nodes are a set of non-joint points of interest (belonging to the mechanism) that are assigned to accomplish certain tasks. Later, we will see how to identify the end-effector nodes in the door mechanism and the skid steer mechanism. We use a solid circle as the kinematic diagram representation of each node.
3. ***Links***: As shown in Fig. 3.1, there are different types of links depending on how many nodes (important points or areas) are located on the link. The binary link has two nodes and we sketch it as a solid line. For all other types of links (ternary, quaternary, and any link with more than two nodes), we will crosshatch the area between the link's nodes as shown in Fig. 3.1.

We are now going to use the kinematic diagram's three major elements to sketch an entire kinematic diagram. The following steps can be used to sketch a kinematic diagram from a 3D model or a real machine.

1. Specify the ground link and the ground nodes.
2. Specify and sketch the type of grounded joints (joints that are connected to the ground link).
3. Specify the unground joints and sketch them.
4. Specify the unground links and their nodes (joint nodes and end-effector nodes). Sketch each link as a solid line or a crosshatched area.

The upcoming examples are for practicing the sketching of kinematic diagrams. For the remainder of the book, we will make the kinematic diagram a must for our analyses.

Table 3.1 Joints' kinematic diagram shapes. Figures in the second column are designed by the author and implemented by both the author and Pranav Dhola. Courtesy of Pranav Dhola

Joint Name	Graphical Shape	Kinematic Diagram Shape
Revolute, Pin, or Hinge Joint between non-ground links.	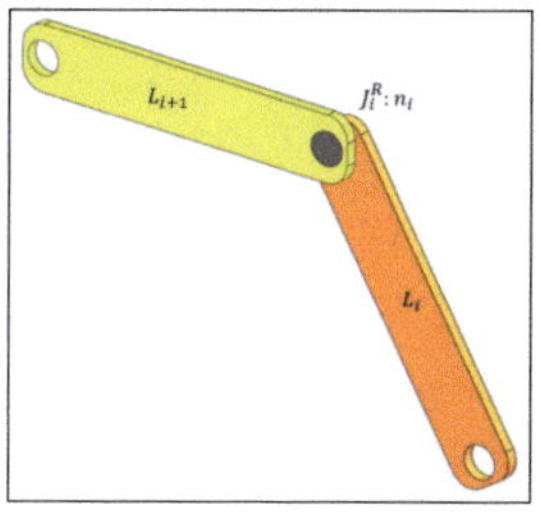	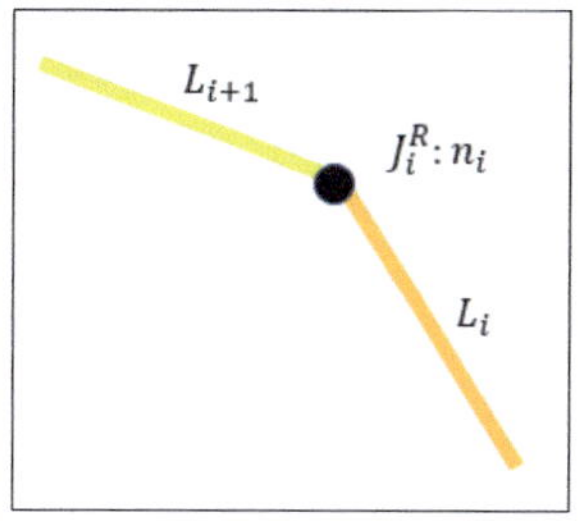
Revolute, Pin, or Hinge Joint between a ground link and a non-ground link.	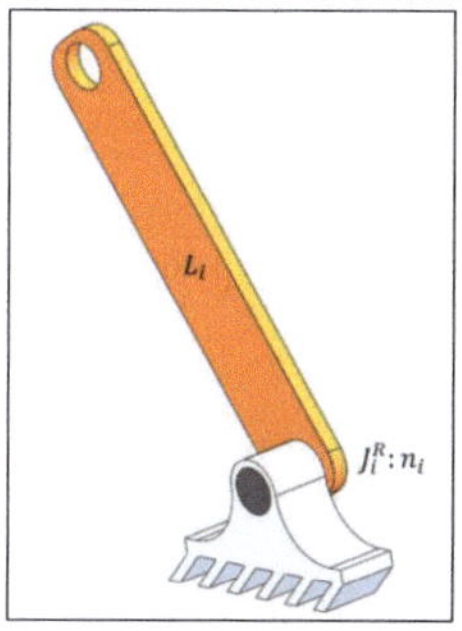	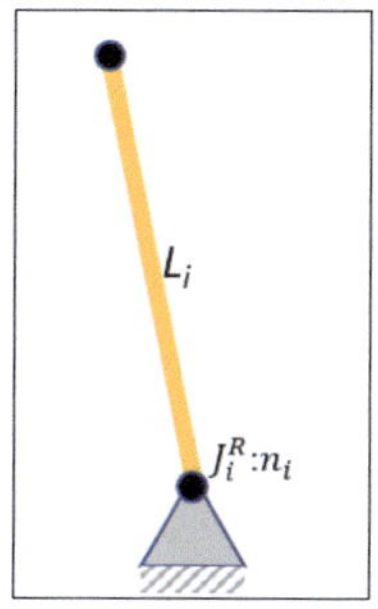
Prismatic, Sliding, or Piston Joint between non-ground links.	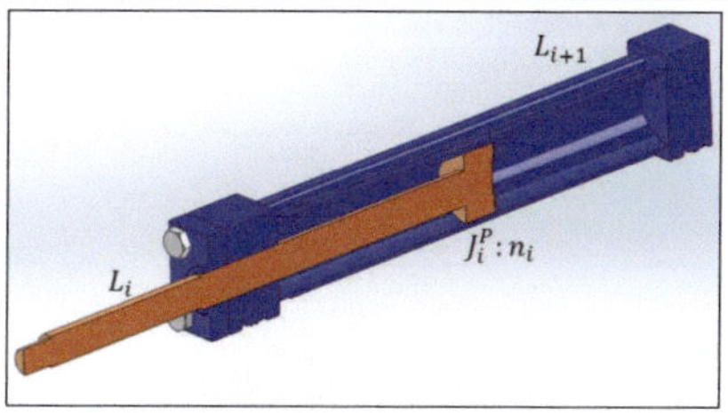	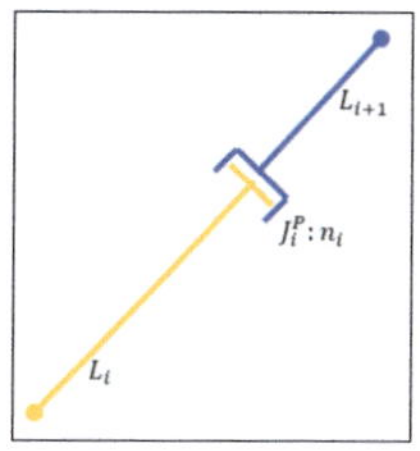
Prismatic, Sliding, or Piston Joint between a ground link and a non-ground link.	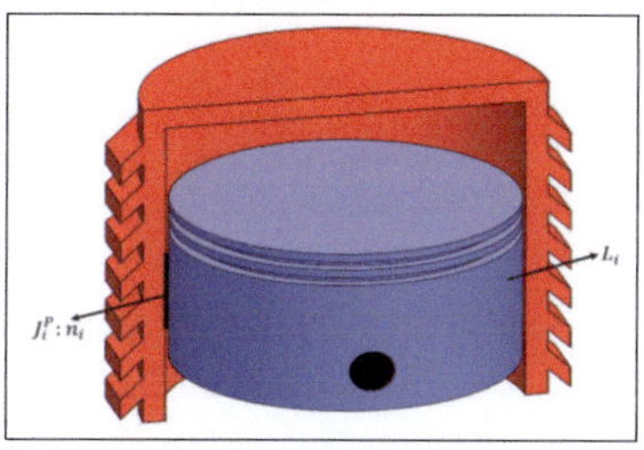	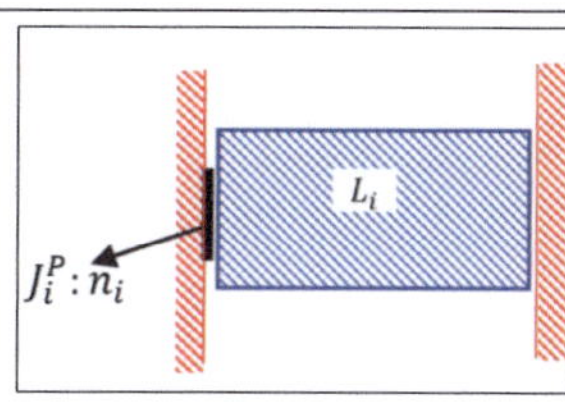
Pin-in-a-slot Joint between one ground link and a non-ground link.	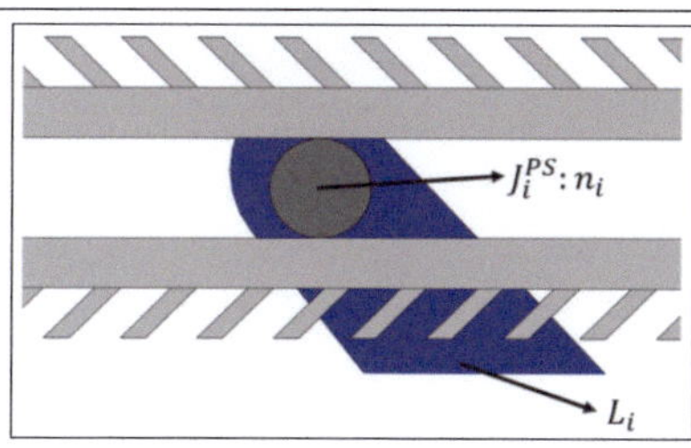	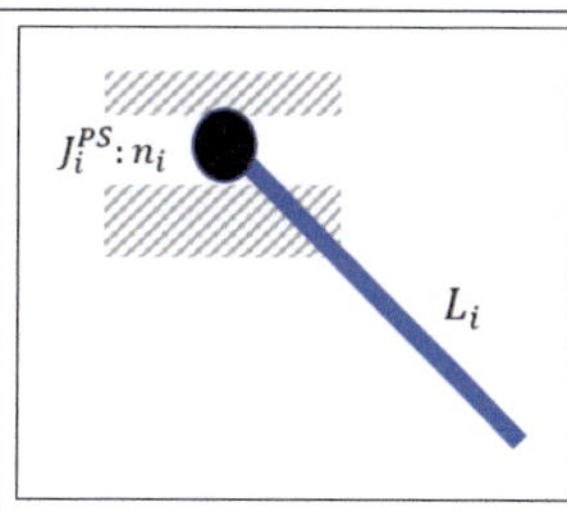

Table 3.1 (continued)

Pin-in-a-slot Joint between non-ground links.	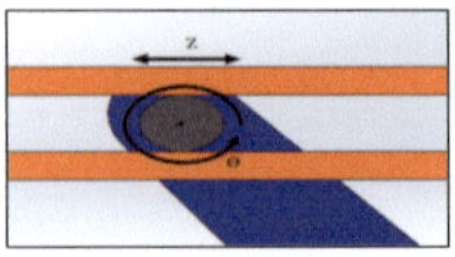	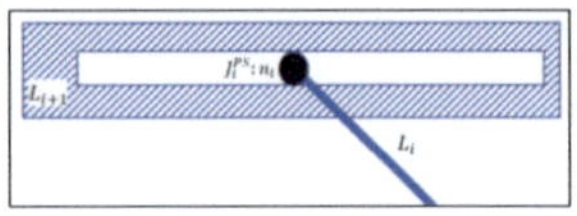
CAM Joint	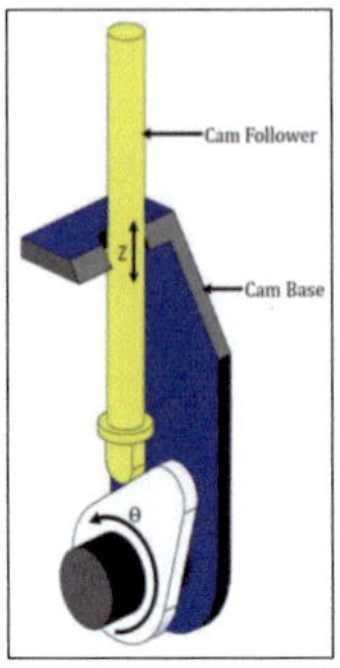	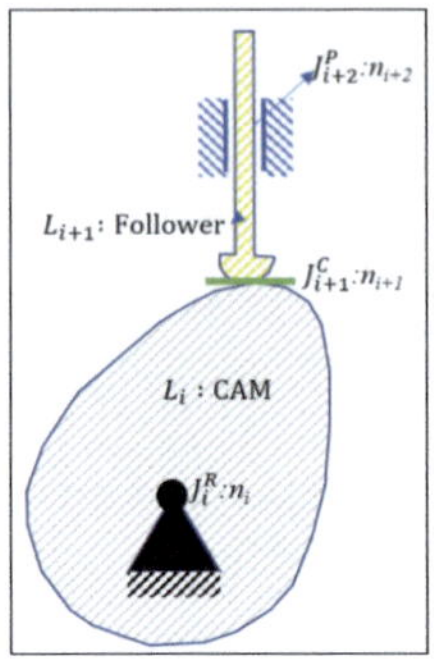
Gear front view joint	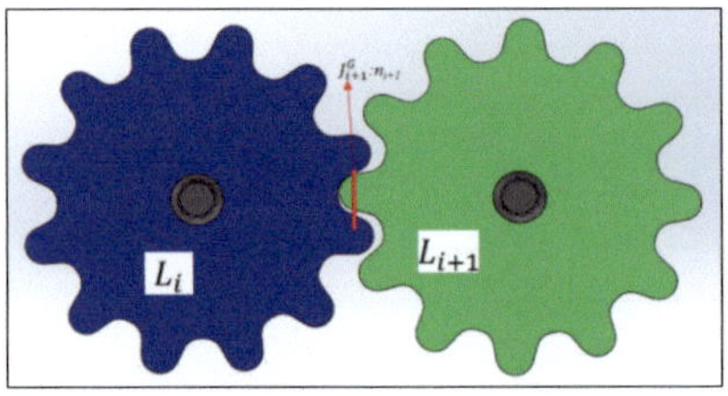	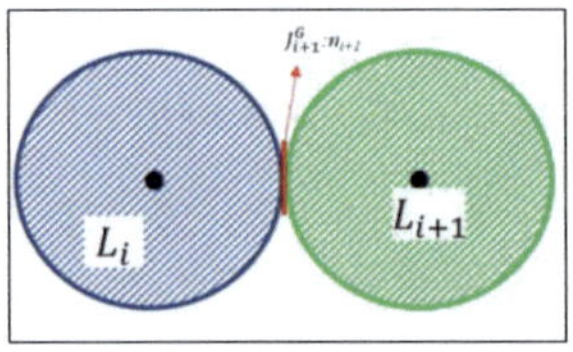
Gear side view joint	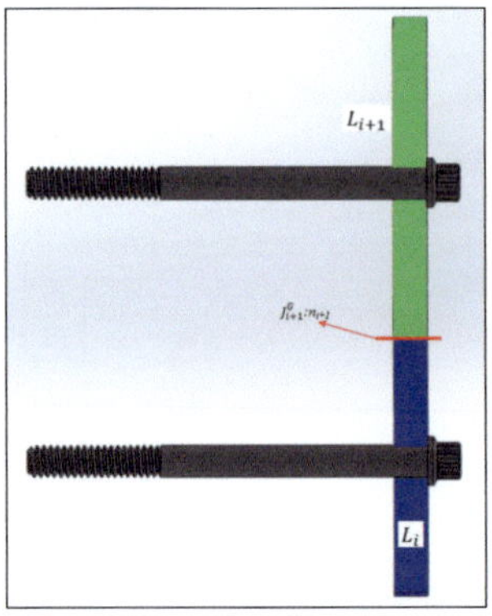	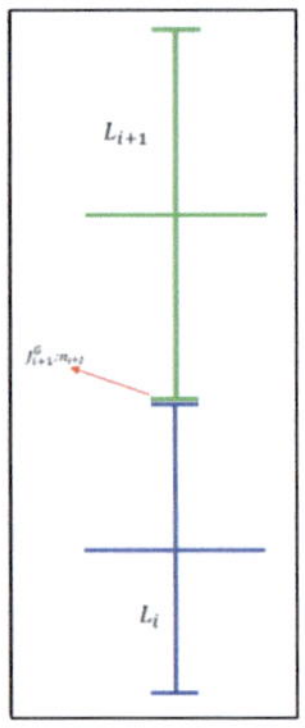

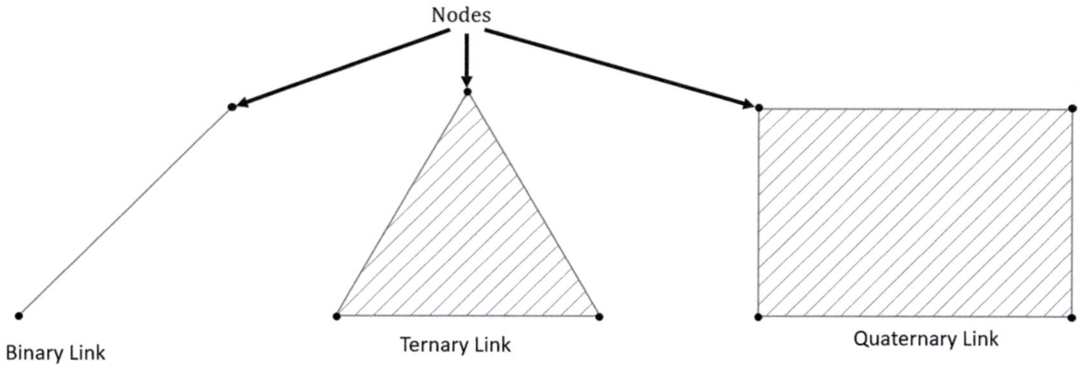

Fig. 3.1 Kinematic diagram of different types of links

Example 3.1

Sketch the kinematic diagram of the door linkage shown in Fig. 3.2.

Fig. 3.2 Door linkage. Designed by the author and implemented by both the author and Pranav Dhola. Courtesy of Pranav Dhola

Solution

From Example 2.1 in Chap. 2, we are going to use the same indices of joints, links, and nodes. The kinematic diagram is shown in Fig. 3.3. Note, we used the following color codes to specify the kinematic diagram steps:

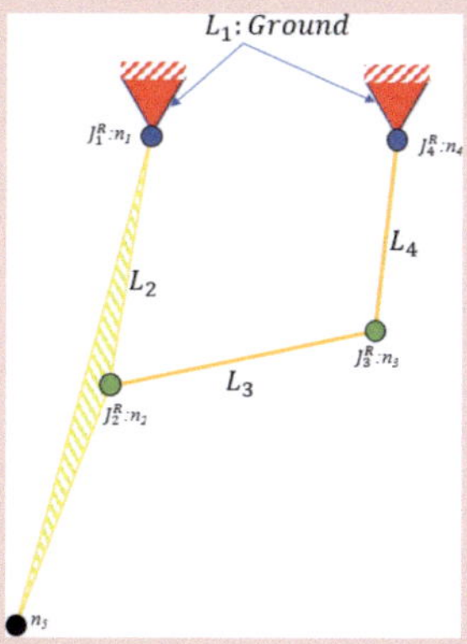

Fig. 3.3 Kinematic diagram of door linkage

Step 1: is colored red.
Step 2: is colored blue.
Step 3: is colored green.
Step 4: is colored black for the end-effector node and orange and yellow for links.

Note, there is no joint at node n_5. The node n_5 is an end-effector node. We encourage the reader to watch the attached animation video of the door linkage.

Example 3.2

Sketch the kinematic diagram of the articulated bucket mechanism in a skid steer vehicle as shown in Fig. 3.4.

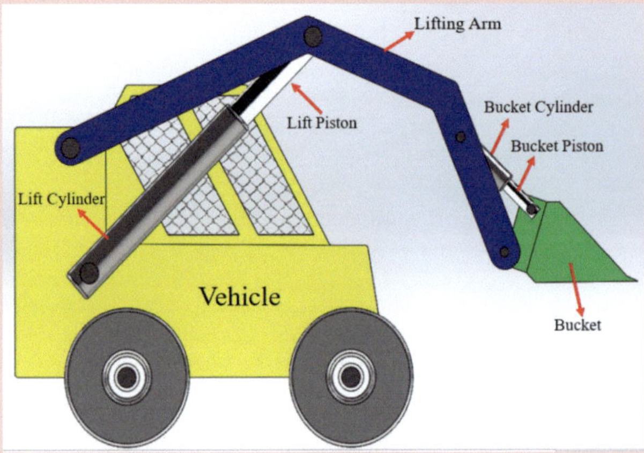

Fig. 3.4 Skid steer articulated bucket mechanism. Figure and associated Skid Steer Articulated Bucket Lift Mechanism.mp4 video designed by the author and implemented by both the author and Pranav Dhola. Both courtesy of Pranav Dhola

Solution

From Example 2.5 in Chap. 2, we are going to use the same indices of joints, links, and nodes. The kinematic diagram is shown in Fig. 3.5. Note that we used the following color codes to specify the kinematic diagram steps:

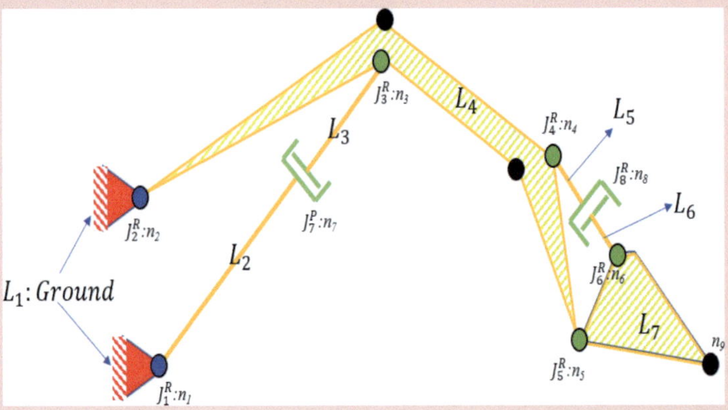

Fig. 3.5 Kinematic diagram of skid steer articulated bucket mechanism

Step 1: is colored red.
Step 2: is colored blue.
Step 3: is colored green.
Step 4: is colored black for the end-effector nodes and yellow for links.

Example 3.3

Sketch the kinematic diagram of the casement window opener as shown in Fig. 3.6.

Fig. 3.6 Casement Window Opener Mechanism. Figure and associated Casement_Window_Mechanism video designed by the author and implemented by both the author and Pranav Dhola. Both courtesy of Pranav Dhola

Solution

Figure 3.7 shows our selected indices of nodes, links, and joints. The mechanism has six links: L_1 (ground), L_2 (L_2^{P1} is the driving gear and L_2^{P2} is the arm), L_3 (L_3^{P1} is the big intermediate gear and L_3^{P2} is the small intermediate gear), L_4 (driven gear with arm), L_5 (window frame), and L_6 (slider). There are eight joints: J_1^R, J_2^R, J_3^R, J_4^R, J_5^R, J_6^P, J_7^G, and J_8^G. There are eight nodes: n_1, n_2, n_3, n_4, n_5, n_6, n_7, and n_8. The kinematic diagram is sketched in Fig. 3.8 with the color code below.

Step 1: is colored red.
Step 2: is colored blue.
Step 3: is colored green.
Step 4: is colored black for the end-effector nodes and yellow for links.

Link L_2 has two parts: the gear part and the link part. These two parts are welded together. We label the gear part L_2^{P1} and the arm L_2^{P2}. We can apply the same rule for Link L_4. To better understand the mechanism, we encourage the reader to watch the attached animation video of the casement window.

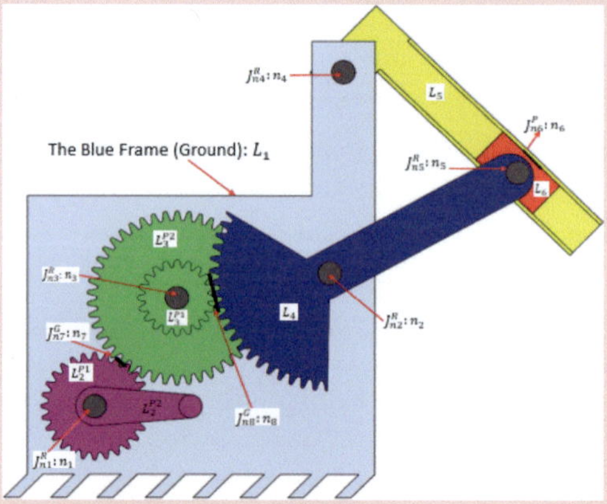

Fig. 3.7 Indices of nodes, links, and joints for a window opener mechanism. Designed by the author and implemented by both the author and Pranav Dhola. Courtesy of Pranav Dhola

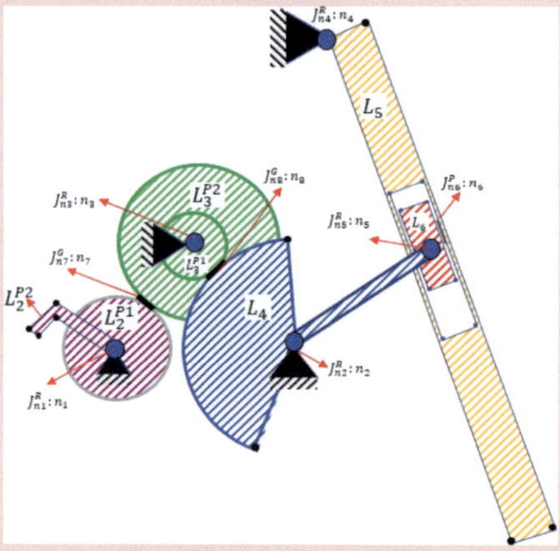

Fig. 3.8 Kinematic diagram of a casement window opener mechanism

Example 3.4

Sketch the kinematic diagram of the piston engine valve opener as shown in Fig. 3.9.

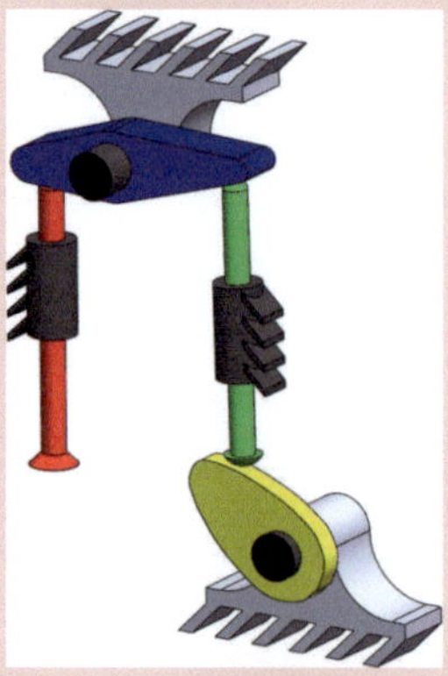

Fig. 3.9 Piston engine valve opener. Designed by the author and implemented by both the author and Pranav Dhola. Courtesy of Pranav Dhola

Solution

Figure 3.10 shows our selected indices of nodes, links, and joints. The mechanism has five links: L_1 (ground), L_2, L_3, L_4, and L_5. There are seven joints: J_1^R, J_2^C, J_3^P, J_4^{SP}, J_5^R, J_6^{SP}, and J_7^R that are considered as 7 nodes: n_1, n_2, n_3, n_4, n_5, n_6, and n_7. The kinematic diagram is sketched in Fig. 3.11 with the color code below.

Step 1: is colored red.
Step 2: is colored blue.
Step 3: is colored green.
Step 4: is colored black for the end-effector nodes and yellow for links.Note, the joints J_4^{SP} and J_6^{SP} can be considered as cam joints or pin-in-a-slot joints. We encourage the reader to watch the attached animation video of a piston engine valve opener.

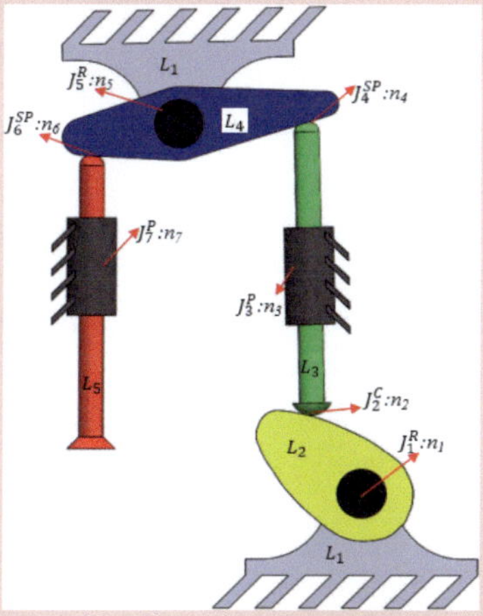

Fig. 3.10 Piston engine valve opener in 2D. Designed by the author and implemented by both the author and Pranav Dhola. Courtesy of Pranav Dhola

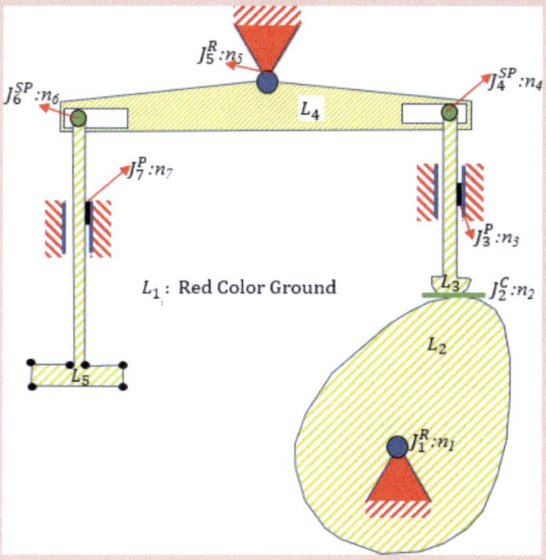

Fig. 3.11 Kinematic diagram of piston engine valve opener

3.3 Four-Bar Mechanism

Definition

Four-Bar Mechanism:

A four-bar linkage, also called a four-bar mechanism, is the simplest movable closed-chain linkage. It consists of four bodies, called bars or links, connected in a loop by four joints. Generally, the joints are configured so the links move in parallel planes, and the assembly is called a planar 2D four-bar linkage [4].

In Chap. 2, we concluded that a four-bar linkage must have a minimum number of 4 links for us to build a mechanism. We cannot build a mechanism with two- and three-bar linkages. Most designers prefer to have a closed-chain machine to avoid flexibility. A four-bar linkage has 1DoF. For this reason, the four-bar linkage needs only one input to move the entire mechanism. For example, the pumpjack mechanism shown in Fig. 3.12 has four links. As an input, a motor is used to rotate the crank that causes the horsehead to move up to pump oil. The horsehead is considered an output for the pumpjack machine. The pumpjack mechanism is converting a rotational input motion from the motor driving the crank into an oscillatory rotational output motion of the walking beam. We encourage the reader to watch the attached animation video of the pumpjack mechanism.

A four-bar linkage is the simplest and most common linkage model. For example, the door mechanism in Fig. 3.2 is a four-bar mechanism. Many other examples can support our statement about how common the four-bar linkage model is. For this reason, we are going to study the four-bar linkage models in more detail during this section.

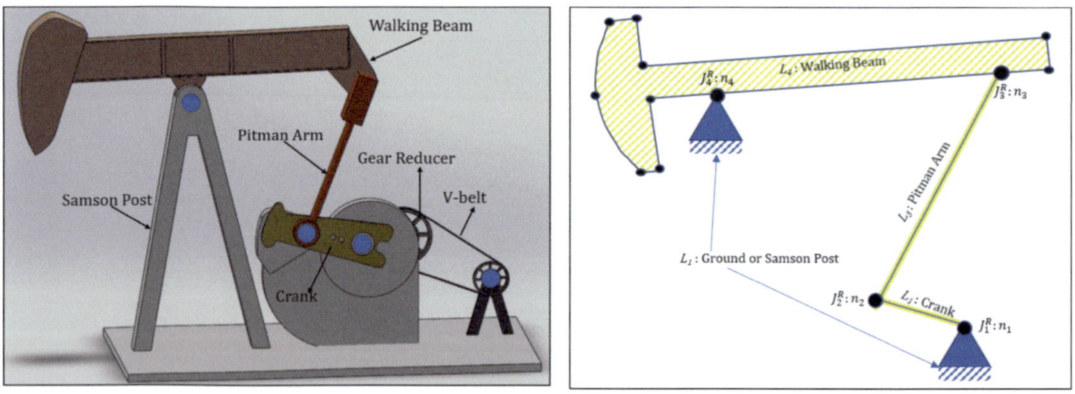

Fig. 3.12 Pumpjack mechanism. Designed by the author and implemented by both the author and Pranav Dhola. Courtesy of Pranav Dhola

3.3.1 Grashof's Law

Grashof's Law (also Known as the Grashof Condition or Grashof Criterion)

Grashof's law is a simple relationship that predicts the rotational behavior, or rotationality, of a four-bar linkage model [1–6].

Grashof's law offers an estimation for a four-bar mechanism's motion from the dimensional ratios of the four links. First, we are going to define:

s—the length of the shortest link.
l—the length of the longest link.
p and q—the lengths of the intermediate length links.

The law states that a four-bar mechanism is called a Grashof mechanism if it satisfies $s+l \leq p+q$. Such a mechanism has at least one crank link or a link that can make a complete 360° cycle. In all other cases, when $s+l>p+q$, all links will be rockers. Such mechanisms are called non-Grashof. In general, Grashof's law estimates the motion of any four-bar mechanism based on the length of each link. In addition, this condition is used to evaluate whether each link is a crank or a rocker. Further discussion and derivation of Grashof's law are listed in Refs. [5, 6].

Engineering Tips—Crank and Rocker

A crank link is a pivoted link that can make a complete 360° rotation, such as a bicycle crank (pedal) or the engine crankshaft.

A rocker link is a pivoted link that rotates but is not able to make a complete 360° rotation such as the pitman arm in a pumpjack mechanism.

Based on Grashof's law, if $s+l>p+q$, no link will be able to complete a full revolution, or the mechanism has no crank link, and all the non-grounded links are rocker links. We call a mechanism with $s+l>p+q$ a *triple rocker mechanism* or GRRR mechanism where GRRR refers to ground–rocker-rocker-rocker (L_1-L_2-L_3-L_4).

There is at least one link that makes a full rotation when $s+l \leq p+q$. The five cases of this condition are:

Case 1 Crank–Rocker with $s+l<p+q$: The shortest link is connected to the ground, or the shortest link is located as L_2 or L_4 as shown in Table 3.2. The shortest link is the only link that can make a full revolution. If the shortest link is L_2, the mechanism is called a "crank–rocker" mechanism or GCCR mechanism where GCCR refers to ground–crank–coupler–rocker (L_1-L_2-L_3-L_4). Or L_1 is the fixed link, L_2 is the crank, L_3 is the coupler, and L_4 is the rocker. If the shortest link is L_4, the mechanism is called rocker-crank or GCRC.

Case 2 Double Crank with $s+l<p+q$: The shortest link is the ground link, or the shortest link is located at L_1. Links L_2 and L_4 will be able to make a full revolution. The mechanism is then called a double crank mechanism. If L_3 is capable of making a rotation, the mechanism is called GCCC;

Table 3.2 Four-bar linkage types

Case	Linkage distribution	Motion	Example of shape
Case 1 Crank–rocker (GCCR)	$s+l<p+q$ s is pivoted to the ground or s is located at L_2 or L_4	The shorter link is the crank link, and it is the only link that can make a full revolution. The other two links are a coupler and a rocker links, GCCR or GRCC	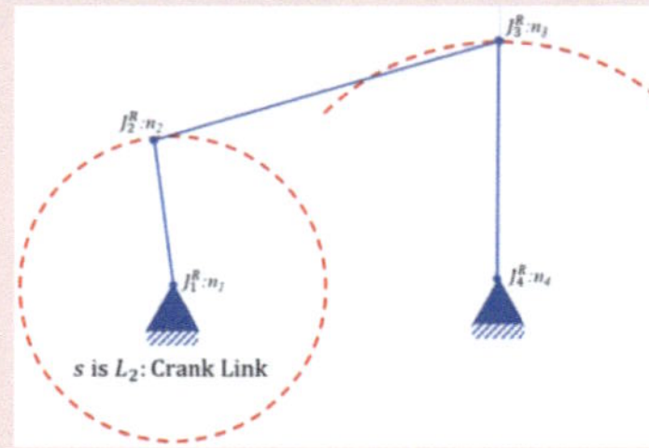
Case 2 Double crank (GCRC)	$s+l<p+q$ s is the ground link or L_1	The two links pivoted to the ground links, or L_2 and L_4 are cranks. GCRC or GCCC	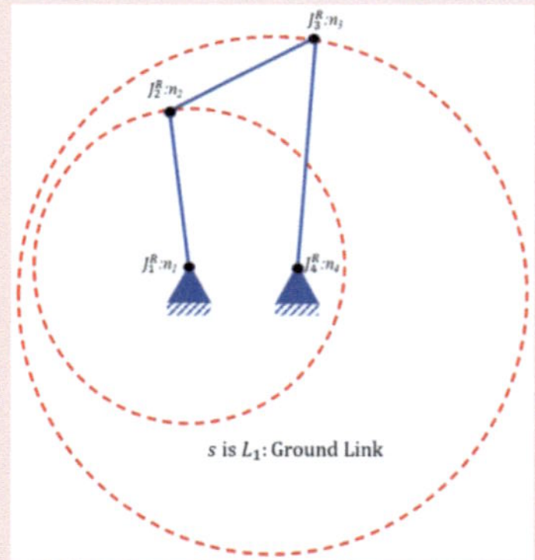
Case 3 Double rocker (GRCR)	$s+l<p+q$ s is not connected to the ground or L_3	The shorter link L_3 is the only crank link	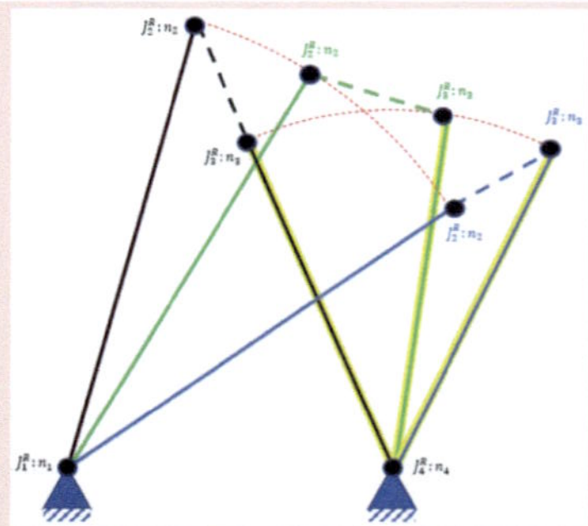

(continued)

Table 3.2 (continued)

Case	Linkage distribution	Motion	Example of shape
Case 4 Collinear parallelogram double crank	$s+l=p+q$ The equal links are opposite to each other or $L_1=L_3$ and $L_2=L_4$	The two links pivoted to the ground links L_2 and L_4 are cranks, GCRC The motion can be a singular configuration when the four links overlap each other as shown	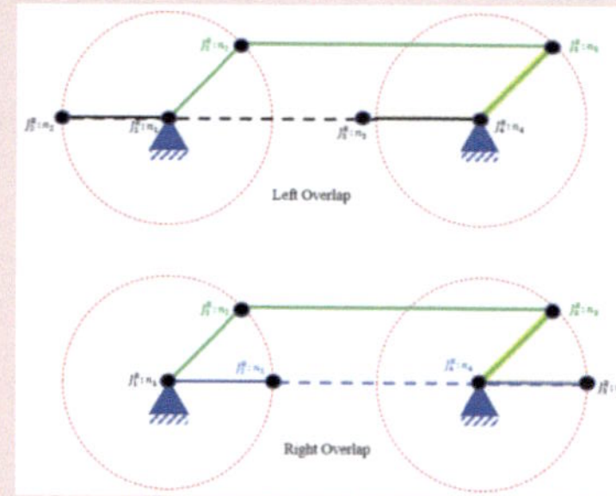
Case 5 Change point mechanism (collinear deltoid crank–rocker)	$s+l=p+q$ The equal links are adjacent to each other which means $L_1=L_2$ and $L_3=L_4$ or $L_2=L_3$ and $L_1=L_4$	The shorter link is the crank link. The other two links are rocker links, GCRR or GRRC The motion can be a singular configuration when the four links overlap each other	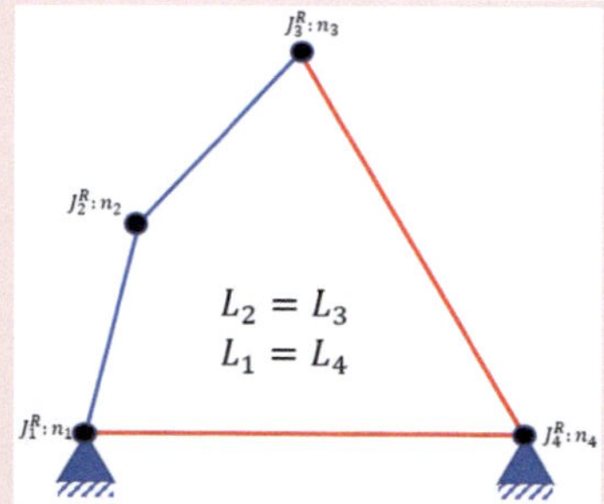

Associated videos: Crank Rocker Mechanism.mp4, Double Crank Mechanism.mp4, Double Rocker Mechanism.mp4, Parallelogram Mechanism.mp4 and Collinear Deltoid Crank Rocker Mechanism.mp4 designed by the author and implemented by both the author and Pranav Dhola. Courtesy of Pranav Dhola otherwise, it is called GCRC, where GCCC refers to ground crank–crank–crank (L_1-L_2-L_3-L_4) and GCRC refers to ground crank–rocker–crank (L_1-L_2-L_3-L_4). An example is listed in Table 3.2.

Case 3 Double Rocker with $s+l<p+q$: The shortest link is not connected to the ground, or the shortest link is located at L_3 and is the only link to make a full revolution, thus being the only crank in the mechanism. The other two links L_2 and L_4 are rocker links. The mechanism is then called a double rocker mechanism, or GRCR. An example is shown in Table 3.2.

Case 4 Collinear Parallelogram Double Crank with $s+l=p+q$: In this case, we have two pairs of equal length. The equal links are *opposite* to each other or $L_1=L_3$ and $L_2=L_4$. We can also call this mechanism a double crank or GCRC. The four links can be in a straight line as shown in Table 3.2.

Case 5 Change Point Mechanism (Collinear Deltoid Crank–Rocker) with $s+l=p+q$: In this case, we have two pairs of equal length. The equal links are *adjacent* to each other so that $L_1=L_2$ and $L_3=L_4$ or $L_2=L_3$ and $L_1=L_4$. We can also call this mechanism a crank–rocker or GCRR. The four links can be in a straight line as shown in Table 3.2. This mechanism's motion is indeterminate—when the mechanism reaches a position where the links are collinear, then friction, elasticity, and dimensional tolerances will cause its motion to be unpredictable.

Note that case 4 and case 5 become the same if $s=l=p=q$. In a four-bar mechanism, if the link L_1 is not the ground, we call the mechanism an inversion. The above five cases are applied to the inversion mechanisms. We encourage the reader to watch the accompanying five animation videos for each Grashof case. In the examples below, we will see how useful Grashof's condition can be to estimate the motion of any four-bar mechanism.

Example 3.5

For the shown four-bar linkage model in Fig. 3.13, specify whether the mechanism is Grashof or non-Grashof. Note, all dimensions are in cm. Specify the Grashof case of this mechanism

Fig. 3.13 Four-bar mechanism

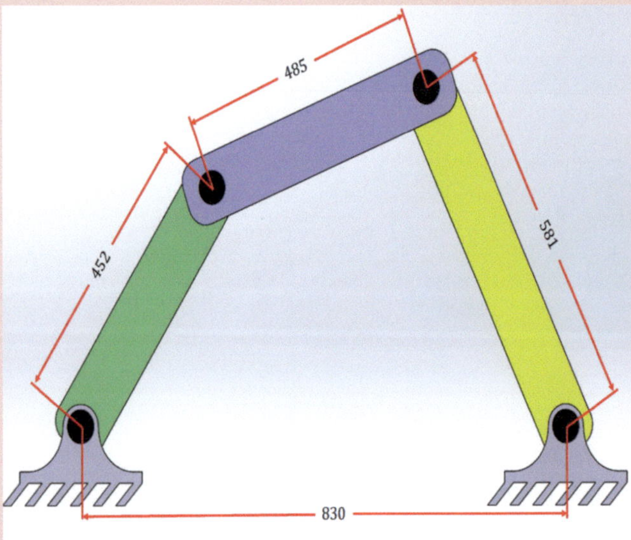

Solution

First, we are going to sketch the kinematic diagram as shown in Fig. 3.14. Based on the labels of the four links: $s=452$ cm, $l=830$ cm, $p=485$ cm, and $q=581$ cm. Since $s+l=1282$ cm and $p+q=1066$ cm, then $s+l>p+q$ and the mechanism is non-Grashof.

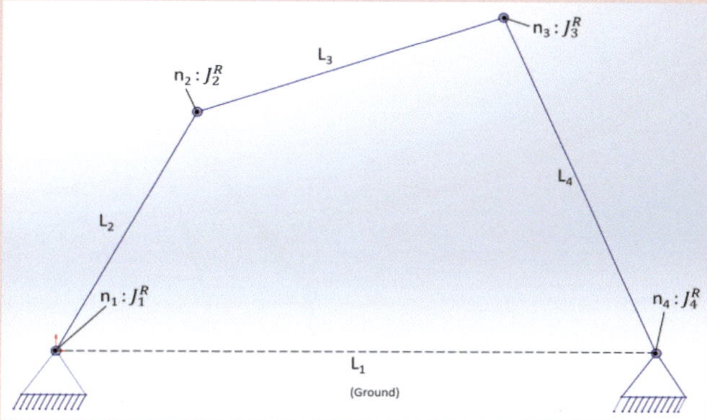

Fig. 3.14 Kinematic diagram of four-bar mechanism

Example 3.6

For the bicycle rider shown here we can come up with the four-bar mechanism in Fig. 3.15. The crank arm length is 10 cm, the distance from the saddle to the chainring center is 62 cm, the length from buttocks to knee is 46 cm, and the length from knee to heel is 45 cm. Prove that this rider is capable of rotating the bicycle crank 360° around the bottom bracket hinge. Determine the Grashof case of this bicycle mechanism

Fig. 3.15 Bicycle rider mechanism. Figure and associated Bicycle_Rider_Mechanism.mp4 video designed by the author and implemented by both the author and Pranav Dhola. Both courtesy of Pranav Dhola

Solution

First, we are going to sketch the kinematic diagram as shown in Fig. 3.16. The bicycle mechanism has four links L_1 (ground), L_2, L_3, and L_4. There are four joints J_1^R, J_2^R, J_3^R, and J_4^R that are considered as four nodes n_1, n_2, n_3, and n_4. The kinematic diagram is sketched in Fig. 3.16 with the color code below.

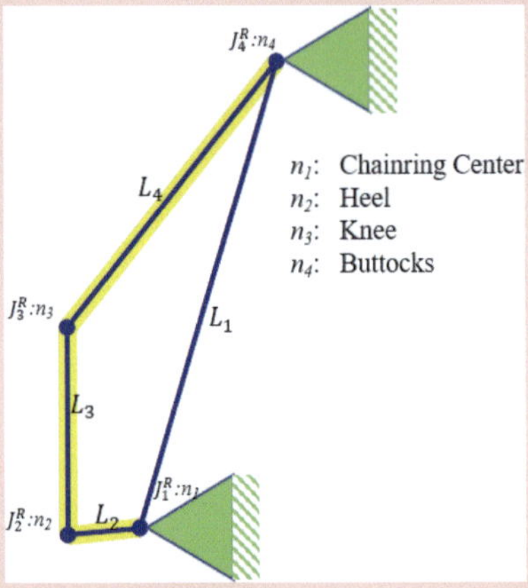

Fig. 3.16 Kinematic diagram of a bicycle rider mechanism

Step 1: is green.
Step 2: is blue.
Step 3: is also blue.
Step 4: is blue-yellow for links.

Based on the labels of the four links: $s = 10$ cm, $l = 62$ cm, $p = 46$ cm, and $q = 45$ cm. We can calculate $s + l = 72$ cm and $p + q = 91$ cm. Since $s + l < p + q$, the mechanism is Grashof and the driver can rotate the bottom bracket 360°.

Since s, or L_2, is connected to the ground and $s + l < p + q$, the mechanism is a crank–rocker mechanism, GCRR. Note that this mechanism can be classified as an inversion. We encourage the reader to watch the animation video of a bicycle rider mechanism.

3.4 Slider–Crank Four-Bar Mechanism

Definition

Slider–Crank Four-Bar Mechanism:

A slider–crank linkage is a four-link mechanism with three revolute joints and one prismatic, or sliding, joint [1–4].

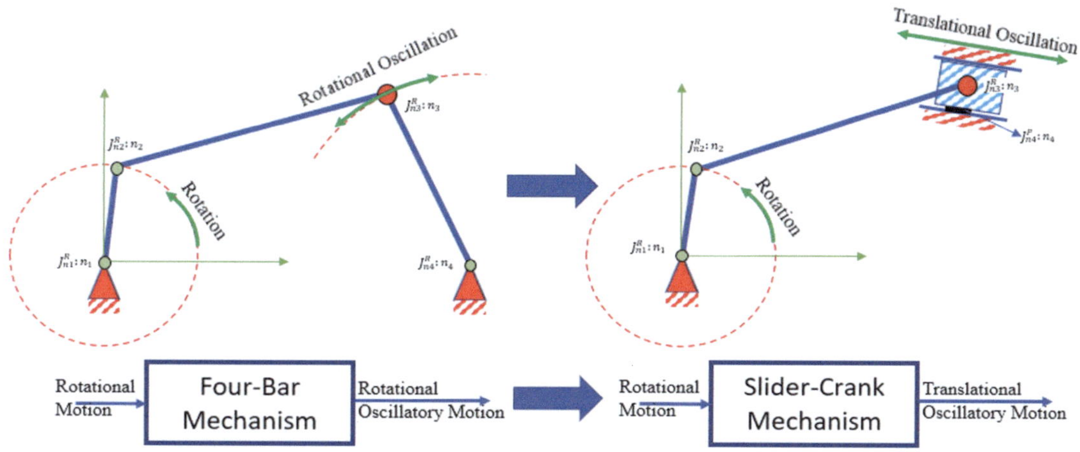

Fig. 3.17 Mapping from four-bar mechanism into slider–crank mechanism

In our analysis so far, the four-bar mechanism converts an input rotational/oscillational motion into an output rotational/oscillational motion depending on the Grashof condition and the inversion configuration. For some engineering designs, there is a need to convert a rotational/oscillational motion into a translational motion or vice versa. This goal can be accomplished by replacing one of the revolute joints from a four-bar mechanism with a prismatic joint as shown in Fig. 3.17. This is called a slider–crank mechanism in which the prismatic joint is a ground node. As we can see in Fig. 3.17, the four-bar mechanism has a rotational motion as an input at node n_1 and a rotational oscillatory motion at node n_3. When it is converted to a slider–crank, the mechanism has a rotational motion as an input at node n_1 and a translational oscillatory motion at node n_3.

The slider–crank shown in Fig. 3.17 has four links connected by three pin joints and one sliding joint. If we apply Grübler's equation, we find that the slider–crank four-bar mechanism has 1DoF; therefore, the four-bar linkage needs only one input to move the entire mechanism. For example, in a piston–cylinder mechanism, the input comes from the expansion of gases caused by the air–fuel reaction pushing the piston down. This causes the connecting rod to move and rotate. Consequently, the crankshaft rotates around the holder. The output is what the mechanism aims to accomplish. For the piston–cylinder, the output is to rotate the crankshaft around the holder as shown in Fig. 3.18. We encourage the reader to watch the attached animation video of a piston–cylinder mechanism.

3.5 Inverted Slider–Crank Four-Bar Mechanism

Definition

Inverted Slider–Crank Four-Bar Mechanism:

An inverted slider–crank mechanism is defined as when the coupler link of a slider–crank mechanism is attached to the ground. So, the inverted slider–crank is a simple inversion of a slider–crank mechanism [1–4, 7, 8].

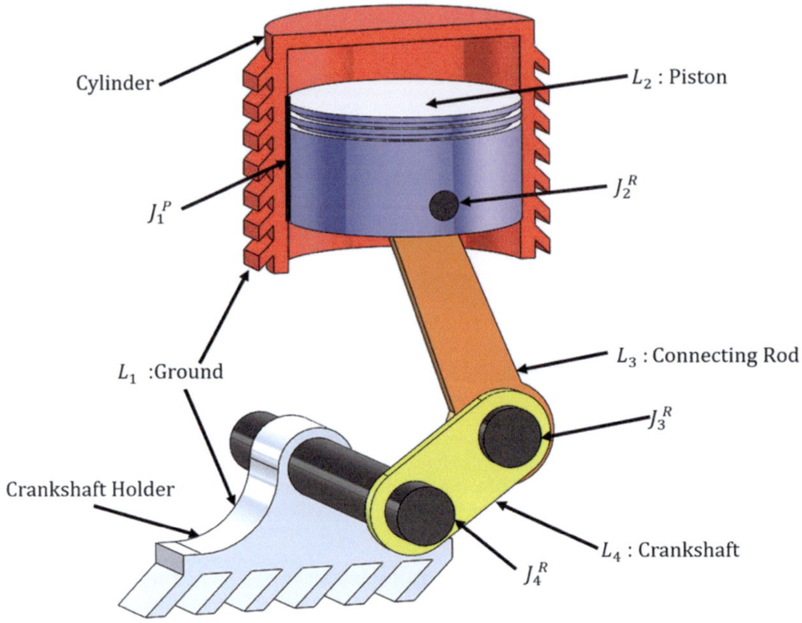

Fig. 3.18 Piston–cylinder mechanism and kinematic diagram. Designed by the author and implemented by both the author and Pranav Dhola. Courtesy of Pranav Dhola

When we replace one of the revolute joints with a prismatic joint, there are two options: a grounded prismatic joint (slider–crank mechanism as shown in Fig. 3.17) or an ungrounded prismatic joint called an inverted slider–crank mechanism as shown in Fig. 3.19. As we can see in Fig. 3.19, the four-bar mechanism has a rotational motion as an input at node n_1 and a rotational oscillatory motion at node n_3. When it is converted to an inverted slider–crank, the mechanism has a rotational motion as an input at node n_1 and a translational oscillatory motion at node n_3.

Similar to a slider–crank, if we apply Grübler's equation, we can find that the inverted slider–crank four-bar mechanism has 1DoF. This means that we need only one input to move the entire mechanism. For example, in the car suspension mechanism, as shown in Fig. 3.20, the input comes from the ground bouncing the wheel up and down. This input excites the hydraulic piston–cylinder or the prismatic joint to move as part of the suspension mechanism. The hydraulic piston–cylinder (or the prismatic joint) is called the *shock absorber*. It keeps the tires in contact with the road when the car goes over any bump. (Note that the spring is a separate component and is not part of the shock absorber.) An animation video of the car suspension mechanism is attached.

The inverted slider–crank can be itemized in four different cases as shown in Fig. 3.21 whereby we are going to call the links L_A, L_B, L_C, and L_D. Link L_B is the crank. Link L_C is the connecting rod. Link L_D is the slider.

Case 1: Link L_A is fixed, or L_A is the ground. As the crank L_B rotates in a cyclic motion, the slider L_D oscillatory moves up and down or vice versa. One example is the piston–cylinder mechanism in Fig. 3.18.

Case 2: Link L_B, the crank, is fixed, or L_B is the ground. As the connecting rod L_C oscillatory rotates, the slider L_D slides back and forth over the rotating link L_A. One example is the skid steer lift mechanism as shown in Fig. 3.4.

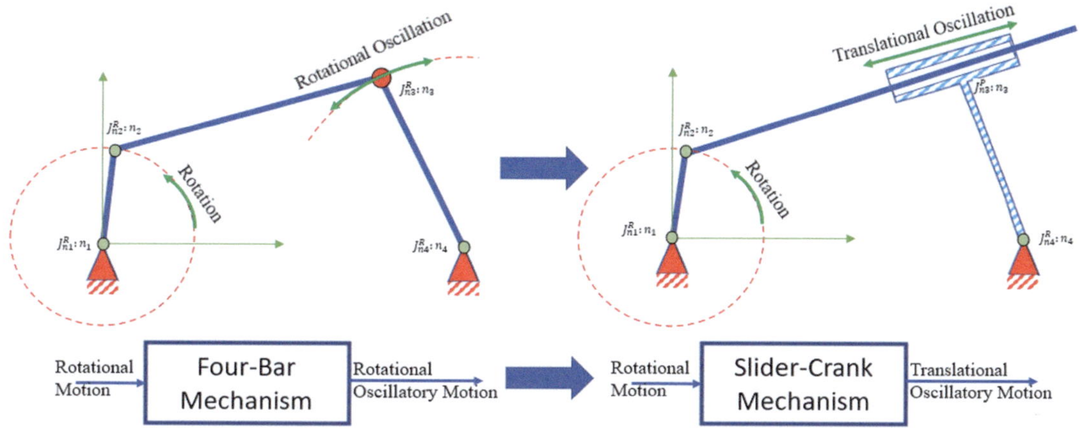

Fig. 3.19 Four-bar mechanism and inverted slider–crank mechanism

Fig. 3.20 Suspension mechanism. Figure and associated Car Suspension Mechanism.mp4 video designed by the author and implemented by both the author and Pranav Dhola. Both courtesy of Pranav Dhola

Case 3: Link L_C, the connecting rod, is fixed, or L_C is the ground. As the crank L_B rotates in a cyclic motion, the link L_D slides back-and-forth link L_A.

Case 4: Link L_D, the slider, is fixed, or $\underline{L_D}$ is the ground. As the crank L_B rotates in a cyclic motion, the link L_A slides in-and-out link L_D.

In general, there are many common examples of the inverted crank–slider. We are going to explore further examples in the solved examples and practice problems. We encourage the reader to watch the attached animated videos of the inverted slider–crank cases.

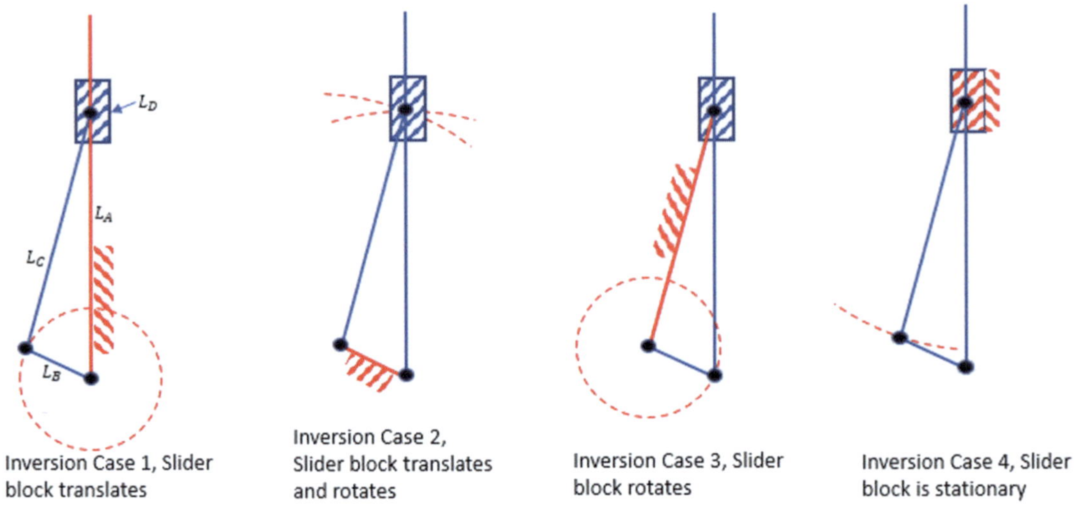

Fig. 3.21 Four different cases of the inverted slider–crank mechanism. Associated Case 1- Case 2- Case 3- and Case-4-Inverted Slider–Crank.Mechanism.mp4 videos designed and implemented by both the author and Pranav Dhola. Courtesy of Pranav Dhola

3.6 Five-Bar Mechanism

Figure 3.22 shows an example of a closed-chain five-bar mechanism that has one ground link L_1 and the remaining links are connected with five revolute joints. If we use Grübler's equation as in Eq. (3.1), the five-bar mechanism has 2DoF.

$$M = \text{DoF} = 3(n_l - 1) - 2n_P - n_h = 3(5 - 1) - 2 * 5 = 2 \tag{3.1}$$

The goal of most applications of five-bar mechanisms is to increase the DoF and avoid flexibility in designing a 2D robot arm. As an example, the SCARA robot arm shown in Fig. 3.23 has two rotational input motions at $J_1^R : n_1$ and $J_5^R : n_5$. The output signals are the x-direction motion and the y-direction motion of the node $J_3^R : n_3$. One common application of a five-bar mechanism is to locate a cutter at node n_3 to cut a piece of glass to a specific shape, or to locate a pin to plot a specific picture. We encourage the reader to watch the attached animation video of the five-bar robot.

3.7 Six-Bar Mechanism

A six-bar linkage is a one-degree-of-freedom mechanism that is constructed from six links and seven joints. One example of a six-bar mechanism is the car hood mechanism as shown in Fig. 3.24. The car hood mechanism has two four-bar submechanisms with six links: L_1 (ground or the car frame), and L_2, L_3, L_4, L_5, and L_6. There are seven joints: J_1^R, J_2^R, J_3^R, J_4^R, J_5^R, J_6^R, and J_7^R that are considered as 7 nodes: n_1, n_2, n_3, n_4, n_5, n_6, and n_7. The degrees of freedom M are calculated from

$$M = \text{DoF} = 3(n_l - 1) - 2n_P - n_h = 3(6 - 1) - 2 * 7 = 1 \tag{3.2}$$

If the six-bar mechanism provides the same degrees of freedom as a four-bar mechanism, why would the designer bring a more sophisticated mechanism to accomplish the same job? The answer

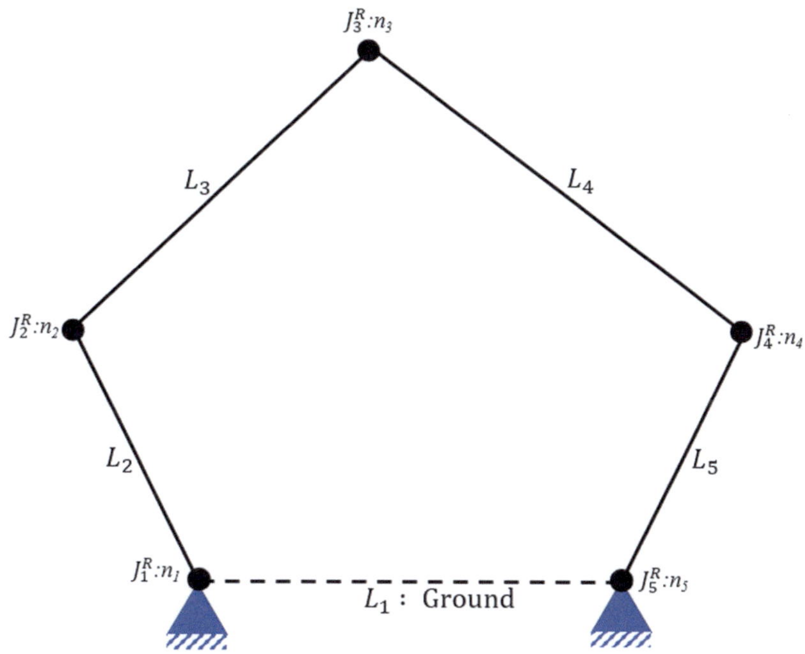

Fig. 3.22 Kinematic diagram of the five-bar mechanism. Associated Five Bar Mechanism.mp4 video designed and implemented by both the author and Pranav Dhola. Courtesy of Pranav Dhola

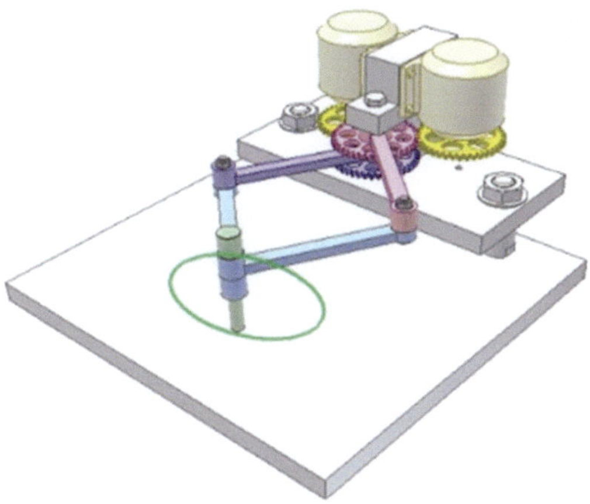

Fig. 3.23 High-speed SCARA parallel robot. Figure and associated Five Bar Mechanism.mp4 video designed by the author and implemented by both the author and Pranav Dhola. Both courtesy of Pranav Dhola

is that six-bar mechanisms have many advantages over four-bar mechanisms. For example, in the car hood mechanism, we can design the same device using a four-bar mechanism instead of a six-bar mechanism, however, the force required to open the hood will be greater in the case of the four-bar mechanism than the six-bar mechanism. In addition, and very importantly, the weight of the hood would cause only one four-bar to fail or unwantedly close. We encourage the reader to watch the attached animation video of the hood mechanism.

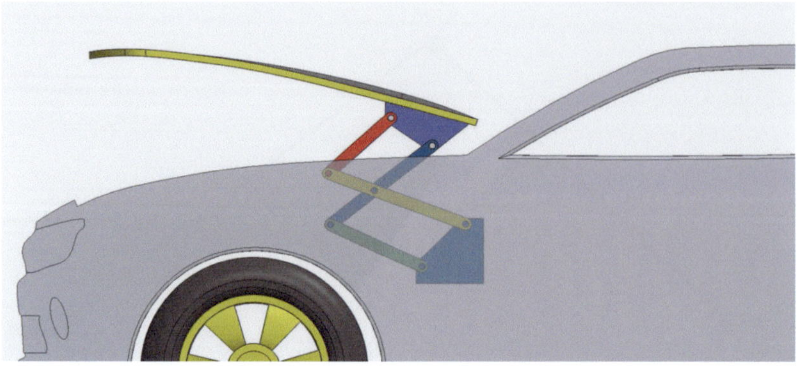

Fig. 3.24 Six-bar hood mechanism. Figure and associated Car_Hood_Mechanism.mp4 video designed by the author and implemented by both the author and Pranav Dhola. Both courtesy of Pranav Dhola

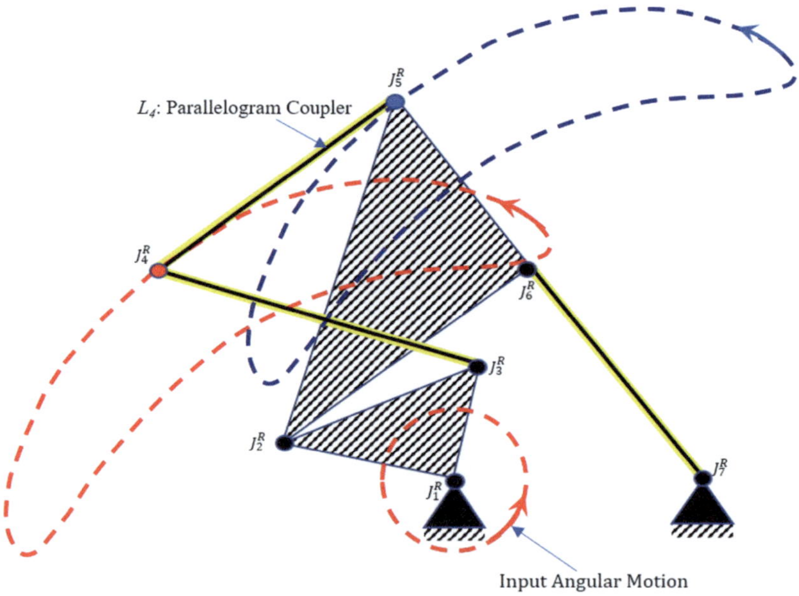

Fig. 3.25 Six-bar with parallelogram coupler motion. Associated Six Bar Mechanism.mp4 video designed and implemented by both the author and Pranav Dhola. Courtesy of Pranav Dhola

Parallelogram coupler motion is one of the six-bar mechanism's main advantages. The six-bar mechanism in Fig. 3.25 has a coupler L_4 that's designed to have no pitch motion, or the relative rotational motion is zero. We encourage the reader to watch the attached animation video of this parallelogram coupler mechanism. As you can see, when the input rotational motion at $J_1^R : n_1$ makes one cycle (rotation from 0° to 360°), the two nodes $J_4^R : n_4$ and $J_5^R : n_5$ create the same motion trajectory with the same velocity vector and the same acceleration vector. The coupler curve traces an approximately straight-line trajectory. This mechanism is considered a parallelogram coupler motion design that is difficult to be accomplished using a four-bar mechanism. Note, the same six-bar parallelogram coupler motion exists in the car hood mechanism in Fig. 3.24 in which the coupler is the hood. The two nodes on the hood have parallel trajectories or the hood (coupler) has no pitch motion. The input motion is activated at node n_7 instead of n_1.

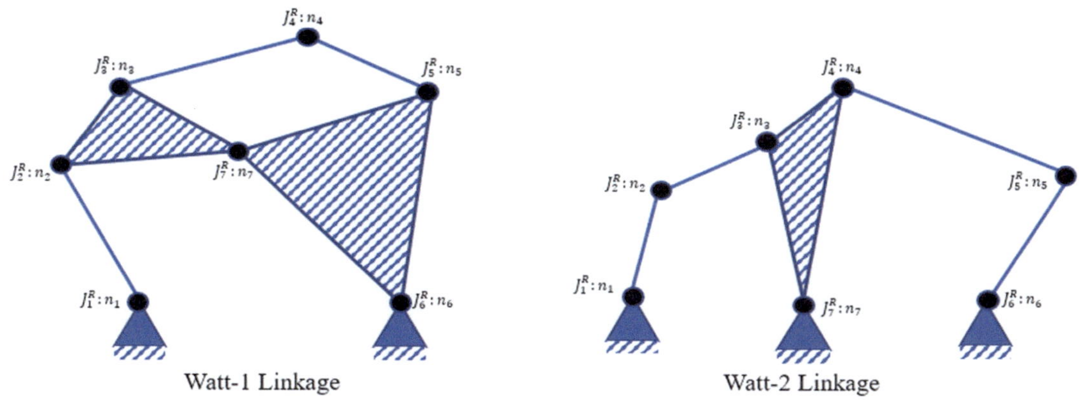

Fig. 3.26 Kinematic diagram of Watt's mechanism

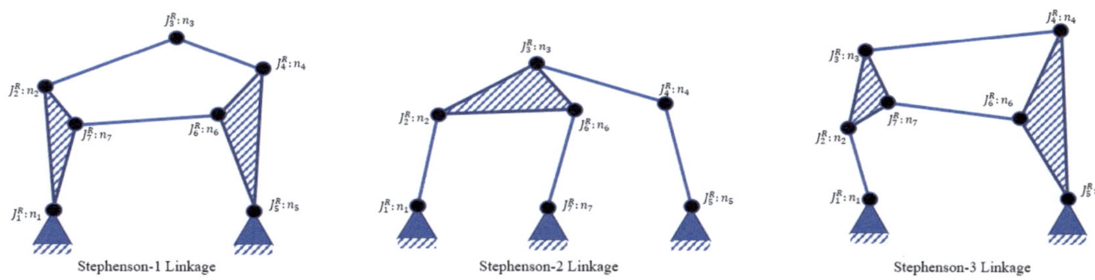

Fig. 3.27 Kinematic diagram of Stephenson's mechanism

There are two well-known six-bar mechanisms: Watt's mechanism as shown in Fig. 3.26 and Stephenson's mechanism as shown in Fig. 3.27. We can describe these six-bar mechanisms as combinations of four-bar linkages, with some links and joints shared. One interesting application for the six-bar mechanism is designing a prosthetic knee joint in Ref. [9].

Engineering Tips—Pitch Motion

Assume a link with *n* distinct nodes. If two of the link's nodes have the same velocity vector, then the link is not rotating at all. It is just translating. All the link's nodes are moving parallel to each other as shown in Fig. 3.25 and the attached video. Further explanation of this motion is in Chaps. 5, 6, and 7. In engineering terms, we say that the link has no pitch motion.

3.8 Special Types of Mechanisms

The four-bar mechanism still proves to be the simplest linkage model that provides 1DoF; however, for special or constrained 1DoF motions, there are a few special types of mechanisms. Some of them are more than 4 links and some are fewer. In the end, each special type has a special or constrained 1DoF.

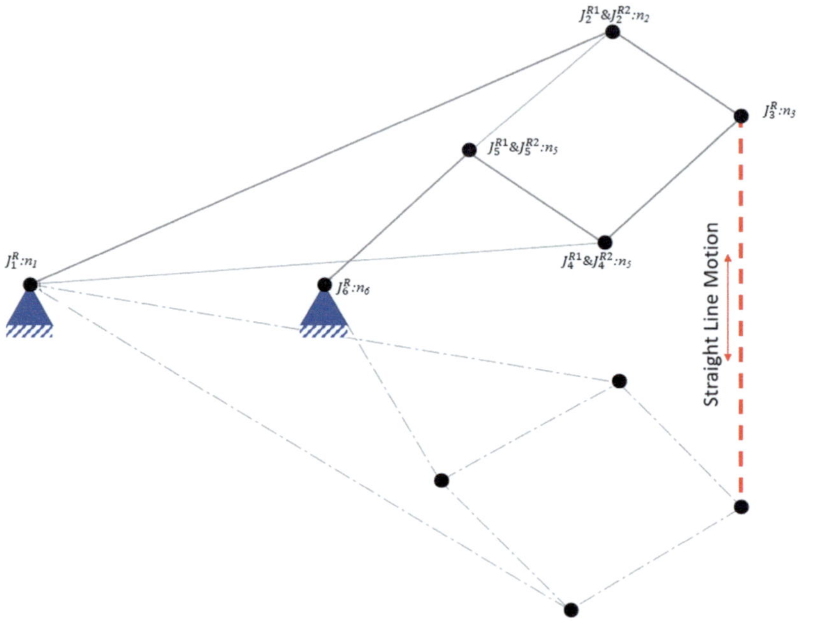

Fig. 3.28 Kinematic diagram of a straight-line mechanism

3.8.1 Straight-Line Mechanism

Figure 3.28 shows a straight-line mechanism that has eight links and ten revolute joints. We can calculate the degrees of freedom as:

$$M = \text{DoF} = 3(n_l - 1) - 2n_P - n_h = 3(8 - 1) - 2 * 10 = 1 \tag{3.3}$$

The straight-line mechanism has eight links with 1DoF but converts rotational motion into straight-line motion and vice versa. One example is the steam engine as shown in Fig. 3.29. The input is the piston's straight-line motion and the output is the wheel's rotational motion. The input starts by heating water until it converts to steam. The increase in the volume from water to steam creates high pressure that pushes the pistons to the drive wheels that operate the locomotive. We encourage the reader to watch the attached video of this mechanism.

3.8.2 Parallelogram Mechanism

The parallelogram mechanism is explained in Table 3.2 where there is motion without any pitch, or the coupler has parallelogram motion without any angular velocity motion—similar to the example in Fig. 3.25. One example is the scissor lift in Fig. 3.30. The operation of the scissor lift mechanism starts when the cylinder(s) with hydraulic fluid or compressed air activates the prismatic joint or the cylinder is pushed outwards and causes the legs to push apart. The top table will be moved up or down as a pure translation without any pitching. We encourage the reader to watch the attached animation video of the scissor lift table. There are additional examples of parallelogram mechanisms that the reader can explore in the problems section.

Fig. 3.29 Steam engine straight-line mechanism. Figure and associated Steam Engine Train Mechanism.mp4 video designed by the author and implemented by both the author and Pranav Dhola. Both courtesy of Pranav Dhola

Fig. 3.30 Scissor lift table mechanism. Figure and associated Scissor Lift Table Mechanism.mp4 video designed by the author and implemented by both the author and Pranav Dhola. Both courtesy of Pranav Dhola

3.8.3 Quick-Return Mechanism

Figure 3.31 shows a kinematic diagram of a Whitworth quick-return mechanism that has five bars with four revolute joints, one prismatic joint, and one pin-in-a-slot joint. We can calculate the degrees of freedom as:

$$M = \text{DoF} = 3(n_l - 1) - 2n_P - n_h = 3(5 - 1) - 2 * 5 - 1 = 1 \tag{3.4}$$

There is a need for only one input to control or to move the 1DoF quick-return machine. One may ask why we have to use five bars to get 1DoF if we could use a simpler four-bar machine. The answer would be that, in a quick-return machine, we add a motor with a counterclockwise constant angular velocity at node n_2. Since n_2 and n_6 are on the same link, node n_6 will rotate with the same constant angular velocity over the red–orange circle as shown in Fig. 3.31. When node n_6 rotates on the red arc from point A to point B, node n_4 will move backward, or in the retraction direction. When node n_6 rotates on the orange arc from point B to point A, node n_4 would move forward or in

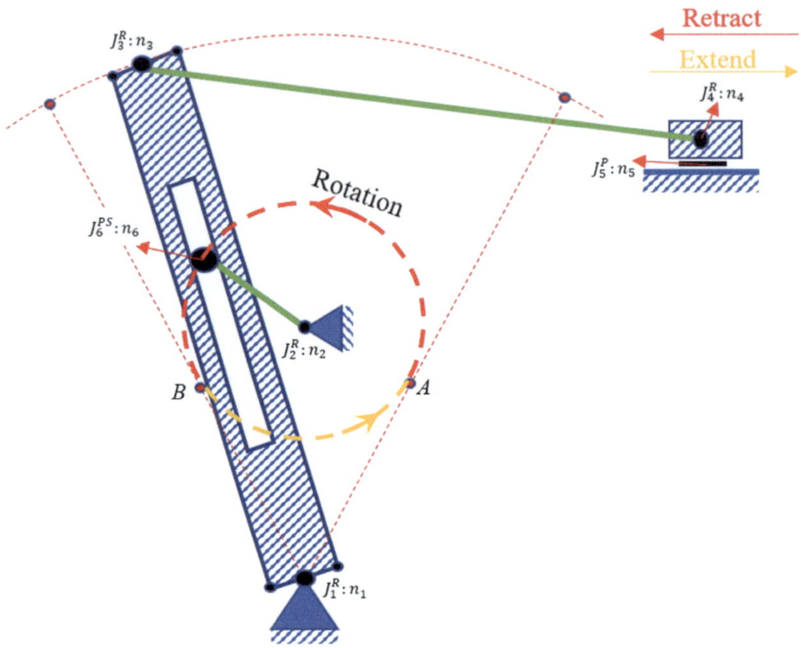

Fig. 3.31 Kinematic diagram of a quick-return mechanism

the extension direction. The red arc is longer than the orange arc; therefore, the time to go over the red arc is longer than the time to go over the orange arc. Moving node n_6 over the red arc means retraction for node n_4. Moving node n_6 over the orange arc means an extension for node n_4. Thus, the extension would take less time than the retraction.

A common example of this quick-return mechanism is the cutting machine as shown in Fig. 3.32. When the smaller gear is rotating with a constant speed as an input, the prismatic motion of the cutting tool has a slow cutting stroke and a fast return stroke. We encourage the reader to watch the attached animation video of the cutting machine.

3.8.4 Scotch Yoke Mechanism

As shown in Fig. 3.33, the Scotch Yoke mechanism has three links with a revolute joint, pin-in-a-slot joint, and prismatic joint. We can calculate the degrees of freedom as:

$$M = \text{DoF} = 3(n_l - 1) - 2n_P - n_h = 3(3 - 1) - 2 * 2 - 1 = 1 \tag{3.5}$$

From Eq. (3.5), there is a need for only one input to control or move the 1DoF for the Scotch Yoke mechanism. The Scotch Yoke mechanism converts rotational motion into translational motion, or vice versa; very similar to the cam joint mechanism. In Fig. 3.33, when a DC motor is connected to node n_1 to rotate, node n_3 would slide back and forth causing an oscillatory translational motion. One example is the water pump as shown in Fig. 3.34. The input is the motor to rotate the crank. The output is the prismatic motion of the slotted yoke arm. We encourage the reader to watch the attached animation video of the water pump to enhance their understanding of the mechanism.

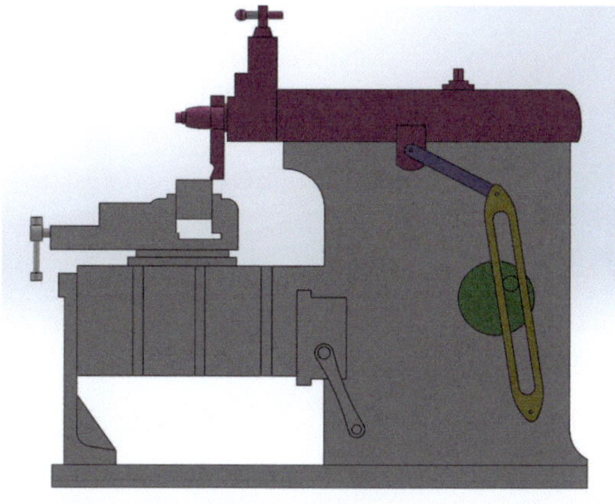

Fig. 3.32 Cutting machine mechanism. Figure and associated Cutting Machine.mp4 video designed by the author and implemented by both the author and Pranav Dhola. Both courtesy of Pranav Dhola

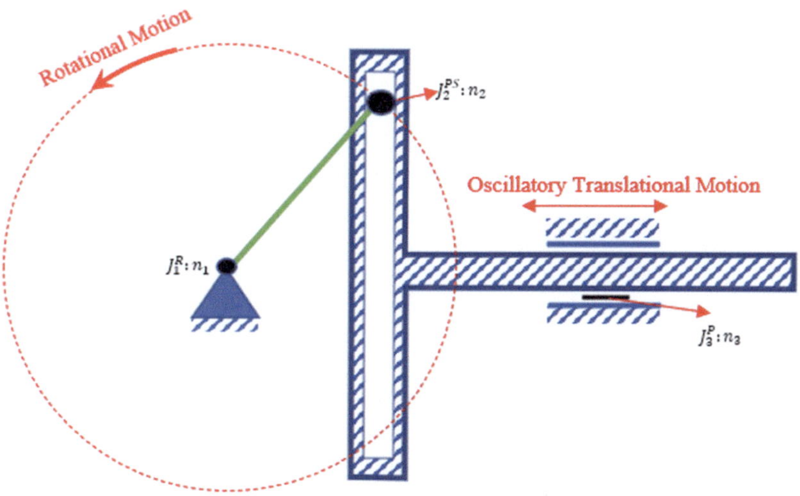

Fig. 3.33 Kinematic diagram of Scotch Yoke mechanism

3.8.5 Intermittent Mechanisms

3.8.5.1 Geneva Mechanism

A Geneva mechanism (Geneva drive or Maltese Cross) is a commonly used mechanism for producing an intermittent rotary motion from a uniform input speed. As shown in Fig. 3.35, the Geneva mechanism consists of:

1. Driving disk that has a locking ring and a drive
2. Driven disk that has driven slots and a locking surface

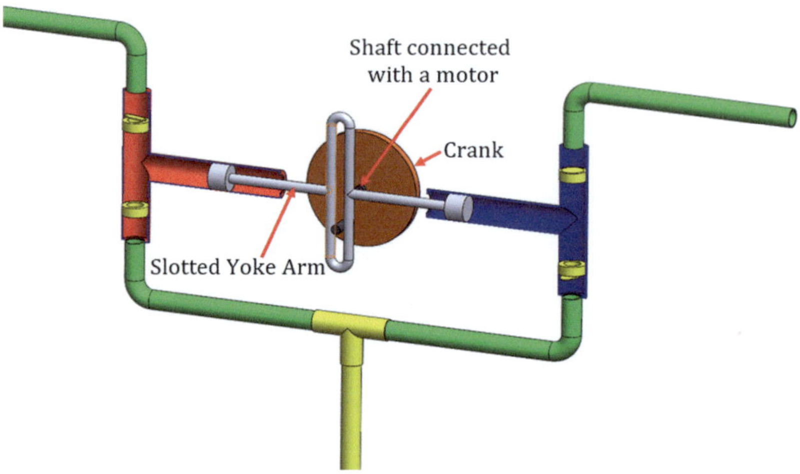

Fig. 3.34 Scotch Yoke for water pump mechanism. Figure and associated Scotch Yoke Pump Mechanism.mp4 video designed by the author and implemented by both the author and Pranav Dhola. Both courtesy of Pranav Dhola

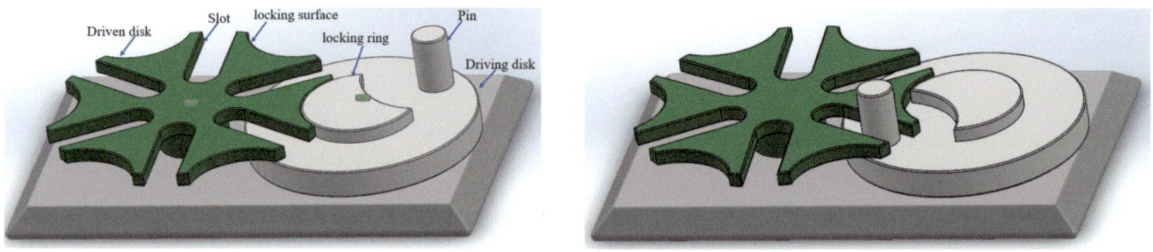

a Geneva Mechanism Configuration, First Situation when the locking ring rotates and slides over the locking surface.

b Second Situation when the pin gets inside one of the drive slots

Fig. 3.35 Rotational Geneva drive mechanism. Figure and associated Rotational Geneva Mechanism.mp4 video designed by the author and implemented by both the author and Pranav Dhola. Both courtesy of Pranav Dhola

When the driving disk rotates, there are two possible cases. The first case is when the locking ring rotates and slides over the locking surface. Because of the sliding motion, the driven disk remains stationary ("dwell period"). The second case is when the pin gets into one of the drive slots. The translational motion of the pin inside the slot will force the driven disk to rotate. In general, the driven disk has a periodic intermittent rotational motion between the pin inside the slot and the locking ring sliding over the locking surface. Note that the number of slots determines the ratio between the stationary (dwell) and motion periods of the driven disk. We encourage the reader to watch the attached animation video of the rotational Geneva mechanism to enhance their understanding of the mechanism.

A Geneva mechanism can also produce a linear intermittent driven motion as shown in Fig. 3.36. When the crank rotates, the pin has two situations. The first situation is when the pin enters one of the slider's slots. This causes the entire slider to translate. The second situation is when the pin is not in any slot. This causes the slider to remain stationary (dwell). The Geneva mechanism has many applications such as the "Geneva Operated Roller Conveyor" as shown in Fig. 3.37 and as listed in Ref. [10]. It is recommended that the reader watches the attached animation video of the "Geneva Operated Roller Conveyor" to enhance their understanding of the mechanism.

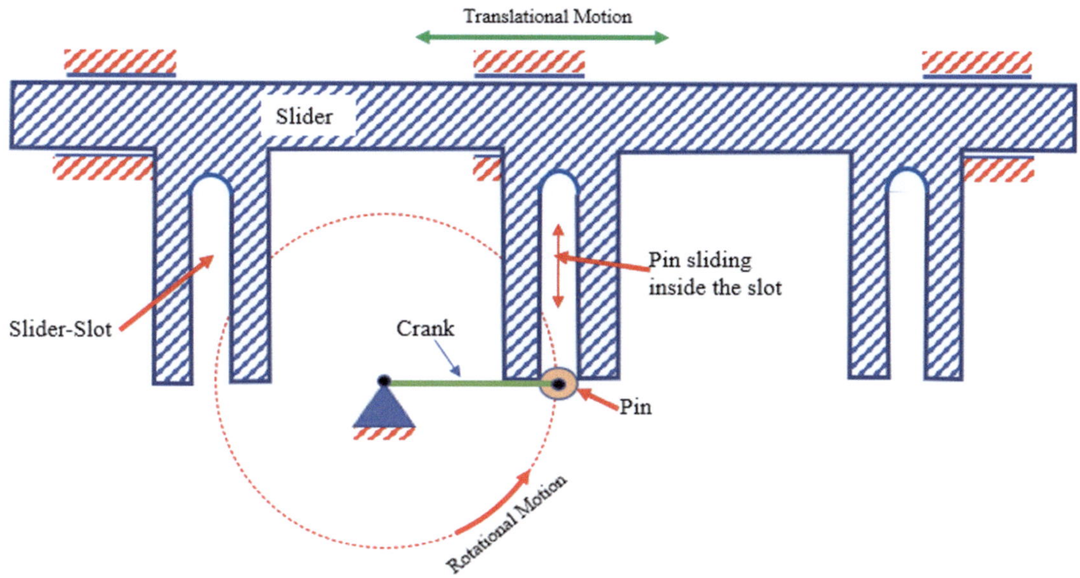

Fig. 3.36 Translational Geneva drive mechanism

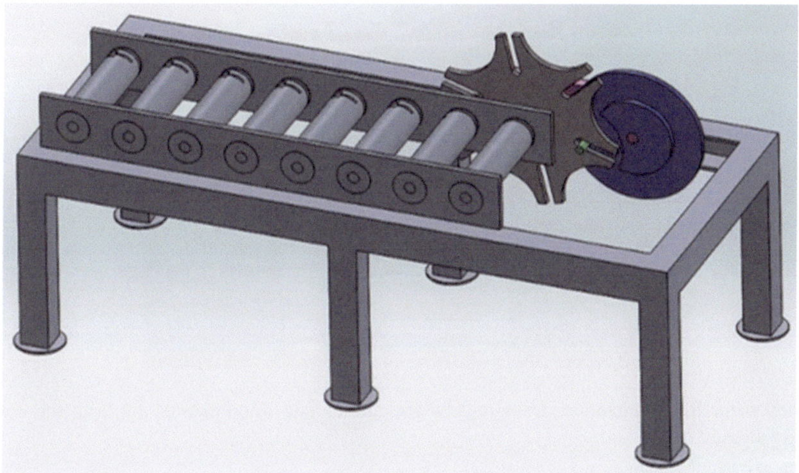

Fig. 3.37 Mechanism of Geneva operated roller conveyor. Figure and associated Geneva Roller Conveyor Mechanism.mp4 video designed by the author and implemented by both the author and Pranav Dhola. Both courtesy of Pranav Dhola

3.8.5.2 Ratchet–Pawl Mechanism

As shown in Fig. 3.38, the ratchet–pawl mechanism consists of a ratchet wheel (a gear), a locking pawl (a pivoting, spring-loaded finger), a driven pawl (a pivoting, spring-loaded finger), and an arm. The gear has suitably shaped teeth, receiving an intermittent circular motion from an oscillating ratchet wheel. When the arm rotates counterclockwise, the driven pawl will force the wheel to rotate. When the driven pawl rotates clockwise, it will slide over the points of the teeth while the wheel remains stationary (dwell) due to the locking pawl.

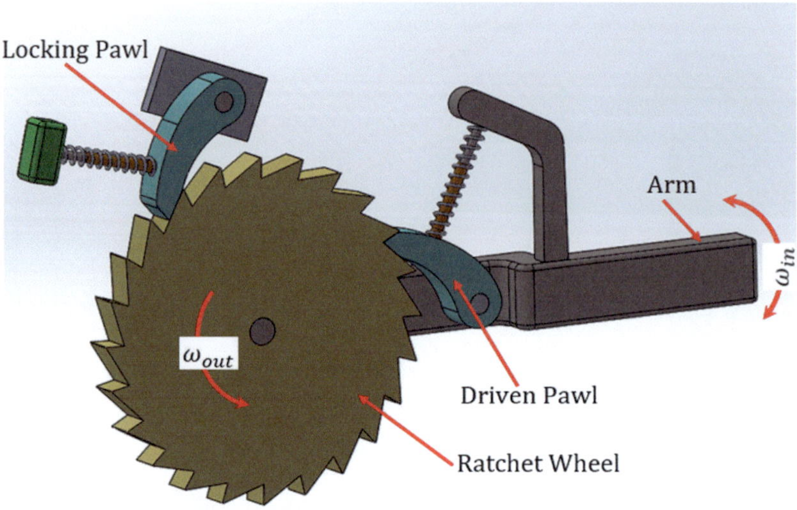

Fig. 3.38 Ratchet–pawl mechanism. Figure and associated Ratchet–Pawl Mechanism.mp4 video designed by the author and implemented by both the author and Pranav Dhola. Both courtesy of Pranav Dhola

Fig. 3.39 Ratchet water lift mechanism. Designed by the author and implemented by both the author and Pranav Dhola. Courtesy of Pranav Dhola

Just as in the case of the Geneva mechanism, the ratchet mechanism can be implemented using rotational and translational motion with many applications such as the grease gun and the loom. One of the earliest applications of the ratchet mechanism was to retrieve water from a well as shown in Fig. 3.39. The person uses this ratchet mechanism to rest by taking his/her hands away from the handle while the pawl falls into the "dip" between the teeth, which will prevent the bucket from falling back into the well. We encourage the reader to watch the attached animated video about the ratchet released mechanism.

3.9 Actuators

Definition

Actuator:

An actuator is a functional element that can be used to control the flow of energy, mass, or volume. The output quantity of an actuator is energy, often in the form of a mechanical working potential "force times displacement" [11].

For a mechanism to function, it needs an energy source that can be in either a rotational form or a translational form. These motion energy sources are called actuators. As you may recall from elementary school, the Law of Conservation of Energy states that energy can be neither created nor destroyed, but can change form. Therefore, actuators are energy converters and are considered the motion energy sources of a mechanism. For example, a motor converts electrical energy into rotational motion energy, and an engine converts combustion energy into rotational motion energy.

Every mechanism that has n DoF needs at least n actuators to activate the motion. These n actuators are located at a set of joints. For example, the engine piston–cylinder mechanism in Fig. 3.18 has 1DoF. Therefore, we need at least one actuator to move the mechanism. The required actuator converts the combustion energy into mechanical or rotational motion energy. When the combustion process occurs in the cylinder, there is a huge increase in pressure that acts as a pushing force on the piston. The motion of the piston will activate the motion of the entire mechanism until the crankshaft rotates. This is the mechanism's output. We call the mechanism's output "mechanical work" not "mechanical energy". Figure 3.40 shows a simplified system block diagram describing the piston–cylinder mechanism.

Some mechanisms are manually actuated, i.e., by hand, leg, or by body pushing. The manual energy motion source is restricted to a human's or animal's capability in terms of the required amount of energy or the required time of work. For example, as shown in Fig. 3.15, a human is capable of moving (riding) a bicycle with his/her legs for a couple of hours. Moving a car by a human is virtually impossible—most humans do not have the required amount of energy to do so. Even if a human could move a car, he/she couldn't do so for a long time. In this case, there is a need for automated actuators that provide a high energy amount and/or provide a continuous energy source for a long time. Even when a manual actuator could be used, in the modern age we often convert manual actuators into automated actuators for reliability and luxury. For example, when you ride a bicycle, your legs on the crank can be implemented by a hydraulic cylinder, motor, pneumatic cylinder, steam engine, or combustion processes in jet, diesel, or gasoline engines.

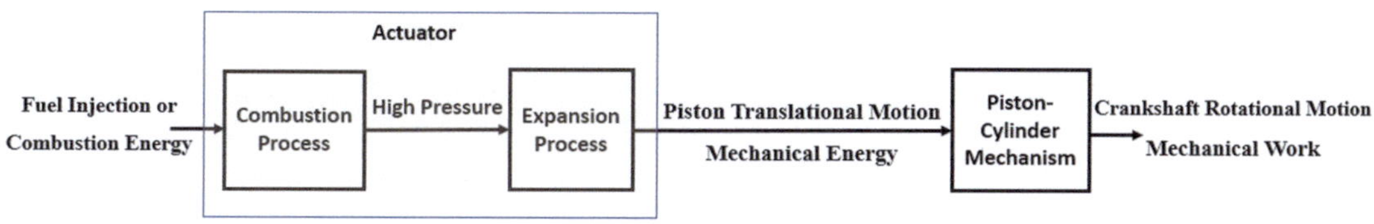

Fig. 3.40 System block diagram of energy transfer for piston–cylinder actuation

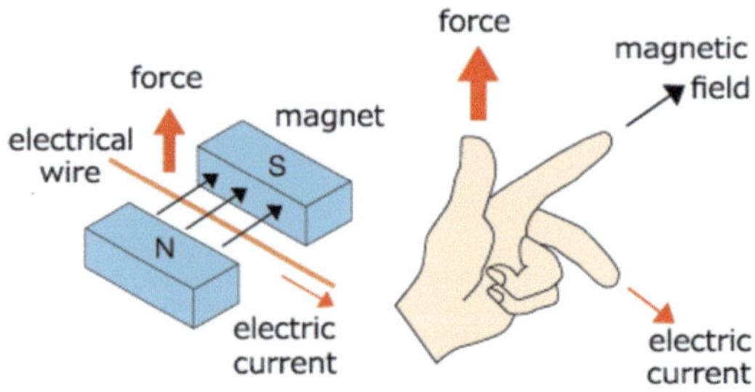

Fig. 3.41 Electromagnetism: Electricity, magnet, and force. Designed by the author and implemented by both the author and Pranav Dhola. Courtesy of Pranav Dhola

In automated actuators, there is an energy conversion from one source to the kinematic sink. For example, in Fig. 3.40, there is a conversion from combustion energy (source) into mechanical energy (kinematic sink). The same concept exists for all automated actuators. In this book, we are going to focus on motor systems and hydraulic systems. One example is the hydraulic actuation in the skid steer bucket lift machine in Fig. 3.4 which has 2DoF with the prismatic motion at the two hydraulic subsystems as inputs.

3.9.1 Electric Motor System

A motor is an electrical machine that converts electrical energy into mechanical energy. The physics behind the motor is electromagnetism. Electric motors generate magnetic fields by means of electric current through a coil. The magnetic field then causes a force with a magnet that causes movement or spinning as shown in Fig. 3.41. Then, we can generate rotational motion or kinetic energy from electrical energy. The directions of electric current, force, and magnetic field follow Fleming's left-hand rule.

The motor can be driven by DC or AC electricity. There are other motor types such as the stepper motor and the universal motor. Most automated motor actuators work in a closed-loop control by providing the current and measuring rotational motion as feedback. This is out of the scope of this book.

Example 3.7

Sketch the system diagram of the energy transfer to actuate the Scotch Yoke water pump mechanism shown in Fig. 3.34.

Solution

Electrical energy (voltage and current) excites the motor to generate the crank rotational motion. The Scotch Yoke mechanism generates the stroke motion as shown in Fig. 3.42.

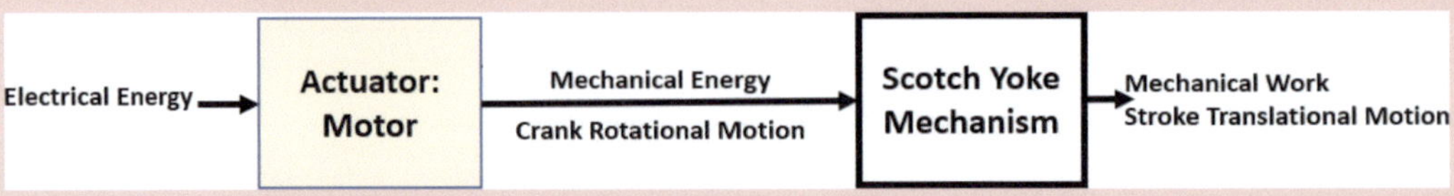

Fig. 3.42 System diagram of energy transfer for water pump machine actuation

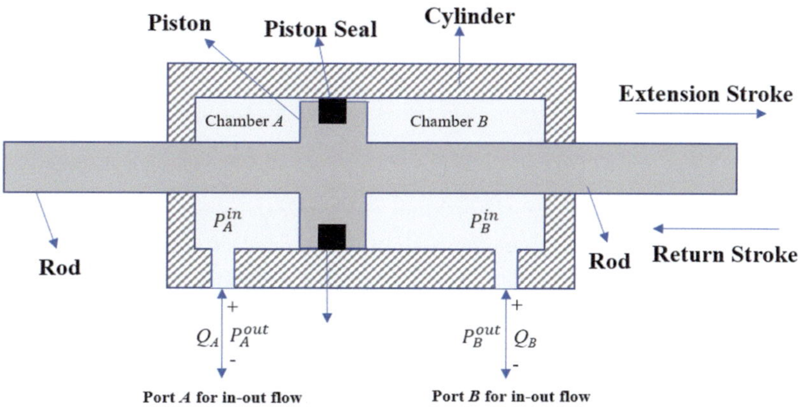

Fig. 3.43 Double-acting piston–cylinder subsystem

3.9.2 Hydraulic System

Hydraulic fluids or oils are (in theory) incompressible flows with almost constant material densities or specific volumes. Based on this concept, a hydraulic piston–cylinder subsystem is used as a linear motion actuator. The hydraulic piston–cylinder subsystem consists of a cylinder barrel, in which a piston connected to a piston rod moves back and forth. The piston divides the inside of the cylinder into two: the bottom chamber (cap end) and the piston rod side chamber (rod end/head end). The fluid enters and leaves the hydraulic cylinder using control valves.

Figure 3.43 shows a double-acting hydraulic piston–cylinder subsystem. The flow in and flow out are driven by control valves, a hydraulic pump, and a container. The flow in and flow out determine the difference in pressure capable of moving the piston. The piston motion activates any mechanism that needs linear motion as an input. For example, in Fig. 3.43, if the pressure P_A^{out} is higher than P_A^{in}, the mass flow rate $+Q_A$ will move through port A to enter chamber A which causes the piston to move in the extension direction. The pressure P_B^{out} is then less than P_B^{in} which causes the flow $-Q_B$ to exit chamber B through port B. More details about how the hydraulic actuator affects the mechanism motion will be introduced in Chap. 8.

There are many examples of using the hydraulic cylinder as an actuator. For example, in the skid steer mechanism in Fig. 3.4, when fuel is injected into the engine, the combustion process will be able to start. High pressure from this combustion process moves the piston. From the piston–cylinder mechanism, the crankshaft will consequently rotate and will start the hydraulic pump. The control valve will control the flow coming from the hydraulic pump. The hydraulic piston–cylinder subsystem moves the pistons of the two hydraulic systems. The skid steer mechanism will then begin to move the bucket.

Example 3.8

Sketch the system diagram of the energy transfer to actuate the skid steer boom mechanism shown in Fig. 3.4.

Solution

Figure 3.44 shows the system diagram of the energy transfer through the skid steer boom actuation process. The energy transfer can be summarized as:

Step 1: is converting combustion energy into mechanical energy. The combustion process uses the injected fuel to generate high pressure which rotates the crankshaft.

Step 2: is converting mechanical energy into hydraulic energy. A hydraulic pump is connected to the crankshaft. The pump pressurizes the flow in the hydraulic subsystem (pipes, valves, and hydraulic cylinders).

Step 3: is converting hydraulic energy into mechanical energy. The controlled flow (supplied from the hydraulic pump) moves the piston inside the hydraulic cylinder.

Step 4: is converting mechanical energy into mechanical work. The boom mechanism converts the piston's translational motion into the boom arm's rotation motion to lift the load.

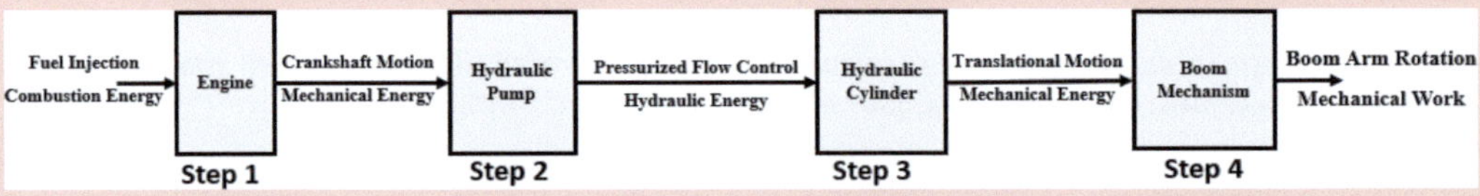

Fig. 3.44 System diagram of energy transfer for skid steer boom mechanism actuation process

In some applications, when a translational motion energy source is required and the hydraulic actuators or the pneumatic actuator are not accessible or allowed, we can use a motor actuator plus a mechanism designed to convert rotary motion from a motor into linear motion such as a slider–crank mechanism as shown in Fig. 3.17, an inverted crank–slider as shown in Fig. 3.19 or the Scotch Yoke mechanism as shown Fig. 3.33. If we locate a motor at $J_1^R : n_1$ to rotate the crank L_l, the link L_3 would have a back-and-forth translational motion.

Another example is the "stroke-doubling mechanism" (as shown in Fig. 3.45 which has four links (crank, rod, and movable rack plus the ground) and four joints (three revolute joints J_1^R, J_2^R, and J_3^R, plus one rack and pinion joint J_4^{RP}). The rack and pinion joint has a circular gear (the pinion) engaging a linear gear (the rack). We can calculate the degrees of freedom as:

$$M = \text{DoF} = 3(n_l - 1) - 2n_P - n_h = 3(4 - 1) - 2 * 4 = 1 \quad (3.6)$$

The rack and pinion joint is capable of translating rotational motion into linear motion as we can see in Fig. 3.45. When the motor rotates the crank, the rod link pushes the movable rack through the

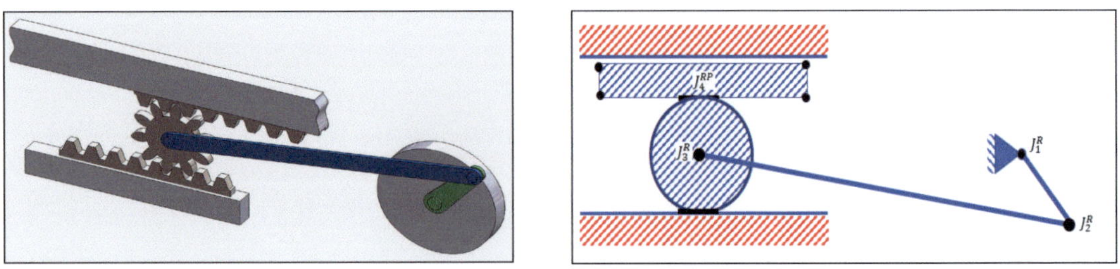

Fig. 3.45 Stroke-doubling mechanism. Figure and associated Stroke Doubling Mechanism.mp4 video designed by the author and implemented by both the author and Pranav Dhola. Both courtesy of Pranav Dhola

gear. The moveable rack reciprocates back and forth. Reciprocation is a repetitive up-and-down or back-and-forth linear motion. We encourage the reader to watch the attached animation video of the stroke-doubling mechanism.

3.10 Chapter Summary

Outcomes

1. Sketching the kinematic diagrams of machines.
2. Using Grashof's law to estimate whether a four-bar mechanism has a crank or not.
3. Understanding that there are 14 different types of the four-bar mechanism.
4. Understanding the difference between the slider–crank four-bar mechanism and the inverted slider–crank mechanism.
5. Understanding there are some special types of mechanisms to accomplish some motion constraints.
6. Breaking down any machine or mechanism into a free body diagram.
7. Sketching the system diagram for the energy transfer through actuators and the mechanisms.
8. Understanding motor and hydraulic actuators.

Problems

For each problem from 1–6, sketch the kinematic diagram showing and name each link and joint.

Problem 1: Folding Chair (Fig. 3.46).

Problem 2: Ironing Table (Fig. 3.47).

Problem 3: Potato Chopper (Fig. 3.48).

Problem 4: Car hood (Fig. 3.49).

Problem 5: Dumpster truck (Fig. 3.50).

Problem 6: Landing Gear (Fig. 3.51).
For each problem from 7–10, for the figure shown, specify whether the mechanism is Grashof or non-Grashof. Specify the Grashof case of each linkage.

Problem 7: Folding chair (Fig. 3.52).

Problem 8: Pumpjack (Fig. 3.53).

Problem 9: Door Closer Mechanism (Fig. 3.54).

Problem 10: Car window wiper (Fig. 3.55).

For each problem from 11–12, sketch the system diagram of the energy transfer to actuate the shown mechanism or machine.

Problem 11: Scissor Lift Table (Fig. 3.56).

Problem 12: Pumpjack (Fig. 3.57).

Fig. 3.46 Folding chair mechanism. Figure and associated Folding_Chair_Mechanism.mp4 video designed by the author and implemented by both the author and Pranav Dhola. Both courtesy of Pranav Dhola

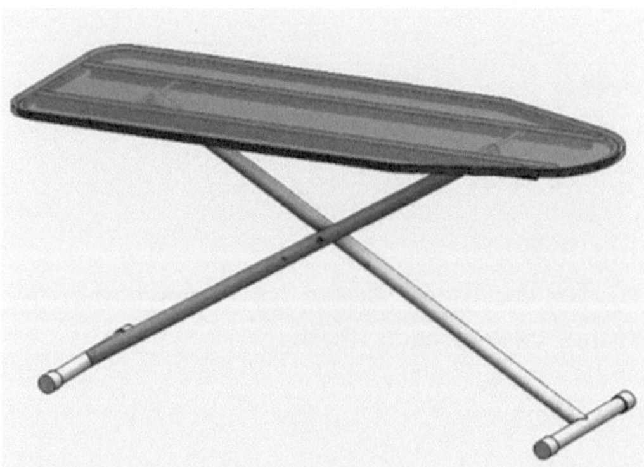

Fig. 3.47 Ironing board mechanism. Figure and associated Ironing_Table_Mechanism.mp4 video designed by the author and implemented by both the author and Pranav Dhola. Both courtesy of Pranav Dhola

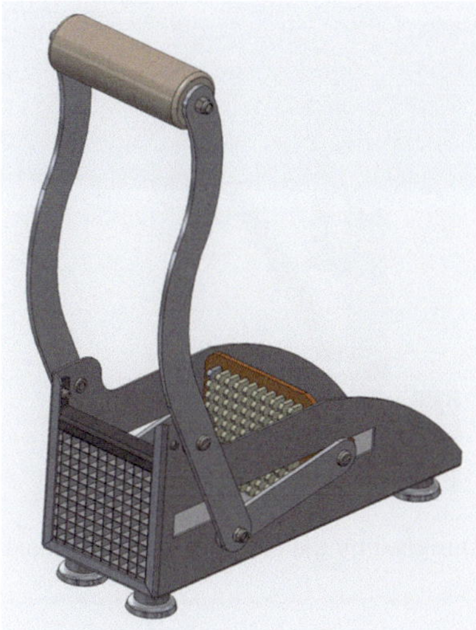

Fig. 3.48 Potato chopper machine. Figure and associated Potato_Chopper_Mechanism.avi video designed by the author and implemented by both the author and Pranav Dhola. Both courtesy of Pranav Dhola

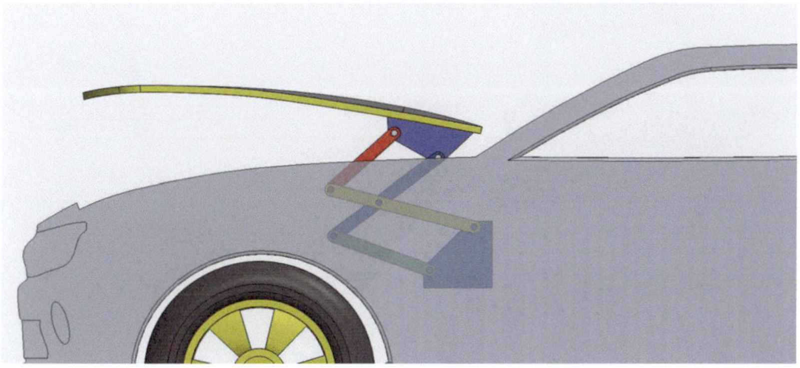

Fig. 3.49 Car hood mechanism. Figure and associated Car_Hood_Mechanism.mp4 video designed by the author and implemented by both the author and Pranav Dhola. Both courtesy of Pranav Dhola

Fig. 3.50 Dumpster truck mechanism. Designed by the author and implemented by both the author and Pranav Dhola. Courtesy of Pranav Dhola

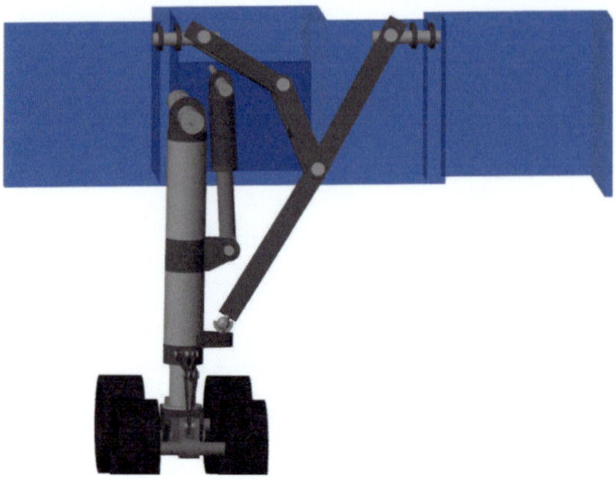

Fig. 3.51 Landing gear mechanism. Designed by the author and implemented by both the author and Pranav Dhola. Courtesy of Pranav Dhola

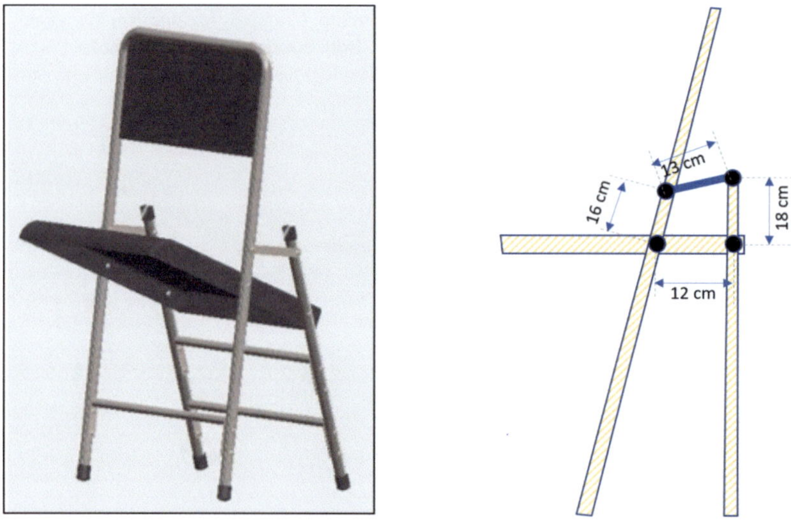

Fig. 3.52 Folding chair mechanism with dimensions. Designed by the author and implemented by both the author and Pranav Dhola. Courtesy of Pranav Dhola

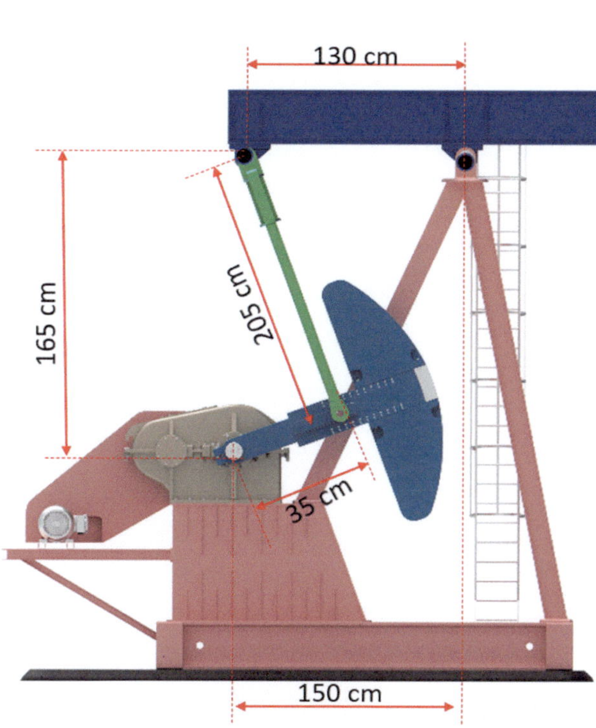

Fig. 3.53 Pumpjack mechanism with dimensions. Designed by the author and implemented by both the author and Pranav Dhola. Courtesy of Pranav Dhola

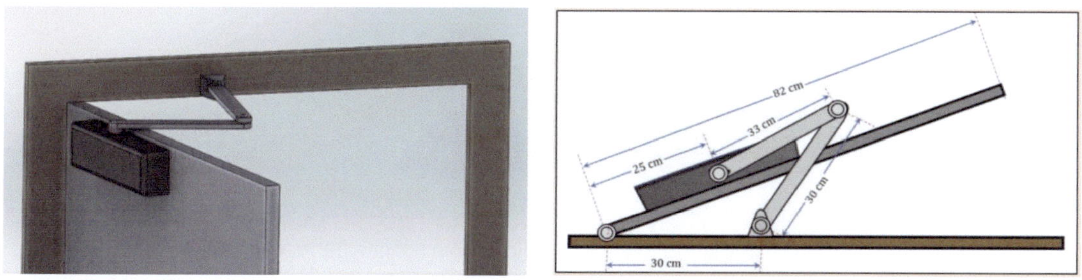

Fig. 3.54 Door holder mechanism with dimensions. Figure and associated Door_4Bar_Mechanism.mp4 video designed by the author and implemented by both the author and Pranav Dhola. Both courtesy of Pranav Dhola

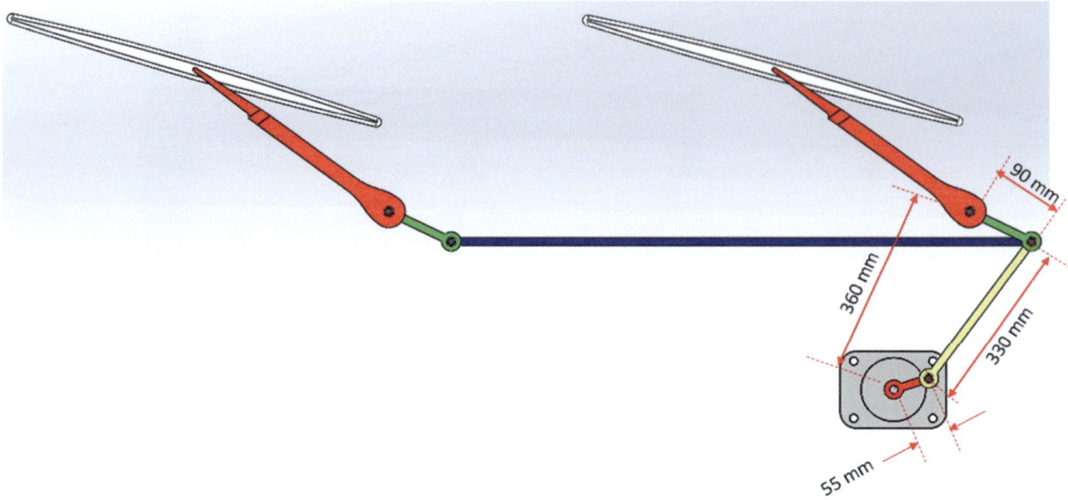

Fig. 3.55 Car wiper mechanism with dimensions. Designed by the author and implemented by both the author and Pranav Dhola. Courtesy of Pranav Dhola

Fig. 3.56 Scissor Lift table mechanism. Figure and associated Scissor Lift Table Mechanism.mp4 video designed by the author and implemented by both the author and Pranav Dhola. Both courtesy of Pranav Dhola

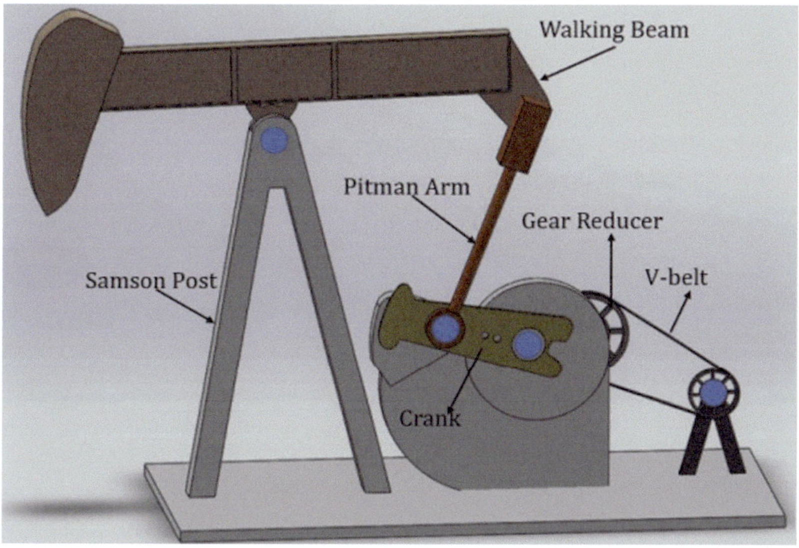

Fig. 3.57 Pumpjack mechanism. Designed by the author and implemented by both the author and Pranav Dhola. Courtesy of Pranav Dhola

References

1. "Design of Machinery," Robert L. Norton, McGraw-Hill; 5th Edition, 2011.
2. "Theory of Machines and Mechanisms," John Uicker, Gordon Pennock and Joseph Shigley, Oxford University Press; 5th Edition, 2016.
3. "Machines & Mechanisms: Applied Kinematic Analysis", David Myszka, Pearson; 4th Edition, 2011.
4. "Kinematic synthesis of linkages," Hartenberg, R.S. & J. Denavit, McGraw-Hill, 1994.
5. "Theoretische Mashinenlehre," Grashof, F., Leipzig, 1883.
6. "A note on Grashof's theorem" Wen-Tung Chang, Chenchou Lin, and Long-Iong Wu, Journal of Marine Science and Technology, Vol. 13, No. 4, pp. 239–248 (2005).
7. "Complete classification of planar four-bar linkage," Baker, C., Journal of Mechanical Design, 1985.
8. "Modelling an inverted slider crank mechanism considering kinematic analysis and multibody aspects, " R.C. Silva, M.A. Nunes, J.P.M. Bento and V.E. Costa, DINAME, Proceedings of the XV international symposium on dynamic problems of mechanics, 2013.
9. "Kinematic and dynamic performance of prosthetic knee joint using six-bar mechanism," Dewen Jin, Ruihong Zhang, HO Dimo, Rencheng Wang, and Jichuan Zhang, Journal of Rehabilitation Research and Development, Volume 40, Number 1, 2003.
10. "Design, Cad Modeling & Fabrication of Geneva Operated Roller Conveyor, " M. V. Ingalkar , A. R. Sonekar , and Y. D. Bansod., International Journal of Engineering Trends and Technology (IJETT) – Volume-46 Number-2 -April 2017.
11. "Actuators, Basics and Applications," Hartmut Janocha, Springer; 2004th Edition, 2004.

4 Kinematics of Machines Using Simulink® and Simscape™ Multibody™

Chapter Objectives

1. Introduction to scripting and functions in MATLAB®
2. Understanding how to build a Simulink® model to simulate a mathematical script
3. Understanding the source library, the sinks library, and the math library in Simulink®
4. Understanding how to use Simscape™ Multibody™
5. Understanding joints, solids, and frames in Simscape™ Multibody™
6. Understanding how to model and simulate a machine in Simscape™ Multibody™

4.1 Introduction

In practice, engineers deal with large-scale problems that have sometimes hundreds of variables. Solving such systems by hand would traditionally take a great deal of time and manpower. To deal with this, programs like MATLAB® exist for the sole purpose of making these large-scale problems easier to deal with in a reasonable amount of time. This chapter aims to introduce the reader to using MATLAB® to model, analyze, and simulate the kinematics of machines. The chapter will start by introducing MATLAB® functions and scripts. After that, we will continue with a brief discussion about the usage of Simulink® to model and simulate a mathematical system. Finally, we will cover descriptive steps for how to model, simulate, and analyze machines in the Simulink®, Simscape™, and Simscape™ Multibody™ toolboxes. This chapter is based on MATLAB® version 2020b. We encourage the reader to watch "Onramp MATLAB®" and "Onramp Simulink®" available on the MathWorks website while working through this chapter.

NOTE: Since MathWorks issues new versions of MATLAB® regularly, the screenshots in the book may differ slightly from what you will see in the program.

Supplementary Information The online version contains supplementary material available at https://doi.org/10.1007/978-3-031-72701-6_4.

A. Omran, *Kinematics and Dynamics of Machines*, https://doi.org/10.1007/978-3-031-72701-6_4

4.2 MATLAB®

Definition[1]

MATLAB®:

MATLAB® combines a desktop environment tuned for iterative analysis and design processes with a programming language that expresses matrix and array mathematics directly. It includes the Live Editor for creating scripts that combine code, output, and formatted text in an executable notebook [1].

MATLAB® (short for Matrix Laboratory) was created by MathWorks. MATLAB® was originally a solution for solving large matrices that advanced hand calculators could not handle. This gradually developed into a program suite that can take massively complex systems and provide solutions and trends. MATLAB® provides a development environment in which the user can write scripts which contain common programming constructs like loops and logical tests while also gaining access to powerful mathematical and logical functions to solve complex problems. The traditional MATLAB® interface is used predominantly for solving systems that are represented mathematically, like functions, matrices, and variables; however, MATLAB® includes another program called Simulink®, which will be the subject of this chapter and future content in this textbook.

Figure 4.1 shows the home screen of MATLAB®, which consists of:

1. Three tabs, (1) HOME tab, (2) PLOTS tab, and (3) APPS tab
2. In the Home tab, there are many options such as opening a new script or m-file.
3. The default view has four windows: Current folder, details, command window, and workspace.

Most engineering schools teach MATLAB® in the first or second year before the students are heading to a Kinematics and Dynamics of Machines course. In this book, we are briefly going over some common functions as listed in Table 4.1 in addition to well-known mathematical operations such as +, −, =, \, *, etc. More details about the scripting and programming of MATLAB® are in [1]. When we use any specific MATLAB® function, a brief explanation will be provided. The reader can also use "help <function name>" in the command window.

4.3 Simulink®

Definition[2]

Simulink®:

Simulink® is a block diagram environment used to design systems with multidomain models, simulate before moving to hardware, and deploy without writing code [1].

[1] Reproduced from www.mathworks.com/products/matlab by MathWorks with permission from MathWorks.

[2] Reproduced from www.mathworks.com/products/simulink by MathWorks with permission from MathWorks.

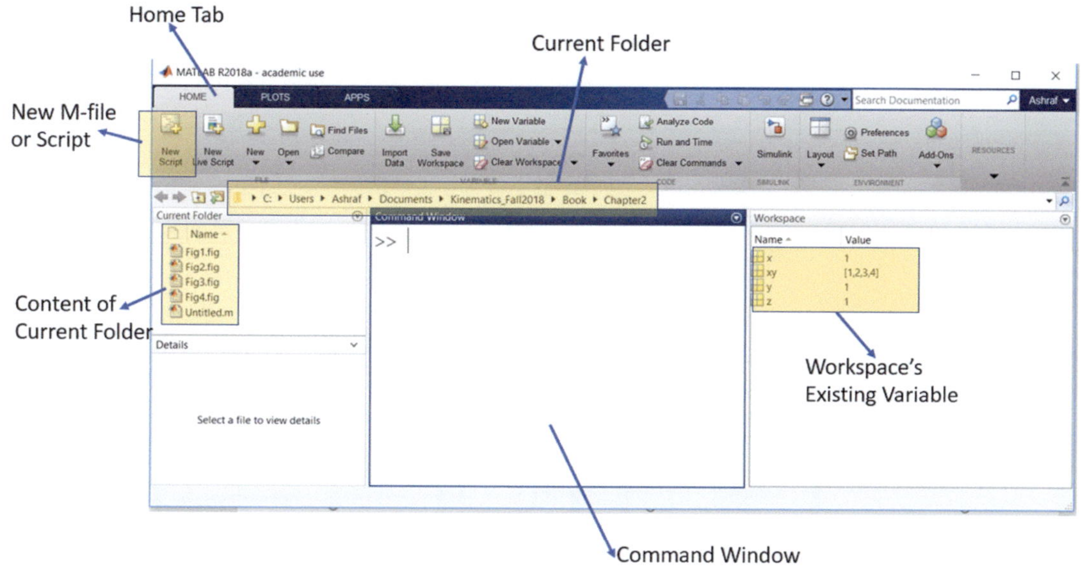

Fig. 4.1 MATLAB® homepage. Designed and implemented by the author. Images used with permission of MathWorks

Table 4.1 Common MATLAB® functions

Category	Function
Trigonometric functions	deg2rad(), rad2deg(), sin(), cos(),...
2D and 3D plotting functions	plot(), area(), loglog(), contour(), surf(), ...
Interpolation functions	interp1(), interp2(), pchip(), meshgrid(), ...
Random number functions	rand(), randn(), ...
Linear algebra functions	expm(), cross(), lu(), eig(), ...

Simulink® can be described as a graphical programming environment for the modeling, analysis, and simulation of systems. The entire concept of Simulink® is to develop subroutines using the so-called block diagram technique. To gain a better understanding, we are going to start with a basic example to help explain how the block diagrams work. Let's assume the following subroutine:

$$\begin{aligned} X &= 3 \\ Y &= 3 \\ Z &= X + Y \end{aligned} \tag{4.1}$$

where X and Y are independent variables and Z is a dependent variable. Think of each element (X, Y, and Z) as its own block that represents that element. These blocks are then arranged and connected to form a diagram that represents how the blocks interact with one another. The independent variables in this example would be classified as *source blocks*. The independent variable $X=3$ is replaced by a block that has 3 as an output with no input because it is a declaration. The same technique is used for $Y=3$. Since the simulation is to express the variation of independent and dependent variables as a function of time, a variable in Simulink® means a function of time. For example, if we declare $X=2$, that means $X(t)=2$ where X is the Simulink® variable as a function of time t. In the block diagram technique, this subroutine can be implemented as shown in the upper part of Fig. 4.2.

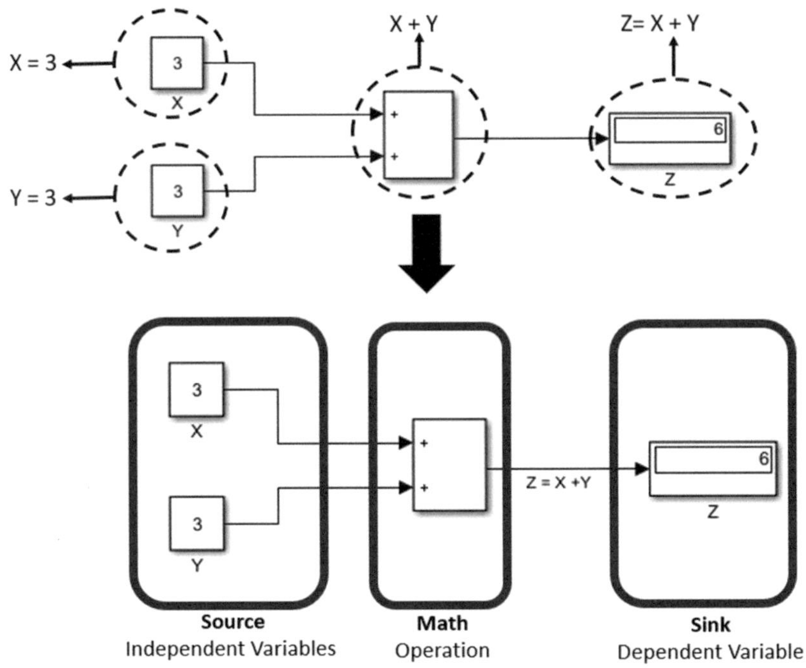

Fig. 4.2 Summation Simulink® model. Designed and implemented by the author. Images used with permission of MathWorks

For the mathematical part, $Z = X + Y$ is a block diagram with two inputs X and Y followed by an output Z. Finally, a sink is used to store the output of the routine (Z = X + Y). . The sink has one input (Z) and the output is the saved variable Z. Scripting is read from left to right.

Simulink® has three types of blocks:

(1) Source blocks (no input because it is a declaration)
(2) Sink blocks (no output because it is a saved variable)
(3) Math (or operation) blocks (with input and output signals).

The generic shape of a block diagram is given in Fig. 4.2. The arrows indicate the flow from input to output.

To run a Simulink® model:

(1) After starting MATLAB®, start Simulink® by either typing "Simulink®" in the command window or by double-clicking on the icon as shown in Fig. 4.3.
(2) Clicking on the Simulink® bar adds an extension for the Simulink® templates. Although other options for specific situations can be selected, the recommended method is to choose "blank model" as shown in Fig. 4.4 which will open a new Simulink® model as shown in Fig. 4.5.
(3) To access all the available libraries in Simulink®, double-click "library browser" as shown in Fig. 4.5.
(4) After building a Simulink® model, you can execute the simulation using the "Run" command. You can specify how long you want to run your simulation as shown in Fig. 4.6. When ready, execute the program by pressing "Start" as shown in Fig. 4.6.

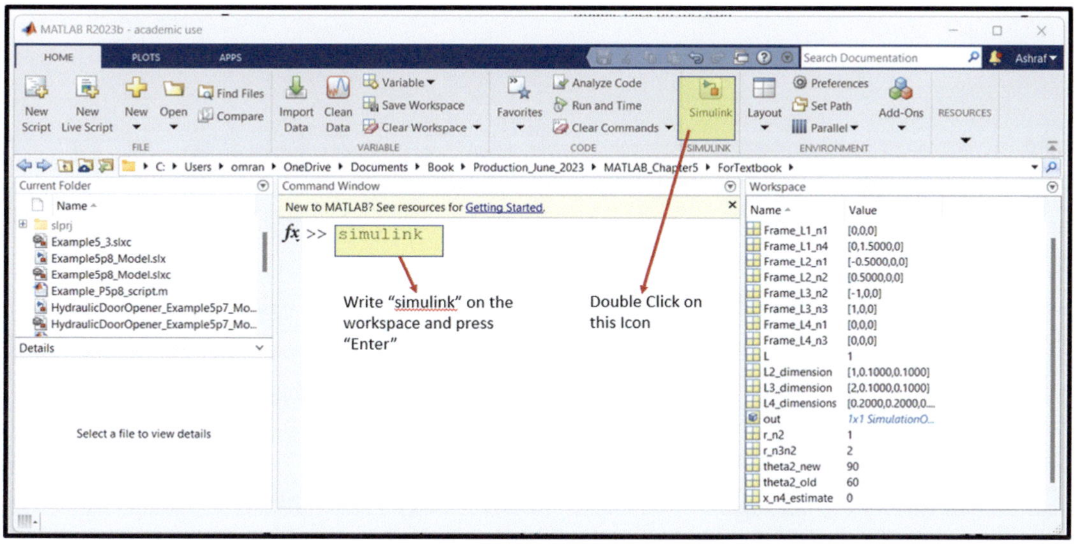

Fig. 4.3 How to start Simulink®. Designed and implemented by the author. Images used with permission of MathWorks

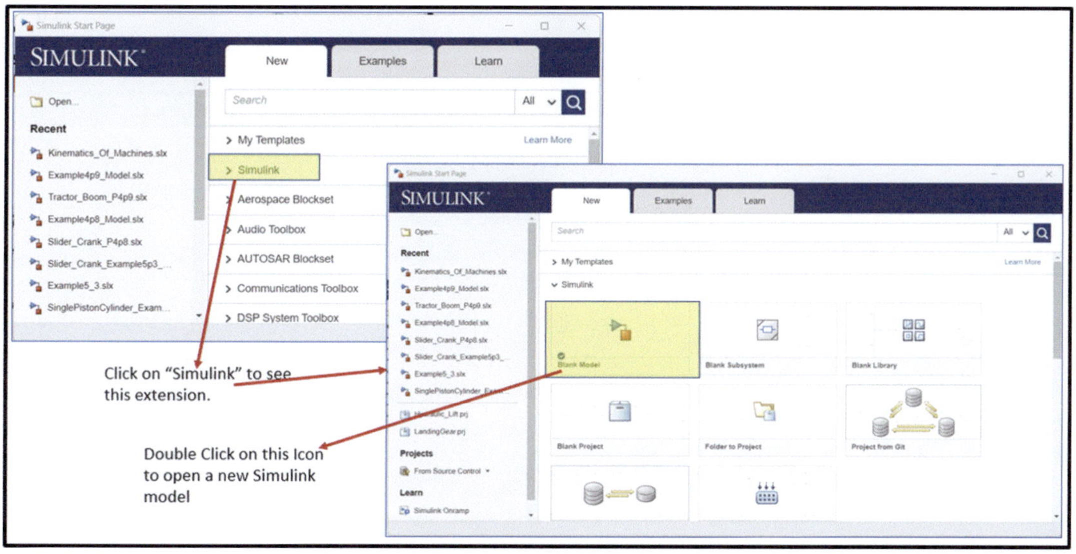

Fig. 4.4 Opening blank model. Designed and implemented by the author. Images used with permission of MathWorks

Note—We will be focusing on 4 libraries as they contain the components for constructing basic sub-routines. For details about other libraries, we recommend referring to the Simulink® tutorial.

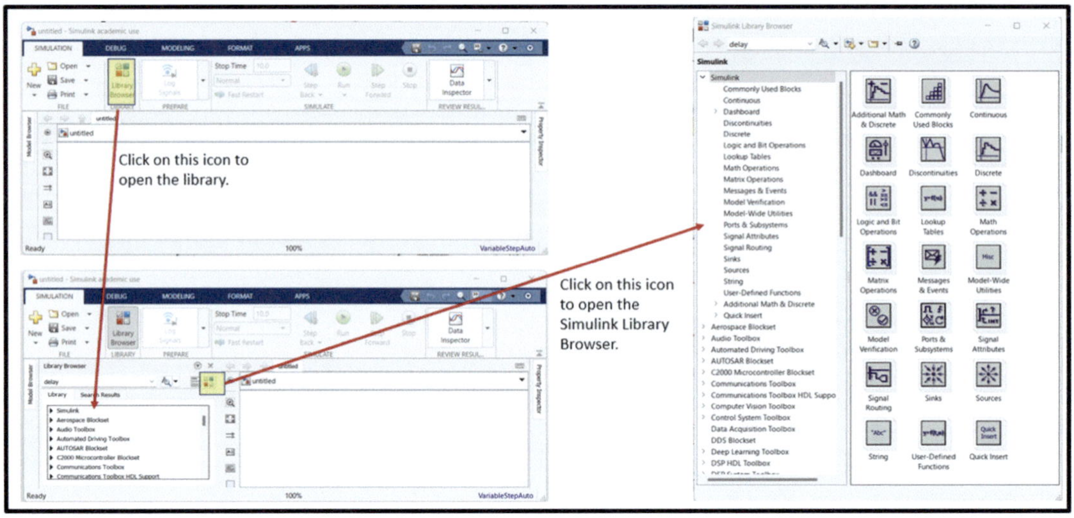

Fig. 4.5 New blank model. Designed and implemented by the author. Images used with permission of MathWorks

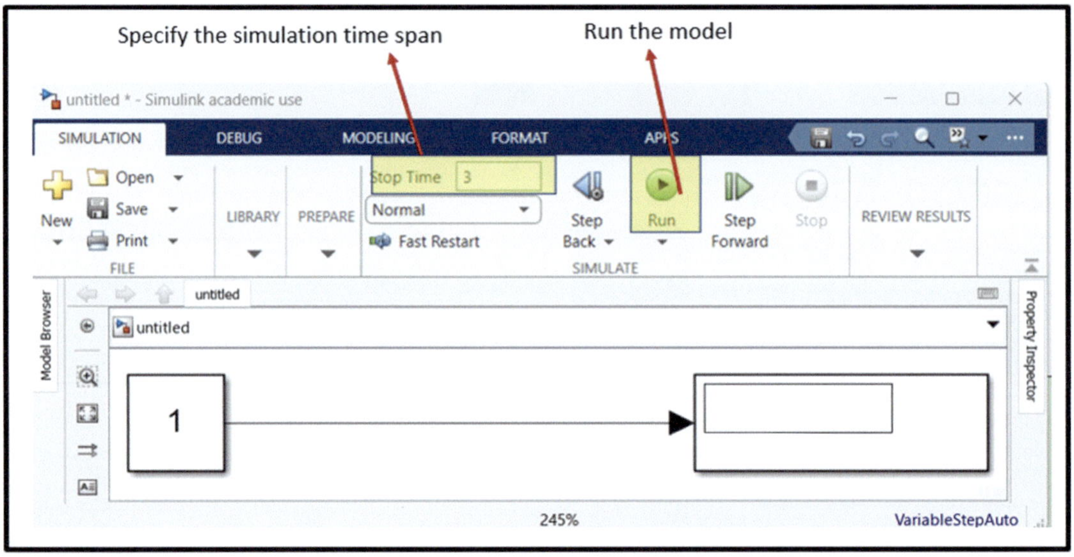

Fig. 4.6 Simulink® model running. Designed and implemented by the author. Images used with permission of MathWorks

4.4 Simulink® Sources Library

Definition[3]

Simulink® Source Library:

The Simulink® source library generates, or imports, signal data blocks such as ***Sine Waves*** and ***From Workspace*** [1].

Figure 4.7 shows the components of the source library. For version 2020b, there are 27 blocks to declare different functions with time. To use any block, drag the block from the "library browser" to the "Simulink® Model Window" or click on any empty space on the "Simulink® Model Window" and write the name of your block. For example, if you want to use the "Ramp" block to declare a variable (signal with time):

(1) Go to any empty area on the "Simulink® Model Window".
(2) Click on the empty screen.
(3) Write "Ramp" and pick the block from the list as shown in Fig. 4.8.

We are only going to cover the source and declaration blocks necessary for the material in this book. The rest of the uncovered source blocks can be found in the Simulink® Help or Simulink® Tutorial in Ref. [1].

4.4.1 Constant Block

Definition[4]

Simulink® Source Constant Block:

The constant block generates a real or complex constant value. The block generates scalar, vector, or matrix output, depending on–

- The dimensionality of the ***Constant value*** parameter
- The setting of the ***Interpret vector parameters as a 1D*** parameter

The block's output has the same dimensions and elements as the ***Constant value*** parameter. If you specify for this parameter a vector that you want the block to interpret as a vector, select the ***Interpret vector parameters as a 1D*** check box. Otherwise, if you specify a vector for the ***Constant value*** parameter, the block treats that vector as a matrix [1].

---

[3] Reproduced from https://www.mathworks.com by MathWorks with permission from MathWorks.

[4] See Footnote 3.

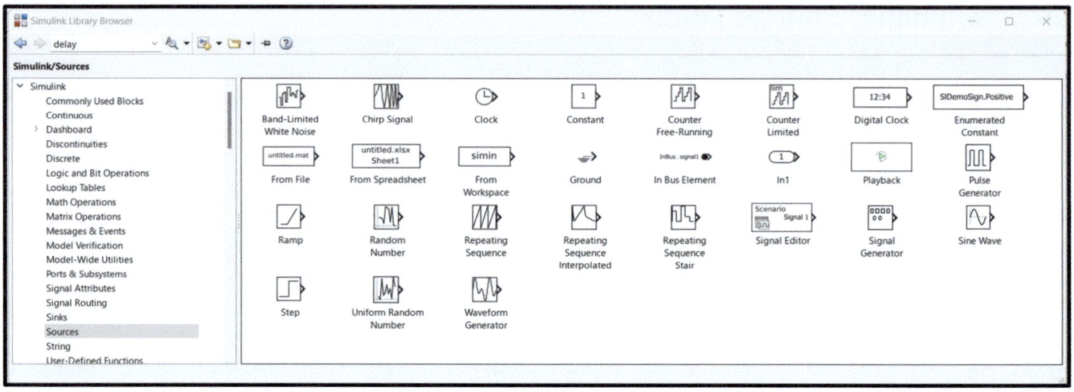

Fig. 4.7 Simulink® source library. Designed and implemented by the author. Images used with permission of MathWorks

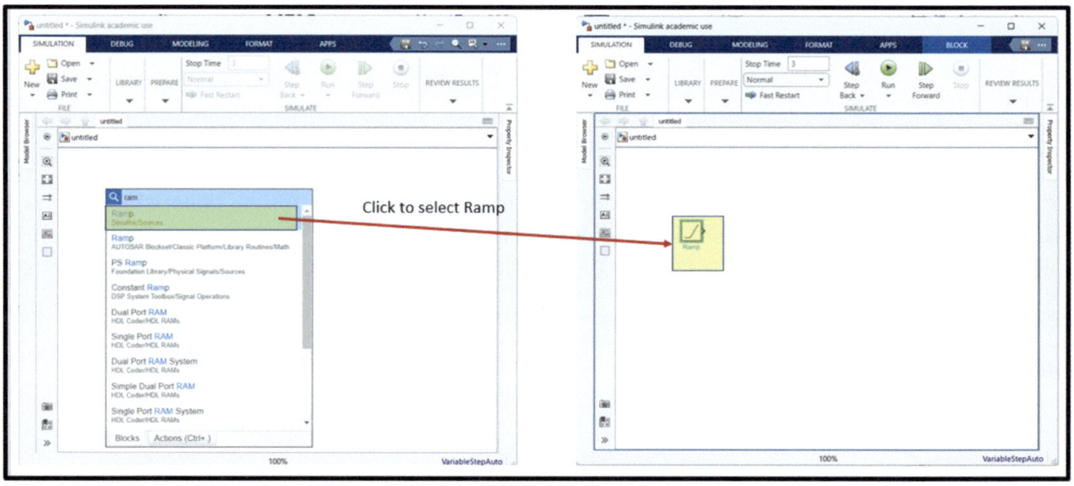

Fig. 4.8 Inserting ramp source block. Designed and implemented by the author. Images used with permission of MathWorks

The constant block provides a "signal" with a ***Constant Value*** "C" with respect to time and can be a single element or a vector. Figure 4.9 shows the constant block parameter and the expected output signal or declared signal from this block. The user can get more information about the constant block in the help link as well as Ref. [1] .

4.4.2 Signal Editor Block

Definition[5]

Signal Editor Block:

The signal editor block allows you to create interchangeable groups of piecewise linear signal sources and use them in a model. You can quickly switch the signal groups in and out of a model to facilitate testing [1].

[5] See Footnote 3.

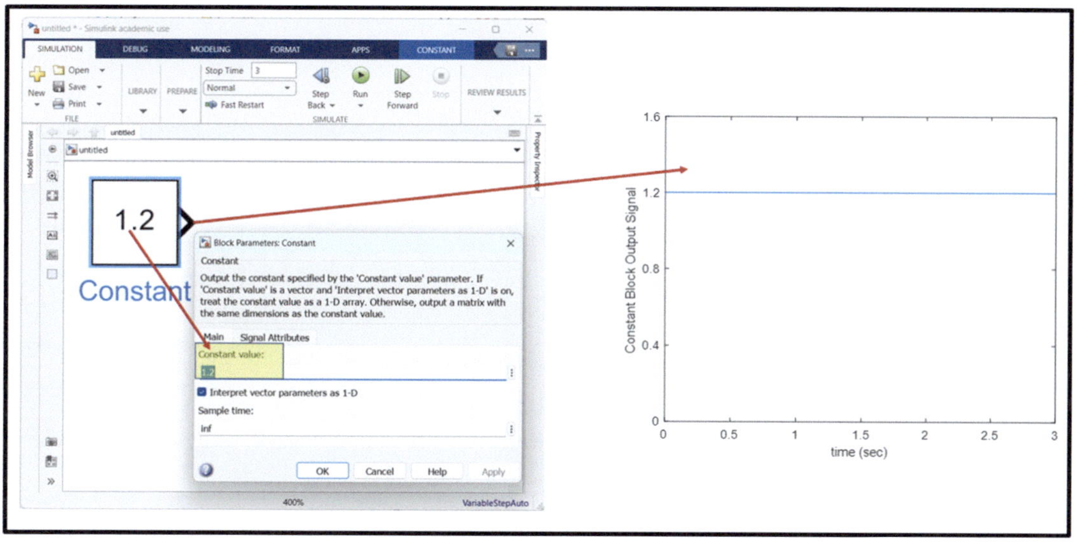

Fig. 4.9 Constant source block. Designed and implemented by the author. Images used with permission of MathWorks

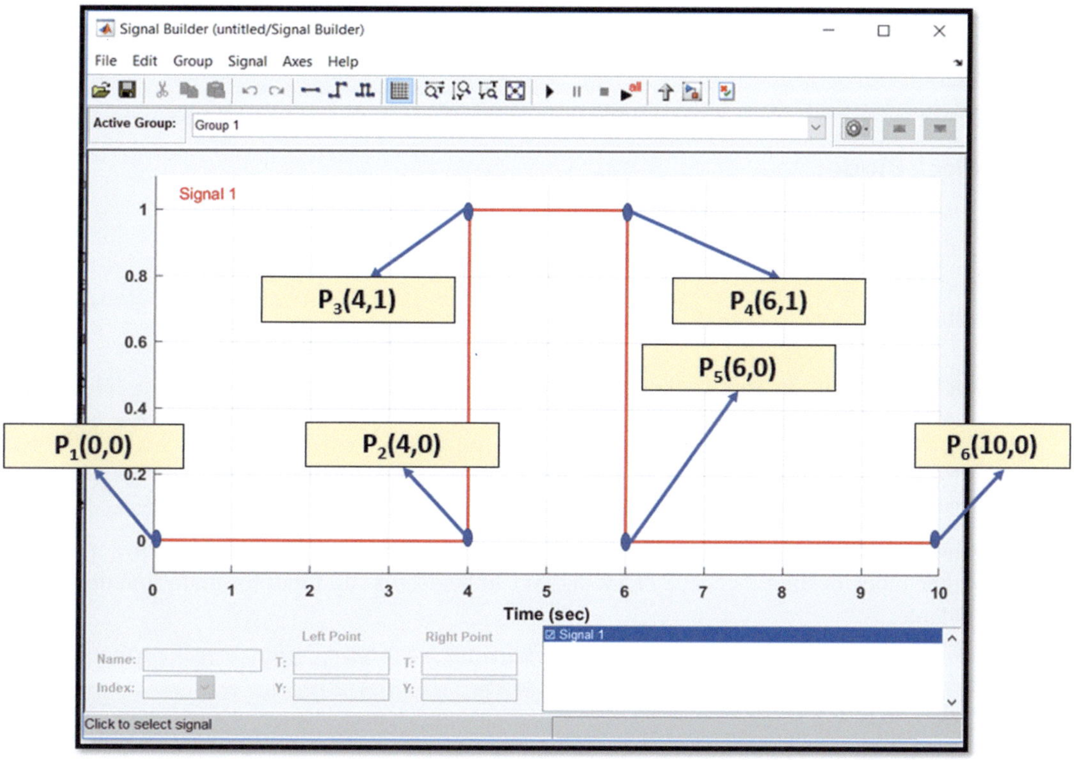

Fig. 4.10 Default signal builder block. Designed and implemented by the author. Images used with permission of MathWorks

The signal editor block provides the signals $y_1, y_2, \ldots, y_n$ as functions of time. The signals $y_1, y_2, \ldots, y_n$ are generated as an interpolation between piecewise points that the user provides. Once you double-click the signal editor block on a model, you will get a pulse signal with six points as shown in Fig. 4.10. You can modify the signal by:

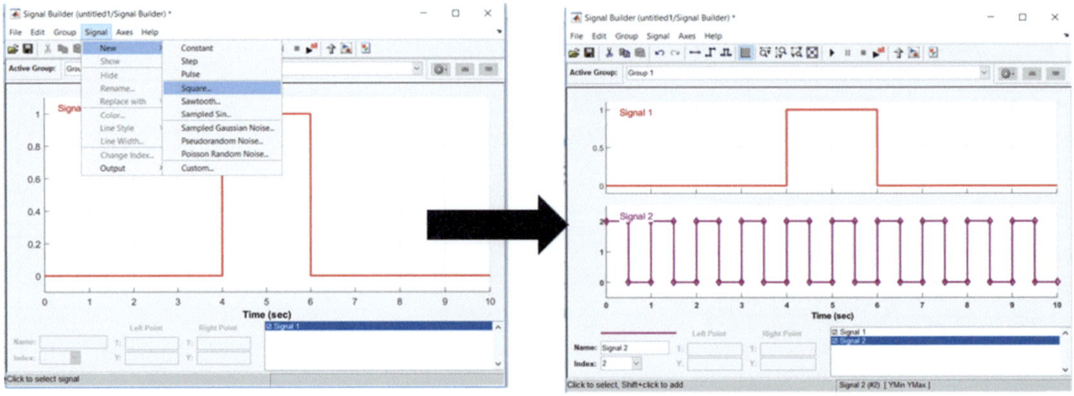

Fig. 4.11 Signal builder block. Designed and implemented by the author. Images used with permission of MathWorks

(1) Adding more signals to the output group by clicking on "signal" and picking the type of signal as shown in Fig. 4.11.
(2) Changing the time value and y-value for each point.
(3) Creating a new point by shift+click on a specific point on the line.

More details about this block are given in Simulink® Help.

4.4.3 Clock Block

Definition[6]

Clock Block:

The ***Clock*** block outputs the current simulation time at each simulation step. This block is useful for other blocks that need time as a value for their input. When you need the current time within a discrete system, use the ***Digital Clock*** block [1].

The clock block provides a signal equivalent to time or output = time. Figure 4.12 shows the output signal from the clock or the timer. The clock signal can be used as an input for math function blocks or a lookup table.

[6] See Footnote 3.

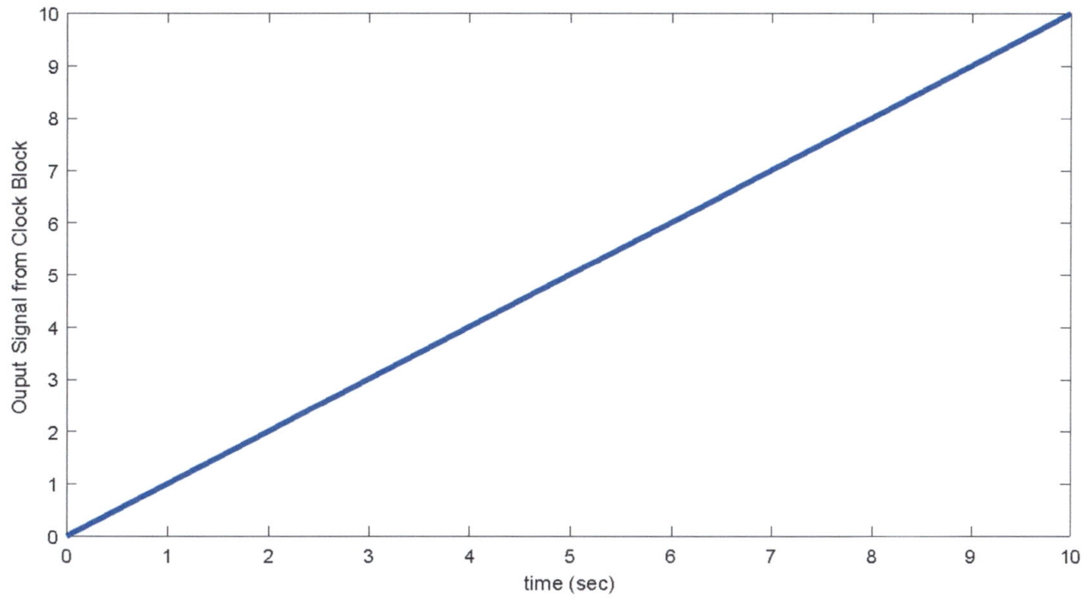

Fig. 4.12 Output of clock block

4.5 Simulink® Sinks Library

Definition[7]

Simulink® Sinks Library:

Display or export signal data blocks such as ***Scope*** and ***To Workspace*** [1].

After running any simulation, we need to save the results for future reference and analysis. Saving results is done through the sinks library using the ***Scope*** block, ***Display*** block, "save to a mat file" block, or by exporting the variable directly to the workspace. Figure 4.13 shows the composition of the ***Sinks*** library. For version 2020b, 10 different blocks can save the signal or discard it. We are going to briefly explain the most important blocks from the sinks library. The rest of the sink blocks are important and reading the Simulink® help in Ref. [1] is recommended.

4.5.1 Scope Block

Definition[8]

Scope Block:

To visualize your simulation results over time, use a ***Scope*** block, ***Time Scope*** block, or ***Time Scope System object*** [1].

[8] See Footnote 3.

[7] See Footnote 3.

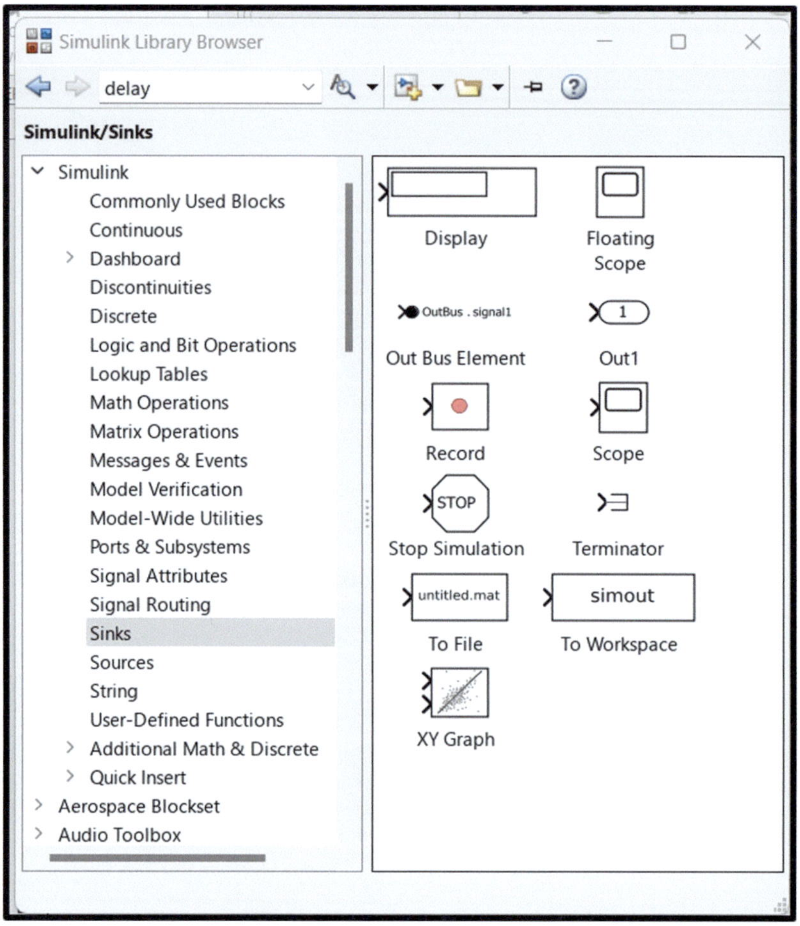

Fig. 4.13 Simulink® sinks library. Designed and implemented by the author. Images used with permission of MathWorks

The scope block is a time scope block that displays time-domain signals. Scope does not automatically save the signal under any variable name and is only used to view a simulation or its results. The scope block provides:

(1) The ability to see the time history of a signal as shown in Fig. 4.14.
(2) Zoom-in and zoom-out options for analysis using the steps below and as shown in Fig. 4.15.
- Press "tools" and pick "zoom in", and
- with the mouse, click on the domain that you want to zoom-in on.

(3) The ability to plot signals on top of one another by use of a mux block that can merge signals to be plotted on the same graph using the steps below and as shown in Fig. 4.16.
- Go to "commonly used blocks",
- drag the mux into the model,
- double-click on the mux block to specify how many signals,
- connect the signals to the input and the scope to the output, and
- execute (run) the simulation.

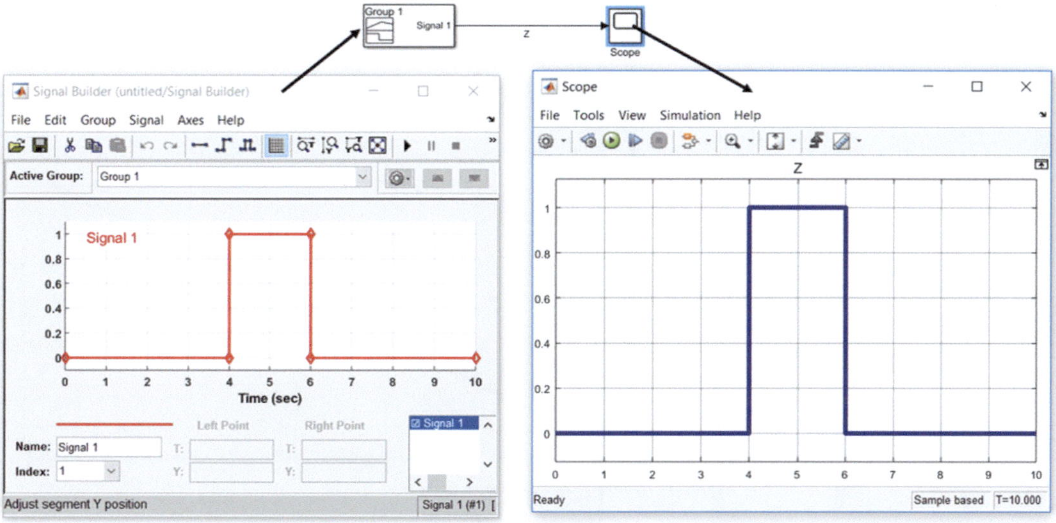

Fig. 4.14 Scope visualization for signal builder. Designed and implemented by the author. Images used with permission of MathWorks

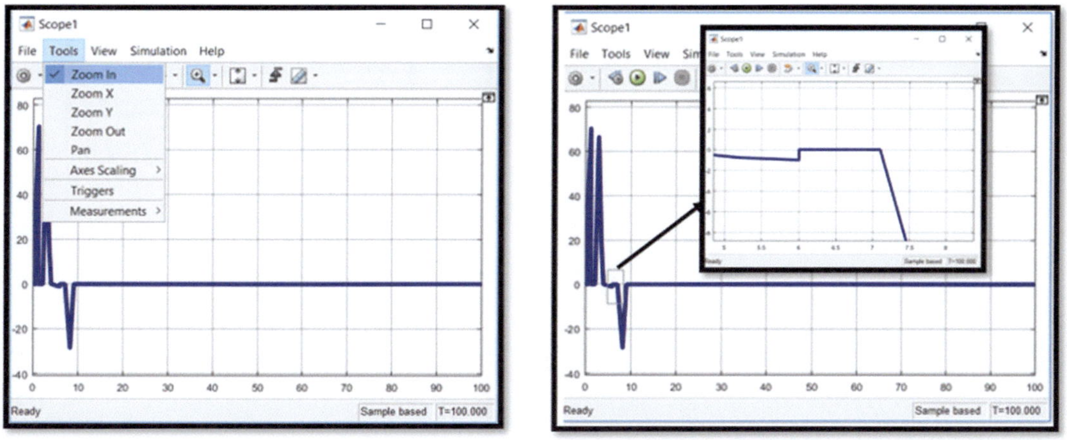

Fig. 4.15 Zoom-in option in scope block. Designed and implemented by the author. Images used with permission of MathWorks

(4) The setting of more than one signal as subplots using the steps below and as shown in Fig. 4.17.

- Drag the scope block to the Simulink® model,
- double-click to get the graphical menu,
- pick "configuration properties" from the view, and
- change the number of the inputs to whatever is needed.

Many other options make the scope a unique tool in Simulink® models. More information can be found in Ref. [1].

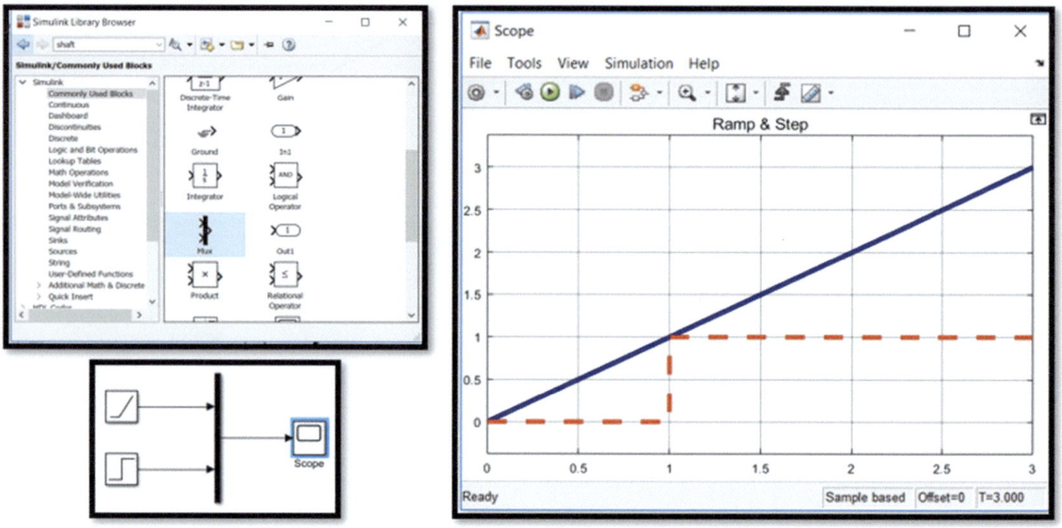

Fig. 4.16 Scope block and mux block. Designed and implemented by the author. Images used with permission of MathWorks

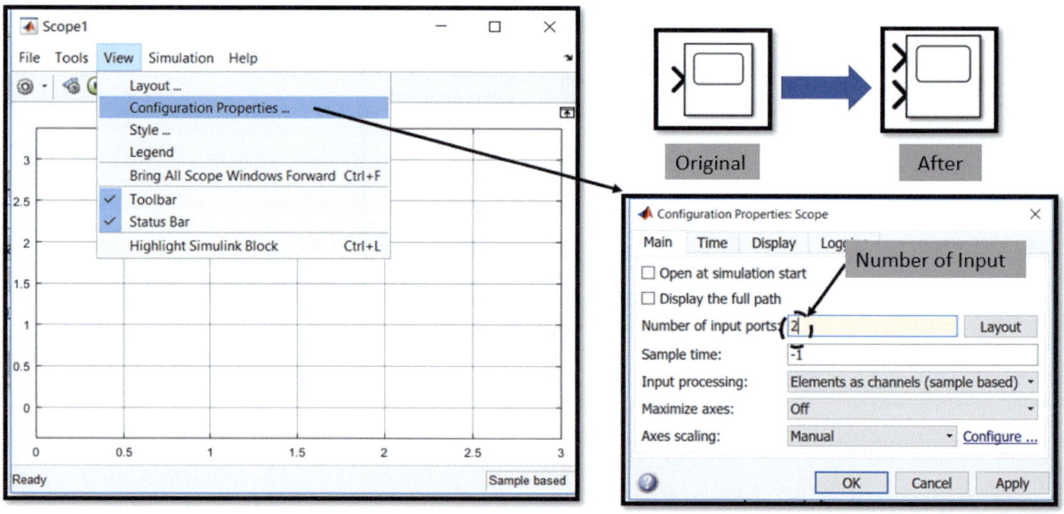

Fig. 4.17 Scope block subplot. Designed and implemented by the author. Images used with permission of MathWorks

4.5.2 Terminator Block

Definition[9]

Terminator Block:

Use the ***Terminator*** block to cap blocks whose output ports do not connect to other blocks. If you run a simulation with blocks having unconnected output ports, Simulink® issues warning messages [1].

[9] See Footnote 3.

When you run or execute the simulation, it is better to terminate the signal if it is not needed to be analyzed or visualized. Figure 4.18 shows some unnecessary outputs that are terminated. Terminating the signal instead of keeping the signal line free saves the computational time needed to run a model.

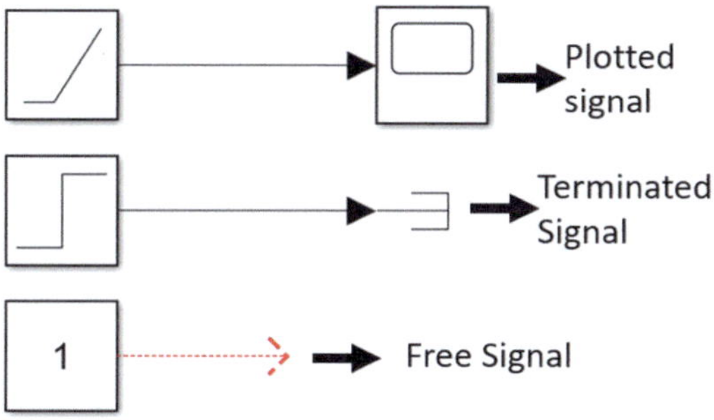

Fig. 4.18 Displaying signal, terminating signal, and make signal free. Designed and implemented by the author. Images used with permission of MathWorks

4.6 Simulink® Math Operations Library

Definition[10]

Math Operations Library:

Contains mathematical function blocks such as ***Gain***, ***Product***, and ***Sum*** [1].

Figure 4.19 shows the components of the math operations library. As stated earlier, the math operation block adds a desired function that will take connected inputs and create new outputs.

- **Gain Block** is a mathematical block that multiplies a user-defined factor to the input and exports the new output.
- **Sum, Add, and Subtract Blocks** are basically equivalent. They are used to add or subtract an *n* number of input signals.
- **Product and Divide Blocks** are used to multiply or divide a set number of input signals.
- **Math Function Block** performs numerous mathematical functions as shown in Table 4.2.

[10] See Footnote 3.

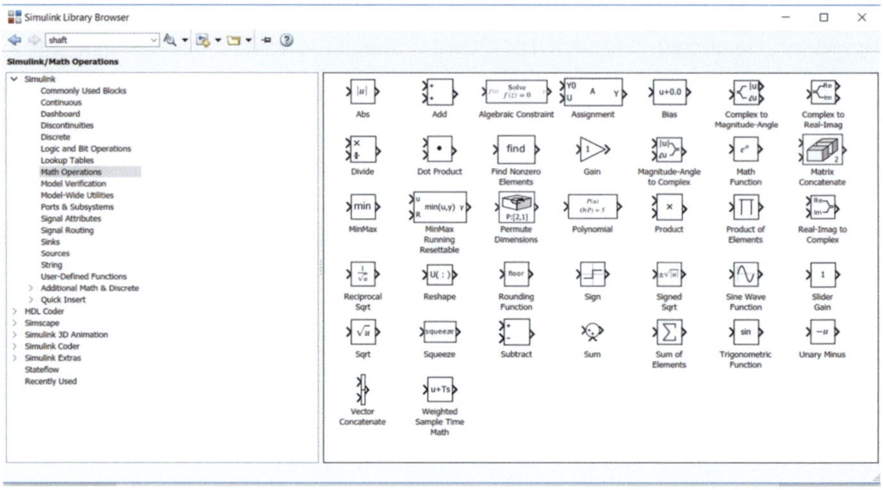

Fig. 4.19 Math operations library. Designed and implemented by the author. Images used with permission of MathWorks

Table 4.2 Mathematical functions of math function block

Functions	Description
exp	Exponential
log	Natural logarithm
10^u	Power of base 10
Log 10	Common (base 10) logarithm
Magnitude ^2	Complex modulus
Square	Power 2
Pow	Power
Conj	Complex conjugate
reciprocal	Reciprocal
hypot	Square root of sum squares
rem	Remainder after division
mod	Modulus after division
transpose	Transpose
hermitian	Complex conjugate transpose

Example 4.1

Implement the following equation in a Simulink® model. Run the model and show the results.

$$Y = 3 * X$$

where X is the input and Y is the output. Assume, X is a ramp source with a slope value of 1 and provides a scope for both the ramp block and as a sink for the output.

Solution

Start by adding a ramp block from the source library. We will rename this block "X" as well as the signal coming from the ramp block. Next, add a gain block with constant 3 just after the

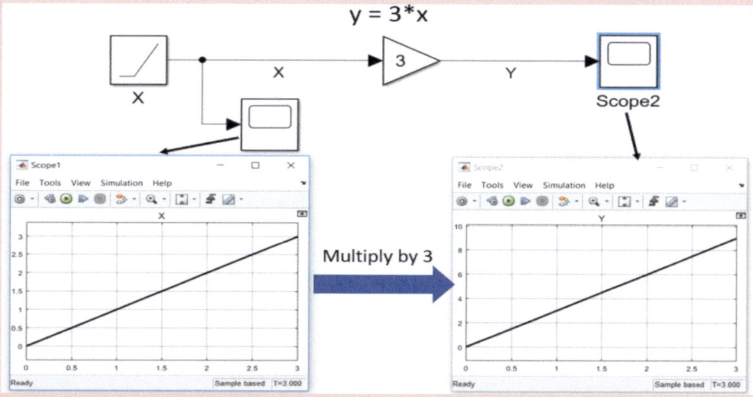

Fig. 4.20 Simulink® model of Example 4.1

ramp block. Connect the ramp with gain, we get "Y" as the output from the gain block. Add scope to both "X" (anywhere in between the ramp block and the gain block) and "Y" signals (connected to the output of the gain block). The visualized time histories of both X and Y are shown in Fig. 4.20.

Example 4.2

Implement the following equation in a Simulink® model. Run the model and show the results.

$$X1 = 3^{*}t$$
$$X2 = 2$$
$$X3 = t$$
$$Y = X1 + X2 - X3$$

Solution

From the equations:

(1) X1 is a ramp source block with a slope of **3**.
(2) X2 is a constant source block with gain **2**.
(3) X3 is a ramp source block with a slope of **1**.

The model can be implemented by:

(1) Dragging the three source blocks and edit their properties
(2) Dragging the sum block and double-clicking to replace "+ +" with "+ +-" and pressing "Ok"
(3) Connecting the output from the sources into the input of the sum block
(4) Dragging a scope and editing the properties to accept 4 inputs in a square shape as shown in Fig. 4.21.

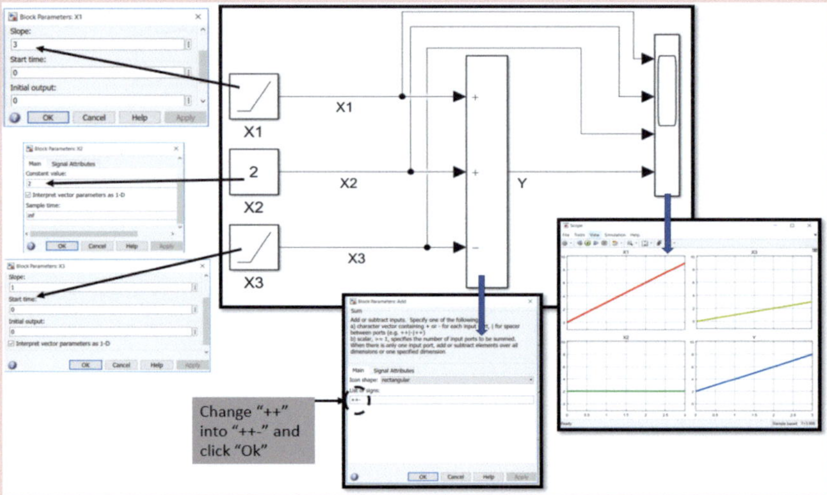

Fig. 4.21 Simulink® model of Example 4.2. Designed and implemented by the author. Images used with permission of MathWorks

Example 4.3

Implement the following equation in a Simulink® model. Run the model and show the results.

$$X1 = 3^{*}t$$
$$X2 = t$$
$$X3 = 5$$
$$Y = X1^{*}X2/X3$$

Solution

From the equations:

(1) X1 is a ramp source block with a slope "3".
(2) X2 is a ramp source block with a slope "1".
(3) X3 is a constant source block with a gain "5".

The model can be implemented by

(1) Dragging the three source blocks and editing their properties
(2) Dragging the product or divide block and double-clicking to replace "2" with "**/" in the case of product and to replace "*/" with "**/". After that, press "Ok".
(3) Connecting the output from the sources into the input to the product or divide block

Drag a scope and edit the properties to accept 4 inputs in a square shape as shown in Fig. 4.22.

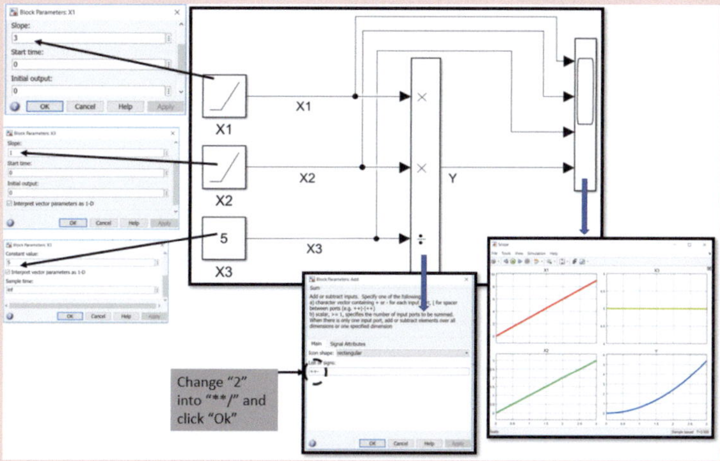

Fig. 4.22 Simulink® model of Example 4.3. Designed and implemented by the author. Images used with permission of MathWorks

Engineering Tips—Simulink® Command Via Keyboard

In the above-explained procedure, you start by opening the library browser. After that, you go all over the library's stack to find the selected block. Finally, you drag and drop the block into the Simulink® model. This process is time-consuming, especially if your computer is resource-constrained or your model is complicated.

You can make it faster simply by bringing your mouse pointer into an empty white spot on the Simulink® model. Using the mouse left button, double-click and write the name of the desired block. You will find a list of the blocks with the closest names to whatever you wrote. Pick the block and it will show up on the model. Figure 4.23 shows an example of bringing in a "Gain" block using the keyboard. When you start typing "ga", a Simulink® pop-up menu will show the list of the closest blocks' names. You pick one of them and it will show up. The keyboard dragging method is faster and easier, especially since Simulink® has the memory intelligence to improve the listed menu.

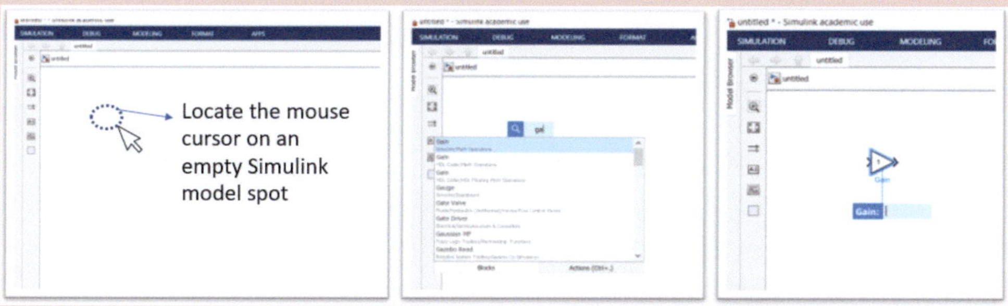

Fig. 4.23 Drop the gain block using the keyboard command. Designed and implemented by the author. Images used with permission of MathWorks

4.7 Simscape™ Multibody™

Definition[11]

Simscape™:

Simscape™ enables you to rapidly create models of physical systems within the Simulink® environment. With Simscape™, you build physical component models based on physical connections that directly integrate with block diagrams and other modeling paradigms. You can model systems such as electric motors, bridge rectifiers, hydraulic actuators, and refrigeration systems by assembling fundamental components into a schematic. Simscape™ add-on products provide more complex components and analysis capabilities [2–5].

Engineers care about modeling physical systems as well as mathematical models. For example, electrical engineers care about modeling electrical systems including physical components. Mechanical engineers care about modeling mechanical physical systems such as machines, thermodynamics, and fluid dynamics that include different physical components such as engines, shafts, gears, and cams. In other words, each engineering field requires different components to accurately model their respective systems. For example, an engineer can be hired to work on designing, modeling, and simulating a kitchen machine as shown in Fig. 4.24. In the beginning, the engineer will

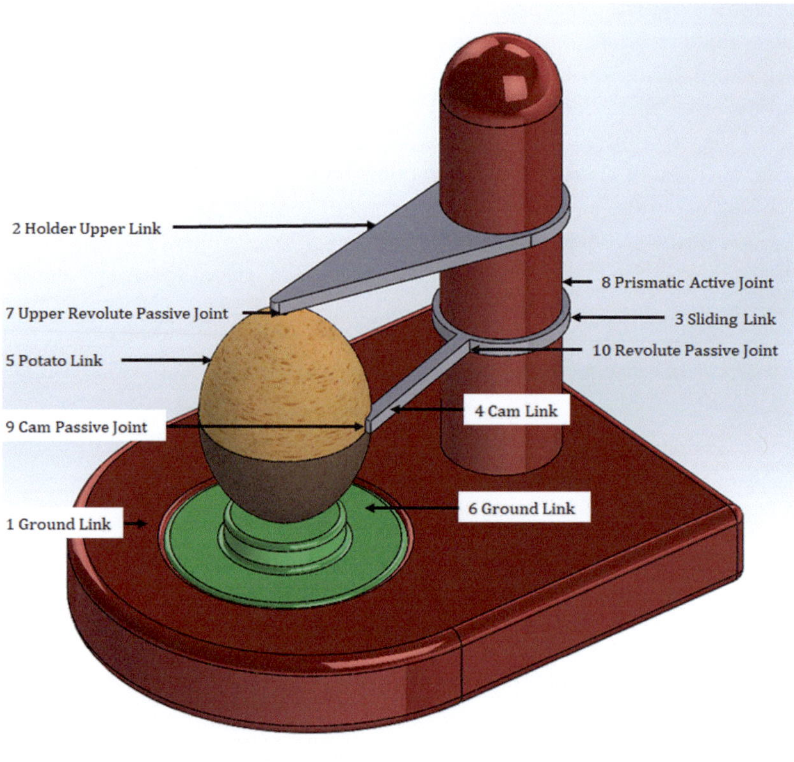

Fig. 4.24 Kitchen peel machine. Figure and associated Potato Peeling Mechanism.mp4 video designed by the author and implemented by both the author and Pranav Dhola. Both courtesy of Pranav Dhola

[11] Reproduced from [2–5] by MathWorks with permission from MathWorks.

break down the machine into links and joints. It will take the engineer a considerable amount of time and work to find the mathematical model and implement it in addition to going back and forth to make sure that his model is correct. Recently, many companies are offering physically based graphical modeling and simulation packages created for mechanical systems. MathWorks offers **Simscape™** as an extension under Simulink® to model and simulate multidomain physical systems.

Now, the engineer who is responsible for designing the kitchen machine will use the Simscape™ language that allows for the physical modeling of components. The kitchen machine can be modeled as:

1. Ground link
2. Holder upper link
3. Sliding link
4. Cam peeler arm follower link
5. Potato cam link
6. Lower revolute active joint
7. Upper revolute passive joint
8. Prismatic active joint
9. Cam passive joint
10. Revolute passive joint

Instead of doing the mathematical model, we are going to introduce how to use the physical modelling technique of Simscape™ to model the kitchen machine in a fraction of the time.

Simscape™ helps to implement models for many physical systems and test system-level performance.

For MATLAB® 2020b, as shown in Fig. 4.25, Simscape™ has the following block libraries:

- **Foundation Library**: Fundamental blocks
- **Electrical**: Electrical components and circuits
- **Gas**: Gas systems models
- **Hydraulic**: Hydraulic components

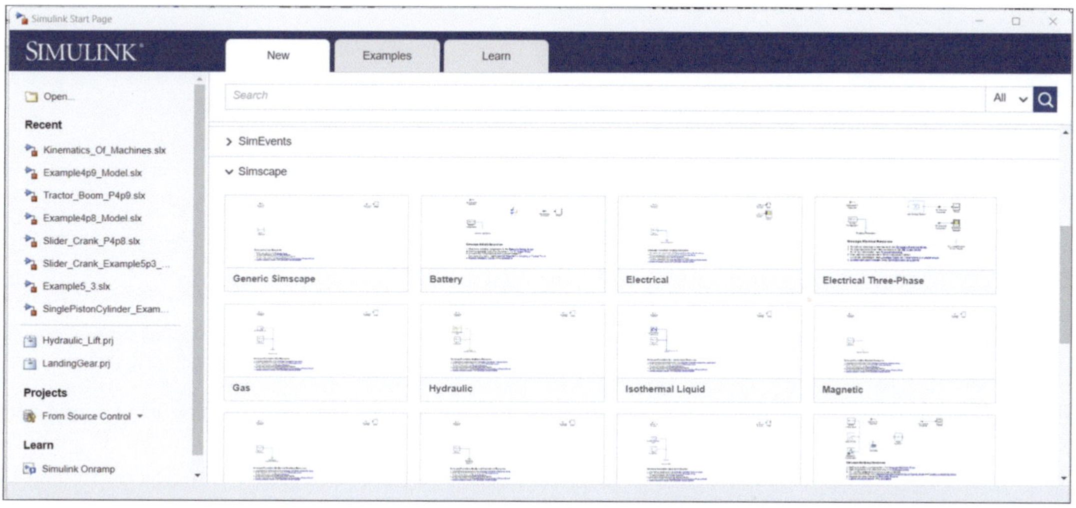

Fig. 4.25 Opening new Simscape™ model. Designed and implemented by the author. Images used with permission of MathWorks

- **Magnetic**: Electromagnetic components
- **Mechanical**: One-dimensional translational and rotational motion components
- **Moist Air**: Two- and three-species gas systems components
- **Multibody**: 3D mechanical systems components
- **Thermal**: Fundamental thermal components
- **Thermal Liquid**: Fundamental thermodynamic components in liquids
- **Two-Phase Fluid**: Fundamental thermodynamic components in working between liquid and vapor

Definition[12]

Simscape™ Multibody™:

Simscape™ Multibody™ (formerly SimMechanics) provides a multibody simulation environment for 3D mechanical systems, such as robots, vehicle suspensions, construction equipment, and aircraft landing gear. You can model multibody systems using blocks representing bodies, joints, constraints, force elements, and sensors. Simscape™ Multibody™ formulates and solves the equations of motion for the complete mechanical system. You can import complete CAD assemblies, including all masses, inertias, joints, constraints, and 3D geometry into your model. An automatically generated 3D animation lets you visualize the system's dynamics [2–5].

For the kinematics of machines, our focus will be on the Multibody toolbox. We will now show you how to use Simscape™ Multibody™ to develop a model for a machine. When you open a Simulink® blank model, open the library browser and click on "Simscape" and then click on "Multibody" as shown in Fig. 4.26.

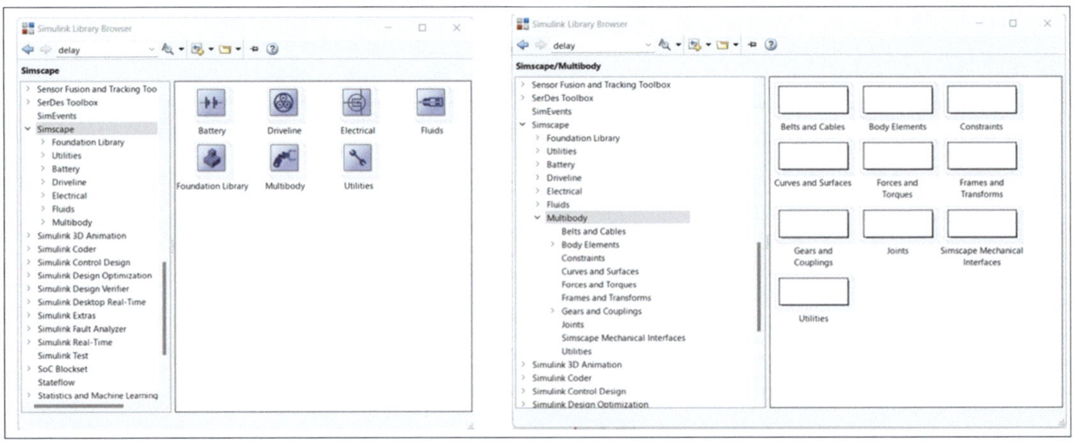

Fig. 4.26 Opening Simscape™ Multibody™ Library. Designed and implemented by the author. Images used with permission of MathWorks

[12] See Footnote 11.

The sublibraries of the multibody are.

1. Belts and cables
2. Body elements
3. Constraints
4. Curves and surfaces
5. Forces and torques
6. Frames and transformers
7. Gears and couplings
8. Joints

Here, we are going to introduce the usage of a few of the above-mentioned multibody blocks used for kinematics. Our focus will be on rigid body (solid) joints, frames, and gears. First, we are going to cover the mathematical foundation for frames and transformation. After that, we will introduce both joints and solids, followed by some examples for assembling machines using the Multibody toolbox.

4.7.1 Simscape™ Multibody™ Joints

Definition[13]

Joint:

Joints connecting bodies provide the primary kinematic constraints that determine how they can move relative to each other. A joint can be a physical connection, such as that between the case and shaft of a linear hydraulic actuator, or a virtual connection, such as that between two-point masses. In Simscape™ Multibody™, you can model both connection types using joint blocks [2–5].

A joint aims to connect two links. The first link is called ***Base***. The second link is called ***Follower***. Each joint has two nodes: Node ***B*** to connect to the base and node ***F*** to connect to the follower. Figure 4.27 shows a revolute joint, prismatic joint, and weld joint block diagrams. We are going to present these joints here as they are the most important ones for the kinematics of machines. Later in Chap. 9, we will introduce the CAM joint; similarly, the gear joint will be introduced in Chap. 10. If a joint connects the ground link with another link, we consider the ground as the base. Also, in the case of the prismatic joint, it is recommended to use the cylinder as the base.

Definition[14]

Revolute Joint:

A ***revolute*** joint block represents a joint with one rotational degree of freedom. One revolute primitive provides the rotational degree of freedom. The base and follower frame origins remain coincident during simulation [2–5].

[13] See Footnote 11.

[14] See Footnote 11.

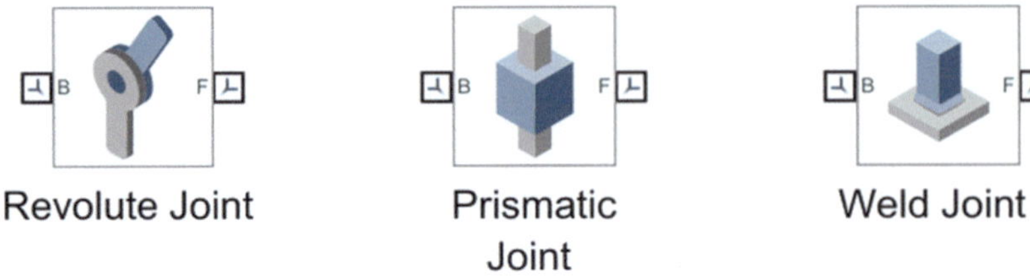

Fig. 4.27 Joints block diagrams. Designed and implemented by the author. Images used with permission of MathWorks

Definition[15]

Prismatic Joint:

A ***Prismatic*** joint block represents a joint with one translational degree of freedom. One prismatic primitive provides the translational degree of freedom. The base and follower frames remain parallel during simulation [2–5].

A revolute joint and prismatic joint have one DoF that can be sensed and/or actuated. If you double-click on the revolute joint block or the prismatic joint block, you can see an option to add Displacement S, Velocity V, and Acceleration A sensors (SVA sensors) or force/torque sensors as output nodes from the joint block as shown in Fig. 4.28. You can click on any sensor and it will show an output node from the block as shown in Fig. 4.29. For the kinematics of machines, we care more about the SVA sensors. Multibody allows for the measurement of position, velocity, and acceleration through these sensors. To measure the jerk rate, a derivative block needs to be added after the acceleration output node. More details about how to use Displacement S, Velocity V, Acceleration A, and Jerk J sensors (SVAJ sensors will be discussed in Chapter 6.)

The actuator of a joint is considered to be an input node. There are two types of actuators: a force/torque actuator and a motion actuator. The force/torque actuator is used for simulating dynamics. The motion actuator is used for simulating kinematic systems. For our interest, we are going to proceed with the motion actuator. Later in Chaps. 11 and 12, we will use the dynamic input and output signals.

The motion actuator is provided in the direction of the joint's DoF. For example, in the case of a revolute joint, we have one DoF given to the follower as rotational motion. Then, the motion actuator is given as an angular displacement for the revolute joint. The motion actuator can be a translational displacement in the case of prismatic joints as shown in Fig. 4.30. In both cases, the input signal must be continuous until the second-order derivative. For example, in the case of driving the revolute joint with a motion actuator, the only signal we should provide is angular displacement $S(t) = \theta(t)$. The first derivative $V(t) = \theta(t)$ must be smooth with no discontinuities; similarly, the second derivative $A(t) = \ddot{\theta}(t)$ must also be smooth with no discontinuities. More details about how to actuate the SVAJ using Multibody will be discussed in Chap. 6.

[15] See Footnote 11.

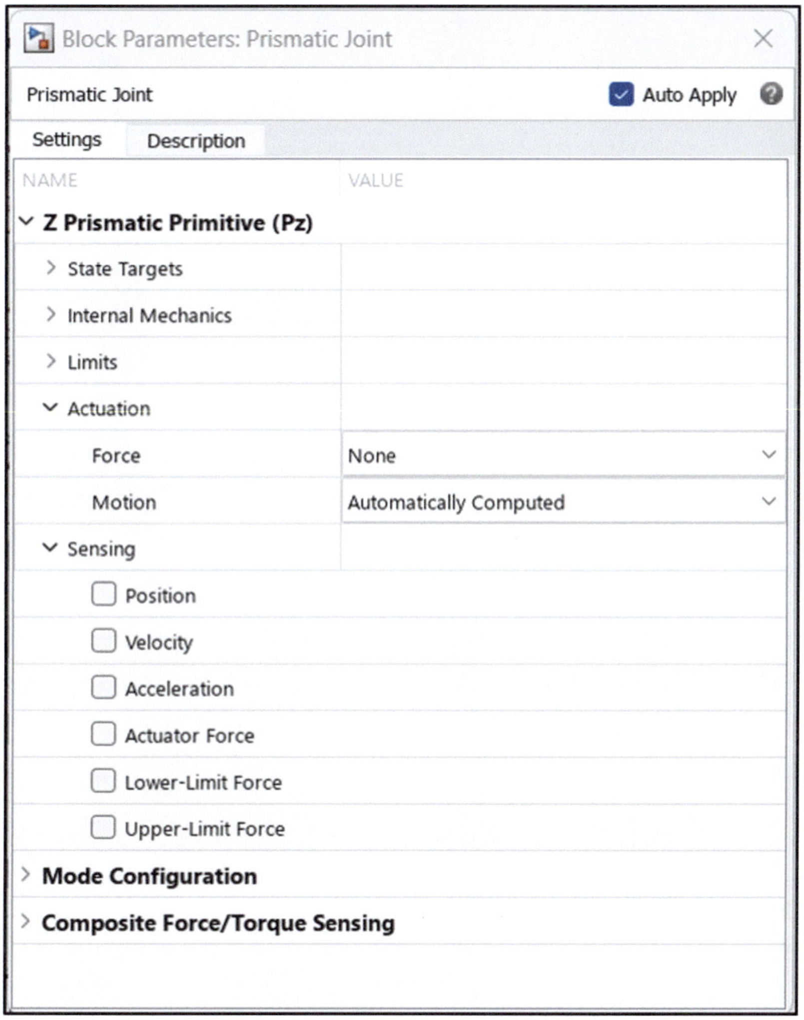

Fig. 4.28 Prismatic joint and revolute joint actuators and sensors options. Designed and implemented by the author. Images used with permission of MathWorks

4.7.2 Simscape™ Multibody™ Rigid Body Links or Solids

Definition[16]

Simscape™ Multibody™ Rigid Body:

Rigid bodies are based on the solid block, the file solid block, or, in special cases, their equivalents of variable mass and geometry. You can find the latter in the **Body Elements>Variable Mass** library. Use the file solid block to import a body from a 3D part file.

[16] See Footnote 11.

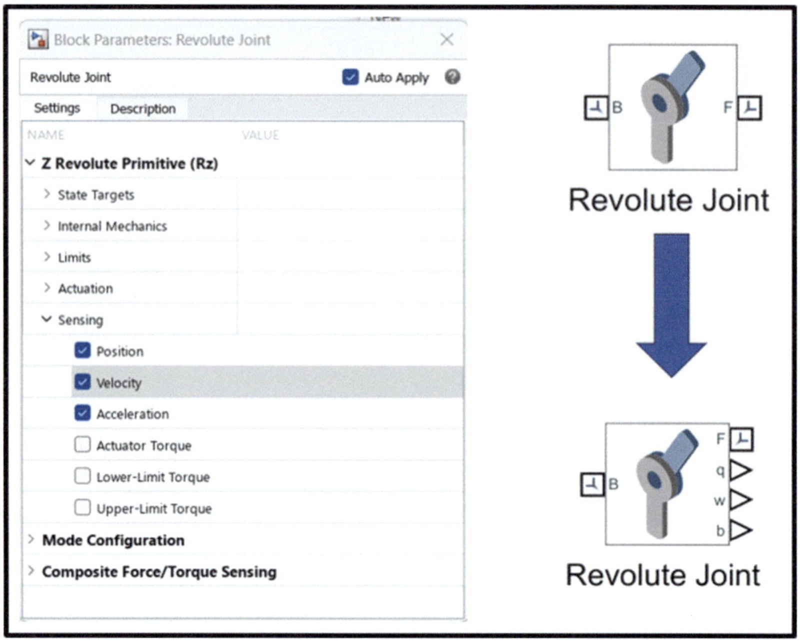

Fig. 4.29 Revolute joint SVA sensors. Designed and implemented by the author. Images used with permission of MathWorks

> The rigid body is composed of simpler body elements fixed to one another. Frame connection lines between the blocks establish the necessary rigid connections between the body elements. Rigid transform blocks, normally inserted in the connection lines, provide the relative positions and orientations required for proper assembly [2–5].

A link is known as a "solid" in the Simscape™ Multibody™ libraries. Figure 4.31 shows the shape of the solid block in a Simscape™ model along with how you can edit the rigid body by double-clicking on the block. As shown in Fig. 4.31, there is one node called "R" as an output from the rigid body block. The symbol "R" refers to the reference frame at the Center of Gravity (C.G.). Later, we are going to show how to add more nodes with extra frames to any solid block.

As shown in Fig. 4.31, we have four editing lists for the solid block:

Geometry: Used to specify the shape of a solid such as ***Cylinder***, ***Sphere***, ***Brick***, ***Ellipsoid***, ***Regular Extrusion***, ***General Extrusion***, ***Revolution,*** or ***From a file***. Table 4.3 shows the corresponding shapes and the needed parameters for each geometry. There is an option called "***From a file***" that is used to import an item as an AutoCAD file. There is an option to import the entire machine assembly from AutoCAD that is out of scope of our textbook.

Inertia: Inertia properties provide the user with the ability to specify: (1) In the case of link, the mass, moment of inertia matrix, center of mass, and product of inertia in the local C.G. reference frame or, (2) just specify the mass (point masses). The inertia properties are not relevant to the kinematics of machines. For this reason, we are not going to go over any more details. In Chap. 12, we will introduce how to calculate the C.G. and the inertia tensor.

Graphics: Used to select the color of the solid. The color can be selected by RGB or simply by picking a color. More details about the graphics options can be found in the MathWorks tutorial in Refs. [2–5].

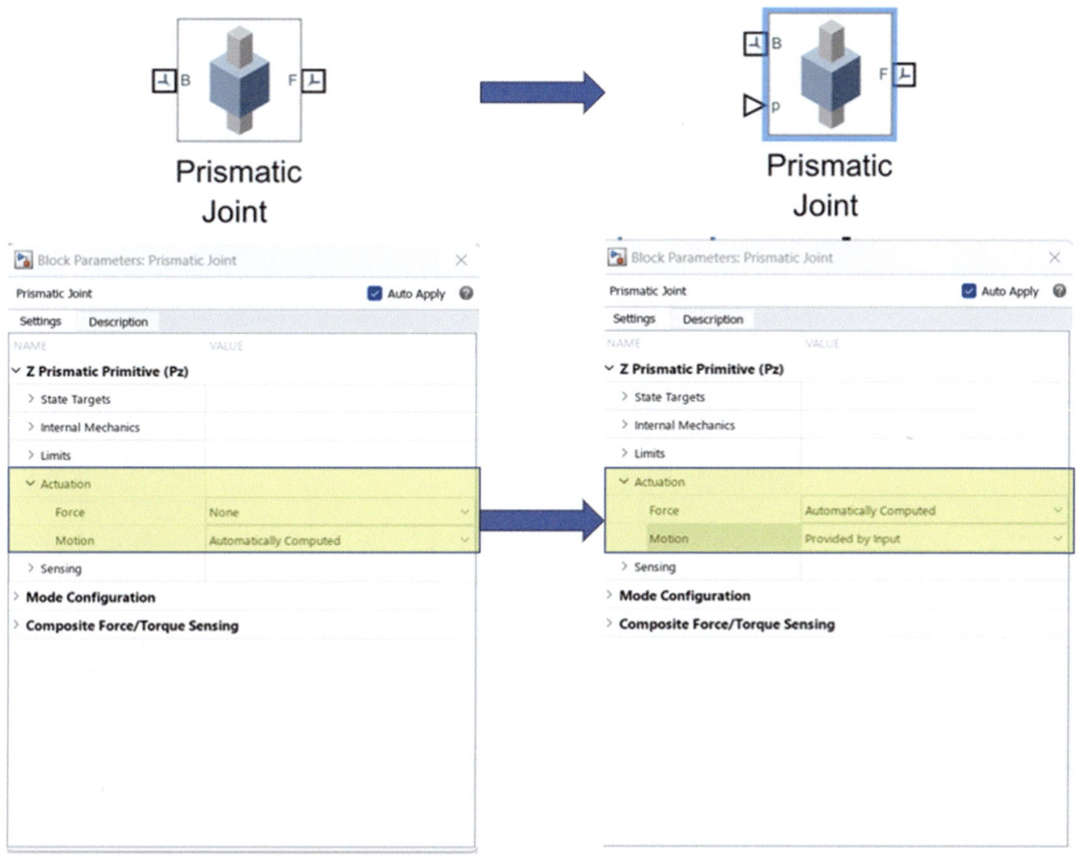

Fig. 4.30 Prismatic joint actuator. Designed and implemented by the author. Images used with permission of MathWorks

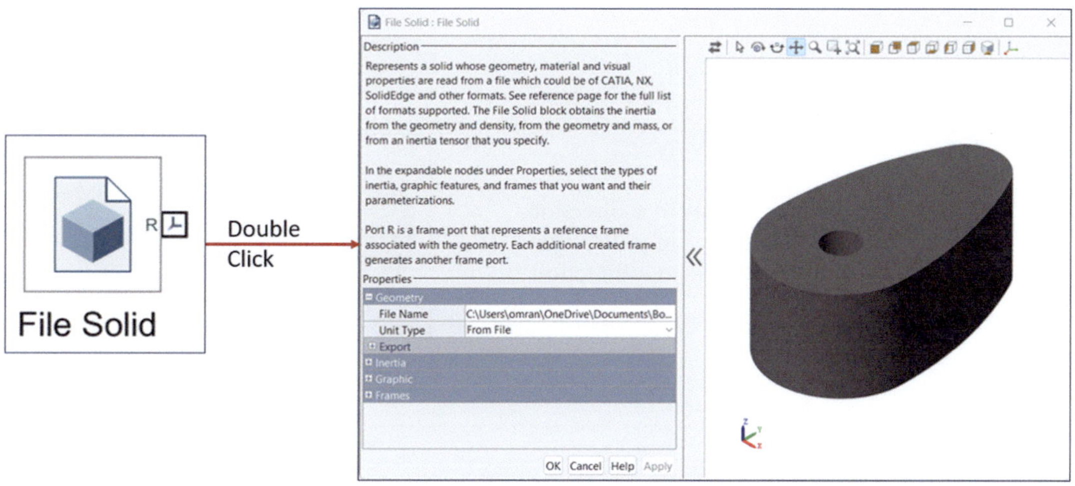

Fig. 4.31 Simscape™ Multibody™ solid block diagram. Designed and implemented by the author. Images used with permission of MathWorks

Table 4.3 Geometry of rigid body block. Figures are Designed and implemented by the author. Images used with permission of MathWorks

Geometry	**Parameters**	**Shape**
Cylinder 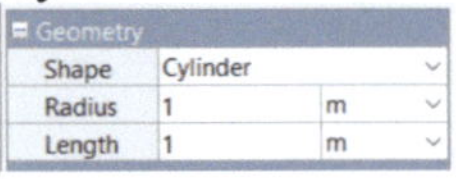	***R:*** Radius ***L:*** Length	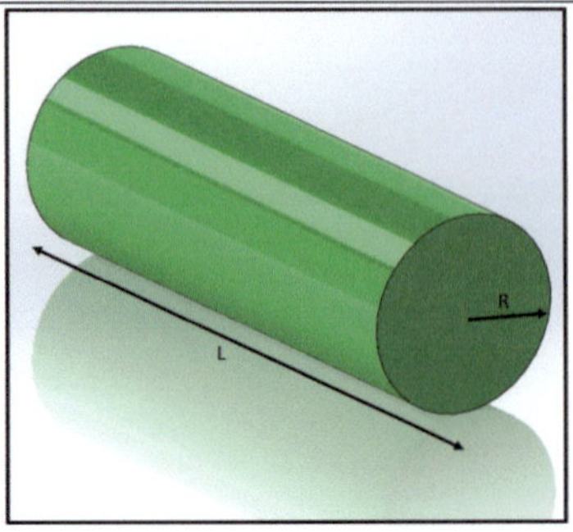
Sphere 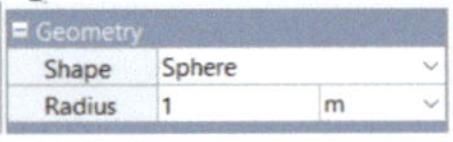	***R:*** Radius	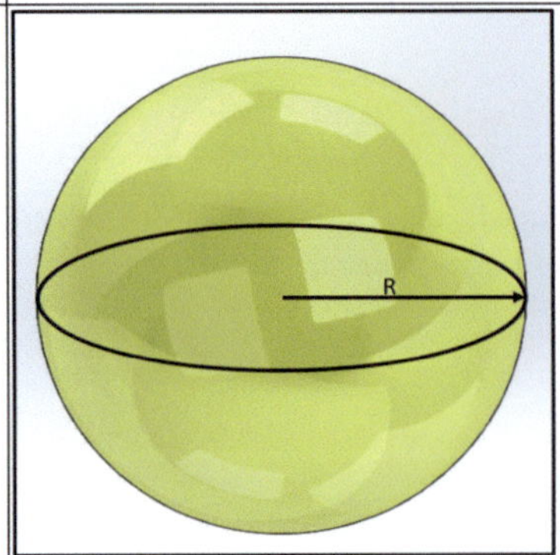
Brick 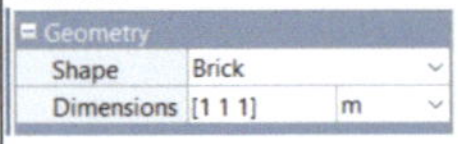	***Dimensions:*** [*Lx Ly Lz*] ***Lx:*** length in x-direction ***Ly:*** length in y-direction ***Lz:*** length in z-direction	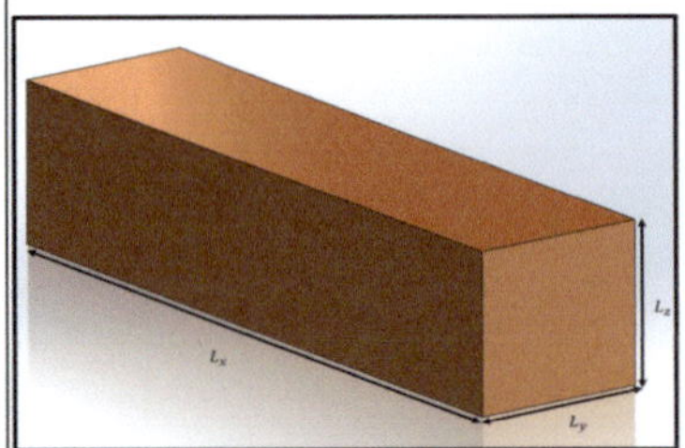
Ellipsoid 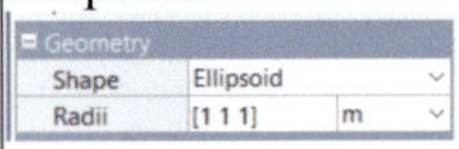	***Radii:*** [*Rx Ry Rz*] ***Rx:*** Radius in x-direction ***Ry:*** Radius in y-direction ***Rz:*** Radius in z-direction	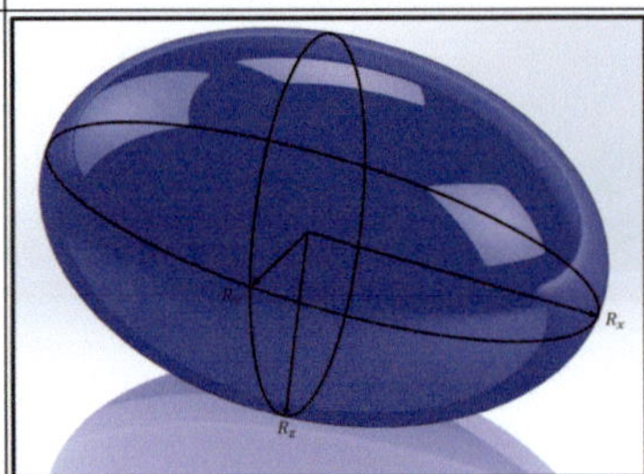

Table 4.3 (continued)

<table>
<tr>
<td>Regular Extrusion
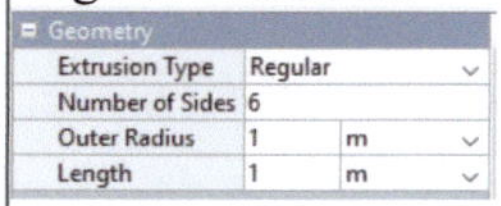
</td>
<td>N: Number of sides
R: Outer radius
L: Length</td>
<td>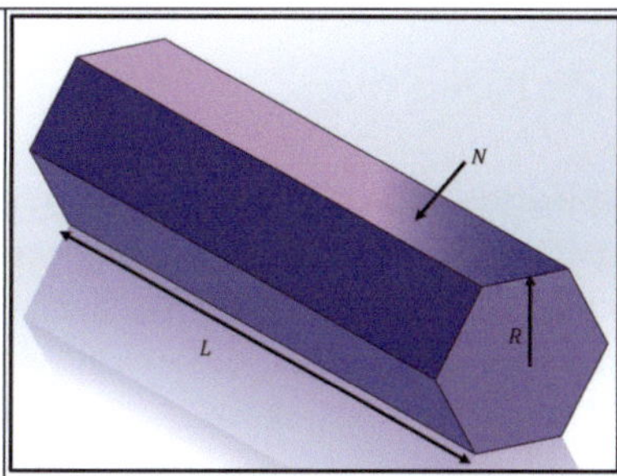
</td>
</tr>
<tr>
<td>General Extrusion
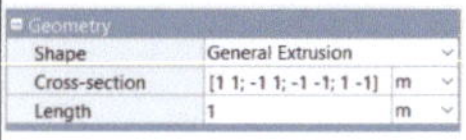
</td>
<td>Cross Section: [x y]
Lz: Length</td>
<td>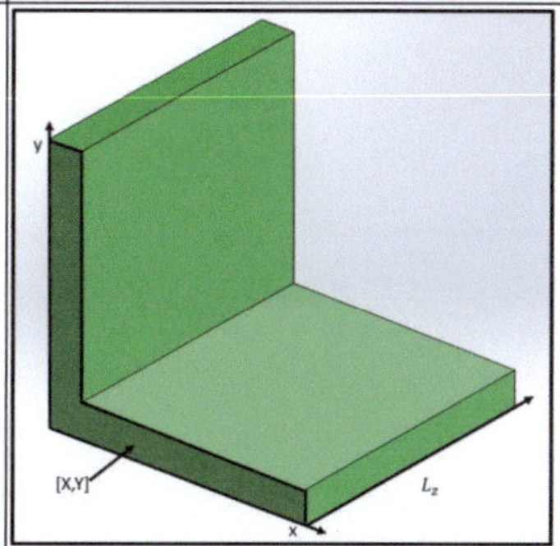
</td>
</tr>
<tr>
<td>Revolution
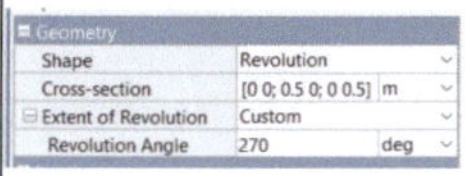
</td>
<td>Cross section: [x z]
β_y: Revolution angle around y-axis.</td>
<td>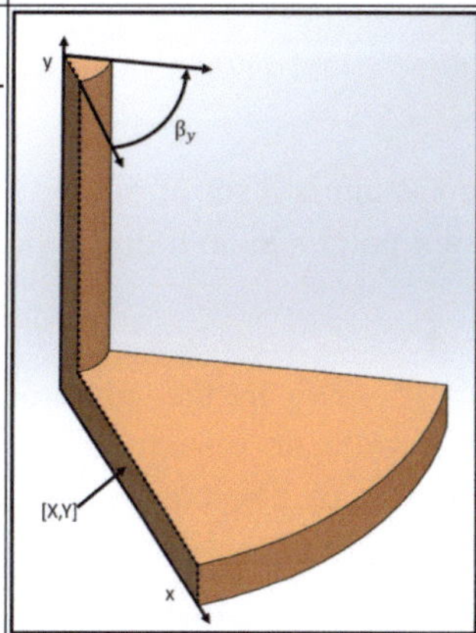
</td>
</tr>
</table>

4.7.3 Simscape™ Multibody™ Frames and Transformation

Definition[17]

Reference Frame Block:

The reference frame block represents a reference frame in which you can define other frames. The reference frame is generally non-inertial. It can accelerate with respect to the world frame. This block is optional in a model [2–5].

[17] See Footnote 11.

Definition[18]

World Frame Block:

The world frame block represents the global reference frame in a model. This frame is inertial and at absolute rest. Rigidly connecting a frame to the world frame makes that frame inertial. Frame axes are orthogonal and arranged according to the right-hand rule.

In a frame network, the world frame is the ultimate reference frame. Directly or indirectly, all other frames are defined with respect to the world frame. If multiple world frame blocks connect to the same frame network, those blocks identify as the same frame. If no world frame block connects to a frame network, a copy of an existing frame, frozen in its initial position and orientation, serves as the world frame [2–5].

Definition[19]

Rigid Transform Block:

The Rigid Transform block applies a time-invariant transformation between two frames. The transformation rotates and translates the follower port frame (F) with respect to the base port frame (B). Connecting the frame ports in reverse causes the transformation itself to reverse. The frames remain fixed with respect to each other during simulation, moving only as a single unit. Combine Rigid Transform and solid blocks to model compound rigid bodies [2–5].

A coordinate system is a set of one or more coordinates that determine the position of a point. The coordinates of a point r in an n-dimensional space are simply an ordered set of n numbers:

$$r = \left[x^{(1)} x^{(2)} \cdots x^{(n)}\right] \tag{4.2}$$

where $x^{(1)}, x^{(2)}, \ldots x^{(n)}$, are the values of each coordinate.

In 3D, any coordinate system has three coordinates: x-coordinate, y-coordinate, and z-coordinate as shown in Fig. 4.32. The coordinates in 3D can be mathematically related to a Cartesian coordinate system by a set of functions:

$$P = P(x, y, z) \tag{4.3}$$

where x, y, z. are the three values of the Cartesian coordinates of the point P.

In 2D, any coordinate system has two coordinates such as the x-coordinate and y-coordinate or the θ-coordinate and r-coordinate as

$$\begin{aligned} P &= P(x, y) \\ P &= P(r, \theta) \end{aligned} \tag{4.4}$$

where x and y are the two Cartesian coordinates of the point while θ and r are the two polar coordinates of point P as shown in Fig. 4.33.

The coordinates of any node can be calculated in different frames. Simscape™ Multibody™ uses two types of frames to configure any node's coordinates: ***world frame*** and ***reference frame***. The connection between any two frames can be facilitated by way of the ***Rigid Transform*** block.

[18] See Footnote 11.
[19] See Footnote 11.

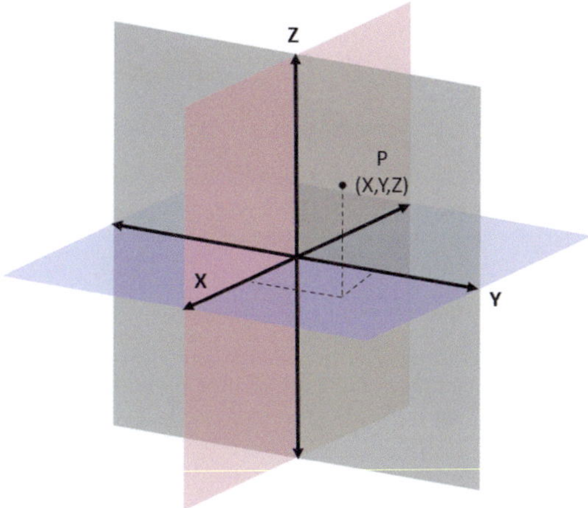

Fig. 4.32 Cartesian coordinates in 3D

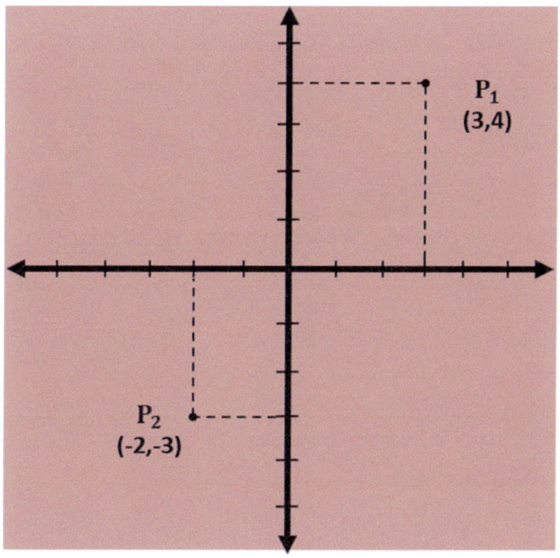

Fig. 4.33 Cartesian coordinates in 2D

World Frame Block

The world frame is a global reference frame that describes the machine's motion. This frame is inertial with zero-translation motion and zero rotational motion and is at absolute rest. The world frame is a coordinate system that consists of axes or a set of points that are fixed in position and orientation with respect to one another. Now, we are going to replace ***point*** with ***node***. For example, Fig. 4.34 shows a tractor machine and Fig. 4.35 shows the kinematic diagram of this tractor machine. For the shown tractor machine or front-end loader, we have two ground nodes: the upper ground node at n_1 and the lower ground node at n_2.

We are going to specify the world frame using one of these two ground nodes. We are going to pick $n_2 : J_2^R$ as the node to specify the Cartesian frame that consists of x-coordinates and y-coordinates for the kinematic diagram in Fig. 4.36. We'll call this lower ground node the ***origin*** of the

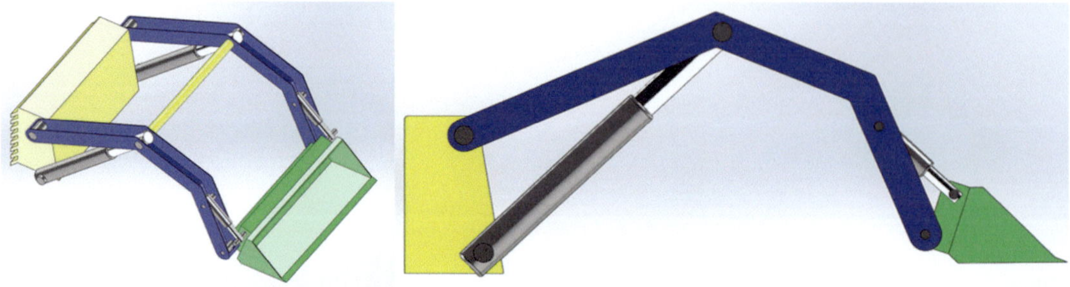

Fig. 4.34 Tractor machine. Designed by the author and implemented by both the author and Pranav Dhola. Courtesy of Pranav Dhola

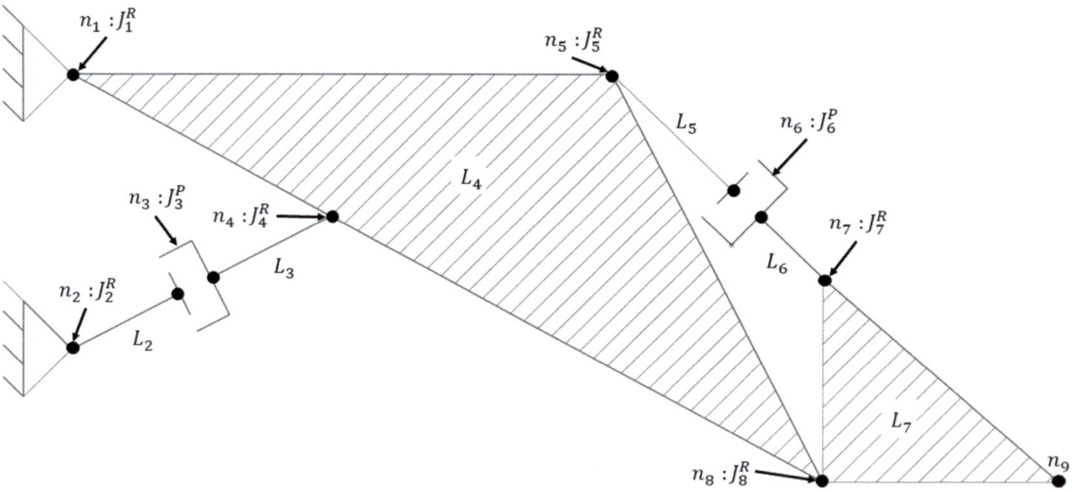

Fig. 4.35 Kinematic diagram of tractor machine

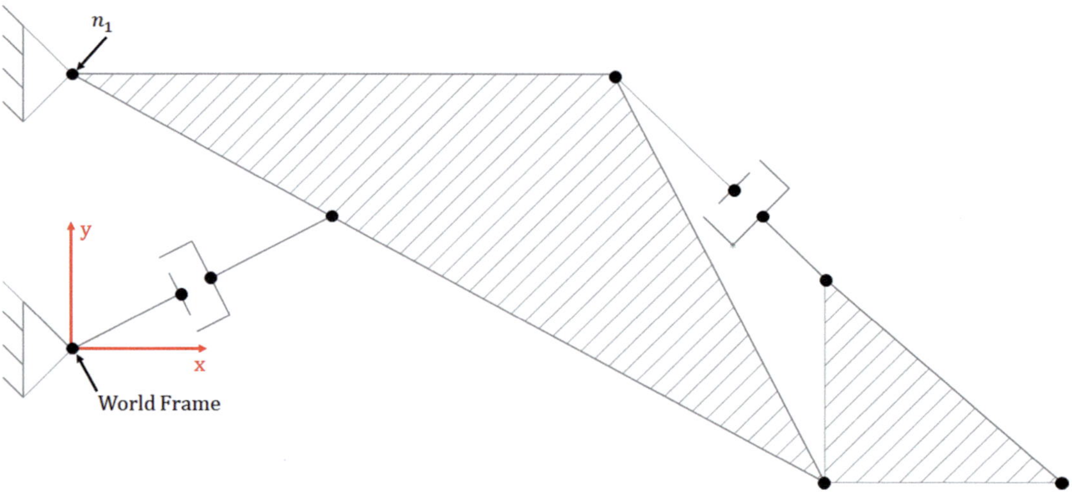

Fig. 4.36 World frame of tractor's kinematic diagram

world frame. The x-axis is defined as a fixed horizontal line that starts from the origin and runs to infinity. The y-axis or y-coordinate is defined as a fixed vertical line that runs from the origin to infinity.

The position of the entire mechanism will be defined with respect to this world frame. Figure 4.37 shows the machine in two different positions. As shown in Fig. 4.37, the coordinate of each node changes from position 1 to position 2. This variation of position is defined as the motion of the machine dictated by the positions of the hydraulic or prismatic joints. The nodes' variation with time is called the trajectory of the machine and can be specified in terms of the SVAJ of each node. During the motion trajectory, shown in Fig. 4.37, it is clear that nodes n_1 and n_2 never change because they are part of the world frame or the "ground link".

If the origin is a ground that is considered as stationary relative to the rest of the components, the frame is called an *inertial frame* or *global frame*. In other words, the inertia frame or global frame of any mechanism has zero translational and rotational velocity with respect to the machine animation. Figure 4.38 shows the block diagram of the world frame. For any multibody model, there is only one world frame. The Cartesian inertial/global frame of any machine is defined using the algorithm below.

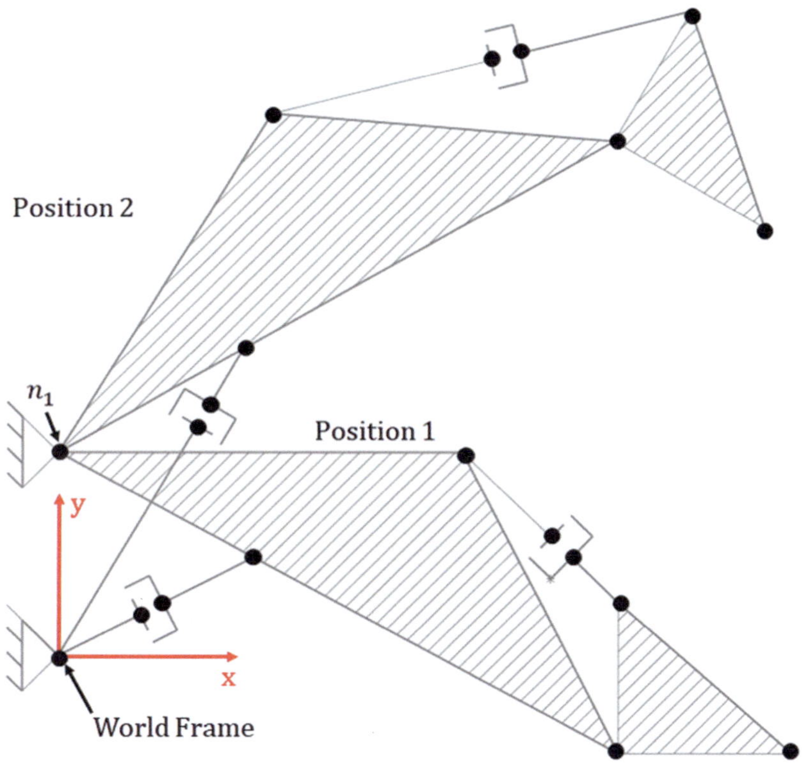

Fig. 4.37 Motion trajectory of the tractor machine

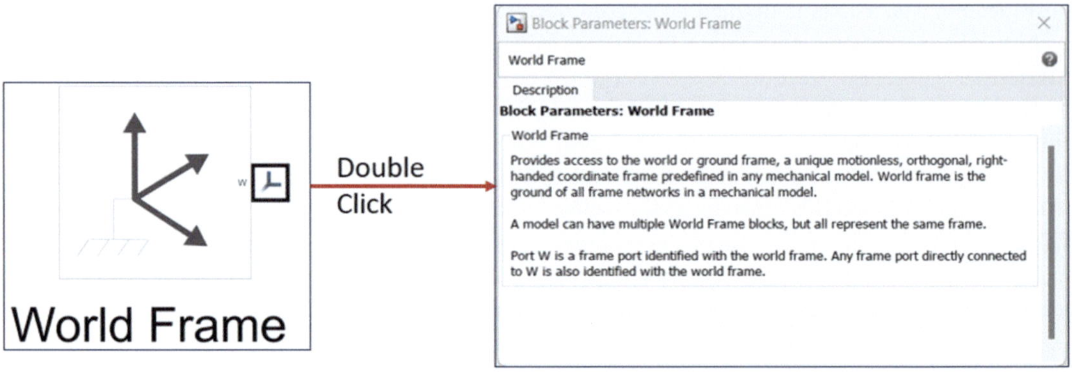

Fig. 4.38 World frame block. Designed and implemented by the author. Images used with permission of MathWorks

(1) Pick one of the ground nodes to be the origin point which is equivalent to *x*-coordinate = 0 and *y*-coordinate = 0 or (0,0).
(2) Extend a horizontal line with an arrow pointing to the infinity direction on the right-hand side. The horizontal line starting from the origin (0,0) ends at infinity on the *x*-axis and 0 on the *y*-axis or (0, ∞). This horizontal line is called the *x*-axis.
(3) Extend a vertical line with an arrow pointing to the infinity direction on the upper side. The vertical line starting from the origin (0,0) ends at infinity on the *y*-axis and 0 on the *x*-axis or (∞, 0). This vertical line is called the *y*-axis.

Example 4.4

Figure 4.39 shows the landing gear machine.

(1) Sketch the kinematic diagram.
(2) Show where we could locate the inertial frame.

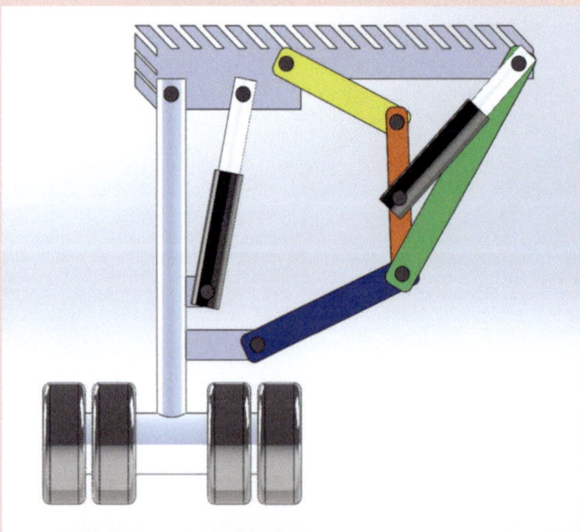

Fig. 4.39 Landing gear machine. Designed by the author and implemented by both the author and Pranav Dhola. Courtesy of Pranav Dhola

Solution

We have four ground nodes n_1, n_2, n_3, and n_4 as shown in Fig. 4.40. We locate the origin at n_1. All n_1, n_2, n_3, and n_4 have the same coordinates regarding any change in the input signals which are the two prismatic joints J_7^P and J_{12}^P with time. The coordinates of the rest of the nodes are changing with the variation of prismatic joints J_7^P and J_{12}^P or with respect to time.

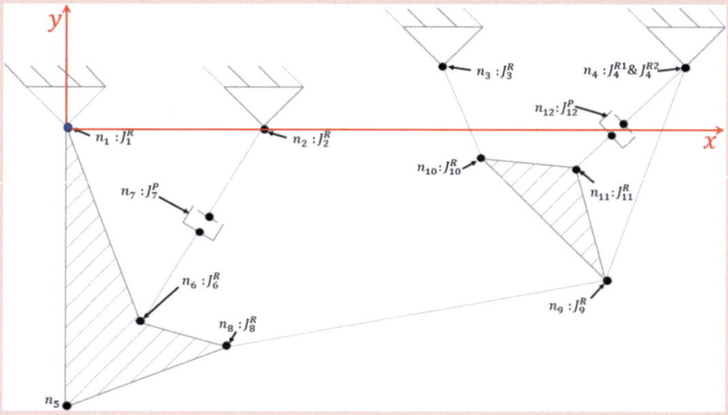

Fig. 4.40 Inertia frame and motion of landing gear machine

Example 4.5

Figure 4.41 shows the radial engine machine.

1. Sketch the kinematic diagram.
2. Show where we could locate the inertial frame.

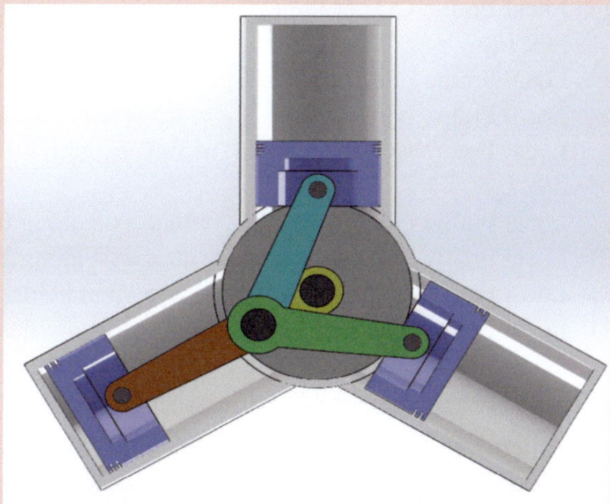

Fig. 4.41 Radial engine machine. Designed by the author and implemented by both the author and Pranav Dhola. Courtesy of Pranav Dhola

Solution

We have four ground nodes n_1, n_6, n_7, and n_8 as shown in Fig. 4.42. We locate the origin at n_1. All n_1, n_6, n_7, and n_8 have the same coordinates regarding any change in the input signal which are the three prismatic joints J_6^P, J_7^P, and J_8^P with time. The coordinates of the remaining nodes are changing with the variation of prismatic joints J_6^P, J_7^P, and J_8^P or with respect to time.

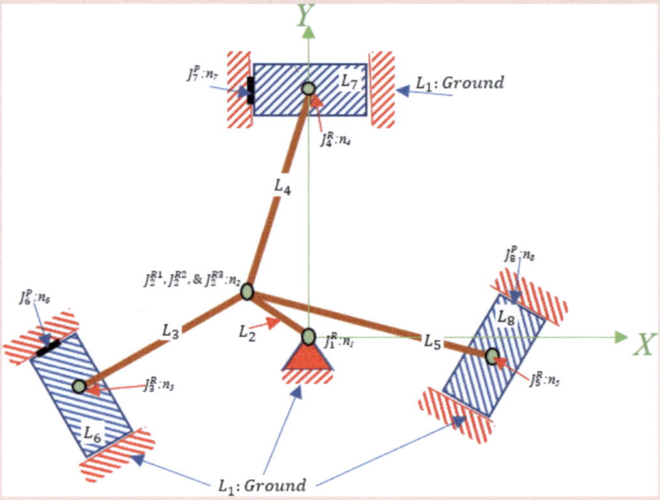

Fig. 4.42 Kinematic diagram of radial engine machine

4.7.3.1 Local Frames on Joints and Links

The local frame is a set of coordinate systems that are relatively fixed as a location and orientation with respect to one another, but this entire coordinate system can move (translation + rotation). For example, Fig. 4.43 shows a four-bar mechanism that has:

(1) One global frame at n_{1L1}, where n_{1L1} is the location of node n_1 on link L_1
(2) Local frame CG_2 at the center of gravity C.G. of link L_2 with x-axis along link L_2 and y-axis perpendicular to link L_2
(3) Local Frame L2n1 at n_1 on L_2 with x-axis along link L_2 and y-axis perpendicular to link L_2
(4) Local Frame L2n2 at n_2 on L_2 with x-axis along link L_2 and y-axis perpendicular to link L_2
(5) Local Frame CG_3 at the C.G. of link L_3 with x-axis along link L_3 and y-axis perpendicular to link L_3
(6) Local Frame L3n2 at n_2 on L_3 with x-axis along link L_3 and y-axis perpendicular to link L_3
(7) Local Frame L3n3 at n_3 on L_3 with x-axis along link L_3 and y-axis perpendicular to link L_3
(8) Local Frame CG_4 at the C.G. of link L_4 with x-axis along link L_4 and y-axis perpendicular to link L_4
(9) Local Frame L4n3 at n_3 on L_4 with x-axis along link L_4 and y-axis perpendicular to link L_4
(10) Local Frame L4n4 at n_4 on L_4 with x-axis along link L_4 and y-axis perpendicular to link L_4
(11) Local Frame 1 at n_1: J_1^R with z-axis perpendicular on the x–y plane
(12) Local Frame 2 at n_2: J_2^R with z-axis perpendicular on the x–y plane
(13) Local Frame 3 at n_3: J_3^R with z-axis perpendicular on the x–y plane
(14) Local Frame 4 at n_4: J_4^R with z-axis perpendicular on the x–y plane

The number of local frames is equal to the total number of joints plus the total number of non-ground links and the number of nodes on each link. There is only one world frame that represents the ground link.

Simscape™ Multibody™ uses the local frames for many reasons but we are going to highlight a couple of them here:

a. To identify the local frame for each joint
b. To identify the nodes on each link
c. To assemble the machine from a set of elements

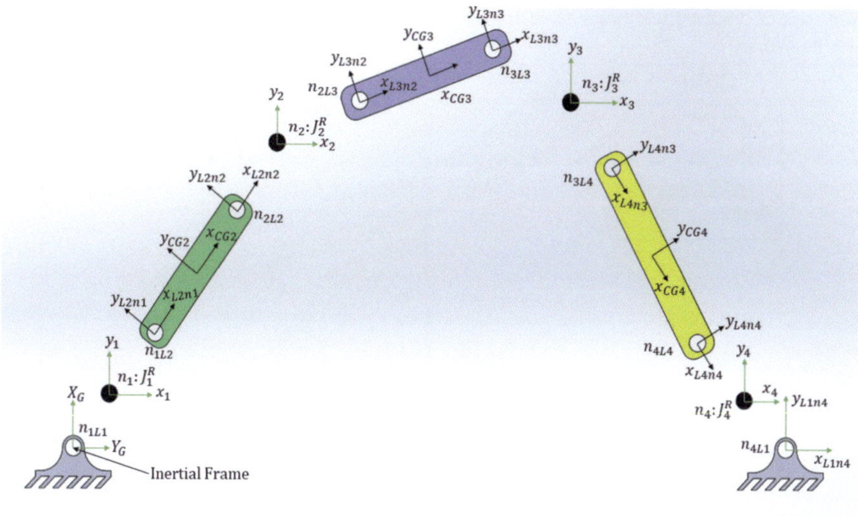

Fig. 4.43 Frames on 4-bar mechanism

4.7.3.1.1 Identify Local Frame for Each Joint

Simscape™ Multibody™ provides an efficient way to specify the direction of DoF for each joint using a standardized local frame for each joint. For example, as shown in Table 4.4, the revolute joint has 1 DoF around one specific axis. The Multibody toolbox defines a fixed local frame with the origin located at the center of the revolute joint. The z-axis of this local frame will be perpendicular to the joint plane as shown in Table 4.4. The case with the prismatic joint is similar, where the local joint frame is defined as the z-axis in the same direction as the allowable translation of the joint. In summary, the z-axis is usually used to define the DoF direction for any specific joint. Table 4.4 shows some examples of local frames of joints.

4.7.3.1.2 To Identify the Nodes on Each Link

For each link, there is a local frame that is attached at the Center of Gravity (C.G.). The origin of this link's local frame is referred to as ***R*** and is one of the link's nodes. Even though the C.G. node is not connected to any other elements (joints or links), Simscape™ Multibody™ establishes other nodes in relation to the C.G. node ***R***. For example, if we have a link with 3 important nodes, we will have to identify three moving/local frames related to the link's C.G. There are two ways to identify the non-C.G. nodes on the link with respect to the C.G. node on the same link.

Method 1: Using an External Rigid Transform

A rigid transform is a transform from one reference frame to another. Figure 4.44 shows the ***Rigid Transform*** block option menu, accessible by double-clicking on the block. There are two major options: Rotational definition and translational definition. Figure 4.45 shows the relation between the two frames. The lower-left-hand side frame is called the ***Base Frame***. The upper-right-hand side frame is called ***Follower Frame***. This block's purpose is to establish the follower frame as a function of the base frame in terms of translational relative motion and rotational relative motion.

Table 4.4 Simscape™ Multibody™ joint local frames. Designed and implemented by the author. Images used with permission of MathWorks

Joint	Local frame
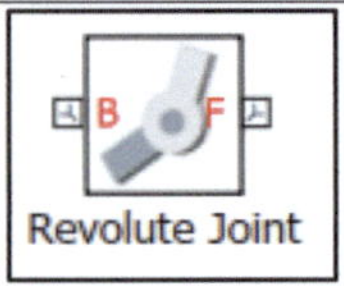	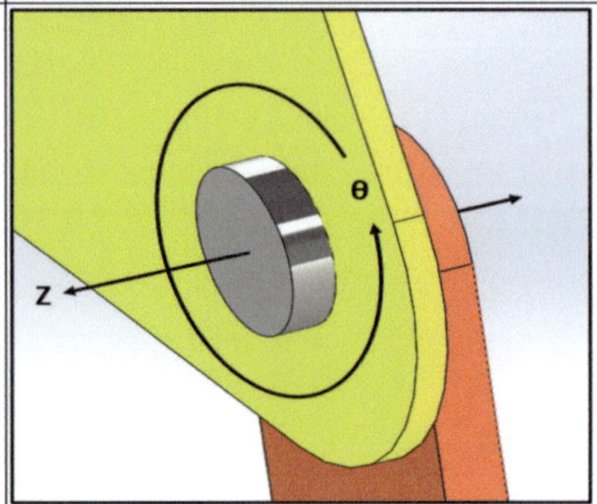
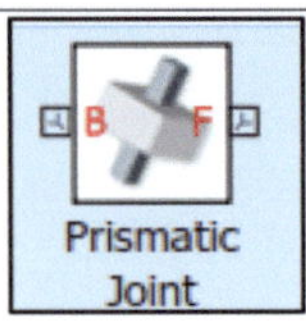	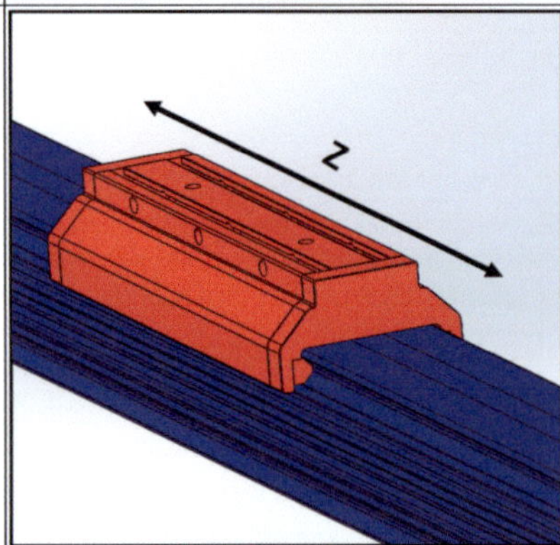
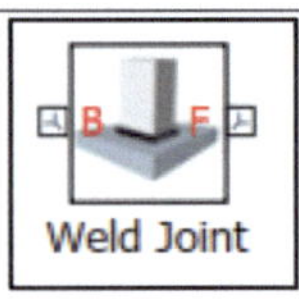	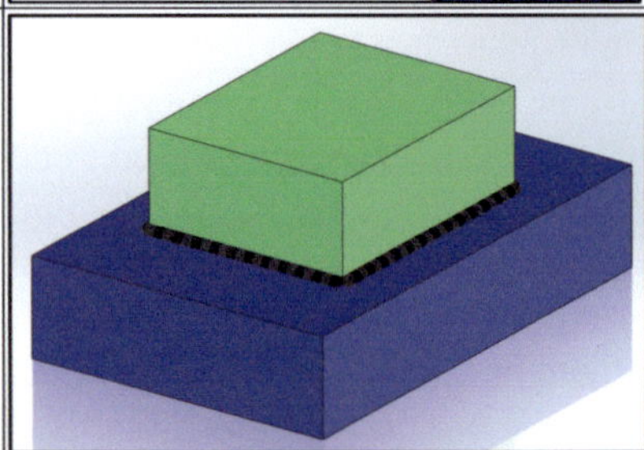
	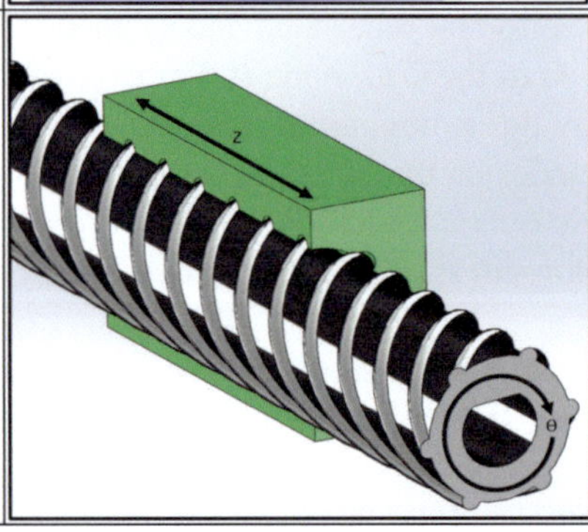

The translation distance between the follower frame and the base frame can be defined using three different methods as listed in Table 4.5. We are going to focus more on the Cartesian method. The translation is defined as

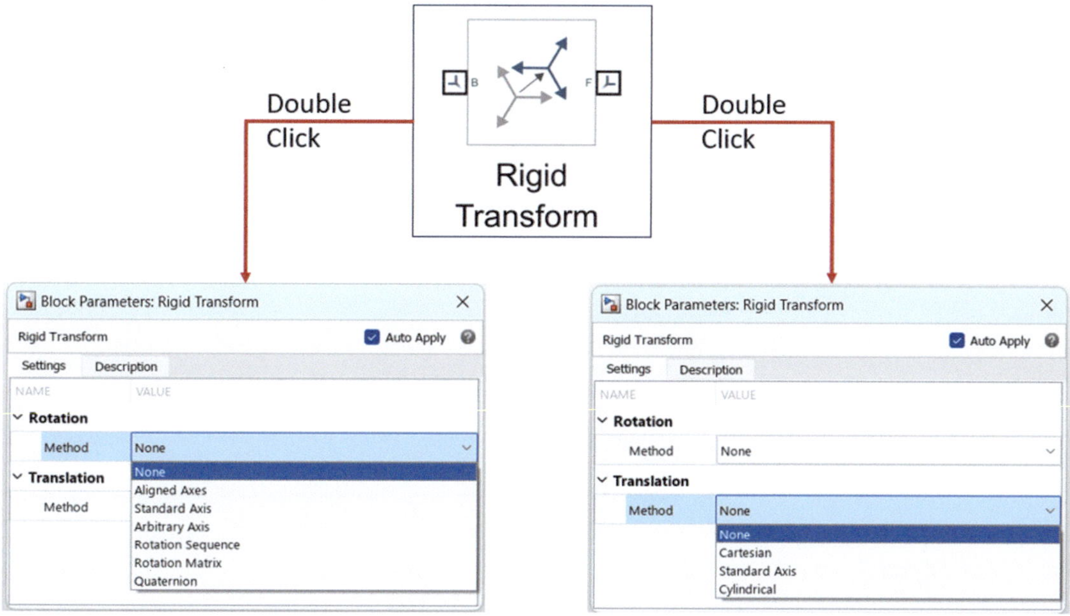

Fig. 4.44 Rigid transform block. Designed and implemented by the author. Images used with permission of MathWorks

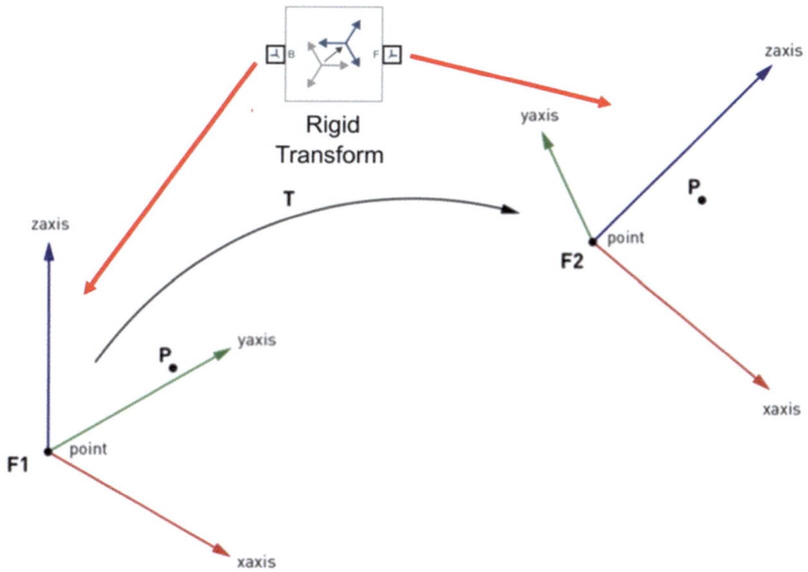

Fig. 4.45 Base frame and follower frame for Rigid Transform. Designed and implemented by the author. Images used with permission of MathWorks

Table 4.5 Translational distance between follower and base for rigid transform

Method	Description
None	Make the base and follower frames coincident. This method requires no parameters
Cartesian	Specify a 3D translation in terms of Cartesian coordinates
Standard axis	Specify a 1D translation along the X-, Y-, or Z-axis
Cylindrical	Specify a 3D translation in terms of cylindrical coordinates

Table 4.6 Rotational between follower and base for Rigid Transform

Method	Description
None	Constrain the base and follower frames to share the same orientation
Aligned axes	Set frame rotation by aligning two follower frame axes with two base frame axes
Standard axis	Specify frame rotation as an angle about a standard axis (x, y, or z)
Arbitrary axis	Specify frame rotation as an angle about a general $[x, y, z]$ axis
Rotation sequence	Specify frame rotation as a sequence of three elementary rotations

$$\Delta P = P_F - P_B = \left[\Delta x \Delta y \Delta z\right] \tag{4.4}$$

where ΔP specifies the **offset** of the follower frame with respect to the base frame in terms of Δx in the x-direction, Δy in the y-direction, and Δz in the z-direction represented in the Cartesian coordinate system in the base frame. Similar descriptions are used for the other three methods but in a different coordinate system. For more details, you can look at Refs. [2–5].

There are five methods to represent the rotation between the frame to the base as listed in Table 4.6. For simplicity, we are going to cover the explanation of two rotational methods: Aligned axes and standard axis.

I-Aligned Axes Rotational Method

The aligned axes rotational method depends on switching the base frame x_B–y_B–z_B into the follower frame x_F–y_F–z_F by an angle "$90 * n$ degree" where $n = \{-4, -3, -2, -1, 0, 1, 2, 3, 4\}$. The user has access to this process by defining two pairs of rotations as shown in Table 4.7, which shows some possible combinations for the aligned axes rotational method with their graphical representation. Axis choices depend on **Pair 1** axis selections. The main disadvantage of the ***alignedaxes rotational*** method is that it is limited to angles that are a function of 90° increments. This means that there is no way to make a rotation with an angle that is not an integer multiple of 90° using this method.

II- Standard Axis Method:

The standard axis method aims to specify the frame rotation as an angle about a standard axis (x, y, or z). The standard axis is selected from one of the three base axes (x_B, $y_{B,}$ or z_B) to be used in relation to the angle. The user can specify the selected axis and the angle of the rotation around this specified axis in degrees as shown in Fig. 4.46.

Table 4.7 Combination of aligned axes rotational method

<table>
<tr><th>Pair 1 & Pair 2</th><th>Graphics</th></tr>
<tr><td>Rotation
Met... Aligned Axes
Pair 1
Foll... +X
Base +Z
Pair 2
Foll... +Y
Base +Y</td><td>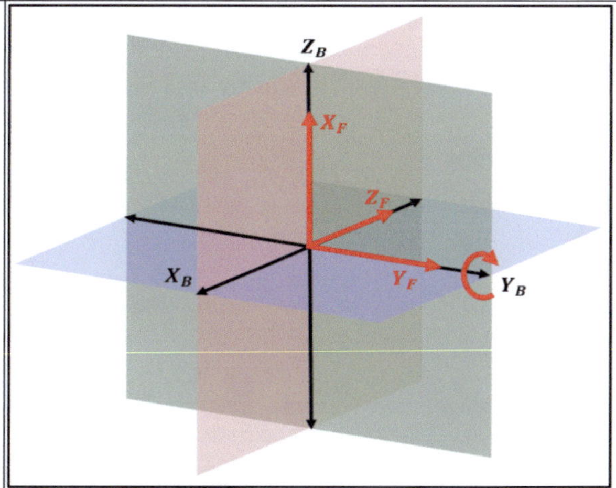
</td></tr>
<tr><td>Rotation
Met... Aligned Axes
Pair 1
Foll... +X
Base -Z
Pair 2
Foll... +Y
Base +Y</td><td>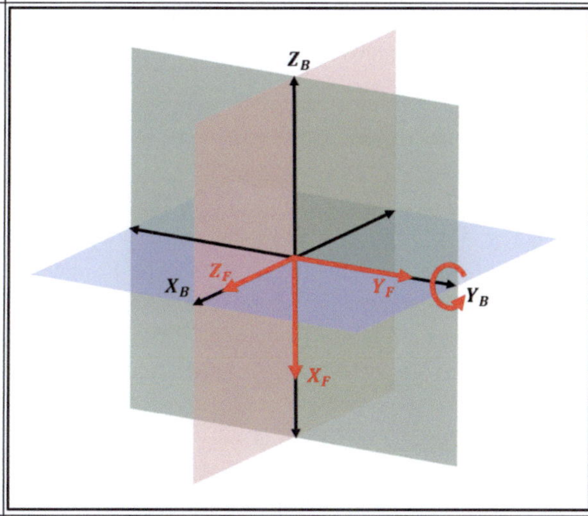
</td></tr>
<tr><td>Rotation
Met... Aligned Axes
Pair 1
Foll... +Y
Base +Z
Pair 2
Foll... +X
Base +X</td><td>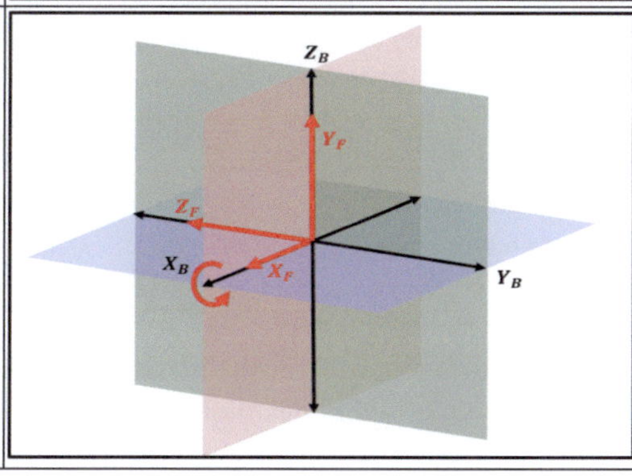
</td></tr>
</table>

Designed and implemented by the author. Images used with permission of MathWorks

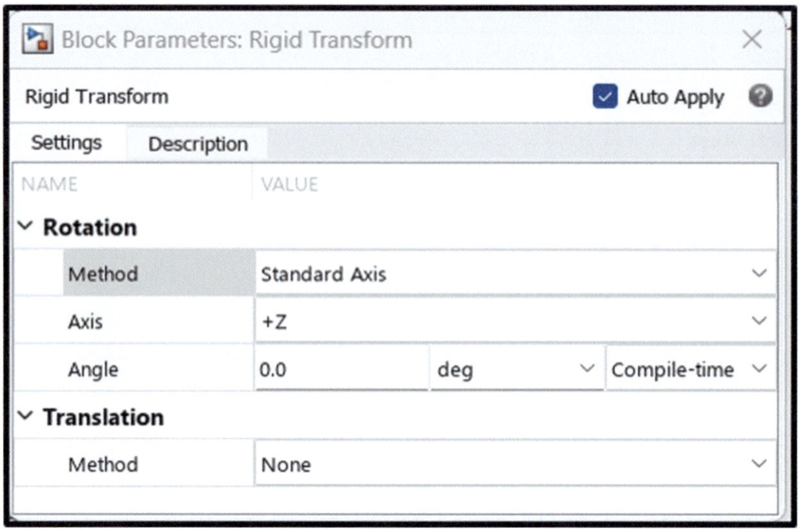

Fig. 4.46 Standard axis rotational method Simscape™ Multibody™. Designed and implemented by the author. Images used with permission of MathWorks

Method 2- Using an Internal Frame Relative to R or C.G. Frame:

Now, let's go back and look at the editing list for the solid block. We are going to focus on the frame list as shown in Fig. 4.47. You can add a new frame by clicking on the "+" sign at "new frame". Underneath the new frame, we have two editing lists: Frame origin and frame axes.

Frame Origin: The frame origin editing list is to define which node has which axis. The origin of the new frame can be defined at the center of mass or the reference frame origin. There is a third option for defining the origin based on geometric features, which is detailed below.

(1) Click on the arrow as shown in Fig. 4.48.
(2) Choose "based on geometric feature" as shown in Fig. 4.48.
(3) Click on the node where you would like to pick the origin as shown in Fig. 4.48.
(4) Click on "use selected feature" and you will see the selected origin on the graph along with a written location of the origin as shown in Fig. 4.48.

Frame Axes: After specifying the frame origin, we need to specify the frame direction or axes. Like the origin, you can specify the new frame to be aligned with the reference frame with Pair 1 and Pair 2. Axes defined by geometric features are detailed below.

(1) Click on the arrow as shown in Fig. 4.49.
(2) Choose "based on geometric feature" for the *z*-axis as shown in Fig. 4.49.
(3) Click on the line between the two nodes that correspond to the desired direction for the z-axis as shown in Fig. 4.49.
(4) Click on "use selected feature" and you will see the selected *z*-axis on the graph along with a written direction of the selected *z*-axis as shown in Fig. 4.49.
(5) Repeat the same step for the *y*-axis.

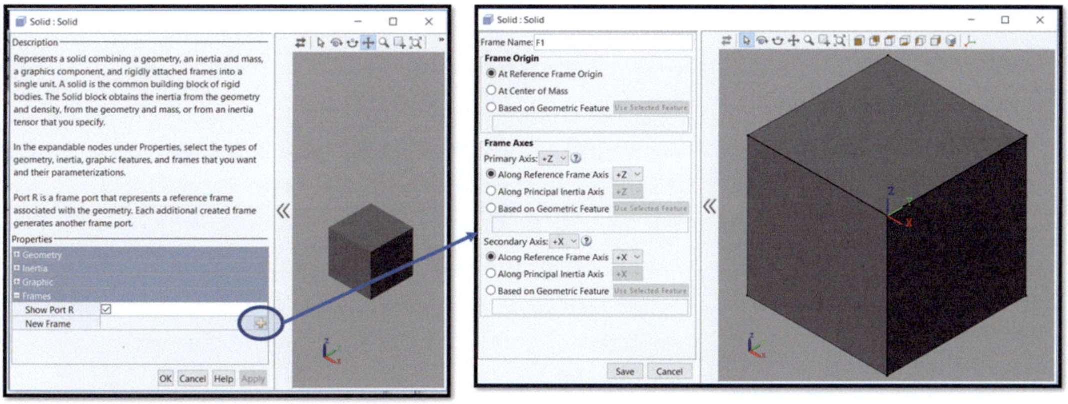

Fig. 4.47 Adding new frame on a link. Designed and implemented by the author. Images used with permission of MathWorks

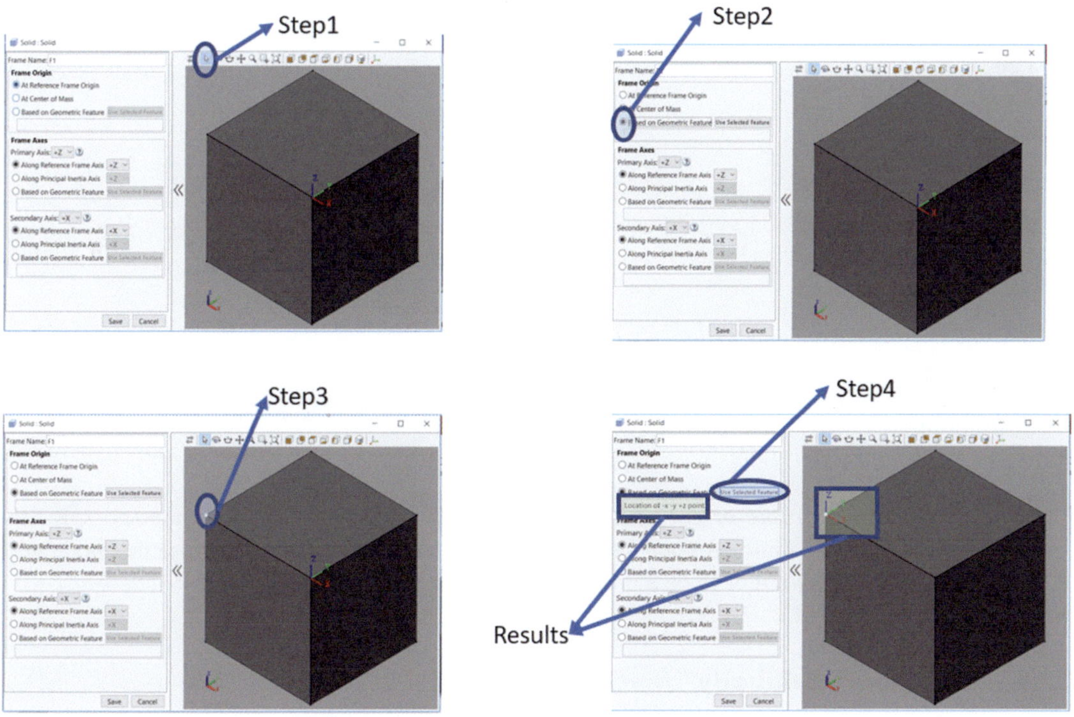

Fig. 4.48 Defining the frame origin using the solid block. Designed and implemented by the author. Images used with permission of MathWorks

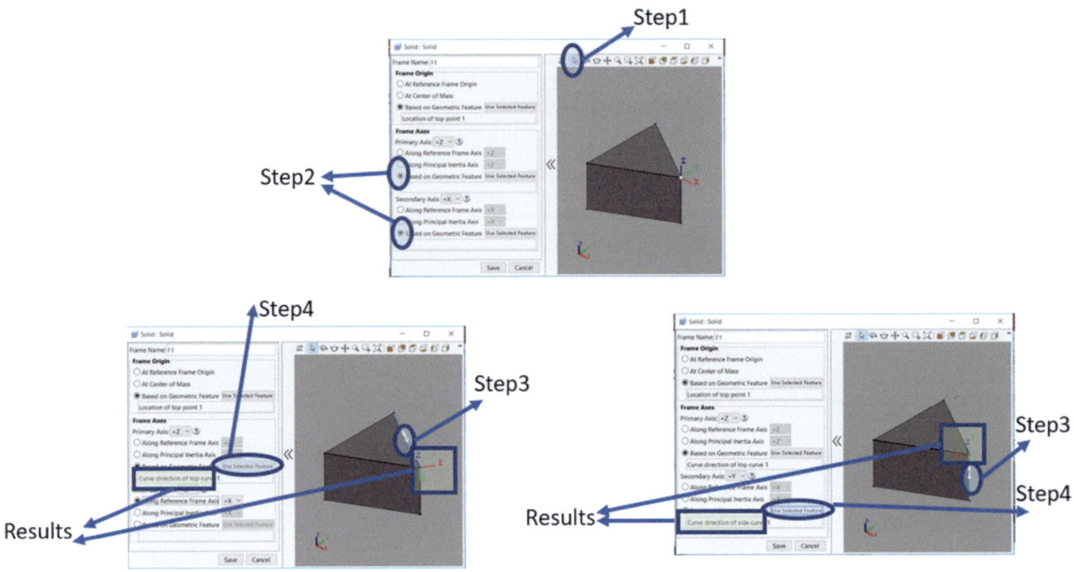

Fig. 4.49 Defining the frame axes using the solid block. Designed and implemented by the author. Images used with permission of MathWorks

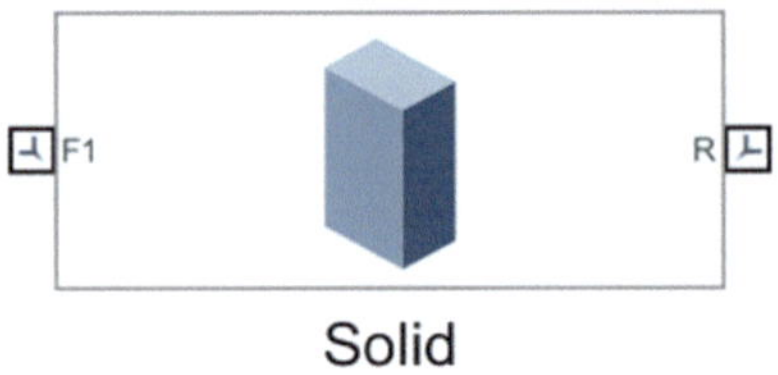

Fig. 4.50 Solid block with reference frame F1. Designed and implemented by the author. Images used with permission of MathWorks

After we are done with ***frame origin*** and ***frame*** axes editing menu, if the user presses "OK" the solid block will have an additional node called F1 as shown in Fig. 4.50. We can use this node with its local frame to attach a joint or to be a point of interest for the link.

4.7.4 Simscape™ Multibody™ Machine Assembly

Now, we are going to use the global frame and the local frames to assemble a mechanism. The steps below will be used.

Assembly Steps:

1. On the physical model, specify how many nodes, joints, and links.
2. Specify each type of joint.
3. Sketch the kinematic diagram of the mechanism.
4. Pick one of the ground nodes to be the origin of your global frame.
5. Specify and sketch the global x-axis, y-axis, and z-axis on the kinematic diagram.
6. List the values of the left-over ground nodes' locations compared to the global frame.
7. Specify the best shape for each link along with its size.
8. Specify the nodes located on each link with respect to the local frame at the link's C.G.
9. For each joint, specify which link will be ***Base*** and which link will be ***Follower***.
10. Sketch the free body diagram (FBD) based on the local frames for each joint and each link.
11. Specify the translational and rotational transformation for each joint (follower) meeting a link (base).
12. Open a new Simscape™ Multibody™ model with the three fixed blocks: (1) World frame (global frame), (2) solver, and (3) configuration.
13. For each ground node add a ***Rigid Transform***. Use the values in step 11 to be added as the translation and rotation between the origin and the left-over ground nodes.
14. Start inserting every joint. Rename each joint with the same name on the kinematic diagram.
15. Add in the links and rename each with the same name on the kinematic diagram. Change the shape and size (step 7) of each link.
16. For each node on a link, add two ***Rigid Transform*** blocks. Use Step 11 to specify the translation and rotation from the local frame at the link's C.G. to this node.
17. Now, arrange the links and joints as expected to be from the kinematic diagram and the FBD.
18. Add a ***Rigid Transform*** between each link and joint to make sure that the local frames are adjusted between two of them.

Example 4.6

For the shown four-bar mechanism in Fig. 4.51, implement a Simscape™ Simulink® model. Note, all the listed lengths are in cm.

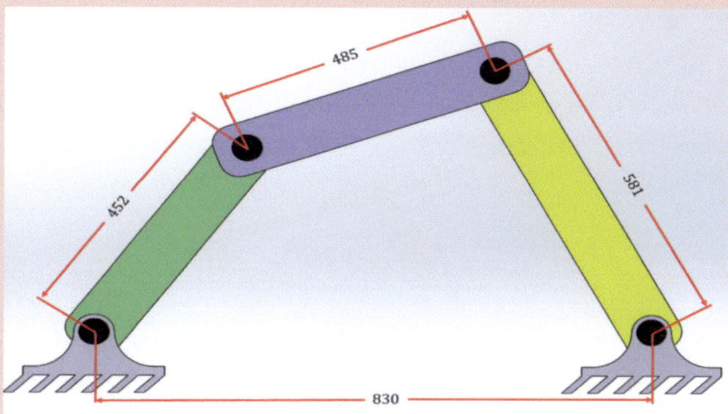

Fig. 4.51 Four-bar mechanism

Solution

Step 1: Number of nodes is 4: n_1, n_2, n_3 and n_4.
Number of joints is 4: J_1^R, J_2^R, J_3^R, and J_4^R.
Number of links is 4: L_1, L_2, L_3 and L_4.
Step 2: The four joints are revolute.
Step 3:: Kinematic diagram as shown in Fig. 4.52
Step 4:: Pick node n_1 as the origin for the global frame.
Step 5: Specify the *x*-axis, *y*-axis, and *z*-axis as shown in Fig. 4.53.
Step 6: $n_1 = n_1(0, 0, 0)$ and $n_4 = n_4(830, 0,0)$ in cm.
Step 7:

- L_1: The ground node by the origin (n_{1L1}) and node (n_{4L1}) as listed in Step 6.
- L_2: Brick with [*dx dy dz*] = [216 25 5] in cm. The length is *dx*. Both *dy* and *dz* are arbitrarily picked up. Color is picked as Red.
- L_3: Brick with [*dx dy dz*] = [485 25 5] in cm. The length is *dx*. Both *dy* and *dz* are arbitrarily picked up. Color is picked as Yellow.
- L_4: Brick with [*dx dy dz*] = [581 25 5] in cm. The length is *dx*. Both *dy* and *dz* are arbitrarily picked up. Color is picked as gray.

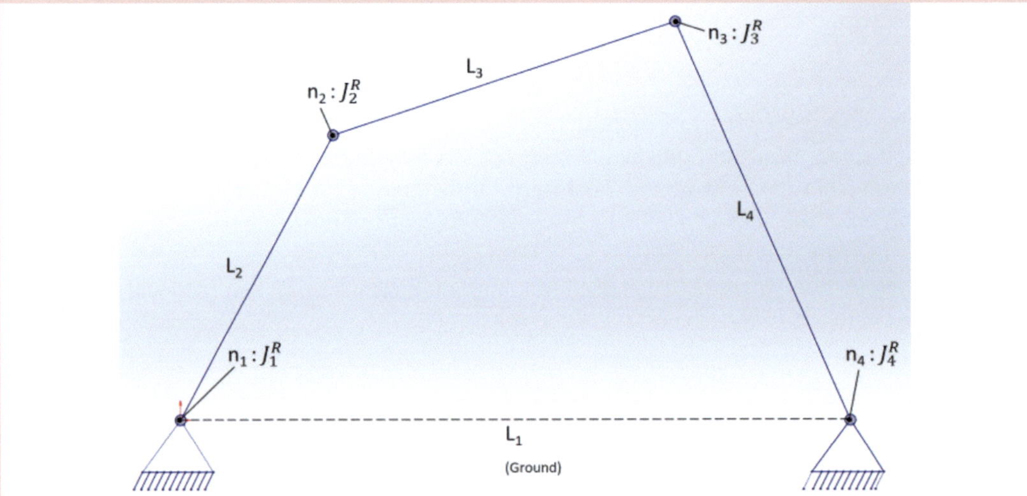

Fig. 4.52 Four-bar mechanism kinematic diagram

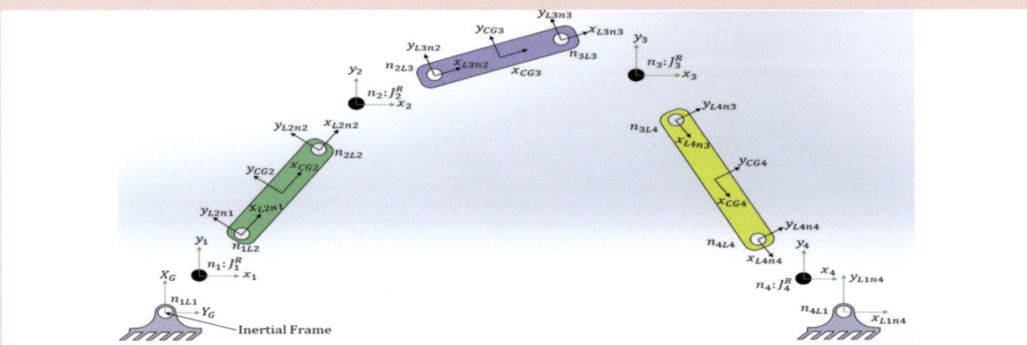

Fig. 4.53 Free body diagram of four-bar mechanism

*Note, we picked $dy = 25$ cm and $dz = 5$ cm same for the links just for better visualization.

Step 8:

Table 4.8 Nodes on each link for four-bar mechanism

Link	First node	Second node
L_1	$n_{1L1} = n_{1L1}$ (0, 0, 0)	$n_{4L1} = n_{4L1}$ (830, 0, 0)
L_2	$n_{1L2} = n_{1L2}$ (−113, 0, 0)	$n_{2L2} = n_{2L2}$ (113, 0, 0)
L_3	$n_{2L3} = n_{2L3}$ (242.5, 0, 0)	$n_{3L3} = n_{3L3}$ (242.5, 0, 0)
L_4	$n_{1L4} = n_{1L4}$ (-290.5, 0, 0)	$n_{4L4} = n_{4L4}$ (290.5, 0, 0)

*Note, n_{iLj} refers to node "*i*" on the link "*j*". For each row in Table 4.8, each node is locally measured from the link's center of gravity

Step 9: See Table 4.9.

Table 4.9 Joints' base and follower

Joint	Base	Follower
J_1^R	L_1	L_2
J_2^R	L_2	L_3
J_3^R	L_3	L_4
J_4^R	L_1	L_4

Step 10: Sketch FBDs.
Step 11: See Table 4.10

Table 4.10 Frames of each link and each node

Frame	Follower and base	Translation	Rotation
Frame_*L1_n1*	J_1^R and L_1	$(0, 0, 0)$	No rotation
Frame_*L2_n1*	J_1^R and L_2	$(-113, 0, 0)$	No rotation
Frame_*L1_n2*	J_2^R and L_2	(113,0,0)	No rotation
Frame_*L3_n2*	J_2^R and L_3	$(-242.5, 0, 0)$	No rotation
Frame_*L3_n3*	J_3^R and L_3	$(242.5, 0, 0)$	No rotation
Frame_*L4_n3*	J_3^R and L_4	$(-290.5, 0, 0)$	No rotation
Frame_*L4_n4*	J_4^R and L_4	(290.5,0,0)	No rotation
Frame_*L1_n4*	J_4^R and L_1	$(830, 0, 0)$	No rotation

Step 12: Open a new Multibody and delete all blocks except the following: **Solver**, **World** (Ground), and **Mechanism Configuration** as shown in Fig. 4.54.

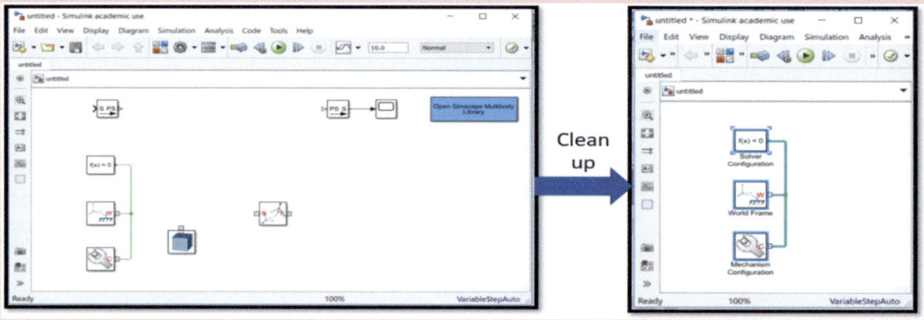

Fig. 4.54 Clean-up model for four-bar mechanism. Designed and implemented by the author. Images used with permission of MathWorks

Step 13: Add two "Rigid Transform" blocks (add one in first then select the block and while holding the **shift** key, click and drag the block to create a copy). Rename them "Frame_L1_n1" and "Frame_L1_n4". There is no translation or rotation for "Frame_L1_n1". Translation is defined as $n_4^1(830, 0, 0)$ for "Frame_L1_n4" as shown in Fig. 4.55.

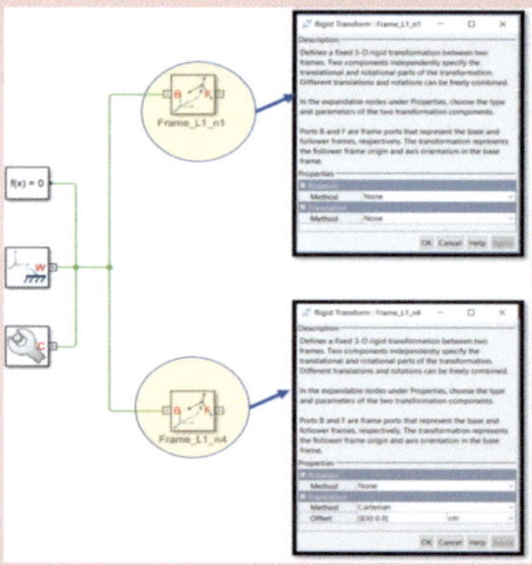

Fig. 4.55 Rigid Transform blocks for 4-bar mechanism. Designed and implemented by the author. Images used with permission of MathWorks

Steps 14–18: The total model is shown in Fig. 4.56. Table 4.11 includes the detailed values for each element. The alignment of each joint and frame is very important. Flipping (Ctrl + i) and rotating (Ctrl + r) are applied on some of the frames, links, and joints.

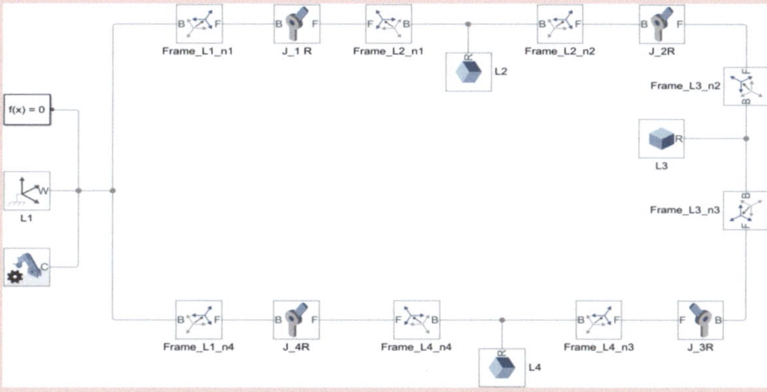

Fig. 4.56 Four-Bar mechanism model Simscape™ Multibody™ model

Table 4.11 Frame Transform editing list of four-bar mechanism. Designed and implemented by the author. Images used with permission of MathWorks

Element	Editing list
Frame_L1_n1	Properties; Rotation: Method None; Translation: Method None
Frame_L2_n1	Properties; Rotation: Method None; Translation: Method Cartesian, Offset [-113 0 0] cm
Frame_L2_n2	Properties; Rotation: Method None; Translation: Method Cartesian, Offset [113 0 0] cm
Frame_L3_n2	Properties; Rotation: Method None; Translation: Method Cartesian, Offset [-242.5 0 0] cm
Frame_L3_n3	Properties; Rotation: Method None; Translation: Method Cartesian, Offset [242.5 0 0] cm
Frame_L4_n3	Properties; Rotation: Method None; Translation: Method Cartesian, Offset [-290.5 0 0] cm

Table 4.11 (continued)

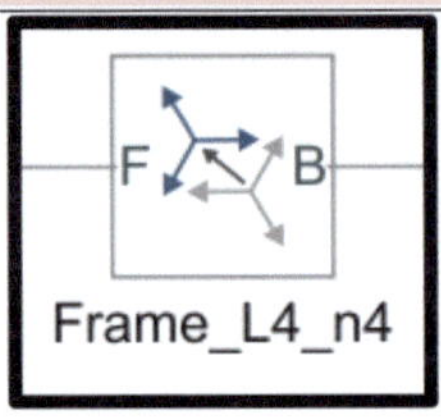	Properties Rotation Method: None Translation Method: Cartesian Offset: [290.5 0 0] cm
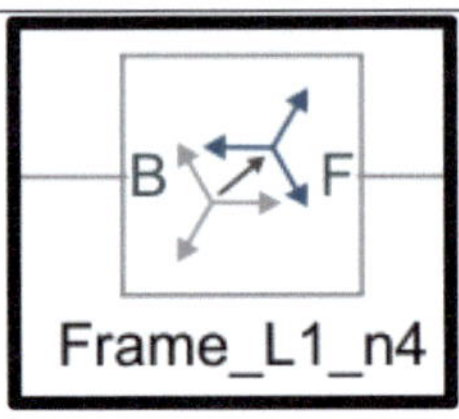	Properties Rotation Method: None Translation Method: Cartesian Offset: [830 0 0] cm

Now, if we run the model, we will get the configuration in Fig. 4.57. This configuration is not what we expected and needs another condition, in this case, an ***initial condition***. To set the configuration at a specific initial condition, we can specify the initial alignment of joint J_1^R to be at 30° deg as shown in Fig. 4.58. After that, re-run the model. In this case, we get our expected configuration as shown in Fig. 4.58. If we do not specify the initial configuration, Multibody will pick up any initial configuration.

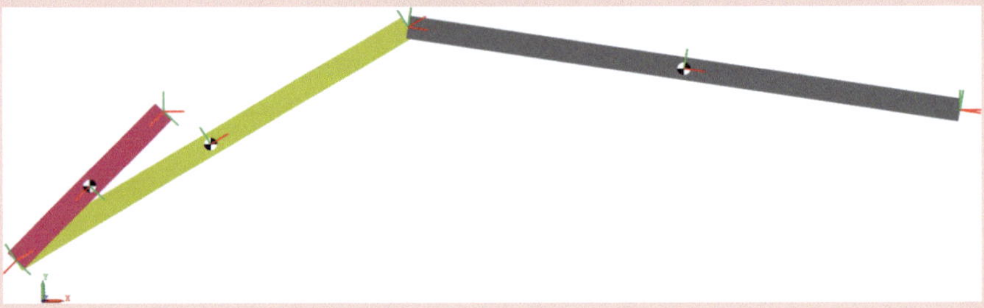

Fig. 4.57 Initial four-bar mechanism configuration

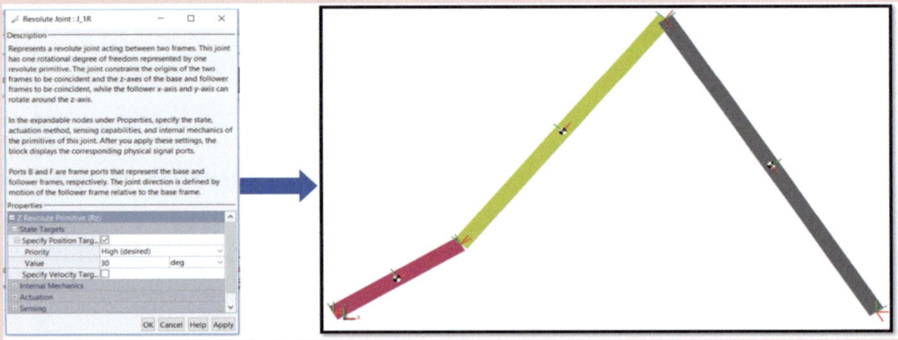

Fig. 4.58 Setting the initial configuration of the four-bar mechanism. Designed and implemented by the author. Images used with permission of MathWorks

4.7.5 Simulink® Subsystems and Masks

Definition[20]

Subsystem Block:

A subsystem is a set of blocks that you group into a single subsystem block. Using subsystems establishes a hierarchical block diagram where a subsystem block is on one layer and the blocks that make up the subsystem are on another, keeps functionally related blocks together, helps reduce the number of blocks displayed in your model window, and establishes an interface with inputs and outputs [1].

Definition[21]

Mask Block:

A mask is a custom user interface for a block. By masking a block, you encapsulate the block diagram to have its own parameter dialog box with its own block description, parameter prompts, and help texts. You can mask an independent custom block that you can reuse as unique blocks like those defined in Simulink® [1].

As we can see from Example 4.6, there is a sort of repeatability between the non-grounded links modeling method. All three non-grounded links have the same shape but are different sizes. For example, L_2, L_3, and L_4 are bricks (links) that are connected to two frames or nodes to be connected with revolute joints. The differences are the size of the bricks (links) and the translations of the frames. To minimize the model complexity and the consumed time to build any complex model, we are going to build a subsystem called "brick with two revolute joints". First, open a new model. As shown in Figs. 4.59 and 4.60, the following steps will be used to build this subsystem and save it as a mask:

Step 1: Connect a brick L with two frames: "Frame_L_First" and "Frame_L_Second".
Step 2: Enter the listed parameters into Table 4.12.
Step 3: Press "Ctrl-A" to select all the blocks and right-click and pick "create subsystem" to build the subsystem as shown in Fig. 4.59. Rename the subsystem "brick with revolute joints".
Step 4: Right-click on the subsystem and select "mask" followed by "create mask". The mask editor window will pop up as shown in Fig. 4.60.
Step 5: As shown in Fig. 4.60, go to the "parameters and dialog" tab. Press "promote" to add: ***BrickDimesions***, ***TranslationCartesianOffsetF***, and ***TranslationCartesianOffsetS*** in the dialog box subwindow. Edit the name and value of the selected parameter in "property editor". After that, click "OK".

When you double-click on the generated mask, you will see the GUI as shown in Fig. 4.61. You can modify the values of the three parameters: Brick dimensions (m) = [L 0.1*L 0.1*L], first revolute joint location (m) = [−0.5*L 0 0], and second revolute joint location (m) = [0.5*L 0 0]. More details about building the mask are given in Ref. [1].

[20] See Footnote 3.
[21] See Footnote 3.

Table 4.12 Subsystem parameters of brick with two revolute joints subsystem. Designed and implemented by the author. Images used with permission of MathWorks

Double-click on *L*, enter "[L 0.1*L 0.1*L]" in the Geometry/Dimensions, and press "OK"	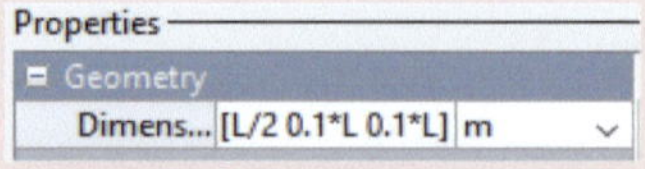
Double-click on the Frame_L_First, enter "[-L/2 0 0]" in the Translation/Cartesian/Offset, and press "OK"	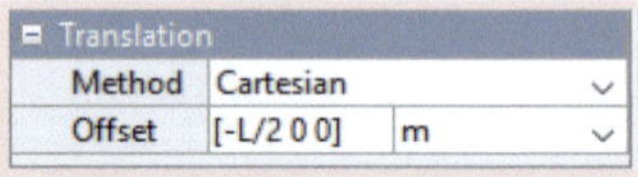
Double-click on the Frame_L_Second, enter "[L/2 0 0]" in the Translation/Cartesian/Offset, and press "OK"	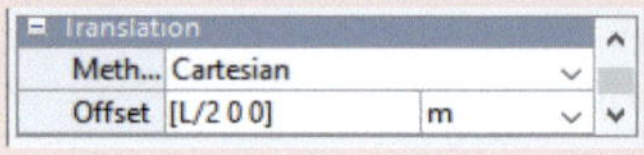

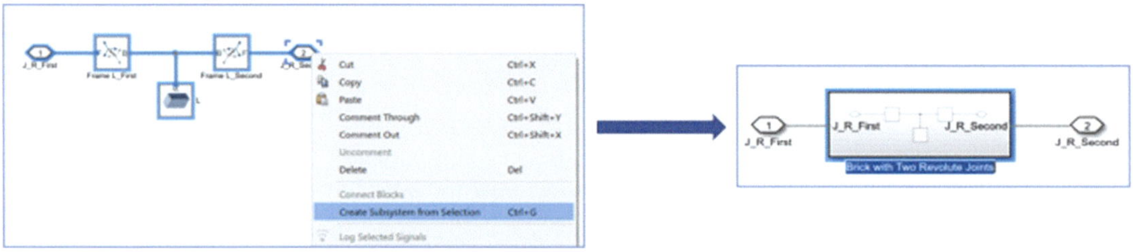

Fig. 4.59 Building brick with two revolute joints subsystem. Designed and implemented by the author. Images used with permission of MathWorks

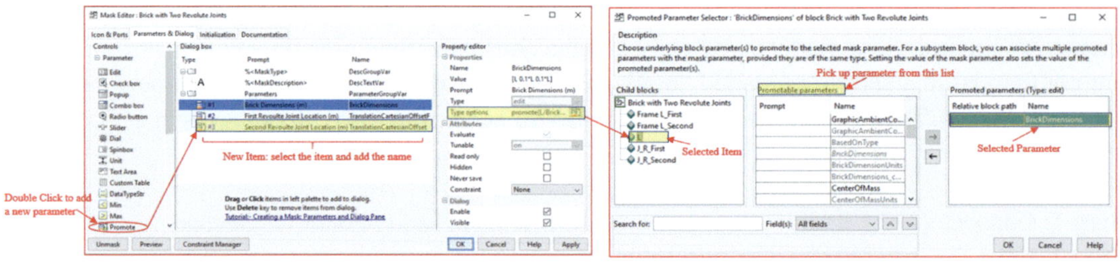

Fig. 4.60 Building mask of link with two revolute joints subsystem. Designed and implemented by the author. Images used with permission of MathWorks

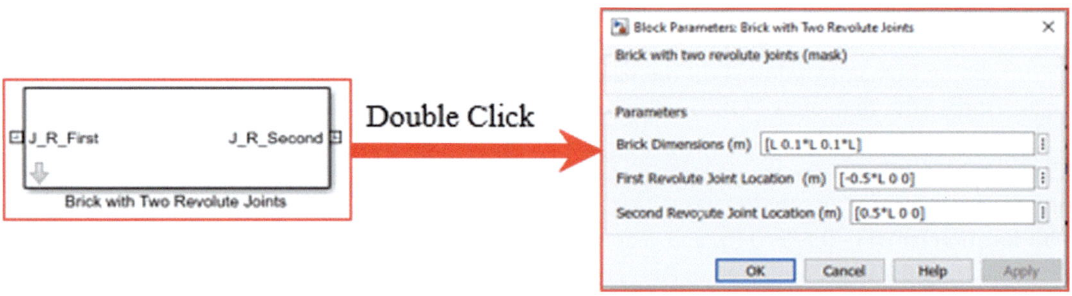

Fig. 4.61 Brick with two revolute joints mask GUI. Designed and implemented by the author. Images used with permission of MathWorks

Using the same method, we can develop many masks. Some of these masks are listed in Table 4.13. From now on, all our developed masks and subsystems are saved in "Kinematics_Of_Machines.slx". During the upcoming chapters, we will build more masks and subsystems inside "Kinematics_Of_Machines.slx" which is provided with the textbook. These masks will be used in

Table 4.13 Multibody developed links masks. Designed and implemented by the author. Images used with permission of MathWorks

<table>
<tr><td>Brick with J_R and J_P</td><td>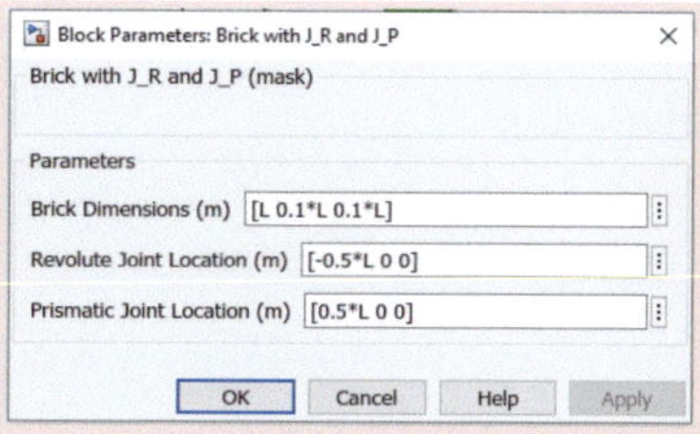
</td></tr>
<tr><td>Ground with two revolute joints</td><td>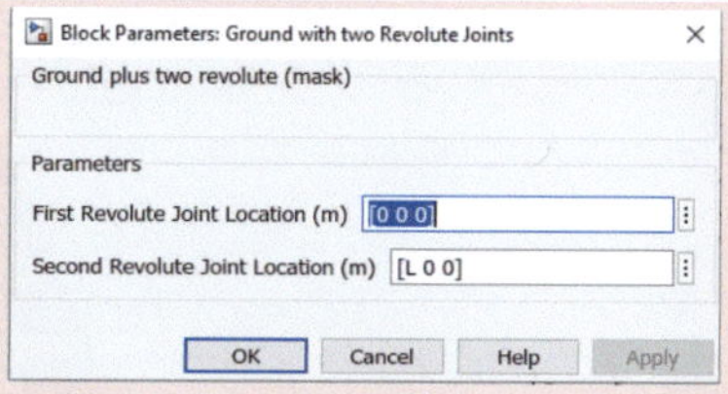
</td></tr>
<tr><td>Ground with J_R and J_P</td><td>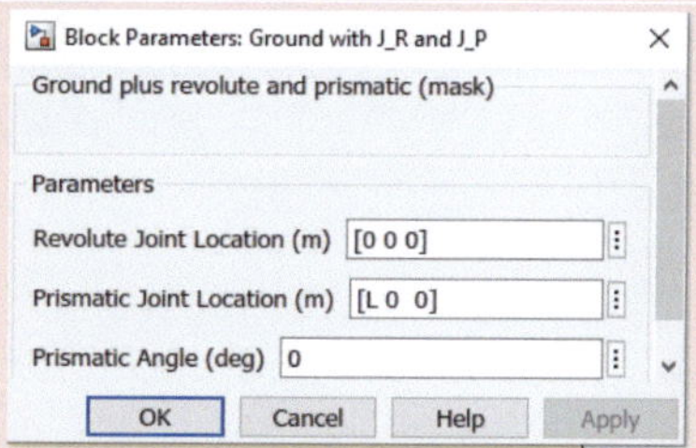
</td></tr>
<tr><td>Extrusion with two revolute joints mask</td><td>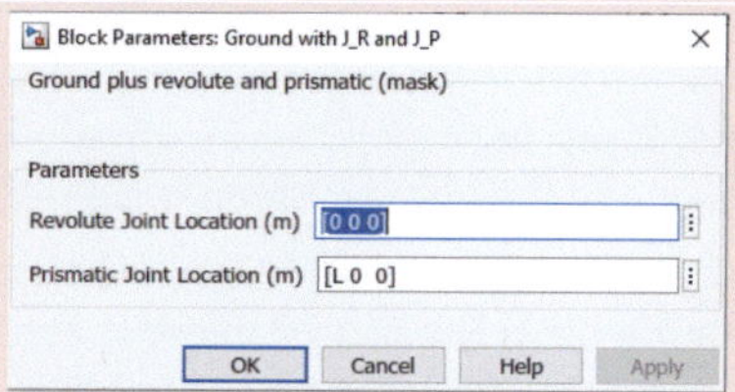
</td></tr>
<tr><td>Extrusion with three revolute joints mask</td><td>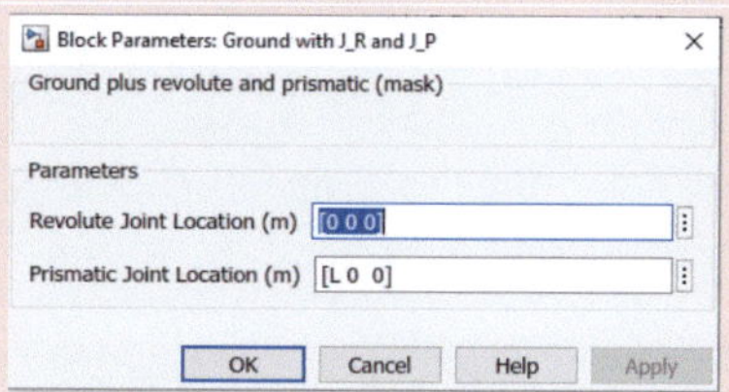
</td></tr>
<tr><td>Extrusion with four revolute joints mask</td><td>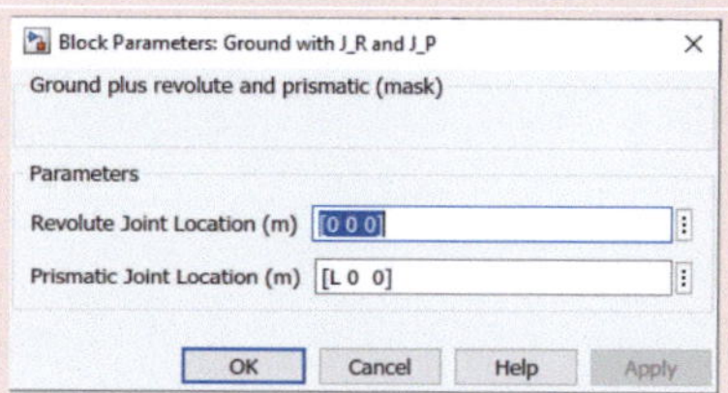
</td></tr>
</table>

solving upcoming problems. In general, the subsystem plus mask are like the function script when the same script is used with different values or inputs. In this case, the mask and subsystem present the link with the two frames. All we must do is drag and drop the mask into the model; then enter the values. Again, this saves time and reduces the model's complexity as a graphic user interface.

Example 4.7

For the mechanism shown in Fig. 4.62, implement a Simscape™, Simulink® model.

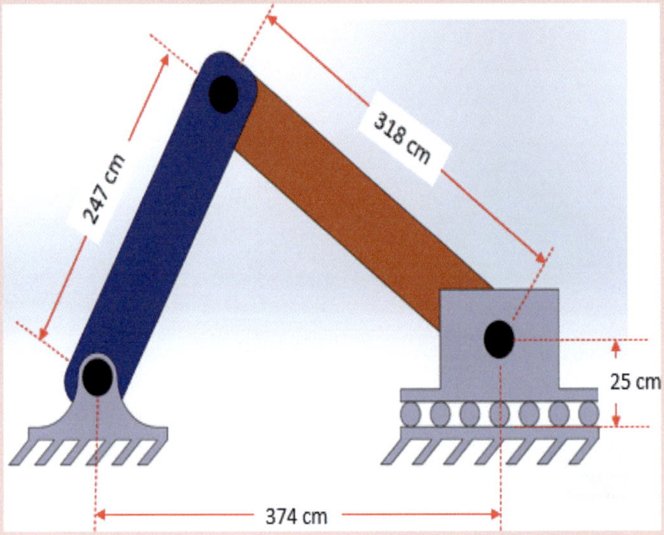

Fig. 4.62 Crank–slider mechanism

Solution

First, we are going to sketch the kinematic diagram as shown in Fig. 4.63. After that, we use the masks technique from our "Kinematics_Of_Machines.slx" to break down the model into.

1. L_1: Ground with J_R and J_P mask
2. L_2: Brick with two revolute joints mask
3. L_3: Brick with two revolute joints mask
4. L_4: Brick with J_R and J_P mask
5. Three revolute joints to connect: L_1 with L_2, L_2 with L_3, and L_3 with L_4
6. One prismatic joint to connect L_4 with L_1

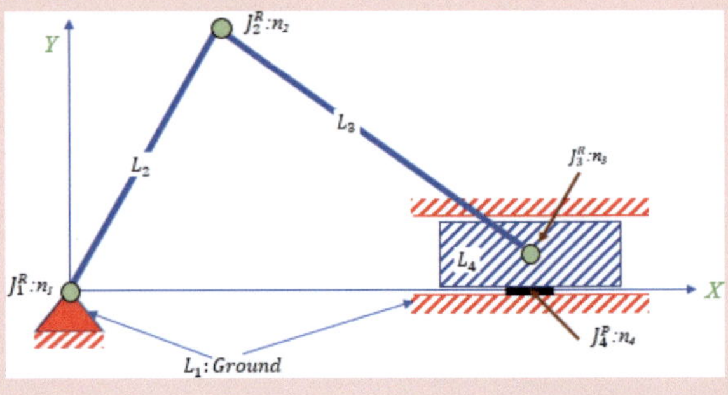

Fig. 4.63 Kinematic diagram of slider–crank mechanism

Figure 4.64 shows the model with the plugged-in parameters' values. After, that you can run the simulation to achieve the configuration in the mechanics explorer as shown in Fig. 4.65.

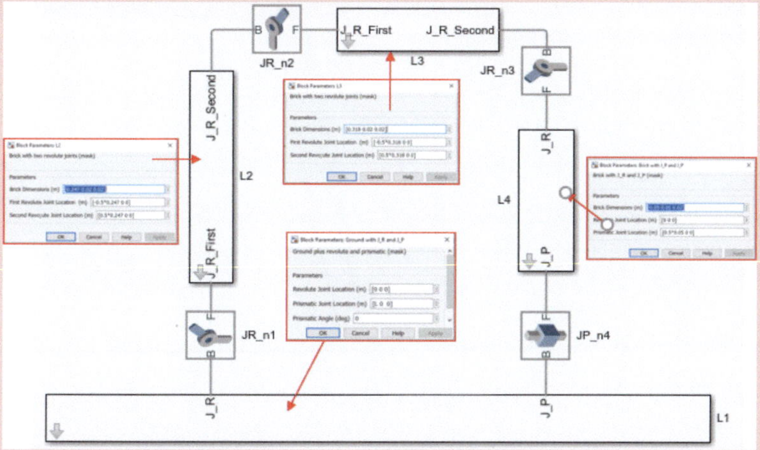

Fig. 4.64 Simscape™ Multibody™ model with plugged-in values for Example 4.7. Designed and implemented by the author. Images used with permission of MathWorks

Fig. 4.65 Simscape™ Multibody™ configuration of slider–crank mechanism

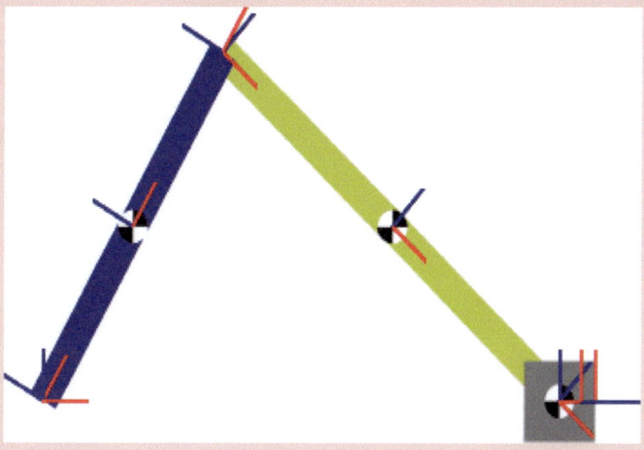

Example 4.8

For the tractor boom mechanism shown in Fig. 4.66, implement a Simscape™, Simulink® model.

Fig. 4.66 Tractor boom mechanism

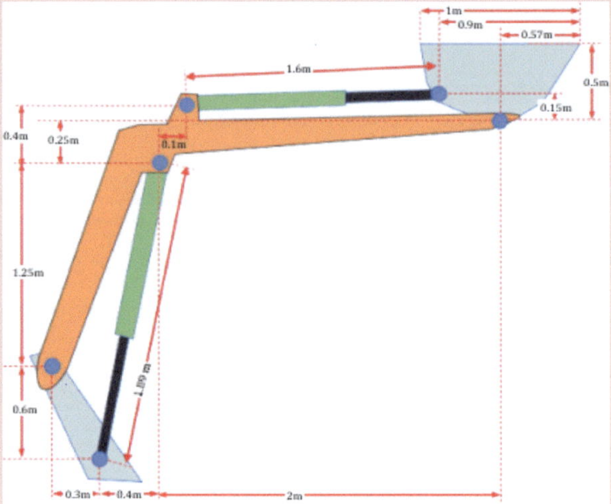

Solution

First, we are going to sketch the kinematic diagram as shown in Fig. 4.67. After that, we use the masks technique to break down the model into.

Fig. 4.67 Tractor boom mechanism kinematic diagram

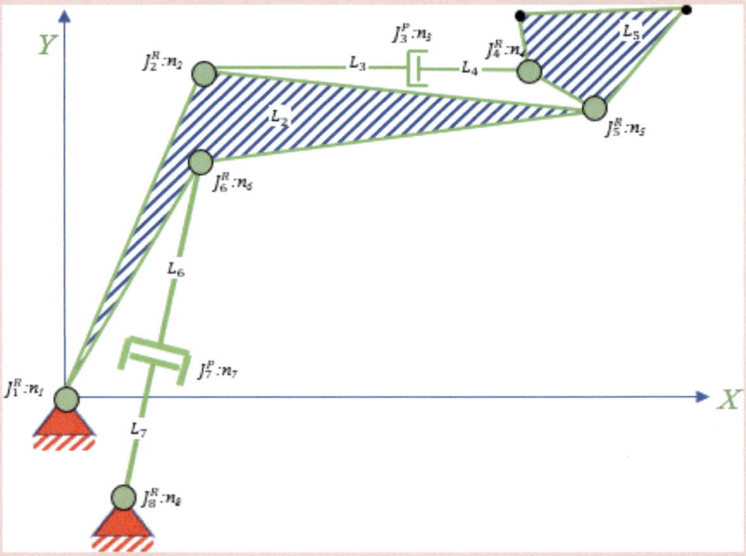

1. L_1: Ground with two revolute joints mask
2. L_2: Extrusion with four revolute joints mask
3. L_3: Brick with J_R and J_P mask
4. L_4: Brick with J_R and J_P mask
5. L_5: Extrusion with two revolute joints mask
6. L_6: Brick with J_R and J_P mask
7. L_7: Brick with J_R and J_P mask
8. Six revolute joints as shown in Fig. 4.67
9. Two prismatic joints as shown in Fig. 4.67

Figure 4.68 shows the model with the plugged-in parameters' values. After, that you can run the simulation to achieve the configuration in the mechanics explorer as shown in Fig. 4.69. We encourage the reader to watch the attached animation video of the tractor.

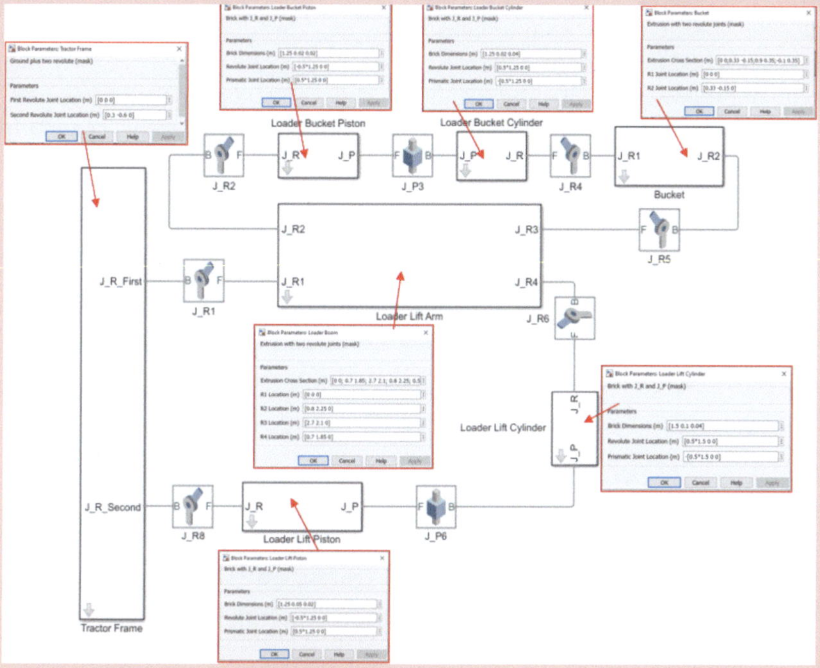

Fig. 4.68 Simscape™ Multibody™ model with plugged-in values for Example 4.8. Designed and implemented by the author. Images used with permission of MathWorks

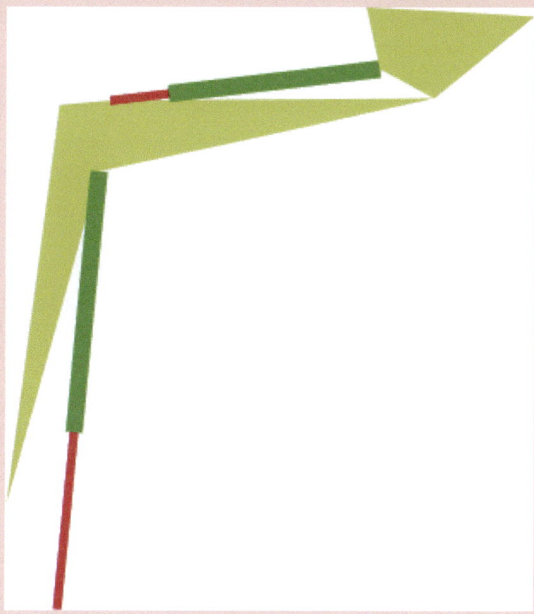

Fig. 4.69 Simscape™ Multibody™ configuration of tractor loader mechanism

Engineering Tips

To show the axes and center of mass, right-click on the mechanics explorer and select "show frames" and "show COMs".

You can change the color of each link inside the mask by double-clicking on the below left arrow to access the model from inside.

For the tractor loader mechanism, it would be better to model the cylinders and pistons as cylinder body blocks. We model the cylinders and pistons as bricks since in 2D both bricks and cylinders have the same side view (rectangular shape). Our analysis in this textbook is restricted to 2D machines and mechanisms.

4.8 Chapter Summary

Outcomes

1. Basic knowledge of MATLAB® and MATLAB® functions
2. Proficiency with the source, sink, and math libraries
3. First-hand experience with Simulink® Mathematical Models
4. Knowing how to use joints, links, and frames in Simscape™ Multibody™
5. Use Simscape™ Multibody™ to model machines and mechanisms

Problems

For each problem from 1–4, implement the listed equation in a Simulink® model. Run the model for 10 s and show the results of x versus t.

Problem 1

$$x(t) = -3 + 5t/10$$

Problem 2

$$x(t) = -3 + 5\left(35\left(\frac{t}{10}\right)^4 - 84\left(\frac{t}{10}\right)^5 + 70\left(\frac{t}{10}\right)^6 - 20\left(\frac{t}{10}\right)^7\right)$$

Problem 3

$$x(t) = -3 + 2.5\left(1 - \cos\left(\frac{\pi(t)}{10}\right)\right)$$

Problem 4

$$x(t) = -3 + 5\left(\frac{t}{10} - \frac{1}{2\pi}\sin\left(\frac{2\pi(t)}{`10}\right)\right)$$

Problem 5
For the shown four-bar-door mechanism in Fig. 4.70:

1. sketch the kinematic diagram,
2. show where we could locate the inertial frame, and
3. implement a Simulink®, Simscape™ Multibody™ model.

Problem 6
For the shown hydraulic door mechanism in Fig. 4.71:

1. sketch the kinematic diagram,
2. show where we could locate the inertial frame, and
3. implement a Simulink®, Simscape™ Multibody™ model.

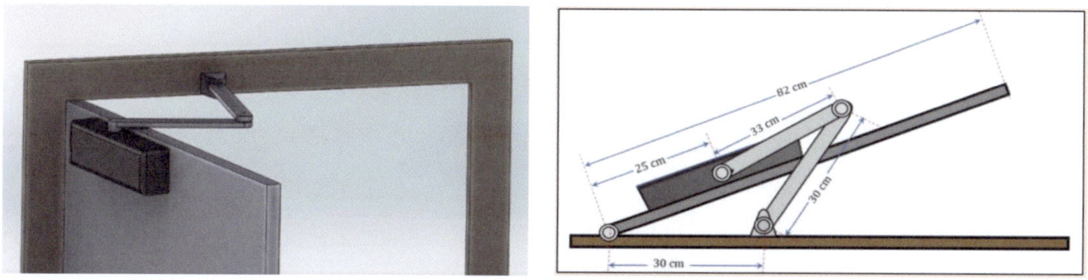

Fig. 4.70 Four-bar door mechanism. Figure and associated Door_4Bar_Mechanism.mp4 video designed by the author and implemented by both the author and Pranav Dhola. Both courtesy of Pranav Dhola

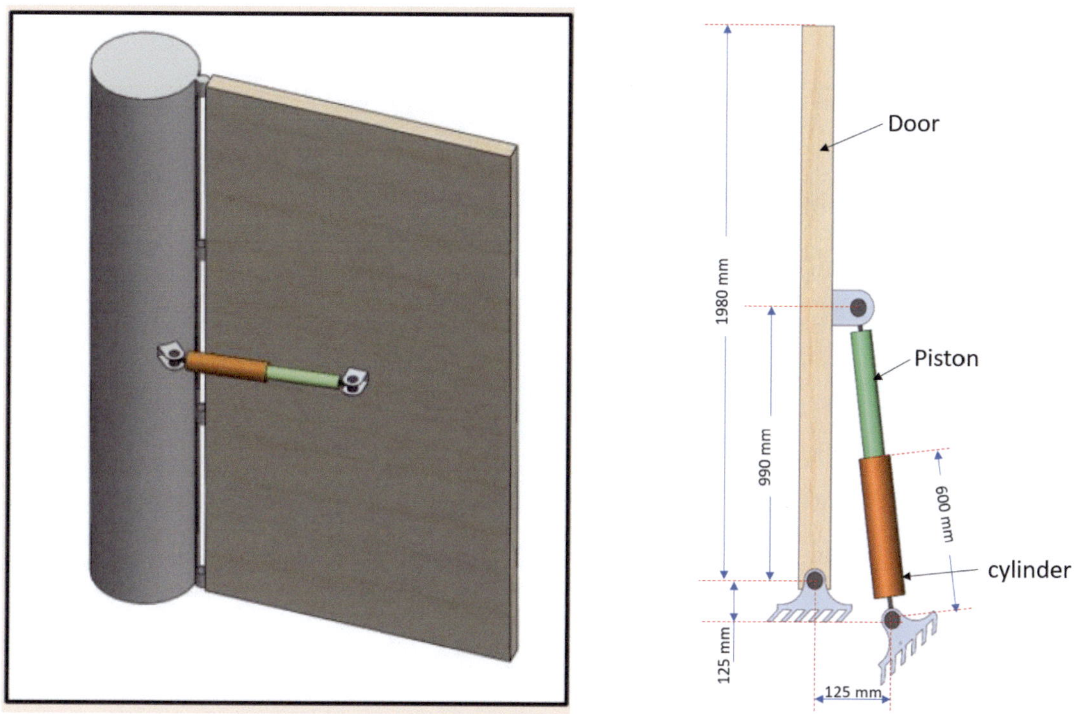

Fig. 4.71 Hydraulic door opener mechanism. Designed by the author and implemented by both the author and Pranav Dhola. Courtesy of Pranav Dhola

Problem 7

For the shown single piston–cylinder mechanism in Fig. 4.72:

1. sketch the kinematic diagram,
2. show where we could locate the inertial frame, and
3. implement a Simulink®, Simscape™ Multibody™ model.

where the crank has a radius of 50 mm, the piston is 35 mm × 25 mm, and the connecting rod has a length of 130 mm.

Problem 8

For the shown dumpster truck mechanism in Fig. 4.73:

1. sketch the kinematic diagram,
2. show where we could locate the inertial frame, and
3. implement a Simulink®, Simscape™ Multibody™ model.

where both cylinder and piston are 1.6 m as length with dimeters of 0.16 m and 0.12 m, respectively.

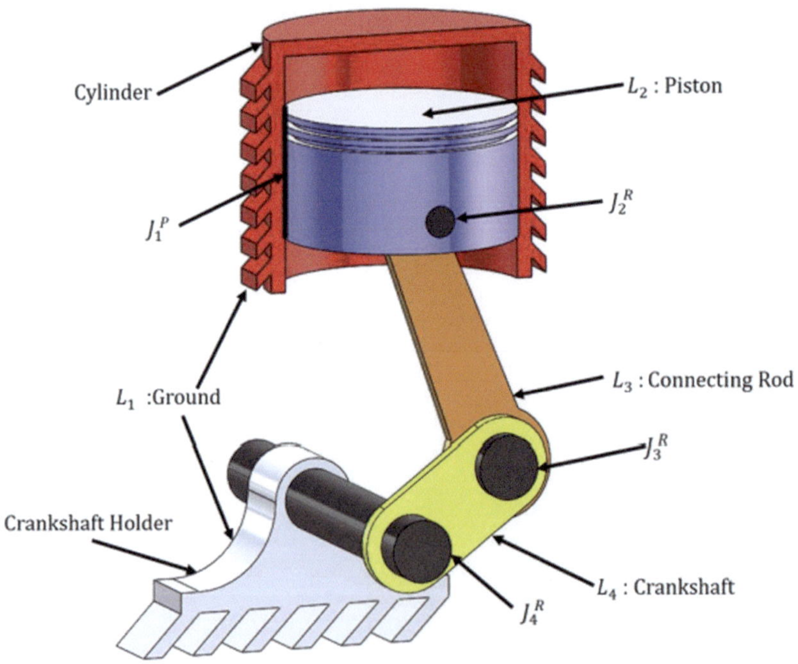

Fig. 4.72 Single piston–cylinder mechanism. Designed by the author and implemented by both the author and Pranav Dhola. Courtesy of Pranav Dhola

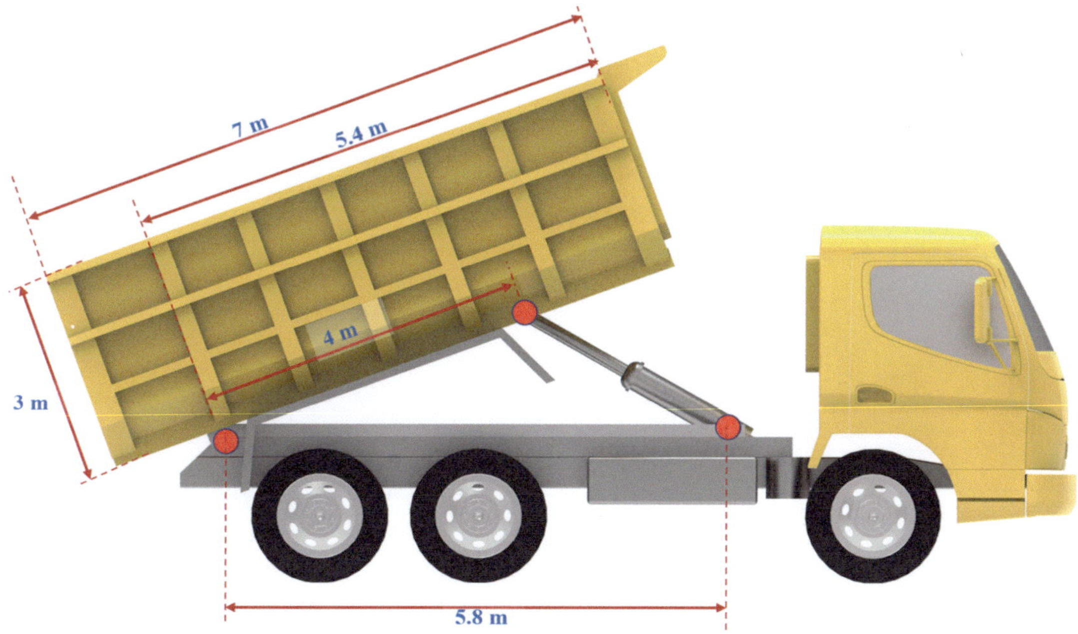

Fig. 4.73 Dumpster truck mechanism. Figure and associated Dumpster Truck Mechanism.mp4 video designed by the author and implemented by both the author and Pranav Dhola. Both courtesy of Pranav Dhola

Problem 9

For the shown landing gear mechanism in Fig. 4.74:

1. sketch the kinematic diagram,
2. show where we could locate the inertial frame, and
3. implement a Simulink®, Simscape™ Multibody™ Model.

Where:
Down lock upper link has length of 0.935 m plus both width and thickness of 0.07.
Down lock upper link has length of 0.98 m plus both width and thickness of 0.07 m.
H-Section link has length of 0.5 m plus both width and thickness of 0.07 m.
Down lock link has length of 0.56 m plus both width and thickness of 0.07 m.
Cylinder has length of 0.45 m and radius of 0.08 m.
Piston has length of 0.5 m and radius of 0.06 m.
In your solution, ignore the lower suspension mechanism and the wheels.

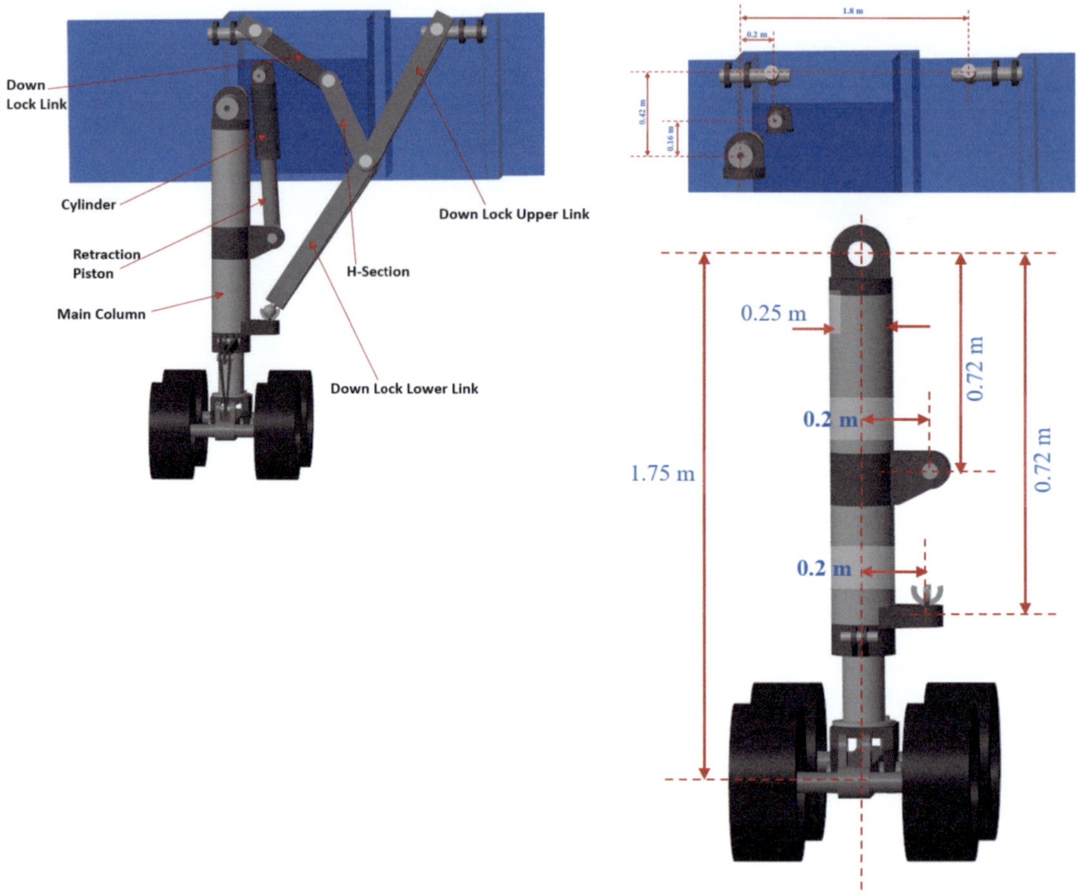

Fig. 4.74 Landing gear mechanism. Designed by the author and implemented by both the author and Pranav Dhola. Courtesy of Pranav Dhola

References

1. https://www.mathworks.com.
2. "Simscape™ Multibody™ Getting Started Guide," MathWorks, 2020.
3. "Simscape™ Multibody™ User's Guide," MathWorks, 2020.
4. "Simscape™ Multibody™ Reference," MathWorks, 2020.
5. "Simscape™ Multibody™ Release Notes," MathWorks, 2020.

Position and Displacement Analysis

5

Chapter Objectives

1. Determining any mechanism node's 2D position and 2D displacement vectors in Cartesian and polar coordinates.
2. Learning how to use the model-based Simulink® and Simscape™ Multibody™ to calculate the position and displacement vectors of any mechanism node.
3. Developing graphical and analytical position models for four-bar, slider–crank, and inverted slider–crank mechanisms.
4. Discussing graphical and analytical position models of the five and beyond bar mechanisms.
5. Developing graphical and analytical position models of any mechanism node of interest.
6. Developing a model for toggle position.
7. Defining and understanding the coupler curve

5.1 Introduction

In this chapter, we explain any mechanism node's position and displacement vectors in both linear and angular terms using mathematical models and model-based Simulink® and Simscape™ Multibody™. After that, we develop three different methods to calculate position and displacement: the graphical method, the analytical method, and the model-based Simulink® and Simscape™ Multibody™, slider–crank, inverted slider–crank mechanisms, inverted slider–crank mechanisms, and five and beyond bar mechanisms. Finally, we discuss the limitations of a machine's movement and the expected coupler curve.

Supplementary Information The online version contains supplementary material available at https://doi.org/10.1007/978-3-031-72701-6_5.

A. Omran, *Kinematics and Dynamics of Machines*, https://doi.org/10.1007/978-3-031-72701-6_5

5.2 Position

Definition

Position:

The position of a point (node) is a position vector that is fixed and locates a point in space relative to another point or the origin of a specific reference axis [1, 2].

Position is a mathematical quantity that is expressed in a vector form to specify how far a specific point or node is from the origin of a reference frame. As an example, to explain the 2D and 3D positions, Fig. 5.1 shows a 3D and 2D view of a living room. First, we are going to locate the origin point "***O***" of our selected frame at the bottom-left point of the door. Using *Fleming's Right-Hand Rule,* the *x*–*y*-*z* axes frame is shown in both 2D and 3D.

Note: The Shorthand (Symbol)+ve Means "Positive". the Shorthand (Symbol) –Ve Means "Negative". You Will See These Symbols Used Throughout This Textbook.

The node "n_1" is located on the table with a distance of 8 m in a+ve *x*-direction, a distance of 3 m in a+ve *y*-direction, and a distance of 1.6 m in the+ve *z*-direction. Instead, we can summarize these distances by saying $n_1\left(x_{n_1}, y_{n_1}, z_{n_1}\right) = (8, 3, 1.6)$. The same concept can be applied to the node $n_2\left(x_{n_2}, y_{n_2}, z_{n_2}\right) = (5, -2, 1.5)$. on the sofa as shown in Fig 5.1a. If we look from the top view, we can see the 2D room plan as shown in Fig. 5.1b. In 2D, there is no existence of a *z*-direction; node $n_1\left(x_{n_1}, y_{n_1}\right) = (8, 3)$ and $n_2\left(x_{n_2}, y_{n_2}\right) = (5, -2)$.. In this book, our analysis and design are restricted to 2D.

The Cartesian coordinate system and polar coordinate system are the most well-known coordinate systems in 2D and are shown in Fig. 5.2. The Cartesian coordinate system specifies any node

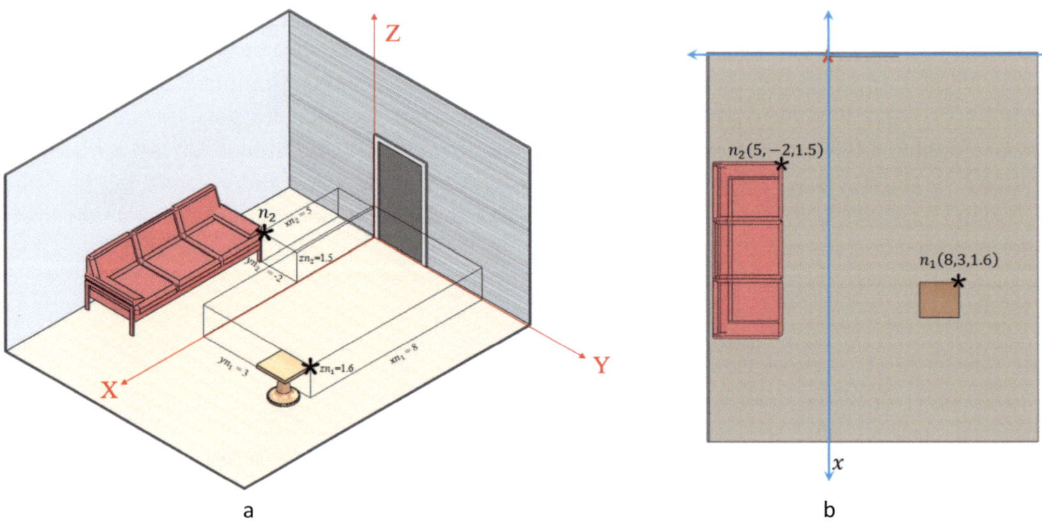

Fig. 5.1 3D view and 2D view of a living room

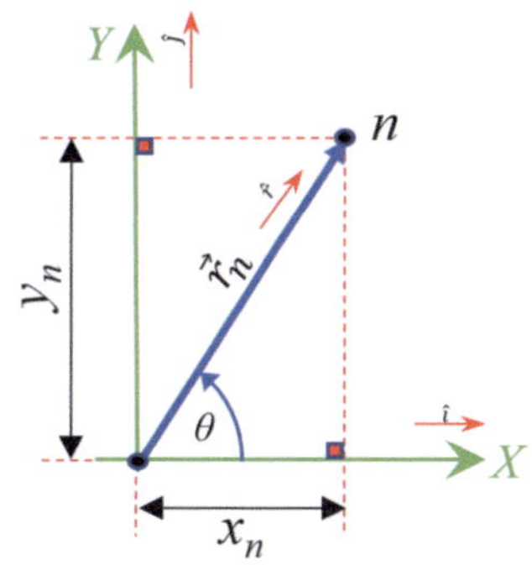

Fig. 5.2 2D Cartesian coordinate and polar coordinate systems

"n" uniquely in a plane by a two-element position vector $n(x_n, y_n)$ that starts from the origin and ends up at the point or the node. The first element represents the perpendicular or shortest distance x_n from the node to *Y-axis*. The second element represents the perpendicular or shortest distance y_n from the node to *X-axis*.

The polar coordinate system uniquely specifies any node n on a plane by a position vector $r_n\angle\theta_n$ where the distance r_n is from the node n to the origin, and the angle $\theta_n is$ measured from +ve X-axis in the counterclockwise direction. The position vector of any node n can be written as:

$$\begin{aligned}\vec{r}_n &= n(x_n, y_n) = x_n\hat{i} + y_n\hat{j}\\ \vec{r}_n &= r_n\angle\theta_n = r_n\hat{r}\end{aligned} \tag{5.1}$$

where unit vector $\hat{i}$ is in x-direction, unit vector $\hat{j}$ is in the y-direction, and unit vector $\hat{r}$ is in the r-direction. EA unit vector is a vector whose magnitude is 1 in a specific direction. From geometry, we can come up with the following polar–Cartesian transformation:

$$\begin{aligned}x_n &= r_n \cos(\theta_n)\\ y_n &= r_n \sin(\theta_n)\\ r_n &= \sqrt{x_n^2 + y_n^2}\\ \theta_n &= \arctan\frac{y_n}{x_n}\end{aligned} \tag{5.2}$$

As shown in Fig. 5.3, the angle θ_n can have four quadrants:

- **First Quadrant**, θ_n changes from 0° to 90° where X is positive (+ve) and Y is positive (+ve)
- **Second Quadrant**, θ_n changes from 90° to 180° where X is negative (−ve) and Y is positive (+ve)
- **Third Quadrant**, θ_n changes from 180° to 270° where X is negative (−ve) and Y is negative (−ve)
- **Fourth Quadrant**, θ_n changes from 270° to 0° where X is positive (+ve) and Y is negative (−ve)

In MATLAB®, we must use "atan2d" instead of "atand" to locate the angle as a degree unit in the correct quadrant. For example, assume a vector $\vec{r} = x\hat{i} + y\hat{j} = -3\hat{i} - 4\hat{j}$ that is in the 3rd quadrant, the angle of the vector can be calculated by atand(-4/-3) = 53° which is not correct. Instead, we must calculate the angle of the vector as atan2d(-4,-3) = -127° or 233° which is correct because the vector

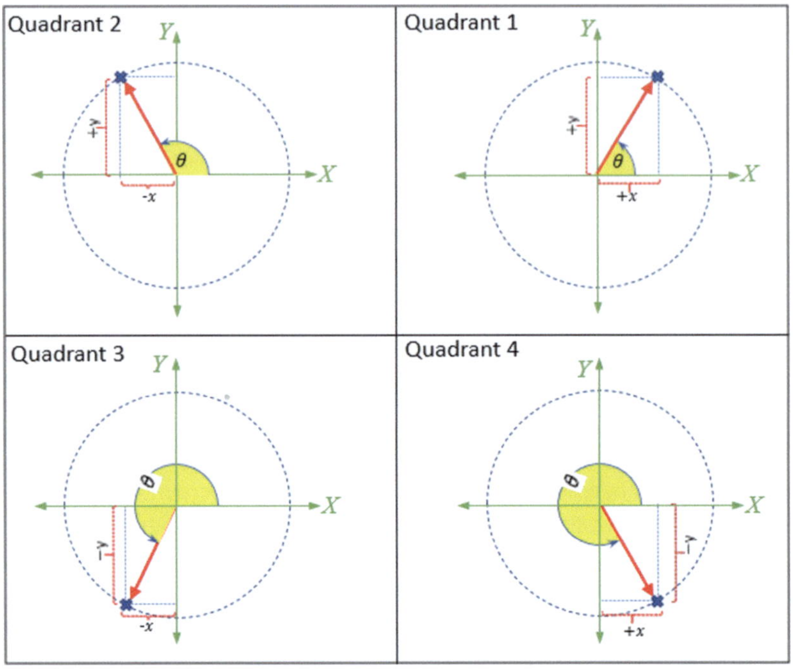

Fig. 5.3 Directional angle

is located in the 3rd quadrant. It is recommended to read the help about atan2d by typing "help atand2d" in the MATLAB® command window.

Vector addition or vector subtraction are commonly used in position vector calculation. For example, let's assume we have two vectors: $\vec{r}_{n2}$ which starts from the origin and ends at node n_2 and $\vec{r}_{n3n2}$ that starts from node n_2 and ends at node n_3. The position vector $\vec{r}_{n3}$ starts at the origin and ends at node n_3 as shown in Fig. 5.4.

To find the position vector $\vec{r}_{n3}$ from the addition of the two vectors $\vec{r}_{n2}$ and $\vec{r}_{n3n2}$, we can follow the sequence in Eq. (5.3).

$$\begin{aligned}\vec{r}_{n2} &= x_{n2}\hat{i} + y_{n2}\hat{j} = r_{n2}\cos(\theta_2)\hat{i} + r_{n2}\sin(\theta_2)\hat{j}\\ \vec{r}_{n3n2} &= x_{n3n2}\hat{i} + y_{n3n2}\hat{j} = r_{n3n2}\cos(\theta_3)\hat{i} + r_{n3n2}\sin(\theta_3)\hat{j}\\ \vec{r}_{n3} &= x_{n3}\hat{i} + y_{n3}\hat{j} = \vec{r}_{n3n2} + \vec{r}_{n2} = (x_{n3n2} + x_{n2})\hat{i} + (y_{n3n2} + y_{n2})\hat{j}\\ &= (r_{n3n2}\cos(\theta_3) + r_{n2}\cos(\theta_2))\hat{i} + (r_{n3n2}\sin(\theta_3) + r_{n2}\sin(\theta_2))\hat{j}\end{aligned} \tag{5.3}$$

Now, we are going to apply the vector operation to calculate the position vectors. For example, Fig. 5.5 shows a robot as an open-chain four-bar mechanism with the kinematic diagram in Fig. 5.6 where L_1 is the ground, L_2 is the blue arm, L_3 is the orange arm, and L_4 is the green arm. The relative position vectors are $\vec{r}_{n2}$, $\vec{r}_{n3n2}$, and $\vec{r}_{n4n3}$ with the angles θ_2, θ_3, and θ_4. The relative position vectors are:

$$\begin{aligned}\vec{r}_{n1} &= 0\hat{i} + 0\hat{j}\\ \vec{r}_{n2} &= x_{n2}\hat{i} + y_{n2}\hat{j} = r_{n2}\cos(\theta_2)\hat{i} + r_{n2}\sin(\theta_2)\hat{j}\\ \vec{r}_{n3n2} &= x_{n3n2}\hat{i} + y_{n3n2}\hat{j} = r_{n3n2}\cos(\theta_3)\hat{i} + r_{n3n2}\sin(\theta_3)\hat{j}\\ \vec{r}_{n4n3} &= r_{n4n3}\cos(\theta_4)\hat{i} + r_{n4n3}\sin(\theta_4)\hat{j}\end{aligned} \tag{5.4}$$

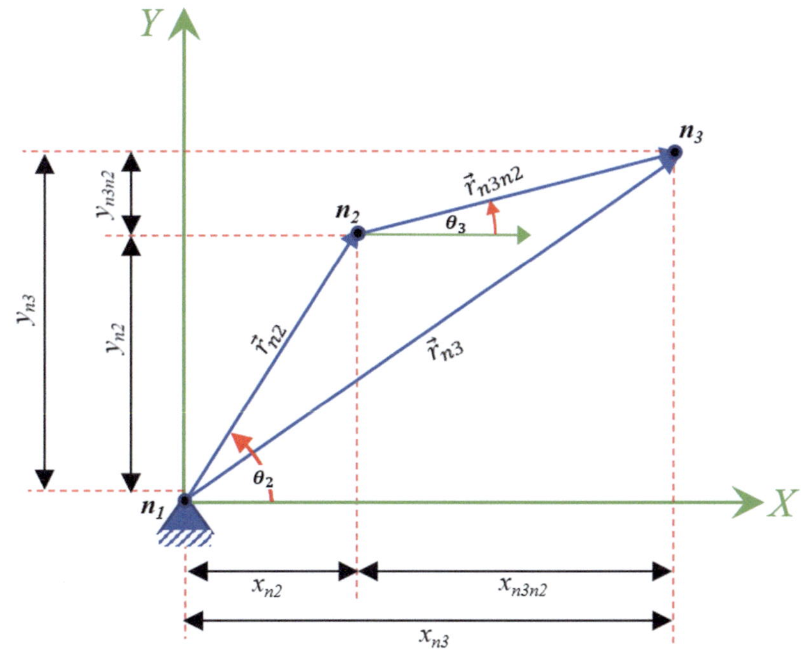

Fig. 5.4 Addition operation of two vectors

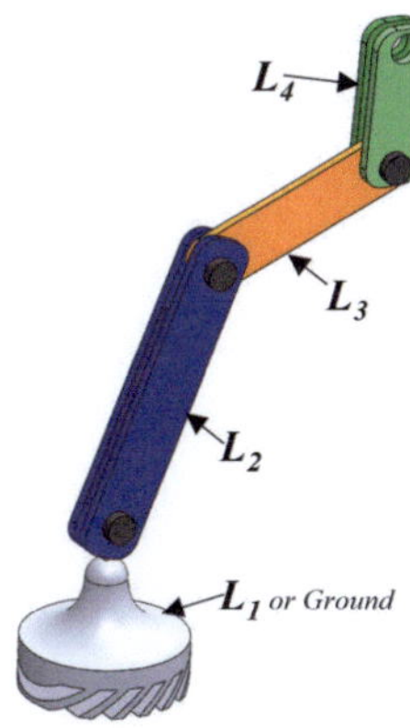

Fig. 5.5 Four-bar mechanism. Designed by the author and implemented by both the author and Pranav Dhola. Courtesy of Pranav Dhola

To calculate the position vector $\vec{r}_{n3}$ or $\vec{r}_{n4}$ with respect to the global frame at n_1, we can use the vector addition rules as:

$$\begin{aligned} \vec{r}_{n3} &= \vec{r}_{n3n2} + \vec{r}_{n2} = r_{n3n2}\cos(\theta_3)\hat{i} + r_{n3n2}\sin(\theta_3)\hat{j} + r_{n2}\cos(\theta_2)\hat{i} + r_{n2}\sin(\theta_2)\hat{j} \\ \vec{r}_{n4} &= \vec{r}_{n4n3} + \vec{r}_{n3} = r_{n4n3}\cos(\theta_4)\hat{i} + r_{n4n3}\sin(\theta_4)\hat{j} + r_{n3n2}\cos(\theta_3)\hat{i} \\ &+ r_{n3n2}\sin(\theta_3)\hat{j} + r_{n2}\cos(\theta_2)\hat{i} + r_{n2}\sin(\theta_2)\hat{j} \end{aligned} \tag{5.5}$$

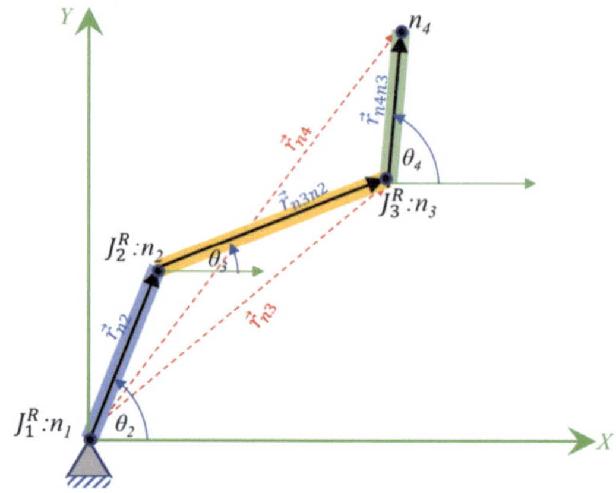

Fig. 5.6 Kinematic diagram and vector diagram of a four-bar open-chain mechanism

Note

The symbol "L" is used for a link. For example, Link i is referred to as L_i.

The ground link is Link 1, L_1.

The symbol "r" refers to a length. For example, the length from node n_1 to node n_2 is referred to as r_{n1n2}.

The symbol "$\vec{r}$" refers to a vector. For example, the vector from node n_1 to node n_2 is referred to as $\vec{r}_{n1n2}$.

In general, $r_{ninj}=r_{njni}$; e.g., $r_{n3n4}=r_{n4n3}$.

We replaced r_{n4n1} with r_{n4} as well as r_{n2n1} with r_{n2} because n_1 is the origin of the global frame.

Example 5.1

Figure 5.7 shows the oil pumpjack mechanism as a four-bar mechanism that includes L_1: the gray arm (Samson post and the base), L_2: the green arm (walking beam and head jack), L_3: the orange arm (Pitman arm), and L4: the red arm (crank and counterweight). We encourage the reader to watch the attached animation video of the pumpjack machine. The dimensions in Fig. 5.7 are for the C80D-133–54 model from Ref. [7].

- Sketch the kinematic diagram of the 4-bar mechanism.
- Find the position vectors of nodes n_1, n_2 and n_3. The selected frame is located in the center of the power shaft (ground node n_1) on the crank.

Solution

The first step is to sketch the kinematic diagram of the pumpjack machine including the walking beam, the crank, and the Pitman arm as shown in Fig. 5.8.

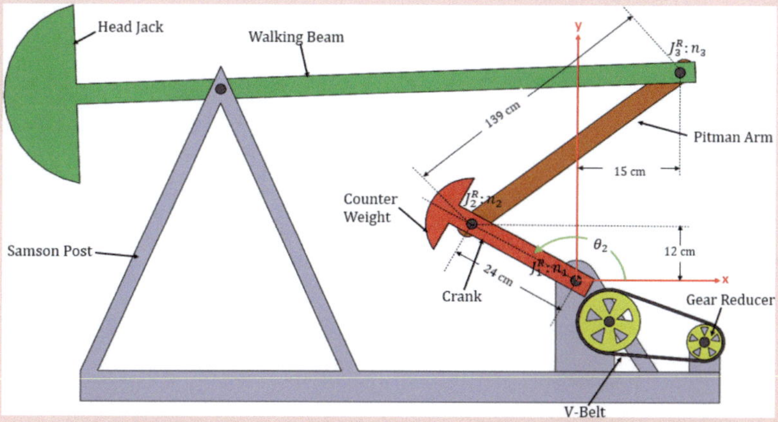

Fig. 5.7 Pumpjack machine. Designed by the author and implemented by both the author and Pranav Dhola. Courtesy of Pranav Dhola

From the geometric configuration as shown in Fig. 5.7, we can find the following parameters:

$$y_{n2} = 12\,\text{cm}$$
$$r_{n2} = 24\,\text{cm}$$
$$r_{n3n2} = 139\,\text{cm}$$

The position vector $\vec{r}_{n1}$ of n_1 is located at the origin and it has the value

$$\vec{r}_{n1} = 0\hat{i} + 0\hat{j}$$

From Eq. (5.2), the angle θ_2 is calculated as

$$\theta_2 = 180^\circ - \sin^{-1}\left(\frac{Y_{n2}}{r_{n2}}\right) = 180^\circ - \sin^{-1}\left(\frac{12}{24}\right) = 150^\circ$$

From Eq. (5.3), the position vector $\vec{r}_{n2}$ of the node n_2 is

$$\vec{r}_{n2} = x_{n2}\hat{i} + y_{n2}\hat{j} = r_{n2}\cos\theta_2\hat{i} + r_{n2}\sin\theta_2\hat{j} = 24\cos 150^\circ\hat{i} + 24\sin 150^\circ\hat{j} = -20.78\hat{i} + 12\hat{j}$$

From the geometry shown in Fig. 5.7

$$x_{n3n2} = x_{n2} + 15 = 20.78 + 15 = 35.78\,\text{cm}$$
$$\theta_3 = \cos^{-1}\left(\frac{x_{n3n2}}{r_{n3n2}}\right) = \cos^{-1}\left(\frac{35.78}{139}\right) = 75.08^\circ$$
$$y_{n3n2} = r_{n3n2}\sin(\theta_3) = 139\sin(75.08^\circ) = 134.3\,\text{cm}$$

From Eq. (5.3), the position vectors $\vec{r}_{n3n2}$ and $\vec{r}_{n3}$ are

$$\vec{r}_{n3n2} = x_{n3n2}\hat{i} + y_{n3n2}\hat{j} = 35.78\hat{i} + 134.3\hat{j}$$
$$\vec{r}_{n3} = \vec{r}_{n3n2} + \vec{r}_{n2} = 35.78\hat{i} + 134.3\hat{j} - 20.78\hat{i} + 12\hat{j} = 15\hat{i} + 156.3\hat{j}$$

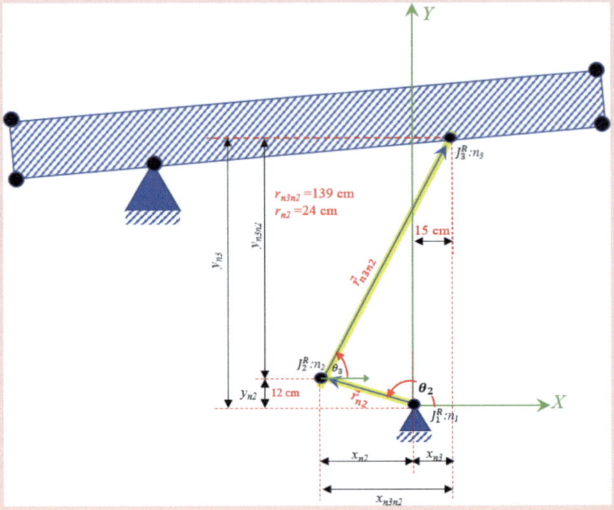

Fig. 5.8 Kinematic and vector diagrams of a pumpjack machine

5.3 Displacements

Definition

Displacement:

Displacement is a vector representing the distance between the starting and ending positions of a point or link. Two types of displacements will be considered: linear and angular [1, 2].

Displacement is the deviation from an old position to a new position. There are two types of displacement: active displacement and passive displacement. Active displacement is manually or automatically actuated with a motor or hydraulic cylinder. Passive displacement is driven by the mechanism without an actuator. For point masses, the displacement can be active or passive; however, the displacement can only be linear. For links, the displacement can be active or passive as a linear displacement and/or angular displacement. In this book, our focus will be on link displacement.

5.3.1 Linear Displacement

Linear displacement is a straight-line distance between a specific node's start and end positions. Figure 5.9 shows an example of a 4-bar mechanism moved from an old position to a new position. The linear displacement of node 3, $\vec{d}_{n3}$, is defined as the difference between the new position $\vec{r}_{n3}^{\,\text{new}}$ and the old position $\vec{r}_{n3}^{\,\text{old}}$ or:

$$\vec{d}_{n3} = \vec{r}_{n3}^{\,\text{new}} - \vec{r}_{n3}^{\,\text{old}} \tag{5.6}$$

If there is a linear actuator at any prismatic joint (node), the joint has an active linear displacement. On the contrary, any node without a linear actuator has a passive linear displacement. The passive linear displacement can be observed on any node located in any specific link.

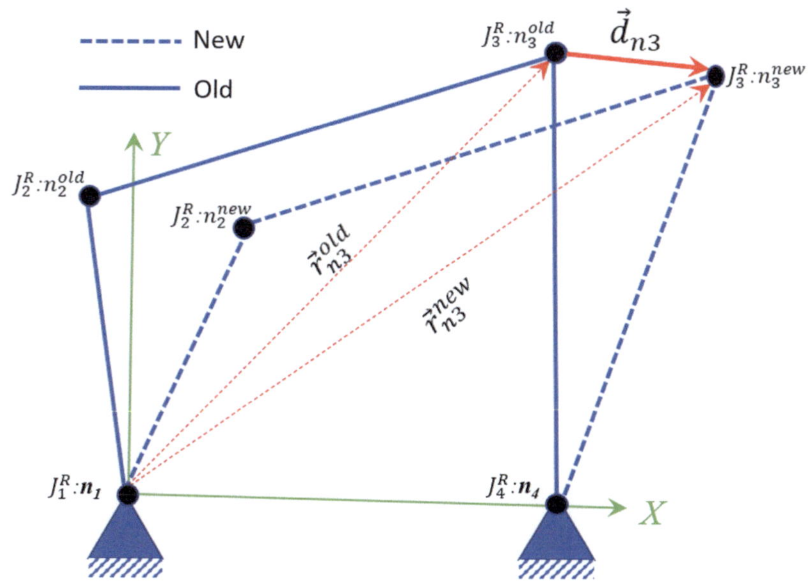

Fig. 5.9 Kinematic diagram of linear displacement of four-bar mechanism

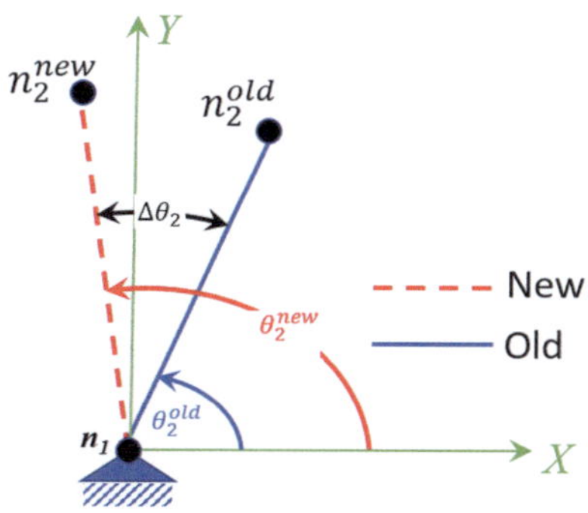

Fig. 5.10 Kinematic diagram of angular displacement of a hinged link

5.3.2 Angular Displacement

Angular displacement is the angular distance between a start angle and an end angle of a link around one node belonging to the link. Figure 5.10 shows an example of the angular displacement $\Delta\theta_2$ starting with θ_2^{old} and ending at θ_2^{new}. In general, the angular displacement is defined as:

$$\Delta\theta = \theta^{\text{new}} - \theta^{\text{old}} \tag{5.7}$$

The concepts of active and passive displacement exist in angular as well as linear displacement. The degrees of freedom of any mechanism represent the minimum number of required actuators to activate a controllable motion. For example, Fig. 5.11 shows a 4-bar mechanism that has 1DoF which means it needs at least one actuator. If we locate a DC motor at joint J_1^R, we call J_1^R an active joint (joint with actuator) and J_2^R, J_3^R, andJ_4^R passive joints (joints without actuator). When the motor rotates or inserts a rotational displacement to n_1, L_1 will start to rotate around n_1 causing the entire mechanism to move. The motor rotation is called the "input-displacement of n_1". The other nodes and links have "state displacement". We could choose angle θ_3 to be our output, or we could use any

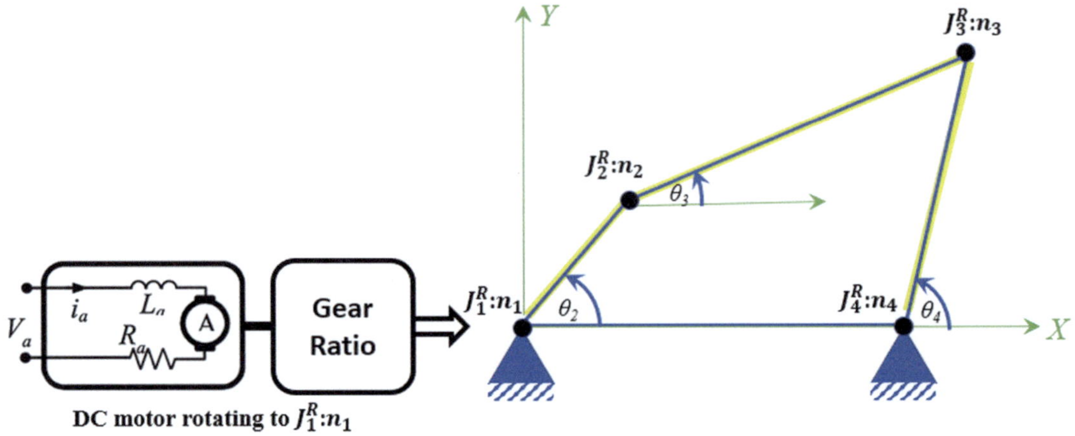

Fig. 5.11 Kinematic diagram of four-bar mechanism with actuator

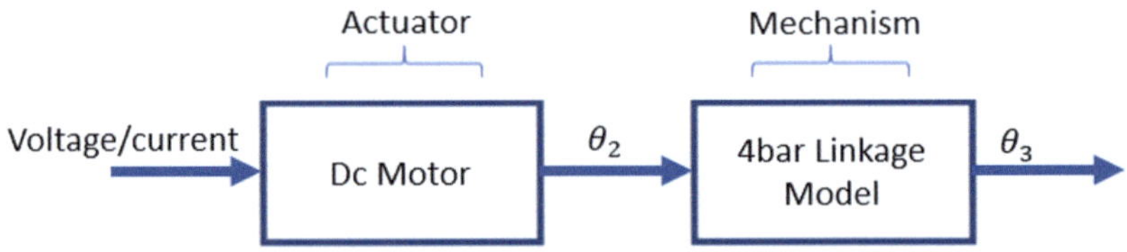

Fig. 5.12 System diagram of four-bar mechanism position

angle that represents the displacement that we are interested in. Figure 5.12 shows the system diagram of the input–output positions for a 4-bar mechanism that is actuated by a DC motor to activate the angular position θ_2 and makes θ_3 an output.

Example 5.2

Refer to the slider–crank 4-bar mechanism shown in Fig. 5.13. In this mechanism, a hydraulic actuator (at node 3) moves the prismatic joint from position 1 to position 2.

1. Calculate the degrees of freedom.
2. Do you think this single hydraulic actuator is enough to control the entire mechanism's motion?
3. Which nodes have active linear displacements?
4. Which nodes have passive linear displacements?
5. Which nodes have active angular displacements?
6. Which nodes have passive angular displacements?
7. Calculate the active linear displacement vectors.
8. Calculate the passive linear displacement vectors.
9. Calculate the active angular displacement.
10. Calculate the passive angular displacement.
11. Sketch the system diagram of the slider–crank 4-bar mechanism where the hydraulic motion is the input and the angular displacement at θ_2 is the output.

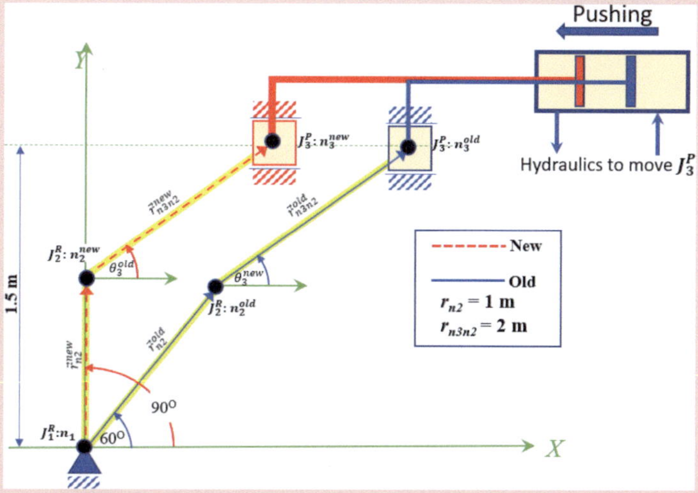

Fig. 5.13 Kinematic diagram and vector diagram of slider–crank mechanism displacement with actuator

Solution

1. The mechanism has four links: L_1, L_2, L_3, and L_4 plus four joints: J_1^R, J_2^R, J_3^P and J_3^R. Substituting into Grübler's equation:

$$M = 3(n_l - 1) - 2n_p - n_h = 3\text{x}(4 - 1) - 2\text{x}4 - 0 = 1$$

The mechanism has 1 DoF.

2. Yes, one actuator (hydraulic system) is enough to control the entire mechanism motion because the mechanism has 1 DoF.
3. The node n_3 or J_3^P has active linear displacement since the actuator is located at n_3.
4. The nodes n_1 and n_2 have passive linear displacements.
5. There is no active angular displacement since the mechanism does not have any rotational actuator.
6. The revolute joints: J_1^R, J_2^R, and J_3^R at nodes n_1, n_2, and n_3 are passive.

The old positions vectors:

$$\theta_2^{\text{old}} = 60^\circ$$

$$\theta_3^{\text{old}} = \sin^{-1}\left(\frac{y_{n3n1} - y_{n3n2}^{\text{old}}}{l_3}\right) = \sin^{-1}\left(\frac{1.5 - 0.866}{2}\right) = 18.46^\circ$$

$$\vec{r}_{n1}^{\text{old}} = 0\hat{i} + 0\hat{j}$$

$$\vec{r}_{n2}^{\text{old}} = r_{n2}^{\text{old}} \cos\left(\theta_2^{\text{old}}\right)\hat{i} + r_{n2}^{\text{old}} \sin\left(\theta_2^{\text{old}}\right)\hat{j}$$

$$\vec{r}_{n3n2}^{\text{old}} = r_{n3n2}^{\text{old}} \times \cos\left(\theta_3^{\text{old}}\right)\hat{i} + r_{n3n2}^{\text{old}} \times \sin\left(\theta_3^{\text{old}}\right)\hat{j}$$

$$\vec{r}_{n3n2}^{\text{old}} = 2 \times \cos(18.46^\circ)\hat{i} + 2 \times \sin\left(18.46^\circ\right)\hat{j} = 1.897\hat{i} + 0.634\hat{j}$$

$$\vec{r}_{n3}^{\text{old}} = \vec{r}_{n3n2}^{\text{old}} + \vec{r}_{n2}^{\text{old}} = 2.397\hat{i} + 1.5\hat{j}$$

The new positions vectors:

$$\theta_2^{\text{new}} = 90°$$

$$\theta_3^{\text{new}} = \sin^{-1}\left(\frac{y_{n3n1} - y_{n3n2}^{\text{new}}}{l_3}\right) = \sin^{-1}\left(\frac{1.5 - 1}{2}\right) = 14.48°$$

$$\vec{r}_{n1}^{\text{new}} = 0\hat{i} + 0\hat{j}$$

$$\vec{r}_{n2}^{\text{new}} = r_{n2}^{\text{new}} \cos\left(\theta_2^{\text{new}}\right)\hat{i} + r_{n2}^{\text{new}} \sin\left(\theta_2^{\text{new}}\right)\hat{j}$$

$$\vec{r}_{n2}^{\text{new}} = 1 \times \cos(90°)\hat{i} + 1 \times \sin(90°)\hat{j} = 0\hat{i} + 1\hat{j}$$

$$\vec{r}_{n3n2}^{\text{new}} = r_{n3n2}^{\text{new}} \times \cos\left(\theta_3^{\text{new}}\right)\hat{i} + r_{n3n2}^{\text{new}} \times \sin\left(\theta_3^{\text{new}}\right)\hat{j}$$

$$\vec{r}_{n3n2}^{\text{new}} = 2 \times \cos(14.48°)\hat{i} + 2 \times \sin(14.48°)\hat{j} = 1.937\hat{i} + 0.5\hat{j}$$

$$\vec{r}_{n3}^{\text{new}} = \vec{r}_{n3n2}^{\text{new}} + \vec{r}_{n2}^{\text{new}} = 1.937\hat{i} + 1.5\hat{j}$$

7. Active linear displacement vector:

$$\vec{d}_{n3} = \vec{r}_{n3}^{\text{new}} - \vec{r}_{n3}^{\text{old}} = 1.937\hat{i} + 1.5\hat{j} - \left(2.397\hat{i} + 1.5\hat{j}\right) = -0.46\hat{i} + 0\hat{j}$$

8. Passive linear displacement vectors:

$$\vec{d}_{n1} = 0\hat{i} + 0\hat{j}$$

$$\vec{d}_{n2} = \vec{r}_{n2}^{\text{new}} - \vec{r}_{n2}^{\text{old}} = 0\hat{i} + 1\hat{j} - \left(0.5\hat{i} + 0.866\hat{j}\right) = -0.5\hat{i} + 0.134\hat{j}$$

9. There is no active angular displacement since there is no angular actuator.
10. There are two passive angular displacements:

$$\Delta\theta_2 = \theta_2^{\text{new}} - \theta_2^{\text{old}} = 90° - 60° = 30°$$

$$\Delta\theta_3 = \theta_3^{\text{new}} - \theta_3^{\text{old}} = 14.48° - 18.46° = -3.98°$$

Note: the calculation shows that the hydraulic system moves the prismatic joint by a distance of 0.46 m in the negative x-direction to change θ_2 from 60° to 90°. There is no displacement at node n_1 because it is a ground node. When the rotation is counterclockwise, the angular displacement is positive such as $\Delta\theta_2$. When the rotation is clockwise, the angular displacement is negative such as $\Delta\theta_3$.

11. Figure 5.14 shows the displacement system diagram of the mechanism. The input is the linear prismatic displacement n_3. The output is the angular displacement $\Delta\theta_2$..

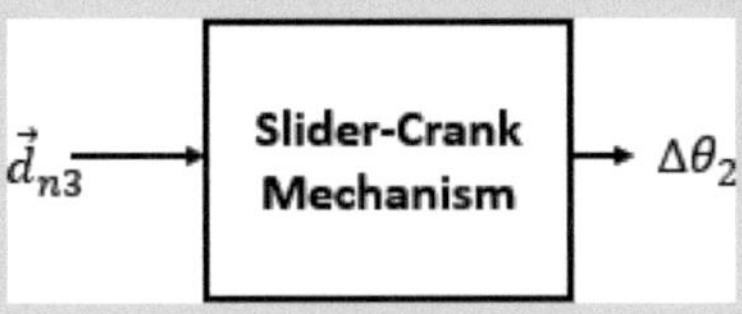

Fig. 5.14 Displacement system diagram of slider–crank mechanism

5.4 Simulink® Simscape™ Multibody™ Motion Sensor and Actuator

Simscape™ Multibody™ provides a sensor block to measure the motion of any node on a mechanism. There is also an actuator block to move the mechanism. Now we are going to explore the sensor block and the actuator block and utilize them to determine the position of any mechanism. We will use the actuator to settle the mechanism at a certain position. We will use a sensor to measure the change in position of any node of interest on the machine.

Definition[1]

Transform Sensor Block:

The transform Sensor block represents a sensor that measures the spatial relationship between two frames. Parameters that this sensor measures include rotational and translational position, velocity, and acceleration. This block provides the broadest motion sensing capability in Simscape™ Multibody™ [3–6].

Simscape™ Multibody™ offers a motion actuation block and a motion sensor block. The motion actuation will be explained in more detail in the upcoming chapters. Here, we are going to focus on the motion sensor block (which Simscape™ Multibody™ calls a "Transformation Sensor") as shown in Fig. 5.15. Port B and port F represent the base and follower frames, respectively. When the user double-clicks on the block, a list of items appears: three rotational motions (rotation, angular velocity, and angular acceleration) and three translational motions (translation, velocity, and acceleration) as shown in Table 5.1. The user can select one of the six items. By expanding the sub-list under each item, the user can pick a specific sensor for a specific motion variable. There would be an output

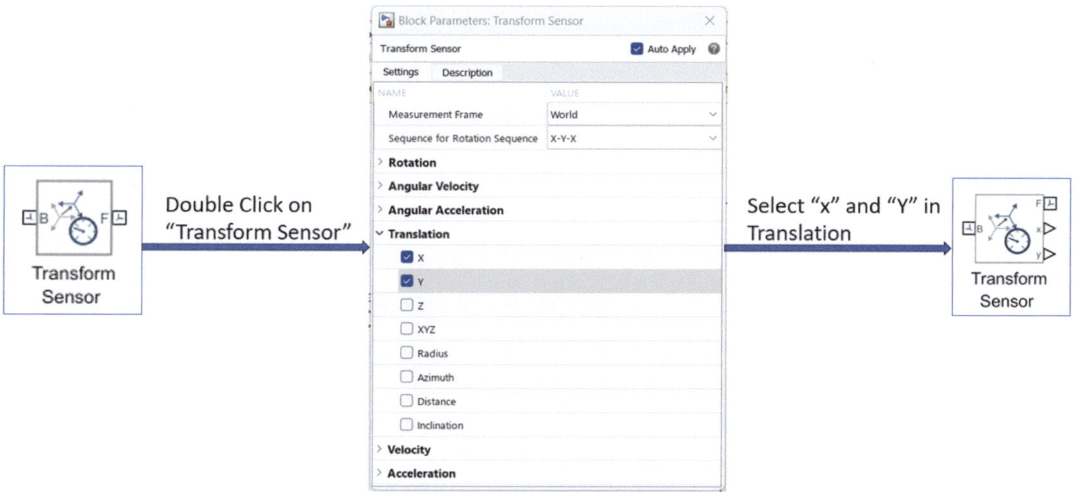

Fig. 5.15 Step to activate displacement sensors from transform sensor block. Designed and implemented by the author. Images used with permission of MathWorks

[1] Reproduced from [3-6] by MathWorks with permission from MathWorks.

Table 5.1 Motion sensor. Designed and implemented by the author. Images used with permission of MathWorks

Sensor motion	Sensor under each motion
Rotation	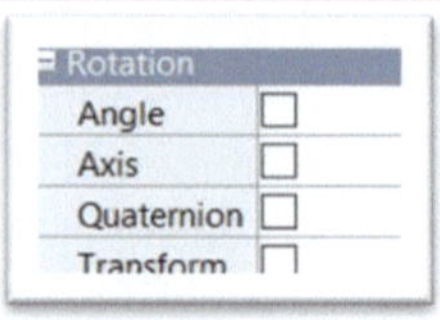
Angular velocity	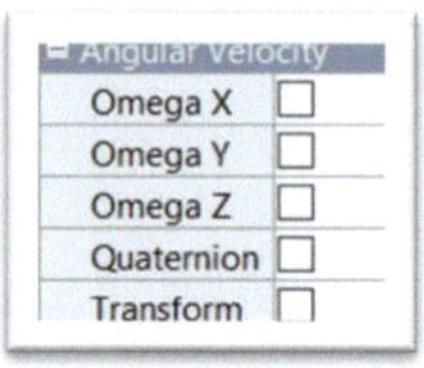
Angular Acceleration	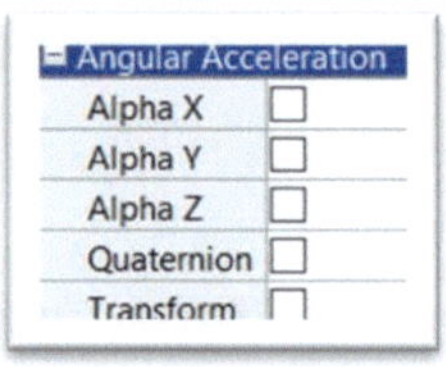
Translation	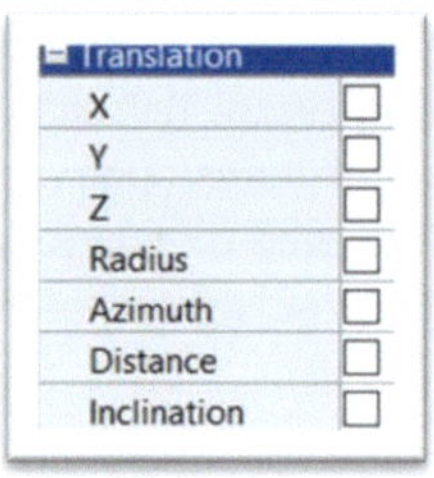
Velocity	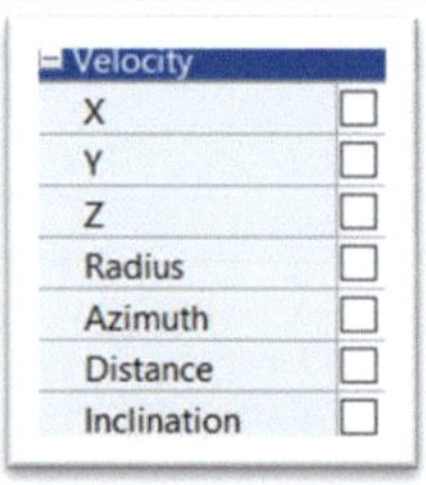
Acceleration	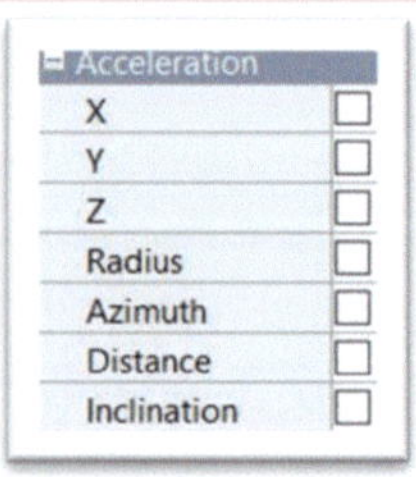

port for each selected sensor. For example, in Fig. 5.15, we select "Translational" motion followed by picking the x-sensor and y-sensor. After pressing "OK", there are two output ports for the x-sensor and y-sensor.

For the displacement, we have the translation motion sensors and the angle motion sensor. In the case of translation motion, we can pick x, y, or z to measure the displacement from the base frame into the follower frame. The same scenario happens in the rotation motion sensor. Port B is going to be connected to the world frame. The follower F port will be connected to the measured node.

Engineering Tips—Sensor as Simulink®Mask

To reduce the complexity of the model visualization, we are going to build and add a set of sensor subsystems in "Kinematics_Of_Machines.slx" file. Figure 5.16 shows an example of the angular position sensor subsystem that consists of the World Frame block, Transform Sensor block, rad_to_deg Gain block, Scope block, and Out block.

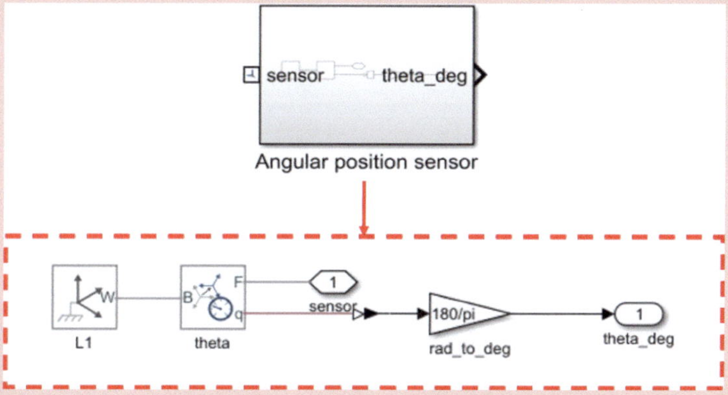

Fig. 5.16 Angular position sensor subsystem. Designed and implemented by the author. Images used with permission of MathWorks

Definition[2]

Motion Actuator:

The motion input of a joint primitive is a timeseries object specifying that primitive's trajectory. For a prismatic primitive, that trajectory is the position coordinate along the primitive axis, given as a function of time [3–6].

In Simscape™ Multibody™, any joint block has an "Actuation" parameter that is divided into Dynamic Actuation (called Force/Torque) and Kinematic Actuation (called Motion). As we previously described in Chap. 1, kinematics deals with motion (position, velocity, and acceleration). In this chapter, we will only discuss the motion actuator since the topic here is position and displacement. The force/torque actuator will be discussed in Chaps. 11 and 12.

[2] Reproduced from [3-6] by MathWorks with permission from MathWorks.

One method used to create angular actuation is a motor with a feedback control loop. The feedback control loop has the motor's motion as an output that is measured using a potentiometer or any motion sensor. The controller applies a mathematical algorithm, such as a PID controller, to change the voltage from the error between the desired motion and the measured motion as shown in Fig. 5.17. The motor torque limit is extremely high compared to the required torque to drive the mechanism so there is no expected torque stall. Due to this control loop, the output signal from the motor or the input signal to the mechanism is the motion, not the torque. The same scenario can be used for translational motion actuation.

To activate a revolute joint with a motion, we can use the following steps:

Step 1: Double-click on the revolute joint and expand the actuation parameter as shown in Fig. 5.18.

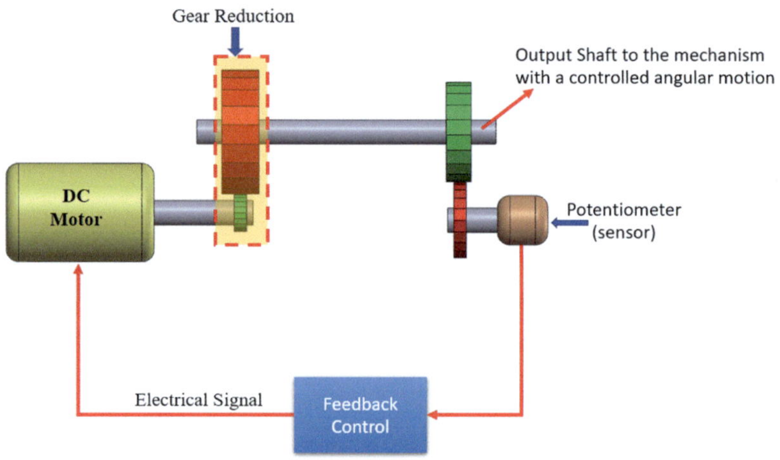

Fig. 5.17 Motion actuator control loop of DC motor. Designed by the author and implemented by both the author and Pranav Dhola. Courtesy of Pranav Dhola

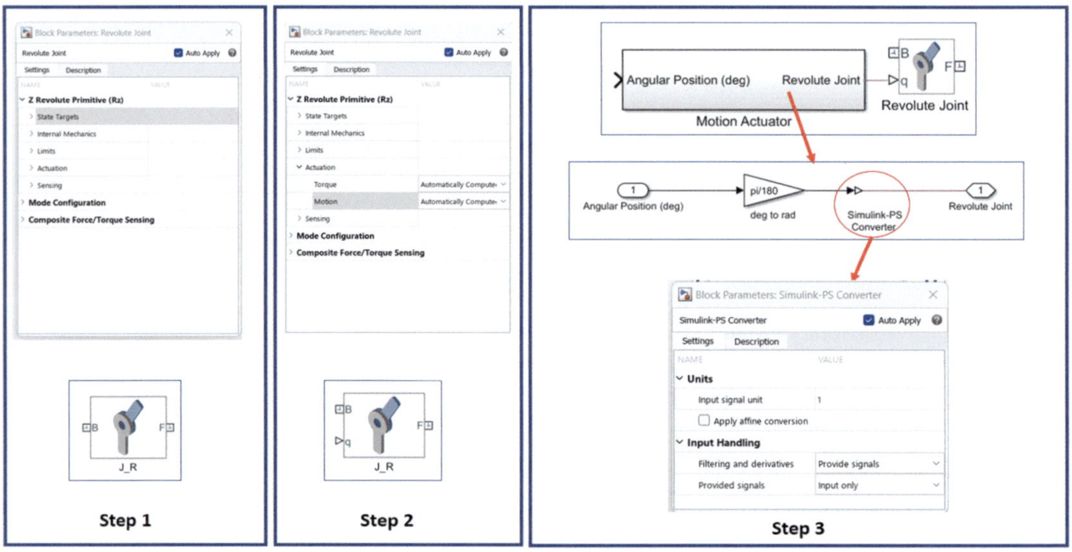

Fig. 5.18 Step for developing motion actuator subsystem. Designed and implemented by the author. Images used with permission of MathWorks

Step 2:

- Select "Provided by Input" under "Motion" then select "Automatically Computed" under "Torque".
- Click "Ok" and you will see an input port "q" generated for the revolute joint as shown in Fig. 5.18.

Step 3:

- Create the "Angular Motion Actuator" subsystem with a "Simulink®-PS Converter" block. This subsystem has one input: angular position plus one output to the revolute joint. There is a gain "pi/180" to convert the degrees to radians.
- The "Input Handling" parameter in the "Simulink®-PS Converter" should be selected to "Provide Signals" and "Input Only" as shown in Fig. 5.18.

Note: The Motion Actuator subsystem is added inside our "Kinematics_Of_Machines.slx" with our developed masks and subsystems. More details about the Motion Actuator can be found in [3–6]. We can use the motion actuator subsystems to set the motion inputs at specific positions as in Example 5.3. Another way to set the position is to double-click on the revolute joint block and go to State Targets > Specify Position Target > Value and add the value as shown in Fig. 5.19. The motion sensor subsystems will be used to measure the position of the nodes of interest.

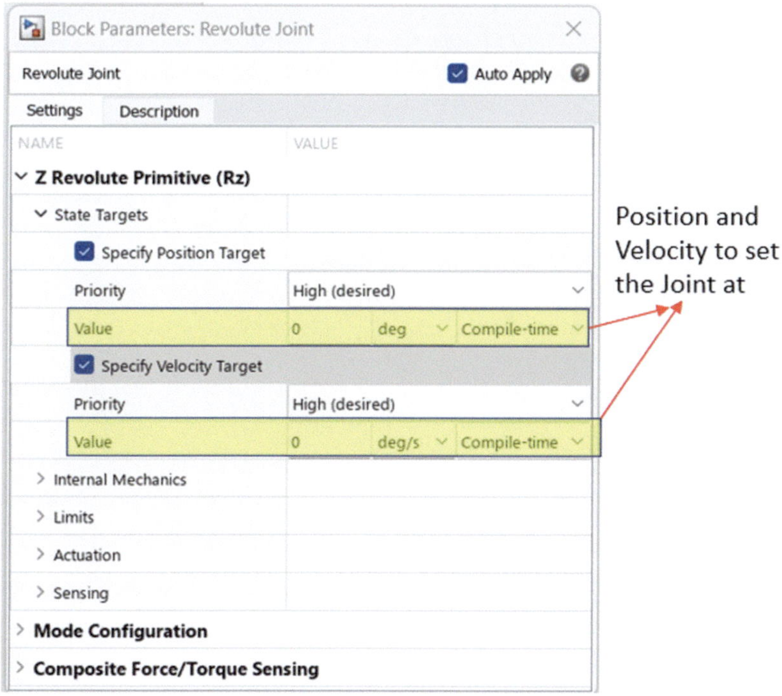

Fig. 5.19 Initial position and velocity setup for revolute joint block. Designed and implemented by the author. Images used with permission of MathWorks

Example 5.3

Build a Simscape™ Multibody™ model of Example 5.2. Use the motion actuator subsystem to activate θ_2 with 60° and 90°. Use the position sensor to measure:

1. Vectors $\vec{r}_{n2}^{new}$, $\vec{r}_{n2}^{old}$, $\vec{r}_{n3}^{new}$, and $\vec{r}_{n3}^{old}$
2. Angles θ_3^{new} and θ_3^{old}

Compare the measured values with the calculated values in Example 5.2.

Solution

The first part is to define the model parameters using a MATLAB® script. Open a new script and save as "Slider_Crank_Example5p3_script.m" that starts by cleaning the variable history. After that, add all the given parameters. Finally, calculate the needed parameters of each link. As we mentioned previously, we encourage the reader to attend the "MATLAB® Onramp" and "Simulink® Onramp".

```
%This file to declare the parameters of Example 5.3
%=================================
%Clear all the variables and close all the figures
%==============================
clc
clear all
close all
%=====================================================================
%Declare the given parameters
%========================
r_n2 = 1; %m
r_n3n2 = 2; %m
theta2_old = 60; %deg
theta2_new = 90; %deg
y_n4 = 1.5; %m
%Calculation of parameters
%======================
x_n4_estimate = 0;
L2_dimension = [1 0.1 0.1]*r_n2; %m
L3_dimension = [r_n3n2 0.1*r_n2 0.1*r_n2]; %m

Frame_L1_n1 = [0 0 0]; %m
Frame_L1_n4 = [x_n4_estimate y_n4 0]; %m

Frame_L2_n1 = [-r_n2/2 0 0]; %m
Frame_L2_n2 = [r_n2/2 0 0]; %m

Frame_L3_n2 = [-r_n3n2/2 0 0]; %m
Frame_L3_n3 = [r_n3n2/2 0 0]; %m

Frame_L4_n1 = [0 0 0]; %m
Frame_L4_n3 = [0 0 0];%m
%=====================================================
%Assumed parameters
%=============
L4_dimensions = [0.2  0.2  0.1]*r_n2; %m
```

Now, we are going to start building our model using Simulink® Environment. Previously in Chap. 4, we developed a set of mask blocks and a set of subsystem blocks that we sorted in our "Kinematics_Of_Machines.slx" file. We are going to use some of these mask blocks to build the slider–crank model. Open a new ".slx" file and save it under "Slider_.

Crank_Example5p3_model.slx".

Step 1: Copy one "Ground plus revolute and prismatic" block from the "Kinematics_Of_ Machines.slx" file and paste this block into the "Slider_Crank_Example5p3_model. slx" file. Rename the block "L1". Double-click on the block and add the parameters of L_1 as shown in Fig. 5.20.

Step 2: Copy the "Brick plus two revolute" block from the "Kinematics_Of_Machines.slx" file and paste this block into the "Slider_Crank_Example5p3.slx" file twice. Rename the blocks "L2" and "L3". Double-click on the first block and add the parameters of L_2. Double-click on the second block and add the parameters of L_3 as shown in Fig. 5.20.

Step 3: Copy the "Brick plus revolute and prismatic" block from the "Kinematics_Of_ Machines.slx" file and paste this block into the "Slider_Crank_Example5p3_model. slx" file. Rename the block "L4". Double-click on this block and add the parameters of L_4 as shown in Fig. 5.20.

Step 4: Drag three revolute joints and one prismatic joint from the Simscape™ Multibody™ Joints library and drop them into the "Slider_Crank_Example5p3_model.slx" file. Connect each joint as shown in Fig. 5.21.

Step 5: Copy an angular position sensor block from the "Kinematics_Of_Machines.slx" file and paste the block into the "Slider_Crank_Example5p3_model.slx" file. Connect this sensor to the follower of Joint J_2R. Connect the sensor to a display as shown in Fig. 5.21.

Step 6: Copy the translational position sensor block twice from the "Kinematics_Of_ Machines.slx" file and paste the blocks into the "Slider_Crank_Example5p3_model. slx" file. Connect the first sensor to the follower of Joint J_2R as the r_n2 sensor.

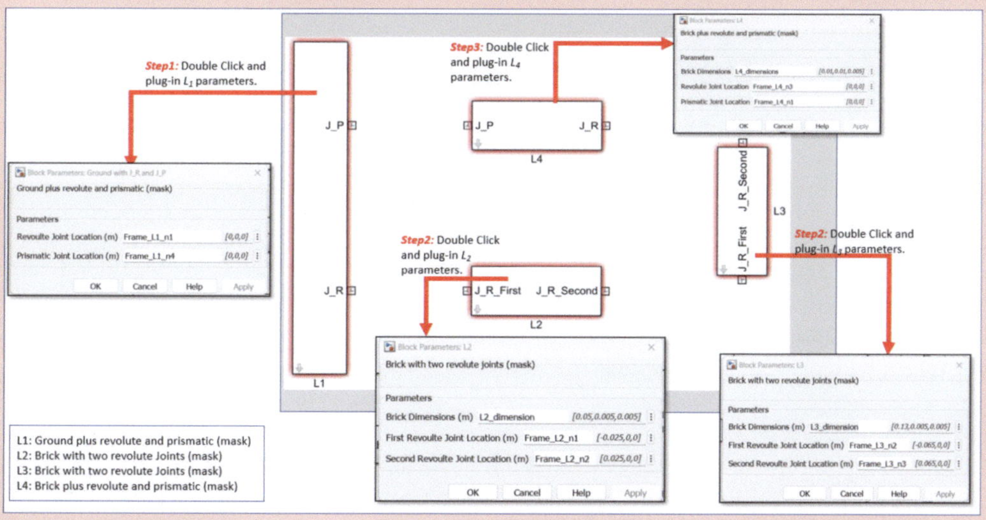

Fig. 5.20 Slider_Crank_Example5p3 model with four links. Designed and implemented by the author. Images used with permission of MathWorks

Connect the second sensor to the follower of Joint J_3P as the r_n3 sensor. Connect the sensors to displays as shown in Fig. 5.21.

Step 7: Double-click on J_1R and go to State Targets>Specify Position Target>Value. Add the variable "theta_2_deg". This process is similar to that in Fig. 5.19.

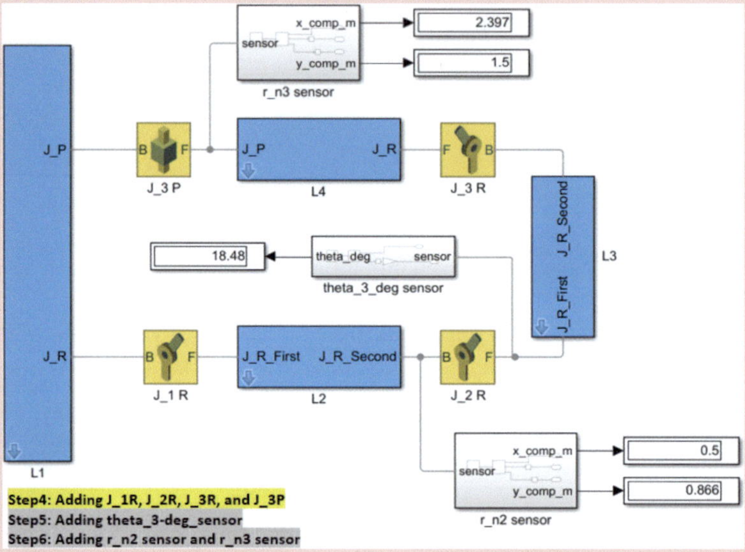

Fig. 5.21 Slider–crank Simscape™ Multibody™ model of Example 5.3

Do not forget to save the files after each step. A copy of the model plus a copy of the script are attached for the reader's reference.

Old position:

After running the "Slider_Crank_Example5p3_script.m" file, write theta2_deg=theta2_old in the command window run the model "Slider_Crank_Example5p3_model.slx" and record the measured values from the displays.

New position:

Write theta2_deg=theta2_new in the command window. Run the model "Slider_Crank_Example5p3_model.slx" and record the measured values from the displays.

Table 5.2 shows the measured values at the old and new positions. Comparing the measured values with our calculated values in Example 5.2, the accuracy is high.

Table 5.2 Old and new position's values

Old position	New position
$\theta_2^{\text{old}} = 60°$	$\theta_2^{\text{new}} = 90°$
$\theta_3^{\text{old}} = 18.48°$	$\theta_3^{\text{new}} = 14.48°$
$\vec{r}_{n2}^{\,\text{old}} = 0.866\hat{i} + 0.5\hat{j}$	$\vec{r}_{n2}^{\,\text{new}} = 0\hat{i} + 1\hat{j}$
$\vec{r}_{n3}^{\,\text{old}} = 2.397\hat{i} + 1.5\hat{j}$	$\vec{r}_{n3}^{\,\text{new}} = 1.936\hat{i} + 1.5\hat{j}$

Fig. 5.22 Old and new positions of slider–crank mechanism from Simscape™ Multibody™ model of Example 5.3

The resultant graphics of the Simscape™ Multibody™ model is shown in Fig. 5.22.

5.5 Displacement Graphical Analysis

Graphical analysis can be done using several methods. To start, as shown in Fig. 5.23, a link L_1 with length r_{n2} has two nodes n_1 and n_2. The link L_1 is connected to the ground at node n_1 through a revolute joint J_1^R. If we plot a circle with radius r_{n2}, we can locate the position of node n_2 as the intersection between Line Angle θ_2 and the Circle Radius r_{n2}.

This concept can be extended to sketch a 4-bar mechanism with an input angle ϑ_2 as follows:

Step 1: Locate the two ground nodes n_1 and n_4 by a distance r_{n4} horizontally and locate the *X-axis* and *Y-axis* origin at n_1 as shown in Fig. 5.24.

Step 2: Plot Circle 1 with a center located at n_1 and radius r_{n2} and plot Circle 2 with a center at n_4 and radius r_{n3n4} as shown in Fig. 5.25.

Step 3: Plot Line 1 starting at n_1 and having an angle θ_2 with a positive X-axis and locate n_2 at the intersection of Line 1 with Circle 1 as shown in Fig. 5.26.

Step 4: Plot Circle 3 with a center at point n_2 with a radius r_{n3n2}. Now, we have two intersections. The upper one is called circuit 1; the lower, is circuit 2. Locate n_{4c1} at the first linkage circuit intersection of Circle 3 with Circle 2 and locate n_{4c2} at the second linkage circuit intersection of Circle 3 with Circle 2 as shown in Fig. 5.27.

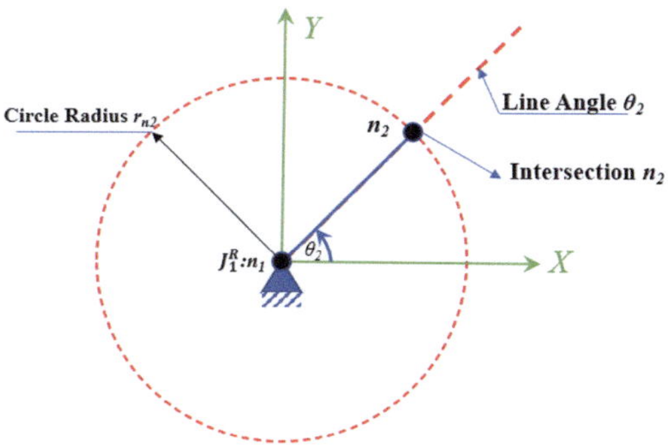

Fig. 5.23 Single link graphical position solution

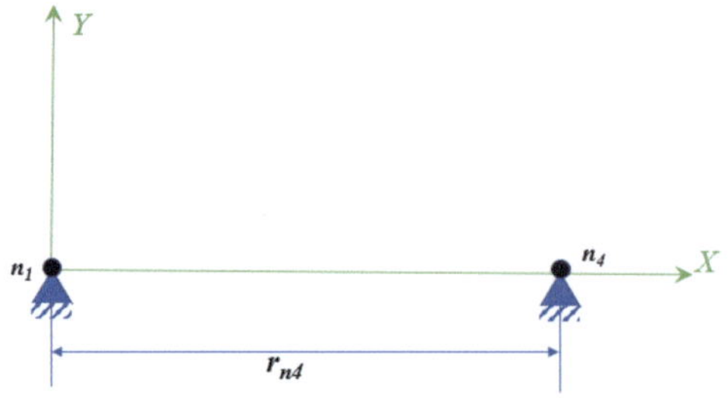

Fig. 5.24 Step 1 of four-bar graphic position solution

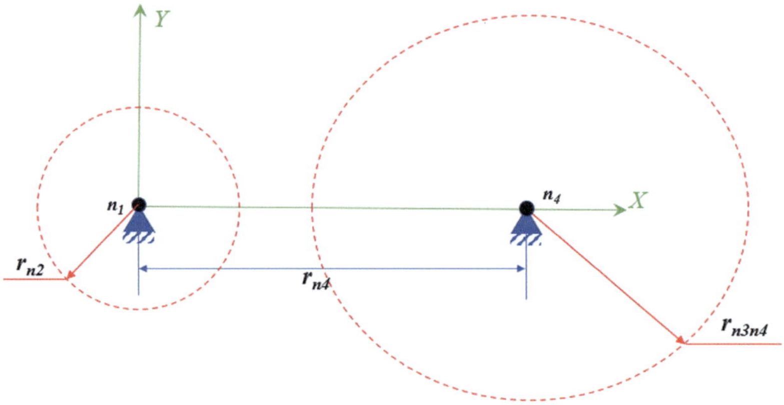

Fig. 5.25 Step 2 of four-bar graphic position solution

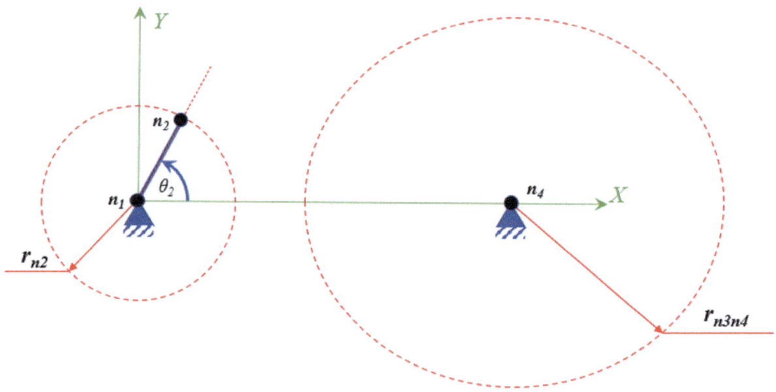

Fig. 5.26 Step 3 of four-bar graphic position solution

Step 5: As shown in Fig. 5.28, we can sketch two linkages from Circuit 1 and Circuit 2. The angles θ_{3c1} and θ_{4c1} in Circuit 1 or the angles θ_{3c2} and θ_{4c2} in Circuit 2 can be measured.

The above algorithm can be applied to any mechanism using the "Combination Drawing Set" or a graphic software such as SolidWorks or AutoCAD. The graphic analysis is more suitable for discrete positions and it takes time and effort to make the sketch in a professional manner.

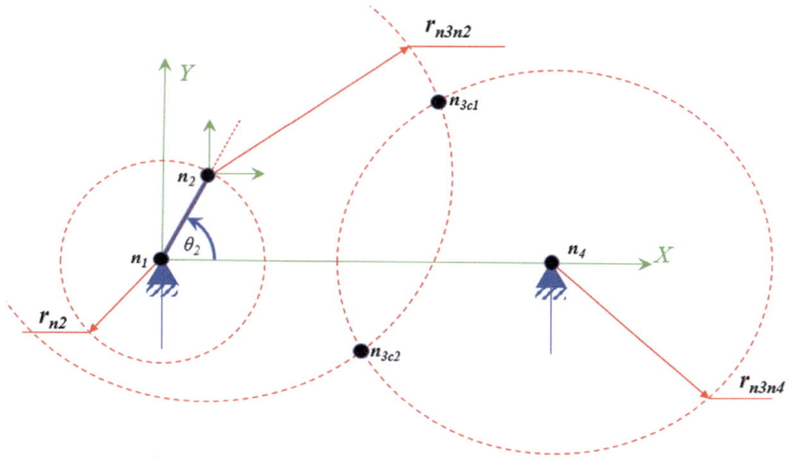

Fig. 5.27 Step 4 of four-bar graphic position solution

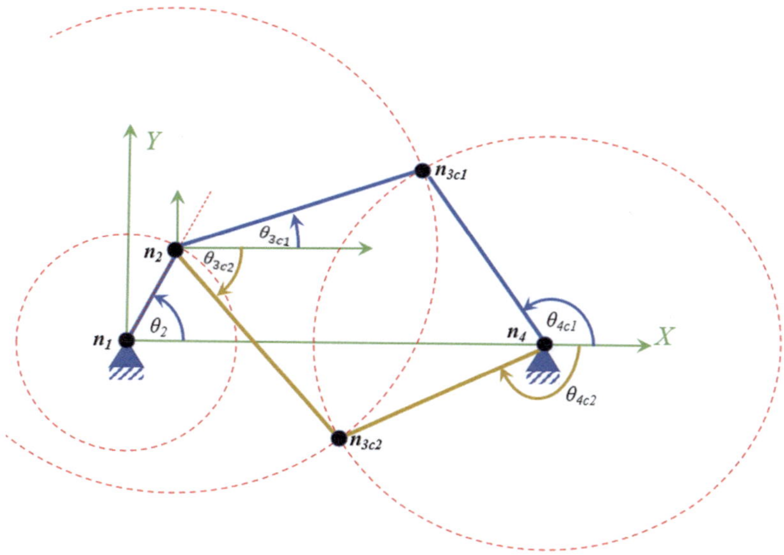

Fig. 5.28 Step 5 of four-bar graphic position solution

Example 5.4

Figure 5.29 shows a piston–cylinder mechanism with a crank radius of 50 mm and connecting rod length of 130 mm. Both the wrist pin (piston-connecting-rod revolute joint) and crankshaft center are horizontally aligned. If the crank angle is at 30° counterclockwise with the horizontal wrist axis, use graphical analysis to:

1. Plot the position of the crank radius and connecting rod.
2. From the plot, measure the angle between the horizontal wrist axis and the connecting rod.
3. From the plot, measure the vertical distance from the wrist pin to the crankshaft center.

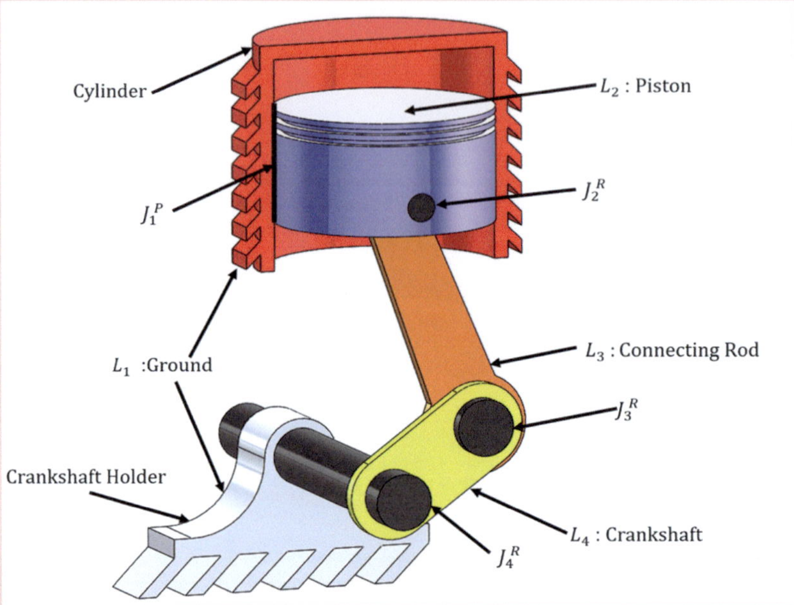

Fig. 5.29 Piston–cylinder mechanism. Designed by the author and implemented by both the author and Pranav Dhola. Courtesy of Pranav Dhola

Solution

Step 1: Locate the *Y-axis* and *X-axis* at the crankshaft center; origin or node n_1 and plot Circle 1 with a radius of 50 mm from n_1 as shown in Fig. 5.30.

Step 2: Draw Line 1 starting from the origin with an angle of 30° from the *X-axis* and locate node n_2 at the intersection of Line 1 and Circle 1 as shown in Fig. 5.31. We are considering the Circuit 1 solution only.

Step 3: Plot Circle 2 with a center at n_2 and radius of 130 mm and locate node n_3 and node n_4 as the intersection between Circle 3 and *X-axis* as shown in Fig. 5.32. We are considering Circuit 1 solution.

Step 4: Plot and connect the links as shown and add the symbols of each node, link, and joint shown in Fig. 5.33.

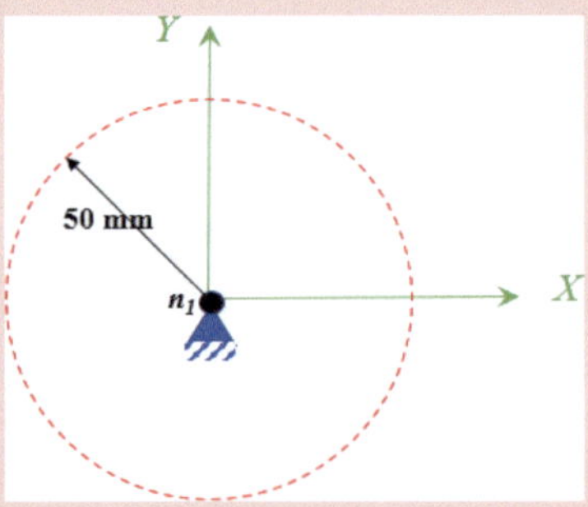

Fig. 5.30 Step 1 of piston–cylinder graphic position solution

From the measurement, we can find $\theta_3 = 169°$ and $r_{n4} = 171$ mm shown in Fig. 5.33. Note, sometimes a reasonable "scale" needs to be established if the dimensions are too big to fit on paper. We encourage the reader to watch the attached animation video of the piston cylinder engine mechanism.

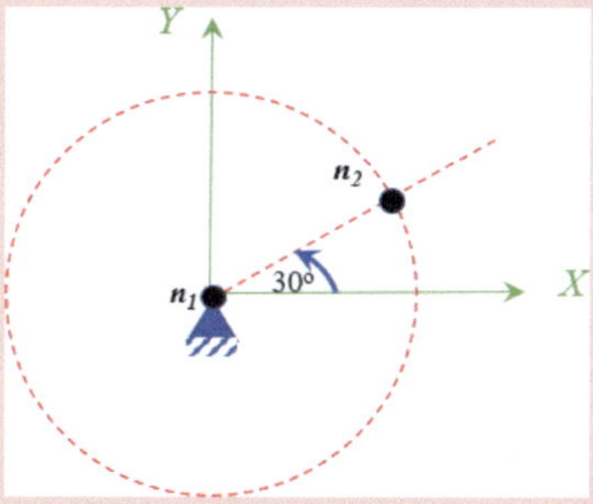

Fig. 5.31 Step 2 of piston–cylinder graphic position solution

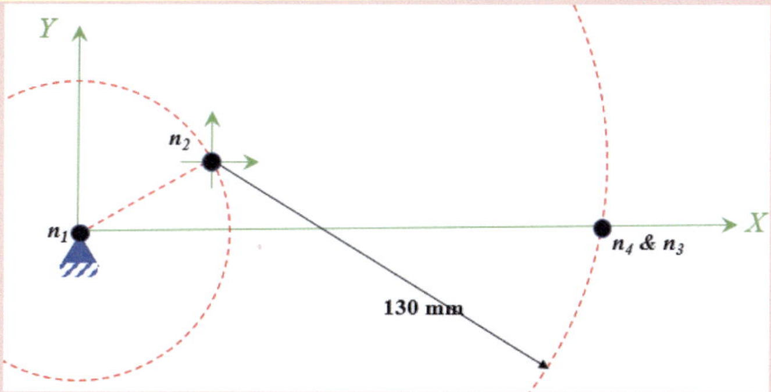

Fig. 5.32 Step 3 of piston–cylinder graphic position solution

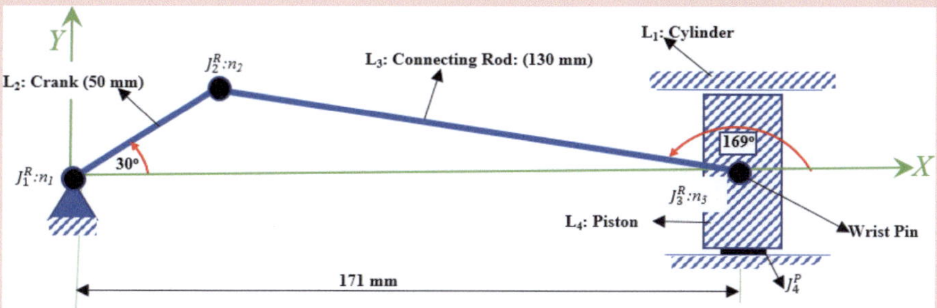

Fig. 5.33 Step 4 of piston–cylinder graphic position solution

5.6 Displacement Analytical Analysis

Analytical solutions provide extremely high accuracy, but they are not easy to develop. There is a need to be an expert and have the time to come up with the closed-form solutions. In this chapter, we are covering closed-loop 4-bar linkage models to develop analytical solutions. The closed-loop four-bar linkage model has one degree of freedom and thus we need only one link position to solve for the position of the other links comprising the mechanism. The best method to develop a closed solution is the "Vector Loop Method". The vector loop closes on itself making the sum of the vectors around the loop zero. The lengths of the vectors are the link lengths. In the upcoming sections, we are going to use the vector loop method.

5.6.1 Four-Bar Mechanism Analytical Analysis

Figure 5.34 shows the kinematic diagram of a 4-bar mechanism with four sketched vectors, one for each link. The loop vector can be written as

$$\vec{r}_{n2} + \vec{r}_{n3n2} = \vec{r}_{n4} + \vec{r}_{n3n4} \tag{5.8}$$

We have three variable angles θ_2, θ_3, and θ_4 plus the constant links' lengths r_{n2}, r_{n3n2}, r_{n3n4}, and r_{n4}. The vector can be expressed as

$$\begin{aligned}
\vec{r}_{n2} &= r_{n2}\cos(\theta_2)\hat{i} + r_{n2}\sin(\theta_2)\hat{j} \\
\vec{r}_{n3n2} &= r_{n3n2}\cos(\theta_3)\hat{i} + r_{n3n2}\sin(\theta_3)\hat{j} \\
\vec{r}_{n3n4} &= r_{n3n4}\cos(\theta_4)\hat{i} + r_{n3n4}\sin(\theta_4)\hat{j} \\
\vec{r}_{n4} &= r_{n4}\hat{i} + 0\hat{j}
\end{aligned} \tag{5.9}$$

Equating the terms in X-direction and Y-direction leads to

$$\begin{aligned}
r_{n2}\cos(\theta_2) + r_{n3n2}\cos(\theta_3) - r_{n3n4}\cos(\theta_4) &= r_{n4} \\
r_{n2}\sin(\theta_2) + r_{n3n2}\sin(\theta_3) - r_{n3n4}\sin(\theta_4) &= 0
\end{aligned} \tag{5.10}$$

The two above equations have three variables θ_2, θ_3, and θ_4. One of the variables must be known to calculate the other two variables. We are going to consider θ_2 as the known variable. The other two variables θ_3 and θ_4 are expressed in terms of θ_2 by:

$$A\tan^2\left(\frac{\theta_4}{2}\right) + B\tan\left(\frac{\theta_4}{2}\right) + C = 0$$

where

$$\begin{aligned}
A &= (1 - K_2)\cos(\theta_2) - K_1 + K_3 \\
B &= -2\sin(\theta_2) \\
C &= K_1 - (K_2 + 1)\cos(\theta_2) + K_3 \\
K_1 &= \frac{r_{n4}}{r_{n2}} \quad K_2 = \frac{r_{n4}}{r_{n3n4}} \quad K_3 = \frac{r_{n2}^2 - r_{n3n2}^2 + r_{n3n4}^2 + r_{n4}^2}{2r_{n2}r_{n3n4}} \\
\theta_3 &= \cos^{-1}\frac{r_{n3n4}\cos(\theta_4) + r_{n4} - r_{n2}\cos(\theta_2)}{r_{n3n2}}
\end{aligned} \tag{5.11}$$

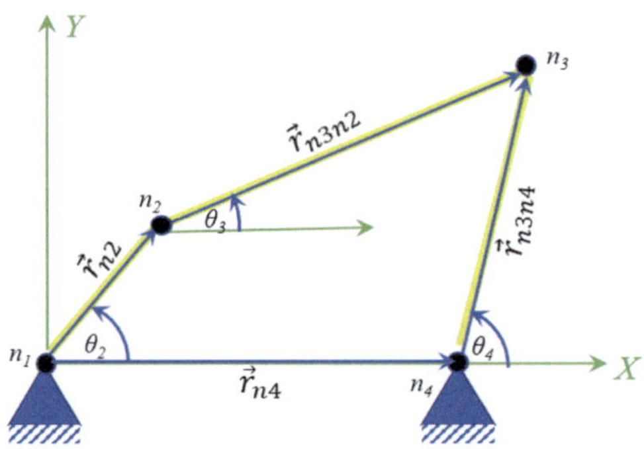

Fig. 5.34 Kinematic diagram and vector loop of four-bar Linkage

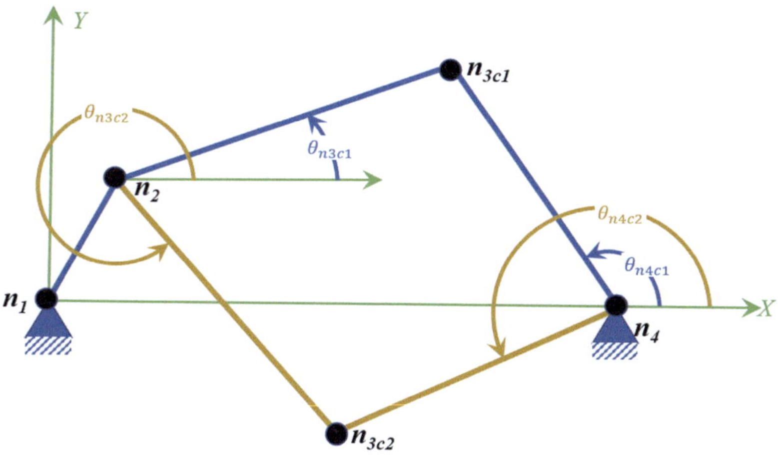

Fig. 5.35 Kinematic diagram of circuit 1 and circuit 2 solutions for four-bar mechanism

The derivation from Eqs. (5.10)–(5.11) is omitted. Please see Ref. [1] for a detailed derivation. Solving Eq. (5.11) provides two solutions: Circuit 1 and Circuit 2 as shown in Fig. 5.35. Simscape™ Multibody™ provides only one solution since it is a numerical simulation.

Example 5.5

For the shown 4-bar linkage model in Fig. 5.36, use the analytic method to find the values of θ_3 and θ_4, if a motor activates $\theta_2 = 60°$. Note, all dimensions are in cm.

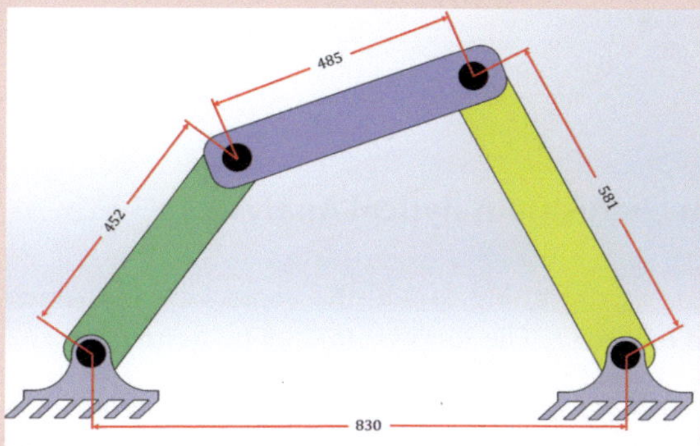

Fig. 5.36 Four-bar mechanism

Solution

Using Eq. (5.11), calculate the ratios K_1, K_2, and K_3:

$$K_1 = \frac{r_{n4}}{r_{n2}} = \frac{830}{452} = 1.836$$

$$K_2 = \frac{r_{n4}}{r_{n3n4}} = \frac{830}{581} = 1.429$$

$$K_3 = \frac{r_{n2}^2 - r_{n3n2}^2 + r_{n3n4}^2 + r_{n4}^2}{2r_{n2}r_{n3n4}} = \frac{452^2 - 485^2 + 581^2 + 830^2}{2 \times 452 \times 581} = 1.896$$

Using the value of $\theta_2 = 60°$, the parameters A, B, and C are:

$$A = (1 - K_2)\cos(\theta_2) - K_1 + K_3 = (1 - 1.429)\cos(60^o) - 1.836 + 1.896 = -0.155$$
$$B = -2\sin(60^o) = -1.732$$
$$C = K_1 - (K_2 + 1)\cos(\theta_2) + K_3 = 2.518$$

Using the values of A, B, and C to find the quadratic equation:

$$-0.155\tan^2\left(\frac{\theta_4}{2}\right) - 1.732\tan\left(\frac{\theta_4}{2}\right) + 2.518 = 0$$

Solving the quadratic equation:

$$\tan\left(\frac{\theta_4}{2}\right) = \frac{1.732 \pm \sqrt{1.732^2 - 4 \times -0.155 \times 2.518}}{2 \times -0.155} = 1.302 \text{ or } -12.468$$

$$\theta_4 = 2 \times \tan^{-1}(1.302) = 104.936° \; or \; \theta_4 = 2 \times \tan^{-1}(-12.468) = -170.829$$

$$\theta_3 = \cos^{-1}\frac{r_{n3n4}\cos(\theta_4) + r_{n4} - r_{n2}\cos(\theta_2)}{r_{n3n2}} = 20.5° \text{ or } 86.4°$$

5.6.2 Slider–Crank Mechanism Analytical Analysis

Similar to the 4-bar linkage, we are going to add the vectors on the slider–crank mechanism kinematic diagram as shown in Fig. 5.37. The loop vector can be written as:

$$\vec{r}_{n4} + \vec{r}_{n3n4} = \vec{r}_{n3n2} + \vec{r}_{n2} \tag{5.12}$$

From the kinematic constraints, $\theta_4 = 90°$. We have three unknowns: θ_2, θ_3, and L_1, plus the links' constant lengths r_{n2}, L_3 and L_4 are used to express the four above vectors as:

$$\begin{aligned} \vec{r}_{n2} &= r_{n2}\cos(\theta_2)\hat{i} + r_{n2}\sin(\theta_2)\hat{j} \\ \vec{r}_{n3n2} &= -r_{n3n2}\cos(\theta_3)\hat{i} - r_{n3n2}\sin(\theta_3)\hat{j} \\ \vec{r}_{n3n4} &= 0\hat{i} + r_{n3n4}\hat{j} \\ \vec{r}_{n4} &= r_{n4}\hat{i} + 0\hat{j} \end{aligned} \tag{5.13}$$

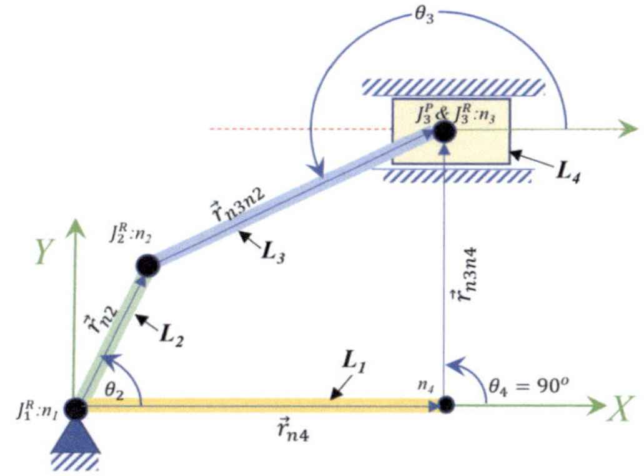

Fig. 5.37 Kinematic diagram and vector loop of slider–crank mechanism

Applying the terms in the x-direction and y-direction leads to:

$$r_{n2}\cos(\theta_2) - r_{n3n2}\cos(\theta_3) = r_{n4}$$
$$r_{n2}\sin(\theta_2) - r_{n3n2}\sin(\theta_3) = r_{n3n4} \tag{5.14}$$

The two above equations have three variables θ_2, θ_3, and r_{n4}. One of the variables must be known to evaluate the other two variables. We are going to consider θ_2 as the known variable. The other two variables θ_3 and L_1 are expressed in terms of θ_2 by:

$$\sin(\theta_3) = \frac{r_{n3n4} - r_{n2}\sin(\theta_2)}{r_{n3n2}} \tag{5.15}$$

The equivalent mathematical model comes to be an easy step to find the closed-loop solution. The term $\sin(\theta_3)$ provides two solutions: The right, or Circuit 1 solution, and the left, or circuit 2 solution (Fig. 5.38).

Example 5.6

In the piston-cylinder mechanism in Example 5.4:

1. Use the slider–crank analytical model to calculate:
   - The angle between the horizontal wrist axis and the connecting rod
   - The vertical distance from the wrist pin to the crankshaft center
2. Use the model-based technique to find:
   - The angle between the horizontal wrist axis and the connecting rod
   - The vertical distance from the wrist pin to the crankshaft center
3. Compare the graphical technique from Example 5.3 with the analytical and model-based technique.

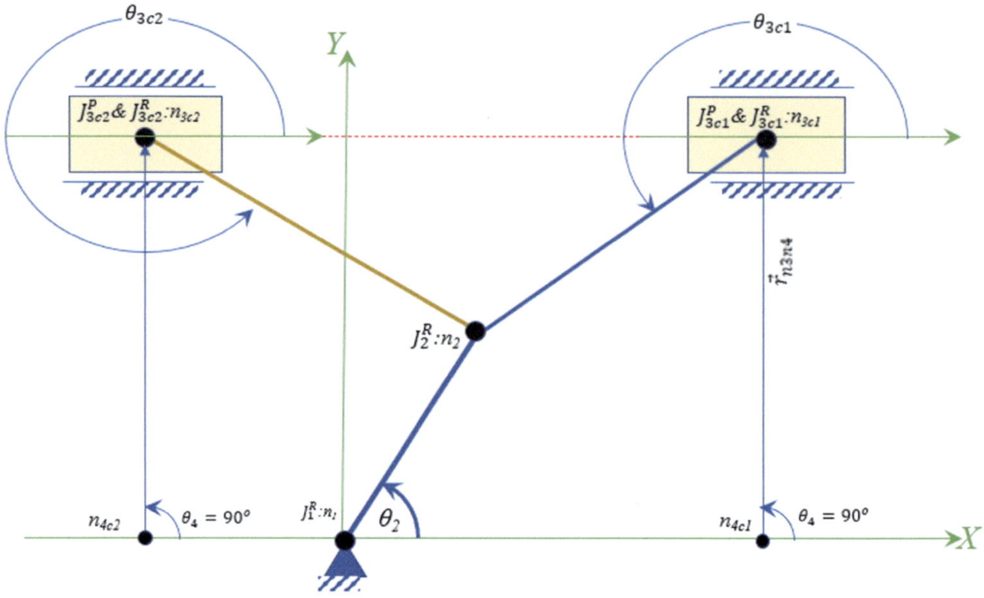

Fig. 5.38 Kinematic diagram of circuit 1 and circuit 2 solutions for slider–crank mechanism

Solution

1. **Analytical Solution**
   Comparing the kinematic diagram in Fig. 5.37 with the piston-cylinder mechanism in Fig. 5.29, we have the following values: $r_{n2} = 50\,\text{mm}$, $r_{n3n2} = 130\,\text{mm}$, $r_{n4n3} = 0\,\text{mm}$ and $\theta_2 = 30°$. The requirements are r_{n4}, and θ_3.
   Using Eq. (5.15):

$$\sin(\theta_3) = \frac{r_{n3n4} - r_{n2}\sin(\theta_2)}{r_{n3n2}} = \frac{0 - 50\sin(30°)}{130} = -0.1923$$

$$\theta_3 = \sin^{-1}(-0.1923) = 168.913° \text{ or } -11.087°$$

   We are going to pick the first circuit solution $\theta_3 = 168.913°$. Using Eq. (5.14):

$$r_{n4} = r_{n2}\cos(\theta_2) - r_{n3n2}\cos(\theta_3) = 50\cos(30°) - 130\cos(168.913°) = 170.87 \text{ mm}$$

2. **Model-Based Simscape™ Multibody™**
   Open "Slider_Crank_Example5p2_model.slx" model in Example 5.3 and save this model as "SinglePistonCylinder_Example5p6_model.slx". Open a new "SinglePistonCylinder_Example5p6_script.m" script file. Write the below commands in the file. In this new script and this new model, we are calling:

   - The ground link (cylinder and the crankshaft holder) L1
   - The crankshaft L2
   - The connecting rod L3
   - The piston L4

```
%This file to declare the parameters of a single piston-cylinder mechanism
%===========================================================================
%Clear all the variables and close all the figures
%============================================================
clc
clear all
close all
%===========================================================================
%Declare the given parameters
%===================================
r_n2 = 50/1000; %mm
r_n3n2 = 130/1000; %mm
theta2_deg = 30; %deg

%Calculation of parameters
%========================
L2_dimension = [1 0.1 0.1]*r_n2; %m
L3_dimension = [r_n3n2 0.1*r_n2 0.1*r_n2]; %m

Frame_L1_n1 = [0 0 0]; %m
Frame_L1_n4 = [0 0 0]; %m, initial assumed value

Frame_L2_n1 = [-r_n2/2 0 0]; %m
Frame_L2_n2 = [r_n2/2 0 0]; %m

Frame_L3_n2 = [-r_n3n2/2 0 0]; %m
Frame_L3_n3 = [r_n3n2/2 0 0]; %m

Frame_L4_n4 = [0 0 0]; %m
Frame_L4_n3 = [0 0 0]; %m

%===========================================================================
%Assumed parameters
%===============================
L4_dimensions = [0.2  0.2  0.1]*r_n2; %m
```

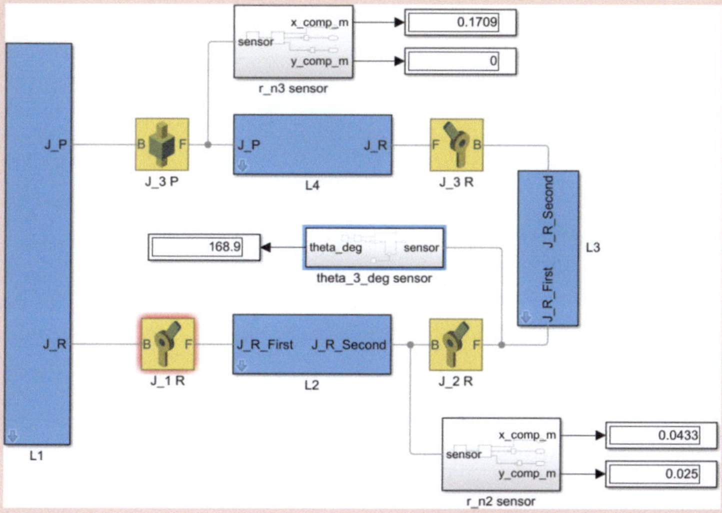

Fig. 5.39 Piston–cylinder engine Simscape™ Multibody™ model

Fig. 5.40 Mechanics explorer of piston–cylinder model from Simscape™ Multibody™ model of Example 5.3 and Example 5.6

After running the script "SinglePistonCylinder_Example5p6_script.m", run the "SinglePistonCylinder_Example5p6_model.slx" model as shown in Fig. 5.39. You will see the piston–cylinder mechanism on Mechanics Explorer as shown in Fig. 5.40. You can record the measured variable from the display.

3. **Comparison**

List the results from the graphical, analytical, and model-based. It is obvious that the three models are providing approximately the same results.

	Graphical	Analytical	Model-based
θ_3(°)	169°	168.9125°	168.9125°
r_{n4}(mm)	171	170.8748	170.8748

5.6.3 Inverted Slider–Crank Mechanism Analytical Analysis

Figure 5.41 shows the vector loop defined on the kinematic diagram of one type of the inverted slider–crank mechanism. The loop vector can be written as:

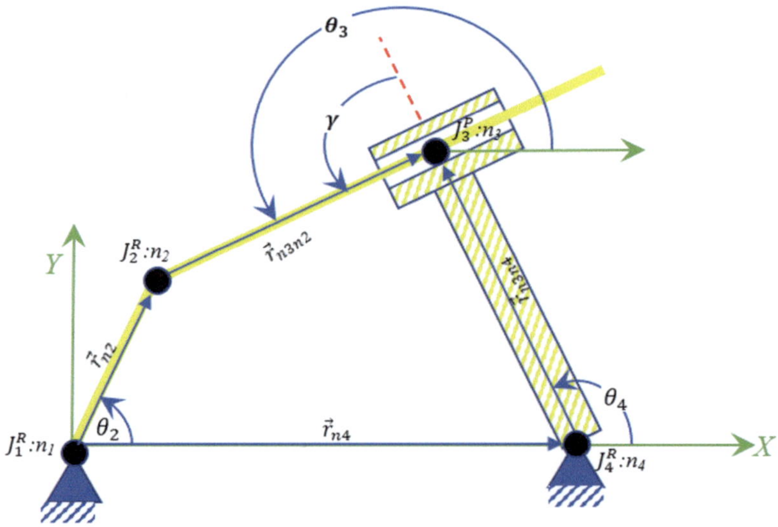

Fig. 5.41 Kinematic diagram and vector loop of inverted slider–crank with fixed rocker linkage

$$\vec{r}_{n4} + \vec{r}_{n3n4} = \vec{r}_{n3n2} + \vec{r}_{n2} \tag{5.16}$$

There is a relationship $\vartheta_3 = \vartheta_4 + \gamma$. We end up with three unknowns: θ_2, γ, and L_3, and the links' constant lengths L_1, r_{n2} and L_4 are used to express the four above vectors as:

$$\begin{aligned}
\vec{r}_{n2} &= r_{n2}\cos(\theta_2)\hat{i} + r_{n2}\sin(\theta_2)\hat{j} \\
\vec{r}_{n3n2} &= -r_{n3n2}\cos(\theta_3)\hat{i} - r_{n3n2}\sin(\theta_3)\hat{j} \\
\vec{r}_{n3n4} &= r_{n3n4}\cos(\theta_4)\hat{i} + r_{n3n4}\sin(\theta_4)\hat{j} \\
\vec{r}_{n4} &= r_{n4}\hat{i} + 0\hat{j} \\
\theta_3 &= \theta_4 + \gamma
\end{aligned} \tag{5.17}$$

Applying the terms in the *x*-direction and *y*-direction leads to:

$$\begin{aligned}
r_{n2}\cos(\theta_2) - r_{n3n2}\cos(\theta_3) &= r_{n3n4}\cos(\theta_4) + r_{n4} \\
r_{n2}\sin(\theta_2) - r_{n3n2}\sin(\theta_3) &= r_{n3n4}\sin(\theta_4)
\end{aligned} \tag{5.18}$$

The two equations above have three variables: θ_2, γ, and L_3. One of the variables must be known to evaluate the other two. We are going to consider θ_2 as the known variable. The other two variables γ and L_3 are expressed in terms of θ_2 by:

$$S\tan^2\left(\frac{\theta_4}{2}\right) + T\tan\left(\frac{\theta_4}{2}\right) + U = 0$$

where

$$\begin{aligned}
&S = R - Q \quad T = 2P \quad U = Q + R \\
&P = r_{n2}\sin(\theta_2)\sin(\gamma) + (r_{n2}\cos(\theta_2) - r_{n4})\cos(\gamma) \\
&Q = -r_{n2}\sin(\theta_2)\cos(\gamma) + (r_{n2}\cos(\theta_2) - r_{n4})\sin(\gamma) \\
&R = r_{n3n4}\sin(\gamma)
\end{aligned} \tag{5.19}$$

The term $\tan(\theta_4/2)$ has two solutions: Circuit 1 and Circuit 2. After that, we can use Eq. (5.18) to find r_{n3n2}.

In addition to Fig. 5.41, there is another configuration called *inverted slider–crank with a fixed crank* as shown in Fig. 5.42. This configuration is commonly used in agriculture, construction, and many other industries. The kinematic diagram shown in Fig. 5.42 has a triangular shape n_1–n_2–n_4. When θ_2 increases in the counterclockwise direction, the length r_{n3n2} ncreases and causes θ_3 to increase as well.

Instead of the vector loop, we are going to use the trigonometric formula applied on the triangle shape n_1–n_2–n_4 as shown in Fig. 5.43 to calculate r_{n3n2} from:

$$\begin{aligned}
(r_{n3n2} + r_{n3n4})^2 &= r_{n2}^2 + r_{n4}^2 - 2r_{n2}r_{n4}\cos(\theta_2 - \delta) \\
r_{n3n2} &= \sqrt{r_{n2}^2 + r_{n4}^2 - 2r_{n2}r_{n4}\cos(\theta_2 - \delta)} - r_{n3n4}
\end{aligned} \tag{5.20}$$

where δ is a constant geometric angle that is defined as shown in Fig. 5.42. Also, we can use the trigonometric relation to find θ_3 from:

$$\begin{aligned}
r_{n2}^2 &= (r_{n3n2} + r_{n3n4})^2 + r_{n4}^2 - 2 \times (r_{n3n2} + r_{n3n4}) \times r_{n4}\cos(180° - \theta_3 + \delta) \\
\theta_3 &= 180° - \cos^{-1}\left(\frac{(r_{n3n2} + r_{n3n4})^2 + r_{n4}^2 - r_{n2}^2}{2 \times (r_{n3n2} + r_{n3n4}) \times r_{n4}}\right) + \delta
\end{aligned} \tag{5.21}$$

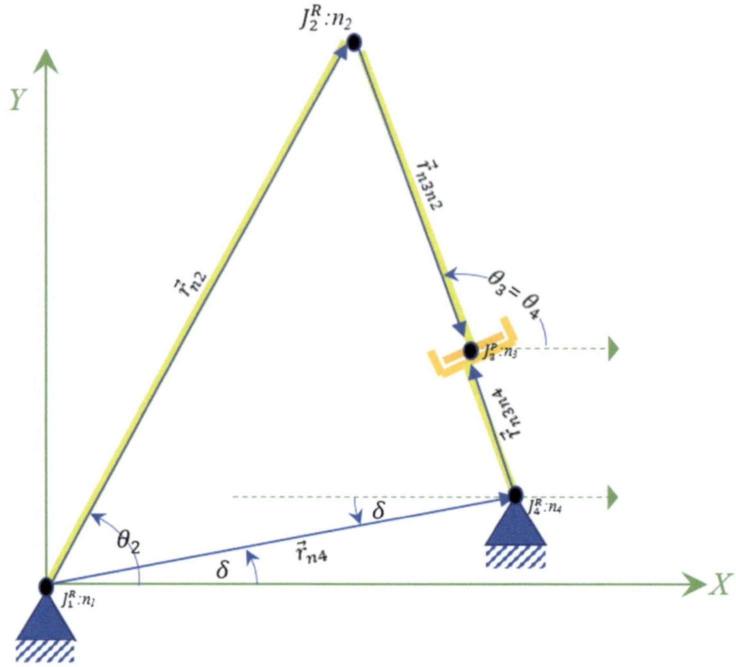

Fig. 5.42 Kinematic diagram and vector loop of inverted slider–crank with a fixed crank

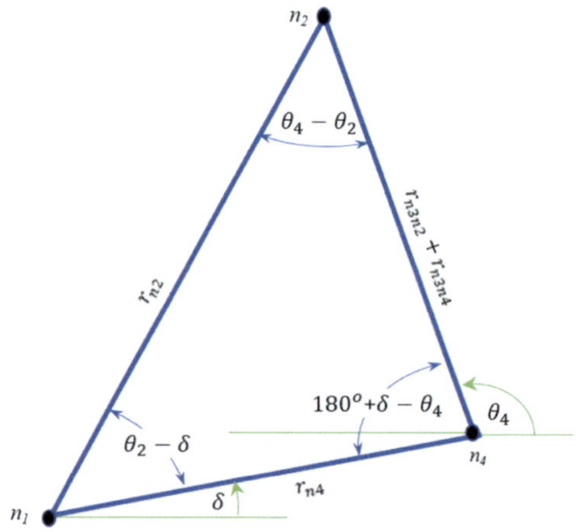

Fig. 5.43 Triangular shape of slider–crank with fixed crank mechanism

Engineering Tips—Triangular Geometry

An angle can be a part of a triangle as shown in Fig. 5.44 where the angle θ is calculated using the law of cosines as

$$O^2 = A^2 + B^2 - 2 \times A \times B \cos(\theta)$$

where the length of the two adjacent sides with lengths ***A*** and ***B*** plus the opposite side with length ***O***.

The law of sines is another rule to be used in triangular geometry as:

$$\frac{\sin(\theta)}{O} = \frac{\sin(\alpha)}{A} = \frac{\sin(\beta)}{B}$$

Note that θ in the triangle does not have a direction as in Fig. 5.44 and it goes from 0° to 180°. The MATLAB® command for inverse cosine in degrees is "acosd" and inverse sine in degrees is "asind". It is recommended to read the MATLAB® help files by typing "help acosd" and/or "help asind" in the MATLAB® command window.

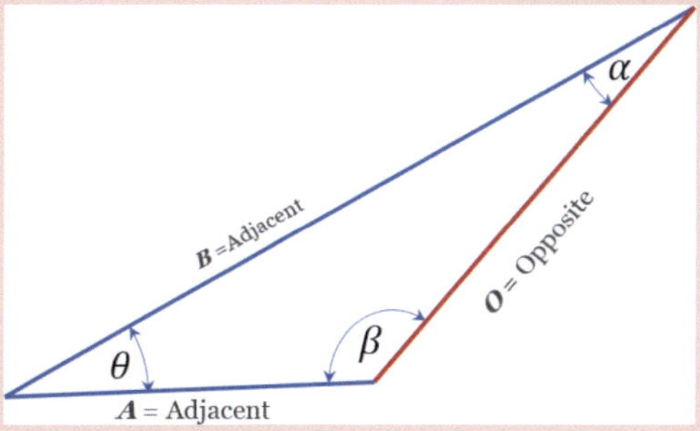

Fig. 5.44 Geometric triangular shape

Example 5.7

For the hydraulic gate opener in Fig. 5.45 with the given dimensional parameters:

1. Sketch the kinematic diagram. If the hydraulic gate opener is treated as an inverted slider–crank, sketch the vector loop on top of the kinematic diagram.
2. Sketch the input–output displacement diagram of the hydraulic door opener mechanism.
3. If the door is opened with a 45° angle, use the analytical equations to find the piston length and the piston angle with respect to the frame.

Solution

1. **Kinematic Diagram and Vector Loop**
   Figure 5.46 shows the kinematic diagram with the vector loops. The frame is L_1, the door is L_2, the piston is L_3, and the cylinder is L_4. As vector loop definition, the door is the yellow triangle with a vector $\vec{r}_{n2}$ from the joint J_1^R to joint J_2^R. The piston is the vector $\vec{r}_{n3n2}$ from the joint J_2^R to joint J_3^P. The cylinder is the vector $\vec{r}_{n3n4}$ from the joint J_4^R to joint J_3^P. The frame is the vector $\vec{r}_{n3n2}$ from joint J_1^R to joint J_4^R.

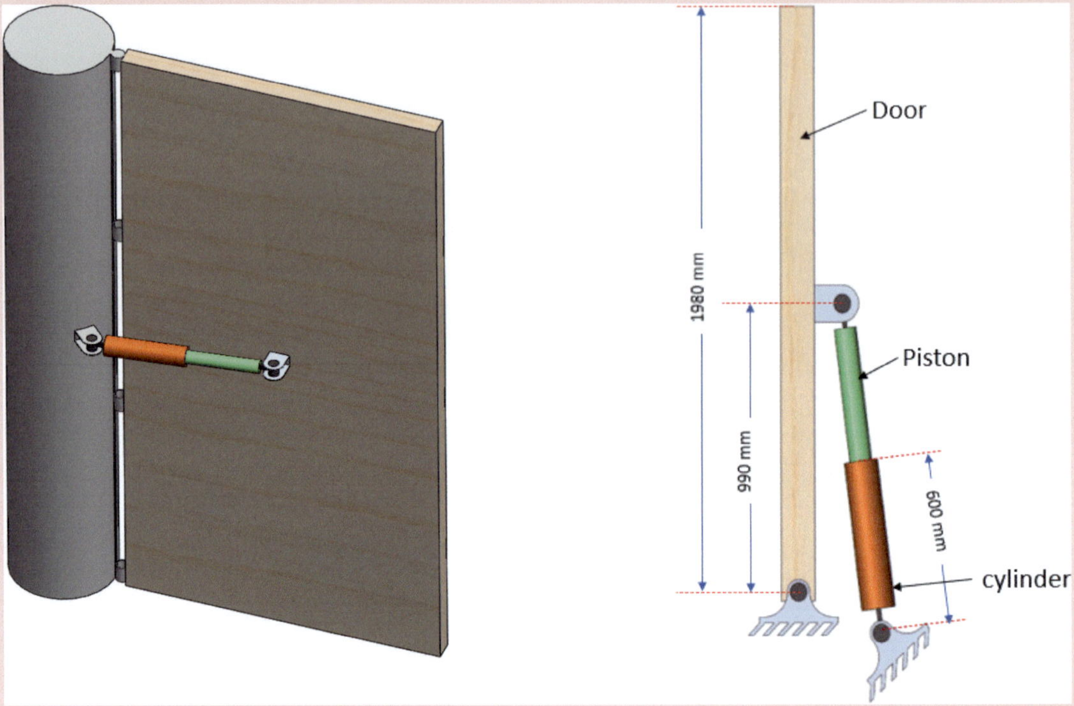

Fig. 5.45 Hydraulic gate opener. Designed by the author and implemented by both the author and Pranav Dhola. Courtesy of Pranav Dhola

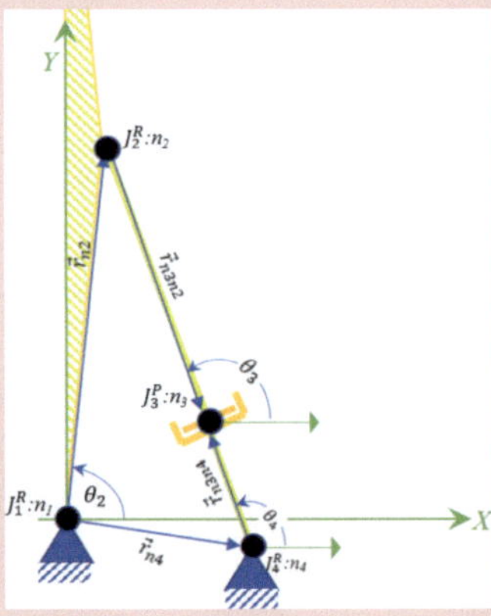

Fig. 5.46 Kinematic diagram with vector loop of hydraulic door opener

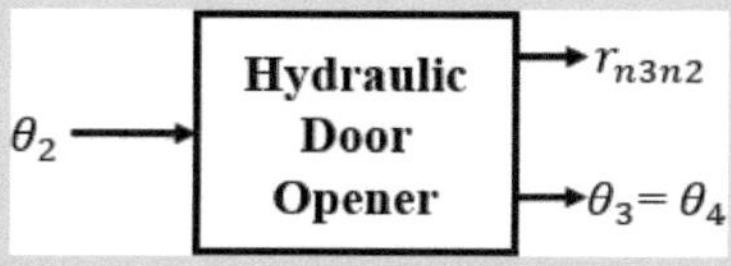

Fig. 5.47 System diagram of hydraulic door opener mechanism position

2. **System Diagram**
   The known parameters are r_{n2}, r_{n3n4}, x_{n4}, and y_{n4}. The known variable (input) is θ_2. The unknown variables (outputs) are r_{n2n3} and θ_3. Figure 5.47 shows the block diagram of the input–output displacement system diagram.

3. **Analytical Solution**
   Use the Eqs. 5.20 and 5.21

$$r_{n4} = \sqrt{x_{n4}^2 + y_{n4}^2} = \sqrt{125^2 + (-125)^2} = 176.77 \text{ mm}$$

$$\delta = \tan^{-1}(y_{n4}/x_{n4}) = \tan^{-1}(-125/125) = -45^\circ$$

$$r_{n3n2} = \sqrt{r_{n2}^2 + r_{n4}^2 - 2 r_{n2} r_n \cos(\theta_2 - \delta)} - r_{n3n4}$$

$$= \sqrt{990^2 + 176.77^2 - 2 \times 990 \times 176.77 \cos(90^\circ)} - 600 = 405.66 \text{ mm}$$

$$\theta_3 = 180^\circ - \cos^{-1}\left(\frac{(r_{n3n2} + r_{n3n4})^2 + r_{n4}^2 - r_{n2}^2}{2 \times (r_{n3n2} + r_{n3n4}) \times r_{n4}}\right) + \delta$$

$$= 180^\circ - \cos^{-1}\left(\frac{(405.66 + 600)^2 + 176.77^2 - 990^2}{2 \times (405.66 + 600) \times 176.77}\right) - 45^\circ = 55.124^\circ$$

4. **Using Model-Based Simscape™ Multibody™ to Calculate r_{n3n2} and θ_3**
   We are going to follow the same steps as in Example 5.3. First, write the script file "HydraulicDoorOpener_Example5p7_script.m". We are going to call:

   - The ground link (the door frame) L_1 that's a ground with two revolute joints mask block.
   - The door L_2 that's an extrusion with two revolute joints mask block.
   - The piston L_3 that's a brick with one prismatic joint and revolute joint mask block.
   - The cylinder L_4 that's a brick with one prismatic joint and revolute joint mask block.

   The door is modeled as a rectangular shape with dimension $2r_{n2} \times 0.05r_{n2}$ with two nodes: n_1 (Frame_L2_n1) and n_2 (Frame_L2_n2).

```
%This file to declare the parameters of a hydraulic door opener mechanism
%==============================================================
%Clear all the variables and close all the figures
%====================================
clc
clear all
close all
%====================================
%Declare the given parameters
%================================
x_n4 = 125/1000; %mm
y_n4 = 125/1000; %m
r_n2 = 990/1000; %m
r_n3n4 = 600/1000; %mm
r_n3n2 = 600/1000; %mm
theta2_deg = 45; %deg

%Calculation of parameters
%=========================
L2_crosssection =[0        0;
            2*r_n2   0;
            2*r_n2   0.05*r_n2;
            0        0.05*r_n2]; %m, this cross section represents the door as 4 points in a counterclockwise
L3_dimension = [r_n3n2 0.05*r_n2 0.1*r_n2]; %m
L4_dimension = [r_n3n4 0.1*r_n2 0.1*r_n2]; %m

Frame_L1_n1 = [0 0 0]; %m
Frame_L1_n4 = [x_n4 -y_n4 0]; %m

Frame_L2_n1 = [0 0 0]; %m
Frame_L2_n2 = [r_n2 0 0]; %m

Frame_L3_n2 = [r_n3n2/2 0 0]; %m
Frame_L3_n3 = [-r_n3n2/2 0 0]; %m

Frame_L4_n4 = [-r_n3n4/2 0 0]; %m
Frame_L4_n3 = [r_n3n4/2 0 0]; %m
```

Secondly, build the model "HydDoorOpenerModel_Example5p7_model.slx" as shown in Fig. 5.48. We plug in the parameter of each block similar to Example 5.3. Finally, go to J_R1 > State Targets > Specify Position Target > Value and add "Theta2_deg" in "deg".

Run the "HydraulicDoorOpener_Example5p7_script.m" script file. Run the "HydraulicDoor.

Opener_Example5p7_model.slx" model. You can see the hydraulic door opener mechanism on Mechanics Explorer as shown in Fig. 5.49. You will get theta3_model_deg = 55.1241 and r_n3n2_model_mm = 405.66.

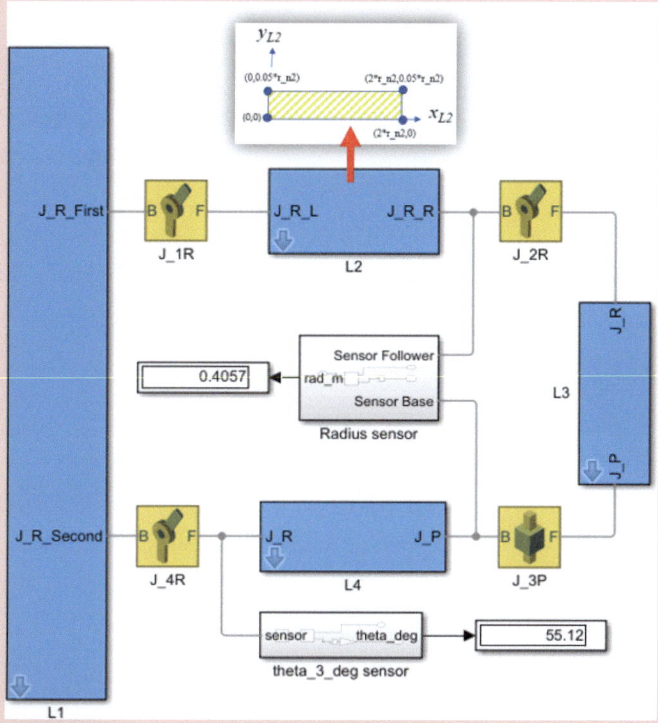

Fig. 5.48 Simscape™ Multibody™ model of hydraulic door opener

Fig. 5.49 Mechanics explorer of hydraulic door opener model

5.6.4 Five and Beyond Bar Mechanisms Analytical Analysis

Engineers can still use the vector loop method for any number of links in a closed-loop configuration. Increasing the number of links leads to more simultaneous equations. Then, obtaining an analytical solution is not an easy job. Many researchers are trying to come up with analytical solutions, but this process continues to be a challenge for engineers.

For example, the five-bar mechanism shown in Fig. 5.50 can be described using the loop vector form as:

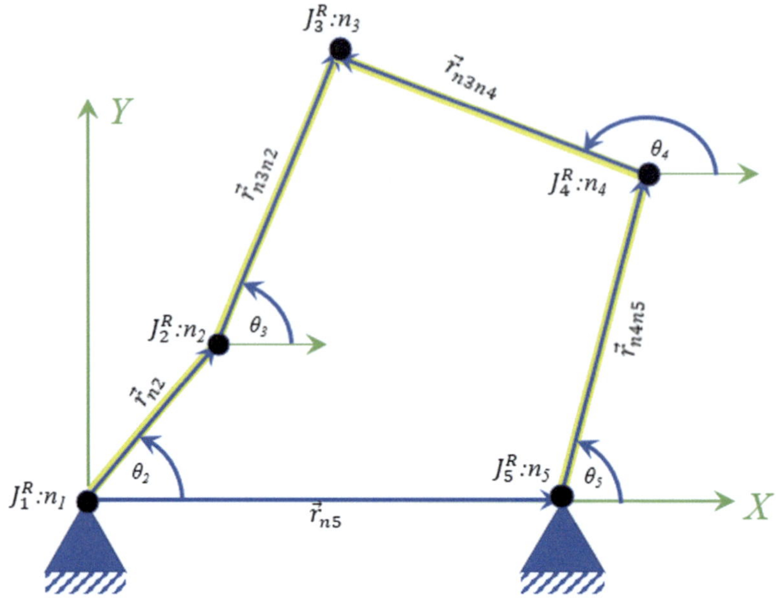

Fig. 5.50 Kinematic diagram of a five-bar mechanism

$$\vec{r}_{n2} + \vec{r}_{n3n2} = \vec{r}_{n5} + \vec{r}_{n4n5} + \vec{r}_{n3n4} \tag{5.22}$$

We have four variable angles θ_2, θ_3, θ_4, and θ_5 plus the constant links' lengths r_{n2}, r_{n3n2}, r_{n3n4}, r_{n4n5}, and r_{n5}. The closed chain 5-bar mechanism vectors are

$$\begin{aligned}
\vec{r}_{n2} &= r_{n2}\cos(\theta_2)\hat{i} + r_{n2}\sin(\theta_2)\hat{j} \\
\vec{r}_{n3n2} &= r_{n3n2}\cos(\theta_3)\hat{i} + r_{n3n2}\sin(\theta_3)\hat{j} \\
\vec{r}_{n3n4} &= r_{n3n4}\cos(\theta_4)\hat{i} + r_{n3n4}\sin(\theta_4)\hat{j} \\
\vec{r}_{n4n5} &= r_{n4n5}\cos(\theta_5)\hat{i} + r_{n4n5}\sin(\theta_5)\hat{j} \\
\vec{r}_{n5} &= r_{n5}\hat{i} + 0\hat{j}
\end{aligned} \tag{5.23}$$

To solve Eq. (5.23), the engineer should first specify the input signals and the output signals. Then a derivation should be conducted to express the outputs as functions of the inputs. Another way to solve Eq. (5.23) is to use a numerical solution. If we build a Simscape™ Multibody™ Model, the model solver will take care of finding the solution. This relieves the engineer from having to bother with the derivation or with the numerical solution steps. They can instead focus on the design and analysis rather than how to solve the equations.

The conclusion from this 5-bar mechanism is that using Simscape™ Multibody™ Model develops a physical-based solution without any sophisticated computational or mathematical models. Similar conclusions can be generalized for any complicated mechanisms or machines.

5.7 Node of Interest Position

A node of interest is any arbitrarily selected point on the linkage model. After developing the solution of the input–output mathematical model, vector calculus (addition and subtraction) is the key to finding the position of any node of interest. The node of interest can be located on the vector loop or outside the vector loop. For example, the 4-bar mechanism in Fig. 5.51 has three nodes of interest at n_2, n_3, and n_5 which can be calculated using vector analysis as:

$$\begin{aligned}\vec{r}_{n2} &= r_{n2}\cos(\theta_2)\hat{i} + r_{n2}\sin(\theta_2)\hat{j}\\ \vec{r}_{n3} &= \vec{r}_{n3n2} + \vec{r}_{n2} = r_{n3n2}\cos(\theta_3)\hat{i} + r_{n3n2}\sin(\theta_3)\hat{j} + r_{n2}\cos(\theta_2)\hat{i} + r_{n2}\sin(\theta_2)\hat{j}\\ \vec{r}_{n5} &= \vec{r}_{n5n2} + \vec{r}_{n2} = r_{n5n2}\cos(\theta_5)\hat{i} + r_{n5n2}\sin(\theta_5)\hat{j} + r_{n2}\cos(\theta_2)\hat{i} + r_{n2}\sin(\theta_2)\hat{j}\end{aligned} \tag{5.24}$$

where

$$\theta_5 = \theta_3 + \gamma$$

Then

$$\begin{aligned}\vec{r}_{n5} =& \vec{r}_{n5n2} + \vec{r}_{n2} = r_{n5n2}\cos(\theta_3+\gamma)\hat{i} + r_{n5n2}\sin(\theta_3+\gamma)\hat{j} + r_{n2}\cos(\theta_2)\hat{i} + r_{n2}\sin(\theta_2)\hat{j}\\ =& \{r_{n5n2}\cos(\theta_3+\gamma) + r_{n2}\cos(\theta_2)\}\hat{i} + \{r_{n5n2}\sin(\theta_3+\gamma) + r_{n2}\sin(\theta_2)\}\hat{j}\end{aligned}$$

The angle γ is a constant since the link L_3 is a rigid body. The nodes n_2 and n_3 are joint points. The node n_5 is an end-effector node.

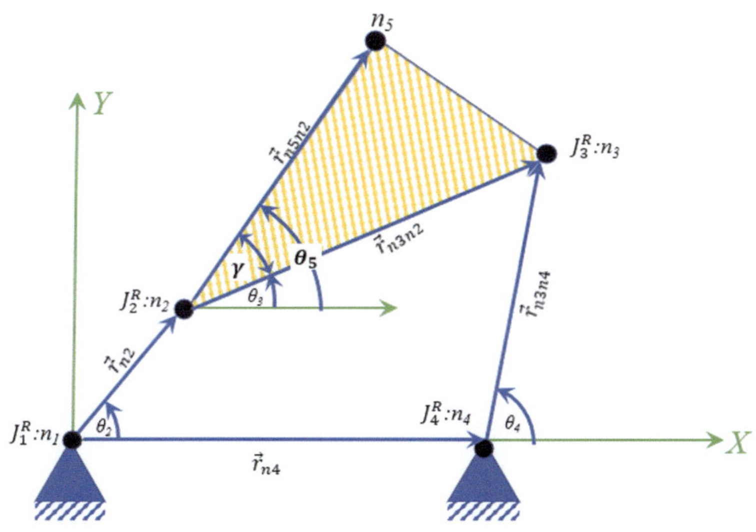

Fig. 5.51 Four-bar mechanism with point of interest

Example 5.8

Revisiting Example 5.3, if the 4-bar linkage model is modified as shown in Fig. 5.52.

1. Use the analytic solution to find the positions of n_2, n_3, and n_5, if a motor activates θ_2 to be at 60°.
2. Develop a Simscape™ Multibody™ model. Then run the model to find the positions of n_2, n_3, and n_5 if a motor activates θ_2 to be at 60°.
3. Compare the results from the Simscape™ Multibody™ model with the analytical solution.
4. Sketch the input–output displacement system diagram if the position of n_5 is considered as the output.

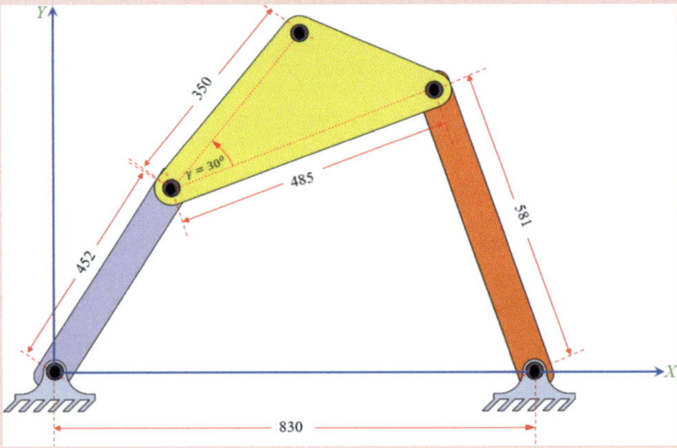

Fig. 5.52 Nodes of interest example

Solution

1. **Analytical Solution**
   Recall the solution of Example 5.3. We can summarize the results as:
   Input: $\theta_2 = 60^\circ$
   Outputs: $\theta_3 = 20.510^\circ$, $\theta_4 = 104.936^\circ$
   Using Eq. (5.24):

$$\vec{r}_{n2} = r_{n2}\cos(\theta_2)\hat{i} + r_{n2}\sin(\theta_2)\hat{j} = 452 \times \cos(60^\circ)\hat{i} + 452 \times \sin(60^\circ)\hat{j} = 226\hat{i} + 391\hat{j}$$

$$\begin{aligned}\vec{r}_{n3} &= r_{n3n2}\cos(\theta_3)\hat{i} + r_{n3n2}\sin(\theta_3)\hat{j} + r_{n2}\cos(\theta_2)\hat{i} + r_{n2}\sin(\theta_2)\hat{j}\\ &= 226\hat{i} + 391\hat{j} + 485\cos(20.51^\circ)\hat{i} + 485\sin(20.51^\circ)\hat{j} = 680\hat{i} + 561\hat{j}\end{aligned}$$

$$\begin{aligned}\vec{r}_{n5} &= \{r_{n5n2}\cos(\theta_3 + \gamma) + r_{n2}\cos(\theta_2)\}\hat{i} + \{r_{n5n2}\sin(\theta_3 + \gamma) + r_{n2}\sin(\theta_2)\}\hat{j}\\ &= \{350 \times \cos(20.51^\circ + 30^\circ) + 226\}\hat{i} + \{350\sin(20.51^\circ + 30^\circ) + 391\}\hat{j} = 448\hat{i} + 661\hat{j}\end{aligned}$$

2. **Model-Based Simscape™ Multibody™ Solution**
   Following Example 5.3, First, write the script file "Example_P5p8_script.m". We are going to call:

- The 830 mm ground link L_1 that's a ground with two revolute joints mask block
- The 452 mm bar L_2 that's a brick with two revolute joints mask block
- The triangular plate L_3 that's an extrusion with two revolute joints mask block
- The 581 mm bar L_4 that's a brick with two revolute joints mask block

The rectangular plate is modeled as a triangular shape with dimensions:

$$\begin{aligned}[n_2 n_3 n_5] &= [00;\ r_{n3n2}\ 0;\ r_{n5n2}\cos(\gamma)\ \ r_{n5n2}\sin(\gamma)] \\ &= [00;\ 0.4850;\ 0.350\ \ \cos d(30^\circ)\quad 0.350\sin d(30^\circ)] \\ &= [00;\ 0.4850;\ 0.3030.107]\mathrm{m}\end{aligned}$$

The rectangular plate has three nodes: n_2, n_3, and n_5.

```
%This file to declare the parameters of example 5.8 mechanism
%===================================================================================
%Clear all the variables and close all the figures
%================================================
clc
clear all
close all
%===================================================================================
%Declare the given parameters
%=========================
r_n2 = 452/1000; %m
r_n3n2 = 485/1000; %m
r_n3n4 = 581/1000; %m
r_n4   = 830/1000; %m
r_n5n2 = 350/1000; %m
gamma_deg = 30; %deg
theta2_deg = 60; %deg

%Calculation of parameters
%=========================
L2_dimension = [1 0.1 0.1]*r_n2; %m
L3_crosssection = [0  0;  r_n3n2  0;r_n3n2*cosd(gamma_deg) r_n3n2*sind(gamma_deg)]; %m corss section as 3 point counter clockwise
L4_dimension =[r_n3n4 0.1*r_n2 0.1*r_n2]; %m

Frame_L1_n1 = [0 0 0]; %m
Frame_L1_n4 = [r_n4 0 0]; %m, intial assumed value

Frame_L2_n1 = [-r_n2/2 0 0]; %m
Frame_L2_n2 = [r_n2/2 0 0]; %m
Frame_L2_n5 = [r_n5n2*cosd(gamma_deg) r_n5n2*sind(gamma_deg)  0];

Frame_L3_n2 = [0 0 0]; %m
Frame_L3_n3 = [r_n3n2 0 0]; %m

Frame_L4_n4 = [-r_n3n4/2 0 0]; %m
Frame_L4_n3 = [r_n3n4/2 0 0];%m
```

Secondly, build the model "Example_P5p8_model.slx" as shown in Fig. 5.53. We plug in the parameter of each block similar to Example 5.3.Finally, go to J_R1 > State Targets > Specify Position Target > Value and add "Theta2_deg" in "deg".

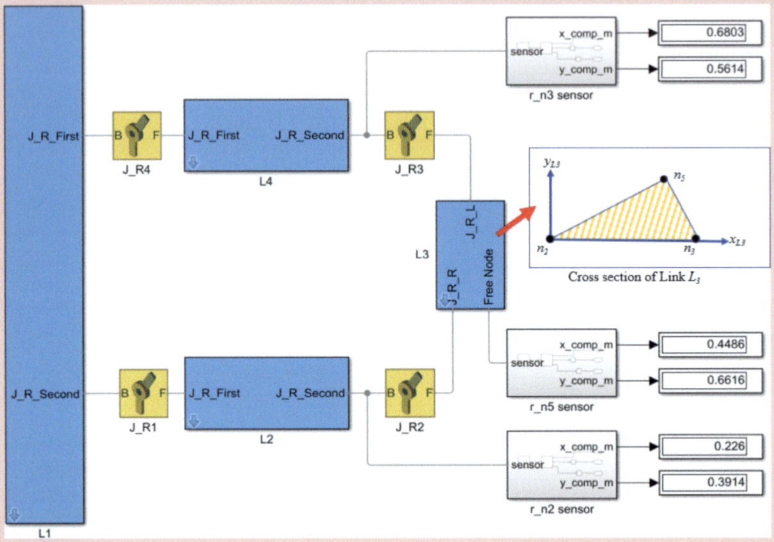

Fig. 5.53 Simscape™ Multibody™ model of Example 5.8

Run the "Example_P5p8_script.m" script file. Run the "Example_P5p7_model.slx" model. You can see Example 5.8 mechanism on Mechanics Explorer as shown in Fig. 5.49. You will get the values listed in Table 5.3 from the displays.

Table 5.3 Simscape™ Multibody™ output variables

$x_{n2} = 226$ cm	$y_{n2} = 391$ cm	$\vec{r}_{n2} = 226\hat{i} + 391\hat{j}$
$x_{n3} = 680$ cm	$y_{n3} = 561$ cm	$\vec{r}_{n3} = 680\hat{i} + 561\hat{j}$
$x_{n5} = 448$ cm	$y_{n5} = 616$ cm	$\vec{r}_{n5} = 448\hat{i} + 661\hat{j}$

3. **Comparison**
   The two methods yield the same answer.

4. **System Block Diagram**
   Figure 5.54 shows the displacement system diagram of the mechanism in Fig. 5.52. The input is the rotational angle displacement θ_2. The outputs are the two displacements x_{n5} and y_{n5}.

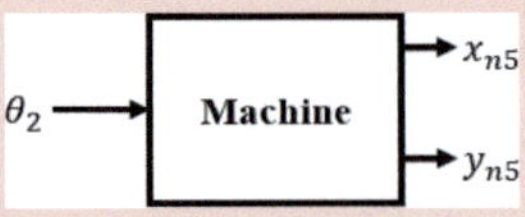

Fig. 5.54 Input–output displacement system diagram Problem 5.5

5.8 Limiting Position Analysis

Definition

Toggle Position:

The toggle positions are determined by the collinearity of two of the moving links. [1]

Recall the 4-bar mechanism as a non-Grashof case as shown in Fig. 5.55-Case I. None of the links can make a 360° rotation, or one cycle. If a motor is connected to rotate Link L_2 counterclockwise, there is a position where Link L_3 and Link L_4 are going to be collinear which means that L_3 and L_4 are acting as one link as shown in Fig. 5.55-Case II. If the motor keeps trying to rotate Link L_2, there are two options: (1) breaking the linkage model as shown in Fig. 5.55-Case III, or (2) stuck in the position as shown in Fig. 5.55-Case II that we call a toggle position. Even in the crank–rocker Grashof mechanism, there is a toggle position.

The toggle position has a triangle shape n_1-n_2-n_4. We can apply a trigonometric relationship to the triangle n_1-n_2-n_4 to calculate of $\theta_{2toggle}$ from:

$$\begin{aligned}(r_{n3n2}+r_{n4})^2 &= r_{n2}^2+r_{n4}^2-2r_{n2}r_{n4}\cos\left(\theta_{2\text{toggle}}\right)\\ \theta_{2\text{toggle}} &= \cos^{-1}\left(\frac{r_{n2}^2+r_{n4}^2-(r_{n3n2}+r_{n4})^2}{2r_{n2}r_{n4}}\right)\end{aligned} \tag{5.25}$$

The issue with the formula in Eq. (5.19) is the restriction of a 4-bar mechanism with θ_2 as an input. Any more complicated mechanisms or any other input will need a new derivation. Instead, using a Simscape™ Multibody™ model provides the toggle position for any mechanism with any input. To do so, simply provide the maximum value to the starting position of the actuated joint and run the simulation. The resultant measured displacement at the actuated joint is the toggle position.

Example 5.9

Calculate the toggle angle of the mechanism in Example 5.5 using the analytical solution and the Simscape™ Multibody™ model.

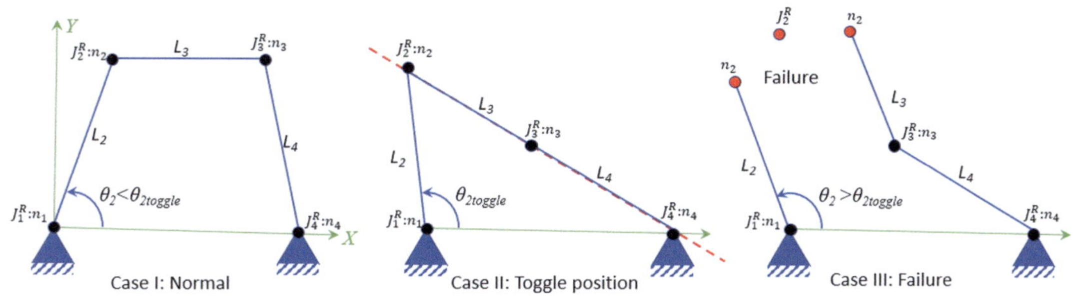

Fig. 5.55 Non-Grashof mechanism

Solution

1. **Analytical Solution**
   From Eq. (5.25)

$$\theta_{2\text{toggle}} = \cos^{-1}\left(\frac{r_{n2}^2 + r_{n4}^2 - (r_{n3n2} + r_{n4})^2}{2r_{n2}r_{n4}}\right)$$

$$= \cos^{-1}\left(\frac{452^2 + 830^2 - (485 + 581)^2}{2 \times 452 \times 830}\right) = 108.9^\circ \text{ or } -108.9^\circ$$

2. **Model-Based**
   - Go to the developed model in Example 5.8.
   - Double-click on the block J_1R and set "Specify Position Target" to "179°".
   - Add a revolute sensor to J_1R.
   - After you run the model, the sensor reading is 108.9°.

Note: Unlike the Analytical Equation, the Model-Based Provides Only One Value.

5.9 Coupler Curve

The Grashof crank–rocker 4-bar mechanism shown in Fig. 5.56 has three common components names: crank Link L_2, coupler Link L_3, and rocker Link L_4. Assume, a 4-bar mechanism with the following ratios $r_{n2}/r_{n4}=0.2$, $r_{n3n2}/r_{n4}=0.5$ and $r_{n3n4}/r_{n4}=0.75$. A DC motor is connected to rotate the crank counterclockwise with a complete cycle 360° from joint J_1^R. The input is the angle θ_2 and the output is the x–y position of node n_5. In Fig. 5.56, n5 is located at $\gamma=30^\circ$ with $r_{n5n2}/r_{n3n2}=1/2$. The shape of the n_5 trajectory is shown in Fig. 5.56 which looks like a "Pseudo Ellipse".

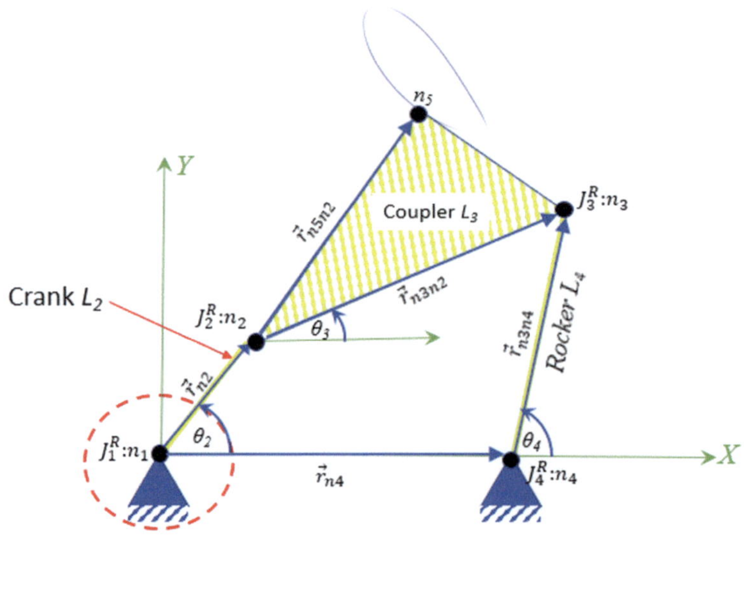

Fig. 5.56 Coupler curve example

The shape of n_5's trajectory changes with the location of n_5 on Link L_3. Figure 5.57 shows the variation of the n_5 trajectory with various values of angle γ and length ratio r_{n5n2}/r_{n3n2}. The most common shapes are described in Ref. 1. We can see the crunode (or, ordinary double point) shape as in the case of $\gamma = 45^o$ and $r_{n5n2}/r_{n3n2} = 1.5$. Also, we can see the cusp as in the case of $\gamma = 0^o$ and $r_{n5n2}/r_{n3n2} = 1.5$.

The shape of the path motions of n_5 in Fig. 5.57 is called coupler curves. The path motion of the coupler curve can be used in many machine design applications such as cutting machines or amusement park game machines. One very important application is the car suspension system (as shown in Fig. 5.58) which is designed as a 4-bar mechanism. The center of gravity of the wheel is located on the coupler. This suspension design aims to keep the wheel's center of gravity coupler curve nearly a straight line, keeping the tire perpendicular to the ground over a small vertical displacement from the road bump. This design provides improved vehicle ride quality.

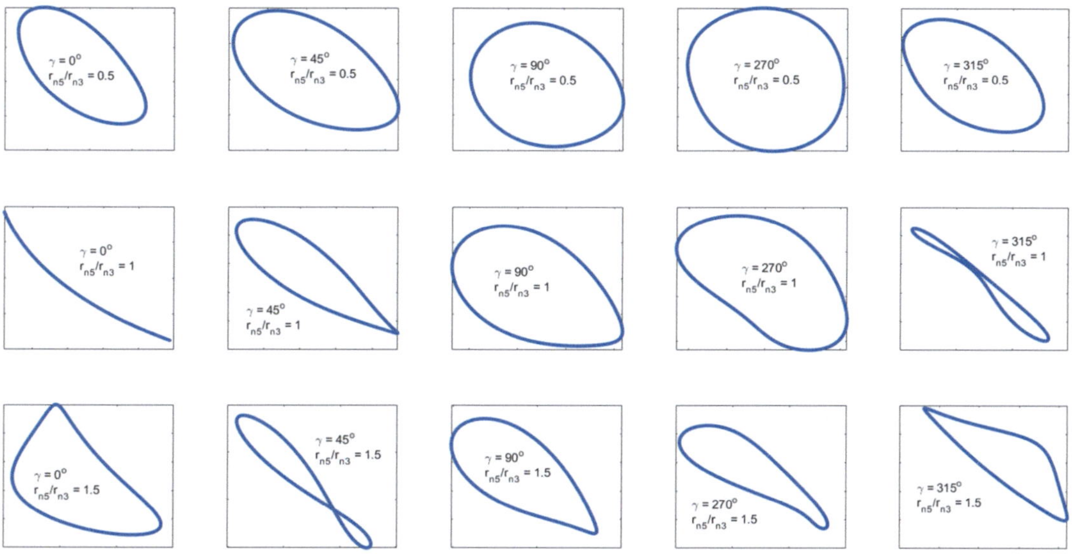

Fig. 5.57 Different coupler curves example

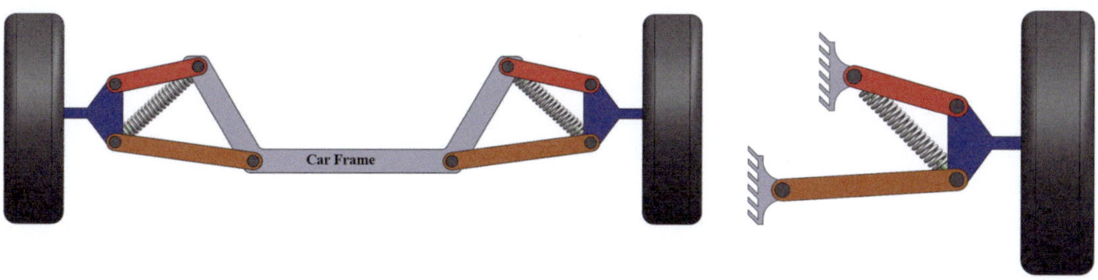

Fig. 5.58 Car suspension as four-bar mechanism. Designed by the author and implemented by both the author and Pranav Dhola. Courtesy of Pranav Dhola

5.10 Chapter Summary

Outcomes

1. Calculating the position vector and the displacement vector of any mechanism node using the graphical method, analytical method, and model-based Simulink® and Simscape™ Multibody™ for 4-bar, slider–crank, and inverted slider–crank mechanisms.
2. Understanding the limitation of developing the analytical model for a higher-order bar mechanism.
3. Understanding the efficiency of the model-based Simulink® and Simscape™ Multibody™ applied to a higher-order bar mechanism.
4. Estimating the toggle position and understanding the applications.
5. Estimating the coupler curve and understanding the applications.

Problems

For each problem from 1–3, for the given dimensions, find the position vectors of nodes n_1, n_2 and n_3.

Problem 1
Use $\theta_2 = 20°$ and $\theta_4 = 40°$

Problem 2
Use $\theta_2 = 20°$ and $\theta_3 = \theta_4 = 150°$.

Problem 3
See Fig. 5.61.

Problem 4
Refer to the shown skid steer bucket mechanism. In this mechanism, a hydraulic actuator (at node 3) moves the prismatic joint (Fig. 5.62).

1. Which one of the joint nodes has a passive linear displacement?
2. Which one of the joint nodes has an active linear displacement?
3. Which one of the joint nodes has a passive angular displacement?
4. Which one of the joint nodes has an active angular displacement?
5. Calculate the degrees of freedom.
6. Do you think this single hydraulic actuator is enough to control the entire mechanism's motion?

Problem 5
Refer to the shown sewing machine, a motor as an actuator (at node 1) rotates the orange-colored roll.

1. Which one of the joint nodes has a passive linear displacement?
2. Which one of the joint nodes has an active linear displacement?

3. Which one of the joint nodes has a passive angular displacement?
4. Which one of the joint nodes has an active angular displacement?
5. Calculate the degrees of freedom.
6. Do you think this single motor actuator is enough to control the entire machine's motion?

Problem 6

For the mechanism shown in Fig. 5.59, use the graphical *Combination Drawing Set* method, when $\theta_2 = 30°$, to measure both vertical distance and horizontal distance between n_1 and n_3.

Problem 7

For the mechanism shown in Fig. 5.60, use the graphical *Combination Drawing Set* method, when $\theta_2 = 30$ deg, to measure.

- the angle $\theta_3 = \theta_4$,
- the piston displacement which we consider as the distance from n_2 to n_3, and
- both vertical distance and horizontal distance between n_1 and n_3.

Problem 8

For the mechanism shown in Fig. 5.60, use the graphical *Combination Drawing Set* method, when the piston is extended by a distance 1.475 m, to measure.

- the angle $\theta_3 = \theta_4$,
- the angle θ_2, and
- both vertical distance and horizontal distance between n_1 and n_3.

Problem 9

For the mechanism shown in Fig. 5.63, when $\theta_2 = 30$ deg, use the graphical *Combination Drawing Set* method to measure the position vector of node n_3 where the distance between n_1 and n_2 is 15 mm, the distance between n_2 and n_3 is 30 mm, and the distance between n_1 and n_4 is 70 mm.

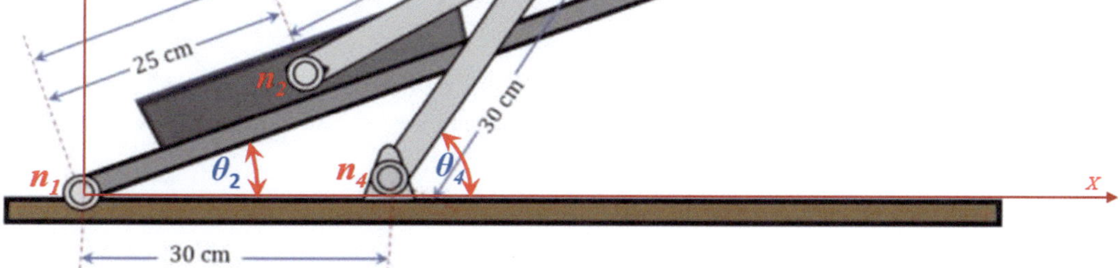

Fig. 5.59 Door closer mechanism

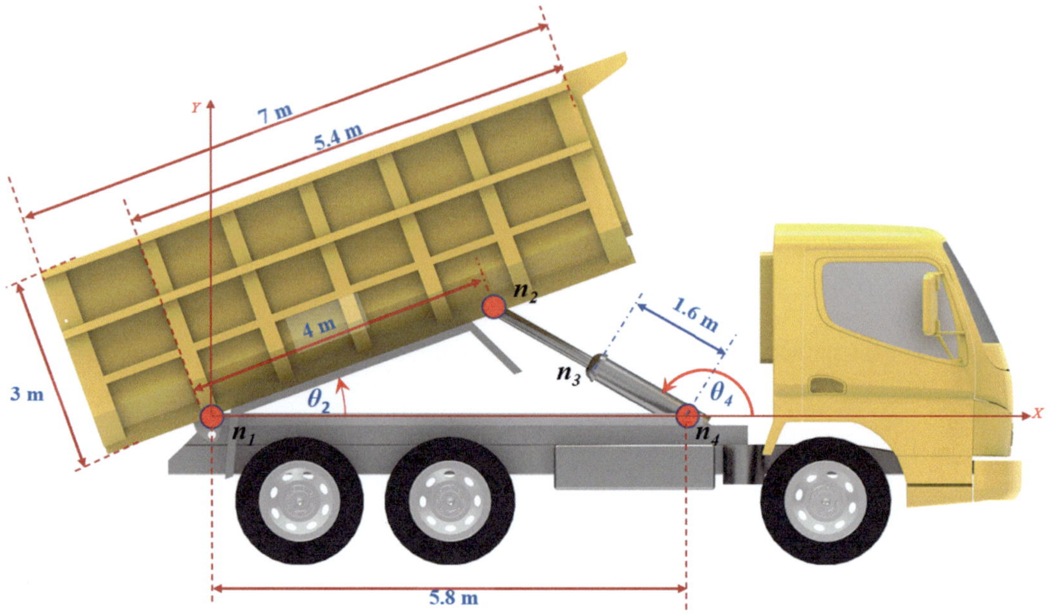

Fig. 5.60 Dumpster truck mechanism. Figure and associated Dumpster Truck Mechanism.mp4 video designed by the author and implemented by both the author and Pranav Dhola. Both courtesy of Pranav Dhola

Problem 10
Build a Simscape™ Multibody™ model for the mechanism shown in Fig. 5.59 Problem 1. Use the motion actuator subsystem to activate θ_2 with 30°. Use the position sensor to measure the position vector of node n_3.

Problem 11
Build a Simscape™ Multibody™ model for the mechanism shown in Fig. 5.60. Use the motion actuator subsystem to extend the piston by a distance 1.475 m. Use the position sensor to measure the position vector of node n_3 and measure both θ_2 and $\theta_4 = \theta_3$.

Problem 12
Build a Simscape™ Multibody™ model for the mechanism in Fig. 5.63. Use the motion actuator subsystem to activate θ_2 with 30°. Use the position sensor to measure the position vectors of node n_3. The distance between n_1 and n_2 is 15 mm, the distance between n_2 and n_3 is 30 mm, and the distance between n_1 and n_4 is 70 mm. The radius of the yellow roller is 20 mm. The length of the blue needle bar is 55 mm.

Problem 13
For the mechanism shown in Fig. 5.59, use the analytic method to find the position vector of node n_3 when $\theta_2 = 30$ deg.

Problem 14
For the mechanism shown in Fig. 5.60, when the piston is extended by a distance 1.475 m, use the analytical method to calculate.

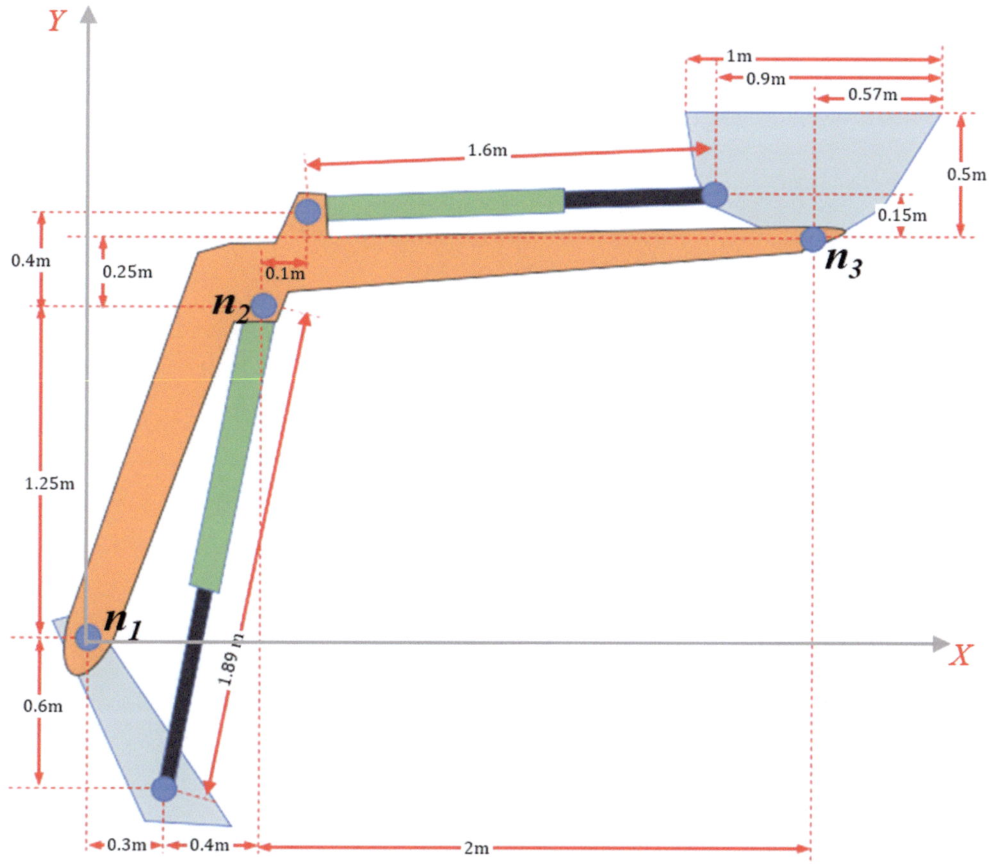

Fig. 5.61 Tractor boom mechanism

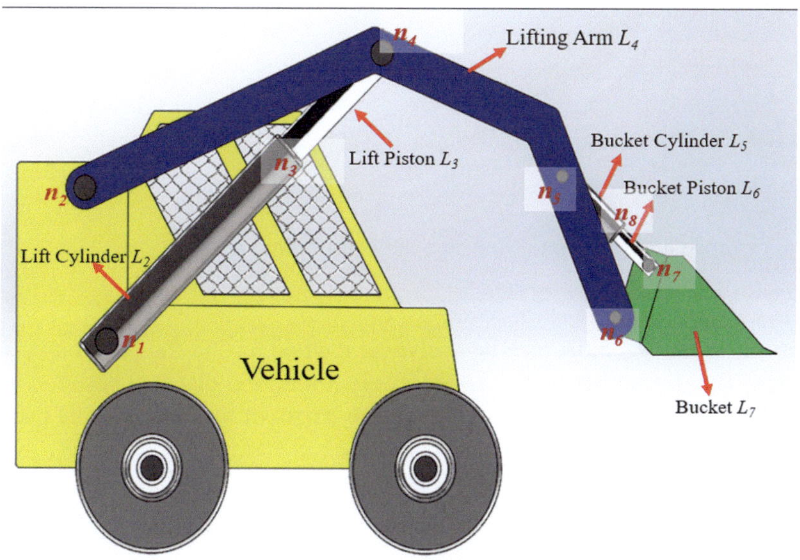

Fig. 5.62 Skid steer articulated bucket mechanism. Figure and associated Skid Steer Articulated Bucket Mechanism. mp4 video designed by the author and implemented by both the author and Pranav Dhola. Both courtesy of Pranav Dhola

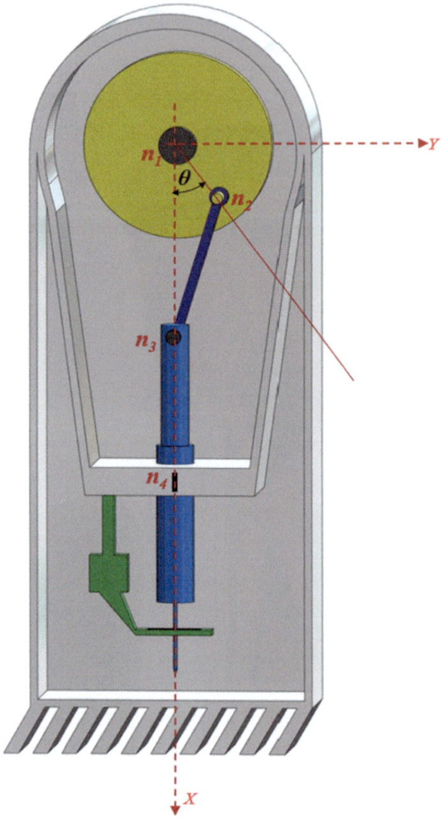

Fig. 5.63 Sewing mechanism. Designed by the author and implemented by both the author and Pranav Dhola. Courtesy of Pranav Dhola

Table 5.4 Parameters of four-bar mechanisms in Problem 17

Mechanism Number	L_1 (m)	L_2 (m)	L_3 (m)	L_4 (m)
1	1	0.5	0.7	1
2	1.2	0.3	0.5	0.8

- the angle $\theta_3 = \theta_4$,
- the angle θ_2, and
- both vertical distance and horizontal distance between n_1 and n_3.

Problem 15

For mechanism shown in Fig. 5.63, use the analytical method to find the position vector of node n_3 when $\theta_2 = 30$ deg. The distance between n_1 and n_2 is 15 mm, the distance between n_2 and n_3 is 30 mm, and the distance between n_1 and n_4 is 70 mm. The length of the entire needle bar (link L_4) is 46 mm.

Problem 16

1. Make a table to compare the results from Problem 6 with Problem 13 and with Problem 10.
2. Make a table to compare the results from problem 7 with problem 8, problem 14, and problem 11.
3. Make a table to compare the results from problem 9 with problem 15 and with problem 12.

Problem 17

Calculate the toggle angle of two 4-bar mechanisms that have the listed parameter in Table 5.4 using the analytical solution and the Simscape™ Multibody™ model. You could use the model that is developed in Example 5.5 by modifying the parameters.

Problem 18

Build a MATLAB® script that uses the analytical model to calculate the angles θ_3, $\theta_{4,}$ and position vectors of nodes n_2, n_3, and n_4 for a four-bar mechanism. The script receives the mechanism parameters as L_1, L_2, L_3, and L_4. The script receives θ_2 as an input. Also, the script calculates the toggle angle.

References

1. "Design of Machinery," Robert L. Norton, McGraw-Hill; 5th Edition, 2011.
2. "Machines & Mechanisms: Applied Kinematic Analysis", David Myszka, Pearson; 4th Edition, 2011.
3. "Simscape™ Multibody™ Getting Started Guide," MathWorks, 2023.
4. "Simscape™ Multibody™ User's Guide," MathWorks, 2023.
5. "Simscape™ Multibody™ Reference," MathWorks, 2023.
6. "Simscape™ Multibody™ Release Notes," MathWorks, 2023.
7. "Sucker Rod Pump Surface Units Catalog", Global Energy Services & Equipment | Schlumberger (slb.com), 2019.

6 Velocity Analysis

Chapter Objectives

1. Learning 2D translational velocity vector and 2D rotational velocity vector of any mechanism node in Cartesian and polar coordinates
2. Learning the translational velocity vector and the rotational velocity vector of a single link connected to the ground link
3. Learning the difference between the absolute velocity vector and relative velocity vector
4. Using the model-based Simulink® and Simscape™ Multibody™ to calculate the translational velocity vector and the angular velocity vector of any mechanism node
5. Determining the instantaneous zero center of velocity of a single link, the instantaneous center of velocity of two links, and Kennedy's theorem for three links
6. Developing and using the graphical velocity model plus the analytical velocity models for four-bar, slider–crank, and inverted slider–crank mechanisms

6.1 Introduction

In this chapter, we explain the translational velocity vector and rotational velocity vector of any mechanism node followed by an explanation of absolute velocity vector and relative velocity vector including sliding and rolling motions. After that, we develop three different methods to calculate the velocities: graphical method, analytical method, and model-based Simulink® and Simscape™ Multibody™ method. These methods will be implemented for four-bar, slider–crank, and inverted slider–crank mechanisms. For the graphical method, we explain the difference among the instantaneous zero center of velocity of a single link, the instantaneous center of velocity of two links, and Kennedy's theorem for three links with a couple of applications.

Supplementary Information The online version contains supplementary material available at https://doi.org/10.1007/978-3-031-72701-6_6.

A. Omran, *Kinematics and Dynamics of Machines*, https://doi.org/10.1007/978-3-031-72701-6_6

6.2 Translational Velocity and Angular Velocity

Definition

Linear Velocity:

The linear velocity of a point is the linear displacement of that point per unit time or the rate of change of position with respect to time [1–3].

Angular Velocity:

Angular velocity is defined as the rate of change of angular displacement with respect to time for an object rotating about a fixed axis [4].

Translational velocity is how fast the body is moving forward in a specific direction. Angular velocity is how fast the body rotates in a specific direction. For example, as shown in Fig. 6.1 If you push a box on a floor, the box will move with a translational speed v without any rotation. If you push a cylinder on a floor, the cylinder will move with a translational velocity v and will rotate with an angular velocity ω as shown in Fig. 6.1. Now, we are going to introduce the mathematical formulas for the translational velocity and rotational velocity plus the combined velocity.

6.2.1 Translational Velocity

If we assume a node that has a 2D position vector $\vec{r}$, measured from the origin, at time t is

$$\vec{r}(t) = x(t)\hat{i} + y(t)\hat{j} \tag{6.1}$$

where $x(t)$ is the position component in X-direction and $y(t)$ is the position component in Y-direction in the world frame, the 2D average velocity vector $\vec{\underline{v}}$ has two components: $\underline{v}_x$ in X-direction and $\underline{v}_y$ in Y-direction. Over a time interval Δt, the average velocity components $\underline{v}_x$ and $\underline{v}_y$ can be defined to be the displacements Δx and Δy divided by the time interval Δt.

$$\vec{\overline{v}}(t, \Delta t) = \overline{v}_x(t, \Delta t)\hat{i}+\overline{v}_y(t, \Delta t)\hat{j} = \frac{\Delta x}{\Delta t}\hat{i}+\frac{\Delta y}{\Delta t}\hat{j} = \frac{x(t+\Delta t)-x(t)}{\Delta t}\hat{i}+\frac{y(t+\Delta t)-y(t)}{\Delta t}\hat{j} \tag{6.2}$$

or

$$\overline{v}_x(t, \Delta t) = \frac{x(t+\Delta t)-x(t)}{\Delta t} \text{and} \overline{v}_y(t, \Delta t) = \frac{y(t+\Delta t)-y(t)}{\Delta t}$$

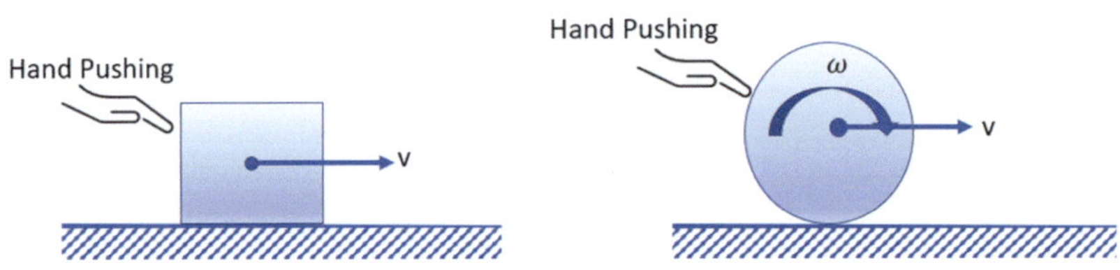

Fig. 6.1 Box and cylinder motion on floor

The instantaneous velocity $\vec{v}(t)$ at time t is the limit of the average velocity $\underline{\vec{v}}$ when Δt tends to zero. This limit is equal to the derivative of the position components as listed in Eq. (6.3).

$$\begin{aligned}\vec{v}(t) = \lim_{\Delta t\to 0}\overline{\vec{v}}(t,\Delta t) &= \lim_{\Delta t\to 0}\overline{v}_x(t,\Delta t)\hat{i} + \lim_{\Delta t\to 0}\overline{v}_y(t,\Delta t)\hat{j}\\ &= \lim_{\Delta t\to 0}\frac{\Delta x(t,\Delta t)}{\Delta t}\hat{i} + \lim_{\Delta t\to 0}\frac{\Delta y(t,\Delta t)}{\Delta t}\hat{j}\\ &= \frac{\mathrm{d}x(t)}{\mathrm{d}t}\hat{i} + \frac{\mathrm{d}y(t)}{\mathrm{d}t}\hat{j} = \dot{x}(t)\hat{i} + \dot{y}(t)\hat{j}\\ &= v_x(t)\hat{i} + v_y(t)\hat{j}\end{aligned} \tag{6.3}$$

or

$$v_x(t) = \frac{\mathrm{d}x(t)}{\mathrm{d}t} \text{ and } v_y(t) = \frac{\mathrm{d}y(t)}{\mathrm{d}t}$$

To express the velocity in polar coordinates, we need to express the Cartesian unit vectors $\hat{i}$ in the X-direction and $\hat{j}$ in the Y-direction as the polar unit vectors $\hat{r}$ in r-direction (or normal direction) and $\hat{\theta}$ in θ-direction (or tangential direction). As shown in Fig. 6.2, the polar unit vectors can be decomposed into the Cartesian as

$$\begin{aligned}\hat{i} &= \cos(\theta)\hat{r} - \sin(\theta)\hat{\theta}\\ \hat{j} &= \sin(\theta)\hat{r} + \cos(\theta)\hat{\theta}\end{aligned} \tag{6.4}$$

Using Eqs. (6.3) and (6.4), the polar velocity components are:

$$\begin{aligned}\vec{v}(t) = \dot{x}(t)\hat{i} + \dot{y}(t)\hat{j} &= \dot{x}(t)\left(\cos(\theta)\hat{r} - \sin(\theta)\hat{\theta}\right) + \dot{y}(t)\left(\sin(\theta)\hat{r} + \cos(\theta)\hat{\theta}\right)\\ &= (\dot{x}(t)\cos(\theta) + \dot{y}(t)\sin(\theta))\hat{r} + (-\dot{x}(t)\sin(\theta) + \dot{y}(t)\cos(\theta))\hat{\theta}\\ &= v_r(t)\hat{r} + v_\theta(t)\hat{\theta}\end{aligned} \tag{6.5}$$

or

$$\begin{aligned}v_r(t) &= \dot{x}(t)\cos(\theta) + \dot{y}(t)\sin(\theta)\\ v_\theta(t) &= -\dot{x}(t)\sin(\theta) + \dot{y}(t)\cos(\theta)\end{aligned}$$

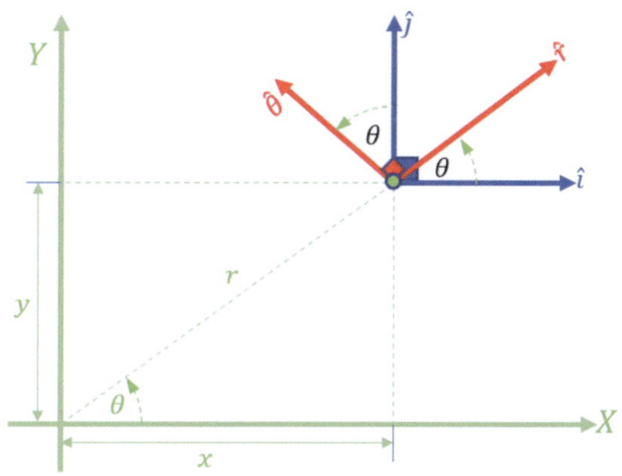

Fig. 6.2 Unit vectors of Cartesian coordinate and polar coordinates

Example 6.1

The shown pitcher water pump mechanism in Fig. 6.3 has four links: pump body (ground), handle, connecting rod, and the piston in addition to four joints: three revolute joints J_1^R, J_2^R, and J_3^R plus one prismatic joint J_4^P. Assume a person pushes the handle down at node n_5. The handle rotates clockwise around at node n_1 and the node n_2 moves up causing the connecting rod to push the piston up with the motion below:

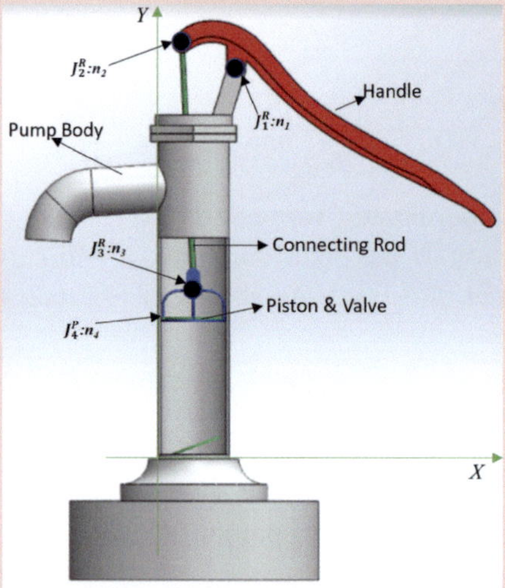

Fig. 6.3 Pitcher water pump. Designed by the author and implemented by both the author and Pranav Dhola. Courtesy of Pranav Dhola

$$y_{\text{piston}}(t) = \frac{A}{2}(1 - \cos(\beta(t))) \text{ where } \beta(t) = \frac{\pi}{T}\left(t - \frac{T}{8\pi}\sin\left(\frac{\pi t}{T}\right)\right) \text{for} 0 \le t \le T \text{s}$$

Consider the amplitude $A = 200$ mm and the time span $T = 1$ s.

1. Calculate the average velocity over the time span.
2. Calculate the instantaneous velocity at $t = 0$ s, $t = 0.5$ s, and $t = 1$ s.
3. Write a MATLAB® script to plot the position displacement $y_{\text{piston}}(t)$ with respect to time and the instantaneous velocity $dy_{\text{piston}}(t)/dt$ with respect to time.

Solution

The position vector is:

$$\vec{r}_{\text{piston}}(t) = y_{\text{piston}}(t)\hat{j}$$

where:

$$y_{\text{piston}}(t) = 100(1 - \cos(\beta(t))) \text{ and } \beta(t) = \pi\left(t - \frac{1}{8\pi}\sin(\pi t)\right)$$

Note that $x_{\text{piston}}(t)$ is zero from the kinematic constraints of the prismatic joint.

Average Velocity

First, we calculate the position at initial time $t = 0$ s and the final time $t = T = 1$ s:

$$\beta(t=0) = \pi\left(t - \frac{1}{4\pi}\sin(2\pi t)\right)\Big|_{t=0} = \pi\left(0 - \frac{1}{4\pi}\sin(\pi \times 0)\right) = 0$$

$$y_{\text{piston}}(t=0) = 100(1-\cos(\beta(t)))|_{t=0} = 100(1-\cos(0)) = 0 \text{ mm}$$

$$\beta(t=1) = \pi\left(t - \frac{1}{4\pi}\sin(2\pi t)\right)\Big|_{t=1} = \pi\left(1 - \frac{1}{8\pi}\sin(\pi \times 1)\right) = \pi$$

$$y_{\text{piston}}(t=1) = 100(1-\cos(\beta(t)))|_{t=1} = 100(1-\cos(\pi)) = 200 \text{ mm}$$

Calculate the average velocity from Eq. (6.2):

$$\bar{\vec{v}}_{\text{piston}}(t=0, \Delta t=1) = \frac{x_{\text{piston}}(t+\Delta t) - x_{\text{piston}}(t)}{\Delta t}\Bigg|_{\substack{t=0\\ \Delta t=1}}\hat{i} + \frac{y_{\text{piston}}(t+\Delta t) - y_{\text{piston}}(t)}{\Delta t}\hat{j}$$

$$= \frac{x_{\text{piston}}(t=1) - x_{\text{piston}}(t=0)}{1-0}\hat{i} + \frac{y_{\text{piston}}(t=1) - y_{\text{piston}}(t=0)}{1-0}\hat{j}$$

$$= 0\hat{i} + 200\hat{j} \text{ mm/s}$$

Instantaneous Velocity

We start with calculating the derivatives of the position vector components with respect to time:

$$x_{\text{piston}}(t) = 0$$

$$\frac{dx_{\text{piston}}(t)}{dt} = 0$$

$$y_{\text{piston}}(t) = 100(1-\cos(\beta(t)))\text{and}\beta(t) = \pi\left(t - \frac{1}{8\pi}\sin(\pi t)\right)$$

$$\frac{dy_{\text{piston}}(t)}{dt} = \frac{dy_{\text{piston}}(t)}{d\beta} \times \frac{d\beta}{dt} = 100\sin(\beta(t))$$

$$\times \pi\left(1 - \frac{1}{8}\cos(\pi t)\right) = 100\pi\left(1 - \frac{1}{8}\cos(\pi t)\right)\sin(\beta(t))$$

The instantaneous velocity is calculated from Eq. (6.3):

$$\vec{v}_{\text{piston}} = \frac{dx_{\text{piston}}}{dt}\hat{i} + \frac{dy_{\text{piston}}}{dt}\hat{j} = 100\pi\left(1 - \frac{1}{8}\cos(\pi t)\right)\sin(\beta(t))\hat{j}$$

Substitute with $t = 0$ s, t = 0.5 s and t = 1 s:

$$\beta(t=0) = \pi\left(t - \frac{1}{8\pi}\sin(\pi t)\right)\Big|_{t=0} = \pi\left(0 - \frac{1}{8\pi}\sin(\pi \times 0)\right) = 0$$

$$\vec{v}_{\text{piston}}(t=0) = 100\pi\left(1 - \frac{1}{8}\cos(\pi t)\right)\sin(\beta(t))\Big|_{t=0}\hat{j}$$

$$= 100\pi\left(1 - \frac{1}{8}\cos(\pi \times 0)\right)\sin(\beta(0))\hat{j} = 0 \text{ mm/s}$$

$$\beta(t=0.5) = \pi\left(t - \frac{1}{8\pi}\sin(\pi t)\right)\Big|_{t=0.5} = \pi\left(0.5 - \frac{1}{8\pi}\sin(\pi \times 0.5)\right) = 1.446$$

$$\vec{v}_{\text{piston}}(t=0.5) = 100\pi\left(1-\frac{1}{8}\cos(\pi t)\right)\sin(\beta(t=0.5))\bigg|_{t=0.5}\hat{j}$$
$$= 100\pi\left(1-\frac{1}{8}\cos(\pi\times 0.5)\right)\sin(\beta(t=0.5))\hat{j} = 311.716\hat{j}\ \text{mm/s}$$
$$\beta(t=1) = \pi\left(t-\frac{1}{8\pi}\sin(\pi t)\right)\bigg|_{t=1} = \pi\left(1-\frac{1}{8\pi}\sin(\pi\times 1)\right) = \pi$$
$$\vec{v}_{\text{piston}}(t=1) = 100\pi\left(1-\frac{1}{8\pi^2}\cos(\pi t)\right)\sin(\beta(t))\bigg|_{t=1}\hat{j}$$
$$= 100\pi\left(1-\frac{1}{8\pi^2}\cos(\pi)\right)\sin(\beta(1))\hat{j} = 0\ \text{mm/s}$$

MATLAB® Plotting

You can write the following commands in MATLAB® script:

```
%This file to plot the instantaneous velocity components of Example 6.1
%=====================================
%Clear all the variables and close all the figures
%=====================================
clc
clear all
close all
%==========================================================================
T = 1; %sec, time span as a given info
A = 200; %mm, amplitude as a given info

%calculating the time
dt = 0.0001; %sec, selected time step
t =[0:dt:T]; %sec, time starts from zero and end at T by dt as time step

%Calculating the position
beta = pi/T*(t-T/8/pi*sin(pi*t/T)); %unit less, given parameter
y = A/2*(1-cos(beta)); %mm, given position
x = 0*y;

%calculating instantaneous velocity components
dbeta_dt = pi/T*(1-1/8*cos(pi*t/T));
dy_dbeta = A/2*sin(beta);
dy_dt = dy_dbeta.*dbeta_dt;

%plot results
[hAx,hLine1,hLine2] = plotyy(t,y,t,dy_dt);
xlabel('t (sec)')
ylabel(hAx(1),'y (mm)') % left Y-axis
ylabel(hAx(2),'dy/dt (mm/sec') % right Y-axis
```

After running the above m file, Fig. 6.4 will be the results.

Note: We did not plot the *X-axis* position component or the *X-axis* velocity component because both have a zero value. We encourage the reader to watch the attached animation video of the pitcher water pump for a better visual understanding.

Fig. 6.4 Position and velocity components in y-axis for pitcher water pump in Example 6.1

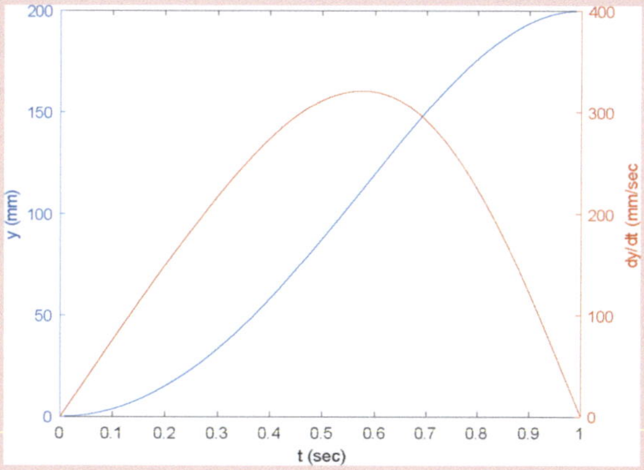

6.2.2 Angular Velocity

Just like translational velocity, the angular velocity has a vector form as well. In 2D, the angular velocity is the rate of change in the rotational angle around any Z-axis (world frame or relative frame). For example, assume a motor is connected to the center of a shaft as shown in Fig. 6.5. If we locate the world frame, the rotational angle is then measured around the Z-axis and it can be mathematically defined as

$$\vec{\Omega}(t) = 0\hat{i} + 0\hat{j} + \omega\hat{k} \tag{6.6}$$

In this textbook, any angular velocity will be defined around the Z-axis or as written in Eq. (6.6) because our analysis is restricted to 2D. Similar to the translational velocity, the angular velocity can be defined as the average angle velocity vector $\underline{\vec{\Omega}}(t, \Delta t)$ and the instantaneous angular velocity vector $\vec{\Omega}(t)$ by

$$\overline{\vec{\Omega}}(t, \Delta t) = \overline{\omega}(t, \Delta t)\hat{k} = \frac{\Delta\theta}{\Delta t}\hat{k} = \frac{\theta(t + \Delta t) - \theta(t)}{\Delta t}\hat{k} \tag{6.7}$$

and

$$\begin{aligned}\vec{\Omega}(t) &= \lim_{\Delta t \to 0} \overline{\vec{\Omega}}(t, \Delta t) = \lim_{\Delta t \to 0} \overline{\omega}(t, \Delta t)\hat{k} \\ &= \lim_{\Delta t \to 0} \frac{\Delta\theta(t, \Delta t)}{\Delta t}\hat{k} = \frac{d\theta(t)}{dt}\hat{k} = \dot{\theta}(t)\hat{k} = \omega(t)\hat{k}\end{aligned}$$

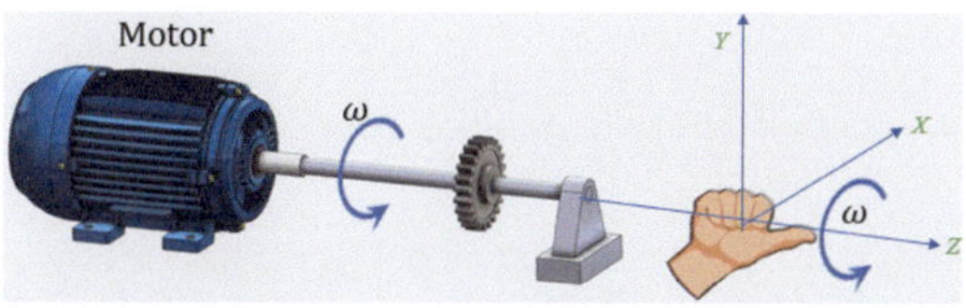

Fig. 6.5 Motor angular velocity around z-axis. Designed by the author and implemented by both the author and Pranav Dhola. Courtesy of Pranav Dhola

Example 6.2

For the car wiper mechanism shown, the motor can rotate at three different angular positions as

$$\theta_2^{\mathrm{Low}}(t) = 1.367\pi t$$
$$\theta_2^{\mathrm{Mid}}(t) = 2\pi t$$
$$\theta_2^{\mathrm{High}}(t) = 3.533\pi t$$

where θ_2 is in radians and t is in seconds.

1. Calculate the required time to go over one cycle from $\theta_2=0°$ to $\theta_2=360°$ for the low-, mid- and high-angular positions.
2. Calculate the average angle velocity in rad/sec and in rpm for the low-, mid-, and high-angular positions.
3. Calculate the instantaneous angular velocity in rpm at $\theta_2=0°$ and at $\theta_2=180°$ for the low-, mid-, and high-angular positions. Figure 6.6

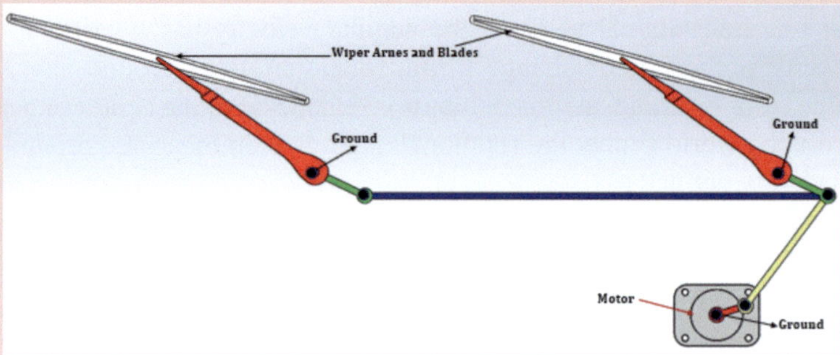

Fig. 6.6 Car wiper mechanism. Designed by the author and implemented by both the author and Pranav Dhola. Courtesy of Pranav Dhola

Solution

One Cycle Time

The cycle time is the time to make one cycle, or to rotate 360°. We can calculate this time as the difference $\Delta t^i = t(\theta_2^i = 360°) - t\left(\theta_2^i = 0°\right)$ where i = low, mid, and high are:

$$\Delta t^{\mathrm{Low}} = t(\theta_2^{\mathrm{low}} = 360°) - t\left(\theta_2^{\mathrm{Low}} = 0°\right) = \frac{\theta_2^{\mathrm{Low}}}{1.367\pi}\Big|_{\theta_2^{\mathrm{Low}}=360°} - \frac{\theta_2^{\mathrm{low}}}{1.367\pi}\Big|_{\theta_2^{\mathrm{Low}}=0°} = \frac{2\pi}{1.367\pi} - \frac{0}{1.367\pi} = 1.463\mathrm{s}$$

$$\Delta t^{\mathrm{Mid}} = t(\theta_2^{\mathrm{Mid}} = 360°) - t\left(\theta_2^{\mathrm{Mid}} = 0°\right) = \frac{\theta_2^{\mathrm{Mid}}}{2\pi}\Big|_{\theta_2^{\mathrm{Mid}}=360°} - \frac{\theta_2^{\mathrm{Mid}}}{2\pi}\Big|_{\theta_2^{\mathrm{Mid}}=0°} = \frac{2\pi}{2\pi} - \frac{0}{2\pi} = 1\ \mathrm{s}$$

$$\Delta t^{\mathrm{High}} = t(\theta_2^{\mathrm{High}} = 360^0) - t\left(\theta_2^{\mathrm{High}} = 0°\right) = \frac{\theta_2^{\mathrm{High}}}{2\pi}\Big|_{\theta_2^{\mathrm{High}}=360°} - \frac{\theta_2^{\mathrm{High}}}{2\pi}\Big|_{\theta_2^{\mathrm{High}}=0°} = \frac{2\pi}{3.533\pi} - \frac{0}{3.533\pi} = 0.5661\ \mathrm{s}$$

Average Angular Velocity

Using Eq. (6.7)

$$\overline{\vec{\Omega}}_2^{\text{Low}}\left(t=0,\Delta t=\Delta t^{\text{Low}}=1.463\right)=\overline{\omega}_2^{\text{Low}}(t=0,\Delta t=1.463)\hat{k}=\frac{\Delta\theta_2^{\text{Low}}}{\Delta t}\hat{k}=\frac{2\pi}{1.463}\hat{k}$$
$$=4.294\hat{k}\frac{\text{rad}}{\text{s}}=4.294\frac{60}{2\pi}\hat{k}\text{ rpm}=41\hat{k}\text{ rpm}$$
$$\overline{\vec{\Omega}}_2^{\text{Mid}}\left(t=0,\Delta t=\Delta t^{\text{Mid}}=1\right)=\overline{\omega}_2^{\text{Mid}}(t=0,\Delta t=1)\hat{k}=\frac{\Delta\theta_2^{\text{Mid}}}{\Delta t}\hat{k}=\frac{2\pi}{1}\hat{k}$$
$$=6.283\hat{k}\frac{\text{rad}}{\text{s}}=6.283\frac{60}{2\pi}\hat{k}\text{ rpm}=60\hat{k}\text{ rpm}$$
$$\overline{\vec{\Omega}}_2^{\text{High}}(t=0,\Delta t=0.566)=\overline{\omega}_2^{\text{High}}(t=0,\Delta t=0.566)\hat{k}=\frac{\Delta\theta_2^{\text{High}}}{\Delta t}\hat{k}=\frac{\theta_2^{\text{High}}(t=0.566)-\theta_2^{\text{High}}(t=0)}{\Delta t}\hat{k}$$
$$=\frac{2\pi-0}{0.566-0}\hat{k}=11.101\frac{\text{rad}}{\text{s}}\hat{k}=11.101\frac{60}{2\pi}\hat{k}=106\hat{k}\text{ rpm}$$

Instantaneous Angular Velocity

Using Eq. (6.7) apply the first derivatives:

$$\vec{\Omega}_2^{\text{Low}}(t)=\omega_2^{\text{Low}}(t)\hat{k}=\frac{d\theta_2^{\text{Low}}(t)}{dt}\hat{k}=1.367\pi\hat{k}\frac{\text{rad}}{\text{s}}=1.367\pi\times\frac{60}{2\pi}\hat{k}=41\hat{k}\text{ rpm}$$
$$\vec{\Omega}_2^{\text{Mid}}(t)=\omega_2^{\text{Mid}}(t)\hat{k}=\frac{d\theta_2^{\text{Mid}}(t)}{dt}\hat{k}=2\pi\hat{k}\frac{\text{rad}}{\text{s}}=2\pi\times\frac{60}{2\pi}\hat{k}=60\hat{k}\text{ rpm}$$
$$\vec{\Omega}_2^{\text{High}}(t)=\omega_2^{\text{High}}(t)\hat{k}=\frac{d\theta_2^{\text{High}}(t)}{dt}\hat{k}=3.533\pi\hat{k}\frac{\text{rad}}{\text{s}}=3.533\pi\times\frac{60}{2\pi}\hat{k}=106\hat{k}\text{ rpm}$$

The instantaneous angular velocity is constant with time, so:

$$\vec{\Omega}_2^{\text{Low}}(t=0)=\vec{\Omega}_2^{\text{Low}}(t=1.367)=41\hat{k}\text{ rpm}$$
$$\vec{\Omega}_2^{\text{Mid}}(t=0)=\vec{\Omega}_2^{\text{Mid}}(t=1)=60\hat{k}\text{ rpm}$$
$$\vec{\Omega}_2^{\text{Mid}}(t=0)=\vec{\Omega}_2^{\text{Mid}}(t=0.566)=106\hat{k}\text{ rpm}$$

We encourage the reader to watch the animation video of the wiper mechanism.

Now we are going to go over some kinematic terminologies to describe the relative translational motion between two in-contact rigid bodies or two in-contact links. The two links connected by a revolute joint J_k^R, where k is the joint index, have zero relative translational motion. For the rest of the joints J_k^l, where k is the joint index and $l=\{P, S, PS, G, C\}$ is the joint type, there is a relative translational motion that can be described as *sliding motion* or/and *rolling motion.*

6.2.3 Sliding Velocity

The sliding motion is a driven motion when one of the two links (L_i or L_j) is pushed by a direct actuator, such as a hydraulic system, or one of the two links is pushed as a consequence of a mechanism's motion. The sliding relative motion is active and deterministic. Note, deterministic motion means that we can estimate the motion by using the kinematics of machines analysis. One example is the scissor lift table mechanism as shown in Fig. 6.7. At $J_3^P : n_3$, the piston L_3 slides in and out of the cylinder L_4 using a direct hydraulic system actuator. The sliding motion at node n_3 causes two non-actuated sliding motions (coming from the mechanism motion). The first non-actuated sliding motion is at n_1 when link L_2 slides on link L_1. The second non-actuated sliding motion is at n_5 when L_5 slides on L_6. The three sliding motions at n_1, n_3, and n_5 are deterministic. We encourage the reader to watch the attached animation video of the scissor lift table.

6.2.4 Rolling Velocity

Rolling contact or rolling motion is where two links (L_i and L_j) are connected at node n_k. The velocity v_k is tangential on the surfaces of L_i and L_j at node n_k. Assume that link L_i is driven by a motor causing L_i to rotate. When the joint at node n_k forces link L_j to slide over L_i with the same velocity v_k, we call this relative motion ***rolling without slipping.*** This means the relative speed is: $\Delta v_k = v_k^{L_i} - v_k^{L_j} = 0$ or $v_k = v_k^{L_j} = v_k^{L_i}$.

For example, in Fig. 6.8, a wheel (L_2) with radius r is driven at the center node n_c with an angular velocity ω. The wheel keeps in contact with the ground (L_1). We are going to consider two extra nodes: the bottom node n_b and the top node n_t, on the wheel. Note, because we are considering the

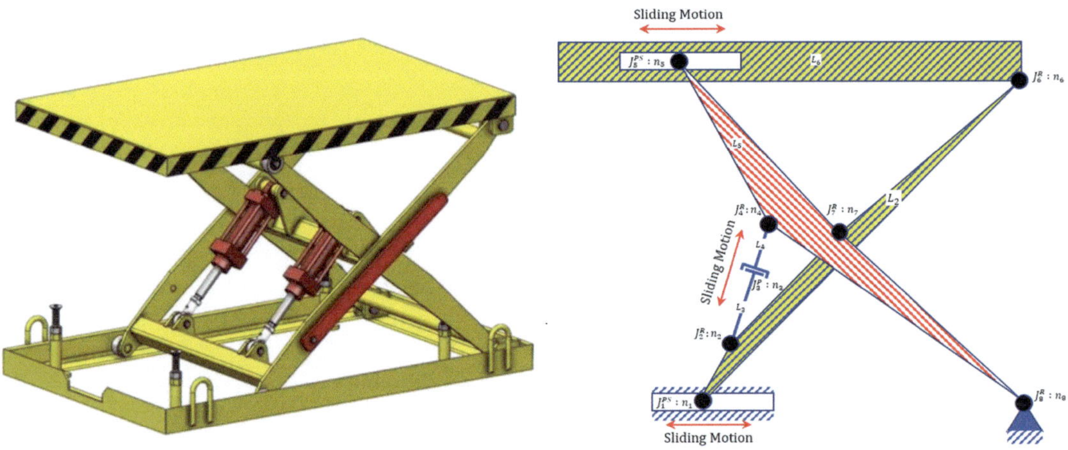

Fig. 6.7 Scissor lift table machine. Figure and associated Scissor Lift Table Mechanism.mp4 video designed by the author and implemented by both the author and Pranav Dhola. Both courtesy of Pranav Dhola

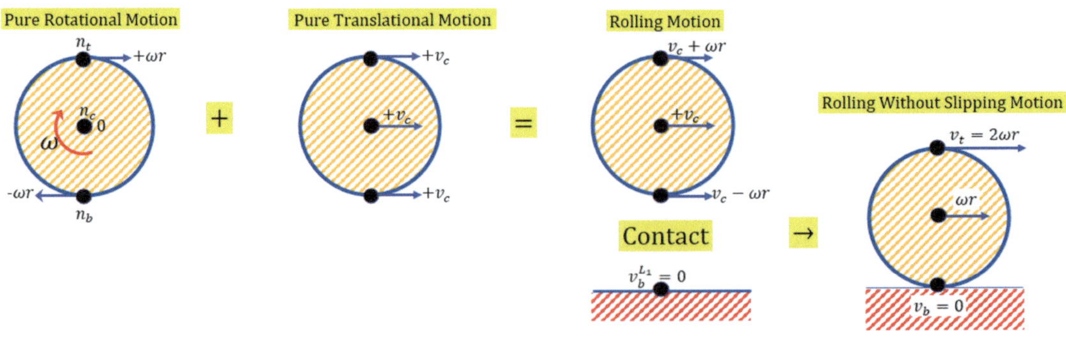

Fig. 6.8 Rolling without slipping motion

wheel as a link and not as a point, the nodes n_b, n_t and n_c have different velocities $v_b \neq v_t \neq v_c$ and they are in X-direction only. To keep the contact between the wheel and the ground, the wheel and ground must be in a rolling motion that has the velocity as a combination of pure translation motion and a pure rotation motion as shown in Fig. 6.8. The rotational angular velocity ω causes the tangential velocities:

$$\begin{aligned} v_b^{\text{Rotational}} &= -\omega r \\ v_t^{\text{Rotational}} &= +\omega r \\ v_c^{\text{Rotational}} &= 0 \end{aligned} \tag{6.8}$$

Note, "+" and "−" are in positive X-*axis* or in negative X-*axis*. The pure translational velocities are the same as v_c or:

$$\begin{aligned} v_b^{\text{Translational}} &= +v_c \\ v_t^{\text{Translational}} &= +v_c \\ v_c^{\text{Translational}} &= +v_c \end{aligned} \tag{6.9}$$

The total velocity at each node is the summation of the rotational velocities and the translational velocities as

$$\begin{aligned} v_b &= v_b^{\text{Rotational}} + v_b^{\text{Translational}} = v_c - \omega r \\ v_c &= v_c^{\text{Rotational}} + v_c^{\text{Translational}} = v_c \\ v_t &= v_t^{\text{Rotational}} + v_t^{\text{Translational}} = v_c + \omega r \end{aligned} \tag{6.10}$$

Note that the speed of any node on the ground is zero. For a rolling without slipping motion between the wheel and the ground, the velocity of the contact node n_b should be zero whereby the wheel is in contact with the ground:

$$\begin{aligned} & v_c = \omega r \\ & v_b^{L_2} = v_b^{L_1} = 0 \\ & \text{or } v_c - \omega r = 0 \end{aligned} \tag{6.11}$$

Then

$$\begin{aligned} v_b &= 0 \\ v_c &= \omega r \\ v_c &= 2\omega r \end{aligned}$$

If the relative velocity is less than zero $\Delta v_k < 0$, the motion is called ***rolling with slipping***. If the relative velocity is higher than zero $\Delta v_k > 0$, the motion is called ***rolling with skidding***. Slipping and skidding are more related to tires during braking, acceleration, cornering, tire blowout, wet road, snowy conditions, and icy conditions. The slipping and skidding motions are not considered in this textbook.

6.3 Single Link Connected to Ground

In general, a free link has a translational motion and a rotational motion or a complex motion. Having the links connected causes constraints on the motion. Now, we are going to study a couple of examples for calculating the velocity vector for a single link connected to the ground.

6.3.1 Hinged to Ground Link

Fig. 6.9 shows a two-node link L_2 attached to the ground at node n_1 (or, the origin) with one revolute joint J_1^R. A motor is attached at node n_1 to rotate the link with an angular position θ_2 and an angular velocity ω_2. The velocity vectors of node n_1 and n_2 are:

$$\begin{aligned}
\vec{r}_{n1} &= 0\hat{i} + 0\hat{j} \\
\vec{v}_{n1} &= \dot{\vec{r}}_{n1} = 0\hat{i} + 0\hat{j} \\
\vec{r}_{n2} &= r_{n2}\cos(\theta_2)\hat{i} + r_{n2}\sin(\theta_2)\hat{j} \\
\vec{v}_{n2} &= \dot{\vec{r}}_{n2} = \underbrace{\frac{dr_{n2}}{dt}}_{=0}\cos(\theta_2)\hat{i} + r_{n2}\underbrace{\frac{d\cos(\theta_2)}{dt}}_{-\dot{\theta}_2\sin(\theta_2)}\hat{i} + \underbrace{\frac{dr_{n2}}{dt}}_{=0}\sin(\theta_2) + r_{n2}\underbrace{\frac{d\sin(\theta_2)}{dt}}_{\dot{\theta}_2\cos(\theta_2)}\hat{j} \\
\vec{v}_{n2} &= -r_{n2}\dot{\theta}_2\sin(\theta_2)\hat{i} + r_{n2}\dot{\theta}_2\cos(\theta_2)\hat{j} = -\omega_2 r_{n2}\sin(\theta_2)\hat{i} + \omega_2 r_{n2}\cos(\theta_2)\hat{j} = \dot{x}_{n2}\hat{i} + \dot{y}_{n2}\hat{j}
\end{aligned} \tag{6.12}$$

or

$$\dot{x}_{n2} = -\omega_2 r_{n2}\sin(\theta_2)\text{and}\dot{y}_{n2} = \omega_2 r_{n2}\cos(\theta_2)$$

Note, in Eq. (6.12), r_{n2} is constant but θ_2 is changing with time with derivative ω_2. Now we are going to use Eq. (6.4) to convert $\vec{v}_{n2}$ into polar coordinates as:

$$\begin{aligned}
v_{rn2} &= \dot{x}_{n2}\cos(\theta) + \dot{y}_{n2}\sin(\theta) = -\omega_2 r_{n2}\sin(\theta)\cos(\theta) + \omega_2 r_{n2}\cos(\theta)\sin(\theta) = 0 \\
v_{\theta n2} &= -\dot{x}(t)\sin(\theta) + \dot{y}(t)\cos(\theta) = \omega_2 r_{n2}\sin(\theta) + \omega_2 r_{n2}\cos(\theta) = \omega_2 r_{n2}
\end{aligned} \tag{6.13}$$

or

$$v_{rn2} = 0 \quad \text{and} \quad v_{\theta n2} = \omega_2 r_{n2}$$

From Eq. (6.13), we can come up with the rule that for any link L_i connected to ground at node n_j by a revolute joint J_j^R that rotates with an angular velocity ω_i, the node k located on the link L_i has only tangential velocity component part $v_{\theta nk} = \omega_i r_k$, where r_k is the radius from node j to node k. Our example in Fig. 6.9 has $i=2, j=1$ and $k=2$.

Assume there is a link with a set of $m+1$ nodes and connected to the ground by a revolute joint at node n_j that has a zero velocity ($v_{\theta nj} = 0$ and $v_{rnj} = 0$). The revolute joint is driven to rotate the link with an angular velocity ω. On the link, any joint n_i only has a tangential velocity $v_{\theta ni} = \omega r_{ni}$, where r_{ni} is the radius from n_j to n_i. The node with the highest radius value from n_j and located on the link will have the highest total velocity that is equal to the tangential velocity.

For example, Fig. 6.10 shows a link L_2 connected to the ground by a revolute joint J_1^R at node n_1. On the link L_2, there are three nodes n_2, n_3, and n_4 that are far from n_1 with radii: r_2, r_3, and r_4 with the ascending order $r_2 < r_4 < r_3$. Note, the symbol "<" means less in a quantitative value. The joint J_1^R is driven to rotate the link with ω_2. The three nodes n_2, n_3, and n_4 have the total velocity equal to the tangential velocity as $v_{\theta n2} = \omega_2 r_2, v_{\theta n4} = \omega_2 r_4$, and $v_{\theta n3} = \omega_2 r_3$ that leads to $v_{\theta n2} < v_{\theta n4} < v_{\theta n3}$.

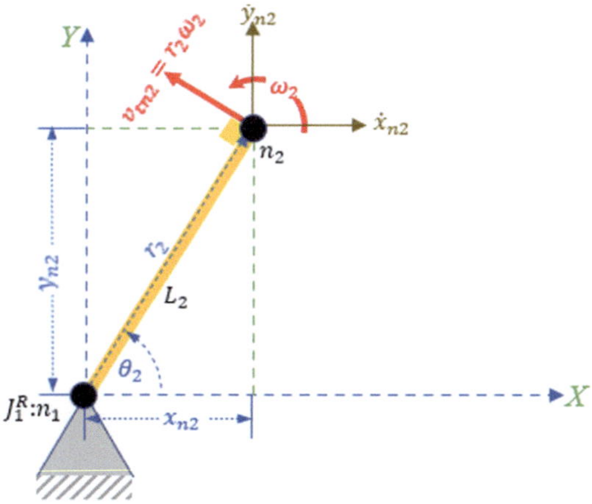

Fig. 6.9 Velocity components of two nodes hinged to ground link

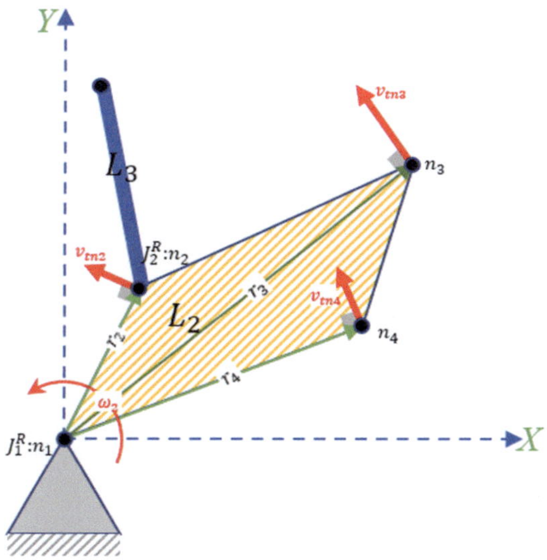

Fig. 6.10 Kinematic diagram of four nodes link with velocity vectors

Example 6.3

The door closer mechanism, in Fig. 6.11, consists of four links: frame (ground), follower, coupler, and the door (the crank) in addition to four revolute joints. If a person opens the door from angle 0° to 85° in two seconds with a constant angular velocity:

- Sketch the kinematic diagram.
- Which door node (stile edge n_5 or door hinged n_2) has a higher velocity? Prove your answer with values.

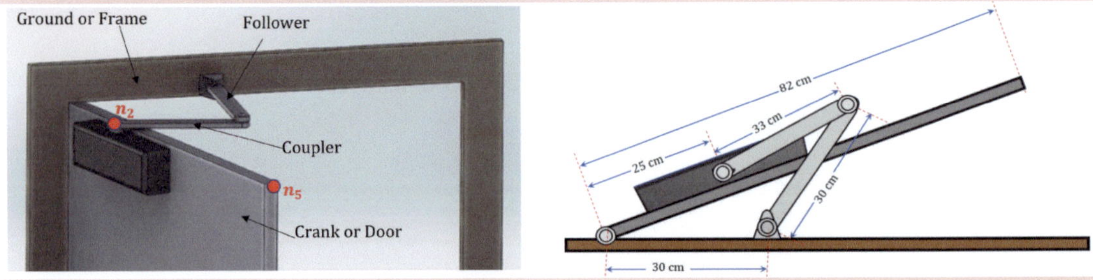

Fig. 6.11 Door closer mechanism. Figure and associated Door_4Bar_Mechanism.mp4 video designed by the author and implemented by both the author and Pranav Dhola. Both courtesy of Pranav Dhola

Solution

Kinematic Diagram

Figure 6.12 shows the kinematic diagram of the door. We selected the frame or the ground link as L_1, the crank or door as L_2, the coupler rod (33 cm) as L_3, and the follower rod (30 cm) as L_4. The mechanism has four revolute joints $J_1^R : n_1, J_2^R : n_2, J_3^R : n_3$, and $J_4^R : n_4$. The door link has an end-effector at n_5.

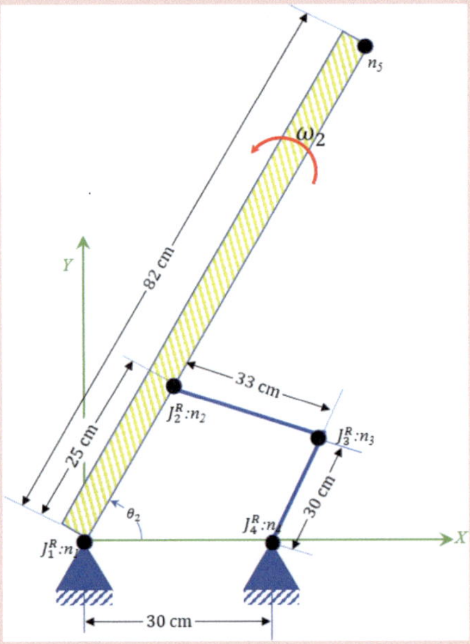

Fig. 6.12 Kinematic diagram of door close mechanism

Which Node Has a Higher Velocity Value?

The two nodes n_2 and n_5 are located on the link L_2. Node n_5 has $r_5 = 82$ cm which is higher than $r_2 = 25$ cm at n_2. Then, $\vec{v}_{n2} < \vec{v}_{n5}$ because $(v_{\theta n2} = \omega_2 r_2) < (v_{\theta n5} = \omega_2 r_5)$.

Numerical Proof

The angular velocity ω_2 can be calculated by dividing 85° over 2 s or:

$$\omega_2 = \frac{85}{2} \times \frac{\pi}{180} = 0.742\,\text{rad/s}$$

The velocities at nodes n_2 and n_5 are tangential only:

$$v_{\theta n2} = \omega_2 r_2 = 0.742 \times 0.25 = 0.1855\frac{\text{m}}{\text{s}}$$

$$v_{\theta n5} = \omega_2 r_5 = 0.742 \times 0.82 = 0.6084\frac{\text{m}}{\text{s}}$$

From the numerical proof, the velocity at node n_5 is higher than the velocity at n_2. We recommend that the reader watches the attached door opener mechanism animation video for further understanding.

6.3.2 Pin-in-a-Slot to Ground

Figure 6.13 shows a two-node link L_k attached to the ground at node n_j with a pin-to-slot joint J_i^{PS} The constant sliding angle γ_{ni} makes the sliding velocity component $\dot{x}_{ni}$ and $\dot{y}_{ni} = \dot{x}_{ni}\tan(\gamma_{ni})$. The link L_k is also rotating with an angular velocity ω_k around the node n_i. The node n_j is located on the link with a radius r_{njni} from node n_i. The velocity of n_j can be broken down into a sliding velocity $\dot{x}_{ni}$ and $\dot{y}_{ni} = \dot{x}_{ni} \tan\tan(\gamma_{ni})$ plus a tangential rotating velocity $\omega_2 r_{n2}$. The velocity vector of nodes n_1 and n_2 are:

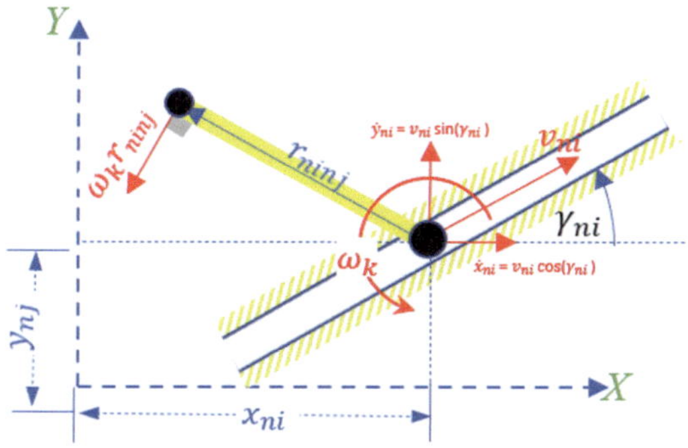

Fig. 6.13 Kinematic diagram of connected link with a pin-slotted joint to ground

$$\begin{aligned}
\vec{r}_{ni} &= x_{ni}\hat{i} + x_{ni}\tan(\gamma_{ni})\hat{j} \\
\vec{v}_{ni} &= \dot{\vec{r}}_{ni} = \dot{x}_{ni}\hat{i} + \dot{x}_{ni}\tan(\gamma_{ni})\hat{j} \\
\vec{r}_{nj} &= \left(x_{ni} + r_{njni}\cos(\theta_k)\right)\hat{i} + \left(x_{ni}\tan(\gamma_{ni}) + r_{njni}\sin(\theta_k)\right)\hat{j} \\
\vec{v}_{nj} &= \dot{\vec{r}}_{nj} = \left(\dot{x}_{ni} + \underbrace{r_{ninj}}_{\text{constant}}\frac{\mathrm{d}\cos(\theta_k)}{\mathrm{d}t}\right)\hat{i} + \left(\dot{x}_{ni}\tan\left(\underbrace{\gamma_{ni}}_{\text{constant}}\right) + \underbrace{r_{ninj}}_{\text{constant}}\frac{\mathrm{d}\sin(\theta_k)}{\mathrm{d}t}\right)\hat{j} \\
\vec{v}_{nj} &= \left(\dot{x}_{ni} - \dot{\theta}_i r_{njni}\sin(\theta_k)\right)\hat{i} + \left(\dot{x}_{ni}\tan(\gamma_{ni}) + \dot{\theta}_k r_{njni}\cos(\theta_k)\right)\hat{j} \\
\vec{v}_{nj} &= \left(\dot{x}_{ni} - \omega_k r_{njni}\sin(\theta_k)\right)\hat{i} + \left(\dot{x}_{ni}\tan(\gamma_{ni}) + \omega_k r_{njni}\cos(\theta_k)\right)\hat{j}
\end{aligned} \tag{6.14}$$

Example 6.4

For the kinematic diagram shown of the piston–cylinder mechanism in Fig. 6.14, at an instantaneous position, the combustion force moves the piston outward with a sliding velocity of 25 m/s that causes the connecting rod to have an angular velocity of 2000 rpm around the wrist rest pin. The 130 mm connecting rod has an angle of 170° with the *X-axis*. Calculate the instantaneous Cartesian velocity vector at the crankpin.

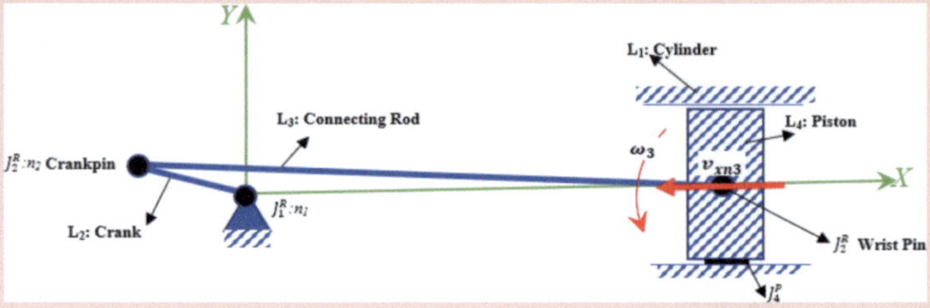

Fig. 6.14 Kinematic diagram of piston–cylinder mechanism

Solution

The connecting rod is the same as the case in Fig. 6.13 where i=3, j=2, k=3, γ_{n2}=0°, r_{n2n3}=r_{n3n2}=130°, $\dot{x}_{n3} = -25$ m/s, and $\dot{y}_{n3} = 0$. Using the free body diagram of the connecting rod as shown in Fig. 6.15 and using Eq. (6.14)

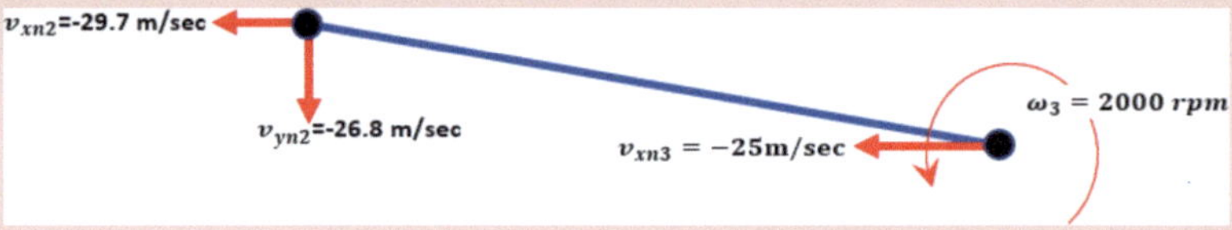

Fig. 6.15 Free body diagram of connecting rod

$$\begin{aligned}\vec{v}_{n2} &= (\dot{x}_{n3} - \omega_3 r_{n3n2}\sin(\theta_3))\hat{i} + \omega_3 r_{n3n2}\cos(\theta_3)\hat{j}\\ &= \left(-25 - 2000 \times \frac{\pi}{30} \times \frac{130}{1000} \times \sin(170°)\right)\hat{i} + 2000 \times \frac{\pi}{30} \times \frac{130}{1000} \times \cos(170°)\hat{j}\\ &= -29.728\hat{i} - 26.814\hat{j}\text{ m/s}\end{aligned}$$

6.4 Absolute and Relative Velocities

In Fig. 6.16, the fixed world or inertial frame *XY* is located at the ground node $n_{A,}$ or the origin. A local rotating and moving frame $X_{nB}Y_{nB}$ are located at node n_B with a position vector $\vec{r}_{nB} = x_{nB}\hat{i} + y_{nB}\hat{j}$. The translational velocity of the local frame $X_{nB}Y_{nB}$ is the absolute velocity of the node n_B with symbol $\vec{v}_{nB} = \dot{x}_{nB}\hat{i} + \dot{y}_{nB}\hat{j} = v_{xnB}\hat{i} + v_{ynB}\hat{j}$. The angular velocity of the local frame $X_{nB}Y_{nB}$ is $\Omega_B = \omega_B\hat{k}$. A node n_C is defined by the absolute position vector $\vec{r}_{nC} = x_{nC}\hat{i} + y_{nC}\hat{j}$. The relative position vector of n_C with respect to n_B is $\vec{r}_{nCnB} = x_{nCnB}\hat{i} + y_{nCnB}\hat{j}$. The absolute velocity vector of n_C, $\vec{v}_{nC}$, can be related to the relative velocity $\vec{v}_{nCnB}^{relative}$ as, Ref. [4],

$$\vec{v}_{nC} = \vec{v}_{nB} + \vec{\Omega}_B \times \vec{r}_{nCnB} + \vec{v}_{nCnB}^{\text{relative}} \tag{6.15}$$

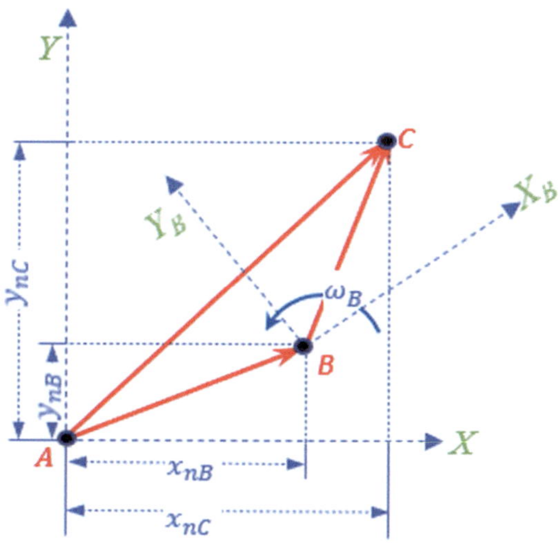

Fig. 6.16 Rotating relative frames

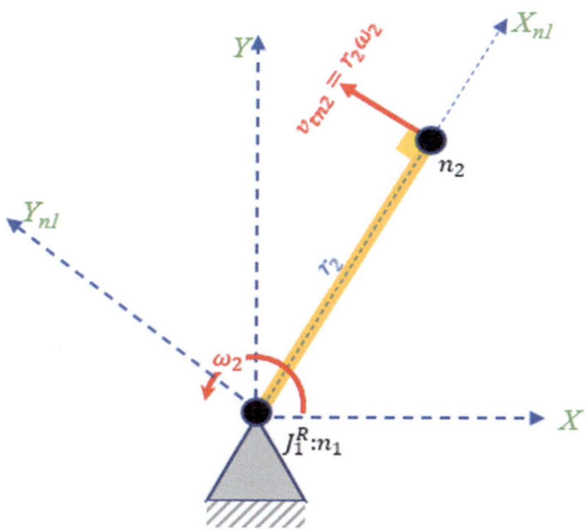

Fig. 6.17 Relative motion of hinged link to ground

The absolute velocity has three components: the absolute velocity vector $\vec{v}_{nB}$ at node n_B, the relative velocity vector $\vec{v}^{relative}_{nCnB}$ of node n_C with respect to node n_B, and the rotational velocity vector $\vec{\Omega}_B \times \vec{r}_{nCnB}$ of the local frame $x_{nB}y_{nB}$. Note, the relation in Eq. (6.15) is a general form so that node n_A, node $n_{B,}$ and node n_C do not have to be on the same link. We are going to use the formula in Eq. (6.15) to calculate the relative and absolute velocities for the rigid bodies of the links.

We can apply Eq. (6.15) to the model in Fig. 6.9; a single link connected to the ground by a revolute joint. Node $n_A = n_B = n_1$ meaning $\vec{v}_{nB} = \vec{0}$. The rotating frame, $X_{nB}Y_{nB} = X_{n1}Y_{n1}$ with an angular velocity $\vec{\Omega}_B = \omega_2 \hat{k}$, is going to be located at node n_1 and X_{n1} is alignment with the link L_2 as shown in Fig. 6.17. The node $n_C = n_2$ has a zero relative velocity with respect to the rotating frame $X_{n1}Y_{n1}$ or $\vec{v}^{\text{relative}}_{nCnB} = \vec{v}^{\text{relative}}_{n2n1} = \vec{0}$. Equation (6.16) will be the same as Eq. (6.13) as listed in Eq. (6.16). A similar technique can be used for the model in Fig. 6.13 using Eq. (6.15).

$$\underbrace{\vec{v}_{nC}}_{\vec{v}_{n2}} = \underbrace{\vec{v}_{nB}}_{\vec{0}} + \underbrace{\vec{\Omega}_B}_{\omega_{2\hat{k}}} \times \underbrace{\vec{r}_{nCnB}}_{r_{n2}\hat{\mathrm{r}}} + \underbrace{\vec{v}^{\text{relative}}_{nCnB}}_{\vec{0}} = \omega_2 r_{n2}\hat{\theta} \tag{6.16}$$

6.5 Simscape™ Multibody™ Velocity Sensor and Actuator Blocks

In Simscape™ Multibody™, the "***Transform Sensor***" block can be used to measure the velocity as well as the displacement. Figure 6.18 shows the "***Transform Sensor***" block that has Port B and Port F which represent the base and follower frames respectively. As an example to explain the functionality of Port B and Port F, if we are measuring the relative velocity of node n_i with respect to node n_j, we will connect Port B (base) to n_j and we will connect Port F (follower) to n_i. You can drag and drop the "***Transform Sensor***" block onto the model.

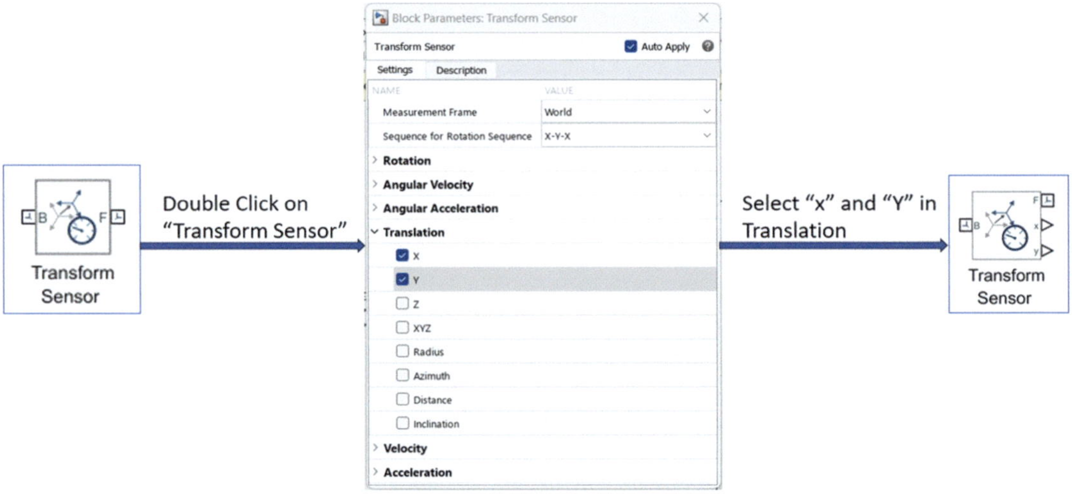

Fig. 6.18 Step to activate displacement sensors from transform sensor block. Designed and implemented by the author. Images used with permission of MathWorks

Table 6.1 Motion sensor. Figures are designed and implemented by the author. Images used with permission of MathWorks

Sensor motion	Sensors under each motion
Angular velocity	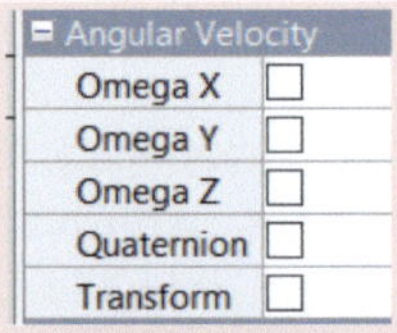
Velocity	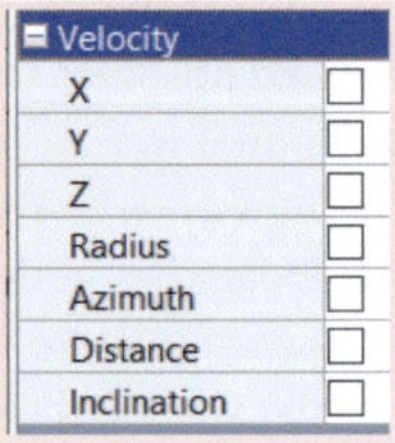

As shown in Fig. 6.18, when you double-click the "***Transform Sensor***" block, there is a list of items: Rotation, angular velocity, angular acceleration, translation, velocity, and acceleration. If you click on any of the listed items, a sublist will show. Table 6.1 shows the sublists of "angular velocity" and "velocity", which are our focus in this chapter. We can pick whatever signal that we would like to measure. For example, if we are interested in measuring the angular velocity around the *z*-axis at a specific node, we select "Omega Z". Immediately, we will get an extra output port from the block, called "WZ". There is further information about the "***Transform Sensor***" block in "Simscape™ Multibody™ Getting Started Guide" and "Simscape™ Multibody™ User's Guide", Refs. [5, 6].

Engineering Tips—Velocity Sensor as Simulink® Subsystem

To reduce the model visualization's complexity, we are going to build a set of velocity sensor subsystems and save them in our "Kinematics_Of_Machines.slx" file. One example is the

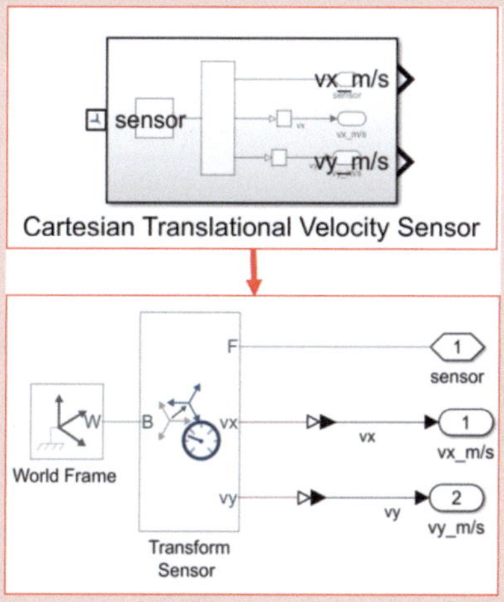

Fig. 6.19 Velocity sensor subsystem. Designed and implemented by the author. Images used with permission of MathWorks

Cartesian translational velocity sensor as shown in Fig. 6.19 that can be built by the following steps:

1. Drag a "***Transform Sensor***" block from the Simscape™ Multibody™ library and drop it into a new model.
2. Double-click on the "***Transform Sensor***" block and select "Velocity". Select X and Y in the velocity sublist. Press "Ok". You will see two more output ports, vx and vy.
3. Connect port F of the "***Transform Sensor***" block to the "***Connection Port***" block. Rename the "***Connection Port***" "sensor". Connect the two ports vx and vy to two "***PS-Simulink® Converter***" followed by two "Out" blocks and call them "vx_m/s" and "vy_m/s".
4. Connect the B port to the world frame.
5. Create a subsystem and rename the subsystem as "***Cartesian Translational Velocity Sensor***".

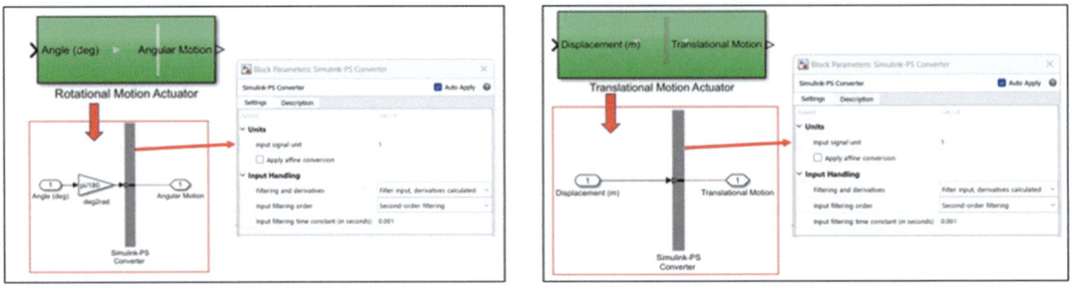

Fig. 6.20 Motion actuators. Designed and implemented by the author. Images used with permission of MathWorks

More velocity sensor subsystems were built and added to our file "Kinematicsof_ Machines.slx". The reader can obtain more details about the velocity sensor subsystems in the provided "Kinematics_Of_Machines.slx" model.

For the actuation block, we are going to use our previously developed subsystems "Rotational Motion Actuator" and "Translational Motion Actuator" from our "Kinematics_Of_Machines. slx" file as shown in Fig. 6.20. We pass the displacement input signal to the subsystem. The Simulink®-PS Converter block makes the derivatives (calculates the velocity and acceleration from the displacement) as we set input handling properties. The "input filtering time constant" is set at 0.001 s. We will discuss how to change this input filtering time constant relative to the input signal. For example, if we have a translational motion input that changes from 0 to 5 m in 2 s with a constant speed, we can use the Signal Builder block to build a signal that changes from 0 to 5 m in 2 s. We will explore this technique in the upcoming examples.

Example 6.5

For the door mechanism in Fig. 6.11, if a person opens the door from angle 0° to 85° in two seconds with a constant angular velocity:

- Build a Simscape™ Multibody™ model to simulate this scenario.
- Measure the speed at the door hinge node and the stile edge node.
- Compare your results with Example 6.3

Solution

Simscape™ Multibody™ model:

Open a new script file and save it as "DoorOpener_P6p5_script.m". Enter the commands below. This script starts by clearing the history of all the variables followed by declaring the input parameters and calculating each link's required parameters.

```
clc
clear all
close all
%===========================================================================
%Declare the given parameters
%==========================
r_n2   = 25/100; %m
r_n5   = 82/100; %m
r_n3n2 = 33/100; %m
r_n3n4 = 30/100; %m
r_n4   = 30/100; %m

theta2_old_deg = 0; %deg
theta2_new_deg = 85; %deg

%Calculation of parameters
%=======================
L2_crosssection = [0 0;r_n2 0;r_n5 0;r_n5 r_n2/10;0 r_n2/10];
L3_dimension   = [r_n3n2 0.1*r_n2 0.1*r_n2]; %m
L4_dimension   = [r_n4 0.1*r_n2 0.1*r_n2]; %m

Frame_L1_n1 = [0 0 0]; %m
Frame_L1_n4 = [r_n4 0 0]; %m
%
Frame_L2_n1 = [0 0 0]; %m
Frame_L2_n2 = [r_n2 0 0]; %m
Frame_L2_n5 = [r_n5 0 0]; %m

Frame_L3_n2 = [-r_n3n2/2 0 0]; %m
Frame_L3_n3 = [r_n3n2/2 0 0]; %m

Frame_L4_n4 = [-r_n3n4/2 0 0]; %m
Frame_L4_n3 = [r_n3n4/2 0 0];%m
```

We are going to use the developed masks and subsystems from "Kinematics_Of_Machines.slx" to build the model as shown in 6.21a

1. Open a new model with the name "DoorOpener_P6p5_model.slx".
2. Drag and drop a "Signal Builder" block from the Simulink®\Source library. Edit the signal $\theta_2(t) = 85/2 \times t$ for $0 \leq t \leq 2$ shown in Fig. 6.21b.
3. Copy and paste the four listed links' masks in Table 6.2 using the declared parameters from "DoorOpener_P6p5_script.m". Note, "w.r.t" is an abbreviation for "with respect to" and C.G is an abbreviation for "Center of Gravity".
4. Drag and drop the four revolute joints and name them J_1R, J_2R, J_3R, and J_4R. After that, connect the joints.
5. Copy two "Tangential Velocity Sensor" subsystems from "Kinematics_Of_Machines.slx" and paste them in "DoorOpener_P6p5_model.slx". Connect the two pasted sensors and name them vt_n2 and vt_n5. Connect the measured velocities vt_n2 and vt_n5 to a scope.
6. Double-click on J_1R and a motion actuator. Copy and paste the "Rotational Motion Actuator" subsystem and connect to the "Signal Builder" block and J_1R block.

Run the script file followed by running the model. Double-click the scope to see the results as shown in Fig. 6.22a. The results are consistent with the analytical solution in Example 6.3 During the simulation, you will see your first animation for the mechanism. Congratulations!

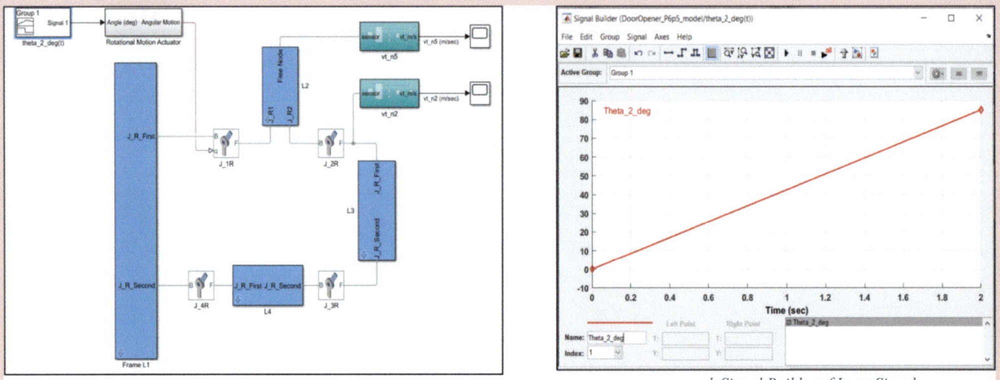

a Simscape Multibody Model *b Signal Builder of Input Signal*

Fig. 6.21 **a** Simscape™ Multibody™ model **b** Signal builder of input signal

Table 6.2 Model parameters of Example 6.5

Link	Selected mask	Plugged-in parameters
Frame, L_1	Ground with two revolute joints	Position of n_1 w.r.t C.G of L_1: Frame_L1_n1 = [0 0 0] Position of n_4 w.r.t C.G of L_1: Frame_L1_n4 = [r_n4 0 0]
Door, L_2	Extrusion with two revolute joints	Cross section: L2_crosssection = [0 0;r_n2 0;r_n5 0;r_n5 r_n2/10;0 r_n2/10] Position of n_1 w.r.t origin of L_2: Frame_L2_n1 = [0 0 0] Position of n_2 w.r.t origin of L_2: Frame_L2_n2 = [r_n2 0 0] Position of n_5 w.r.t origin of L_2: Frame_L2_n5 = [r_n5 0 0]
Coupler, L_3	Brick with two revolute joints	Dimension: L3_dimension = [r_n3n2 0.1*r_n2 0.1*r_n2] Position of n_2 w.r.t C.G of L_3: Frame_L3_n2 = [-r_n3n2/2 0 0] Position of n_3 w.r.t C.G of L_3: Frame_L3_n3 = [r_n3n2/2 0 0]
Follower, L_4	Brick with two revolute joints	Dimension: L4_dimension = [r_n3n4 0.1*r_n2 0.1*r_n2] Position of n_4 w.r.t C.G of L_3: Frame_L4_n4 = [-r_n3n4/2 0 0] Position of n_3 w.r.t C.G of L_3: Frame_L4_n3 = [r_n3n4/2 0 0]

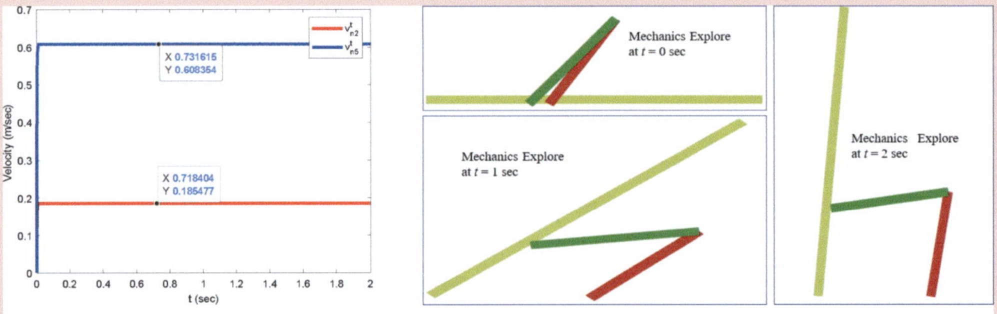

a Velocities at n_2 and n_5 *b Mechanics Explorer at Different Times*

Fig. 6.22 Results of Simscape™ Multibody™ model for door mechanism

We encourage the reader to watch the attached animated video from this model as well as to look at the attached model to provide a further understanding of the problem.

6.6 Graphical Velocity Analysis

Graphical velocity analysis is a necessary method that every engineer should know to visualize the velocity components even if there is an analytical or a numerical solution available. When you discuss the design analysis with your coworker or your technical manager, you will need to use this graphical velocity analysis tool.

We use two coordinate systems, Cartesian and polar, to visualize the velocity components of any node on a link. First, we will cover some basic analyses of the Multibody™ kinematics: Rigid Body Instantaneous Center of Zero Velocity, Kinematic Pair Instantaneous Centers of Velocity, and Kennedy's Theorem. We will then apply these analyses to the four-bar mechanism, the slider–crank mechanism, and beyond. Additionally, we will demonstrate how to use the Multibody™ kinematic analysis in estimating the velocity at any mechanism node.

6.6.1 Rigid Body Instantaneous Center of Zero Velocity

Definition

Rigid Body Instantaneous Center of Zero Velocity:

The instant center of rotation, also called the instantaneous center of zero velocity, is the point fixed to a body undergoing planar movement that has zero velocity at a particular instant of time, Ref. [7].

For each rigid body, there is an existing instantaneous zero velocity point or, as we refer to it in this textbook, "instantaneous ground node". As shown in Fig. 6.23, a rigid body or link L has two nodes n_A and n_B with two instantaneous velocities $\vec{v}_A(t)$ and $\vec{v}_B(t)$. There an instantaneous zero velocity point or an instantaneous ground node $IC(t)$ that is defined as the intersection of the two perpendicular lines on the two velocities vectors $\vec{v}_A(t)$ and $\vec{v}_B(t)$. Another way to think about $IC(t)$ is that if we were to extend the rigid body or link to the $IC(t)$ location, then we could model the link as a hinged link to the instantaneous ground as shown in Fig. 6.23. In general, any rigid body can be analyzed as a hinged link to the ground as listed in Sect. 6.3.1 where the ground is the instantaneous center of zero velocity.

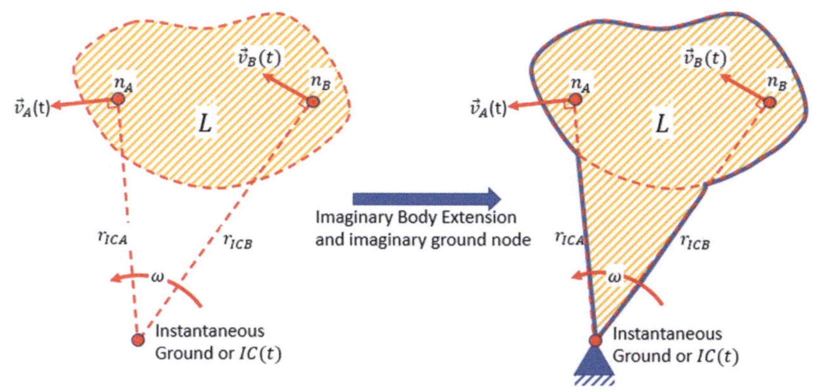

Fig. 6.23 Instantaneous center of zero velocity and instant ground

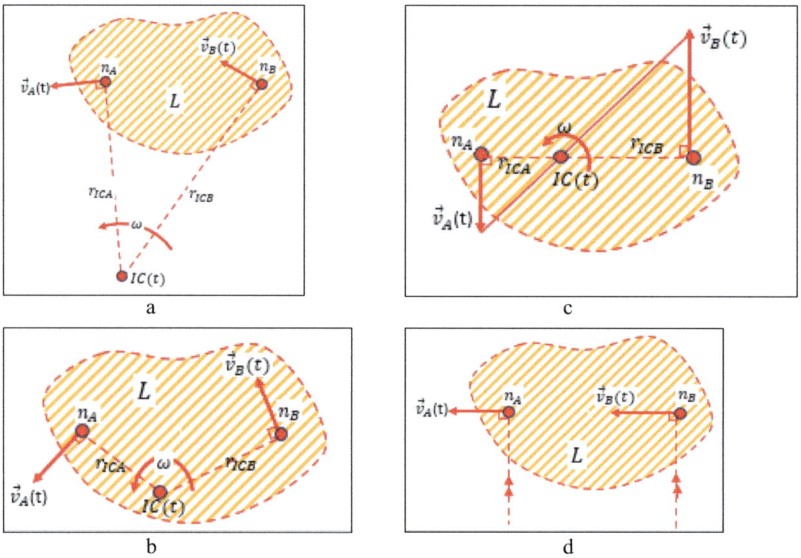

Fig. 6.24 Different locations of single link IC

The *IC* can be stationary or does not change with time if one of the link's nodes is connected to the ground. Otherwise, there are three cases of the *IC* location:

1. The two perpendicular lines on the velocity vectors $\vec{v}_A(t)$ and $\vec{v}_B(t)$ have a single intersection point that could be inside the actual link as shown in Fig. 6.24a or outside the link as shown in Fig. 6.24b. The link is rotating around this $IC(t)$ node. The magnitude of the two velocities $\vec{v}_A(t)$ and $\vec{v}_B(t)$ can be expressed as:

$$\begin{aligned} v_A &= \omega r_{ICA} \\ v_B &= \omega r_{ICB} \end{aligned} \tag{6.17}$$

2. The two perpendicular lines on the velocity vectors $\vec{v}_A(t)$ and $\vec{v}_B(t)$ are on the top of each other with an infinite number of intersection points as shown in Fig. 6.24c. In this case, the *IC*(t) node is located somewhere in the middle between nodes n_A and n_B. Equation (6.17) can be used to calculate the magnitude of the two velocities $\vec{v}_A(t)$ and $\vec{v}_B(t)$.
3. The two perpendicular lines on the velocity vectors $\vec{v}_A(t)$ and $\vec{v}_B(t)$ are parallel to each other with no intersection point or $IC(t)=\infty$ as shown i n Fig. 6.24d. In this case, the body has no rotational speed or $\omega=0$ rad/s.

6.6.2 Kinematic Pair Instantaneous Centers of Velocity

Definition

Kinematic Pair Instantaneous Center of Velocity:

The instantaneous center of velocity is defined as the instantaneous location of a pair of coincident points of two different rigid bodies in planar motion for which the absolute velocities of the two points are equal. The instantaneous center of velocity may also be defined as the location of a pair of coincident points of two different rigid bodies in planar motion for which the apparent velocity of one of the points is zero, as seen by an observer on the other body [2].

For each kinematic pair (two links connected by a joint as listed in Chap. 2), there is an instantaneous center of velocity or velocity pole at which the two links have the same velocity. In other words, assume a kinematic pair of two links L_i and L_j are connected by a joint $J_k^l : n_k$ where i, j, and k are integral values and $l = \{ R, C, G, P, S, PS\}$. There is an instantaneous center of velocity pole (or node) I_{ij} that belongs to the two links and has the same velocity. For example, assume link L_3 is connected to link L_2 at a revolute joint $J_5^R : n_5$. Node n_5 belongs to both L_2 and L_3. Since the joint is revolute, there is no slipping motion, and node n_5 has the same velocity with respect to L_2 and L_3. Then, n_5 is the instantaneous center of velocity I_{23}.

The same concept of a revolute joint kinematic pair, that has two links connected through a revolute joint, can be generalized for any kinematic pair. In general, we have two types of kinematic pairs to describe the instantaneous centers of velocity: non-slipping or rolling-without-sliding kinematic pairs (revolute, gear, and cam) and sliding kinematic pairs (prismatic, pin-in-a-slot, and screw). Now, we are going to discuss how to estimate the instantaneous center of velocity for each type.

No-slipping and no-sliding kinematic pair: A link L_i is connected to the link L_j by a joint J_k^l that does not allow sliding motion between the two links and the joint J_k^l does not have a slipping motion between the two links. Note, k is the joint index, and $l = \{R, G\}$ is the joint type. The node n_k is called the instantaneous center of velocity for L_i and L_j since the joint node is the only point that has the same absolute velocity (no-slipping and no-sliding) for both L_i and L_j. We refer to this instantaneous center of velocity with the symbol I_{ij} in addition to n_k. Figure 6.25 shows two different examples of the instantaneous center of velocity for no-slipping or no-sliding kinematic pairs connected by revolute joints and gear joints.

Sliding kinematic pair: If a link L_i is connected to the link L_j by a sliding joint J_k^l, where k is the joint index and $l = \{P, S, PS\}$ is joint type, the two links slide on each other and there is no node with the same velocity belonging to these two links. In this case, we say that the instantaneous center of velocity is located at infinity perpendicular to the sliding motion. We refer to this as the instantaneous center of velocity with the symbol I_{ij}. Fig. 6.26 shows the instantaneous center of velocity for a sliding kinematic pair connected by a prismatic joint. Note, there is only a specific direction of I_{ij}; there is no specific position of I_{ij}.

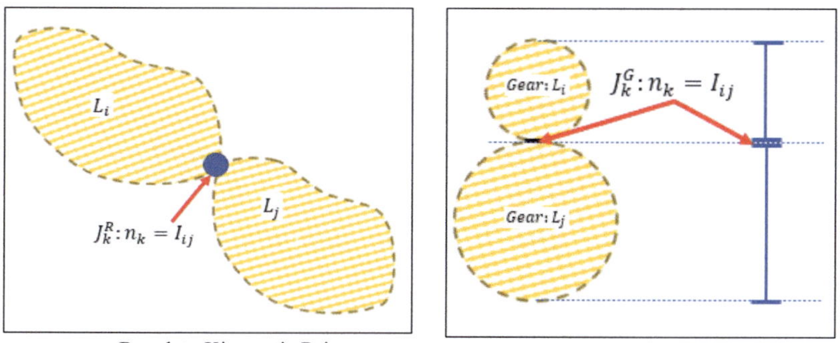

a Revolute Kinematic Pair

b Gear Kinematic Pair

Fig. 6.25 Instantaneous centers of no-sliding or no-slipping kinematic pairs

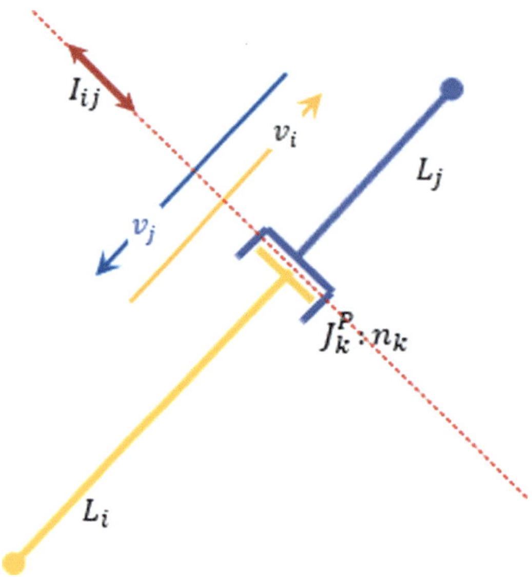

Fig. 6.26 Instantaneous center of a kinematic pair connected by a prismatic joint

The idea of locating the instantaneous center of velocity for a kinematic pair can be extended to the instantaneous centers of velocity for a mechanism. For any mechanism with n number of links, there are N instantaneous centers of velocity for this mechanism where, from Ref. [2],

$$N = \frac{n(n-1)}{2} \tag{6.18}$$

For example, if we have a mechanism with five links, regardless of the types of joints, there are $N = n(n-1)/2 = 5 \times 4/2 = 10$ instantaneous centers of velocity. To identify the N number of instantaneous centers of velocity and their location, we are going to introduce Kennedy's circle and Kennedy's theorem.

6.6.3 Kennedy's Theorem

Rule

Kennedy's Theorem

Three instant centers shared by ***three*** rigid bodies in motion with respect to one another (whether or not they are connected) all lie on the same straight line [2].

Kennedy's theorem states that if three planar links have motion relative to one another, their instantaneous centers should lie in a straight line. For example, the three links L_i, L_j, and L_k are connected with two no-sliding and no-slipping joints J_m and J_n at nodes n_n and n_m as shown in Fig. 6.27. The two instant centers I_{ij} and I_{jk} are located at nodes n_n and n_m since the joints J_m and J_n do not have a sliding motion or a slipping motion. We call these two instant centers I_{ij} and I_{jk} "primary instantaneous". Kennedy's theorem tells us that the third instant center I_{ik} (Secondary Instantaneous) should be located on the line of I_{ij} and I_{jk}.

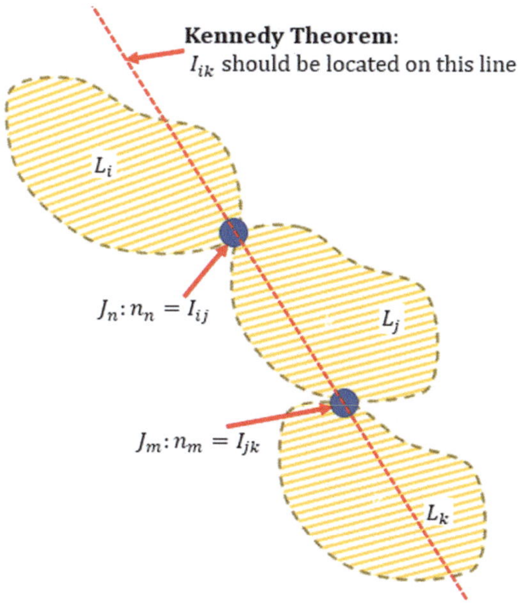

Fig. 6.27 Applying Kennedy's theorem to three links with two no-slipping and no-sliding joints

By using Kennedy's theorem, we can determine that there are two types of instantaneous centers of velocity in any mechanism: primary and secondary. Primary instantaneous centers of velocity come from the kinematic pairs (sliding joint or non-sliding joint). The locations of the primary instantaneous centers can be calculated as shown in Fig. 6.25 or Fig. 6.26. Secondary instantaneous centers of velocity come from non-kinematic pairs. The locations of these secondary centers are calculated using Kennedy's theorem.

Kennedy's theorem proposes the so-called Kennedy Circle to identify and determine the instantaneous centers of any mechanism using the following steps:

1. Locate the primary instantaneous centers of velocity connections using the joint types between each kinematic pair.
2. Draw a circle and mark evenly spaced ticks to represent each link in the mechanism.
3. Draw solid lines identifying each of the known primary instant centers for each kinematic pair.
4. Draw dashed lines between each point (or non-kinematic pair links) to represent the remaining secondary instant centers of velocity to be found.
5. Apply Kennedy's theorem to find the unknown secondary instantaneous centers of velocity.

The instantaneous center of zero velocity of a single rigid body or a single link along with the instantaneous centers of velocity of a mechanism is extremely useful for calculating the angular velocities and velocity vectors at any node on the mechanism.

Now, we are going to apply our knowledge of the zero center of velocity plus Kennedy's theorem to a four-bar mechanism, a slider–crank mechanism, and an inverted slider–crank mechanism to develop a graphical technique to estimate the velocity at each node and estimate the output-to-input velocity ratio.

6.6.4 Four-Bar Mechanism Graphical Velocity Analysis

The four-bar mechanism has four links L_1, L_2, $L_{3,}$ and L_4 that have $N = n(n-1)/2 = 4 \times 3/2 = 6$ instantaneous centers of velocity. Assume ω_2 is the input in CCW as the input signal. Now, we follow the algorithm of drawing the Kennedy Circle as:

- As shown in Fig. 6.28, locate the four primary instantaneous centers of velocity connections using the joint types between each kinematic pair.
  (1) Pair 1, L_1-L_2, connected through a non-slipping and non-sliding revolute joint J_1^R that makes I_{12} located at node n_1.
  (2) Pair 1, L_2–L_3, connected through a non-slipping and non-sliding revolute joint J_2^R that makes I_{23} located at node n_2.
  (3) Pair 1, L_3–L_4, connected through a non-slipping and non-sliding revolute joint J_3^R that makes I_{34} located at node n_3.
  (4) Pair 1, L_1-L_4, connected through a non-slipping and non-sliding revolute joint J_4^R that makes I_{14} located at node n_4.
- Draw a circle and mark four evenly spaced tick marks to represent each link in the mechanism as shown in Fig. 6.28a.
- Draw solid Line 12, Line 23, Line 34, and Line 14 to represent the four primary instant centers for each kinematic pair as Fig. 6.28b.
- Draw dashed Line 13 and Line 24 to represent the remaining secondary instant centers of velocity for each non-kinematic pair Fig. 6.28c.

Now, we have six instantaneous centers of velocity. To estimate their location, start with Fig. 6.29a where we draw the four primary instantaneous centers of velocity I_{12}, I_{23}, I_{34}, and I_{14} that are located at each joint J_1^R, J_2^R, J_3^R, and J_4^R. The first secondary center I_{13} is calculated using Kennedy's theorem as

- ***Line 1:*** The three centers I_{12}, $I_{23,}$ and I_{13} must lie on the same straight line.
- ***Line 2:*** The three centers I_{14}, $I_{34,}$ and I_{13} must lie on the same straight line.

The intersection of ***Line 1*** and ***Line 2*** is the instantaneous center I_{13} as shown in Fig. 6.29b. Using the same reasoning, we are going to locate the second secondary center I_{24} as shown in Fig. 6.29c to be the intersection of Line I_{14}–I_{12}–I_{24} and Line I_{34}–I_{23}–I_{24}.

Observation 1: The node I_{13} is instantaneously located on both L_1 and L_3 with the same velocity v_{I13}. Any node on the ground link L_1 has zero velocity including I_{13} or $v_{I13} = 0$. The node I_{13} is then considered as link L_3 instantaneous zero center of velocity IC_3. This observation can be used

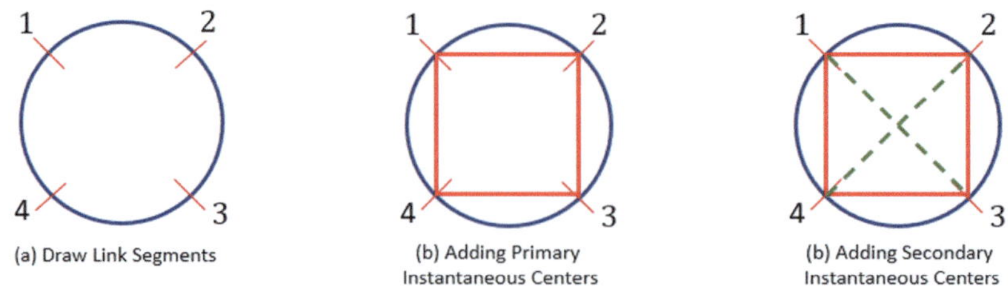

Fig. 6.28 Kennedy circle of four-bar mechanism

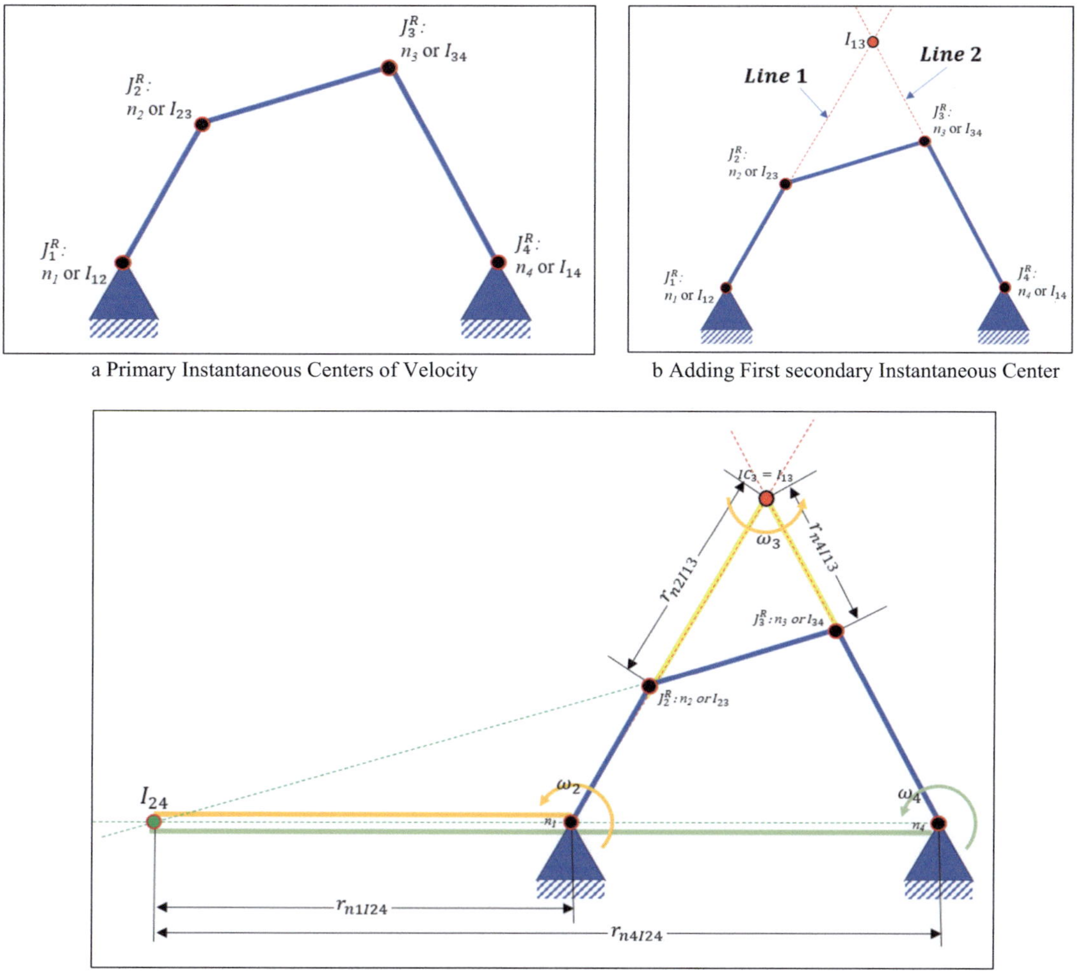

a Primary Instantaneous Centers of Velocity

b Adding First secondary Instantaneous Center

c Adding Second Secondary Instantaneous Center

Fig. 6.29 Instantaneous centers of velocity of a four-bar mechanism

to calculate the velocity at any node on link L_3 from the location of $I_{13}=I_{c3}$. For example, the two nodes n_2 and n_3 belong to L_3. The two velocities v_{n2} and v_{n3} can be calculated as:

$$v_{n2} = \omega_3 r_{n2IC3} = \omega_3 r_{n2I13}$$
$$v_{n3} = \omega_3 r_{n3IC3} = \omega_3 r_{n3I13} \tag{6.19}$$

where $r_{n2IC3} = r_{n2I13}$ is the radius from node n_2 to node $IC_3=I_{13}$ and $r_{n3IC3} = r_{n3I13}$ is the radius from node n_3 to node $IC_3=I_{13}$ as shown in Fig. 6.29c.

Observation 2: The node I_{24} is instantaneously located on both L_2 and L_4 with the same velocity v_{I24} that can be also calculated from

$$v_{I24} = \omega_2 r_{n2I24}$$
$$v_{I24} = \omega_4 r_{n2I24} \tag{6.20}$$

where r_{n2I24} is the radius from node n_2 to node I_{24} and r_{n4I24} is the radius from node n_4 to node I_{24} as shown in Fig. 6.29c.

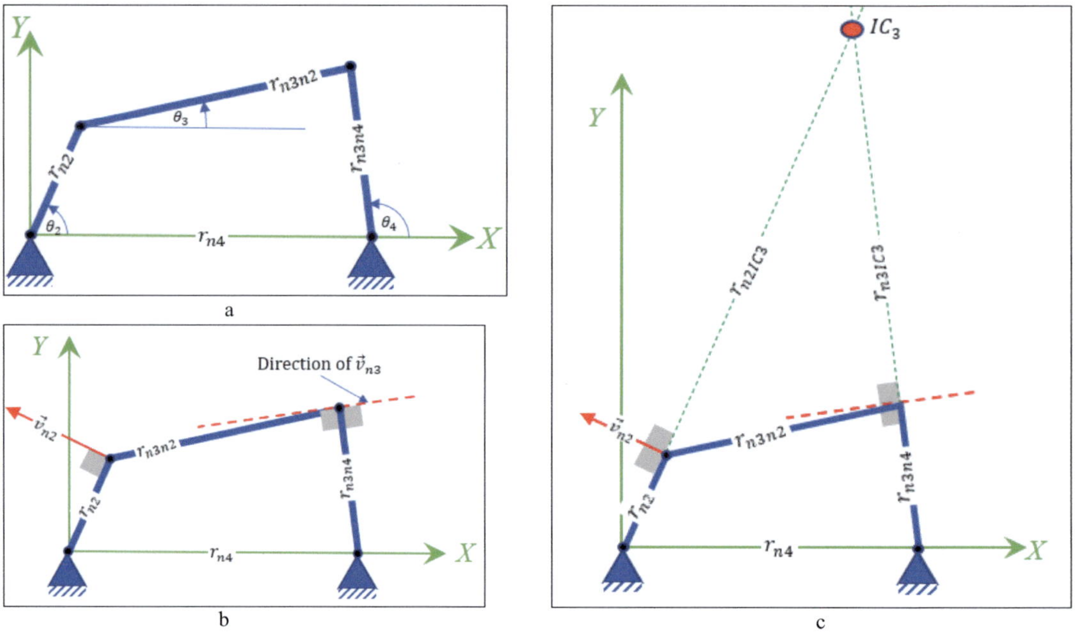

Fig. 6.30 Kinematic diagrams of graphical velocity analysis to determine four-bar velocity

Now, we are going to use the instantaneous zero velocity and the instantaneous centers of velocity in the graphical technique to estimate the velocity vector at any node on the four-bar mechanism.

Step 1: Use the graphical algorithm in Chap. 5 to plot the entire mechanism according to the given parameters ($r_{n2}, r_{n2n3}, r_{n3n4}, andr_{n4}$) and the given input signal (θ_2). Measure the output signals (θ_3 and θ_4) as shown in Fig. 6.30a. Note, sometimes a reasonable "scale" needs to be established if the dimensions are too big to fit on paper.

Step 2: Plot the velocity vector $v_{n2} = \omega_2 r_{n2}$ at node n_2 and perpendicular to r_{n2} as shown in Fig. 6.30b.

Step 3: Plot the direction v_{n3} at node n_3 to be perpendicular to r_{n3n4} as shown in Fig. 6.30b.

Step 4: Locate $IC_3 = I_{13}$ as the intersection of the two lines perpendicular to the velocities v_{n3} *and* v_{n2} or the intersection of the two extended lines n_1–n_2 and n_4–n_3 as shown in Fig. 6.30c.

Step 5: Measure r_{n2IC3} and calculate:

$$\omega_3 = v_{n2} / r_{n2IC3} \tag{6.21}$$

Step 6: Measure $r_{n3IC3.}$ and calculate:

$$v_{n3} = \omega_3 r_{n3IC3} \tag{6.22}$$

Step 7: Use the value of v_{n3} to calculate:

$$\omega_4 = v_{n3} / r_{n3n4} \tag{6.23}$$

Observation 3: The velocity vector of any node on link L_3, can be calculated using the location of $I_{C3} = I_{13}$ and ω_3.The velocity vector of any node on link L_2, can be calculated using the location of $n_2 = I_{C2} = I_{12}$ and ω_2. The velocity vector of any node on link L_4 can be calculated using the location of $n_4 = IC_4 = I_{14}$ and ω_4.

The second application of the instantaneous zero velocity and the instantaneous centers of velocity is to identify the torque ratio between two links or the force ratio between two links. Most of the time, we are interested in knowing the output-to-input torque ratio. For example, in the four-bar mechanism, we assume a motor as a source rotates link L_2 at joint J_2^R with an angular velocity ω_2 and a torque T_2 to deliver an input power $P_{in} = T_2 \times \omega_2$. Assume there is no power loss in the mechanism and there is no transformation of energy into less useful forms via friction, producing thermal energy, or sound and vibration energy. The input power is the same as the output power or:

$$P_{\text{in}} = P_{\text{out}} \tag{6.24}$$

where P_{out} is the output power. Assume the output power is the rotational motion of link L_4 or $P_{out} = T_4\omega_4$, where T_4 is output torque. Equating the input power with the output power will lead us to

$$P_{\text{in}} = T_2\omega_2 = P_{\text{out}} = T_4\omega_4 \tag{6.25}$$

or

$$\frac{T_4}{T_2} = \frac{\omega_2}{\omega_4}$$

From Eq. (6.25) the output-to-input torque ratio is proportional to the input-to-output angular velocity ratio. This means that to increase the output torque, we need to increase the input angular velocity. One advantage of most mechanisms is to magnify the output torque compared to the input torque or to make the output-input torque ratio T_4/T_2 high. To calculate the angular velocity ratio ω_2/ω_4, we can use the velocity values of v_{I24} from Eq. (6.26) as:

$$v_{I24} = \omega_2 r_{n2I24} = \omega_4 r_{n2I24} \tag{6.26}$$

or

$$\frac{T_4}{T_2} = \frac{\omega_2}{\omega_4} = \frac{r_{n2I24}}{r_{n2I24}}$$

Example 6.6

For the door opener in Fig. 6.11, assume a person opens the door from angle 0° to 85° in two seconds with a constant angular velocity.

1. Use the graphical method to find the angular velocity of the coupler and the angular velocity of the follower when $\theta_2 = 0°$.
2. Use the graphical method to calculate the output-to-input torque ratio, where the input torque is applied torque to open the door or link L_2 and the output torque is the torque on the link L_4.

Solution

Follow the graphical method steps for a four-bar mechanism as shown in Fig. 6.31. You can use any drawing scale that you prefer. From, we have $\omega_2 = 0.742$ rad/s. The location of IC3 is node n4. That leads to $r_{n2IC3} = r_{n4} - r_{n2} = 0.3 - 0.25 = 0.05$ m and $r_{n3IC3} = r_{n3n4} = 0.33$ m. From Eqs. (6.22) and (6.23), we can calculate the angular velocities as:

Fig. 6.31 Kinematic diagram of a door opener at $\theta_2 = 0°$

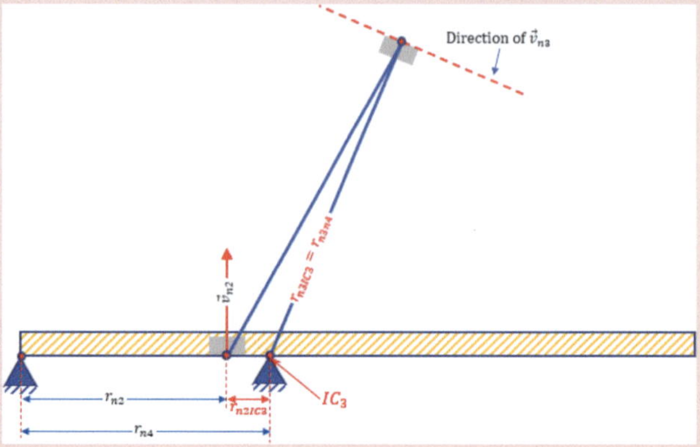

$$\omega_3 = -\frac{r_{n2}}{r_{n2IC3}}\omega_2 = \frac{0.25}{0.05} \times 0.742 = -3.71\,\text{rad/s}$$

$$\omega_4 = \frac{v_{n3}}{r_{n3n4}} = \frac{r_{n3IC3}}{r_{n3n4}}\omega_3 = -\frac{r_{n3IC3}}{r_{n3n4}}\frac{r_{n2}}{r_{n2IC3}}$$

$$\omega_2 = \frac{0.33}{0.33} \times \frac{0.25}{0.05} \times 0.742 = -3.71\,\text{rad/s}$$

The velocity direction of v_{n2} provides a clockwise angular velocity around IC_3 that causes ω_3 and ω_4 to be negative as listed above. Note that at a different value of angle, there will be different values for $\omega_3 and \omega_4$.

To calculate the torque ratio, we are going to use Eq. (6.25):

$$\frac{T_4}{T_2} = \frac{\omega_2}{\omega_4} = \frac{0.742}{-3.71} = -0.2$$

This means the output torque T_4 is 20% of the input torque T_2. We could use Eq. (6.26) but that would require measuring r_{n2I24} and r_{n2I24}.

6.6.5 Slider–Crank Mechanism Graphical Velocity Analysis

Similar to the four-bar mechanism, the slider–crank mechanism has four links with 6 instantaneous centers of velocity as in Fig. 6.28. To estimate the location of these six instantaneous centers of velocity, we start with Fig. 6.29 where we draw the four primary instantaneous centers of velocity I_{12}, I_{23}, and I_{34} located at the three revolute joints $J_1^R : n_1, J_2^R : n_2$, and $J_3^R : n_3$. The instantaneous center of velocity I_{14} is located vertically at infinity in the perpendicular to the horizontal sliding motion. The first secondary center I_{13} is calculated using Kennedy's theorem as: Fig. 6.32

- ***Line 1***: The three centers I_{12}, $I_{23,}$ and I_{13} must lie on the same straight line.
- ***Line 2***: The three centers I_{14}, $I_{34,}$ and I_{13} must lie on the same straight line.

The intersection of ***Line 1*** and ***Line 2*** is the instantaneous center I_{13}. Using the same method, we can locate the second secondary center I_{24} to be the intersection of Line I_{14}–I_{12}–I_{24} and Line I_{34}–I_{23}–I_{24}.

Observation 1: Like the four-bar mechanism, node I_{13} is instantaneously located on both L_1 and L_3 with the same velocity $v_{I13}=0$ or $IC_3=I_{13}$. Using this fact, the two velocities v_{n2} and v_{n3} are calculated using Eq. (6.19).

Observation 2: Like the four-bar mechanism, node I_{24} is instantaneously located on both L_2 and L_4 with the same velocity v_{I24} as listed in Eq. (6.20).

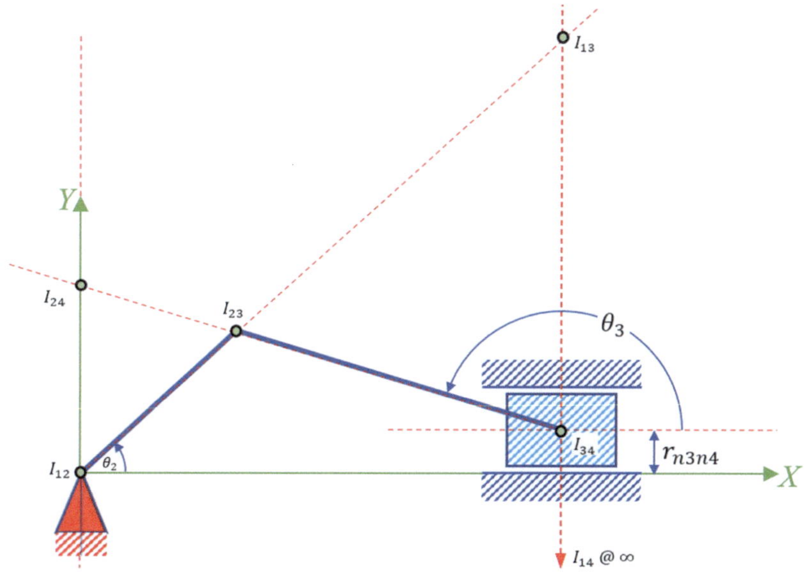

Fig. 6.32 Instantaneous centers of velocity of a slider–crank mechanism

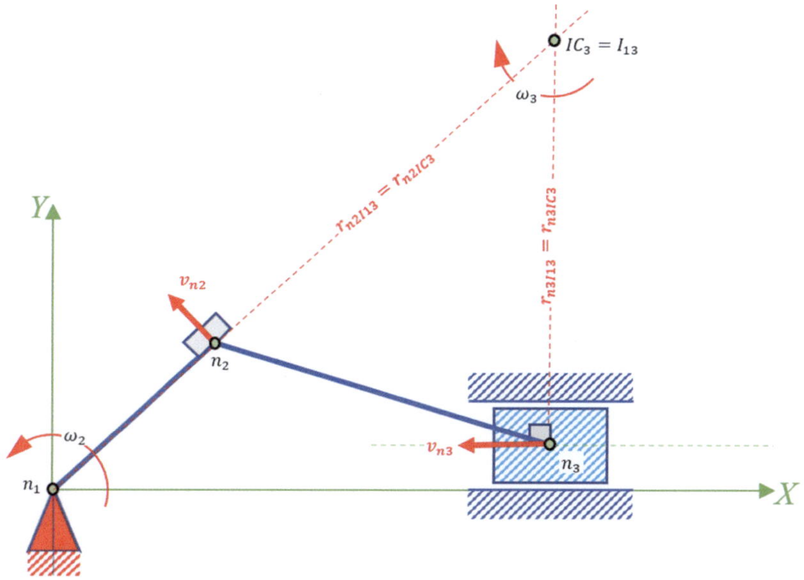

Fig. 6.33 Kinematic diagram of graphical velocity analysis to determine slider–crank velocity

Now, we are going to use ***Observation 1*** and ***Observation 2*** to graphically calculate the velocity vectors.

Step 1: Use the graphical algorithm from Chap. 5 to plot the entire mechanism according to the given parameters $(r_{n2}, r_{n2n3}, and r_{n3n4})$and the given input signal (θ_2). Measure the output signals $(\theta_3 and r_{n4})$. Note, sometimes a reasonable "scale" needs to be established if the dimensions are too big to fit on paper.

Step 2: Plot the velocity vector $v_{n2} = \omega_2 r_{n2}$ at node n_2 and perpendicular to r_{n2}, Fig. 6.33.

Step 3: Plot the direction v_{n3} at node n_3 to be -ve *X-axis* Fig. 6.33.

Step 4: Locate $IC_3 = I_{13}$ as the intersection of the two lines perpendicular to the velocities v_{n3} and $v_{n2,}$ Fig. 6.33.

Step 5: Measure r_{n2IC3} and calculate:

$$\omega_3 = v_{n2}/r_{n2IC3} \tag{6.27}$$

Step 6: Measure $r_{n3IC3.}$ and calculate:

$$v_{n3} = \omega_3 r_{n3IC3} \tag{6.28}$$

Example 6.7

Figure 6.34 shows a piston–cylinder mechanism with a crank radius of 50 mm and connecting rod length of 130 mm. Both the wrist pin (piston-connecting-rod revolute joint) and crank shaft center are horizontally aligned. At an instantaneous position, the combustion force moves the piston outward with a sliding velocity of 25 m/s. The crank angle is at 30° counterclockwise to the horizontal wrist axis.

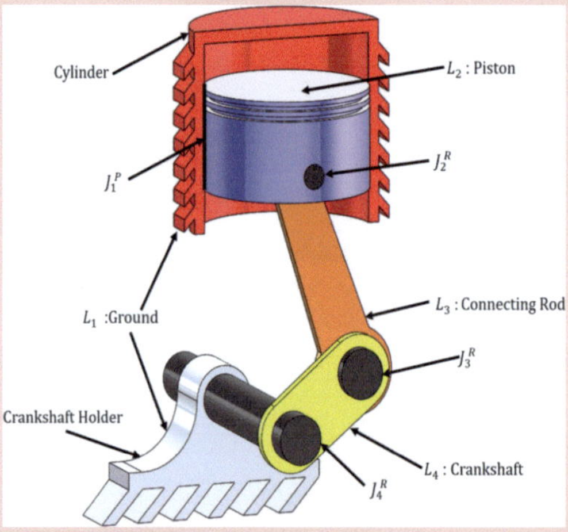

Fig. 6.34 Piston–cylinder mechanism. Designed by the author and implemented by both the author and Pranav Dhola. Courtesy of Pranav Dhola

1. Use the graphical velocity analysis to calculate the crank and the connecting rod angular velocities.
2. Compare your results with the simulation of the piston–cylinder mechanism model-based.

Solution

Graphical Method:

Follow the steps of the graphical method for the slider–crank mechanism as shown in Fig. 6.35. You can use any drawing scale that you prefer. From Example 5.4, we have $\theta_3 = 169$ deg and $r_{n4} = 171$ mm. From the location of IC_3, we can measure $r_{n2IC3} = 147.3$ mm and $r_{n3IC3} = 98.6$ mm From Eq. (6.28), we can calculate the angular velocity as:

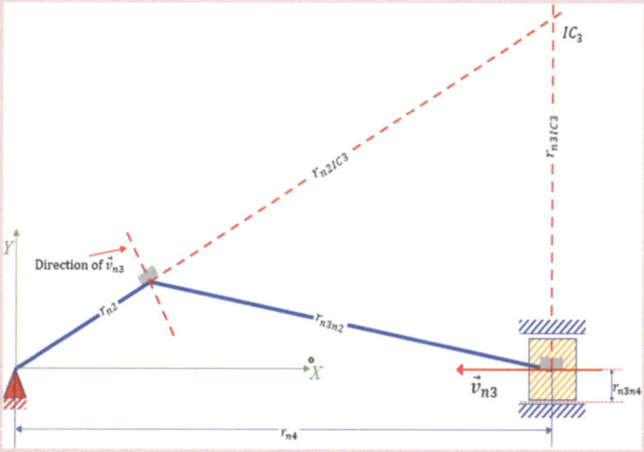

Fig. 6.35 Kinematic diagram with velocities of single piston–cylinder at $\theta_2 = 30°$

$$\omega_3 = \frac{v_{n3}}{r_{n3IC3}} = \frac{25}{0.0986} = 253.4\,\text{rad/s}$$

Using the formula $v_{n2} = \omega_2 r_{n2}$ and Eq. (6.27):

$$\omega_2 = \frac{r_{n2IC3}}{r_{n2}}\omega_3 = \frac{r_{n2IC3}}{r_{n2}}\omega_3 = \frac{147.35}{50}253.4 = 747.5\,\text{rad/s}$$

Note at different values of angle θ_2, there will be different values of ω_3 and ω_4.

Model-Based Method

For the model-based method:

1. Copy and rename "SinglePistonCylinder_P5p6_model.slx" to "SinglePistonCylinder_P6p7_model.slx" and "SinglePistonCylinder_P5p6_script.m" to "SinglePistonCylinder_P6p7_script.m". Run the script file.
2. If we look at the attached animation video of the piston–cylinder mechanism, we can conclude that the piston would move with the velocity $v_{xn3} = -25$ m/s from the fully extended position:

$$x_{n3}^{\text{Full Extr}} = r_{n2} + r_{n3n2} = 0.05 + 0.13 = 0.18\text{m and}\theta_2 = 0°$$

to the fully retracted position:

$$x_{n3}^{\text{Full Retr}} = -r_{n2} + r_{n3n2} = -0.05 + 0.13 = 0.8\text{m and}\,\theta_2 = 180°$$

The required time to make this motion is:

$$\Delta t = \left(x_{n3}^{\text{Full Extr}} - x_{n3}^{\text{Full Retr}}\right)/v_{xn3} = (0.18 - 0.8)/25 = 0.0032\text{ s}$$

To model this motion, we are going to drag and drop a "Signal Builder" block. Change the input signal from 0.18 m to 0.8 m in 0.0032 sec and constant after that as shown in Fig. 6.36. The simulation stop time should be $t_{\text{stop}} = 0.005$ s.

3. Double-click on J_3P and select the actuator. Select the motion actuator as an input and the force to be automatically computed.
4. Insert the "Translational Motion Actuator" block from "Kinematics_Of_Machines.slx". Connect this block to the "Signal Builder" block and the "J_3P" block as shown in Fig. 6.36.
5. Add a "Rotational Position Sensor" and an "Angular Velocity Sensor" at node n_2 and an "Angular Velocity Sensor" at node n_3. Connect the two sensor blocks to a scope block.

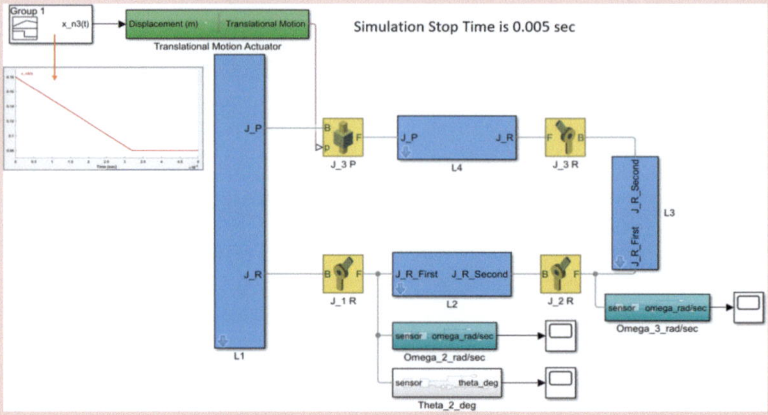

Fig. 6.36 Piston cylinder Simscape™ Multibody™ model

Run the model and examine the results from the scope as shown in Fig. 6.37a. We can determine from the figure that at $\theta_2(t = 0.0003725) = 30°, \omega_2(t = 0.0003725) = 748.5$ rad/s, and $\omega_3(t = 0.0003725) = -254$ rad/s. During the simulation, you can see the animation for the mechanism. Some screenshots are provided in Fig. 6.37b. We encourage the reader to watch the attached animated video from this model as well as to look at the attached model.

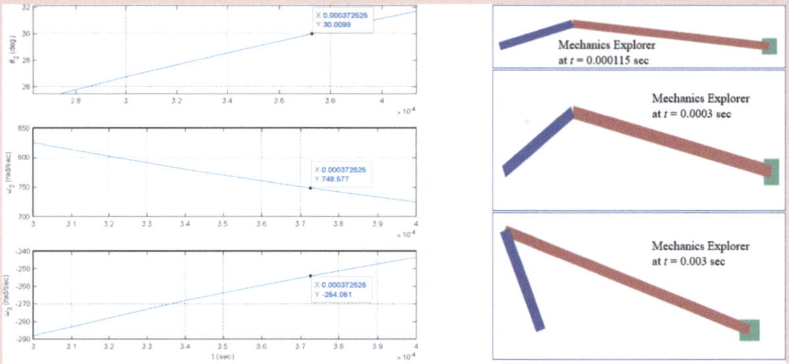

a Position and Velocities Signals

b Mechanics Explorer at Different Time

Fig. 6.37 Results of Simscape™ Multibody™ model for piston–cylinder mechanism

6.6.6 Inverted Slider–Crank Mechanism Graphical Velocity Analysis

Similar to the four-bar mechanism and the slider–crank mechanism, the inverted slider–crank mechanism has four links with six instantaneous centers of velocity as in Fig. 6.28. To estimate the location of these six instantaneous centers of velocity in Fig. 6.29, we draw the four primary instantaneous centers of velocity I_{12}, I_{23}, and I_{14} that are located at the three revolute joints $J_1^R : n_1, J_2^R : n_2$, and $J_4^R : n_4$. The instantaneous center of velocity I_{34} is located at infinity in the perpendicular on the sliding motion as shown in Fig. 6.38a. The first secondary center I_{13} is calculated using Kennedy's theorem as:

- ***Line 1***: The three centers I_{12}, $I_{23,}$ and I_{13} must lie on the same straight line.
- ***Line 2***: The three centers I_{14}, $I_{34,}$ and I_{13} must lie on the same straight line.

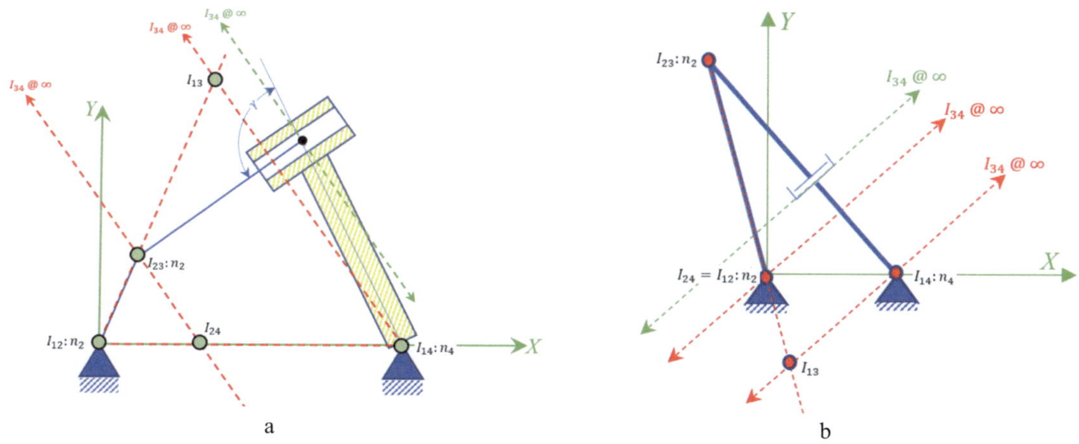

Fig. 6.38 Instantaneous centers of velocity of an inverted slider–crank mechanism

The intersection of ***Line 1*** and ***Line 2*** is the instantaneous center I_{13}. Using the same method, we are going to locate the second secondary center I_{24} to be the intersection of Line I_{14}-I_{12}-I_{24} and Line I_{34}–I_{23}–I_{24}. The angle $\gamma = \theta_3 - \theta_4$ is a constant geometric angle. One important special case is when $\gamma = 0$ or $\theta_3 = \theta_4$. The instantaneous centers of velocity for $\gamma = 0$ are shown in Fig. 6.38b.

Observation 1: Like the four-bar mechanism, node I_{13} is instantaneously located on both L_1 and L_3 with the same velocity v_{I13}=0 or IC_3=I_{13}. Using this fact, the two velocities v_{n2} and v_{n3} are calculated using Eq. (6.19).

Observation 2: Like the four-bar mechanism, node I_{24} is instantaneously located on both L_2 and L_4 with the same velocity v_{I24} as listed in Eq. (6.20).

Now, we are going to use ***Observation 1*** and ***Observation 2*** to graphically calculate the velocity vectors as

Step 1: Use the graphical algorithm from Chap. 5 to plot the entire mechanism according to the given parameters ($r_{n2}, r_{n4},$ and r_{n3n4})and the given input signal (θ_2). Measure the output signals ($\theta_3 and r_{n3n2}$). Note, sometimes a reasonable "scale" needs to be established if the dimensions are too big to fit on paper.

Step 2: Plot Line 1 of I_{12}, I_{23} and I_{13}. Measure r_{n2I13} and r_{n3I13}.

Step 3: Calculate:

$$\omega_3 = v_{n2}/r_{n2I13} \tag{6.29}$$

$$v_{n3} = \omega_3 r_{n3IC3} \tag{6.30}$$

Example 6.8

Figure 6.39 shows a hydraulic gate opener mechanism. Assume the door is open from the initial position at $\theta_2 = 0°$ to the final position $\theta_2 = 50°$ with an angular velocity ω_2 =1 rad/sec. At the position when $\theta_2 = 45°, \theta_3 = 55°$ and $r_{n3n2} = 405.67$ mm:

1. Use graphical velocity analysis to calculate the cylinder angular velocity and the piston translational velocity.
2. Compare your results with the simulation of the previously developed hydraulic gate opener mechanism model-based.

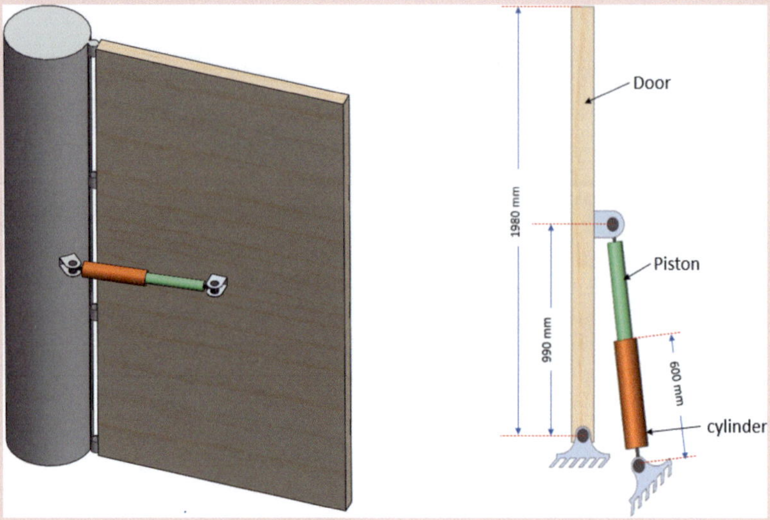

Fig. 6.39 Hydraulic gate opener. Designed by the author and implemented by both the author and Pranav Dhola. Courtesy of Pranav Dhola

Solution

First, draw the kinematic diagram using any scale that you prefer. As shown in Fig. 6.40, the node $n_2 = I_{23}$ is located on both L_2 and L_3 with the same velocity $\vec{v}_{n2}$ that is perpendicular to the position vector $\vec{r}_{n2}$ and $\vec{v}_{n2}$ and can be calculated by:

$$v_{n2} = \omega_2 r_{n2} = 1 \times 0.99 = 0.99\,\text{m/s}$$

The velocity $\vec{v}_{n2}$ has two components, sliding velocity component $v_{\text{sn2n4}} = v_{\text{sn3n4}}$ in the direction of $\vec{r}_{n2n4}$ and rotational velocity components $v_{\theta n2n4} = \omega_4 r_{n2n4}$, where $r_{n2n4} = r_{n2n3} + r_{n3n4}$. Using the graphical sketch, we can calculate the sliding velocity component:

$$v_{sn3n4} = v_{sn2n4} = v_{n2}\sin(\theta_4 - \theta_2) = 0.99 \times \sin(10^\circ) = 0.172\,\text{m/s}$$

Also, we can calculate a rotational velocity component:

$$v_{\theta n2n4} = v_{n2}\cos(\theta_4 - \theta_2) = 0.99 \times \cos\cos(10^\circ) = 0.975\,\text{m/s}$$

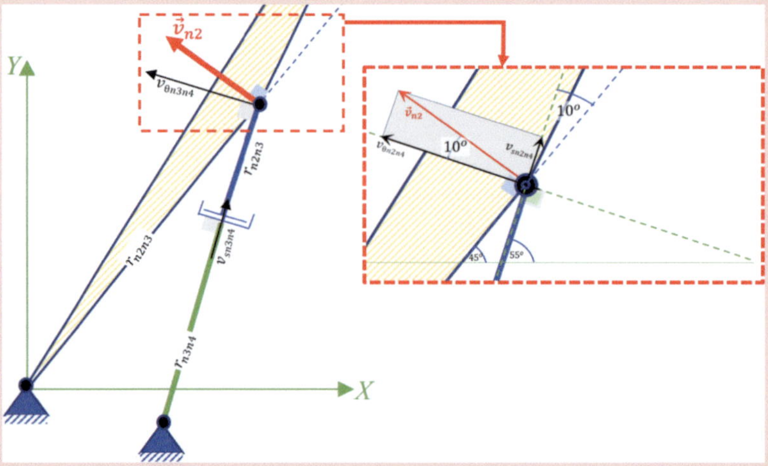

Fig. 6.40 Kinematic diagram with velocities of hydraulic door opener at $\theta_2 = 45^\circ$

Using the fact that $v_{\theta n2n4} = \omega_4 r_{n2n4}$, we can calculate the angular velocity:

$$\omega_4 = \frac{v_{\theta n2n4}}{r_{n2n4}} = \frac{0.975}{0.6 + 0.405} = 0.97 \text{ rad/s}$$

The aspect ratio in Fig. 6.40 has been adjusted to show the details in the figure. Note, another way to calculate the velocity is to use Eqs. (6.29) and (6.30). We encourage the reader to do the other graphical method and confirm the results.

For the Model-Based:

We are going to use our previously developed "HydraulicDoorOpener_P5p7_model.slx" and "HydraulicDoorOpener_P5p7_script.m" to build the new model as:

1. Copy and rename "HydraulicDoorOpener_P5p7_script.m" to "HydraulicDoorOpener_P6p8_script.m" and "HydraulicDoorOpener_P5p7_model.slx" to "HydraulicDoorOpener_P6p8_model.slx". Run the script file.
2. Drag and drop the "Signal Builder" block. Edit the signal of this block as θ_2 that changes from 0° to 50° with an angular velocity $\omega_2 = 1\text{rad/s s} = 1 \times 180/\pi = 57.298°\text{deg/s}$. This signal can be built by three points $\theta_2 = \left[0 \frac{180}{\pi} \frac{180}{\pi}\right]$ and $t = [011.5]$, where θ_2 is in deg and t is in sec, Fig. 6.41. The simulation stop time should be $t_{\text{stop}} = 1.5$ s.
3 Double-click on the J_1R block and select the actuator. Select the motion actuator as an input and select the torque to be automatically computed.
4. Bring "Rotational Motion Actuator" block from "Kinematics_Of_Machines.slx". Connect this block to the "Signal Builder" block and the "J_1R" block, Fig. 6.41.

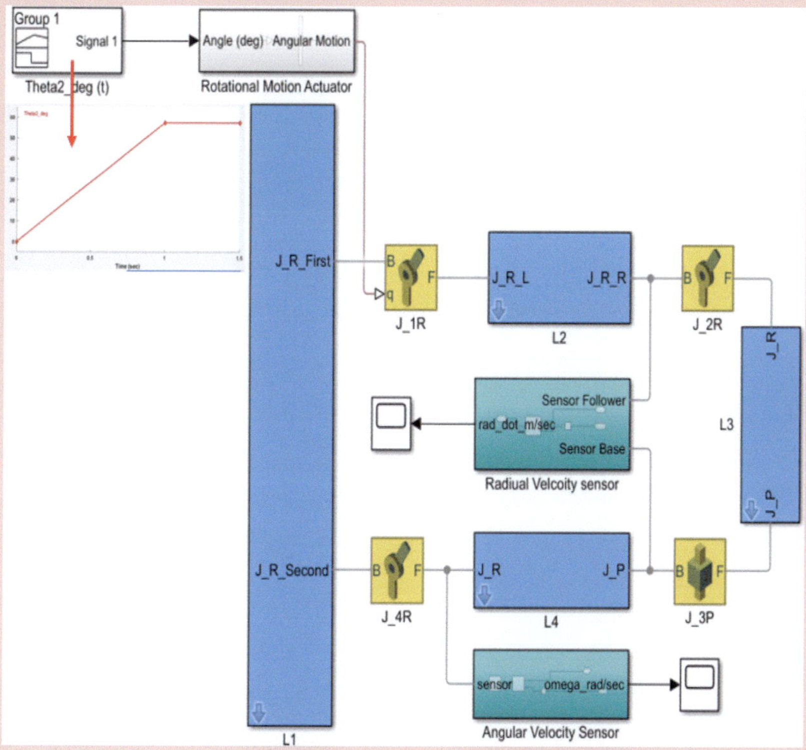

Fig. 6.41 Hydraulic door opener Simscape™ Multibody™ model

5. Add "Rotational Velocity Sensor" subsystem and "Radial Velocity Sensor" subsystem from "Kinematics_Of_Machines.slx", Fig. 6.41.

Run the model and examine the results from the scope as shown in Fig. 6.42. Because $\omega_2 = 57.298°$ deg/s, we can calculate when $\theta_2 = 45°$, $t = \frac{45°}{57.298°} = 0.785$ s. Looking at the two scopes, we can measure $\omega_4 = 0.174$ m/s and $v_{sn3n4} = 0.97$ m/s that is consistent with the graphical analysis. During the simulation, you can see the animation for the mechanism. We encourage the reader to watch the attached animated video from this model as well as look at the attached model plus the attached animation video.

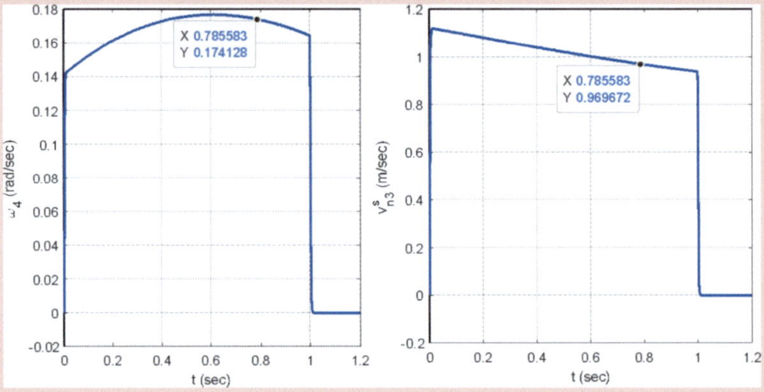

Fig. 6.42 Results of Simscape™ Multibody™ model for hydraulic door opener mechanism

Engineering Tips—Log Data to Workspace

In the previous examples, we used the scope to determine output values at certain times. This process can be bothersome and inaccurate. To get more accurate values, we recommend the reader log the data to the workspace using the below steps:

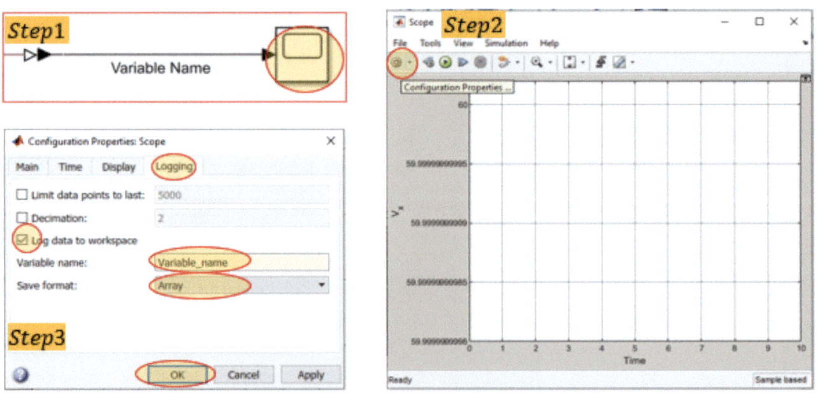

Fig. 6.43 Log data to workspace. Designed and implemented by the author. Images used with permission of MathWorks

Step 1: Connect the signal with a scope block. Double-click on the scope block to open the scope window as shown in Fig. 6.43.
Step 2: When the scope window opens, click on the "Configuration Properties" icon to open the "Configuration Properties: Scope" window as shown in Fig. 6.43.
Step 3: Select "Log data to workspace". Enter the name of your variable in "Variable Name". Save format array. Press "OK" as shown in Fig. 6.43.

After running the model. You can use the variable name to plot the signal or to make any calculation.

6.6.7 General Mechanism Graphical Velocity Analysis

The same concept can be applied to any mechanism. The general algorithm for the graphical velocity analysis can be summarized as:

Step 1: Use the graphical algorithm from Chap. 5 to plot the entire mechanism using the given parameters and input signals. Measure the output signals.
Step 2: Use the velocity input signals (translational velocity or/and angular velocity) along with the kinematic constraints at each input node to plot the velocity vectors.
Step 3: Use the kinematic constraints at each node to plot the velocity directions.
Step 4: For each link with two known velocity directions, find the *IC* or I_{ij}.**Step 5**: Use the equations of the kinematic velocity model to estimate the unknown velocity magnitudes.

Graphical velocity analysis is a unique tool for understanding a mechanism, but it is time-consuming. Also, it depends on the position or the time. For example, the door mechanism in, if we simulate opening the door from $\theta_2(t=0)=0°$ to $\theta_2(t=2)=85°$ with a time step $\Delta t=0.01$, we will have to come up with 200 sketches! This is very time-consuming, especially if we are changing the design parameters for optimization. On the other hand, we can try to find an analytical solution or we can use model-based techniques. In the next section, we are going to introduce the analytical solution of simplified mechanisms.

6.7 Analytical Velocity Analysis

Analytical solutions are exact solutions, if we can obtain them. However, this is virtually impossible, especially if the mechanism in question has many links and joints. To develop these analytical models, we must be experts. Also, understanding the physical meaning of each component in the solution is a challenging task. This is the purpose of master's theses and doctoral dissertations! We still need the analytical solution to justify the numerical solutions that are provided by software such as Simulink® and Simscape™ Multibody™. In this chapter, we cover four-bar linkage models to develop analytical velocity solutions. These analytical solutions enable us to understand the nature of motion. In addition, we can these analytical models to justify the model-based solution.

6.7.1 Four-Bar Mechanism Analytical Velocity Analysis

Figure 6.44 shows the kinematic diagram of a four-bar mechanism plus the system diagram of the position and velocity analytical solution. From our position analytical solution in Chap. 5, the

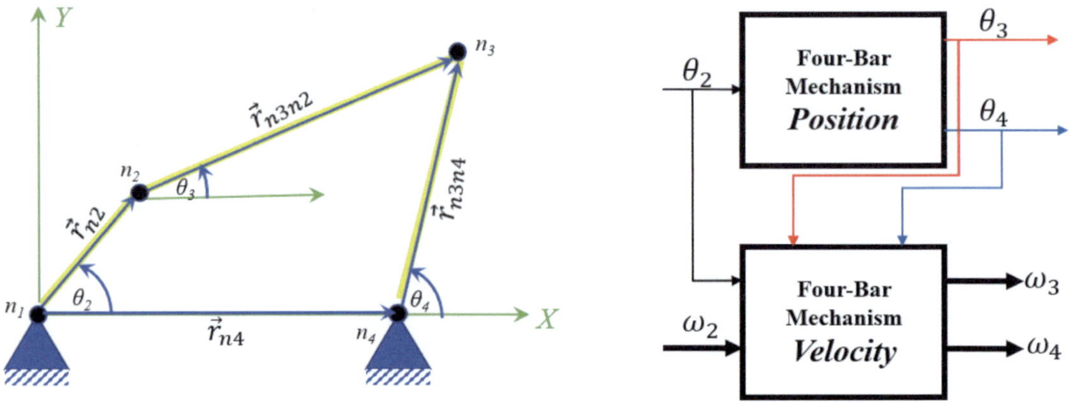

Fig. 6.44 Kinematic diagram and system diagram of four-bar mechanism

four-bar mechanism has three variable angles θ_2, $\theta_{3,}$ and θ_4 plus the constant links' lengths r_{n2}, r_{n3n2}, $r_{n3n4,}$ and r_{n4}. We are going to call the variable angles "signals". The block diagram of the position analytical solution has the input signal θ_2 and the output signals θ_3 and θ_4 in addition to any node position vector. The analytical velocity block diagram analytical solution has the input signal ω_2 and the output signals $\omega_3\ and \omega_4$ for the velocity analytical solution.

Recall the vector loop equations of the four-bar mechanism from Chap. 5:

$$\begin{aligned}\vec{r}_{n2} &= r_{n2}\cos(\theta_2)\hat{i} + r_{n2}\sin(\theta_2)\hat{j}\\ \vec{r}_{n3n2} &= r_{n3n2}\cos(\theta_3)\hat{i} + r_{n3n2}\sin(\theta_3)\hat{j}\\ \vec{r}_{n3n4} &= r_{n3n4}\cos(\theta_4)\hat{i} + r_{n3n4}\sin(\theta_4)\hat{j}\\ \vec{r}_{n4} &= r_{n4}\hat{i} + 0\hat{j}\end{aligned} \tag{6.31}$$

Equating the terms in the *X*-direction and *Y*-direction leads to:

$$\begin{aligned} r_{n2}\cos(\theta_2) + r_{n3n2}\cos(\theta_3) - r_{n3n4}\cos(\theta_4) &= r_{n4}\\ r_{n2}\sin(\theta_2) + r_{n3n2}\sin(\theta_3) - r_{n3n4}\sin(\theta_4) &= 0\end{aligned} \tag{6.32}$$

Differentiate Eq. (6.32) with respect to the time:

$$\begin{aligned} -r_{n2}\omega_2\sin(\theta_2) - r_{n3n2}\omega_3\sin(\theta_3) + r_{n3n4}\omega_4\sin(\theta_4) &= 0\\ r_{n2}\omega_2\cos(\theta_2) + r_{n3n2}\omega_3\cos(\theta_3) - r_{n3n4}\omega_4\cos(\theta_4) &= 0\end{aligned} \tag{6.33}$$

Solving the Eq. (6.33) to find $\omega_3$and ω_4 in terms of ω_2 as:

$$\begin{aligned}\frac{\omega_3}{\omega_2} &= \frac{r_{n2}}{r_{n2n3}} \times \frac{\sin(\theta_4 - \theta_2)}{\sin(\theta_4 - \theta_3)}\\ \frac{\omega_4}{\omega_2} &= \frac{r_{n2}}{r_{n3n4}} \times \frac{\sin(\theta_3 - \theta_2)}{\sin(\theta_4 - \theta_3)}\end{aligned} \tag{6.34}$$

In Eq. (6.34), when $\theta_2 = \theta_4$ or the link L_2 is parallel to L_4, the link L_3 has zero angular velocity, $\omega_3 = 0$ rad/sec. This scenario happens in the parallelogram mechanism such as the folding staircase as shown in Fig. 6.45 where the steps do not rotate when the staircase is folded or unfolded. We call

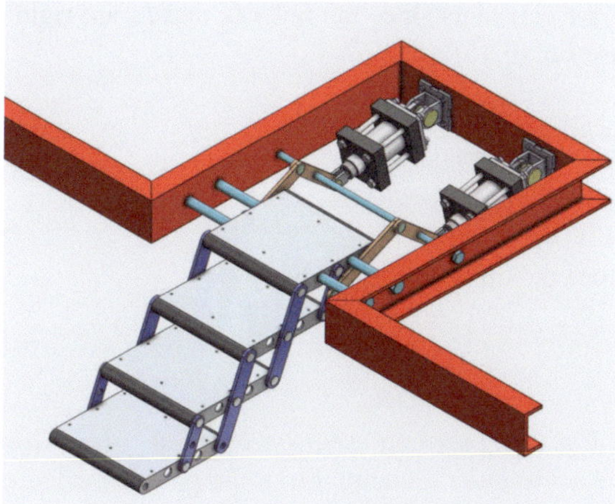

Fig. 6.45 Folding staircase mechanism. Figure and associated Folding Staircase Mechanism.mp4 video designed by the author and implemented by both the author and Pranav Dhola. Both courtesy of Pranav Dhola

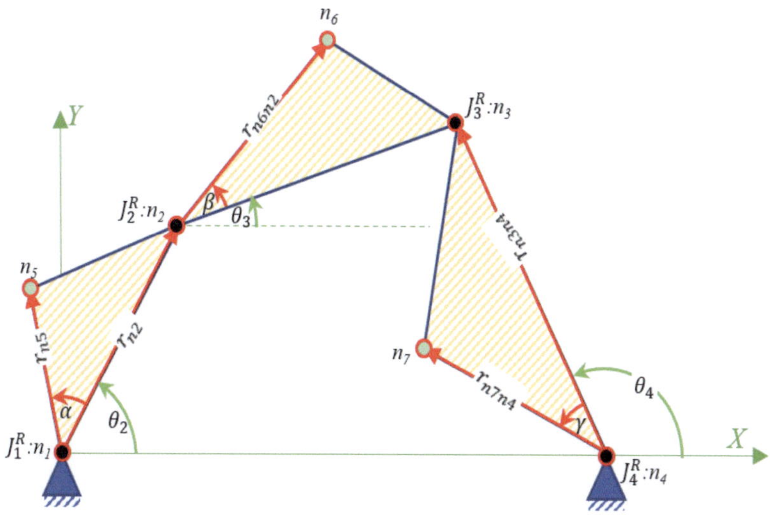

Fig. 6.46 Four-*bar mechanism with extra points* of *interest*

this parallelogram motion a motion without any pitch. We recommend the reader look at the attached video of the folding staircase. Another example is the car wiper mechanism as shown in Fig. 6.6.

Once the signals ω_2, ω_3, and ω_4 are known, we can calculate the velocity vector of any node belonging to the mechanism. Figure 6.46 shows a four-bar mechanism that has five nodes of interest at n_2, n_3, n_5, n_6, and n_7. Using Eq. (6.13), we can calculate the velocities at these nodes:

$$\begin{aligned}
\vec{v}_{n2} &= \omega_2 r_{n2}\hat{\theta} = -\omega_2 r_{n2}\sin(\theta_2)\hat{i} + \omega_2 r_{n2}\cos(\theta_2)\hat{j} \\
\vec{v}_{n5} &= \omega_2 r_{n5}\hat{\theta} = -\omega_2 r_{n5}\sin(\theta_2+\alpha)\hat{i} + \omega_2 r_{n5}\cos(\theta_2+\alpha)\hat{j} \\
\vec{v}_{n3} &= -\omega_4 r_{n3n4}\sin(\theta_4)\hat{i} + \omega_4 r_{n3n4}\cos(\theta_4)\hat{j} \\
\vec{v}_{n7} &= -\omega_4 r_{n7n4}\sin(\theta_4+\gamma)\hat{i} + \omega_4 r_{n7n4}\cos(\theta_4+\gamma)\hat{j}
\end{aligned} \tag{6.35}$$

The two angles α and γ are constants since the links L_2 and L_4 are rigid bodies. For the velocity vector at node n_6, we can use Eq. (6.15) as:

$$\begin{aligned}\underbrace{\vec{v}_{nC}}_{\vec{v}_{n6}} &= \underbrace{\vec{v}_{nB}}_{\vec{v}_{n2}} + \underbrace{\vec{\Omega}_B}_{\omega_3\hat{k}} \times \underbrace{\vec{r}_{nCnB}}_{r_{n6n2}\hat{r}} + \underbrace{\vec{v}^{relative}_{nCnB}}_{\vec{0}} \\ &= -\omega_2 r_{n2}\sin(\theta_2)\hat{i} + \omega_2 r_{n2}\cos(\theta_2)\hat{j} - \omega_3 r_{n6n2}\sin(\theta_3+\beta)\hat{i} + \omega_3 r_{n6n2}\cos(\theta_3+\beta)\hat{j}\end{aligned} \tag{6.36}$$

The angle β is a constant since the link L_3 is a rigid body.

Example 6.9

For the door opener in Fig. 6.11, assume a person opens the door from angle 0° to 85° in two seconds with a constant angular velocity. From the position calculation, if $\theta_2 = 0^o$, then $\theta_3 = -130.428^o and \theta_4 = 56.861^o$.

- Use the analytical method to find the angular velocity of the coupler and the follower when $\theta_2 = 0^o$.
- Use the analytical method to find the velocity vector at the stile edge and door hinge.
- Use the analytical method to evaluate the model-based method and the graphical method.

Solution

Using Eq. (6.34):

$$\omega_3 = \frac{r_{n2}}{r_{n2n3}} \times \frac{\sin(\theta_4-\theta_2)}{\sin(\theta_4-\theta_3)} \times \omega_2 = \frac{0.25}{0.33} \times \frac{\sin(56.861° - 0°)}{\sin(56.861° + 130.428°)} \times 0.742 = -3.71\,\text{rad/s}$$

$$\omega_4 = \frac{r_{n2}}{r_{n3n4}} \times \frac{\sin(\theta_3-\theta_2)}{\sin(\theta_4-\theta_3)} \times \omega_2 = \frac{0.25}{0.3} \times \frac{\sin(-130.428° - 0°)}{\sin(56.861° + 130.428°)} \times 0.742 = -3.71\,\text{rad/s}$$

Using Eq. (6.35):

$$\vec{v}_{n2} = \omega_2 r_{n2}\hat{e}_t = 0.742 \times 0.25\hat{e}_t = 0.189\hat{e}_t\text{m/s s} = 0.189\hat{j}\ \text{m/s}$$

$$\vec{v}_{n5} = \omega_2 r_{n5}\hat{e}_t = 0.742 \times 0.82\hat{e}_t = 0.608\hat{e}_t\text{m/s s} = 0.608\hat{j}\text{m/s}$$

The analytical values match the model-based method in Example 6.5 and the graphical method in Example 6.6

6.7.2 Slider–Crank Mechanism Analytical Velocity Analysis

Figure 6.47 shows the kinematic diagram of a slider–crank mechanism plus the system diagram of the position and velocity analytical solution. From our position analytical solution in Chap. 5, the slider–crank mechanism has three variable angles θ_2, $\theta_{3,}$ and r_{n4} plus the constant links' lengths r_{n2}, r_{n3n2}, and r_{n3n4} while $\theta_3 = 90°$. The system diagram of the position analytical solution has the input signal θ_2 and output signals θ_3 and r_{n4} in addition to any node position vector. The velocity system diagram analytical solution has the input signal $\omega_2$and the output signals ω_3 and $\dot{r}_{n4}$ for the velocity analytical solution.

Recall the vector loop equations of the slider–crank mechanism from Chap. 5:

$$\begin{aligned} r_{n2}\cos(\theta_2) - r_{n3n2}\cos(\theta_3) &= r_{n4} \\ r_{n2}\sin(\theta_2) + r_{n3n2}\sin(\theta_3) &= r_{n3n4} \end{aligned} \tag{6.37}$$

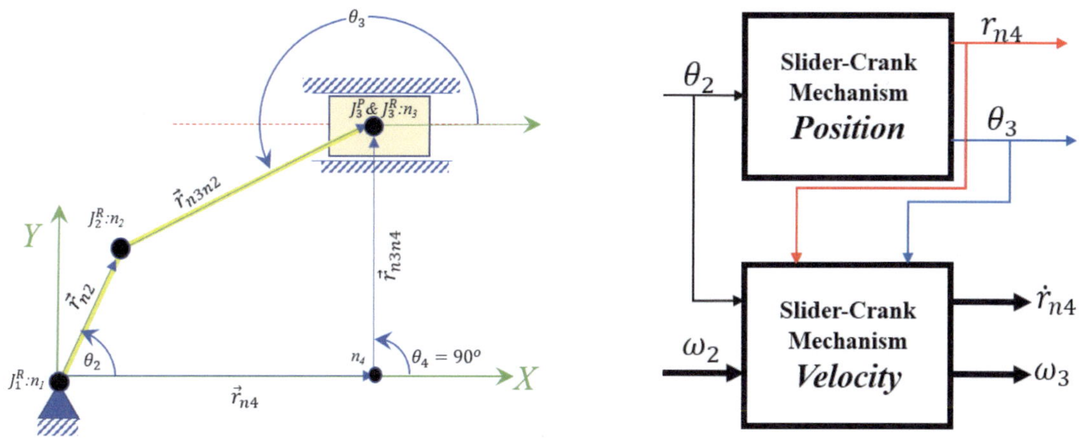

Fig. 6.47 Kinematic diagram and velocity system diagram of slider–crank mechanism

Differentiate Eq. (6.37) with respect to the time:

$$\begin{aligned} -r_{n2}\omega_2 \sin(\theta_2) - r_{n3n2}\omega_3 \sin(\theta_3) &= \dot{r}_{n4} \\ r_{n2}\omega_2 \cos(\theta_2) - r_{n3n2}\omega_3 \cos(\theta_3) &= 0 \end{aligned} \tag{6.38}$$

Equation (6.38) can be used to find ω_3 and ω_2 in terms of $\dot{r}_{n4}$ or getting ω_3 in terms of ω_2 as:

$$\begin{aligned} \frac{\omega_3}{\omega_2} &= \frac{r_{n2}}{r_{n2n3}} \times \frac{\cos(\theta_2)}{\cos(\theta_3)} \\ \dot{r}_{n4} &= -r_{n2}\omega_2 \sin(\theta_2) + r_{n2}\omega_2 \cos(\theta_2)\tan(\theta_3) \\ \omega_2 &= \frac{\dot{r}_{n4}}{r_{n2}(\cos(\theta_2)\tan(\theta_3) - \sin(\theta_2))} \end{aligned} \tag{6.39}$$

The velocity vector at each node on the mechanism can be calculated using Eqs. (6.13) and (6.15) similar to the four-bar mechanism case.

6.7.3 Inverted Slider–Crank Mechanism Analytical Velocity Analysis

Figure 6.48 shows the kinematic diagram of one type of the inverted slider–crank mechanism. Recall the inverted slider–crank vector equation from Chap. 5.

$$\begin{aligned} r_{n2}\cos(\theta_2) - r_{n3n2}\cos(\theta_3) &= r_{n3n4}\cos(\theta_4) + r_{n4} \\ r_{n2}\sin(\theta_2) - r_{n3n2}\sin(\theta_3) &= r_{n3n4}\sin(\theta_4) \end{aligned} \tag{6.40}$$

where $\theta_4 = \theta_3 - \gamma$. Differentiate Eq. (6.40) with respect to the time:

$$\begin{aligned} -r_{n2}\omega_2 \sin(\theta_2) + r_{n3n2}\omega_3 \sin(\theta_3) - \dot{r}_{n3n2}\cos(\theta_3) &= -r_{n3n4}\omega_3 \sin(\theta_3 - \gamma) \\ r_{n2}\omega_2 \cos(\theta_2) - r_{n3n2}\omega_3 \cos(\theta_3) - \dot{r}_{n3n2}\sin(\theta_3) &= r_{n3n4}\omega_3 \cos(\theta_3 - \gamma) \end{aligned} \tag{6.41}$$

Equation (6.41) can be used to find ω_3 and ω_2 in terms of $\dot{r}_{n3n2}$ or getting ω_3 and $\dot{r}_{n3n2}$ in terms of ω_2 as:

$$\begin{aligned} \frac{\omega_3}{\omega_2} &= -\frac{r_{n2}\cos(\theta_2 - \theta_3)}{r_{n2n3} + r_{n3n4}\cos(\theta_4 - \theta_3)} \\ \dot{r}_{n2n3} &= \frac{-r_{n2}\omega_2 \sin(\theta_2) + \omega_4(r_{n2n3}\sin(\theta_3) + r_{n3n4}\sin(\theta_4))}{\cos(\theta_3)} \end{aligned} \tag{6.42}$$

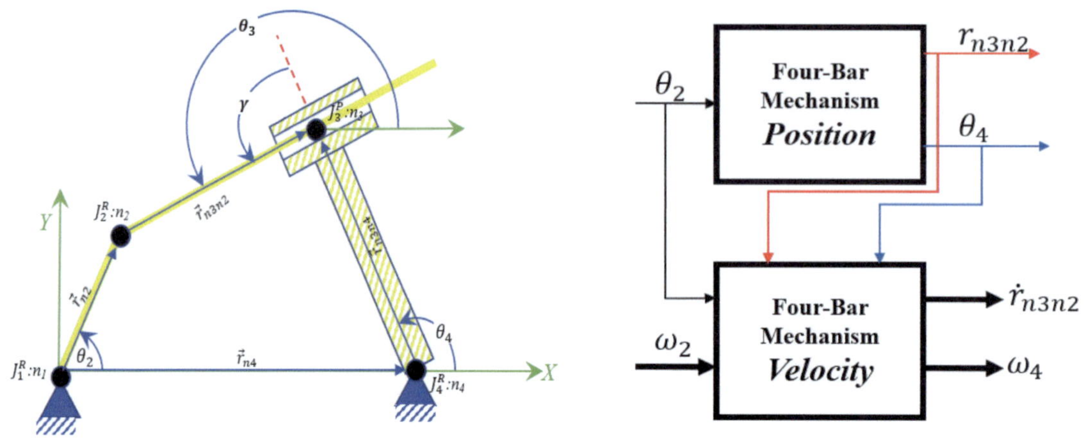

Fig. 6.48 Kinematic diagram and system diagram of an inverted slider–crank linkage

Fig. 6.49 Kinematic diagram inverted slider crank with $\gamma = 0°$

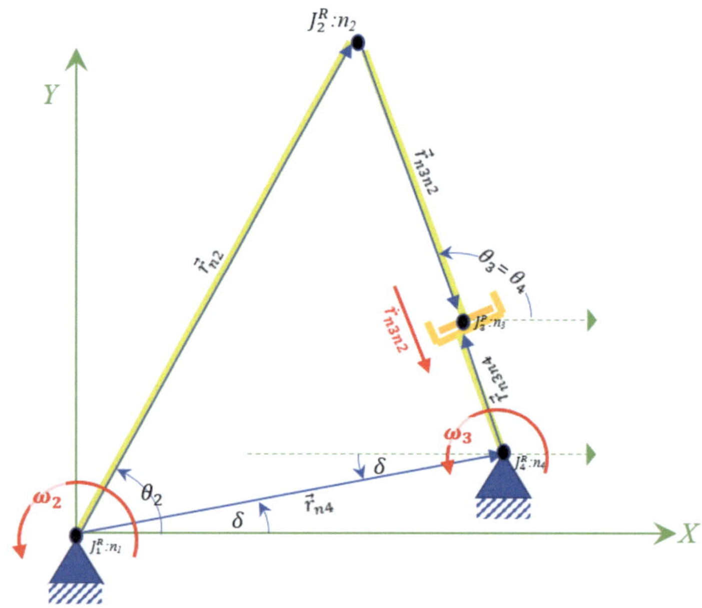

In addition to Fig. 6.48, there is another configuration called the inverted slider crank with fixed crank as shown in Fig. 6.49 that is commonly used in agriculture, construction machines, and others. This configuration is a special case from the configuration when $\gamma = 0°$. Thus, when θ_2 increases in counterclockwise, the length r_{n3n2} increases and causes $\theta_3 = \theta_4$ to increase as well and Eq. (6.42) will be:

$$\frac{\omega_3}{\omega_2} = \frac{r_{n2}\cos(\theta_2 - \theta_3)}{r_{n2n4}}$$
$$\dot{r}_{n2n3} = \frac{\omega_2 r_{n2}\sin(\theta_2) + \omega_4 r_{n2n4}\sin(\theta_4)}{cos(\theta_3)} \tag{6.43}$$

Example 6.10

For the hydraulic gate opener in Fig. 6.39. Assume the door is open with an angular velocity $\omega_2 = 1$ rad/s from the initial position $\theta_2 = 0°$ to final position $\theta_2 = 50°$. At the position $\theta_2 = 45°, \theta_3 = 55°$ and $r_{n3n2} = 405.67$ mm:

- Use the analytical velocity method to calculate the cylinder angular velocity and the piston translational velocity.
- Use the analytical method to evaluate the model-based method and the graphical method.

Solution

Figure 6.50 shows the kinematic diagram with the vector loop. The mechanism has four links: the door frame or the ground L_1, the door L_2, the piston $L_{3,}$ and the cylinder L_4 that are connected by $J_1^R : n_1, J_2^R : n_2, J_3^P : n_3, and J_4^R : n_4$, respectively. The vector loop is shown in Fig. 6.50.

Use Eq. (6.43):

$$\omega_4 = \omega_3 = \frac{r_{n2}\cos(\theta_2 - \theta_3)}{r_{n2n4}} \times \omega_2 = \frac{0.99\cos(45° - 55°)}{0.405 + 0.600} \times 1 = 0.970\,\text{rad/s}$$

$$\dot{r}_{n2n3} = \frac{-\omega_2 r_{n2}\sin(\theta_2) + \omega_4 r_{n2n4}\sin(\theta_4)}{\cos(\theta_3)} = \frac{-1 \times 0.99 \times \sin(45°) + 0.97 \times (0.405 + 0.600) \times \sin(55°)}{\cos(55^o)}$$

$$= 0.172\ \text{m/s}$$

The model-based solution and graphical solution come very close to the analytical solution.

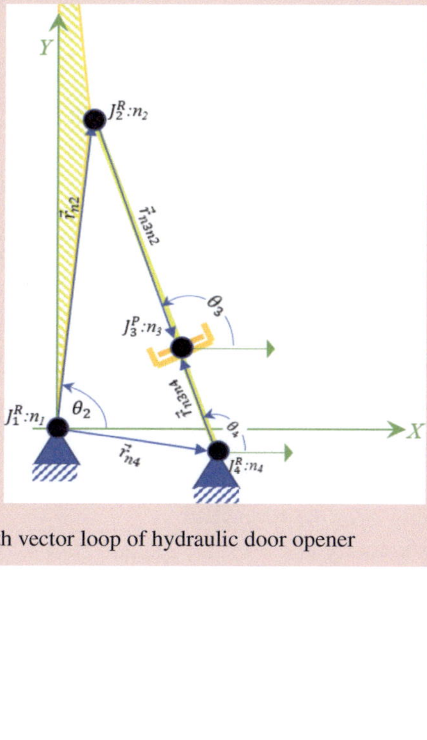

Fig. 6.50 Kinematic diagram with vector loop of hydraulic door opener

6.8 Chapter Summary

Outcomes

1. Calculating the translational velocity vector and the rotational velocity vector of any mechanism node using the graphical method, analytical method, and model-based Simulink® and Simscape™ Multibody™ for a four-bar mechanism, slider–crank mechanism, and inverted slider–crank mechanism
2. Understanding the usage of Kennedy's theorem to estimate the velocity at any node of the mechanism and to estimate the torque ratio between the input motion and output motion of any mechanism
3. Understanding the efficiency of the model-based Simulink® and Simscape™ Multibody™ applied to a higher-order bar mechanism
4. Understanding the difference between sliding motion and rolling motion

Problems

Problem 1

For the shown sewing machine in Fig. 6.51, assume the motor pushes the needle down at node n_5 with the motion below:

$$x_{\text{needle}}(t) = \frac{-A}{2}\left(1 - \cos\left(\frac{\pi t}{T}\right)\right) \text{ for } 0 \le t \le T\,\text{s}$$

Consider the amplitude $A = 32$ mm and the time span $T = 0.05$ s.

1. Calculate the average velocity over the time span.
2. Calculate the instantaneous velocity at $t = 0$ s, $t = 0.025$ s, and $t = 0.05$ s.
3. Write a MATLAB® script to plot the position displacement $x_{\text{needle}}(t)$ with respect to time and the instantaneous velocity $\mathrm{d}x_{\text{needle}}(t)/\mathrm{d}t$ with respect to time.

Problem 2

For the shown potato chopper machine in Fig. 6.52, assume a person rotates the handle so that the plate moves following the motion below:

$$x_{\text{plate}}(t) = A - \frac{Bt}{T}\text{for} 0 \le t \le T\,\text{s}$$

Consider the amplitude $A = 50$ cm, B $= -40$ cm/s, and $T = 4$ s.

1. Calculate the average velocity over the time span.
2. Calculate the instantaneous velocity at $t = 0$ s, $t = 2$ s, and $t = 4$ s.
3. Write a MATLAB® script to plot the position displacement $x_{\text{plate}}(t)$ with respect to time and the instantaneous velocity $\mathrm{d}x_{\text{plate}}(t)/\mathrm{d}t$ with respect to time.

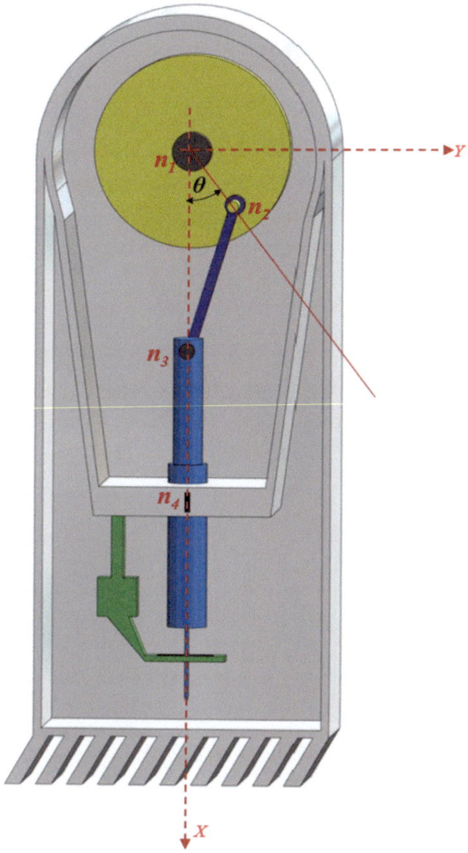

Fig. 6.51 Sewing mechanism. Designed by the author and implemented by both the author and Pranav Dhola. Courtesy of Pranav Dhola

Problem 3

For the pumpjack mechanism shown in Fig. 6.53, the motor is set to rotate the crank with

where θ_2 is in radians and t is in seconds.

$$\theta_2(t) = \frac{2\pi}{3}t$$

1. Calculate the required time, T, to go over one cycle from $\theta_2=0°$ to $\theta_2=360°$.
2. Calculate the average angle velocity over one cycle in rad/sec and in rpm.
3. Calculate the instantaneous angular velocity in rpm at $\theta_2=0°$ and at $\theta_2=180°$.

Problem 4

For the sewing mechanism shown in Fig. 6.51, the motor is set to rotate at

$$\theta(t) = \frac{50\pi}{3}t$$

where θ is in radians and t is in seconds.

1. Calculate the required time to go over one cycle from $\theta = 0°$ to $\theta = 360°$.
2. Calculate the average angle velocity in rad/sec and in rpm.
3. Calculate the instantaneous angular velocity in rpm at $\theta = 0°$ and at $\theta = 180°$.

Problem 5

For the shown dumpster truck mechanism in Fig. 6.54, the hydraulic actuator moves the piston to keep the bed rotating around node n_1 with a constant angular velocity $\omega_2 = 5$ deg/s. Prove with values that node n_5 has more absolute velocity than node n_2.

Problem 6

For the shown potato chopper machine in Fig. 6.52, if a person pushes the handle at node n_5 to rotate around the ground node n_1 with a constant angular velocity $\omega_2 = 15$ rpm, prove with values that node n_5 has more absolute velocity than node n_2. The distance between n_1 and n_5 is 38.8 cm while the distance between n_1 and n_2 is 27.2 cm.

Problem 7

For the shown dumpster truck mechanism in Fig. 6.54, at an instantaneous position, the hydraulic actuator moves the piston outward with 0.4 m and a sliding velocity of 0.1 m/s that causes the cylinder–piston part to rotate with 0.0817 rad/s around n_4. Calculate the instantaneous total velocity vector at node n_2.

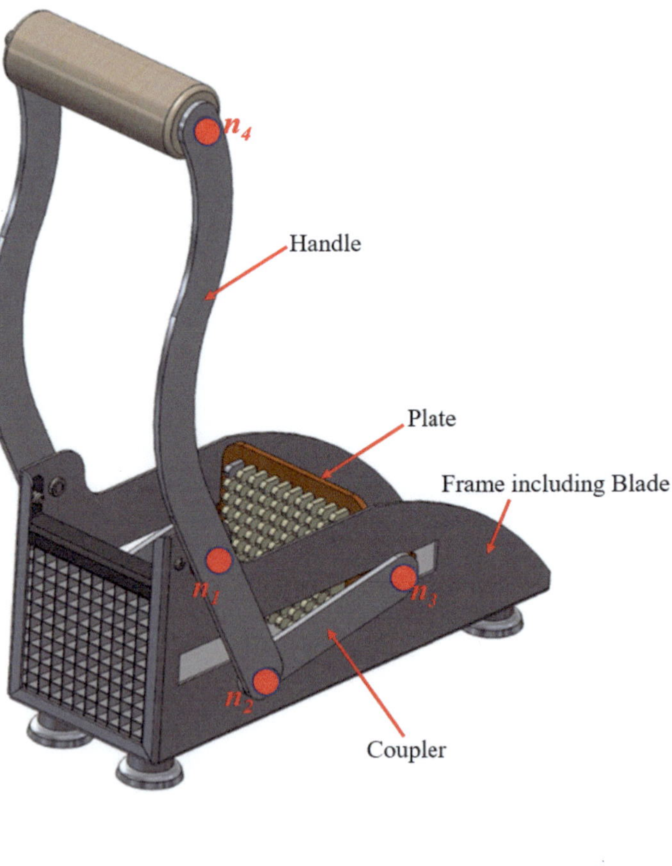

Fig. 6.52 Potato chopper machine. Figure and associated Potato_Chopper_Mechanism.avi video designed by the author and implemented by both the author and Pranav Dhola. Both courtesy of Pranav Dhola

Problem 8

For the shown potato chopper mechanism in Fig. 6.52, if a person pushes the handle at node n_5 to rotate around the ground node n_1. At an instantaneous position, the rocker has an angle 213.6° with the positive X-axis and is rotating with an angular velocity 0.201 rad/s while the node n_3 slides inward with a velocity 0.1 m/s. The rocker has a length of 29 cm. Calculate the instantaneous total velocity vector at node n_2.

Problem 9

For the pumpjack mechanism shown in Fig. 6.53, assume the motor rotates the crank with an angular velocity $\omega_2 = 0.5$ rad/sec. Draw the kinematic diagram when $\theta_2 = 20°$ and

1. Use the graphical method to find the angular velocity of the Pitman arm and the angular velocity of the walking beam.
2. Use the graphical method to calculate the output-to-input torque ratio, where the input torque is applied torque to crank (link L_2) and the output torque is the torque on the walking beam (link L_4).

Problem 10

For the sewing mechanism shown in Fig. 6.51, assume the motor rotates the roll with an angular velocity $\omega_2 = 500$ rpm. Draw the kinematic diagram when $\theta_2 = 30°$ and use the graphical method to find the angular velocity of the coupler plus the sliding velocity of node n_4. The distance between n_1 and n_2 is 15 cm, the distance between n_2 and n_3 is 30 cm, and the distance between n_1 and n_4 is 70 cm. The radius of the yellow roller is 20 cm. The length of the blue needle bar is 55 cm.

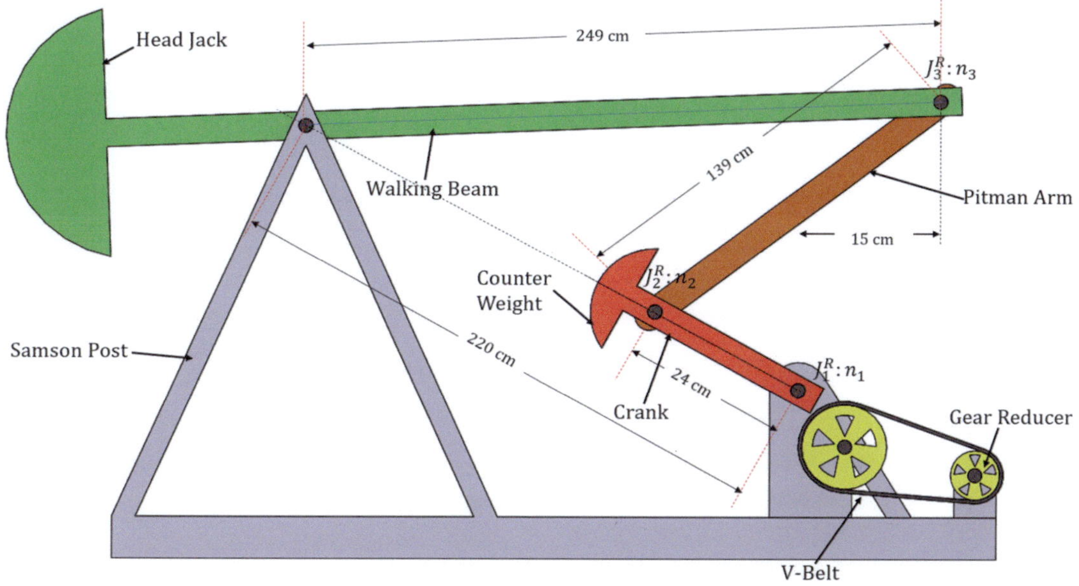

Fig. 6.53 Pumpjack machine. Designed by the author and implemented by both the author and Pranav Dhola. Courtesy of Pranav Dhola

Problem 11

For the dumpster truck mechanism shown in Fig. 6.54, assume the hydraulic actuator moves the piston with a speed 0.1 m/s. Draw the kinematic diagram when the piston is extended by 0.4 m and use the graphical method to find the cylinder angular velocity and the bed angular velocity.

Problem 12

Build a Simscape™ Multibody™ model for the mechanism shown in Fig. 6.51. Use the motion actuator subsystem to move θ_2 from 0° to 90° with the angular velocity 500 rpm. Use the velocity sensors to measure the variation of the angular velocity of the coupler plus the sliding velocity of node n_4 with respect to the time.

Problem 13

Build a Simscape™ Multibody™ model for the mechanism shown in Fig. 6.54. Use the motion actuator subsystem to move the piston by 0.4 m with a speed 0.1 m/s. Use the velocity sensors to measure the variation of both the cylinder angular velocity and the bed angular velocity with time.

Problem 14

Build a Simscape™ Multibody™ model for the mechanism in Fig. 6.53. Use the motion actuator subsystem to rotate θ_2 for one cycle with an angular velocity of 10 rad/sec. Use the velocity sensors to measure the variation of both the angular velocity of the Pitman arm and the angular velocity of the walking beam with time.

Problem 15

For the pumpjack mechanism shown in Fig. 6.53, assume the motors rotate the crank with an angular velocity $\omega_2 = 0.5$ rad/sec. When $\theta_2 = 20°$.

1. Use the analytical method to find both the angle θ_3 and the angle θ_4.
2. Use the analytical method to find both the angular velocity ω_3 and the angular velocity ω_4.

Problem 16

For the shown dumpster truck mechanism in Fig. 6.54, at an instantaneous position, the hydraulic actuator moves the piston outward with 0.4 m and a sliding velocity of 0.1 m/s. Use the analytical method to calculate.

- The angles $\theta_3 = \theta_4$ and θ_2
- The angular velocities $\omega_3 = \omega_4$ and ω_2

Problem 17

For the sewing mechanism shown in Fig. 6.51, assume the motors rotate the roll with an angular velocity $\omega_2 = 500$ rpm. When $\theta_2 = 30°$, use the analytical method to find:

- The angle θ_3 and the X-component of the position vector of node n_3
- The angular velocity of the coupler plus the sliding velocity of node n_4

The distance between n_1 and n_2 is 15 mm, the distance between n_2 and n_3 is 30 cm, and the distance between n_1 and n_4 is 70 mm. The radius of the yellow roller is 20 mm. The length of the blue needle bar is 55 mm.

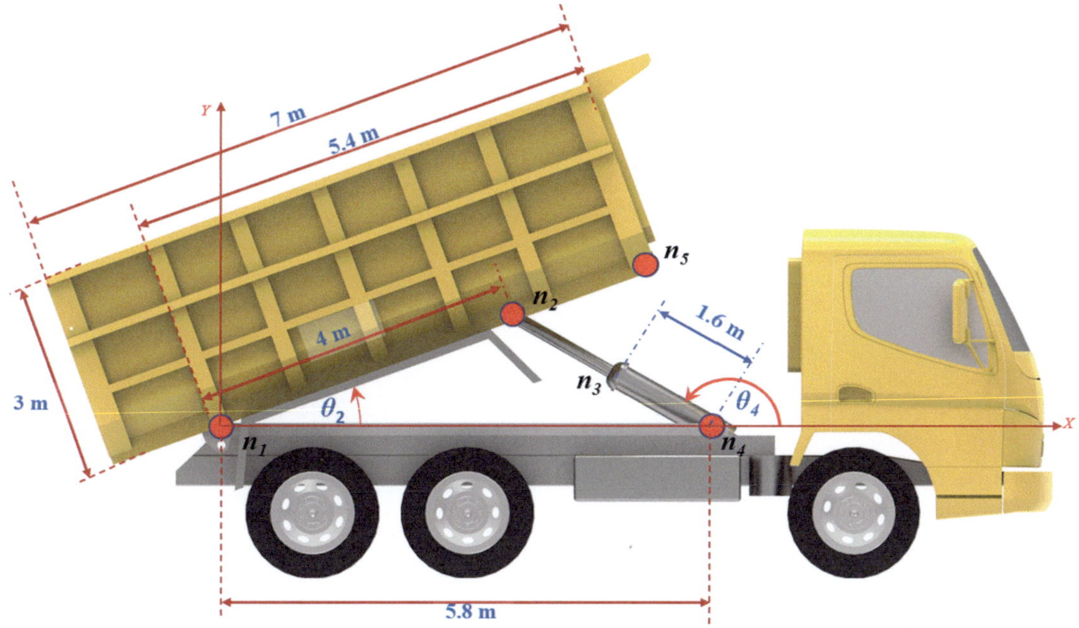

Fig. 6.54 Dumpster truck mechanism. Figure and associated Dumpster Truck Mechanism.mp4 video designed by the author and implemented by both the author and Pranav Dhola. Both courtesy of Pranav Dhola

Problem 18

Build a MATLAB® script that uses the analytical model to calculate θ_3, θ_4, ω_3, ω_4 and position vectors of nodes n_2, n_3, and n_4 for a four-bar mechanism. The script receives the mechanism parameters as L_1, L_2, L_3, and L_4. The script receives θ_2 as an input. Also, the script calculates the toggle angle.

References

1. "Design of Machinery," Robert L. Norton, McGraw-Hill; 5th Edition, 2011.
2. "Theory of Machines and Mechanisms," John Uicker, Gordon Pennock and Joseph Shigley, Oxford University Press; 5th Edition, 2016.
3. "Machines & Mechanisms: Applied Kinematic Analysis", David Myszka, Pearson; 4th edition, 2011.
4. "Newton Engineering and Physical Science Pocket Book", John Bird, 2nd Edition, 1996.
5. "Simscape™ Multibody™ Getting Started Guide" MathWorks, 2021.
6. "Simscape™ Multibody™ User's Guide", MathWorks, 2021.
7. "Engineering Mechanics: Dynamics", Russel Hibbeler, 14th Edition, 2016.
8. "Kinematic synthesis of linkages," Hartenberg, R.S. & J. Denavit, McGraw-Hill, 1994.
9. "Complete classification of planar four-bar linkage," Baker, C., Journal of Mechanical Design, 1985.
10. "Modelling an inverted slider crank mechanism considering kinematic analysis and multibody aspects, " R.C. Silva, M.A. Nunes, J.P.M. Bento and V.E. Costa, DINAME, Proceedings of the XV international symposium on dynamic problems of mechanics, 2013.

Acceleration and Jerk Analysis

7

Chapter Objectives

1. Learning 2D translational acceleration vectors and 2D angular acceleration vectors of any single node (point) in both Cartesian and polar coordinates.
2. Learning the 2D acceleration vector diagram.
3. Learning the difference between the absolute acceleration vector and the relative acceleration vector.
4. Learning the difference between sliding, normal, tangential, and Coriolis acceleration components.
5. Learning that the translational acceleration vector and the angular acceleration vector of any node belong to a link connected to the ground link.
6. Using the model-based Simulink® and Simscape™ Multibody™ to calculate the translational acceleration vector and the angular acceleration vector of any node belonging to a mechanism.
7. Developing and using the graphical acceleration models plus the analytical velocity models for four-bar, slider–crank, and inverted slider–crank mechanisms.
8. Introducing jerk analysis and the influence of the jerk's magnitude value on mechanism flexibility.

7.1 Introduction

In this chapter, we explain the translational acceleration and angular acceleration vectors of any node followed by an explanation of absolute acceleration vectors and relative acceleration vectors. We include the definition of sliding, normal, tangential, and Coriolis acceleration components. The acceleration vector diagram is introduced. After that, we develop three different methods to calculate the acceleration of any mechanism's node: model-based Simulink® and Simscape™ Multibody™ method, graphical method, and analytical method. These methods will be implemented for the four-bar mechanisms, the slider–crank mechanisms, and the inverted slider–crank mechanisms. Finally, jerk analysis is covered plus an introduction to the flexible Simulink® and Simscape™ Multibody™ model.

Supplementary Information The online version contains supplementary material available at https://doi.org/10.1007/978-3-031-72701-6_7.

A. Omran, *Kinematics and Dynamics of Machines*, https://doi.org/10.1007/978-3-031-72701-6_7

7.2 Translational Acceleration and Angular Acceleration

Definition

Linear Acceleration:

The translational acceleration of a node (point) is the linear velocity of that point per unit of time. Alternatively, linear acceleration is the rate of change of linear velocity with respect to time [1–3].

Angular Acceleration:

Angular acceleration is defined as the rate of change of angular velocity with respect to time for an object rotating about a fixed axis [4].

Just as the total velocity vector of any node on a mechanism includes a translational velocity vector $\vec{v}$ and an angular velocity vector $\vec{\Omega}$, so the total acceleration vector includes a translational acceleration vector $\vec{a}$ and an angular acceleration vector $\vec{\alpha}$. The translational acceleration is the time rate of change for the node's translational velocity in a specific direction. The angular acceleration is the time rate of the node's angular velocity in a specific direction. Now, we shall introduce the mathematical formulas for the translational acceleration vector and angular acceleration vector.

7.2.1 Translational Acceleration

In a 2D domain, if we assume a node with a translational velocity vector $\vec{v}$ at time t

$$\vec{v}(t) = v_x(t)\hat{i} + v_y(t)\hat{j} \tag{7.1}$$

where $v_x(t)$ is the translational velocity component in the x-direction and $v_y(t)$ is the translational velocity component in the Y-direction in the world frame or a non-rotating frame. The 2D average translational acceleration vector $\vec{\overline{a}}$ has two components: $\overline{a}_x$ in X-direction and $\overline{a}_y$ in Y-direction. Over a time interval Δt, the average translational acceleration components $\overline{a}_x$ and $\overline{a}_y$ can be defined to be the translational velocity displacements Δv_x and Δv_y divided by the time interval Δt as

$$\begin{aligned}\vec{\overline{a}}(t, \Delta t) &= \overline{a}_x(t, \Delta t)\hat{i} + \overline{a}_y(t, \Delta t)\hat{j} = \frac{\Delta v_x}{\Delta t}\hat{i} + \frac{\Delta v_y}{\Delta t}\hat{j} \\ &= \frac{v_x(t+\Delta t) - v_x(t)}{\Delta t}\hat{i} + \frac{v_y(t+\Delta t) - v_y(t)}{\Delta t}\hat{j}\end{aligned}$$

or

$$\overline{a}_x(t, \Delta t) = \frac{v_x(t+\Delta t) - v_x(t)}{\Delta t} \text{ and } \overline{a}_y(t, \Delta t) = \frac{v_y(t+\Delta t) - v_y(t)}{\Delta t} \tag{7.2}$$

The instantaneous translational acceleration $\vec{a}(t)$ at time t is the limit of the average translational acceleration $\vec{\overline{a}}$ when Δt tends to zero. This limit is equal to the time derivative of the translational velocity components as listed in Eq. (7.3).

$$\begin{aligned}\vec{a}(t) &= \lim_{\Delta t\to 0}\overline{\vec{a}}(t,\Delta t) = \lim_{\Delta t\to 0}\overline{a}_x(t,\Delta t)\hat{i} + \lim_{\Delta t\to 0}\overline{a}_y(t,\Delta t)\hat{j}\\ &= \lim_{\Delta t\to 0}\frac{\Delta v_x(t,\Delta t)}{\Delta t}\hat{i} + \lim_{\Delta t\to 0}\frac{\Delta v_y(t,\Delta t)}{\Delta t}\hat{j}\\ &= \frac{\mathrm{d}v_x(t)}{\mathrm{d}t}\hat{i} + \frac{\mathrm{d}v_y(t)}{\mathrm{d}t}\hat{j} = \frac{\mathrm{d}^2x(t)}{\mathrm{d}t^2}\hat{i} + \frac{\mathrm{d}^2y(t)}{\mathrm{d}t^2}\hat{j}\\ &= \ddot{x}(t)\hat{i} + \ddot{y}(t)\hat{j} = a_x(t)\hat{i} + a_y(t)\hat{j}\end{aligned} \tag{7.3}$$

or

$$a_x(t) = \frac{\mathrm{d}v_x(t)}{\mathrm{d}t} = \dot{v}_x(t) = \frac{\mathrm{d}^2x(t)}{\mathrm{d}t^2} = \ddot{x}(t) \text{ and}$$
$$a_y(t) = \frac{\mathrm{d}v_x(t)}{\mathrm{d}t} = \dot{v}_y(t) = \frac{\mathrm{d}^2x(t)}{\mathrm{d}t^2} = \ddot{y}(t)$$

Example 7.1

Previously in Example 6.1, we had the water pump (Fig. 7.1) with the following piston position and velocity profiles:

$$y_{\text{piston}}(t) = 100(1-\cos(\beta(t)))$$
$$\dot{y}_{\text{piston}}(t) = 100\pi\left(1-\frac{1}{8}\cos(\pi t)\right)\sin(\beta(t))$$
$$\text{where } \beta(t) = \pi\left(t-\frac{1}{8\pi}\sin(\pi t)\right) \text{ for } 0 \le t \le 1\text{ s}$$

1. Calculate the average translational acceleration over the time span, $T_{\text{span}} = 1$ s.
2. Calculate the instantaneous translational acceleration at $t=0$ s, $t=0.5$ s, and $t=1$ s.
3. Write MATLAB® script to plot the position displacement $y_{\text{piston}}(t)$ with respect to time, the instantaneous velocity $\dot{y}_{\text{piston}}(t)$ with respect to time, and the instantaneous acceleration $\ddot{y}_{\text{piston}}(t)$ with respect to time.

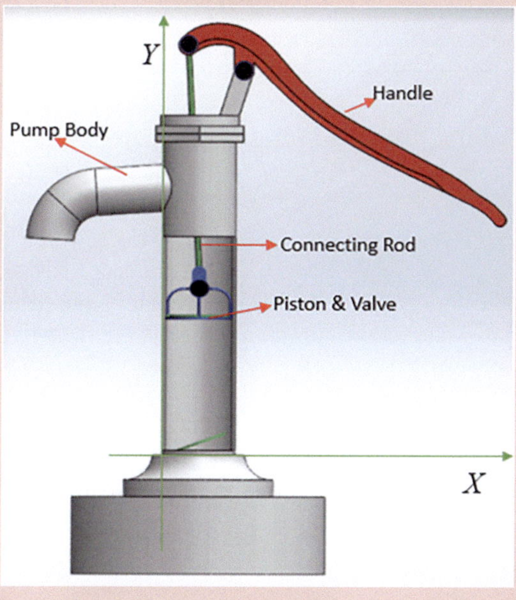

Fig. 7.1 Pitcher water pump. Designed by the author and implemented by both the author and Pranav Dhola. Courtesy of Pranav Dhola

Solution

The velocity vector is

$$\vec{r}_{\text{piston}}(t) = \dot{y}_{\text{piston}}(t)\hat{j}$$

where

$$\dot{y}_{\text{piston}}(t) = 100\pi\left(1 - \frac{1}{8}\cos(\pi t)\right)\sin(\beta(t)) \text{ and } \beta(t) = \pi\left(t - \frac{1}{8\pi}\sin(\pi t)\right)$$

Note that $\dot{x}_{\text{piston}}(t)$ is zero from the kinematic constraints of the prismatic joint. Also, the frame is an inertia frame or not a rotating frame.

Average Acceleration

First, we calculate the velocity at the initial time $t=0$ s and the final time $t=1$ s

$$\left.\begin{array}{l}\beta(t=0) = \pi\left(t - \frac{1}{4\pi}\sin(2\pi t)\right)\Big|_{t=0} = \pi\left(0 - \frac{1}{4\pi}\sin(\pi \times 0)\right) = 0 \\ \dot{y}_{\text{piston}}(t=0) = 100\pi\left(1 - \frac{1}{8}\cos(\pi t)\right)\sin(\beta(t))\end{array}\right|_{t=0} = 0\,\text{mm/s}$$

$$\beta(t=1) = \pi\left(t - \frac{1}{4\pi}\sin(2\pi t)\right)\Big|_{t=1} = \pi\left(1 - \frac{1}{8\pi}\sin(\pi \times 1)\right) = \pi$$

$$\dot{y}_{\text{piston}}(t=1) = 100\pi\left(1 - \frac{1}{8}\cos(\pi t)\right)\sin(\beta(t))\Big|_{t=1} = 0\,\text{mm/s}$$

Calculate the average acceleration from Eq. (7.2)

$$\begin{aligned}\bar{\vec{a}}_{\text{piston}}(t=0, \Delta t=1) &= \left.\frac{\dot{x}_{\text{piston}}(t+\Delta t) - \dot{x}_{\text{piston}}(t)}{\Delta t}\right|_{\substack{t=0\\ \Delta t=1}} \hat{i} \\ &\quad + \frac{\dot{y}_{\text{piston}}(t+\Delta t) - \dot{y}_{\text{piston}}(t)}{\Delta t}\hat{j} \\ &= \frac{\dot{x}_{\text{piston}}(t=1) - \dot{x}_{\text{piston}}(t=0)}{1-0}\hat{i} \\ &\quad + \frac{\dot{y}_{\text{piston}}(t=1) - \dot{y}_{\text{piston}}(t=0)}{1-0}\hat{j} \\ &= 0\hat{i} + 0\hat{j}\,\text{mm/s}\end{aligned}$$

Instantaneous Velocity

We start with calculating the derivatives of the velocity vector components with respect to time

$$\begin{aligned}v_{x\text{piston}}(t) &= \frac{\mathrm{d}x_{\text{piston}}(t)}{\mathrm{d}t} = 0 \\ a_{x\text{piston}}(t) &= \frac{\mathrm{d}v_{x\text{piston}}(t)}{\mathrm{d}t} = \frac{\mathrm{d}^2 x_{\text{piston}}(t)}{\mathrm{d}t^2} = 0 \\ v_{y\text{piston}}(t) &= \frac{\mathrm{d}y_{\text{piston}}(t)}{\mathrm{d}t} = 100\pi\left(1 - \frac{1}{8}\cos(\pi t)\right)\sin(\beta(t)) \\ a_{y\text{piston}}(t) &= \frac{\mathrm{d}v_{y\text{piston}}(t)}{\mathrm{d}t} = \frac{\mathrm{d}^2 y_{\text{piston}}(t)}{\mathrm{d}t^2} \\ &= \frac{25\pi^2}{2}\sin(\pi t)\sin(\beta(t)) + 100\pi^2\left(1 - \frac{1}{8}\cos(\pi t)\right)^2\cos(\beta(t))\end{aligned}$$

The instantaneous acceleration vector is calculated from Eq. (7.3)

$$\vec{a}_{\text{piston}}(t) = a_{x\text{piston}}(t)\hat{i} + a_{y\text{piston}}(t)\hat{j} = \frac{\mathrm{d}^2 x_{\text{piston}}(t)}{\mathrm{d}t^2}\hat{i} + \frac{\mathrm{d}^2 y(t)}{\mathrm{d}t^2}\hat{j}$$
$$= \left(\frac{25\pi^2}{2}\sin(\pi t)\sin(\beta(t)) + 100\pi^2\left(1 - \frac{1}{8}\cos(\pi t)\right)^2\cos(\beta(t))\right)\hat{j}$$

Substitute with $t=0$ s, $t=0.5$ s and $t=1$ s

$$\beta(t=0) = \pi\left(t - \frac{1}{8\pi}\sin(\pi t)\right)\Big|_{t=0} = \pi\left(0 - \frac{1}{8\pi}\sin(\pi\times 0)\right) = 0$$

$$\vec{a}_{\text{piston}}(t=0) = \frac{25\pi^2}{2}\sin(\pi t)\sin(\beta(t)) + 100\pi^2\left(1 - \frac{1}{8}\cos(\pi t)\right)^2\cos(\beta(t))\Big|_{t=0}\hat{j}$$
$$= \left(\frac{25\pi^2}{2}\sin(0)\sin(\beta(0)) + 100\pi^2\left(1 - \frac{1}{8}\cos(0)\right)^2\cos(\beta(0))\right)\hat{j}$$
$$= \vec{0}\,\text{mm/s}^2$$

$$\beta(t=0.5) = \pi\left(t - \frac{1}{8\pi}\sin(\pi t)\right)\Big|_{t=0.5} = \pi\left(0.5 - \frac{1}{8\pi}\sin(\pi\times 0.5)\right) = 1.446$$

$$\vec{a}_{\text{piston}}(t=0.5) = \frac{25\pi^2}{2}\sin(\pi t)\sin(\beta(t)) + 100\pi^2\left(1 - \frac{1}{8}\cos(\pi t)\right)^2\cos(\beta(t))\Big|_{t=0.5}\hat{j}$$
$$= \left(\frac{25\pi^2}{2}\sin(0.5\pi)\sin(\beta(0.5)) + 100\pi^2\left(1 - \frac{1}{8}\cos(0.5\pi)\right)^2\cos(\beta(0.5))\right)\hat{j}$$
$$= \left(\frac{25\pi^2}{2}\sin(0.5\pi)\sin(1.446) + 100\pi^2\left(1 - \frac{1}{8}\cos(0.5\pi)\right)^2\cos(1.446)\right)\hat{j}$$
$$= 245.26\hat{j}\,\text{mm/s}^2$$

$$\beta(t=1) = \pi\left(t - \frac{1}{8\pi}\sin(\pi t)\right)\Big|_{t=1} = \pi\left(1 - \frac{1}{8\pi}\sin(\pi\times 1)\right) = \pi$$

$$\vec{a}_{\text{piston}}(t=1) = \frac{25\pi^2}{2}\sin(\pi t)\sin(\beta(t)) + 100\pi^2\left(1 - \frac{1}{8}\cos(\pi t)\right)^2\cos(\beta(t))\Big|_{t=0}\hat{j}$$
$$= \left(\frac{25\pi^2}{2}\sin(\pi)\sin(\beta(1)) + 100\pi^2\left(1 - \frac{1}{8}\cos(\pi)\right)^2\cos(\beta(1))\right)\hat{j}$$
$$= \left(\frac{25\pi^2}{2}\sin(\pi)\sin(\pi) + 100\pi^2\left(1 - \frac{1}{8}\cos(\pi)\right)^2\cos(\pi)\right)\hat{j}$$
$$= -1249.1\hat{j}\,\text{mm/s}^2$$

MATLAB® Plotting

You can write the commands below in MATLAB® script. After running the m file below, Fig. 7.2 will be the results. Note: We did not plot the *X*-axis position component, *X*-axis velocity component, or the *X*-axis acceleration component because the three of them have a zero value. We encourage the reader to watch the attached animation video of the pitcher water pump. We also encourage the reader to watch the MATLAB® onramp tutorial to help in understanding the script below. Also, we encourage the reader to watch the animation video of the water pump mechanism.

```
% Example 7.1
%===================================
%Clear all the variables and close all the figures
%===============================
clc
clear all
close all
%===================================================================

%calculating the time
dt = 0.0001; %sec, selected time step
t =[0:dt:T]; %sec, time starts from zero and end at T by dt as time step

%Calculating the position
Beta   = pi*(t-1/8/pi*sin(pi*t));
y_piston =  100*(1-cos(Beta));

%calculating instantaneous velocity
vy_piston = 100*pi*(1-1/8*cos(pi*t)).*sin(Beta);

%calculating instantaneous acceleration
ay_piston = (25*pi^2)/2*sin(pi*t).*sin(Beta)+100*pi^2*((1-1/8*cos(pi*t)).^2).*cos(Beta);

%Plot the signals
%===============
subplot(3,1,1)
plot(t,y_piston,'LineWidth',2);
xlabel('t (sec)');
ylabel('y_p_i_s_t_o_n (mm)');
grid on;
subplot(3,1,2)
plot(t,vy_piston,'LineWidth',2);
xlabel('t (sec)');
ylabel('v_y_p_i_s_t_o_n (mm/sec)');
grid on;
subplot(3,1,3)
plot(t,ay_piston,'LineWidth',2);
xlabel('t (sec)');
ylabel('a_y_p_i_s_t_o_n (mm/sec^2)');
grid on;
```

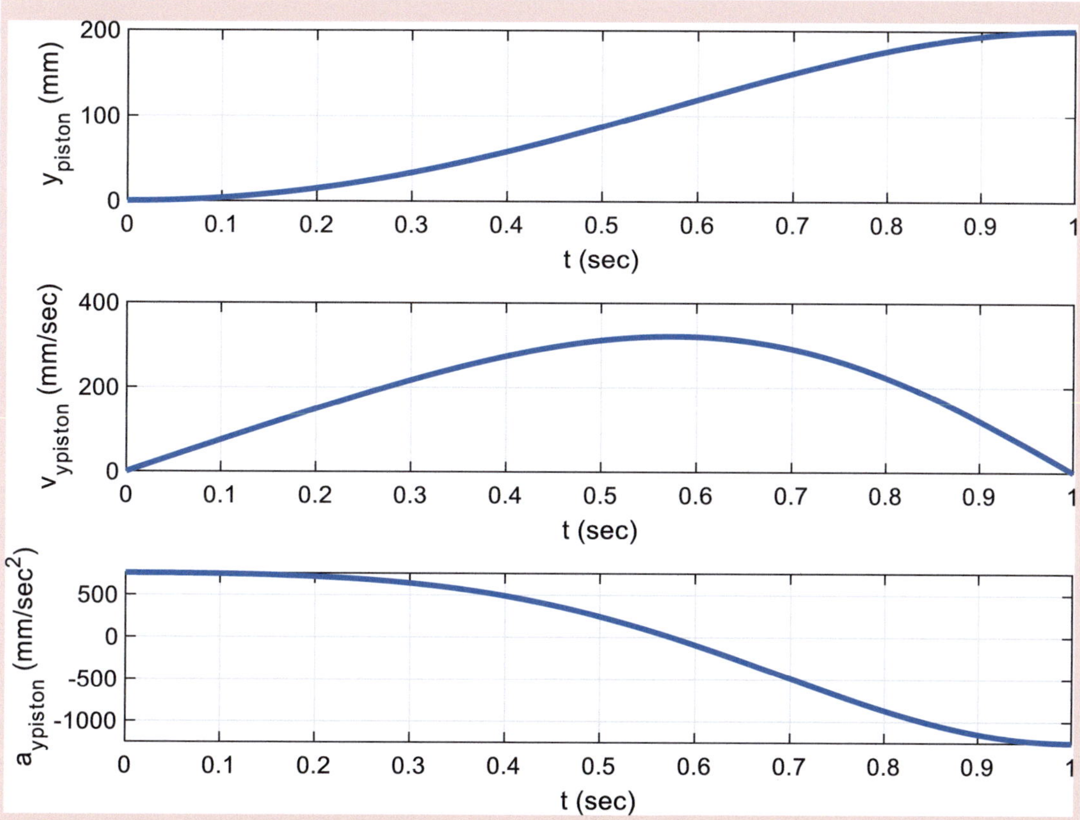

Fig. 7.2 Position, velocity, and acceleration in *y*-axis for pitcher water pump in Example 7.1

One important technique for understanding the translational acceleration vector is to sketch the so-called "***Acceleration Vector Diagram***" that describes the magnitude and the direction of any acceleration quantity by a vector arrow. For example, assume a node n_i with a translational acceleration vector $\vec{a}_{ni}(t) = a_{xni}(t)\hat{i} + a_{yni}(t)\hat{j}$ has a 2D acceleration vector diagram at a specific time t that can be visualized at a new 2D acceleration *xy*-plane using the following steps:

- Plot *XY*-plane with an origin *O*.
- Move $a_{xni}(t)$ in *X*-direction and plot a vertical dashed line parallel to the *Y*-axis.
- Move $a_{yni}(t)$ in *Y*-direction and plot a horizontal dashed line parallel to the *X*-axis.
- Define point *A* as the intersection between the vertical dashed line and the horizontal dashed line.
- Plot a vector (line with an arrow) headed from the origin *O* to point *A* to define $\overrightarrow{OA} = \vec{a}_{n2}$.

Figure 7.3 shows the translational acceleration vector diagram of node n_i. Note that the acceleration vector diagram is visualized on a different *XY*-plane from what we use in the kinematic diagram. We can use an acceleration vector diagram to visualize the translational acceleration vectors of any number of nodes. For example, if we have a mechanism with three important nodes, we can plot the three acceleration vectors on one acceleration vector diagram.

7.2.2 Angular Acceleration

The angular acceleration can be expressed as a vector quantity in the same way as the translational acceleration. In 2D, the angular acceleration is the rate of change in the angular speed around any

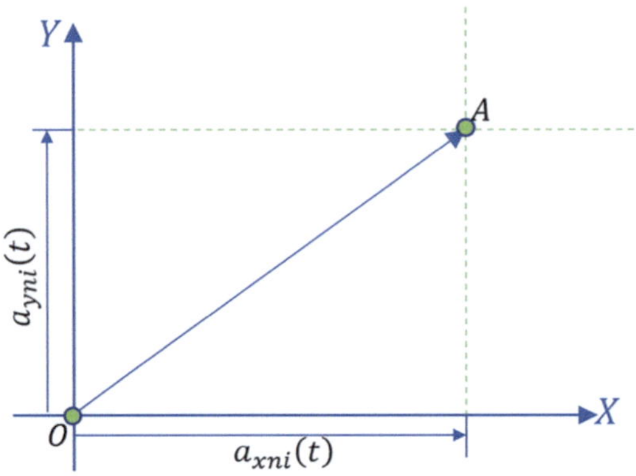

Fig. 7.3 Acceleration vector diagram of node n_i

Z-axis (world frame or relative frame). For example, assume that a motor is connected to the center of a shaft as shown in Fig. 7.4. The angular velocity and the angular acceleration are then measured around the *Z*-axis, and they can be mathematically defined as

$$\begin{aligned}\vec{\Omega}(t) &= 0\hat{i} + 0\hat{j} + \omega\hat{k} \\ \vec{\alpha}(t) &= 0\hat{i} + 0\hat{j} + \alpha\hat{k}\end{aligned} \tag{7.4}$$

In this textbook, any angular acceleration will be defined around the *Z*-axis or as written in Eq. (7.4) because our analysis is restricted to 2D. The angular acceleration is expressed as the average angular acceleration vector $\vec{\overline{\alpha}}(t, \Delta t)$ and the instantaneous angular acceleration vector $\vec{\alpha}(t)$ whereby the average angular acceleration vector $\vec{\overline{\alpha}}(t, \Delta t)$ has one component $\overline{\alpha}$ around *Z*-axis and instantaneous angular acceleration vector $\vec{\alpha}(t)$ has one acceleration component α around *Z*-axis. Over a time interval Δt, the average angular acceleration component $\overline{\alpha}$ is the angular velocity displacements $\Delta\omega$ per time Δt. The instantaneous angular acceleration component is where Δt tends to zero or the time derivative of ω.

$$\vec{\overline{\alpha}}(t, \Delta t) = \overline{\alpha}(t, \Delta t)\hat{k} = \frac{\Delta\omega}{\Delta t}\hat{k} = \frac{\omega(t + \Delta t) - \omega(t)}{\Delta t}\hat{k} \tag{7.5}$$

and

$$\begin{aligned}\vec{\alpha}(t) &= \lim_{\Delta t \to 0} \vec{\overline{\alpha}}(t, \Delta t) = \lim_{\Delta t \to 0} \overline{\alpha}(t, \Delta t)\hat{k} \\ &= \lim_{\Delta t \to 0} \frac{\Delta\omega(t, \Delta t)}{\Delta t}\hat{k} = \frac{d\omega(t)}{dt}\hat{k} = \dot{\omega}(t)\hat{k} = \alpha(t)\hat{k}\end{aligned}$$

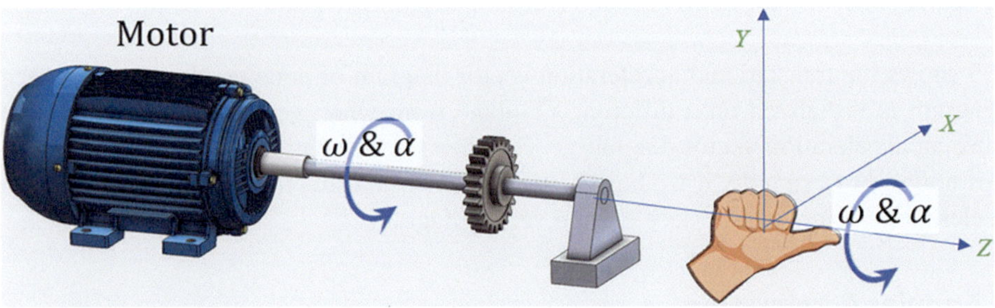

Fig. 7.4 Motor angular velocity and angular acceleration around the *Z*-axis. Designed by the author and implemented by both the author and Pranav Dhola. Courtesy of Pranav Dhola

Example 7.2

Figure 7.5 shows the car wiper four-bar mechanism. The motor can rotate the crank per one cycle using the below equation
where θ_2 is in deg and t is in sec.

$$\theta_2(t) = A\left(1 - \cos\left(\frac{\pi t}{T_{\text{cycle}}}\right)\right), T_{\text{cycle}} = 1\ \text{s} \quad \text{and } A = 180^\circ$$

$$\omega_2(t) = \frac{A\pi}{T_{\text{cycle}}}\sin\left(\frac{\pi t}{T_{\text{cycle}}}\right)$$

1. Calculate the instantaneous angular velocity in deg/s at $t=0$ s and $t=T_{\text{cycle}}$.
2. Calculate the average angular acceleration in deg/s^2 over one cycle.
3. Calculate the instantaneous angular acceleration in deg/s^2 as a function of time.
4. Write a MATLAB® script to plot the angular position, angular velocity, and angular acceleration w.r.t. time over one cycle.

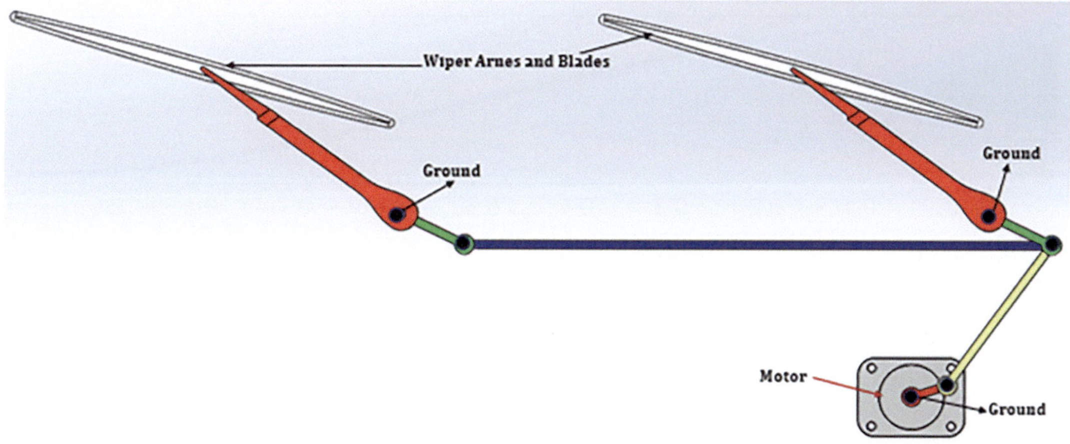

Fig. 7.5 Car wiper mechanism. Designed by the author and implemented by both the author and Pranav Dhola. Courtesy of Pranav Dhola

Solution

1. **Angular velocity at $t=0$ s and at $t = T_{\text{cycle}} = 1$ s**

$$\vec{\Omega}_2(t=0) = \omega_2(t=0)\hat{k} = \frac{A\pi}{T_{\text{cycle}}}\sin\left(\frac{\pi t}{T_{\text{cycle}}}\right)\Big|_{t=0}$$

$$\hat{k} = \frac{180^\circ\pi}{1}\sin\left(\frac{\pi \times 0}{1}\right)\hat{k} = 0\hat{k}$$

$$\vec{\Omega}_2(t=1) = \omega_2(t=1)\hat{k} = \frac{A\pi}{T_{\text{cycle}}}\sin\left(\frac{\pi t}{T_{\text{cycle}}}\right)\Big|_{t=1}$$

$$\hat{k} = \frac{180^\circ\pi}{1}\sin\left(\frac{\pi \times 1}{1}\right)\hat{k} = 0\hat{k}$$

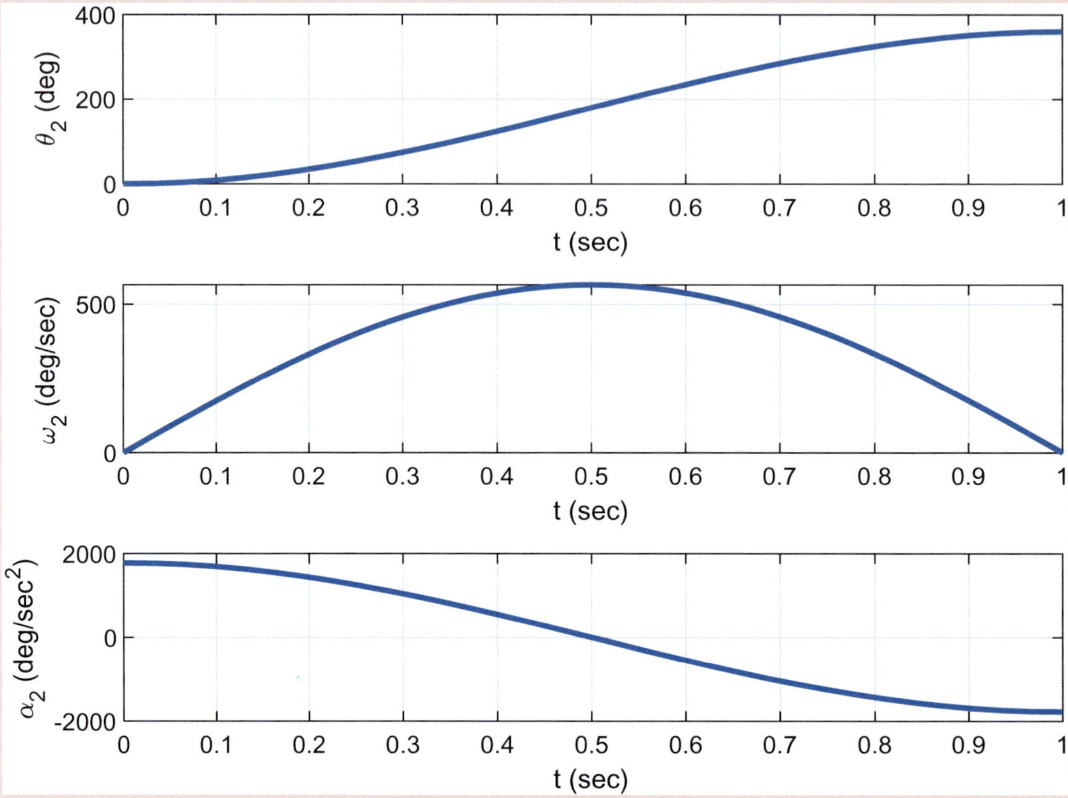

Fig. 7.6 Angular position, angular velocity, and angular acceleration of motor in car wiper mechanism

2. **Average angular acceleration over one cycle**
Using Eq. (7.5)

$$\vec{\overline{\alpha}}_2\left(t=0, \Delta t = T_{cyce} = 1\right) = \overline{\alpha}_2(t, \Delta t)\Big|_{\substack{t=0\\ \Delta t=1}} \hat{k} = \frac{\Delta\omega_2}{\Delta t}\Big|_{\substack{t=0\\ \Delta t=1}}$$

$$\hat{k} = \frac{\omega_2(t+\Delta t) - \omega_2(t)}{\Delta t}\Big|_{\substack{t=0\\ \Delta t=1}} \hat{k}$$

$$= \frac{\omega_2(0+1) - \omega_2(0)}{1}\hat{k} = 0\hat{k}\ \mathrm{deg/s^2}$$

3. **Instantaneous angular acceleration as a function of time**
Using Eq. (7.5)

$$\vec{\alpha}_2(t) = \alpha_2(t)\hat{k} = \frac{\mathrm{d}\omega_2(t)}{\mathrm{d}t}\hat{k} = \frac{A\pi^2}{T_{\text{cycle}}^2}\cos\left(\frac{\pi t}{T_{\text{cycle}}}\right)\hat{k}$$

4. **MATLAB® Plotting**

You can write the commands below in MATLAB® script. After running the m file below, Fig. 7.6 will be the results. We encourage the reader to watch the attached animation video of the car wiper mechanism. Also, we encourage the reader to watch the MATLAB® onramp tutorial to help in understanding the script below.

```
%Example 7.2
%=====================================
%Clear all the variables and close all the figures
%=================================
clc
clear all
close all
%====================================================================================
%plugging in the parameters
%=========================
T_cycle = 1; %sec
A = 180; %deg

%Calculating the signals
%========================
t = [0:0.001:T_cycle]; %sec
theta_2 = A*(1-cos(pi*t/T_cycle)); %deg
omega_2 = A*pi/T_cycle*sin(pi*t/T_cycle); %deg/sec
alpha_2 = A*(pi/T_cycle)^2*cos(pi*t/T_cycle); %deg/sec^2

%Plot the signals
%=================
subplot(3,1,1)
plot(t,theta_2,'LineWidth',2);
xlabel('t (sec)');
ylabel('\theta_2 (deg)');
grid on;
subplot(3,1,2)
plot(t,omega_2,'LineWidth',2);
xlabel('t (sec)');
ylabel('\omega_2 (deg/sec)');
grid on;
subplot(3,1,3)
plot(t,alpha_2,'LineWidth',2);
xlabel('t (sec)');
ylabel('\alpha_2 (deg/sec^2)');
grid on;
```

7.2.3 Absolute Polar Acceleration Components

The absolute polar acceleration components are defined with respect to a frame that is located at the instantaneous center of the rotation and not from any other frames. For example, assume we have the node n_i belonging to the link L_k as shown in Fig. 7.7. The absolute polar acceleration components of n_i must be measured from the link L_k instantaneous center of the rotation IC_k where the unit vector$\hat{r}$ is in the direction of the vector r_{niICK}. The unit vector $\hat{\theta}$ is perpendicular on the unit vector $\hat{r}$. The frame $\hat{r} - \hat{\theta} - \hat{k}$ follows the right-hand rule as well as the frame $\hat{i} - \hat{j} - \hat{k}$.

To map the Cartesian acceleration components onto the polar acceleration components or vice versa, we are going to use an acceleration vector diagram. Assume that node n_i has the absolute transitional acceleration vector.

$$\begin{aligned} \vec{a}_{ni} &= a_{xni}\hat{i} + a_{yni}\hat{j} \\ \vec{a}_{ni} &= a_{rni}\hat{r} + a_{\theta ni}\hat{\theta} \end{aligned} \tag{7.6}$$

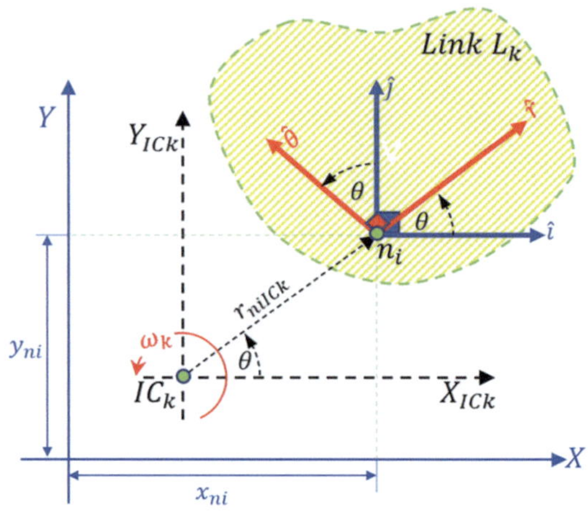

Fig. 7.7 Unit vectors of Cartesian coordinate and polar coordinates

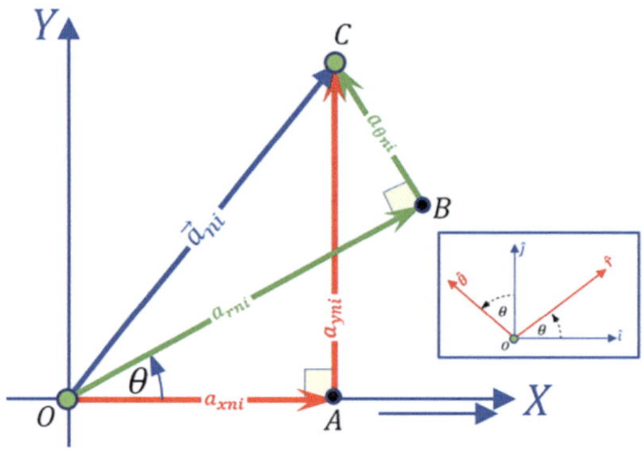

Fig. 7.8 Acceleration vector diagram for acceleration Cartesian components and acceleration polar components

In Eq. (7.6), the same vector $\vec{a}_{ni}$ is decomposed as Cartesian components and as polar components. From the primary vector courses, we can add vectors in the acceleration vector diagram. We can plot $\vec{a}_{ni}$ as a summation of $\vec{a}_{xni}$ in *X*-direction and $\vec{a}_{yni}$ in *Y*-direction. We can also plot $\vec{a}_{ni}$ as a summation of $\vec{a}_{rni}$ in *r*-direction and $\vec{a}_{\theta ni}$ in θ-direction using the steps below.

- Plot *XY*-plane with an origin *O*.
- Plot vector $\overrightarrow{OA}$ with a length of a_{xni} starting from point *O* and ending at point *A* on the *X*-axis.
- Plot vector $\overrightarrow{AC}$ with a length of a_{yni} starting from *A* point and ending at *B* point in a parallel direction to the *Y*-axis.
- Plot vector $\overrightarrow{OB}$ with a length of a_{rni} starting from *O* point and ending at *B* point on the *r*-axis.
- Plot vector $\overrightarrow{BC}$ with a length of a_{yni} starting from *B* point and ending at *C* point in a parallel direction to the θ-axis
- Plot the vector $\overrightarrow{OC} = \vec{a}_{ni}$by connecting point *O* with point *C*.

Figure 7.8 shows the results from following the above steps. Now, we can apply some geometric calculations to find:

$$\begin{aligned}
a_r(t) &= a_x(t)\cos(\theta) + a_y(t)\sin(\theta)\\
a_\theta(t) &= -a_x(t)\sin(\theta) + a_y(t)\cos(\theta)\\
&\text{or}\\
a_x(t) &= a_r(t)\cos(\theta) - a_\theta(t)\sin(\theta)\\
a_y(t) &= a_r(t)\sin(\theta) + a_\theta(t)\cos(\theta)
\end{aligned} \tag{7.7}$$

where a_r is the acceleration component in r-direction and a_θ is the acceleration component in θ-direction as originally defined in Fig. 7.7.

Engineering Tips—Polar Coordinate Angle and Polar Coordinate Radius

From our previous definitions: The ***instant center of rotation***, also called the ***instantaneous center of zero velocity,*** is the point fixed to a body undergoing planar movement that has zero velocity at a particular instant of time, Ref. [7]. Using this definition, for any node n_i belonging to link L_k, we could measure the absolute acceleration polar coordinate components of this node by using $\theta = \theta_{niICk}$ and the polar radius $r = r_{niICk}$. The absolute acceleration polar coordinate components are the normal acceleration component a_{rni} and the tangential acceleration component $a_{\theta ni}$.

For example, if one examines the four-bar mechanism in Fig. 7.9, to measure the polar acceleration components of node n_2 belonging to link L_2, we must measure θ_{n2IC2} and r_{n2IC2} from the IC_2 which happens to be stationary at n_1, the origin. That leads us to use $\theta_{n2IC2} = \theta_2$ and Eq. (7.7) becomes

$$\begin{aligned}
a_{rn2} &= a_{xn2}\cos(\theta_{n2IC2}) + a_{yn2}\sin(\theta_{n2IC2})\\
&= a_{xn2}\cos(\theta_2) + a_{yn2}\sin(\theta_2)\\
a_{\theta n2} &= -a_{xn2}\sin(\theta_{n2IC2}) + a_{yn2}\cos(\theta_{n2IC2})\\
&= -a_{xn2}\sin(\theta_2) + a_{yn2}\cos(\theta_2)
\end{aligned}$$

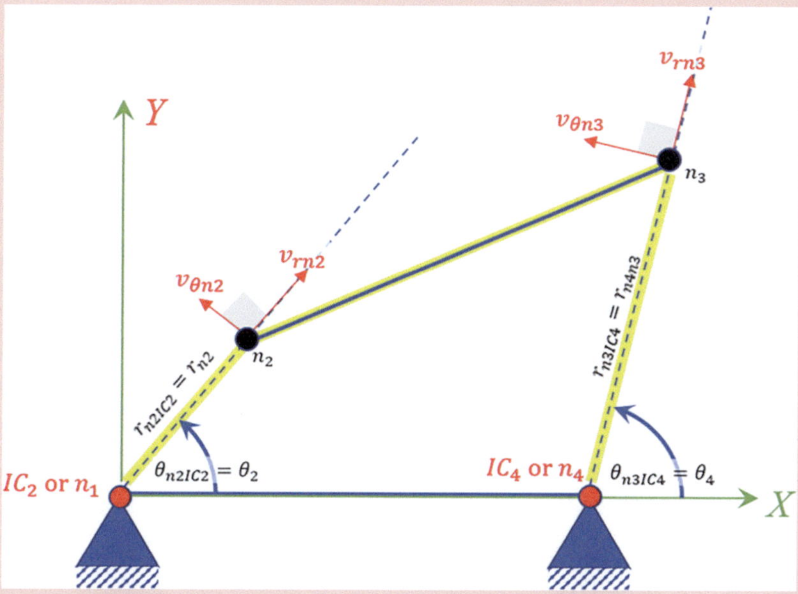

Fig. 7.9 Polar angle and polar radius of n_2 using L_2 center of rotation and n_3 using L_4 center of rotation

We can use the same method to measure the polar acceleration components of node n_3 belonging to link L_4 using IC_4 which happens to be stationary at n_4. That leads us to use $\theta_{n3IC4} = \theta_4$ and $r_{n3IC4} = r_{n3n4}$. Equation (7.7) becomes

$$\begin{aligned} a_{rn3} &= a_{xn3}\cos(\theta_{n3IC4}) + a_{yn3}\sin(\theta_{n3IC4}) \\ &= a_{xn3}\cos(\theta_4) + a_{yn3}\sin(\theta_4) \\ a_{\theta n3} &= -a_{xn3}\sin(\theta_{n3IC4}) + a_{yn3}\cos(\theta_{n3IC4}) \\ &= -a_{xn3}\sin(\theta_4) + a_{yn3}\cos(\theta_4) \end{aligned}$$

If we decide to calculate the polar acceleration components of n_2 using L_3, we should measure θ_{n2IC3} and r_{n2IC3} from the IC_3 that is not stationary like IC_2 or IC_4. Figure 7.10 shows two different positions of IC_3 for a four-bar mechanism. If we decide to measure the polar acceleration components of node n_2 using L_3, Eq. (7.7) becomes

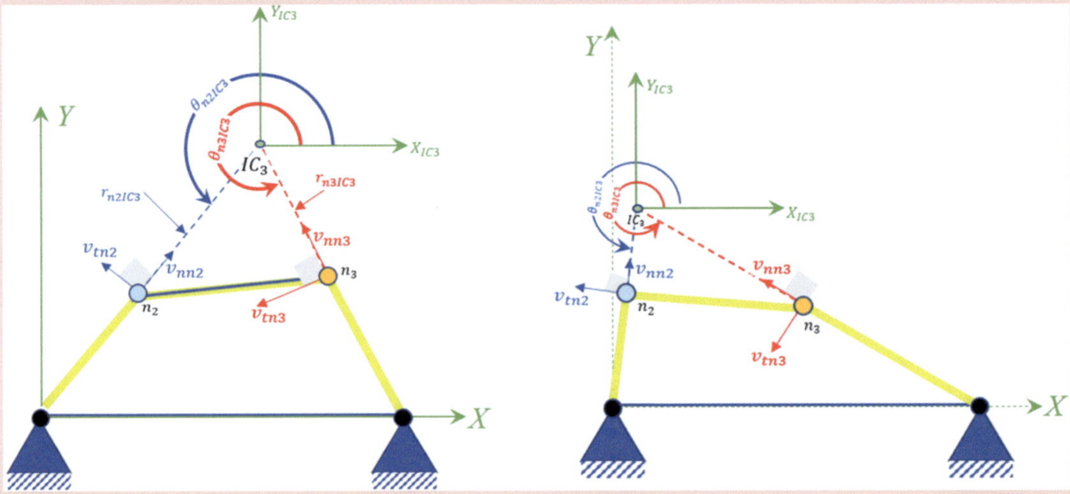

Fig. 7.10 Polar angle of n_2 and n_2 using L_3 center of rotation at two different times

$$\begin{aligned} a_{rn2} &= a_{xn2}\cos(\theta_{n2IC3}) + a_{yn2}\sin(\theta_{n2IC3}) \\ a_{\theta n2} &= -a_{xn2}\sin(\theta_{n2IC3}) + a_{yn2}\cos(\theta_{n2IC3}) \end{aligned}$$

Similarly, if we decide to measure the polar acceleration components of node n_3 using L_3, Eq. (7.7) becomes

$$\begin{aligned} a_{rn3} &= a_{xn3}\cos(\theta_{n3IC3}) + a_{yn3}\sin(\theta_{n3IC3}) \\ a_{\theta n3} &= -a_{xn3}\sin(\theta_{n3IC3}) + a_{yn3}\cos(\theta_{n3IC3}) \end{aligned}$$

After finishing our analysis to understand the motion of a single node, we are now going to talk about the relative motion between two moving objects or two moving nodes (points).

7.3 Absolute and Relative Acceleration

In Fig. 7.11, the fixed world or inertia frame XY is located at the ground node n_1 or the origin. A local rotating and moving frame $X_{ni}Y_{ni}$ is located at node n_i that has an absolute translational position vector $\vec{r}_{ni}$, an absolute translational velocity vector $\vec{v}_{ni}$, and an absolute translational acceleration vector $\vec{a}_{ni}$:

$$\begin{aligned} \vec{r}_{ni} &= x_{ni}\hat{\imath} + y_{ni}\hat{\jmath} \\ \vec{v}_{ni} &= v_{xni}\hat{\imath} + v_{yni\hat{\jmath}} = \dot{x}_{ni}\hat{\imath} + \dot{y}_{ni}\hat{\jmath} \\ \vec{a}_{ni} &= a_{xni}\hat{\imath} + a_{yni}\hat{\jmath} = \dot{v}_{xni}\hat{\imath} + \dot{v}_{yni}\hat{\jmath} = \ddot{x}_{ni}\hat{\imath} + \ddot{y}_{ni}\hat{\jmath} \end{aligned} \tag{7.8}$$

The $X_{ni}Y_{ni}$ frame is rotating with an angle θ_i, an angular velocity vector $\vec{\Omega}_i$, and an angular acceleration vector $\vec{\alpha}_i$ as

$$\begin{aligned} \vec{\Omega}_i &= \omega_i\hat{k} \\ \vec{\alpha}_i &= \alpha_i\hat{k} \end{aligned} \tag{7.9}$$

The position vector of node n_j can be measured as the absolute position vector $\vec{r}_{nj}$ from the origin or it can be measured as the relative position vector $\vec{r}_{njni}$ from node n_i.

$$\begin{aligned} \vec{r}_{nj} &= x_{nj}\hat{\imath} + y_{nj}\hat{\jmath} \\ \vec{r}_{njni} &= x_{njni}\hat{\imath} + y_{njni}\hat{\jmath} \end{aligned} \tag{7.10}$$

The node n_j has a relative velocity $\vec{v}_{njni}^{\text{relative}}$ and a relative acceleration $\vec{a}_{njni}^{\text{relative}}$ with respect to the node n_i. The absolute acceleration vector $\vec{a}_{nj}$ is, Refs. [1–8].

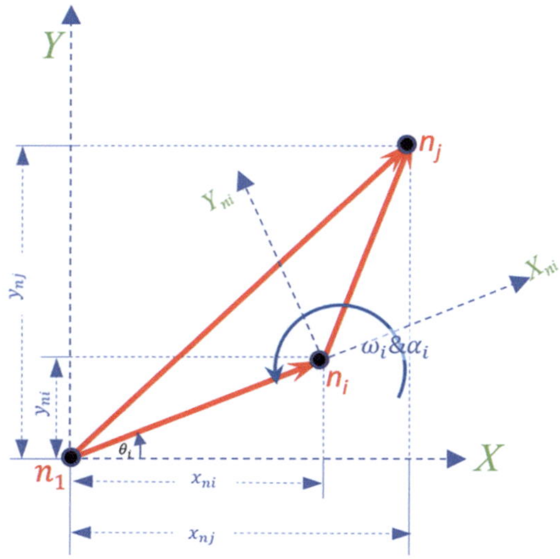

Fig. 7.11 Rotating relative frames

$$\vec{a}_{nj} = \vec{a}_{ni} + \vec{a}_{njni} = \vec{a}_{ni} + \underbrace{\vec{a}^{\text{relative}}_{njni}}_{\textit{Sliding}} + \underbrace{\vec{\alpha}_i \times \vec{r}_{ninj}}_{\textbf{Tangential}} + \underbrace{\vec{\Omega}_i \times \left(\vec{\Omega}_i \times \vec{r}_{njni}\right)}_{\textbf{Normal}} + \underbrace{2\vec{\Omega}_i \times \vec{v}^{\text{relative}}_{njni}}_{\textbf{Coriolis}} \tag{7.11}$$

The derivation of Eq. (7.11) is provided in many textbooks and References [1–8]. We omitted the derivation, and we are going to focus more on the meaning of each component and how we can use Eq. (7.11) in our kinematics of machines analysis.

In Eq. (7.11), besides that the node n_i has an acceleration vector $\vec{a}_{ni}$, there are four relative acceleration terms. The first acceleration term is the translational or sliding relative acceleration component $\vec{a}^{\text{relative}}_{njni}$ that happens when there is a speeding up or slowing down during a sliding motion on a rotating link. The other three acceleration components are a mix between the angular motion terms and the translational motion terms as explained below.

Tangential Acceleration Component $\vec{\boldsymbol{a}}_{Tnjni} = \vec{\boldsymbol{\alpha}}_{ni} \times \vec{\boldsymbol{r}}_{njni}$: As shown in Fig. 7.12a, if there is a moving node on a curved trajectory with a radius r and angular acceleration α, there will be an instantaneously tangential acceleration component on the curved path or perpendicular on the radius r in 2D plane with the value:

$$\begin{aligned} \vec{a}_T &= \vec{\alpha} \times \vec{r} = \alpha\hat{k} \times r\hat{r} = \alpha r\hat{\theta} \\ a_T &= \alpha r \end{aligned} \tag{7.12}$$

Note that in Eq. (7.12), if the angle acceleration α is counterclockwise, the tangential acceleration component a_T will be in the direction as shown in Fig. 7.12a or vice versa. The tangential acceleration component a_T is like linear acceleration. The only difference is that linear acceleration is on a straight trajectory while a_T is on a curved trajectory. For example, imagine a car is driving along a straight road. To increase the speed, there is a need for a linear acceleration. If the car is driving on a curved road, to increase the speed there is a need for tangential acceleration a_T.

Centripetal or Normal Acceleration Component $\vec{\boldsymbol{a}}_{Nnjni} = \vec{\boldsymbol{\Omega}}_i \times \left(\vec{\boldsymbol{\Omega}}_i \times \vec{\boldsymbol{r}}_{njni}\right)$: As shown in Fig. 7.12b, in the 2D plane, if there is a moving node on a curved trajectory with a radius $r\,\hat{e}_r$ and angular velocity $\omega\hat{k}$, there will be an instantaneously normal acceleration component toward the center of the curve with the value

$$\begin{aligned} \vec{a}_N &= \vec{\Omega} \times \left(\vec{\Omega} \times \vec{r}\right) = \omega\hat{k} \times \left(\omega\hat{k} \times r\hat{r}\right) \\ &= \omega\hat{k} \times \left(\omega\hat{k} \times r\hat{r}\right) = \omega^2 r\hat{k} \times \hat{\theta} = -\omega^2 r\hat{r} \\ a_N &= -\omega^2 r \end{aligned} \tag{7.13}$$

Note in Eq. (7.13), regardless of if the angular velocity ω is counterclockwise or clockwise, the centripetal acceleration component a_N will be always in the normal direction toward the center of the rotation as shown in Fig. 7.12b. The centripetal acceleration component represents the normal force that acts on a rotating node, even when the node moves at a constant angular speed with a zero angular acceleration. The centripetal force or acceleration must point toward the center of the circle, and therefore perpendicular to the tangential velocity, to maintain the circular motion.

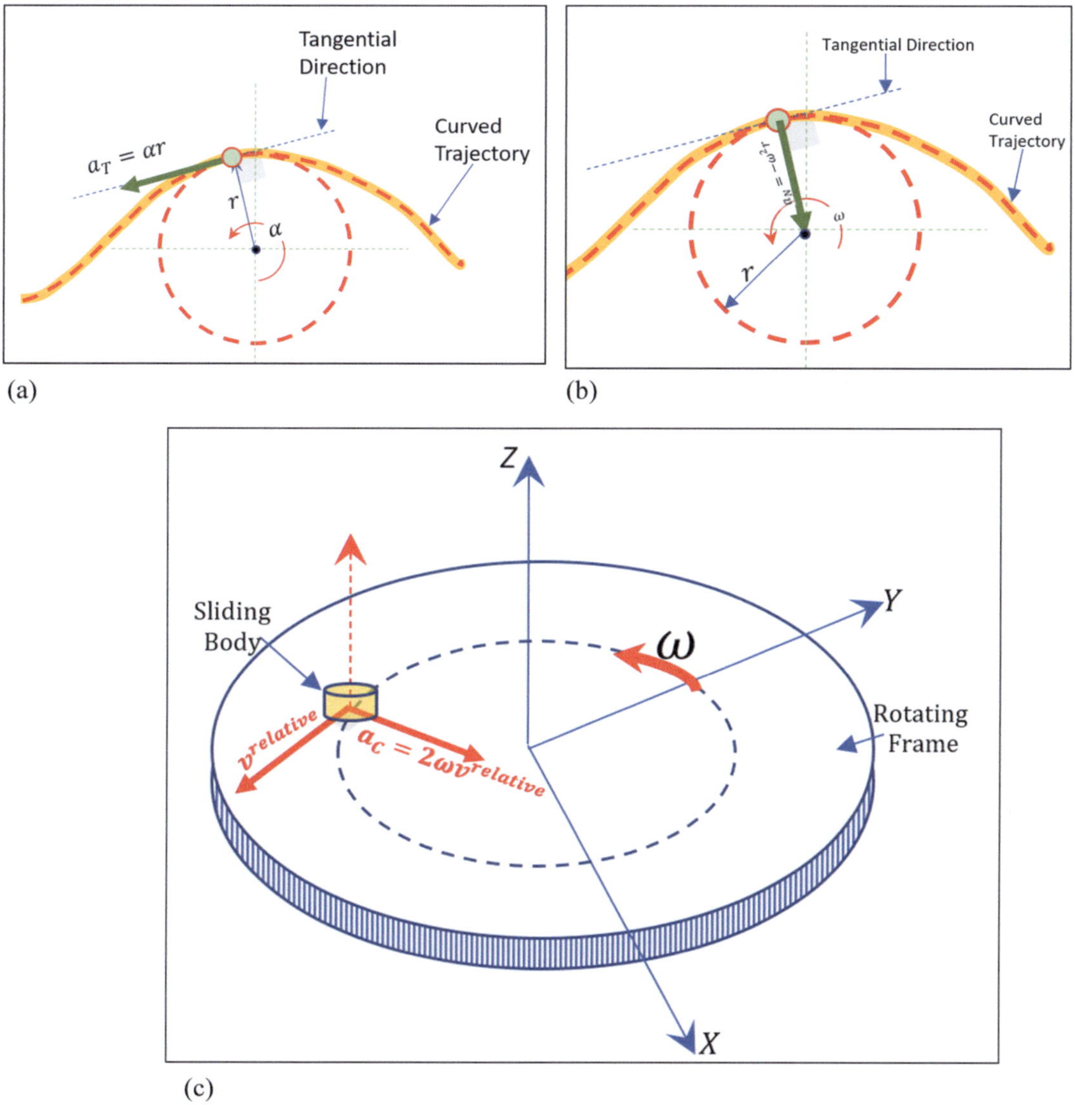

Fig. 7.12 Relative acceleration components

Coriolis Acceleration Component $\vec{a}_{Cnjni} = 2\vec{\Omega}_i \times \vec{v}_{njni}^{relative}$: As shown in Fig. 7.12c, in a 2D plane, if there is a moving node with a sliding relative velocity v^{relative} on a rotating frame with an angular velocity $\vec{\Omega} = \omega\widehat{k}$, there will instantaneously be a Coriolis acceleration component.

$$\begin{aligned} &\vec{a}_C = 2\vec{\Omega} \times \vec{r} = 2\omega\hat{k} \times v^{\text{relative}}\hat{e}_v = 2\omega v^{\text{relative}}\hat{e}_{pv} \\ &a_C = 2\omega v^{\text{relative}} \end{aligned} \tag{7.14}$$

where $\hat{e}_v$ is the unit vector in the direction of v^{relative} and $\hat{e}_{pv}$ is the unit vector perpendicular on $\hat{e}_v$ such that $\hat{e}_v - \hat{e}_{pv} - \hat{k}$ follows the right-hand rule. Note in Eq. (7.14), if the angle velocity ω is

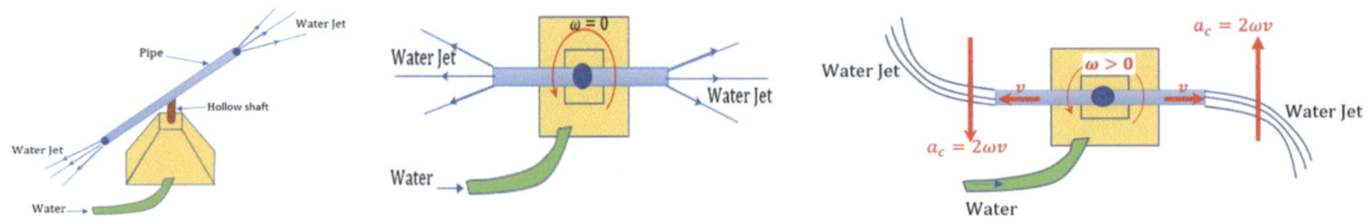

(a) (b) (c)

Fig. 7.13 Coriolis acceleration in rotary lawn sprinkler

counterclockwise, the Coriolis acceleration component a_C will be in the direction as shown in Fig. 7.12c or vice versa.

One everyday example to explain the Coriolis acceleration components is a rotary lawn sprinkler as shown in Fig. 7.13a. This sprinkler is made of a pipe that is fixed to a hollow shaft at its middle and has attached nozzles at both ends. The water runs in the pipe with a constant velocity v. There is a motor to rotate the sprinkler around the shaft. If this motor is not running or $\omega = 0$, the water will exit the nozzle with a constant velocity v as shown in Fig. 7.13b. However, if the motor is rotating with a constant angular velocity ω, there will be an additional Coriolis acceleration component that moves the water a bit tangential as shown in Fig. 7.13c.

In the kinematics of machines, the Coriolis acceleration component occurs when a link is sliding over a rotating link such as the inverted slider–crank mechanism that will be explained later in this chapter.

Engineering Tips—Cross Product

The cross product of the vectors $\vec{a} = a_x\hat{i} + a_y\hat{j} + a_z\hat{k}$ and $\vec{b} = b_x\hat{i} + b_y\hat{j} + b_z\hat{k}$ is

$$\vec{c} = \vec{a} \times \vec{b} = -\vec{b} \times \vec{a} = \det(A)$$

$$\text{where} A = \begin{bmatrix} i & j & k \\ a_x & a_y & a_z \\ b_x & b_y & b_z \end{bmatrix}$$

Note that the symbol det(A) is used to refer to the determinate of matrix A. The vector $\vec{c}$ is perpendicular to both $\vec{a}$ and $\vec{b}$, with a direction given by the right-hand rule as shown in Fig. 7.14. Take your index finger (on your right hand) and point straight ahead, next take your middle finger and point it to the left so it is perpendicular to your index finger. Stick your thumb straight up as if you were giving a thumbs up to a fellow engineer. If your index finger is the $\vec{a}$ *direction* and your middle finger is the $\vec{b}$ *direction*, then your thumb is in the direction of the $\vec{c}$ direction.

The cross product of two unit vectors is a unit vector that is perpendicular on both with a direction given by the right-hand rule. For example, $\hat{i} \times \hat{j} = \hat{k}$ since $\hat{i}$-$\hat{j}$-$\widehat{k}$, follow the right rule and $\hat{j} \times \hat{i} = -\hat{k}$ because $\hat{j}$-$\hat{i}$-"$-\hat{k}$" follow the right rule as well.

Another way to obtain the cross product of two unit vectors is to use the unit-vector circle $\hat{i}$ -$\hat{j}$-$\hat{k}$ or $\hat{\theta}$-$\hat{r}$-$\hat{k}$ as shown in Fig. 7.15. If the two vectors are on a counterclockwise rotation, then the cross product is +ve the third vector. For example, to calculate $\hat{i} \times \hat{j}$, the rotation from $\hat{i}$ to $\hat{j}$ is counterclockwise and the $\hat{i} \times \hat{j} = +\hat{k}$. If the two vectors are on a clockwise rotation, then the cross product is the -ve the third vector. For example, to calculate $\hat{j} \times \hat{i}$, the rotation from $\hat{j}$ to $\hat{i}$ is clockwise and the $\hat{i} \times \hat{j} = -\hat{k}$.

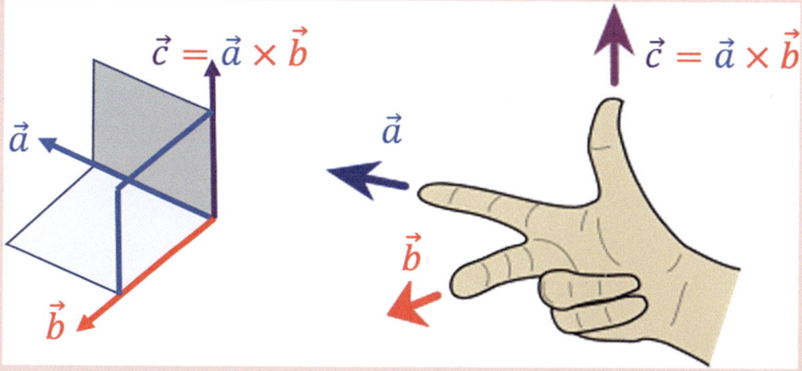

Fig. 7.14 Cross product directions. Designed by the author and implemented by both the author and Pranav Dhola. Courtesy of Pranav Dhola

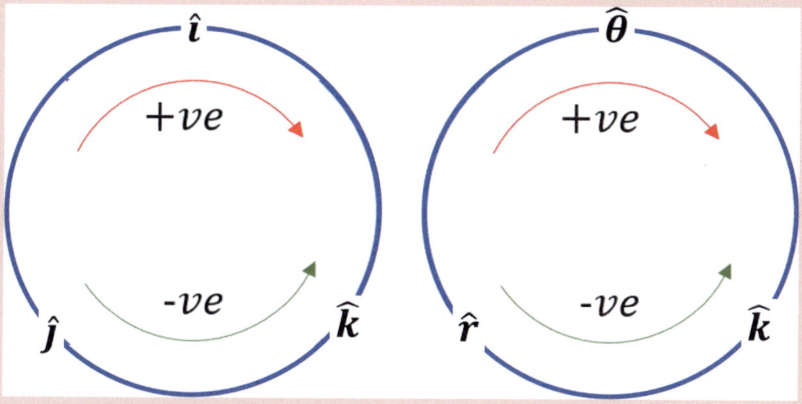

Fig. 7.15 Unit vector circle and cross product direction

From our previous analysis for the four acceleration components, we can rewrite Eq. (7.11) in 2D plane as

$$\vec{a}_{nj} = \vec{a}_{ni} + \vec{a}_{njni} = \vec{a}_{ni} + \underbrace{\alpha_i r_{njni} \hat{\theta}}_{\textit{Tangential}} + \underbrace{-\omega_i^2 r_{nCnB} \hat{r}}_{\text{Normal}} + \underbrace{2\omega_i v_{njni}^{\text{relative}} \hat{e}_{pv}}_{\textit{Coriolis}} + \underbrace{\vec{a}_{njni}^{\text{relative}}}_{\textit{Sliding}} \tag{7.15}$$

We can use the acceleration vector diagram to plot both the relative acceleration vector and the absolute acceleration vector as shown in Fig. 7.16 using the following vector definitions:

$$\begin{aligned}
\overrightarrow{OA} &= \vec{a}_{ni} \\
\overrightarrow{AB} &= \alpha_i r_{njni} \hat{\theta} \\
\overrightarrow{BC} &= -\omega_i^2 r_{njni} \hat{r} \\
\overrightarrow{CD} &= 2\omega_i v_{njni}^{\text{relative}} \hat{e}_{pv} \\
\overrightarrow{DE} &= \vec{a}_{njni}^{\text{relative}} \\
\overrightarrow{OE} &= \vec{a}_{nj}
\end{aligned} \tag{7.16}$$

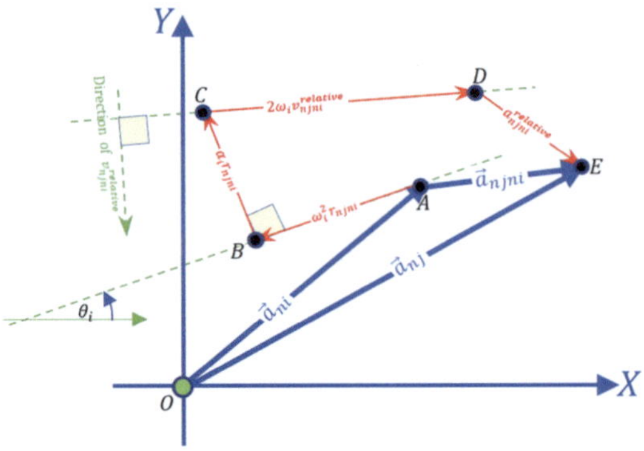

Fig. 7.16 Acceleration vector diagram of absolute and relative acceleration

Example 7.3

Figure 7.17 shows a hydraulic gate opener mechanism. The door is rotating with a constant angular velocity of 1 rad/s. At instantaneous time t, while the door is rotating at a position of an angle 45°, the piston is extracted to the length 0.413 m with a sliding velocity of 0.174 m/s and a sliding acceleration of – 0.0298 m/s^2. The cylinder is at an angle of 55.1° and rotating with an angular velocity of 0.969 rad/s and an angular acceleration of – 0.162 rad/s^2. At the piston–door hinge joint node $J_2^R : n_2$, using the piston link to calculate:

- Tangential acceleration component
- Centripetal acceleration component
- Coriolis acceleration component
- Sliding acceleration component
- Total acceleration in polar coordinate
- Plot the acceleration vector diagram
- Total acceleration in Cartesian coordinate.

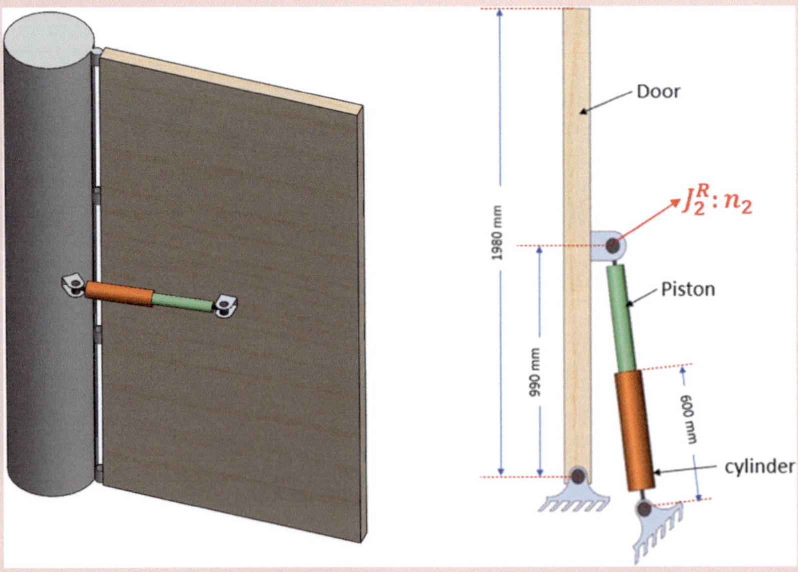

Fig. 7.17 Hydraulic gate opener. Designed by the author and implemented by both the author and Pranav Dhola. Courtesy of Pranav Dhola

Solution

First, we sketch the kinematic diagram with the frame or the ground link as L_1, the door as L_2, the piston as L_3, and the cylinder as L_4 as shown in Fig. 7.18. The mechanism has three revolute joints $J_1^R : n_1, J_2^R : n_2$, and $J_4^R : n_4$ plus a prismatic joint $J_3^R : n_3$.

The relative acceleration components of the node n_2 w.r.t. node n_4 are normal, tangential, Coriolis, and sliding.

The normal component and tangential component are coming from the cylinder rotational motion with $\theta_{\text{cylinder}} = \theta_4 = 55.1°, \omega_{\text{cylinder}} = \omega_4 = 0.969 \frac{\text{rad}}{\text{s}}$, and $\alpha_{\text{cylinder}} = \alpha_4 = -0.162 \frac{\text{rad}}{\text{s}^2}$.

The Coriolis component and the sliding component are coming from the piston slides with $r_{n2n3} = 0.406\,\text{m}, v_{\text{piston}}^{\text{relative}} = v_{n3n4}^{\text{relative}} = v_{n2n4}^{\text{relative}} = 0.174 \frac{\text{m}}{\text{s}}$ and $a_{\text{piston}}^{\text{relative}} = a_{n3n4}^{\text{relative}} = a_{n2n4}^{\text{relative}} = -0.0298 \frac{\text{m}}{\text{s}^2}$ inside the rotating cylinder L_4. The direction of each acceleration component plus the directions of $v_{\text{piston}}^{\text{relative}}, a_{\text{piston}}^{\text{relative}}, \omega_4$ and α_4 are shown in Fig. 7.18.

Using Eqs. (7.12)–(7.14), we can calculate the acceleration components as:

Normal acceleration component: $a_{Nn2n4} = -\omega_4^2 r_{n2n4} = -0.969^2 \times (0.6 + 0.413) = -0.9512\,\text{m/s}^2$.

Tangential acceleration component: $a_{Tn2n4} = \alpha_4 r_{n2n4} = -0.162 \times (0.6 + 0.413) = -0.1641\,\text{m/s}^2$.

Coriolis acceleration component: $a_{Cn2n4} = 2\omega_4 v_{n2n4}^{\text{relative}} = 2 \times 0.969 \times 0.174 = 0.3372\,\text{m/s}^2$.

Sliding acceleration component: $a_{n2n4}^{\text{relative}} = -0.0298\,\text{m/s}^2$.

The polar total acceleration of n_2 is calculated using Eq. (7.15) as

$$\vec{a}_{n2} = \underbrace{\vec{a}_{n4}}_{=\vec{0}} + \underbrace{\alpha_4 r_{n2n4}\hat{\theta}}_{\text{Tangential}} + \underbrace{-\omega_4^2 r_{n2n4}\hat{r}}_{\text{Centripetal}} + \underbrace{2\omega_4 v_{n2n4}^{\text{relative}}\hat{\theta}}_{\textit{Coriolis}} + \underbrace{\vec{a}_{n2n4}^{\text{relative}}\hat{r}}_{\text{Sliding}}$$
$$= \vec{0} - 0.1641\hat{\theta} - 0.9512\hat{r} + 0.3372\hat{\theta} - 0.0298\hat{r} = -0.981\hat{r} + 0.174\hat{\theta}$$

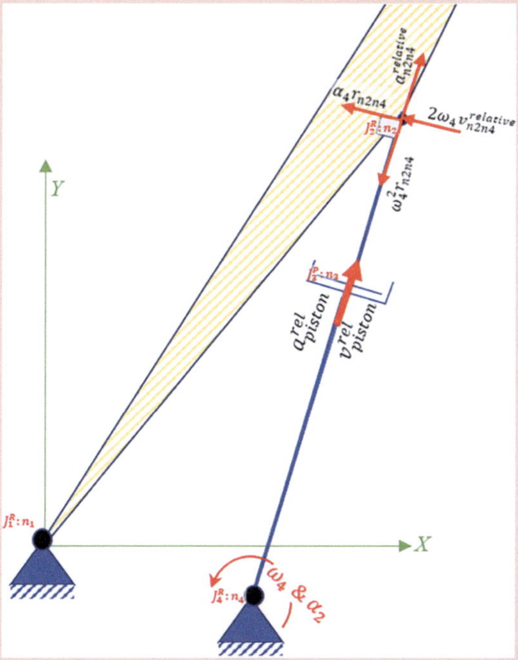

Fig. 7.18 Kinematic diagram with acceleration components of n_2 w.r.t. n_4

The acceleration vector diagram, as shown in Fig. 7.19, can be achieved by:

- Plot vector $\overrightarrow{OA} = \vec{a}_{n4} = \vec{0}$ or locate point A at point O.
- Plot vector $\overrightarrow{AB} = \alpha_4 r_{n2n4}\hat{\theta} = -0.1641\hat{\theta}$.
- Plot vector $\overrightarrow{BC} = -\omega_4^2 r_{n2n4}\hat{r} = -0.9512\hat{r}$.
- Plot vector $\overrightarrow{CD} = 2\omega_4 v_{n2n4}^{\text{relative}}\hat{\theta} = 0.3372\hat{\theta}$.
- Plot vector $\overrightarrow{DE} = \vec{a}_{n2n4}^{\text{relative}}\hat{r} = -0.0298\hat{r}$.
- Plot vector $\overrightarrow{OE} = \vec{a}_{n2}$.

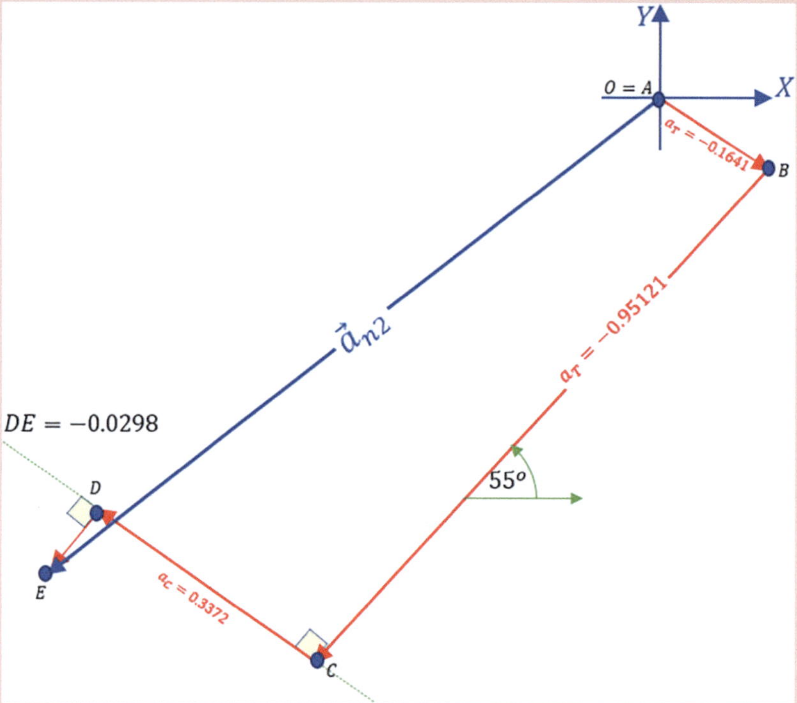

Fig. 7.19 Acceleration vector diagram of n_2

Understanding each component as a value and as a direction helps in estimating the total acceleration of each node on a mechanism. We will go through a couple of examples to apply Eq. (7.15): single link hinged to ground and single link pin-in-a-slot to ground.

7.3.1 Single Link Hinged to Ground

Assume that a single link L_2 is connected to the ground by a revolute joint $J_1^R : n_1$ as shown in Fig. 7.20. The link L_2 has three non-ground nodes, n_2, n_3, and n_4 with radii $r_2 < r_3 < r_4$. A motor is attached at $J_1^R : n_1$ to rotate L_2 with an angular position θ_2, an angular velocity ω_2, and an angular acceleration α_2.

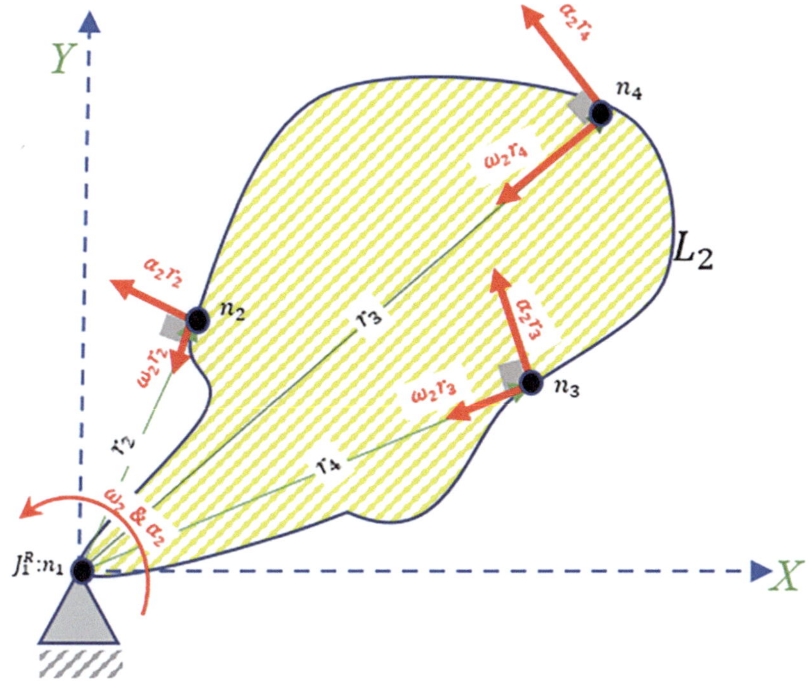

Fig. 7.20 Relative motion of hinged link to ground

To calculate the absolute acceleration at node n_2, we can apply Eq. (7.15) as $n_1 = n_i$ and $n_2 = n_j$. The acceleration $\vec{a}_{n1} = 0$ since n_1 belongs to the ground and J_1^R has no slipping or sliding motion. That also leads to no relative translational velocity ($v_{nCnB}^{\text{relative}} = 0$) and there is no relative translational acceleration ($\vec{a}_{nCnB}^{\text{relative}} = 0$) or there is no Coriolis acceleration component and there is no sliding acceleration component between nodes n_2 and n_1. There is only a tangential acceleration component $\alpha_2 r_2 \hat{e}_t$ plus centripetal acceleration component $-\omega_2^2 r_2 \hat{e}_n$ or

$$\vec{a}_{n2} = \underbrace{\alpha_2 r_2 \hat{\theta}}_{\text{Tangential}} + \underbrace{-\omega_2^2 r_2 \hat{r}}_{\text{Centripetal}} \tag{7.17}$$

Similarly, we can calculate the absolute acceleration at nodes n_3 and n_4.

$$\begin{aligned} \vec{a}_{n3} &= \alpha_2 r_3 \hat{\theta} - \omega_2^2 r_3 \hat{r} \\ \vec{a}_{n4} &= \alpha_2 r_4 \hat{\theta} - \omega_2^2 r_4 \hat{r} \end{aligned} \tag{7.18}$$

Figure 7.20 shows the acceleration components of each node. The three nodes n_2, n_3, and n_4 have the tangential acceleration components $\alpha_2 r_2 < \alpha_2 r_3 < \alpha_2 r_4$ in different directions and the centripetal acceleration components $\omega_2^2 r_2 < \omega_2^2 r_3 < \omega_2^2 r_4$ in different directions. The total absolute values of the accelerations are

$$a_{n2} = r_2 \sqrt{\omega_2^4 + \alpha_2^2} < a_{n3} = r_3 \sqrt{\omega_2^4 + \alpha_2^2} < a_{n4} = r_4 \sqrt{\omega_2^4 + \alpha_2^2} \tag{7.19}$$

In general, we can state that for any node n_j belonging to a link L_k hinged to the ground at node n_i with angular velocity ω_k and angular acceleration α_k, the absolute acceleration vector of node n_j is

$$\vec{a}_{nj} = \alpha_k r_{njni}\hat{\theta} - \omega_k^2 r_{njni}\hat{r} \tag{7.20}$$

where r_{njni} is the radius from the ground node n_i to the node n_j. The tangential direction $\hat{\theta}$ and the normal direction $\hat{r}$ are measured from the node n_i which is not necessarily the origin. In general, the node n_i is the instantaneous center of rotation of the link L_k. The node n_i belongs to the link L_k. The absolute of the tangential acceleration component a_{tninj}, the absolute of the centripetal component is a_{nninj}, and the absolute of the total acceleration a_{ni} increases as the radius r_{ninj} increases.

Example 7.4

For the gate opener shown in Fig. 7.21, if a person opens the door from angle 0°–85° with the following trajectory:

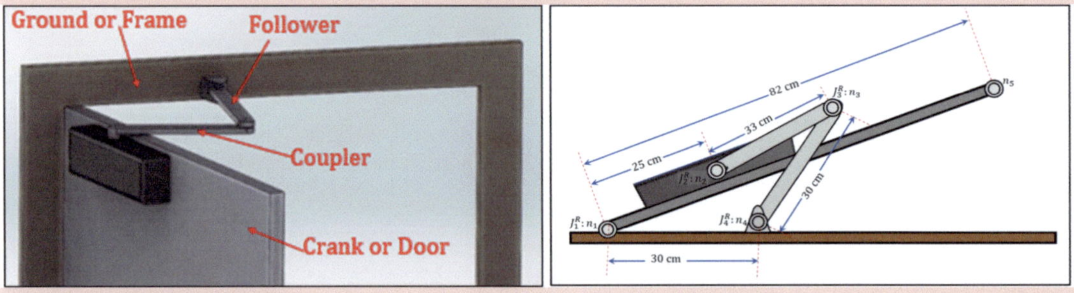

Fig. 7.21 Door mechanism. Figure and associated Door_4Bar_Mechanism.mp4 video designed by the author and implemented by both the author and Pranav Dhola. Both courtesy of Pranav Dhola

$$\theta_2(t) = A\left(1 - \cos\left(\frac{\pi t}{T_{cycle}}\right)\right), T_{cycle} = 2\,\text{s} \quad \text{and } A = 42.5°$$

$$\omega_2(t) = \frac{A\pi}{T_{cycle}} \sin\left(\frac{\pi t}{T_{cycle}}\right)$$

- Find using hand calculation and plot using MATLAB® the absolute acceleration vectors as a function of time at the stile edge node n_5 and the node n_2 where the pivoted joint between the door and the coupler is located.
- Calculate the absolute acceleration vectors at n_2 and n_5 when $t = 1$ s.
- Use the acceleration vector diagram to plot the acceleration components and the absolute acceleration of the stile edge node and the node where the pivoted joint between the door and the coupler is located at $t = 1$ s.

Solution

We start first by sketching the kinematic diagram of the door mechanism as shown in Fig. 7.22 where the frame or the ground link is L_1, the crank or door is L_2, the coupler rod is L_3, and the follower rod is L_4. The mechanism has four revolute joints $J_1^R : n_1, J_2^R : n_2, J_3^R : n_3$, and $J_4^R : n_4$. The door link L_1 has an end-effector at n_5.

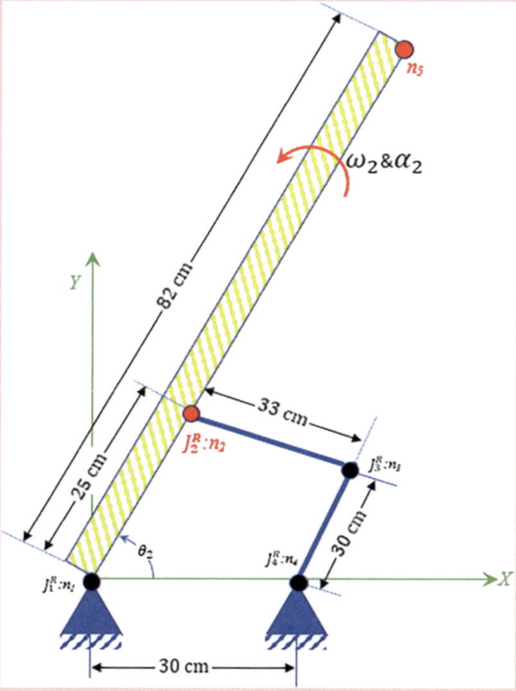

Fig. 7.22 Kinematic diagram of door mechanism

Now, we calculate the angular acceleration using Eq. (7.5)

$$\omega_2(t) = \frac{A\pi}{T_{cycle}} \sin\left(\frac{\pi t}{T_{cycle}}\right)\Bigg|_{\substack{A = 42.5 \\ T_{cycle} = 2}} = 66.759 \sin(1.571t) \text{ deg/s}$$

$$\vec{\alpha}(t) = \alpha_2(t)\hat{k} = \dot{\omega}_2(t)\hat{k}$$

or

$$\alpha_2(t) = \frac{\mathrm{d}\omega_2(t)}{\mathrm{d}t} = \frac{\mathrm{d}}{\mathrm{d}t}\left(\frac{A\pi}{T_{\text{cycle}}} \sin\left(\frac{\pi t}{T_{\text{cycle}}}\right)\right)\Bigg|_{\substack{A = 42.5 \\ T_{\text{cycle}} = 2}}$$

$$= \frac{A\pi^2}{T_{\text{cycle}}^2} \cos\left(\frac{\pi t}{T_{\text{cycle}}}\right)\Bigg|_{\substack{A = 42.5 \\ T_{\text{cycle}} = 2}}$$

$$= 104.865 \cos(1.571t)\frac{\text{deg}}{\text{s}^2}$$

Using Eq. (7.20) for nodes n_2 and n_5

$$\begin{aligned}\vec{a}_{n2} &= \alpha_2 r_{n2}\hat{\theta} - \omega_2^2 r_{n2}\hat{r} \\ &= \pi/180 \times 66.759 \times \cos(1.571t) \times 0.25 \times \hat{\theta} \\ &\quad - (\pi/180 \times 104.865 \times \sin(1.571t))^2 \times 0.25 \times \hat{r} \\ &= 0.458 \times \cos(1.571t)\hat{\theta} - 0.3394 \times \sin^2(1.571t) \times \hat{r} \text{ m/s}^2\end{aligned}$$

$$\begin{aligned}\vec{a}_{n5} &= \alpha_2 r_{n5}\hat{\theta} - \omega_2^2 r_{n5}\hat{r} \\ &= \pi/180 \times 66.759 \times \cos(1.571t) \times 0.82 \times \hat{\theta} \\ &\quad - (\pi/180 \times 104.865 \times \sin(1.571t))^2 \times 0.82 \times \hat{r} \\ &= 1.5 \times \cos(1.571t)\hat{\theta} - 1.113 \times \sin^2(1.571t) \times \hat{r}\ \text{m/s}^2\end{aligned}$$

Note, the value $\pi/180$ is to convert from deg/s into rad/s or from deg/s^2 into rad/s^2. To plot the acceleration components versus time, we write the script below (Fig. 7.23).

```
%Example 7.4
%=====================================
%Clear all the variables and close all the figures
%=================================
clc
clear all
close all
%=========================================================================
%plugging in the parameters
%==========================
T_cycle = 2; %sec
A = 42.5; %deg
rn2 = 0.25; %m
rn5 = 0.82; %m
deg2rad = pi/180; %convert from deg to rad

%Calculating the signals
%========================
t = [0:0.001:T_cycle]; %sec
theta_2 = A*(1-cos(pi*t/T_cycle)); %deg
omega_2 = A*pi/T_cycle*sin(pi*t/T_cycle); %deg/sec
alpha_2 = A*(pi/T_cycle)^2*cos(pi*t/T_cycle); %deg/sec^2

a_n2_tang = alpha_2*rn2*deg2rad; %m/sec^2
a_n2_norm = -omega_2.^2*rn2*deg2rad^2; %m/sec^2

a_n5_tang = alpha_2*rn5*deg2rad; %m/sec^2
a_n5_norm = -omega_2.^2*rn5*deg2rad^2; %m/sec^2

%Plot the signals
%=================
subplot(2,1,1)
plot(t,a_n2_tang,t,a_n2_norm,'LineWidth',2);
xlabel('t (sec)');
ylabel('Node n_2 Acceleration (m/sec^2)');
grid on;
legend('a_t_n_2 (m/sec^2)','a_n_n_2 (m/sec^2)')
subplot(2,1,2)
plot(t,a_n5_tang,t,a_n5_norm,'LineWidth',2);
xlabel('t (sec)');
ylabel('Node n_5 Acceleration (m/sec^2)');
grid on;
```

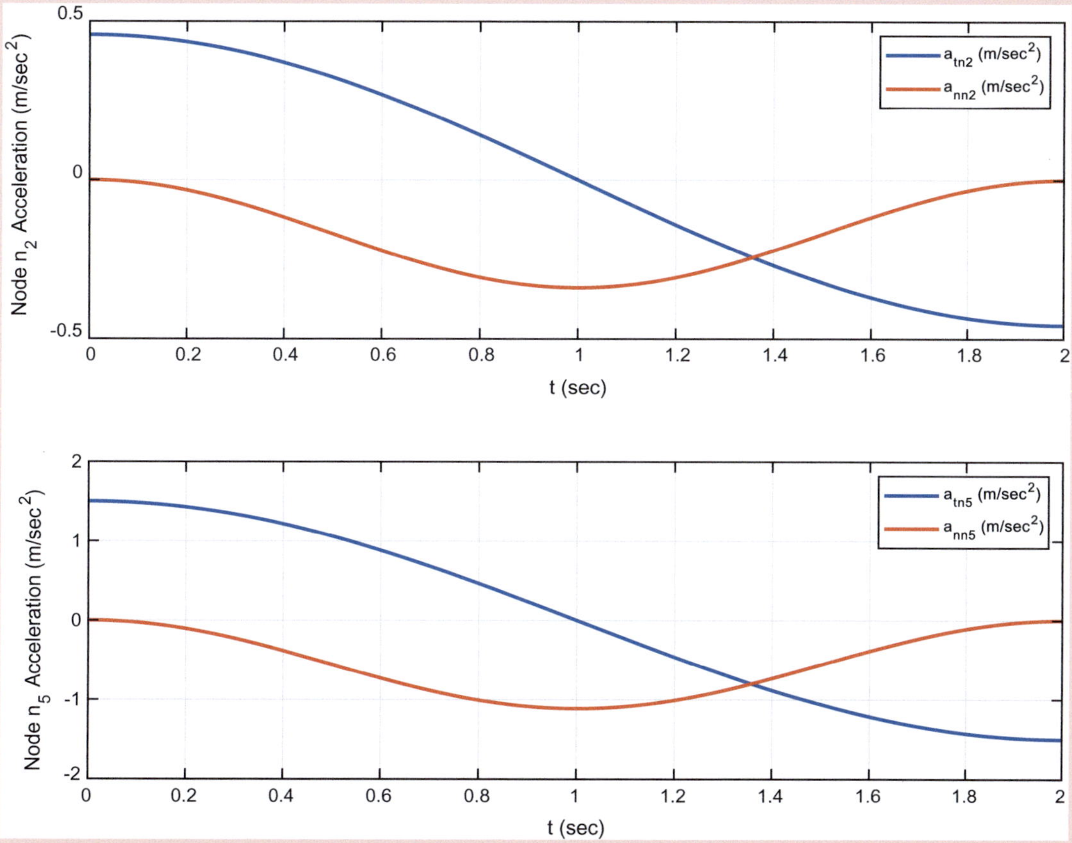

Fig. 7.23 Acceleration components of n_2 and n_5 for door mechanism

At time $t = 1$ s:

$$\vec{a}_{n2}(t=1) = 0.458 \times \cos(1.571t)\hat{\theta} - 0.3394 \times \sin^2(1.571t) \times \hat{r}\Big|_{t=1}$$
$$= -0.3394\hat{r} \text{ m/s}^2$$
$$\vec{a}_{n5} = 1.5 \times \cos(1.571t)\hat{\theta} - 1.113 \times \sin^2(1.571t) \times \hat{r}\Big|_{t=1}$$
$$= -1.113\hat{r} \text{ m/s}^2$$

The acceleration vector diagram, as shown in Fig. 7.24, can be achieved by:

- Plot vector $\overrightarrow{OB} = -\omega_2^2 r_{n2}\hat{r} = -0.3394\hat{r}$.
- Plot vector $\overrightarrow{OC} = -\omega_2^2 r_{n5}\hat{r} = -1.113\hat{r}$.
- Relabel $\overrightarrow{OB} = \vec{a}_{n2} = -\omega_2^2 r_{n2}\hat{r} = -0.3394\hat{r}$.
- Relabel $\overrightarrow{OC} = \vec{a}_{n5} = -\omega_2^2 r_{n5}\hat{r} = -1.113\hat{r}$.

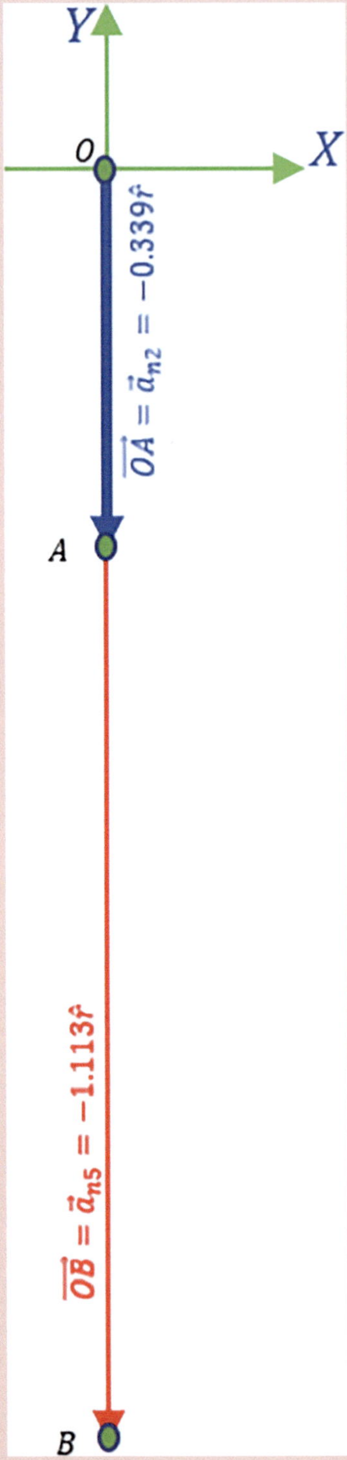

Fig. 7.24 Acceleration vector diagram of n_2 and n_5 at $t = 1$ s for door opener mechanism

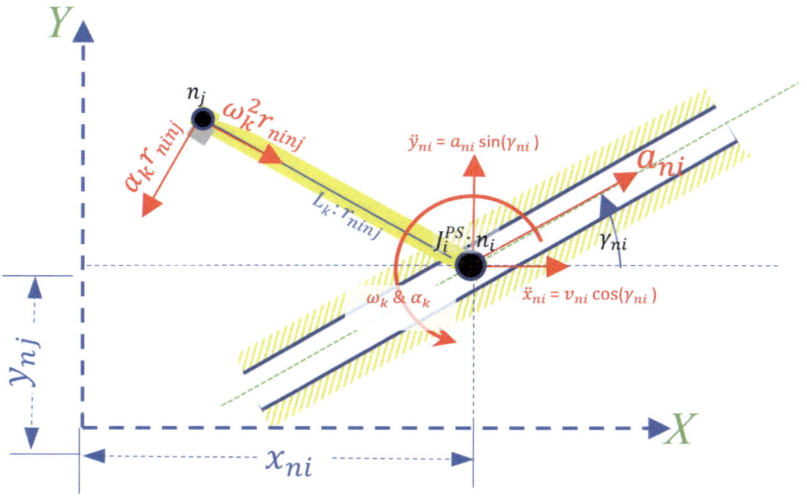

Fig. 7.25 Kinematic diagram of connected link with a pin-slotted joint to ground

7.3.2 Single Link Pin-in-a-Slot to Ground

As shown in Fig. 7.25, the link L_k is attached to the ground at node n_i with a pin-to-slot joint J_i^{PS}. The constant sliding angle γ_{ni} makes the sliding acceleration components as $\ddot{x}_{ni}$ and $\ddot{y}_{ni} = \ddot{x}_{ni} \tan(\gamma_{ni})$. The link L_k is also rotating with an angular velocity ω_k and angular acceleration α_k . The node n_j is located on the link with a radius r_{njni} from node n_i. We can define the ***relative*** polar acceleration components of node n_j w.r.t. node n_i as the relative tangential component $\alpha_k r_{njni}$ and the relative centripetal component $-\omega_k^2 r_{njni}$. Recall Eq. (7.15) and considering $\omega_i = \omega_k$ and $\alpha_i = \alpha_k$. The $v_{nCnB}^{\text{relative}} = 0$ and $\vec{a}_{nCnB}^{\text{relative}} = 0$ since there is no sliding or slipping motion between nodes i and j. Using these substitutions

$$\vec{a}_{nj} = \underbrace{\vec{a}_{ni}}_{\ddot{x}_{ni}\hat{i} + \ddot{x}_{ni}\tan(\gamma_{ni})\hat{j}} + \alpha_k r_{njni}\hat{\theta}_i - \omega_k^2 r_{njni}\hat{r}_i \tag{7.21}$$

Note that $\hat{\theta}_i$ and $\hat{r}_i$ are non-rotating local polar frames at node n_i. Keep in mind that the absolute polar acceleration components of node n_j must be measured from the instantaneous center of rotation IC_K where n_j belongs to the link L_k.

Example 7.5

For the shown kinematic diagram of the piston–cylinder mechanism in Fig. 7.26 where the length of the crank is 76.2 mm and the length of the connecting rod is 203.2 mm. Assume that the combustion force moves the piston with a sliding motion.

$$x_p(t) = 0.2032 + 0.0494 \cos(263.158t) 0 < t < 0.012 \text{ s}$$

where $x_p(t)$ is the piston sliding position in mm. This sliding motion causes the connecting rod to rotate and consequently, the crankshaft rotates. Instantaneously at t=0.006 s, the measured angular velocity and angular acceleration of the connecting rod are − 11.751 rad/s and 9993.58 rad/s^2, where $\theta_2 = 79.62°$ and $\theta_3 = 158.53°$. At this instantaneous time t=0.006 s, calculate

1. The relative acceleration components of the crankpin node w.r.t. the wrist pin node.
2. The absolute acceleration of the crankpin node.

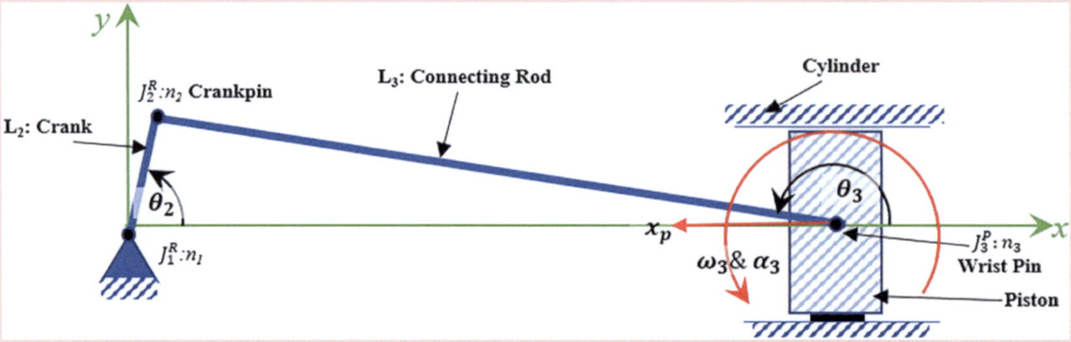

Fig. 7.26 Kinematic diagram of piston–cylinder mechanism

Solution

The connecting rod in the piston–cylinder mechanism is the same as in Fig. 7.25 where $i=3$, $j=2$ and $k=3$. The given information can be summarized as $r_{n2} = 0.0762\text{m}, r_{n3n2} = 0.2032\text{m}, \ \omega_3(t = 0.006) = -11.751 \frac{\text{rad}}{\text{s}}, \ \alpha_3(t = 0.006) = 9993.58 \frac{\text{rad}}{\text{s}^2}$, $\theta_2(t = 0.006) = 79.62°$ and $\theta_3(t = 0.006) = 158.53°$. We can calculate the piston sliding velocity and acceleration as:

$$x_p(t) = 0.203 + 0.0494 \cos(263.158t)$$

$$v_p(t) = \frac{\mathrm{d}x_p}{\mathrm{d}t} = -13 \sin(263.158t)$$

$$a_p(t) = \frac{\mathrm{d}v_p}{\mathrm{d}t} = 3421 \cos(263.158t)$$

At $t=0.006$ s:

$$x_p(t = 0.006) = 0.203 + 0.0494 \cos(263.158 \times 0.006) = 0.2026\,\text{m}$$

$$v_p(t = 0.006) = \frac{dx_p}{dt} = -13 \sin(263.158 \times 0.006) = -12.999\,\text{m/s}$$

$$a_p(t = 0.006) = \frac{dv_p}{dt} = -2421 \cos(263.158 \times 0.006) = -27.885\,\text{m/s}^2$$

As shown in Fig. 7.27, in addition to the sliding acceleration component a_p, there is a relative tangential acceleration component $a_{Tn3n2} = \alpha_3 r_{n3n2}$ and a relative centripetal acceleration component $a_{Nn3n2} = \omega_3^2 r_{n3n2}$ that can be calculated at $t=0.006$ s as:

$$a_{Tn3n2} = \alpha_3 r_{n3n2} = 9993.58 \times 0.2032 = 2024.7\,\text{m/s}^2$$

$$a_{Nn3n2} = \omega_3^2 r_{n3n2} = (-11.751)^2 \times 0.2032 = 27.976\,\text{m/s}^2$$

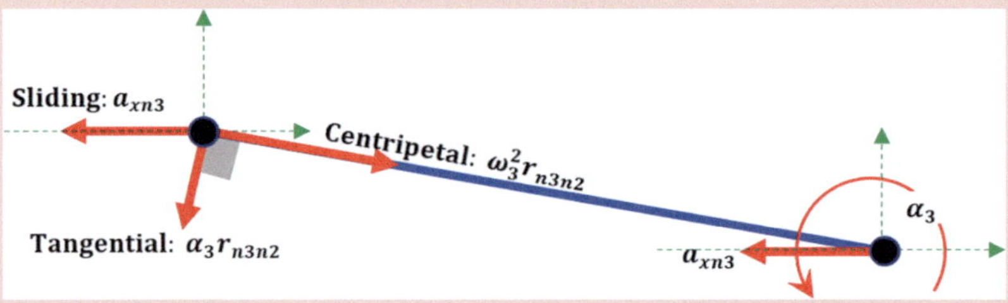

Fig. 7.27 Acceleration components on connecting rod-free body diagram

Using Eq. (7.21), the absolute acceleration at node n_2 is the summation of the piston sliding acceleration a_p and the relative tangential acceleration component a_{Tn3n2} and the relative centripetal acceleration component a_{Nn3n2} as

$$a_{n2}(t = 0.006) = a_p\hat{i} + \alpha_3 r_{n3n2}\hat{\theta} - \omega_3^2 r_{n3n2}\hat{r}\Big|_{t=0.006} = -27.885\hat{i} + 2024.7\hat{\theta} - 27.976\hat{r}$$

Using Eq. (7.6)

$$\hat{r} = \cos(\theta_3)\hat{i} + \sin(\theta_3)\hat{j}$$
$$\hat{\theta} = -\sin(\theta_3)\hat{i} + \cos(\theta_3)\hat{j}$$

$$a_{n2}(t = 0.006) = -27.885\hat{i} + 2024.7 \times \left(-\sin(158.53°)\hat{i} + \cos(158.53°)\hat{j}\right) - 27.976 \times \left(\cos(158.53°)\hat{i} + \sin(158.53°)\hat{j}\right) = -742.92\hat{i} - 1894.4\hat{j}$$

One important tool in the relative acceleration analysis is to know how to find the Cartesian coordinates of the relative acceleration components. The relative acceleration $\vec{a}_{njni}$ of a non-sliding node n_j, that belongs to the hinged link L_l at node n_i with an angle θ_k, angular velocity ω_k, and the angular acceleration α_k, is

$$\vec{a}_{njni} = \vec{a}_{Nnjni} + \vec{a}_{Tnjni} \tag{7.22}$$

where $\vec{a}_{Nnjni}$ and $\vec{a}_{Tnjni}$ are the normal relative acceleration component and the tangential relative acceleration component, respectively. Using geometry, we can compose these two components in an X-direction and a Y-direction as

$$\begin{aligned} \vec{a}_{Nnjni} &= -\omega_k^2 r_{njni}\cos(180° + \theta_k) - \omega_k^2 r_{njni}\sin(180° + \theta_k) \\ \vec{a}_{Tnjni} &= \alpha_k r_{njni}\cos(\theta_k + 90°) + \alpha_k r_{njni}\sin(\theta_k + 90°) \end{aligned} \tag{7.23}$$

7.4 Simscape™ Multibody™ Acceleration Sensor and Actuator Blocks

In Simscape™ Multibody™, the "***Transform Sensor***" block can be used to measure the acceleration as well as the velocity and the displacement. Figure 7.28 shows the "***Transform Sensor***" block that has Port B and Port F which represent the base and follower frames, respectively. As an example, to explain the functionality of Port B and Port F, if we are measuring the relative acceleration of node n_j with respect to node n_i, we will connect Port B (base) to n_i and we will connect Port F (follower) to n_j. You can drag and drop the "***Transform Sensor***" block to the model. After that, when you double-click on the "***Transform Sensor***" block, there will be a list of items. If you click on any of the listed items, a sublist will show. Table 7.1 shows the sublists of "Angular Acceleration" and "Acceleration", which are our focus in this chapter. We can choose whatever signal that we would like to measure. For example, if we are interested in measuring the angular acceleration around the Z-axis at a specific node, we would select "Alpha Z". We would immediately get an extra output port from the block, labeled "bZ". There is more information about the "***Transform Sensor***" block in "Simscape™ Multibody™ Getting Started Guide" and "Simscape™ Multibody™ User's Guide," Refs. [5, 6].

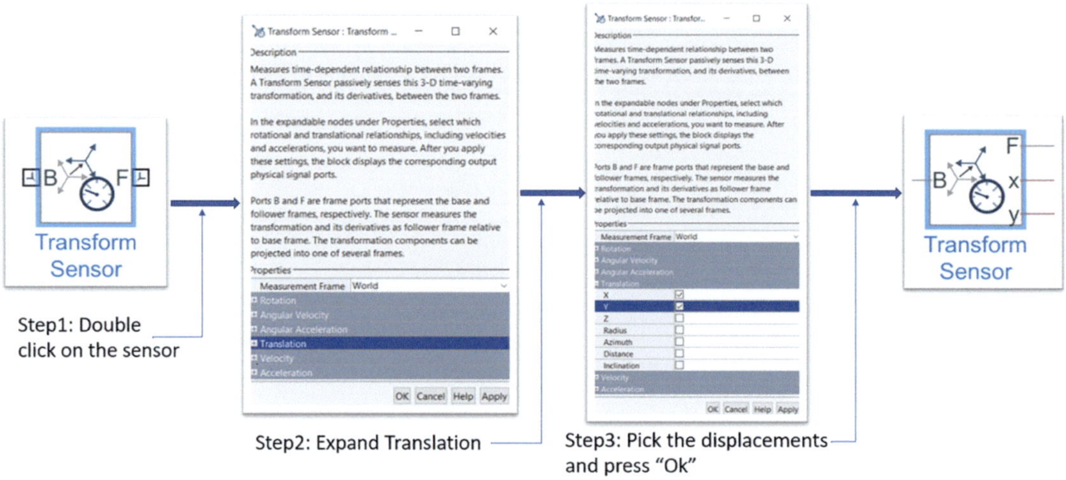

Fig. 7.28 Step to activate displacement sensors from transform sensor block. Designed and implemented by the author. Images used with permission of MathWorks

Table 7.1 Motion sensor. Figures in the second column are designed and implemented by the author. Images used with permission of MathWorks

Sensor motion	Sensors under each motion
Angular velocity	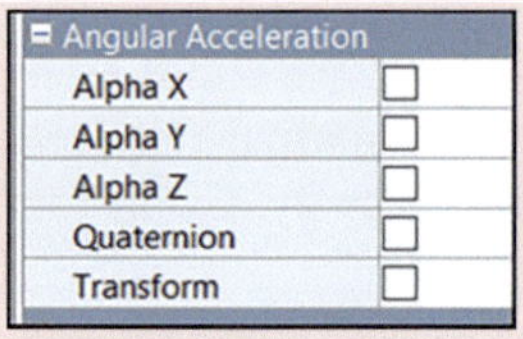
Velocity	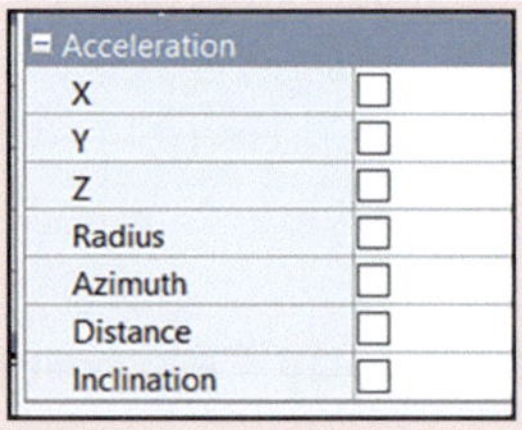

Engineering Tips—Acceleration Sensor as Simulink® Subsystem

To reduce the complexity of the model visualization, we are going to build a set of acceleration sensor subsystems and save them in our "Kinematics_Of_Machines.slx" file. One example is the Cartesian translational acceleration sensor as shown in Fig. 7.29 that can be built by following the steps below:

1. Drag a "***Transform Sensor***" block from the Simscape™ Multibody™ library and drop it into a Kinematics_Of_Machines.slx model.
2. Double click on the "***Transform Sensor***" block and select "Acceleration." Select X and Y in the acceleration sublist. Press "Ok." You will see two more output ports—ax and ay.
3. Connect Port F of "***Transform Sensor***" block to "***Connection Port***" block. Rename the "***Connection Port***" as sensor. Connect the two ports ax and ay to two "***PS-Simulink® Converter***" followed by two "Out" blocks and call them "ax_m/s^2" and "ay_m/s^2".

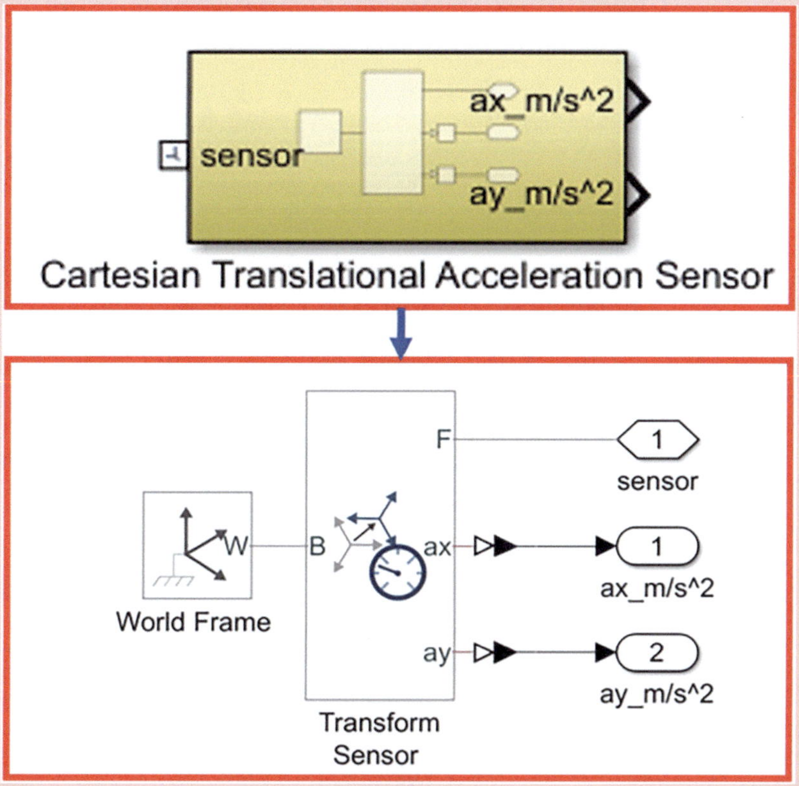

Fig. 7.29 Cartesian translational acceleration sensor subsystem

4. Connect Port B to the world frame.
5. Create a subsystem and rename the subsystem "***Cartesian Translational Acceleration Sensor***".

We are going to build more acceleration sensors subsystems and add them to our file "Kinematics_Of_Machines.slx".

Engineering Tips—Polar Coordinate Acceleration Sensor Simulink® Subsystem

From our previous definitions: The ***instant center of rotation***, also called the ***instantaneous center of zero velocity,*** is the point fixed to a body undergoing planar movement that has zero velocity at a particular instant of time, Ref. [7]. For any link L_i (rigid body), to measure the absolute acceleration polar coordinate components (normal acceleration component a_N and tangential acceleration component a_T), we must measure the polar angle and the polar radius from the ***instant center of rotation*** IC_i.

For the actuation block, we are going to pass the three signals "position, velocity, and acceleration" as shown in Fig. 7.30. In the case of the rotational actuation, we call the subsystem with the three inputs "Triple Inputs Rotational Motion Actuator" that consists of "PS-Simulink® Converter" block, Mux block, and three gain blocks. In the case of the translational actuation, we call the subsystem with the three inputs "Triple Inputs Transitional Motion Actuator" that consists of "PS-Simulink® Converter" block, and Mux block.

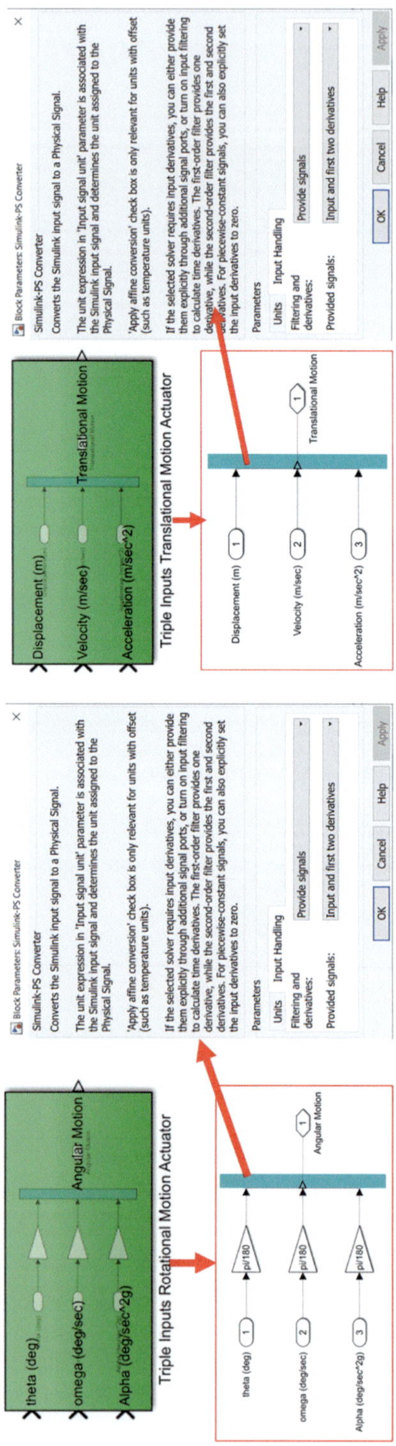

Fig. 7.30 Acceleration and three inputs motion actuators. Designed and implemented by the author. Images used with permission of MathWorks

Example 7.6

For the door mechanism shown in Fig. 7.21, if a person opens the door from angle 0°–85° in two seconds with a constant angular velocity:

- Build a Simscape™ Multibody™ model to simulate this scenario.
- Measure the angular accelerations of L_3 and L_4.
- Measure the acceleration vectors at n_2, n_3, and n_5 in Cartesian coordinates.

Solution

Simscape™ Multibody™ model:
Open a new script file saved as "DoorOpener_E7p6_script.m" and write the command below. The script starts cleaning the history of all the variables followed by declaring the input parameters and calculating each link's required parameters.

```
clc
clear all
close all
%==========================================================================
%Declare the given parameters
%============================
r_n2   = 25/100; %m
r_n5   = 82/100; %m
r_n3n2 = 33/100; %m
r_n3n4 = 30/100; %m
r_n4   = 30/100; %m
T_cycle = 2; %sec
theta_2_max = 85; %deg
omega_2 = theta_2_max/2; %deg/sec
alpha_2 = 0 %deg/sec^2

%Calculation of parameters
%=========================
L2_crosssection = [0 0;r_n2 0;r_n5 0;r_n5 r_n2/10;0 r_n2/10];
L2_dimension    = [1 0.1 0.1]*r_n2; %m
L3_dimension    = [r_n3n2 0.1*r_n2 0.1*r_n2]; %m
L4_dimension    = [r_n4 0.1*r_n2 0.1*r_n2]; %m
Frame_L1_n1 = [0 0 0]; %m
Frame_L1_n4 = [r_n4 0 0]; %m
Frame_L2_n1 = [0 0 0]; %m
Frame_L2_n2 = [r_n2 0 0]; %m
Frame_L2_n5 = [r_n5 0 0]; %m
Frame_L3_n2 = [-r_n3n2/2 0 0]; %m
Frame_L3_n3 = [r_n3n2/2 0 0]; %m
Frame_L4_n4 = [-r_n3n4/2 0 0]; %m
Frame_L4_n3 = [r_n3n4/2 0 0]; %m
```

We are going to bring the model from Example 6.5 and save it as "DoorOpener_E7p6_model.slx". We are going to follow the steps below:

1. Replace the "Rotational Motion Actuator" subsystem with "Triple Inputs Rotational Motion Actuator" from subsystem "Kinematics_Of_Machines.slx" file as shown in Fig. 7.31. Rename this "Triple Inputs Rotational Motion Actuator" by "theta_2 triple input".

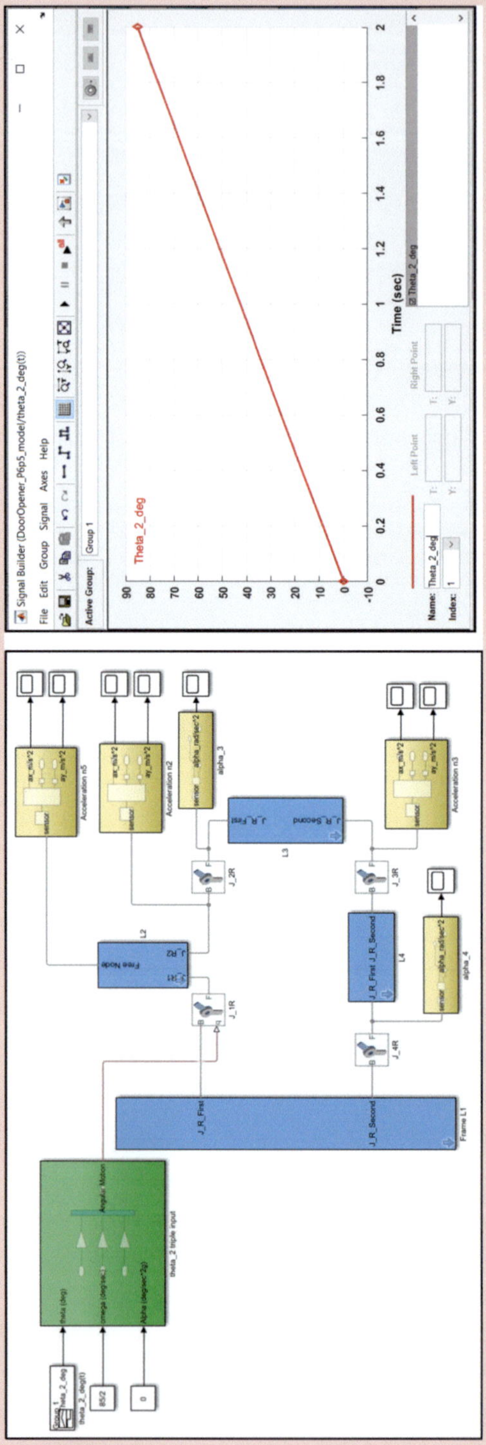

Fig. 7.31 Simscape™ Multibody™ model details of door mechanism

2. Connect the input signals to (1) the "theta_2_deg" signal builder block, (2) a constant block with a value "85/2", and (3) a constant block with a value "0".
3. Bring two "Angular Acceleration Sensor" subsystems from the "Kinematics_Of_Machines.slx" file and rename them "alpha_3" and "alpha_4". Connect these two subsystems as shown in Fig. 7.31. Add a scope block at each output port of these two subsystems. Now, we can measure the angular acceleration α_3 of L_3 from the connected scope to the "alpha_3" subsystem. Also, we can measure the angular acceleration α_4 of L_4 from the connected scope to the "alpha_4" subsystem
4. Bring three "Cartesian Translational Acceleration Sensor" subsystems from the "Kinematics_Of_Machines.slx" file and rename them "Acceleration n2", "Acceleration n3", and "Acceleration n5". Connect these three subsystems at nodes n_2, n_3, and n_5 as shown in Fig. 7.31. Connect the output ports for each of these three subsystems to scope blocks to measure the Cartesian acceleration components:a_{n2x}, a_{n2y}, a_{n3x}, a_{n3y}, a_{n5x}, and a_{n5y}.

Run the script file followed by running the model. Double click on the scopes to see the results as shown in Fig. 7.32. During the simulation, you can see the animation for the mechanism. We encourage the reader to watch the attached animated video from this model as well as look at the attached model.

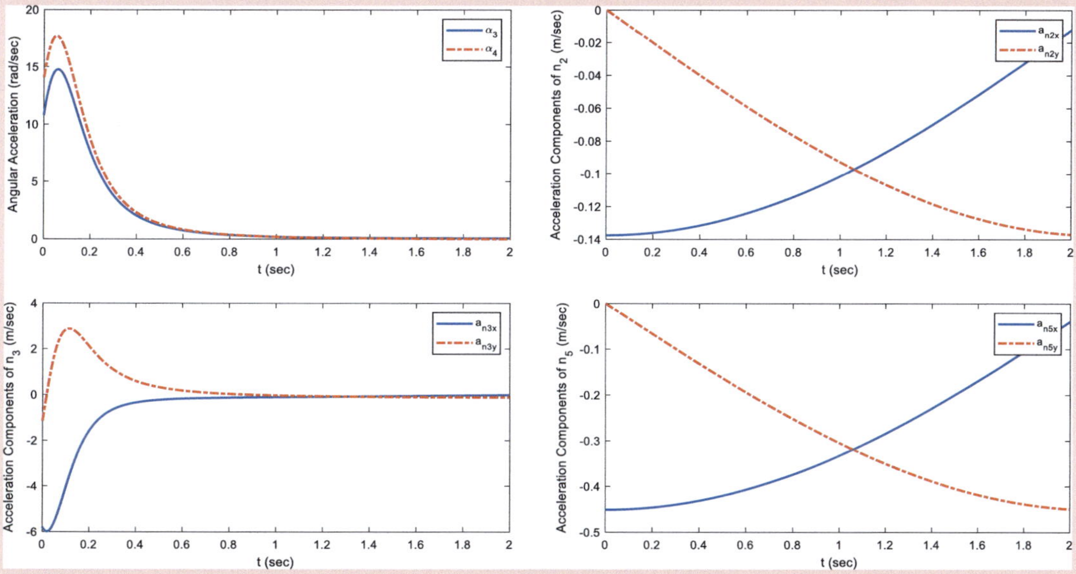

Fig. 7.32 Results of Simscape™ Multibody™ model for door mechanism

7.5 Graphical Acceleration Analysis

The graphical acceleration analysis is as important as the graphical velocity analysis and the graphical position analysis. The main foundation of graphic acceleration is to understand the relative motion between two nodes n_i and n_j. Assume the node n_i with an absolute acceleration vector $\vec{a}_{ni}$. The relative acceleration vector $\vec{a}_{njni}$ is the observed acceleration of n_j w.r.t. n_i. The absolute acceleration vector of n_j can be calculated by Eq. (7.24) or by graphically adding the two vectors $\vec{a}_{ni}$ and $\vec{a}_{njni}$ as shown in Fig. 7.33 whereby $\overrightarrow{OA} = \vec{a}_{ni}$, $\overrightarrow{AB} = \vec{a}_{njni}$, and $\overrightarrow{OB} = \vec{a}_{nj}$.

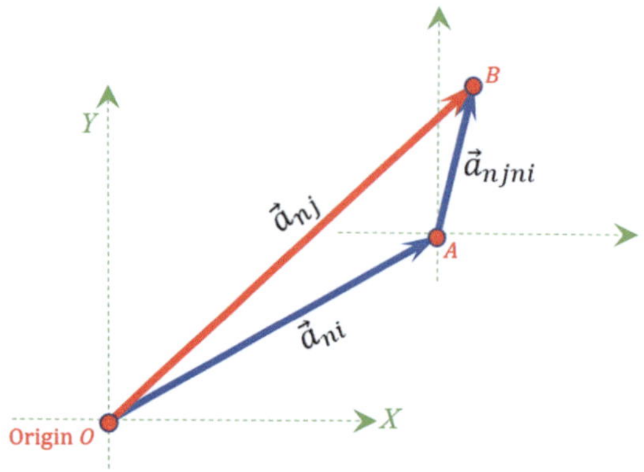

Fig. 7.33 Relative and absolute accelerations between two nodes

$$\vec{a}_{nj} = \vec{a}_{ni} + \vec{a}_{njni} \tag{7.24}$$

Assume that a vector $\vec{r}_{njni}$ starts from n_i and ends at n_j with length r_{njni} that rotates with an angular velocity ω_k and and angular acceleration α_k around the *Z*-axis as shown in Fig. 7.34. The positive ω_k and α_k are counterclockwise. As we previously discussed in Sect. 7.3, the relative acceleration vector $\vec{a}_{njni}$ has four acceleration components: centripetal, tangential, Coriolis, and sliding. The relative centripetal acceleration component $-\omega_k^2 r_{njni}$ is in the negative $\vec{r}_{njni}$ direction. The relative tangential acceleration component $\alpha_k r_{njni}$ is in the perpendicular direction on the vector $\vec{r}_{njni}$. The relative Coriolis acceleration component $2\omega_k v_{nCnB}^{\text{relative}}$ is in the perpendicular direction on the relative velocity vector $\vec{v}_{njni}^{\text{relative}}$. The direction of the relative sliding acceleration component $\vec{a}_{ninj}^{\text{relative}}$ as well as the direction of the relative sliding velocity component can be determined from a kinematic constraint such as the example of a prismatic joint. The total relative acceleration component $\vec{a}_{ninj}$ is shown graphically in Fig. 7.34.

We are now going to use our knowledge of the directions and magnitude of the relative acceleration components to establish a set of graphical algorithms to calculate the acceleration vector at any node or the angular acceleration of any link existing in four-bar mechanism, slide-crank mechanism, and the inverted slide-crank mechanism as listed in the upcoming subsections.

7.5.1 Four-Bar Mechanism Graphical Acceleration Analysis

Assume the four-bar mechanism shown in Fig. 7.35 has $\theta_2, \omega_2,$ and α_2 as a motor input signal at $J_1^R : n_1$. From our previous analyses in Chaps. 5 and 6, we can graphically calculate the position vectors and the velocity vectors of any node located in this four-bar mechanism. That brings us to setting up the graphical acceleration algorithm which has the given signals: $\theta_2, \theta_3, \theta_4, \omega_2, \omega_3, \omega_4$ and α_2 and the required signals: $\alpha_3, \alpha_4, \vec{a}_{n2},$ and $\vec{a}_{n3}$. The four-bar mechanism graphical acceleration algorithm can be summarized as follows:

Step 1: Use the graphical algorithm in Chap. 5 to plot the entire mechanism according to the given parameters ($r_{n2}, r_{n2n3}, r_{n3n4},$ and r_{n4}) and the given input signal (θ_2) as shown in Fig. 7.35. Measure the output signals (θ_3 and θ_4). Note, sometimes a reasonable "scale" needs to be established if the dimensions are too big to fit on paper. We can use any CAD software to do so.

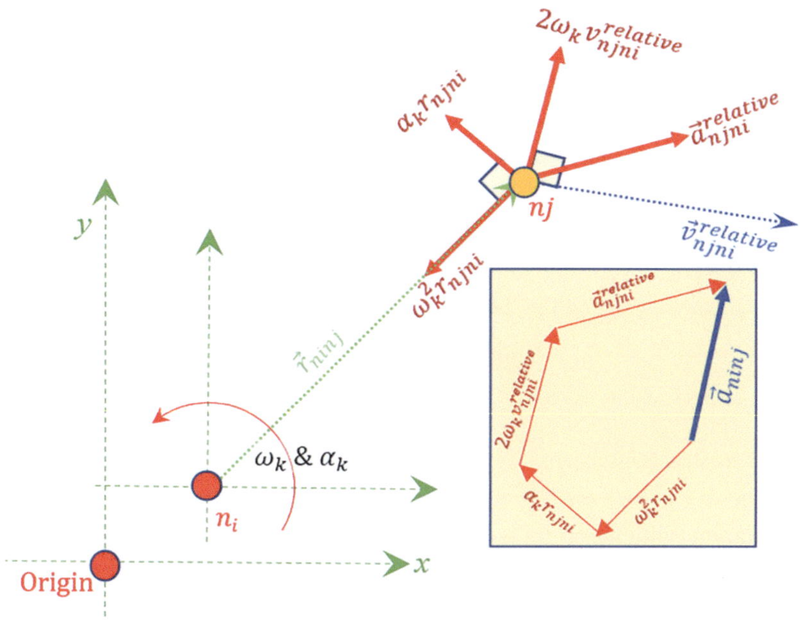

Fig. 7.34 Relative acceleration components and total acceleration component

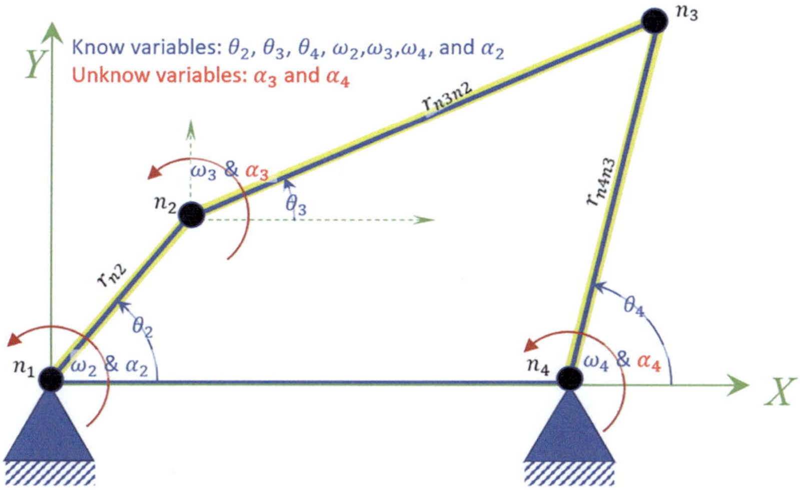

Fig. 7.35 Kinematic diagram of four-bar mechanism

Step 2: Use the graphical algorithm in Chap. 6 to calculate the angular velocities ω_3 and ω_4 from the known independent variables $\theta_2, \theta_3, \theta_4$, and ω_2. After that, we have seven known independent variables: $\theta_2, \theta_3, \theta_4, \omega_2, \omega_3, \omega_4$, and α_2 and two unknown independent variables α_3 and α_4 that will be obtained using the graphical acceleration algorithm.

Step 3: Now we are going to graphically calculate the acceleration $\vec{a}_{n2}$ using the known independent variables θ_2, ω_2, and α_2 and the known parameter r_{n2}. Assume α_2 is positive or in a counterclockwise direction. On the kinematic diagram, plot the normal acceleration component $\vec{a}_{Nn2}$ as a solid line vector that starts from node n_2 and heads in the negative direction of $\vec{r}_{n2}$. After that, plot the tangential acceleration component $\vec{a}_{Tn2}$ as a solid line vector that starts from node n_2 and is perpendicular to $\vec{r}_{n2}$, making an angle $90° + \theta_2$ with the horizontal line passing by node n_2 as shown in Fig. 7.36a.

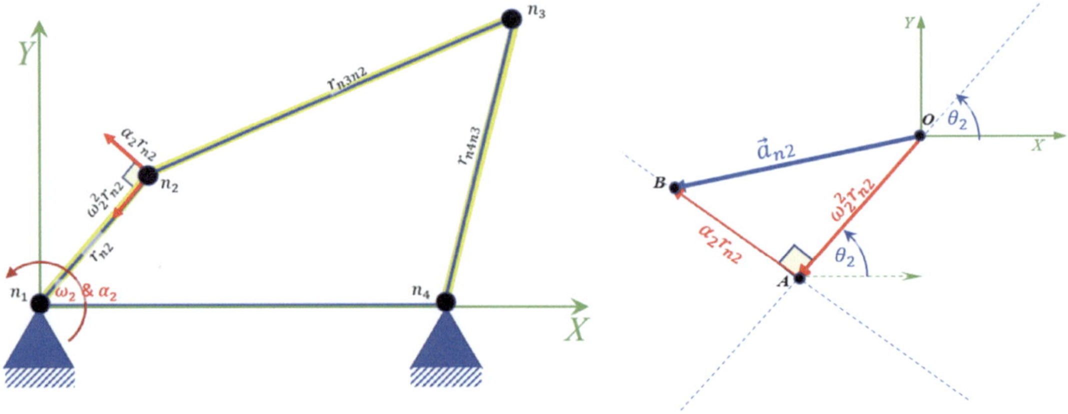

Fig. 7.36 Acceleration components of n_2

Use the vector diagram (with an origin point O) to plot the normal acceleration component $\vec{a}_{Nn2}$ as a solid line vector $\overrightarrow{OA}$ from point O to point A with a length $a_{Nn2} = \omega_2^2 r_{n2}$, making an angle $180° + \theta_2$ as shown in Fig. 7.36b. Plot the tangential acceleration component $\vec{a}_{Tn2}$ as a solid line vector $\overrightarrow{AB}$ that starts from point A to point B with a length $a_{Nn2} = \alpha_2 r_{n2}$ and perpendicular to $\overrightarrow{OA}$, making an angle $90° + \theta_2$ with the X-axis. Plot the total acceleration vector $\vec{a}_{n2}$ as a solid line vector $\overrightarrow{OB}$ to be the summation of both $\overrightarrow{OA}$ and $\overrightarrow{AB}$. Note, if the angular acceleration α_2 is in the counterclockwise direction, the direction of $\vec{a}_{Tn2} = \overrightarrow{AB}$ would have $90° + \theta_2$ with the X-axis or as shown in Fig. 7.36a, b. If the angular acceleration α_2 is counterclockwise, the direction of $\vec{a}_{Tn2}$ is $-90° + \theta_2$ with the X-axis.

Step 4: To graphically calculate the acceleration $\vec{a}_{n3}$, we can use Eq. (7.24) using $i=2$ and $j=3$ as:

$$\vec{a}_{n3} = \vec{a}_{n2} + \vec{a}_{n3n2} \tag{7.25}$$

From Step 3, we know that $\vec{a}_{n2}$. Now, we need to calculate $\vec{a}_{n3n2}$. On the kinematic diagram, plot the relative normal acceleration component $\vec{a}_{Nn3n2}$ as a solid line vector that starts from node n_3 and heads in the negative $\vec{r}_{n3n2}$-direction. Also, plot the direction of the relative tangential acceleration component $\vec{a}_{Tn3n2}$ as a perpendicular dashed line on $\vec{r}_{n3n2}$ and passing by node n_3. Note, we use the dashed line to plot a known vector's direction whereby the vector's magnitude is unknown, and we use the solid line for a known vector (direction and magnitude). On the acceleration vector diagram, add the vector $\overrightarrow{BC}$ that goes from point B to point C with a length $a_{Nn3n2} = \omega_3^2 r_{n3n2}$ and an angle $180° + \theta_3$ with the horizontal line passing by point B as shown in Fig. 7.37b. Plot the dashed line that passes by point C and perpendicular on $\overrightarrow{BC}$ vector as shown in Fig. 7.37b. This dashed line represents the direction of the relative tangential acceleration component $\vec{a}_{Tn3n2}$.

Step 5: Another way to calculate the acceleration vector $\vec{a}_{n3}$ is to use node n_4 as the center of rotation of link L_4. The node n_3 belongs to link L_4. First, use the kinematic diagram to start from node n_3 and plot the normal acceleration component $a_{Nn3} = \omega_4^2 r_{n3n4}$ as a solid vector in the negative direction of $\vec{r}_{n3n4}$. Plot the direction of the tangential acceleration component $a_{Tn3} = \alpha_4 r_{n3n4}$ as a perpendicular dashed line on $\vec{r}_{n3n4}$ passing by node n_3 as shown in Fig. 7.38a. On the acceleration vector diagram, add the vector $\overrightarrow{OD}$ from the origin O to point D with a length $a_{Nn3} = \omega_4^2 r_{n3n4}$ and an angle $180° + \theta_4$ with X-axis as shown in Fig. 7.38b. Plot the direction of the relative acceleration component $a_{Tn3} = \alpha_4 r_{n3n4}$ at point D and perpendicular to $\overrightarrow{AD}$ vector as shown in Fig. 7.38b.

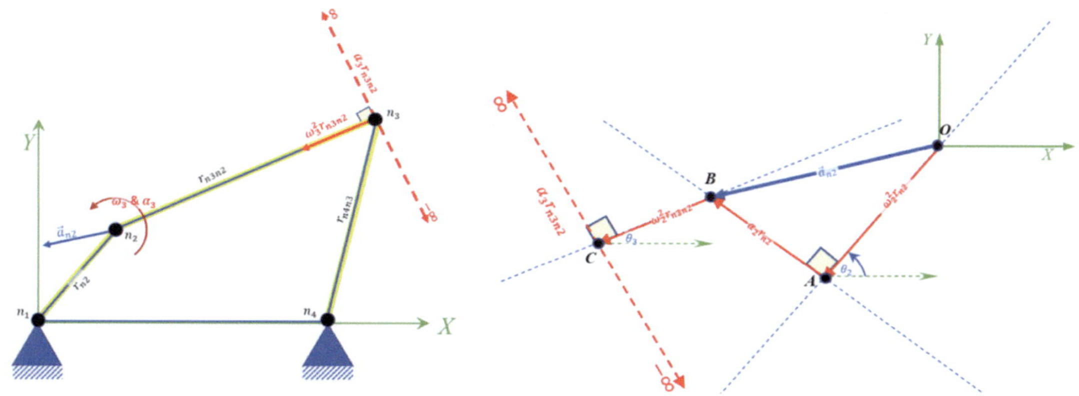

Fig. 7.37 Relative acceleration components of *an*3*n*2

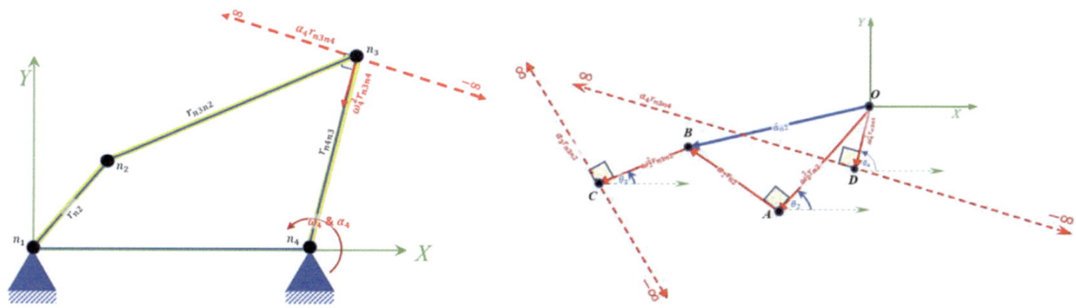

Fig. 7.38 Total acceleration of *an*3

Step 6: Extend the two dashed lines (directions of the two tangential accelerations $a_{Tn3} = \alpha_4 r_{n3n4}$ and $a_{Tn3n2} = \alpha_3 r_{n3n2}$) to intersect at point E as shown in Fig. 7.39. Plot the vector $\overrightarrow{OE}$. Now, we can use the point of intersection E to calculate $\vec{a}_{n3}$ by

$$\vec{a}_{n3} = \vec{a}_{n3} = \vec{a}_{n2} + \vec{a}_{n3n2} \quad \text{or} \quad \overrightarrow{OE} = \overrightarrow{OB} + \overrightarrow{BE} \tag{7.26}$$

From Eq. (7.26), we can calculate $\vec{a}_{n3}$ by measuring the length and angle of $\overrightarrow{OE}$. Also, we can calculate the two independent unknown variables $\alpha_3$and α_4 by measuring the length of $\overrightarrow{CE}$ and $\overrightarrow{DE}$ and use:

$$\begin{aligned} \alpha_3 &= \frac{CE}{r_{n3n2}} \\ \alpha_4 &= \frac{DE}{r_{n3n4}} \end{aligned} \tag{7.27}$$

The above algorithm is based on knowing the variables $\theta_2, \omega_2,$ and α_2 as the motor input signals. The algorithm can be modified if we place the motor at joint $J_2^R : n_2$ ($\theta_3, \omega_3,$ and $\alpha_3$are known) or if we place the motor at joint $J_4^R : n_4$ ($\theta_4, \omega_4,$ and $\alpha_4$are known). We are not going over the new algorithms that are coming from placing the motor $J_3^R : n_3$ or $J_4^R : n_4$. The reader can develop these algorithms using a similar method to the steps above.

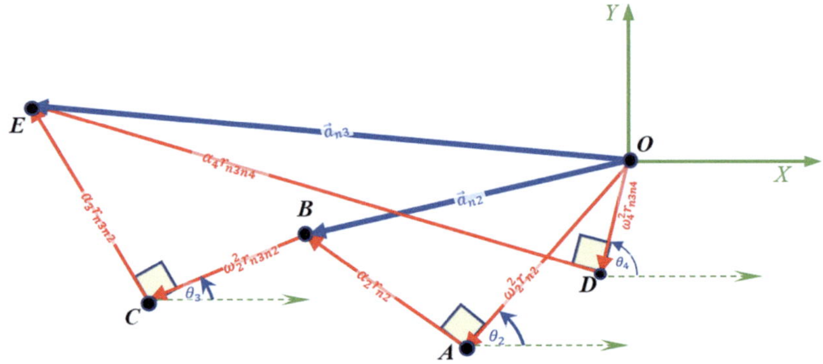

Fig. 7.39 Vector diagram of four-bar mechanism accelerations

Example 7.7

For the door opener mechanism in Fig. 7.21 and the described simulation scenario in Example 7.6, when $t=0.2$ s, the measured variables

$\theta_2 = 8.5°, \theta_3 = 22.15°, \theta_4 = 32.5°$,

$\omega_2 = 0.742\,\frac{\text{rad}}{\text{s}}, \omega_3 = -1.269\,\frac{\text{rad}}{\text{s}}$, and $\omega_4 = -0.809\,\frac{\text{rad}}{\text{s}}$:

1. Use the vector diagram to graphically plot the acceleration components and the total acceleration of node n_2 and node n_3.
2. Use the vector diagram to graphically estimate the values of the angular acceleration of the door coupler and the angular acceleration of the door follower (α_3 and α_4).
3. Check the fidelity of your graphical estimation in step 1 and step 2 with the model-based simulation in Example 7.6

Solution

Follow the steps of the graphical method for a four-bar mechanism. You can use any drawing scale that you prefer. Using Eq. (7.23), plot the two vectors:

$$\begin{aligned}\overrightarrow{OA} = \vec{a}_{Nn2} &= \omega_2^2 r_{n2}\left(\cos(180° + \theta_2)\hat{i} + \sin(180° + \theta_2)\hat{j}\right)\\ &= 0.742^2 \times 0.25 \times \left(\cos(180° + 8.5°)\hat{i} + \sin(180° + 8.5°)\hat{j}\right)\\ &= -0.136\hat{i} - 0.02\hat{j}\end{aligned}$$

$$\begin{aligned}\overrightarrow{OD} = \vec{a}_{Nn3} &= \omega_4^2 r_{n4n3}\left(\cos(180° + \theta_4)\hat{i} + \sin(180° + \theta_4)\hat{j}\right)\\ &= -0.809^2 \times 0.3 \times \left(\cos(180° + 32.5°)\hat{i} + \sin(180° + 32.5°)\hat{j}\right)\\ &= -0.166\hat{i} - 0.106\hat{j}\end{aligned}$$

Note that the door is opened at a constant angular velocity or $\alpha_2 = 0$ rad/s^2, so:

$$\vec{a}_{n2} = \overrightarrow{OB} = \overrightarrow{OA} + \overrightarrow{AB} = \vec{a}_{Nn2} + \underbrace{\vec{a}_{Tn2}}_{\substack{=0 \\ \text{since}\alpha_2=0}} = \overrightarrow{OA} + \vec{0} = -0.136\hat{i} - 0.02\hat{j}$$

Points A and B are located at the same point as shown in Fig. 7.40. Stand at point B and plot the vector $\overrightarrow{BC}$:

$$\begin{aligned}\overrightarrow{BC} = \vec{a}_{Nn3n2} &= \omega_3^2 r_{n3n2}\left(\cos(180^\circ + \theta_3)\hat{i} + \sin(180^\circ + \theta_3)\hat{j}\right) \\ &= -1.269^2 \times 0.33 \times \left(\cos(180^\circ + 22.15^\circ)\hat{i} + \sin(180^\circ + 22.15^\circ)\hat{j}\right) \\ &= -0.492\hat{i} - 0.2004\hat{j}\end{aligned}$$

Stand at point D and plot the direction of the vector $\overrightarrow{DE} = \vec{a}_{Tn3}$ as perpendicular on $\overrightarrow{OD} = \vec{a}_{Nn3}$ in a dashed-line format. Stand at point C and plot the direction of the vector $\overrightarrow{CE} = \vec{a}_{Tn3n2}$ as perpendicular on $\overrightarrow{BC} = \vec{a}_{Nn3n2}$ in a dashed-line format. Define the point E as the intersection of $\overrightarrow{DE}$ and $\overrightarrow{CE}$. Highlight the two lines $\overrightarrow{DE}$ and $\overrightarrow{CE}$ with a solid-line format as shown in Fig. 7.40.

Connect the line $\overrightarrow{OE} = \vec{a}_{n3}$ and measure the X-component and Y-component to get:

$$\overrightarrow{OC} = \vec{a}_{n3} = -1.577\hat{i} + 2.11\hat{j}$$

Measure the length of CE and DE to be 2.572 m/s^2 and 2.627 m/s^2 consecutively. Apply Eq. (7.27) as:

$$\alpha_3 = \frac{CE}{r_{n3n2}} = \frac{2.5165}{0.33} = 7.63\,\text{rad/s}^2$$

$$\alpha_4 = \frac{DE}{r_{n3n4}} = \frac{2.627}{0.3} = 8.77\,\text{rad/s}^2$$

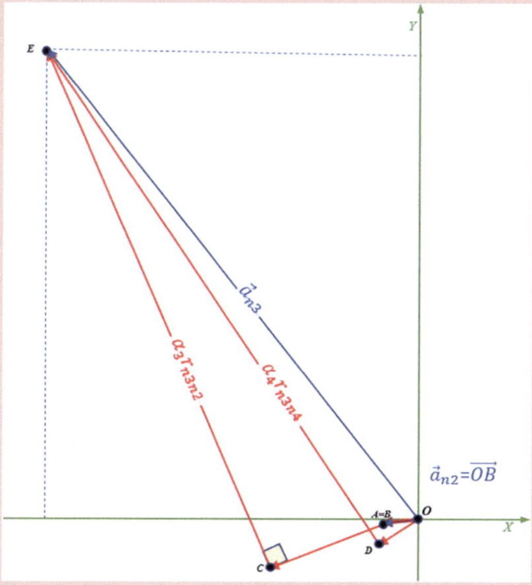

Fig. 7.40 Acceleration vector diagram of a door opener at $t=0.2$ s

If we compare the values from the graphical solution with the model-based solution at $t = 0.2$ s, we can see a very high match.

7.5.2 Slider–Crank Mechanism Graphical Velocity Analysis

Assume the shown slider–crank mechanism in Fig. 7.41 has $\theta_2, \omega_2,$ and α_2 as a motor input signals at $J_1^R : n_1$. From our previous analyses in Chaps. 5 and 6, we can graphically calculate the position vectors and the velocity vectors of any node located in this slider–crank mechanism. That brings us now to set up the graphical acceleration algorithm which has the given signals:$\theta_2, \theta_3, r_{n4}, \omega_2, \omega_3, \dot{r}_{n4}$ and α_2 and the required signals: $\alpha_3, \ddot{r}_{n4}, \vec{a}_{n2}$ and $\vec{a}_{n3}$. The slider–crank mechanism graphical acceleration algorithm can be summarized as follows:

Step 1: Use the graphical algorithm in Chap. 5 to plot the entire mechanism according to the given parameters ($r_{n2}, r_{n2n3},$ and r_{n3n4}) and the given input signal (θ_2) as shown in Fig. 7.41. Measure the output signals (θ_3 and r_{n4}). Note, sometimes a reasonable "scale" needs to be established if the dimensions are too big to fit on paper. We can use any CAD software to do so.

Step 2: Use the graphical algorithm in Chap. 6 to calculate the angular velocities ω_3 and the transitional velocity $\dot{r}_{n4}$ from the known independent variables $\theta_2, \theta_3, r_{n4},$ and ω_2. After finalizing the graphical position analysis and the graphical velocity analysis, we now have seven known independent variables: $\theta_2, \theta_3, r_{n4}, \omega_2, \omega_3, \dot{r}_{n4},$ and α_2 and two unknown independent variables α_3 and $\ddot{r}_{n4}$ as shown in Fig. 7.41.

Step 3: We are now going to graphically calculate the acceleration $\vec{a}_{n2}$ using the known independent variables $\theta_2, \omega_2,$ and α_2 and the known parameter r_{n2}. Assume that α_2 is positive or in the

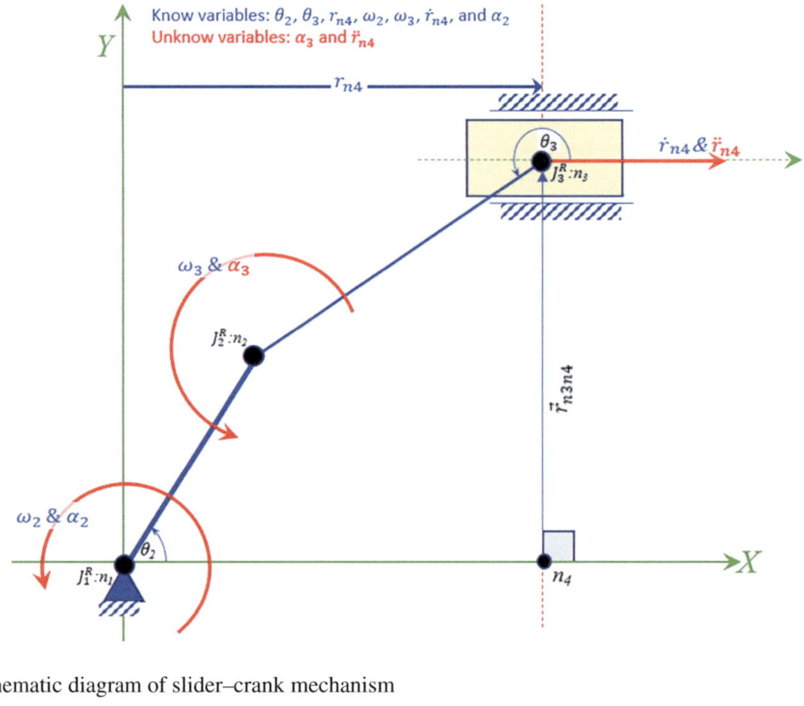

Fig. 7.41 Kinematic diagram of slider–crank mechanism

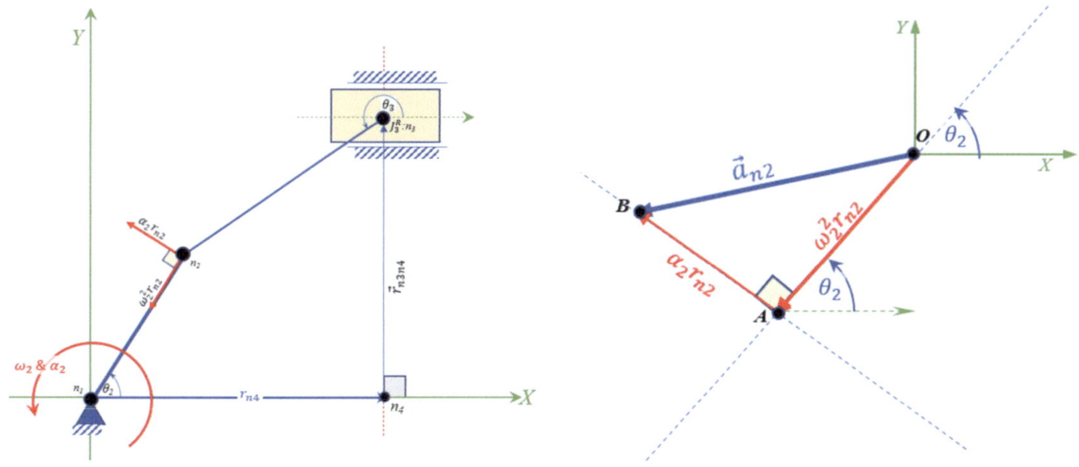

Fig. 7.42 Acceleration components of n_2

counterclockwise direction. On the kinematic diagram, plot the normal acceleration component $\vec{a}_{Nn2}$ as a solid line vector that starts from node n_2 and heads in the negative direction of $\vec{r}_{n2}$. After that, plot the tangential acceleration component $\vec{a}_{Tn2}$ as a solid line vector that starts from node n_2 and is perpendicular to $\vec{r}_{n2}$, making an angle $90° + \theta_2$ with the horizontal line passing by node n_2 as shown in Fig. 7.42a. Use the vector diagram (with an origin point O) to plot the normal acceleration component $\vec{a}_{Nn2}$ as a solid line vector $\overrightarrow{OA}$ from point O to point A with a length $a_{Nn2} = \omega_2^2 r_{n2}$, making an angle $180° + \theta_2$ as shown in Fig. 7.42b. Plot the tangential acceleration component $\vec{a}_{Tn2}$ as a solid line vector $\overrightarrow{AB}$ that starts from point A to point B with a length $a_{Nn2} = \alpha_2 r_{n2}$ and perpendicular on, making an angle $90° + \theta_2$ with the *X-axis* because we assumed the angular $\overrightarrow{OA}$ acceleration in counterclockwise. Plot the total α_2 acceleration vector $\vec{a}_{n2}$ as a solid line vector $\overrightarrow{OB}$ to be the summation of both $\overrightarrow{OA}$ and $\overrightarrow{AB}$.

Step 4: To graphically calculate the acceleration $\vec{a}_{n3}$, we can use Eq. (7.24) using $i=2$ and $j=3$ as:

$$\vec{a}_{n3} = \vec{a}_{n2} + \vec{a}_{n3n2} \tag{7.28}$$

From Step 3, we know $\vec{a}_{n2}$. Now, we need to calculate $\vec{a}_{n3n2}$. On the kinematic diagram, plot the relative normal acceleration component $\vec{a}_{Nn3n2}$ as a solid line vector that starts from node n_3 and heads in the negative $\vec{r}_{n3n2}$-direction. Also, plot the direction of relative tangential acceleration component $\vec{a}_{Tn3n2}$ as a perpendicular dashed line on $\vec{r}_{n3n2}$ and passing by node n_3. On the acceleration vector diagram, add the vector $\overrightarrow{BC}$ that goes from point B to point C with a length $a_{Nn3n2} = \omega_3^2 r_{n3n2}$ and an angle $180° + \theta_3$ with the horizontal line passing by point B as shown in Fig. 7.43b. Plot the dashed line that passes by point C and perpendicular on $\overrightarrow{BC}$ vector as shown in Fig. 7.43b. This dashed line presents the direction of the relative tangential acceleration component $\vec{a}_{Tn3n2}$.

Step 5: From the kinematic constraint of the prismatic joint J_4^P, we know that the direction of the total acceleration $\vec{a}_{n3}$ must be horizontal and equal to $\ddot{r}_{n4}\hat{i}$. We plot the direction of $\vec{a}_{n3}$ as a dashed line as shown in Fig. 7.44a. Also on the acceleration vector diagram, we add a dashed horizontal line passing by point O to present the direction of $\vec{a}_{n3}$ as shown in Fig. 7.44b.

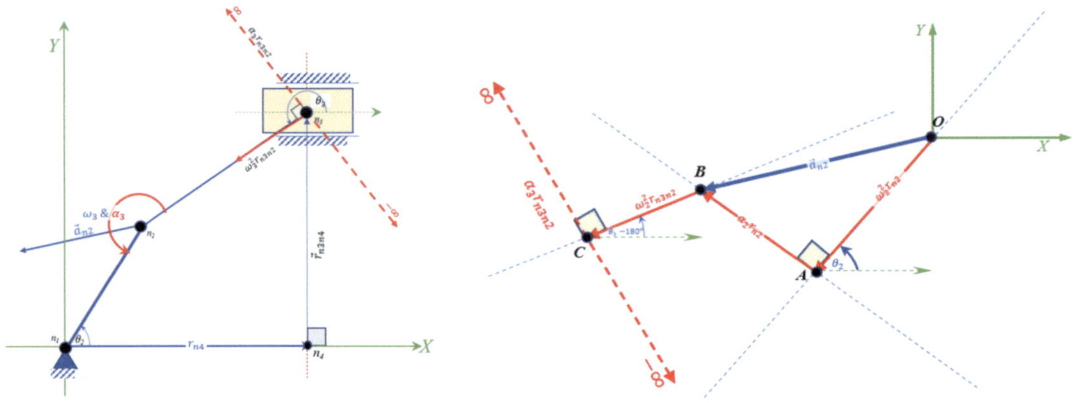

Fig. 7.43 Relative acceleration components of a_{n3n2}

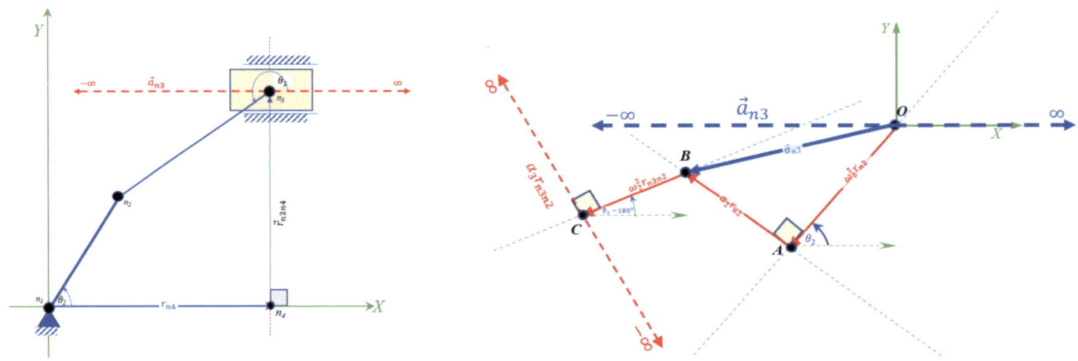

Fig. 7.44 Total acceleration of a_{n3}

Step 6: Extend the two dashed lines (directions of the tangential acceleration $a_{Tn3n2} = \alpha_3 r_{n3n2}$ and the direction of $\vec{a}_{n3}$) to intersect at point D as shown in Fig. 7.45. Now, we can use the point of intersection D to calculate $\vec{a}_{n3}$ by:

$$\vec{a}_{n3} = \ddot{r}_{n4}\hat{i} = \vec{a}_{n2} + \vec{a}_{n3n2} \text{ or } \overrightarrow{OD} = \overrightarrow{OB} + \overrightarrow{BD} \tag{7.29}$$

From Eq. (7.29), we can calculate $\vec{a}_{n3}$ by measuring the length and the angle of $\overrightarrow{OD}$ that is equal to $\ddot{r}_{n4}$. Also, we can calculate the independent unknown α_3 by measuring the length of $\overrightarrow{BD}$ and use:

$$\alpha_3 = \frac{BD}{r_{n3n2}} \tag{7.30}$$

The above algorithm is based on knowing $\theta_2, \omega_2,$ and α_2 as the motor input signals. The algorithm can be modified if we place the motor at joint $J_2^R : n_2$ ($\theta_3, \omega_3,$ and $\alpha_3$are known) or if we attach a hydraulic actuator at joint $J_3^R : n_3$ ($r_{n4}, \dot{r}_{n4},$ and $\ddot{r}_{n4}$ are known). We are not going over the new algorithms that are coming from changing the actuator's location. The reader can develop these algorithms using similar steps to the above.

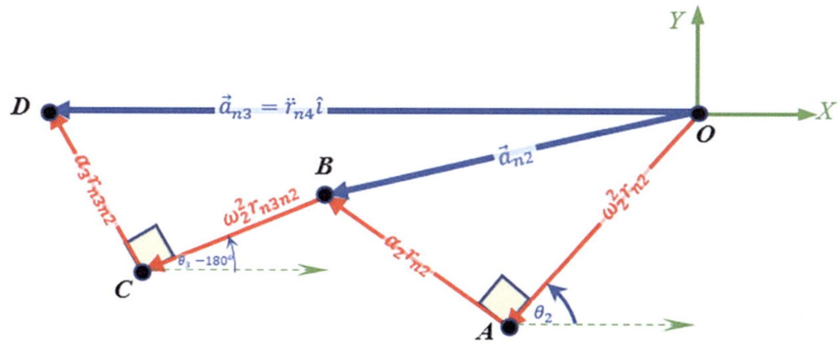

Fig. 7.45 Vector diagram of slider–crank mechanism accelerations

Example 7.8

For the shown kinematic diagram of the piston–cylinder mechanism in Fig. 7.26 where the length of the crank is 76.2 mm and the length of the connecting rod is 203.2 mm. Assume that the combustion force moves the piston toward the crankshaft with a sliding motion:

$$x_p(t) = 0.2032 + 0.0494\cos(263.158t)0 < t < 0.012\,\text{s}$$

where $x_p(t)$ is the piston sliding position in *m*. This sliding motion causes the connecting rod to rotate and consequently, the crankshaft rotates.

1. Build a model-based simulation of the piston–cylinder mechanism and plot:
   - The time history of the crank angular acceleration, α_2
   - The time history of the connecting rod angular acceleration, α_3
   - The time history of the Cartesian components of the absolute acceleration at the crankpin
2. At $t=0.006$ s, we have the measured variables $\theta_2 = 79.05°, \omega_2 = 161.4\,\text{rad/s}$, $\theta_3 = 158.4°, \omega_3 = -12.37\,\text{rad/s}$, $r_{n4} = 0.2034\,\text{m}$, and $\dot{r}_{n4} = -13\,\text{m/s}$ in addition to the input acceleration $\ddot{r}_{n4} = -27.89\,\text{m/s}^2$. Use the acceleration vector diagram to graphically calculate:
   - The acceleration components of the crankpin
   - The absolute Cartesian components of the crankpin
   - The crank angular acceleration
   - The connecting rod angular acceleration
3. Use the results from the model-based simulation to measure the angular accelerations of the crank and the connecting rod at $t=0.006$ s. Compare your measurement with the graphical solution.

Solution

Model-Based Method

1. Copy and rename "SinglePistonCylinder_P6p7_model.slx" to "SinglePistonCylinder_P7p8_model.slx" and "SinglePistonCylinder_P6p7_script.m" to "SinglePistonCylinder_P7p8_script.m".
2. In the "SinglePistonCylinder_P7p8_model.slx", replace the "Translational Motion Actuator" subsystem with "Triple Inputs Translational Motion Actuator" subsystem from

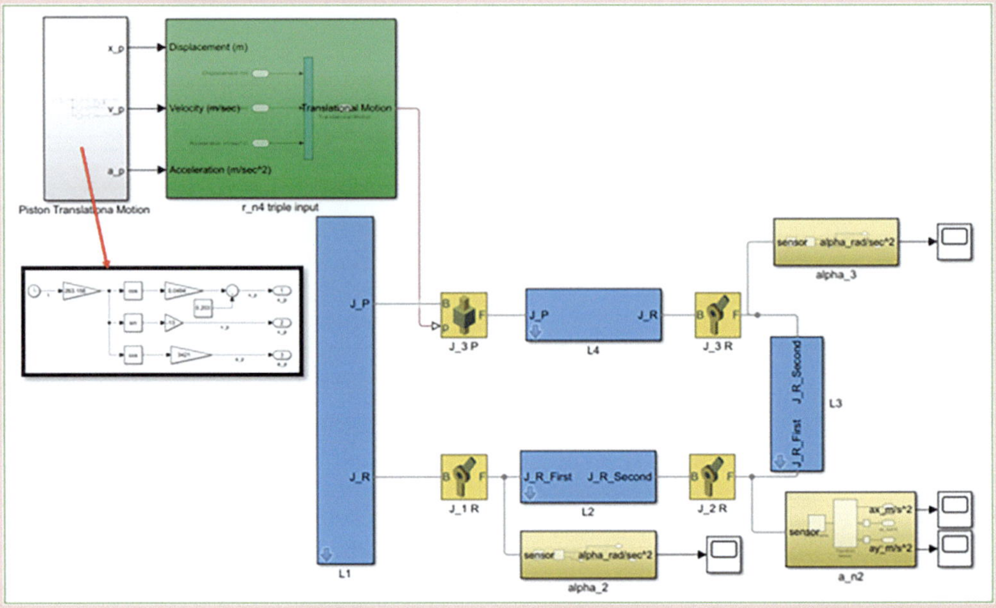

Fig. 7.46 Piston–cylinder Simscape™ Multibody™ model

"Kinematics_Of_Machines.slx" file as shown in Fig. 7.46. Rename this "Triple Inputs Rotational Motion Actuator" by "r_n4 triple input".

3. Implement the "Piston Translational Motion" subsystem that has the position "x_p" Outport, the velocity "v_p" Outport, and the acceleration "a_p" Outport as shown in Fig. 7.46 using the equations:

$$x_p(t) = 0.203 + 0.0494 \cos(263.158t)$$
$$v_p(t) = \frac{dx_p}{dt} = -13 \sin(263.158t)$$
$$a_p(t) = \frac{dv_p}{dt} = 3421 \cos(263.158t)$$

4. Connect the input signals of the "r_n4 triple input" subsystem to (1) "x_p" Outport, (2) "v_p" Outport, and (3) "a_p" Outport from "Piston Translational Motion" subsystem.
5. Bring two "Angular Acceleration Sensor" subsystems from the "Kinematics_Of_
6. Machines.slx" file and rename them "alpha_2" and "alpha_3". Connect these two subsystems as shown in Fig. 7.46. Add scope block at each output port of these two sensor subsystems.
7. Bring one "Cartesian Translational Acceleration Sensor" subsystem from the "Kinematics_Of_Machines.slx" file and rename this subsystem as "an_2". Connect this subsystem as shown in Fig. 7.46. Add scope block at each output port of this sensor subsystem.

Run the script file followed by running the model file with Stop Time = 0.012. Look at the results from the scopes as shown in Fig. 7.47. We encourage the reader to watch the attached animated video from this model as well as look at the attached model.

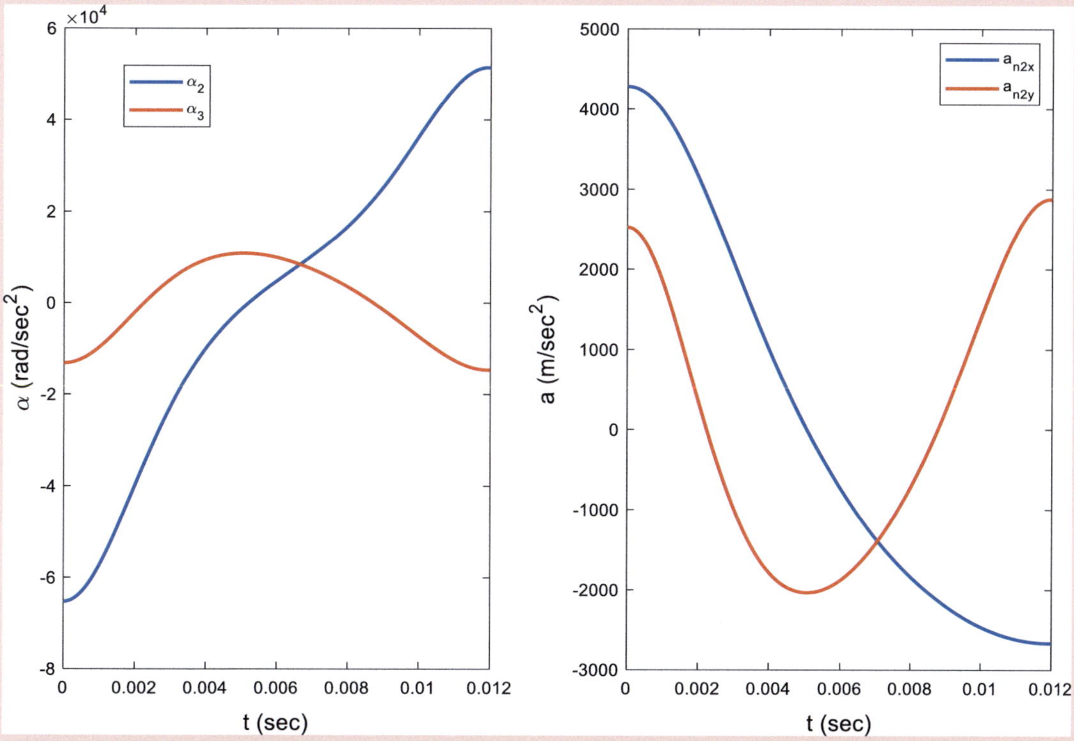

Fig. 7.47 Results of Simscape™ Multibody™ model for piston–cylinder mechanism

Graphical Method

Follow the steps of the graphical method for a slider–crank mechanism. You can use any drawing scale that you prefer. In a solid line format and using Eq. (7.23), plot the three vectors:

$$\begin{aligned}\overrightarrow{OA} = \vec{a}_{Nn2} &= \omega_2^2 r_{n2}\left(\cos(180^\circ + \theta_2)\hat{i} + \sin(180^\circ + \theta_2)\hat{j}\right)\\ &= 161.4^2 \times 0.0762 \times \left(\cos(180^\circ + 79.05^\circ)\hat{i} + \sin(180^\circ + 79.05^\circ)\hat{j}\right)\\ &= -377.06\hat{i} - 1948.9\hat{j}\end{aligned}$$

$$\begin{aligned}\overrightarrow{AB} = \vec{a}_{Nn3n2} &= \omega_3^2 r_{n3n2}\left(\cos(180^\circ + \theta_3 - 180^\circ)\hat{i} + \sin(180^\circ + \theta_3 - 180^\circ)\hat{j}\right)\\ &= -12.37^2 \times 0.2032 \times \left(\cos(180^\circ + 158.4^\circ - 180^\circ)\hat{i} + \sin(180^\circ + 158.4^\circ - 180^\circ)\hat{j}\right)\\ &= -28.91\hat{i} + 11.45\hat{j}\end{aligned}$$

$$\overrightarrow{OD} = \vec{a}_{n3} = -\ddot{r}_{n4}\hat{i} = -27.89\hat{i}$$

Stand at point B and plot the dashed direction line of the vector $\overrightarrow{BC} = \vec{a}_{Tn2}$ as perpendicular on $\overrightarrow{OA} = \vec{a}_{Nn2}$. Stand at point D and plot the dashed direction line of the vector $\overrightarrow{CD} = \vec{a}_{Tn3n2}$ as perpendicular on $\overrightarrow{AB} = \vec{a}_{Nn3n2}$. Define the point C as the intersection of the direction of $\overrightarrow{BC}$ and the direction of $\overrightarrow{CD}$. Highlight the two vectors $\overrightarrow{BC}$ and $\overrightarrow{CD}$ with a solid line format as shown in Fig. 7.48a.

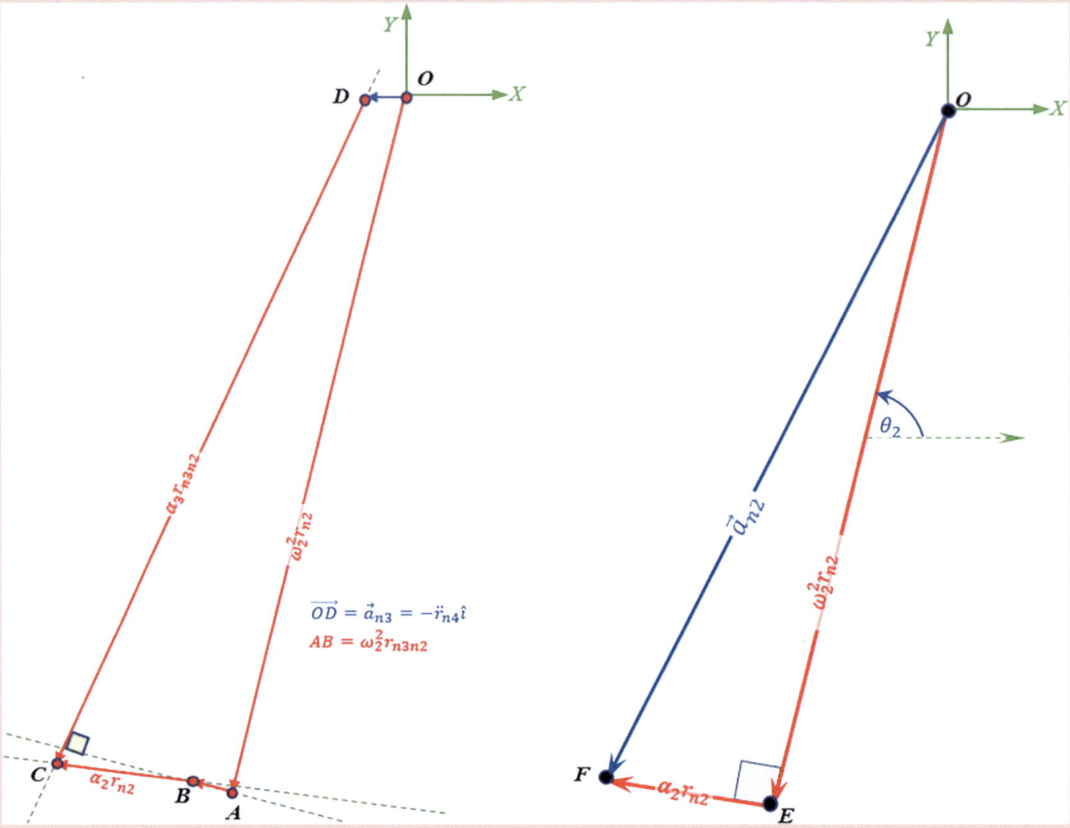

Fig. 7.48 Acceleration vector diagram of single piston–cylinder mechanism

Measure the length of $\overrightarrow{BC}$ and $\overrightarrow{CD}$ to be 368.013 and 2008.5 m/s^2 consecutively. Using the definition of $BC = a_{Tn2} = \alpha_2 r_{n2}$ and $CD = a_{Tn3n2} = \alpha_3 r_{n3n2}$ to calculate:

$$\alpha_2 = \frac{BC}{r_{n2}} = \frac{368.013}{0.0672} = 4829.6\,\text{rad/s}^2$$

$$\alpha_3 = \frac{CD}{r_{n3n4}} = \frac{2008.5}{0.2032} = 9884.4\,\text{rad/s}^2$$

To calculate the absolute acceleration vector $\vec{a}_{n2}$, we sketch a new acceleration vector diagram of the two vectors $\overrightarrow{OE} = \vec{a}_{Nn2}$ and $\overrightarrow{EF} = \vec{a}_{Tn2}$ in Fig. 7.48b using Eq. (7.23):

$$\begin{aligned}\overrightarrow{OE} = \vec{a}_{Nn2} &= \omega_2^2 r_{n2}\left(\cos(180^\circ + \theta_2)\hat{\imath} + \sin(180^\circ + \theta_2)\hat{\jmath}\right)\\ &= 161.4^2 \times 0.0762 \times \left(\cos(180^\circ + 79.05^\circ)\hat{\imath} + \sin(180^\circ + 79.05^\circ)\hat{\jmath}\right)\\ &= -377.06\hat{\imath} - 1948.9\hat{\jmath}\end{aligned}$$

$$\begin{aligned}\overrightarrow{EF} = \vec{a}_{Tn2} &= \alpha_2 r_{n2}\left(\cos(\theta_2 + 90^\circ)\hat{\imath} + \sin(\theta_2 + 90^\circ)\hat{\jmath}\right)\\ &= 4830.5 \times 0.0762 \times \left(\cos(79.05^\circ + 90^\circ)\hat{\imath} + \sin(79.05^\circ + 90^\circ)\hat{\jmath}\right)\\ &= -361.3\hat{\imath} + 69.96\hat{\jmath}\end{aligned}$$

Connect the vector $\overrightarrow{OF} = \vec{a}_{n2}$. Measure the X-component and Y-component of $\overrightarrow{OF}$ to get

$$\overrightarrow{OF} = \vec{a}_{n2} = -738.4\hat{i} - 1878.9\hat{j}$$

Now we are going to compare the model-based results with the graphical analysis results as listed in Table 7.2 that shows remarkably close values.

Table 7.2 Comparing model-based versus graphical analysis for single piston–cylinder mechanism

Quantity	Model-based	Graphical analysis
α_2 (rad/s^2)	4830.5	4829.6
α_3 (rad/s^2)	9884.2	9884.4
a_{n2x} (m/s^2)	– 738.5	– 738.4
a_{n2y} (m/s^2)	– 1877.8	– 1878.9

* Note, the implementations of the piston's position, velocity, and acceleration are a simple collection of "clock" block, "Gain" blocks, "cos" blocks, "sin" block, "Sum" block, "Constant" block, and "Outport" blocks as shown in Fig. 7.46

Engineering Tips—Vector Addition Commutative Property

For a set of three vectors $\vec{V}_1$, $\vec{V}_2$, and $\vec{V}_3$, because of the vector addition commutative property, the summation vector can be written in six forms. Here are three of these six forms:

$$\begin{gathered}\vec{V}_{\text{sum1}} = \vec{V}_1 + \vec{V}_2 + \vec{V}_3 \\ \text{or} \\ \vec{V}_{\text{sum2}} = \vec{V}_2 + \vec{V}_1 + \vec{V}_3 \\ \text{or} \\ \vec{V}_{\text{sum3}} = \vec{V}_3 + \vec{V}_2 + \vec{V}_1 \\ \text{and} \\ \vec{V}_{\text{sum}} = \vec{V}_{\text{sum1}} = \vec{V}_{\text{sum2}} = \vec{V}_{\text{sum3}}\end{gathered} \tag{7.31}$$

Equation (7.31) shows that the order of the elements in the vector summation process does not affect the results. Assume the three arbitrary vectors are graphically presented as shown in Fig. 7.49. Each vector $\vec{V}_i$ starts from point a_i and ends at point b_i, where $i = 1$, 2, or 3. We can graphically add the three vectors and calculate $\vec{V}_{\text{sum}}$ in six different ways. In Fig. 7.50, you can see the three summation forms as listed in Eq. (7.31).

We can use the vector addition commutative property when we plot the total acceleration vector on the acceleration vector diagram. For example, assume we have a link L_k that rotates with an angular velocity ω_k and an angular acceleration α_k. The nodes n_i and n_j belong to link L_k with a fixed distance of r_{njni} (no slipping or sliding). The absolute acceleration of node n_i is known to be $\vec{a}_{ni}$. The relative acceleration, of node n_j relative to node n_i, $\vec{a}_{njni}$ has two components: normal $\vec{a}_{Nnjni}$ and $\vec{a}_{Tnjni}$ and then:

$$\vec{a}_{nj} = \vec{a}_{ni} + \vec{a}_{njni} = \vec{a}_{ni} + \underbrace{\vec{a}_{Nnjni} + \vec{a}_{Tjni}}_{\text{form1}} = \vec{a}_{ni} + \underbrace{\vec{a}_{Tjni} + \vec{a}_{Nnjni}}_{\text{form2}} \tag{7.32}$$

where $\vec{a}_{nj}$ is the total acceleration at node n_j. Equation (7.32) has two forms to calculate the relative acceleration $\vec{a}_{njni}$ or the total acceleration $\vec{a}_{nj}$. Figure 7.51 shows the acceleration

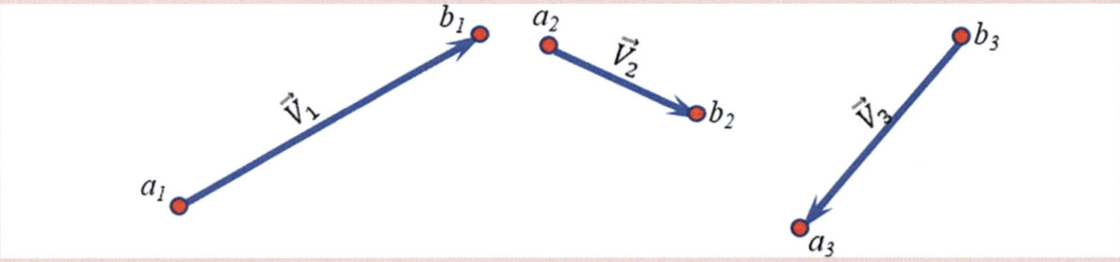

Fig. 7.49 Three arbitrary 2D vectors

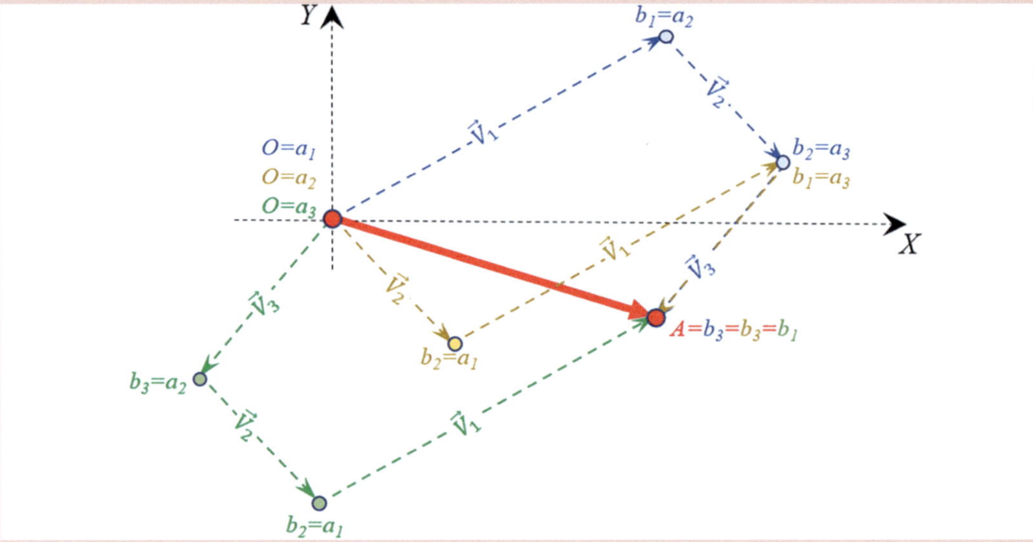

Fig. 7.50 Applying the vector additional commutative property for three vectors

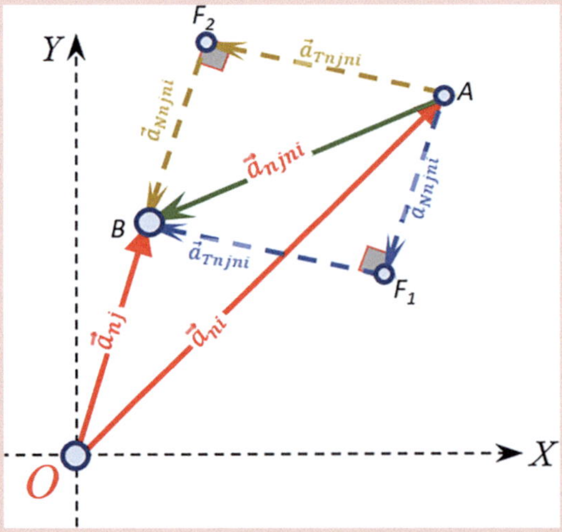

Fig. 7.51 Absolute acceleration vectors of two nodes using two different forms without sliding or slipping

vector $\vec{a}_{nj}$ using the two forms. In both forms, we plot the vector $\overrightarrow{OA} = \vec{a}_{ni}$. In the first form, we plot the relative normal acceleration vector component $\overrightarrow{AF_1} = \vec{a}_{Nnjni}$ followed by plotting the relative tangential acceleration component $\overrightarrow{F_1B} = \vec{a}_{Tnjni}$. In the second form, we plot the relative tangential acceleration vector component $\overrightarrow{AF_2} = \vec{a}_{Tnjni}$ followed by plotting the relative normal acceleration component $\overrightarrow{F_2B} = \vec{a}_{Nnjni}$. Using any of the two forms, we will end up with $\overrightarrow{OB} = \vec{a}_{nj}$. When we have relative acceleration components with known components (known magnitude and known direction) and unknown components (unknown magnitude and known direction), it is recommended to plot the known components before the unknown components' directions.

7.5.3 Inverted Slider–Crank Mechanism Graphical Velocity Analysis

The inverted slider–crank mechanism has four links L_1, L_2, L_3, and L_4. Assume θ_2, ω_2, and α_2 are the motor input signals to the mechanism at $J_1^R : n_1$. Going through our previous graphical position in Chap. 5 and graphical velocity analyses in Chap. 6, we should know the position and the velocity of each link and each node in the inverted slider–crank mechanism. That brings us now to set up the graphical acceleration algorithm which has the given signals: $\theta_2, \theta_3, \theta_4 = \theta_3 - \gamma, r_{n3n2}, \omega_2, \omega_3 = \omega_4, \dot{r}_{n3n2}$ and α_2 and the required signals: $\alpha_3 = \alpha_4, \ddot{r}_{n3n2}, \vec{a}_{n2}$, and $\vec{a}_{n3}$. For simplicity, we are going to cover the case where $\gamma = 0$ (as shown in Fig. 7.52). We are now going to follow a similar algorithm as in the four-bar mechanism case and in the slider–crank case to graphically calculate the required signals.

Step 1: Use the graphical algorithm in Chap. 5 to plot the entire mechanism according to the given parameters ($r_{n2}, \gamma = 0, \delta$, and r_{n4}) and the given input signal (θ_2) as shown in Fig. 7.52.

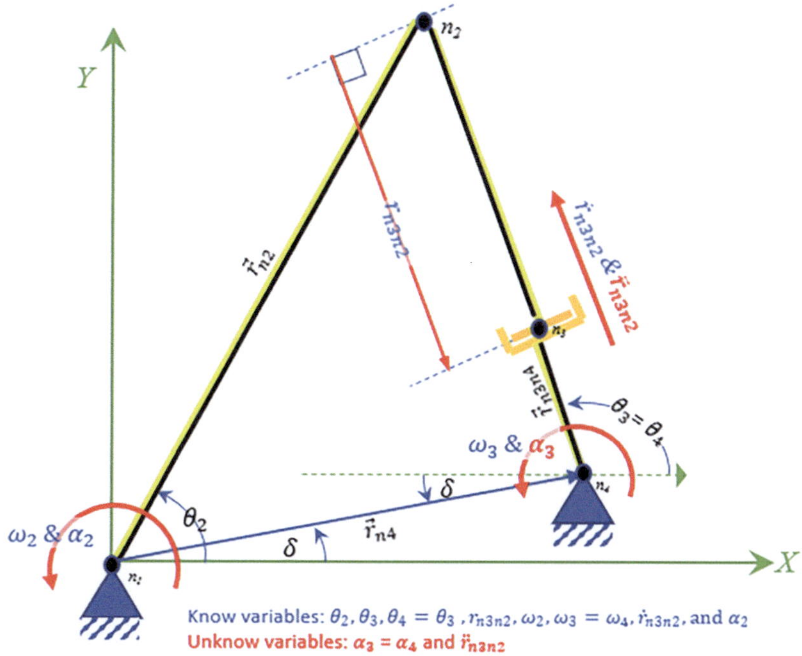

Fig. 7.52 Kinematic diagram of inverted slider–crank mechanism

Measure the output signals ($\theta_3, \theta_4 = \theta_3$, and r_{n3n2}). Note that sometimes a reasonable "scale" needs to be established if the dimensions are too big to fit on paper. We can use any CAD software to do so.

Step 2: Use the graphical algorithm in Chap. 6 to calculate the angular velocities ω_3 and the transitional velocity $\dot{r}_{n3n2}$ from the know independent variables $\theta_2, \theta_3, r_{n3n2}$, and ω_2. After finalizing the graphical position analysis and the graphical velocity analysis, we now have seven known independent variables: $\theta_2, \theta_3, \theta_4 = \theta_3, r_{n3n2}, \omega_2, \omega_3 = \omega_4, \dot{r}_{n3n2}$, and α_2 and two unknown independent variables $\alpha_3 = \alpha_4$ and $\ddot{r}_{n3n2}$ as shown in Fig. 7.52.

Step 3: We are now going to graphically calculate the acceleration $\vec{a}_{n2}$ using the known independent variables θ_2, ω_2, and α_2 and the known parameter r_{n2}. Assume, α_2 is positive or in the counterclockwise direction. On the kinematic diagram, plot the normal acceleration component $\vec{a}_{Nn2}$ as a solid line vector that starts from node n_2 and heads in the negative direction of $\vec{r}_{n2}$. After that, plot the tangential acceleration component $\vec{a}_{Tn2}$ as a solid line vector that starts from node n_2 and perpendicular to $\vec{r}_{n2}$, making an angle $90° + \theta_2$ with the horizontal line passing by node n_2 as shown in Fig. 7.53a. Use the vector diagram (with an origin point O) to plot the normal acceleration component $\vec{a}_{Nn2}$ as a solid line vector $\overrightarrow{OA}$ from point O to point A with a length $a_{Nn2} = \omega_2^2 r_{n2}$, making an angle $180° + \theta_2$ as shown in Fig. 7.53b. Plot the tangential acceleration component $\vec{a}_{Tn2}$ as a solid line vector $\overrightarrow{AB}$ that starts from point A to point B with a length $a_{Nn2} = \alpha_2 r_{n2}$ and perpendicular on $\overrightarrow{OA}$, making an angle $90° + \theta_2$ with the *X-axis* because we assumed the angular acceleration α_2 in counterclockwise. Plot the total acceleration vector $\vec{a}_{n2}$ as a solid line vector $\overrightarrow{OB}$ to be the summation of both $\overrightarrow{OA}$ and $\overrightarrow{AB}$.

Step 4: Another way to graphically calculate the acceleration $\vec{a}_{n2}$, is to use (7.24) using $i=4$ and $j=2$ as:

$$\vec{a}_{n2} = \vec{a}_{n4} + \vec{a}_{n2n4} \tag{7.33}$$

Since n_4 belongs to the ground, then $\vec{a}_{n4} = 0$. Now, we need to calculate $\vec{a}_{n2n4}$. Use the kinematic diagram to plot the normal acceleration component $a_{Nn2n4} = \omega_3^2 r_{n2n4}$ as a solid line, the Coriolis acceleration component $a_{Cn2n4} = 2\omega_3 \dot{r}_{n2n4} = 2\omega_3 \dot{r}_{n3n2}$ as a solid line, sliding acceleration component $a_{Sn2n4} = \ddot{r}_{n2n4} = \ddot{r}_{n3n2}$ as a dashed line, and the tangential acceleration component $a_{Tn2n4} = \alpha_3 r_{n2n4}$ as a dashed line as shown in Fig. 7.54a where the direction of each acceleration component is shown. The solid line means that the acceleration component is known as a direction and as a value. The dashed line means that the acceleration component is known as a direction

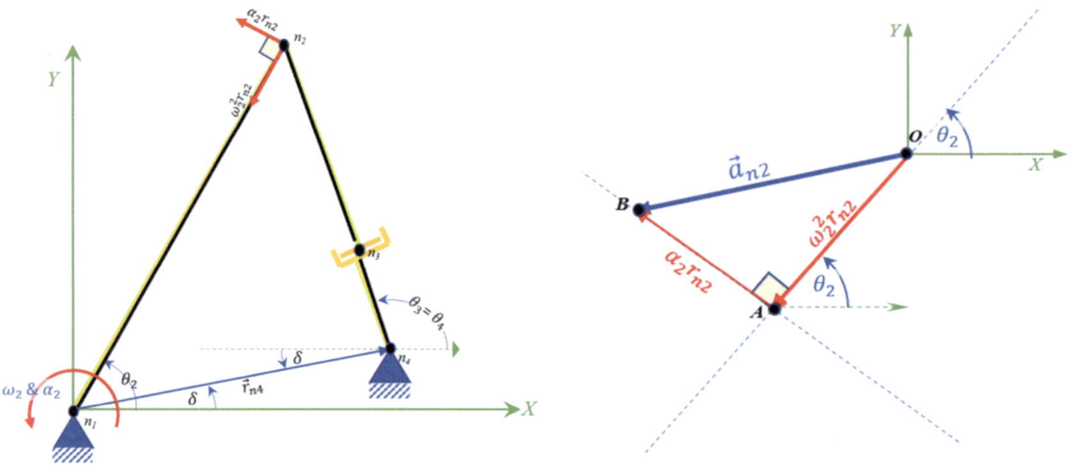

Fig. 7.53 Acceleration components of n_2

but not as a value. Add the vector $\overrightarrow{OC}$ from point O to point C with a length $a_{Nn2n4} = \omega_3^2 r_{n2n4}$ and an angle θ_3 as shown in Fig. 7.54b. Add the vector $\overrightarrow{CD}$ from point C to point D with a length $a_{Cn2n4} = 2\omega_3 \dot{r}_{n3n2}$ and perpendicular on $\overrightarrow{OB}$. Plot the summation vector $\overrightarrow{OD}$. From point D, plot both the directions of $\overrightarrow{DE} = a_{Tn3n2} = \alpha_3 r_{n3n2}$. From point B, plot the direction of $\overrightarrow{EB} = a_{Sn2n4} = \ddot{r}_{n3n2}$ Fig. 7.54b. Note, we know the directions of a_{Sn2n4} and a_{Tn3n2} but we do not know the values.

Step 5: Now, we have the intersection point E between the new location of line $a_{Sn2n4} = \ddot{r}_{n3n2}$ with the line $a_{Tn3n2} = \alpha_3 r_{n3n2}$. Convert the dashed lines into solid lines as shown in Fig. 7.55. We measure the length of $\overrightarrow{EB}$ that is equal to $\ddot{r}_{n3n2}$. We calculate the independent unknown α_3 by measuring the length of $\overrightarrow{DE}$ and use:

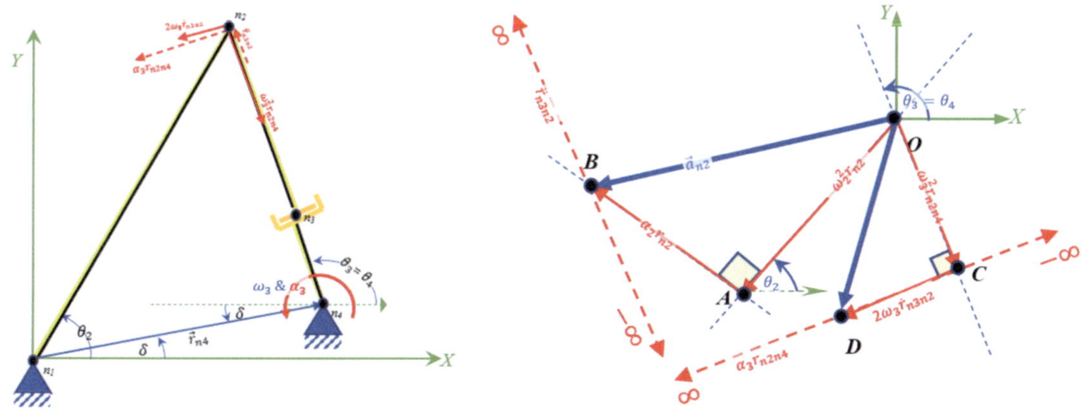

Fig. 7.54 Relative acceleration components of a_{n3n2}

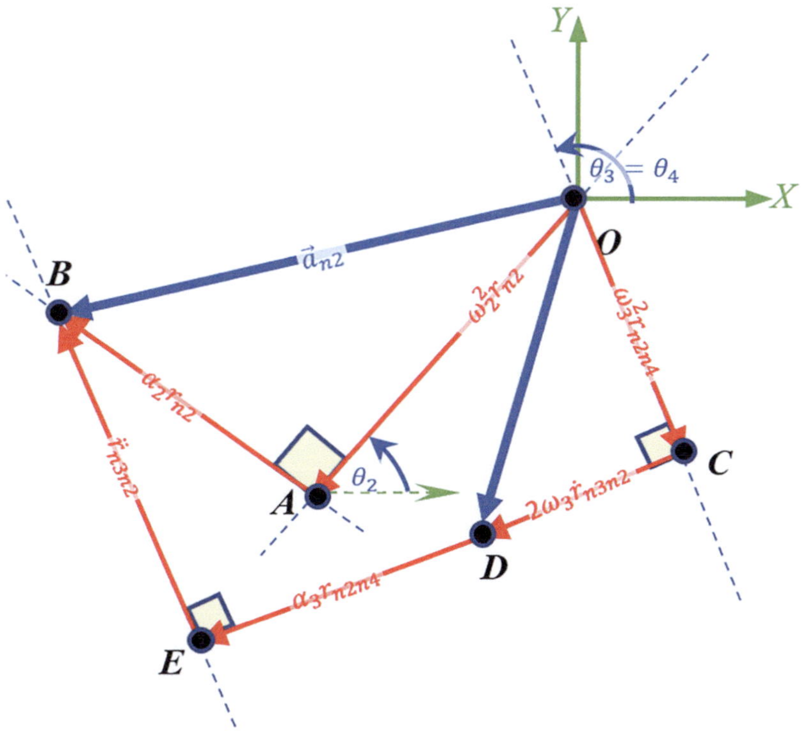

Fig. 7.55 Vector diagram of slider–crank mechanism accelerations

$$\alpha_3 = \frac{DE}{r_{n2n4}} \tag{7.34}$$

The above algorithm is based on knowing $\theta_2, \omega_2,$ and α_2 as the motor input signals. The algorithm can be modified if we have the prismatic joint as the active joint or $r_{n3n2}, \dot{r}_{n3n2},$ and r_{n3n2} are known. We are not going to cover the new algorithms that are coming from changing the actuator's location. The reader can develop these algorithms using steps similar to those above.

Example 7.9

For the shown hydraulic gate opener mechanism in Fig. 7.17. If the door is opened from the initial position at $\theta_2=0°$ with a constant angular velocity $\omega_2=1$ rad/s for 1 s time duration,

Build a model-based simulation of the hydraulic gate opener mechanism and plot:

- The time history of the cylinder angular acceleration, $\alpha_3 = \alpha_4$
- The Cartesian components of the absolute acceleration at node n_2
- At $\theta_2=45°$, from our previous position graphical analysis and velocity graphical analysis, we have $\omega_2 = 1\,\frac{\text{rad}}{\text{s}}, \theta_3 = \theta_4 = 55°, \omega_3 = \omega_4 = 0.97\,\frac{\text{rad}}{\text{s}}$, $r_{n3n2} = 0.413\,\text{m}$, and $\dot{r}_{n3n2} = 0.172\,\text{m/s}$ in addition to the angular acceleration $\alpha_2 = 0\,\text{m/s}^2$. Use the acceleration vector diagram to graphically calculate:
  - The acceleration components of node n_2 with respect to node n_1
  - The acceleration components of node n_2 with respect to node n_4
  - The Cartesian components of the absolute acceleration at node n_2
  - The cylinder angular acceleration

Use the results from the model-based simulation to measure the angular accelerations of the crank and the connecting rod at $\theta_2=45°$. Compare your measurement with the graphical solution.

Solution

For the Model-Based Method

We are going to use our previously developed "HydraulicDoorOpener_P6p8_model.slx" and "HydraulicDoorOpener_P6p8_script.m" to build the new model as:

1. Copy and rename "HydraulicDoorOpener_P6p8_script.m" to "HydraulicDoorOpener_P7p9_script.m" and copy and rename "HydraulicDoorOpener_P6p8_model.slx" to "HydraulicDoorOpener_P7p9_model.slx". Run the script file.
2. Replace the "Rotational Motion Actuator" subsystem with "Triple Inputs Rotational Motion Actuator" subsystem from the "Kinematics_Of_Machines.slx" file as shown in Fig. 7.56.
3. Implement the "Rotational Motion" subsystem that has "theta2_deg" Outport number 1, "Omega2_deg/s" Outport number 2, and "Alpha2_deg/s^2" Outport number 3. Inside "Rotational Motion", we have two Signal Builder blocks. The first Signal Builder block implements "Theta2_deg(t)" and the second Signal Builder block implements "Omega2_deg/s(t)" as shown in Fig. 7.56. The "Alpha2_deg/s^2(t)" Constant block has a zero value since the door is opened at constant angular velocity.

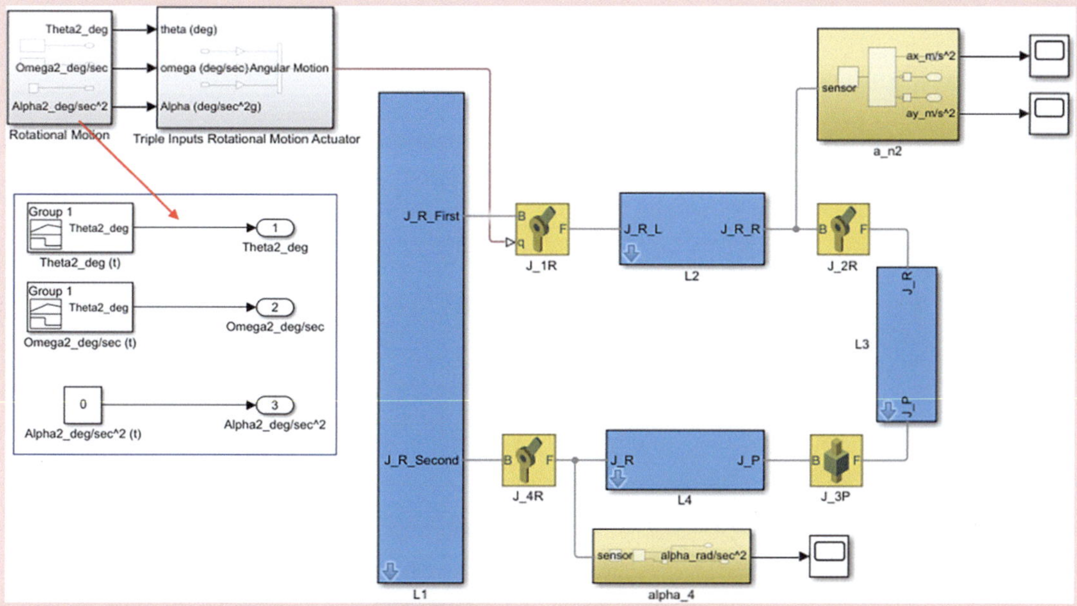

Fig. 7.56 Hydraulic door opener Simscape™ Multibody™ model

4. Add the "Angular Acceleration Sensor" subsystem and the "Cartesian Translational Acceleration Sensor" subsystem from "Kinematics_Of_Machines.slx". Connect the two sensor subsystems as shown in Fig. 7.56.
5. Connect the two sensor subsystems to two scopes as shown in Fig. 7.56.

Run the model and look at the results from the scope as shown in Fig. 7.57. During the simulation, you will see the animation for the mechanism. We encourage the reader to watch the attached animated video from this model as well as look at the attached model.

For the graphical method

Follow the steps of the graphical method for the inverted slider–crank mechanism. You can use any drawing scale that you prefer. In a solid line format and using Eq. (7.23), as shown in Fig. 7.58, plot the four know vectors:

$$\begin{aligned}\overrightarrow{OA} &= \vec{a}_{Nn2n1} = \omega_2^2 r_{n2}\left(-\cos(\theta_2)\hat{i} - \sin(\theta_2)\hat{j}\right)\\ &= 1^2 \times 0.99 \times \left(-\cos(45°)\hat{i} - \sin(45°)\hat{j}\right) = -0.7\hat{i} - 0.7\hat{j}\end{aligned}$$

$$\overrightarrow{AB} = \vec{a}_{Tn2n1} = \underbrace{\alpha_2}_{0}\, r_{n2}\left(\cos(\theta_2 + 90°)\hat{i} + \sin(\theta_2 + 90°)\hat{j}\right) = 0\hat{i} - 0\hat{j}$$

$$\begin{aligned}\overrightarrow{OC} &= \vec{a}_{Nn2n4} = \omega_3^2 r_{n2n4}\left(-\cos(\theta_4)\hat{i} - \sin(\theta_4)\hat{j}\right)\\ &= \omega_3^2 (r_{n3n2} + r_{n3n4})\left(-\cos(\theta_4)\hat{i} - \sin(\theta_4)\hat{j}\right)\\ &= 0.97^2 \times (0.406 + 0.6) \times \left(-\cos(55°)\hat{i} - \sin(55°)\hat{j}\right)\\ &= -0.5429\hat{i} - 0.7754\hat{j}\end{aligned}$$

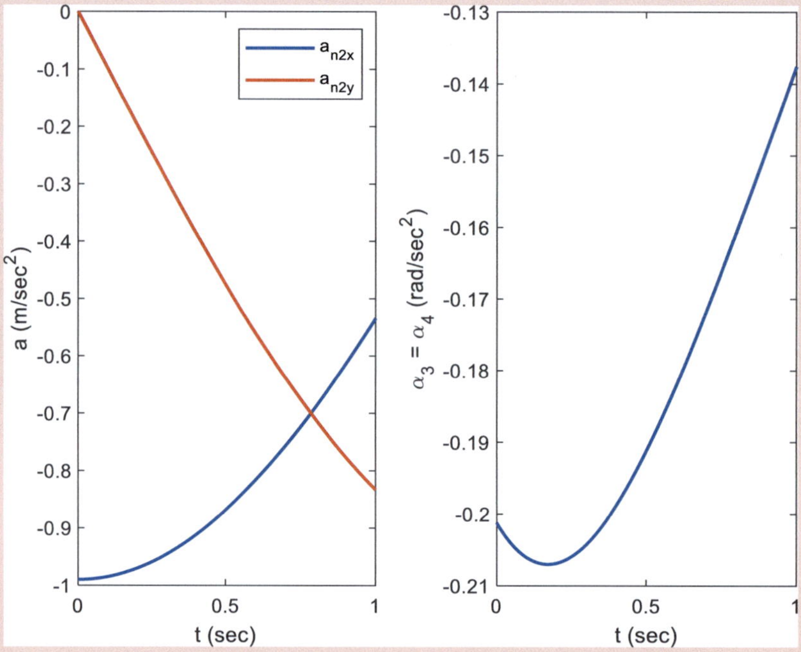

Fig. 7.57 Results of Simscape™ Multibody™ model for hydraulic door opener mechanism

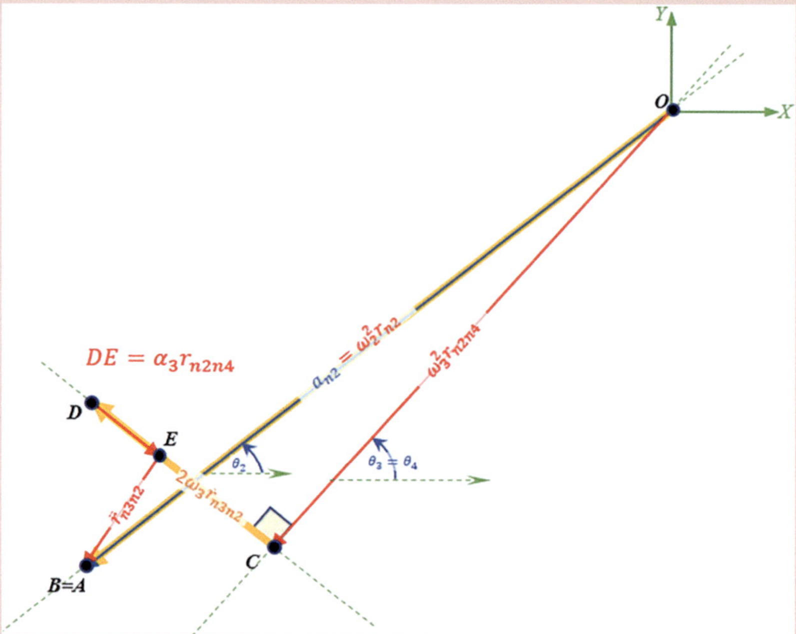

Fig. 7.58 Acceleration vector diagram of hydraulic door opener mechanism

$$\overrightarrow{CD} = \vec{a}_{Cn2n4} = 2\omega_3 \dot{r}_{n3n2}\left(\cos(\theta_4 + 90^\circ)\hat{i} + \sin(\theta_4 + 90^\circ)\hat{j}\right)$$
$$= 2 \times 0.97 \times 0.172 \times \left(\cos(55^\circ + 90^\circ)\hat{i} + \sin(55^\circ + 90^\circ)\hat{j}\right)$$
$$= -0.2733\hat{i} + 0.1914\hat{j}$$

Note, point A and point B are at the same location since $\overrightarrow{AB} = \vec{0}$ because the door is opened at a constant angular velocity or $\alpha_2 = 0$. The total velocity vector of node $n2$ is

$$\overrightarrow{OB} = \overrightarrow{OA} + \overrightarrow{AB} = \vec{a}_{n2} = \vec{a}_{Nn2n1} + \vec{a}_{Tn2n1}$$
$$= -0.7\hat{i} - 0.7\hat{j} + 0\hat{i} - 0\hat{j} = -0.7\hat{i} - 0.7\hat{j}$$

Stand at point B and plot the direction line of the vector $\overrightarrow{EB} = \vec{a}_{Sn2n4}$ as perpendicular on $\overrightarrow{CD} = \vec{a}_{Nn2}$ in a dashed-line format. Stand at point D and plot the direction line of the vector $\overrightarrow{DE} = \vec{a}_{Tn2n4}$ as perpendicular on $\overrightarrow{OC} = \vec{a}_{Nn2n4}$ in a dashed-line format. Define the point E as the intersection of $\overrightarrow{EB}$ and $\overrightarrow{DE}$. Highlight the two lines $\overrightarrow{EB}$ and $\overrightarrow{DE}$ with a solid line format as shown in Fig. 7.58.

Measure the length of EB and DE to be – 0.0283 and – 0.1635 m/s^2, respectively. Note, the negative values are coming in the opposite directions as shown in Fig. 7.58. Using the definition of EB $= a_{Sn2n4} = \ddot{r}_{n3n2}$ and $DE = a_{Tn2n4} = \alpha_4 r_{n2n4}$ to calculate

$$\ddot{r}_{n3n2} = -0.0283\,\text{m/s}^2$$
$$\alpha_4 = \frac{\text{DE}}{r_{n2n4}} = \frac{-0.1635}{1.013} = -0.1614\,\text{rad/s}^2$$

We are now going to compare the model-based results with the graphical analysis results as listed in Table 7.3 that shows remarkably close values.

Table 7.3 Comparing model-based versus graphical analysis for hydraulic door opener mechanism

Quantity	Model-based	Graphical analysis
α_3 (rad/s^2)	– 0.1613	– 0.1614
a_{n2x} (m/s^2)	– 0.7	– 0.7
a_{n2y} (m/s^2)	– 0.7	– 0.7

7.5.4 General Mechanism Graphical Acceleration Analysis

The same concept can be applied to any mechanism. The general algorithm for the Graphical Acceleration Analysis can be summarized as:

Step 1: Use the graphical algorithms from Chaps. 5 and 6 to plot the entire mechanism according to the given parameters and the given input signals. Measure the output signals as positions and as velocities.

Step 2: Use the kinematic constraints at each node to plot the expected acceleration directions on the acceleration vector diagram.

Step 3: Use the known velocity signals (translational velocity and/or angular velocity) to plot the normal acceleration components and the Coriolis acceleration components on the acceleration vector diagram.
Step 4: Plot the expected acceleration components' directions on the acceleration vector diagram.
Step 5: Find the intersection points and measure the unknown acceleration components and the unknown absolute acceleration vectors.

The graphical acceleration analysis is a unique tool to understand the mechanism, but it is time-consuming. Also, it depends on the time. For example, the door mechanism in Fig. 7.21. If we simulate opening the door from $\theta_2(t = 0) = 0°$ to $\theta_2(t = 2) = 85°$ with a time step $\Delta t = 0.01$, we will have to come up with 200 sketches which would be very time-consuming, especially if we are changing the design parameters for optimization. On the other hand, we can try to find an analytical solution, or we can use model-based techniques. In the next section, we are going to introduce the analytical solution of simplified mechanisms.

7.6 Analytical Acceleration Analysis

The analytical acceleration solution, similar to the analytical position solution as well as the analytical velocity solution, provides a closed-form expression to substitute in. This substitution process is easier than developing a numerical, graphical, or even model-based solution that may encounter numerical error propagation or inaccuracy in measuring the graphical magnitudes and the graphical angles. Also, the analytical acceleration solution provides a particularly good tool for tracing the acceleration components or the system response as a function of the design parameters. Developing the analytical solution is a challenge. For example, as you will see at the end of the upcoming subsections, the analytical expressions are in exceptionally long equations that need experts to develop and/or to explain. Also, understanding the physical meaning of each component in the solution is not an easy task. In this chapter, we are covering the four-bar mechanism, slider–crank mechanism, and inverted slider–crank mechanism to develop analytical acceleration solutions. These analytical solutions will provide us with more understanding of the nature of motion, plus we can use these analytical models to justify the model-based solution and the graphical acceleration model.

7.6.1 Four-Bar Mechanism Analytical Acceleration Analysis

Figure 7.59 shows the kinematic diagram of the four-bar mechanism plus the system diagram of the position, velocity, and acceleration analytical solutions. From our position analytical solution in Chap. 5 and the velocity analytical solution in Chap. 6, the four-bar mechanism has six signals or variables $\theta_2(t), \theta_3(t), \theta_4(t), \omega_2(t), \omega_3(t)$ and $\omega_4(t)$ plus the parameters or the constant links' lengths r_{n2}, r_{n3n2}, $r_{n3n4,}$ and r_{n4}. The block diagram of the position analytical solution has the input signal $\theta_2(t)$ and output signals $\theta_3(t)$ and $\theta_4(t)$ in addition to any node position vector. The velocity block diagram analytical solution has the input signal $\omega_2(t)$ and the output signals $\omega_3(t)$ and $\omega_4(t)$ for the velocity analytical solution. Now, we are going to add the acceleration block diagram analytical solution that has $\alpha_2(t)$ as an input signal and $\omega_3(t)$ and $\omega_4(t)$ as output signals.

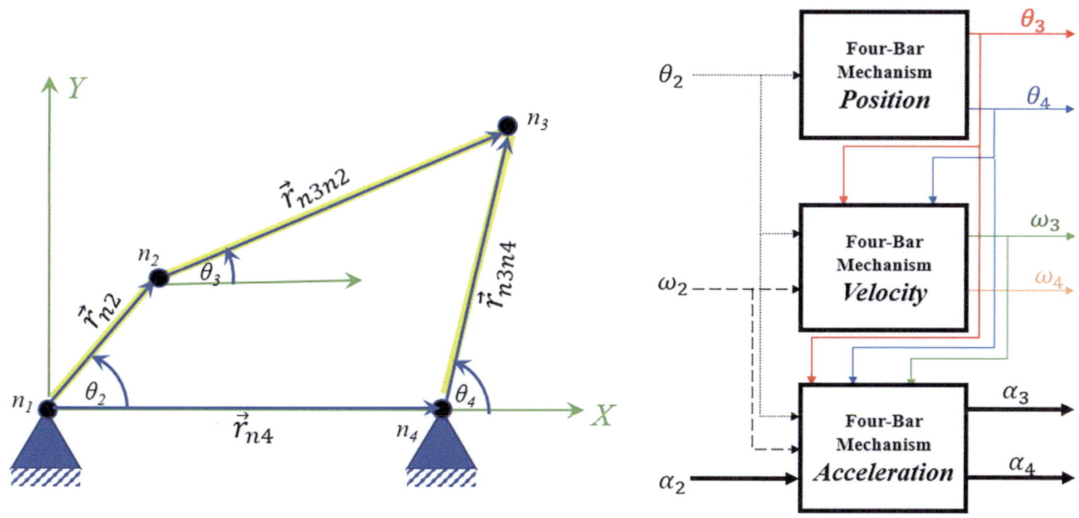

Fig. 7.59 Kinematic diagram and system diagram of four-bar mechanism

Recall the vector loop differentiation equations of the four-bar mechanism from Chap. 6:

$$\begin{aligned} -r_{n2}\omega_2 \sin(\theta_2) - r_{n3n2}\omega_3 \sin(\theta_3) + r_{n3n4}\omega_4 \sin(\theta_4) &= 0 \\ r_{n2}\omega_2 \cos(\theta_2) + r_{n3n2}\omega_3 \cos(\theta_3) - r_{n3n4}\omega_4 \cos(\theta_4) &= 0 \end{aligned} \tag{7.35}$$

Differentiate Equation (7.35):

$$\begin{aligned} &- r_{n2}\alpha_2 \sin(\theta_2) - r_{n2}\omega_2^2 \cos(\theta_2) - r_{n3n2}\alpha_3 \sin(\theta_3) \\ &\quad - r_{n3n2}\omega_3^2 \cos(\theta_3) + r_{n3n4}\alpha_4 \sin(\theta_4) + r_{n3n4}\omega_4^2 \cos(\theta_4) = 0 \\ &r_{n2}\alpha_2 \cos(\theta_2) - r_{n2}\omega_2^2 \sin(\theta_2) + r_{n3n2}\alpha_3 \cos(\theta_3) \\ &\quad - r_{n3n2}\omega_3^2 \sin(\theta_3) - r_{n3n4}\alpha_4 \cos(\theta_4) + r_{n3n4}\omega_4^2 \sin(\theta_4) = 0 \end{aligned} \tag{7.36}$$

Solving the Eq. (7.36) to find α_3 and α_4 as:

$$\begin{aligned} \alpha_3 &= \frac{B \times F - D \times E}{B \times C - A \times D} \\ \alpha_4 &= \frac{C \times E - A \times F}{B \times C - A \times D} \end{aligned} \tag{7.37}$$

Or

$$\begin{aligned} A &= -r_{n3n2}\sin(\theta_3) \\ B &= r_{n3n4}\sin(\theta_4) \\ C &= r_{n3n2}\cos(\theta_3) \\ D &= -r_{n3n4}\cos(\theta_4) \end{aligned}$$

$$\begin{aligned} E &= -r_{n2}\alpha_2\sin(\theta_2) + r_{n2}\omega_2^2 \cos(\theta_2) \\ &\quad + r_{n3n2}\omega_3^2 \cos(\theta_3) - r_{n3n4}\omega_4^2 \cos(\theta_4) \\ F &= -r_{n2}\alpha_2\cos(\theta_2) - r_{n2}\omega_2^2 \sin(\theta_2) \\ &\quad - r_{n3n2}\omega_3^2 \sin(\theta_3) - r_{n3n4}\omega_4^2 \sin(\theta_4) \end{aligned}$$

$$\alpha_3 = -\frac{r_{n2}\omega_2^2\cos(\theta_2-\theta_4)+r_{n3n2}\omega_3^2\cos(\theta_3-\theta_4)-r_{n3n4}\omega_4^2}{r_{n3n2}\sin(\theta_3-\theta_4)}$$
$$\alpha_4 = -\frac{r_{n2}\omega_2^2\cos(\theta_2-\theta_3)-r_{n3n4}\omega_4^2\cos(\theta_3-\theta_4)+r_{n3n2}\omega_3^2}{r_{n3n4}\sin(\theta_3-\theta_4)}$$

After calculating the unknown signals α_3 and α_4, the acceleration at any node can be calculated using the relative acceleration Eq. (7.20) or Eq. (7.21). For example, using Eq. (7.20), any node n_i belongs to link L_2 has an acceleration vector:

$$\begin{aligned} a_{rni} &= -\omega_2^2 r_{ni} \\ a_{\theta ni} &= \alpha_2 r_{ni} \\ \vec{a}_{ni} &= a_{\theta ni}\hat{\theta} + a_{rni}\hat{r} = \alpha_2 r_{ni}\hat{\theta} - \omega_2^2 r_{ni}\hat{r} \\ &\text{where the node } n_i \text{ belongs to } L_2 \end{aligned} \tag{7.38}$$

We can then use the transformation from polar to Cartesian coordinate using Eq. (7.7) as:

$$\begin{aligned} a_{xi} &= -\omega_2^2 r_{ni}\cos(\theta_i) - \alpha_2 r_{ni}\sin(\theta_i) \\ a_{yi} &= -\omega_2^2 r_{ni}\sin(\theta_i) + \alpha_2 r_{ni}\cos(\theta_i) \\ \vec{a}_{ni} &= a_{xi}\hat{i} + a_{yi}\hat{j} \quad \text{where the node } n_i \text{ belongs to } L_2 \end{aligned} \tag{7.39}$$

Similarly, we can use Eqs. (7.20) and (7.7) to calculate the acceleration vector of any node n_j belonging to link L_4 in both polar coordinates and Cartesian coordinates as:

$$\begin{aligned} a_{rj} &= -\omega_4^2 r_{njn4} \\ a_{\theta j} &= \alpha_4 r_{njn4} \\ \vec{a}_{nj} &= \alpha_4 r_{njn4}\hat{\theta} - \omega_4^2 r_{njn4}\hat{r} \quad \text{where the node } n_j \text{ belongs to } L_4 \end{aligned} \tag{7.40}$$

$$\begin{aligned} a_{xj} &= -\omega_4^2 r_{njn4}\cos(\theta_{njn4}) - \alpha_4 r_{njn4}\sin(\theta_{njn4}) \\ a_{yj} &= -\omega_4^2 r_{nin4}\sin(\theta_{njn4}) + \alpha_4 r_{njn4}\cos(\theta_{njn4}) \\ \vec{a}_{nj} &= a_{xj}\hat{i} + a_{yj}\hat{j} \quad \text{where the node } n_j \text{ belongs to } L_4 \end{aligned} \tag{7.41}$$

Note in the case of L_2, we have n_1 as the IC_2 that is stationary. Also, we have the stationary location n_4 as IC_4 of L_4. The angle θ_{njn4} is measured from the non-rotating stationary local frame X_4Y_4 as shown in Fig. 7.60. For L_3, we do not have a stationary IC_3. For this reason, we are not going to calculate the acceleration polar coordinate components. We are going to stick with Cartesian coordinate components only. Using Eq. (7.21), we can calculate the acceleration vector at any node n_k belongs to link L_3 as:

$$\vec{a}_{nk} = \vec{a}_{n2} + \vec{a}_{nkn2} \quad \text{where the node } n_k \text{ belongs to } L_3 \tag{7.42}$$

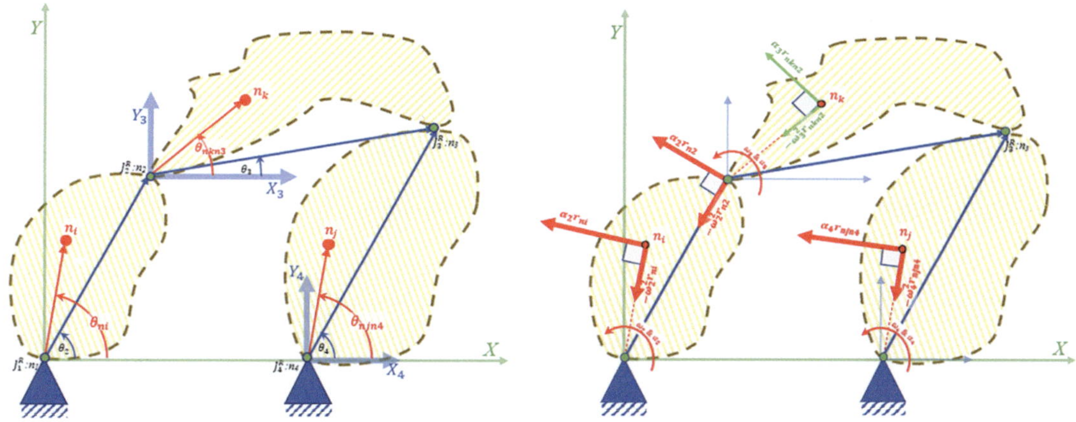

Fig. 7.60 Acceleration components of four-bar mechanism with extra points of interest

where

$$\vec{a}_{n2} = \left(-\omega_2^2 r_{ni}\cos(\theta_2) - \alpha_2 r_{ni}\sin(\theta_2)\right)\hat{i} + \left(-\omega_2^2 r_{ni}\sin(\theta_2) + \alpha_2 r_{ni}\cos(\theta_2)\right)\hat{j}$$

$$\vec{a}_{nkn2} = -\omega_3^2 r_{nkn2}\cos(\theta_3)\hat{r}_2 + \alpha_3 r_{nkn2}\cos(\theta_3)\hat{\theta}_2$$

$$\vec{a}_{nkn2} = \left(-\omega_3^2 r_{nkn2}\cos(\theta_3) - \alpha_3 r_{nkn2}\sin(\theta_3)\right)\hat{i} + \left(-\omega_3^2 r_{nkn2}\sin(\theta_3) + \alpha_3 r_{nkn2}\cos(\theta_3)\right)\hat{j}$$

where $\hat{r}_2$ and $\hat{\theta}_2$ are non-rotating local polar frames located at node n_2 as shown in Figs. 7.59 and 7.60.

Example 7.10

For the door opener in Fig. 7.61, a person opens the door from angle 0°–85° in two seconds with a constant angular velocity $\omega_2 = 0.742$ rad/s. At instantaneous time, if $\theta_2 = 0°$, the displacement analysis provides $\theta_3 = 49.57°$ and $\theta_4 = 56.86°$ and the velocity analysis provides $\omega_3 = \omega_4 = -3.71$ rad/s.

- Use the analytical method to find the angular acceleration of the coupler and the follower when $\theta_2 = 0°$.
- Use the analytical method to find the acceleration vector in the Cartesian coordinate at n_2, n_3, and n_5.
- Use the analytical method to evaluate the model-based method and the graphical method.

Solution

Using Eq. (7.37)

$$A = -r_{n3n2}\sin(\theta_3) = -0.33 \times \sin(49.57°) = -0.2512$$

$$B = r_{n3n4}\sin(\theta_4) = 0.3 \times \sin(56.86°) = 0.2512$$

$$C = r_{n3n2}\cos(\theta_3) = 0.33 \times \cos(49.57°) = 0.214$$

$$D = -r_{n3n4}\cos(\theta_4) = -0.3 \times \cos(56.86°) = -0.164$$

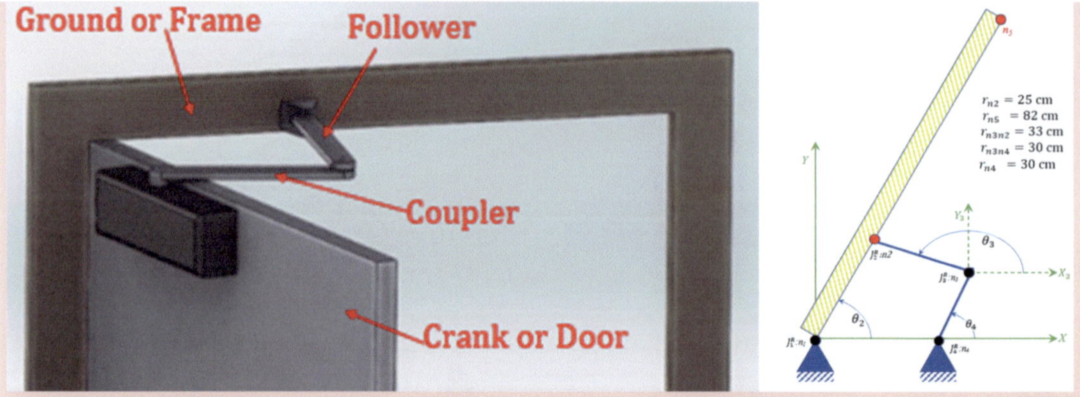

Fig. 7.61 Door mechanism. Figure and associated Door_4Bar_Mechanism.mp4 video designed by the author and implemented by both the author and Pranav Dhola. Both courtesy of Pranav Dhola

$$\begin{aligned}E &= -r_{n2}\alpha_2\sin(\theta_2) + r_{n2}\omega_2^2\cos(\theta_2) + r_{n3n2}\omega_3^2\cos(\theta_3)\\&\quad - r_{n3n4}\omega_4^2\cos(\theta_4) = -0.25\times 0\times\sin(\theta_2)\\&\quad + 0.25\times(0.742)^2\times\cos(0°) + 0.33\times(-3.71)^2\\&\quad \times\cos(49.57°) - 0.3\times(-3.71)^2\times\cos(56.86°) = 0.825\end{aligned}$$

$$\begin{aligned}F &= -r_{n2}\alpha_2\cos(\theta_2) - r_{n2}\omega_2^2\sin(\theta_2) - r_{n3n2}\omega_3^2\sin(\theta_3)\\&\quad - r_{n3n4}\omega_4^2\sin(\theta_4) = -0.25\times 0\times\cos(0°) - 0.25\\&\quad \times(0.742)^2\times\sin(0°) - 0.33\times(-3.71)^2\times\sin(49.57°)\\&\quad - 0.3\times(-3.71)^2\times\sin(56.86°) = -0.1854\end{aligned}$$

$$\alpha_3 = \frac{B\times F - D\times E}{B\times C - A\times D} = \frac{0.2512\times -0.1854 - 0.164\times 0.825}{0.2512\times 0.214 - 0.2512\times -0.164} = 10.776\,\text{rad/s}$$

$$\alpha_4 = \frac{C\times E - A\times F}{B\times C - A\times D} = \frac{0.214\times 0.825 - 0.2512\times F}{0.2512\times 0.214 - 0.2512\times -0.164} = 14.062\,\text{rad/s}$$

Using Eq. (7.39) at n_2

$$\begin{aligned}a_{x2} &= -\omega_2^2 r_{n2}\cos(\theta_2) - \alpha_2 r_{n2}\sin(\theta_2) = -0.742^2\times 0.25\times\cos(0°)\\&\quad - 0\times 0.25\times\sin(0°) = -0.138\,\text{m/s}^2\\a_{y2} &= -\omega_2^2 r_{n2}\sin(\theta_2) + \alpha_2 r_{n2}\cos(\theta_2) = -0.742^2\times 0.25\times\sin(0°)\\&\quad + 0\times 0.25\times\cos(0°) = 0\,\text{m/s}^2\\\vec{a}_{n2} &= a_{x2}\hat{i} + a_{y2}\hat{j} = -0.138\hat{i}\end{aligned}$$

Using Eq. (7.41) at n_3, with $\theta_{n3n4} = \theta_4$:

$$\begin{aligned}a_{x3} &= -\omega_4^2 r_{n3n4}\cos(\theta_4) - \alpha_4 r_{n3n4}\sin(\theta_4)\\&= -(-3.71)^2\times 0.3\times\cos(56.86°) - 14.062\times 0.3\times\sin(56.86°)\\&= -5.790\,\text{m/s}^2\\a_{y3} &= -\omega_4^2 r_{n2}\sin(\theta_4) + \alpha_4 r_{n3n4}\cos(\theta_4)\\&= -(-3.71)^2\times 0.3\times\sin(56.86°) + 14.062\times 0.3\times\cos(56.86°)\\&= -1.151\,\text{m/s}^2\\\vec{a}_{n2} &= a_{x2}\hat{i} + a_{y2}\hat{j} = -5.79\hat{i} - 1.151\hat{j}\end{aligned}$$

Using Eq. (7.39) at n_5

$$a_{x5} = -\omega_2^2 r_{n5}\cos(\theta_2) - \alpha_2 r_{n5}\sin(\theta_2) = -0.742^2 \times 0.82 \times \cos(0°)$$
$$- 0 \times 0.82 \times \sin(0°) = -0452\,\text{m/s}^2$$
$$a_{y5} = -\omega_2^2 r_{n5}\sin(\theta_2) + \alpha_2 r_{n5}\cos(\theta_2) = -0.742^2 \times 0.82 \times \sin(0°)$$
$$+ 0 \times 0.82 \times \cos(0°) = 0\,\text{m/s}^2$$
$$\vec{a}_{n5} = a_{x5}\hat{i} + a_{y5}\hat{j} = -0.452\hat{i}$$

The analytical values match the model-based method in Example 7.6 and the graphical method in Example 7.7.

7.6.2 Slider–Crank Mechanism Analytical Velocity Analysis

Figure 7.62 shows the kinematic diagram of the slider–crank mechanism plus the system diagram of the position, velocity, and acceleration analytical solution. From our position analytical solution and velocity analytical solution in Chaps. 5 and 6, the slider–crank mechanism has six variable angles θ_2, θ_3, ω_2, ω_3, r_{n4}, and $\dot{r}_{n4}$ plus the constant links' lengths r_{n2}, r_{n3n2}, and r_{n3n4} while $\theta_4 = 90°$ and $\omega_3 = 0$ rad/s. The system diagram of the position analytical solution has the input signal θ_2 and output signals θ_3 and r_{n4} in addition to any node position vector. The velocity system diagram analytical solution has the input signal ω_2 and the output signals ω_3 and $\dot{r}_{n4}$ for the velocity analytical solution. Now, we are going to add the acceleration analytical solution with an input signal α_2 and the output signals α_3 and $\ddot{r}_{n4}$ and any additional node acceleration. That leads to increasing the total number of variables to nine: θ_2, θ_3, ω_2, ω_3,α_2, α_3 r_{n4}, $\dot{r}_{n4}$, and $\ddot{r}_{n4}$.

Recall the vector loop equations of the slider–crank mechanism from Chap. 5:

$$\begin{aligned} r_{n2}\cos(\theta_2) - r_{n3n2}\cos(\theta_3) &= r_{n4} \\ r_{n2}\sin(\theta_2) - r_{n3n2}\sin(\theta_3) &= r_{n3n4} \end{aligned} \tag{7.43}$$

Differentiate Equation (7.43) with respect to the time:

$$\begin{aligned} -r_{n2}\omega_2\sin(\theta_2) + r_{n3n2}\omega_3\sin(\theta_3) &= \dot{r}_{n4} \\ r_{n2}\omega_2\cos(\theta_2) - r_{n3n2}\omega_3\cos(\theta_3) &= 0 \end{aligned} \tag{7.44}$$

Differentiate Equation (7.44) with respect to the time:

$$\begin{aligned} &- r_{n2}\alpha_2\sin(\theta_2) - r_{n2}\omega_2^2\cos(\theta_2) + r_{n3n2}\alpha_3\sin(\theta_3) \\ &\quad + r_{n3n2}\omega_3^2\cos(\theta_3) = \ddot{r}_{n4} \\ &r_{n2}\alpha_2\cos(\theta_2) - r_{n2}\omega_2^2\sin(\theta_2) - r_{n3n2}\alpha_3\cos(\theta_3) \\ &\quad + r_{n3n2}\omega_3^2\sin(\theta_3) = 0 \end{aligned} \tag{7.45}$$

Equation (7.45) can be used to find α_3 and α_2 in terms of $\ddot{r}_{n4}$ as:

$$\begin{aligned} \alpha_2 &= \frac{\left(r_{n2}\omega_2^2\cos(\theta_3 - \theta_2) - r_{n3n2}\omega_3^2 + \ddot{r}_{n4}\cos(\theta_3)\right)}{r_{n2}\sin(\theta_3 - \theta_2)} \\ \alpha_3 &= \frac{\left(r_{n2}\omega_2^2 - r_{n3n2}\omega_3^2\cos(\theta_3 - \theta_2) + \ddot{r}_{n4}\cos(\theta_2)\right)}{r_{n2}\sin(\theta_3 - \theta_2)} \end{aligned} \tag{7.46}$$

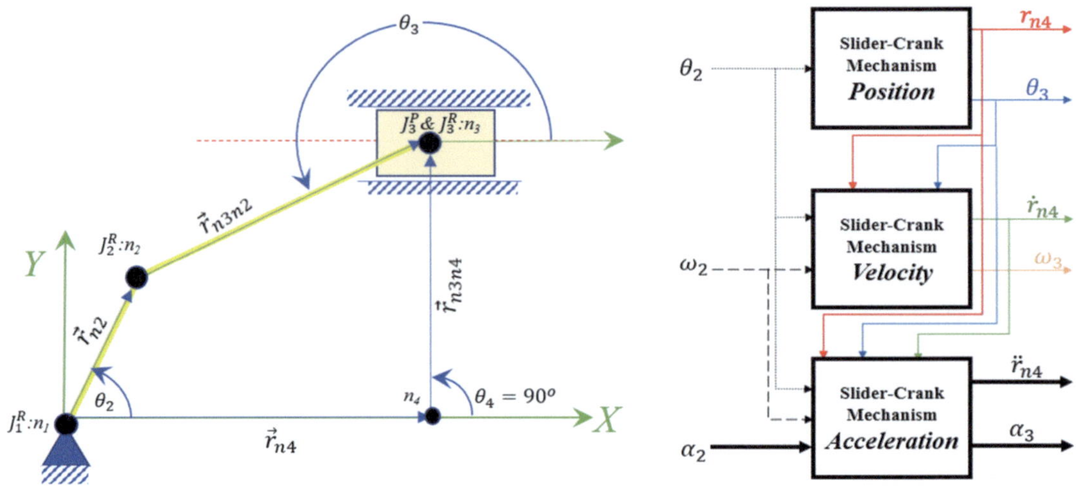

Fig. 7.62 Kinematic diagram and position–velocity–acceleration system diagram of slider–crank mechanism

or to find α_3 and $\ddot{r}_{n4}$ in terms of α_2 as:

$$\alpha_3 = \frac{r_{n2}\alpha_2 \cos(\theta_2) - r_{n2}\omega_2^2 \sin(\theta_2) + r_{n3n2}\omega_3^2 \sin(\theta_3)}{r_{n3n2}\cos(\theta_3)}$$
$$\ddot{r}_{n4} = -r_{n2}\alpha_2 \sin(\theta_2) - r_{n2}\omega_2^2 \cos(\theta_2) + r_{n3n2}\alpha_3 \sin(\theta_3) + r_{n3n2}\omega_3^2 \cos(\theta_3) \quad (7.47)$$

Note, we may use the fact that $\ddot{r}_{n4} = a_{x4} = a_{x3}$ where a_{x3} is the X-component of the acceleration of $a_{\mathrm{n}3}$ and a_{x4} is the X-component of the acceleration of $a_{\mathrm{n}4}$. The solution in Eq. (7.46) and the one in Eq. (7.47) represent a pair of solutions because of the circuit 1 solution and circuit 2 solutions that are discussed in Chap. 5. For example, in the case of the solution in Eq. (7.47), recall the position solution of the slider–crank mechanism from Chap. 5 which is represented by:

$$\sin(\theta_3) = \frac{r_{n3n4} + r_{n2}\sin(\theta_2)}{r_{n3n2}} \quad (7.48)$$

The solution in Eq. (7.48) has two values or circuit 1 solution θ_{3c1} and circuit 2 solution θ_{3c2}. That leads to two angular velocity solutions (ω_{3c1} and ω_{3c2}), two angular acceleration solutions (α_{3c1} and α_{3c2}), two transitional positions (r_{n4c1} and r_{n4c2}), two trinational velocities ($\dot{r}_{n4c1}$ and $\dot{r}_{n4c2}$), and two translational accelerations ($\ddot{r}_{n4c1}$ and $\ddot{r}_{n4c2}$) as shown in Fig. 7.63.

Similar to the four-bar mechanism, we can use the relative acceleration Eq. (7.20) or Eq. (7.21) to calculate the acceleration components at any node on the slider–crank mechanism as shown in Fig. 7.64. For example, using Eq. (7.20), any node n_i that belongs to link L_2 has an acceleration vector

$$a_{rni} = -\omega_2^2 r_{ni}$$
$$a_{\theta ni} = \alpha_2 r_{ni}$$
$$\vec{a}_{ni} = a_{\theta ni}\hat{\theta} + a_{rni}\hat{r} = \alpha_2 r_{ni}\hat{\theta} - \omega_2^2 r_{ni}\hat{r}$$
$$\text{where the node } n_i \text{ belongs to } L_2 \quad (7.49)$$

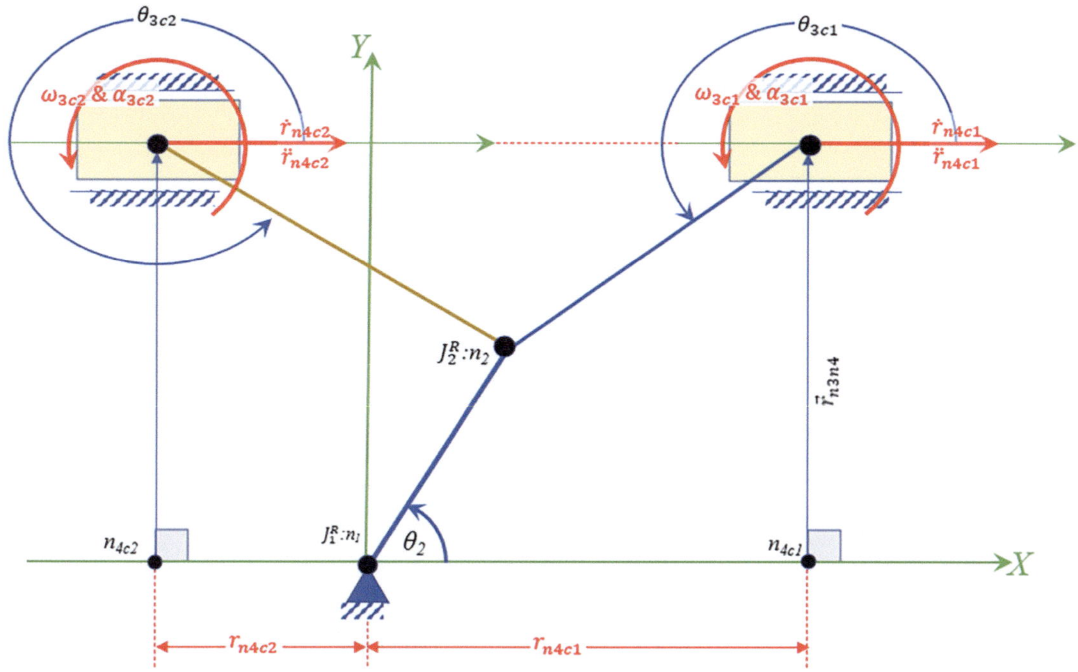

Fig. 7.63 Kinematic diagram of circuit 1 and circuit 2 solutions for four-bar slider–crank mechanism

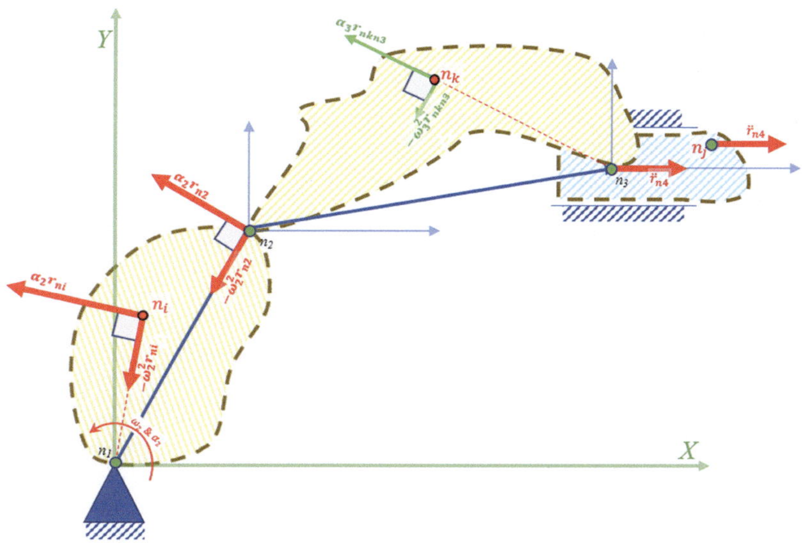

Fig. 7.64 Acceleration components of Slider–Crank mechanism with extra points of interest

We can then use the transformation from polar to Cartesian coordinate using Eq. (7.7) as

$$\begin{aligned} a_{xi} &= -\omega_2^2 r_{ni}\cos(\theta_i) - \alpha_2 r_{ni}\sin(\theta_i) \\ a_{yi} &= -\omega_2^2 r_{ni}\sin(\theta_i) + \alpha_2 r_{ni}\cos(\theta_i) \\ \vec{a}_{ni} &= a_{xi}\hat{i} + a_{yi}\hat{j} \quad \text{where the node } n_i \text{ belongs to } L_2 \end{aligned} \tag{7.50}$$

From the kinematic constraint of the prismatic joint, any node n_j that belongs to the link L_4 has the same acceleration as n_3 which is

$$\vec{a}_{nj} = \ddot{r}_{n4}\hat{i} \quad \text{where } n_j \text{ belongs to } L_4 \tag{7.51}$$

For L_3, using Eq. (7.21), we can calculate the acceleration vector at any node n_k that belongs to link L_3 as

$$\vec{a}_{nk} = \vec{a}_{n3} + \vec{a}_{nkn3} \quad \text{where } n_k \text{ belongs to } L_3 \tag{7.52}$$

where

$$\begin{aligned} \vec{a}_{n3} &= \ddot{r}_{n4}\hat{i} \\ \vec{a}_{nkn3} &= -\omega_3^2 r_{nkn3}\cos(\theta_3)\hat{r}_2 + \alpha_3 r_{nkn3}\cos(\theta_3)\hat{\theta}_2 \\ \vec{a}_{nkn3} &= \left(-\omega_3^2 r_{nkn3}\cos(\theta_3) - \alpha_3 r_{nkn3}\sin(\theta_3)\right)\hat{i} \\ &\quad + \left(-\omega_3^2 r_{nkn3}\sin(\theta_3) + \alpha_3 r_{nkn3}\cos(\theta_3)\right)\hat{j} \end{aligned}$$

Example 7.11

For the shown kinematic diagram of the piston–cylinder mechanism in Fig. 7.26 where the length of the crank is 76.2 mm and the length of connecting rod is 203.2 mm, assume that the combustion force moves the piston outward with a sliding motion

$$x_p(t) = 0.2032 + 0.0494\cos(263.158t) \quad 0 < t < 0.012\,\text{s}$$

where $x_p(t)$ is the piston sliding position in m. This sliding motion causes the connecting rod to rotate and consequently, the crankshaft rotates. Instantaneously at $t=0.006$ s, the measured angular velocity of the connecting rod sec is – 11.751 rad/s, and the angular velocity of the crank is 161.685 rad/s, where $\theta_2 = 79.62°$ and $\theta_3 = 158.53°$. At this instantaneous time $t=0.006$ s, calculate

1. The crank and the connecting rod angular accelerations using the analytical acceleration analysis.
2. Compare your results with the simulation of the piston–cylinder mechanism model-based.

Solution

The given parameters are: $r_{n2} = 0.0762$m and $r_{n3n2} = 0.2032$m. The given piston translational motion or r_{n4} translational motion is:

$$\begin{aligned} r_{n4}(t) &= x_p(t) = 0.2032 + 0.0494\cos(263.158t) \quad 0 < t < 0.012\,\text{s} \\ \dot{r}_{n4}(t) &= \frac{\mathrm{d}r_{n4}(t)}{\mathrm{d}t} = -13\sin(263.158t) \quad 0 < t < 0.012\,\text{s} \\ \ddot{r}_{n4}(t) &= \frac{\mathrm{d}\dot{r}_{n4}(t)}{\mathrm{d}t} = -3421.105\cos(263.158t) \quad 0 < t < 0.012\,\text{s} \end{aligned}$$

At $t = 0.006$ s:

$$r_{n4}(t = 0.006) = 203.2 + 0.0494 \cos(263.158 \times 0.006) = 0.202279\,\text{m}$$
$$\dot{r}_{n4}(t = 0.006) = -13 \sin(263.158 \times 0.006) = -12.999568\,\text{m/s}$$
$$\ddot{r}_{n4}(t = 0.006) = -3421.105 \cos(263.158 \times 0.006) = 27.887017\,\text{m/s}^2$$

Also, at $t = 0.006\,\text{s}$, we have: $\theta_2(t = 0.006) = 79.62°$, $\theta_3(t = 0.006) = 158.53°$, $\omega_2(t = 0.006) = 161.685\,\frac{\text{rad}}{\text{s}}$, $\omega_3(t = 0.006) = -11.751\,\frac{\text{rad}}{\text{s}}$, and $\omega_3(t = 0.006) = -11.751\,\frac{\text{rad}}{\text{s}}$.

We can calculate the required variables $\alpha_2(t = 0.006)$ and$\alpha_3(t = 0.006)$ using Eq. (7.46)

$$\alpha_2(t = 0.006) = \frac{\left(r_{n2}\omega_2^2\cos(\theta_3 - \theta_2) - r_{n3n2}\omega_3^2 + \ddot{r}_{n4}\cos(\theta_3)\right)}{r_{n2}\sin(\theta_3 - \theta_2)}$$
$$= \frac{\begin{pmatrix} 0.0762 \times 161.685^2 \times \cos(158.53 - 79.62) - 0.2032 \\ \times - 11.751^2 + 27.887017 \times \cos(158.53) \end{pmatrix}}{0.0762 \times \sin(158.53 - 79.62)}$$
$$= 4405.603\,\text{rad/s}^2$$
$$\alpha_3 = \frac{\left(r_{n2}\omega_2^2 - r_{n3n2}\omega_3^2\cos(\theta_3 - \theta_2) + \ddot{r}_{n4}\cos(\theta_2)\right)}{r_{n2}\sin(\theta_3 - \theta_2)}$$
$$= \frac{\begin{pmatrix} 0.0762 \times 161.685^2 - 0.2032 \times -11.751^2 \\ \times\cos(158.53 - 79.62) + 27.887017 \times \cos(79.62) \end{pmatrix}}{0.0762 \times \sin(158.53 - 79.62)}$$
$$= 9988.178\,\text{rad/s}^2$$

The analytical values match the model-based method with a very small percentage of error in. Note that this error may happen due to the hand calculation in the analytical model or the numerical integration in the model-based solution.

7.6.3 Inverted Slider–Crank Mechanism Analytical Velocity Analysis

Figure 7.65 shows the kinematic diagram of one of the inverted slider–crank mechanisms plus the system diagram of the position, velocity, and acceleration analytical solution. From our position analytical solution and velocity analytical solution in Chaps. 5 and 6, the inverted slider–crank mechanism has six variable angles θ_2, θ_3, ω_2, ω_3, r_{n3n2}, and $\dot{r}_{n3n2}$ plus the constant links' lengths r_{n2}, r_{n4}, and r_{n3n4} while $\gamma = \theta_3 - \theta_4$ is a constant. The system diagram of the position analytical solution has the input signal θ_2 and output signals θ_3 and r_{n3n2} in addition to any node position vector. The velocity system diagram analytical solution has the input signal ω_2 and the output signals $\omega_3 = \omega_4$ and $\dot{r}_{n3n2}$ for the velocity analytical solution. Now, we are going to add the acceleration analytical solution with an input signal α_2 and the output signals $\alpha_3 = \alpha_4$ and $\ddot{r}_{n3n2}$ and any additional node acceleration. That leads to increasing the total number of variables to nine: θ_2, θ_3, ω_2, ω_3,α_2, $\alpha_3 r_{n4}$, $\dot{r}_{n3n2}$, and $\ddot{r}_{n3n2}$.

Recall the vector loop equations of the slider–crank mechanism from Chap. 5

$$r_{n2}\cos(\theta_2) - r_{n3n2}\cos(\theta_3) = r_{n3n4}\cos(\theta_4) + r_{n4}$$
$$r_{n2}\sin(\theta_2) - r_{n3n2}\sin(\theta_3) = r_{n3n4}\sin(\theta_4) \tag{7.53}$$

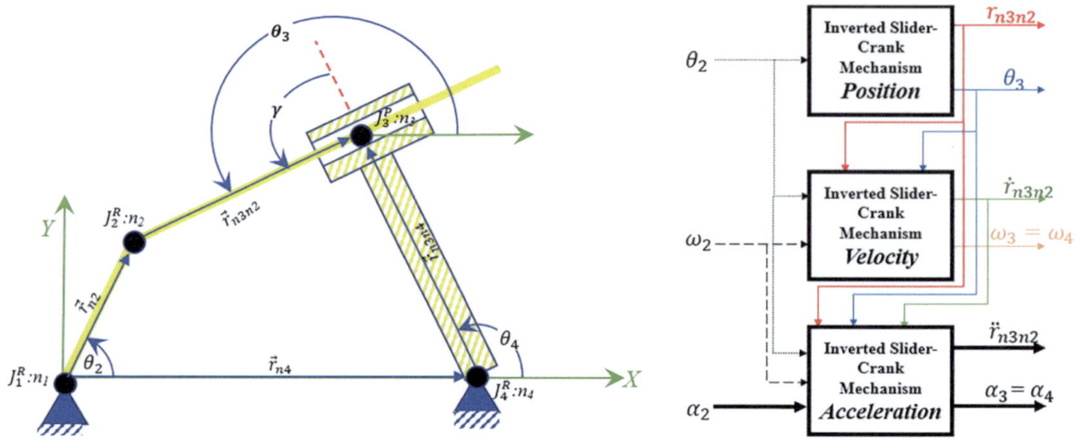

Fig. 7.65 Kinematic diagram and system diagram of inverted slider–crank linkage

Using the fact that $\theta_4 = \theta_3 - \gamma$ and differentiate Equation (7.53) with respect to the time

$$
\begin{aligned}
-r_{n2}\omega_2 \sin(\theta_2) + r_{n3n2}\omega_3 \sin(\theta_3) - \dot{r}_{n3n2}\cos(\theta_3) &= -r_{n3n4}\omega_3 \sin(\theta_3 - \gamma) \\
r_{n2}\omega_2 \cos(\theta_2) - r_{n3n2}\omega_3 \cos(\theta_3) - \dot{r}_{n3n2}\sin(\theta_3) &= r_{n3n4}\omega_3 \cos(\theta_3 - \gamma)
\end{aligned}
\tag{7.54}
$$

Differentiate Equation (7.54) with respect to the time

$$
\begin{aligned}
& r_{n2}\alpha_2 \cos(\theta_2) - r_{n2}\omega_2^2 \sin(\theta_2) - r_{n3n2}\alpha_3 \cos(\theta_3) \\
& + r_{n3n2}\omega_3^2 \sin(\theta_3) - \dot{r}_{n3n2}\omega_3 \cos(\theta_3) - \ddot{r}_{n3n2}\sin(\theta_3) \\
& - \dot{r}_{n3n2}\omega_3^2 \cos(\theta_3) = r_{n3n4}\alpha_3 \cos(\theta_3 - \gamma) - r_{n3n4}\omega_3^2 \sin(\theta_3 - \gamma) \\
& \quad - r_{n2}\alpha_2 \sin(\theta_2) - r_{n2}\omega_2^2 \cos(\theta_2) + r_{n3n2}\alpha_3 \sin(\theta_3) \\
& \quad + r_{n3n2}\omega_3^2 \cos(\theta_3) + \dot{r}_{n3n2}\omega_3 \sin(\theta_3) \\
& \quad - \ddot{r}_{n3n2}\cos(\theta_3) + \dot{r}_{n3n2}\omega_3 \sin(\theta_3) = -r_{n3n4}\alpha_3 \sin(\theta_3 - \gamma) \\
& \quad - r_{n3n4}\omega_3^2 \cos(\theta_3 - \gamma)
\end{aligned}
\tag{7.55}
$$

Equation (7.55) can be used to find α_3 and α_2 in terms of $\ddot{r}_{n4}$ as

$$
\begin{aligned}
G = & -r_{n2}r_{n3n2}\omega_2^2 \cos(\theta_3 - \theta_2) - r_{n2}r_{n3n4}\omega_2^2 \cos(\theta_4 - \theta_2) \\
& - \left(r_{n3n2}^2 + r_{n3n4}^2\right)\omega_3^2 - 2r_{n3n2}r_{n3n4}\omega_3^2 \cos(\gamma) \\
& - 2\dot{r}_{n3n2}r_{n3n4}\omega_3 \sin(\gamma) + \ddot{r}_{n3n2}(r_{n3n2} + r_{n3n4}\cos(\gamma))
\end{aligned}
\tag{7.56}
$$

$$
\begin{aligned}
H = & \,\omega_2^2 r_{n2}^2 + \ddot{r}_{n3n2}r_{n2}\cos(\theta_3 - \theta_2) - \omega_3^2 r_{n2}r_{n3n4}\cos(\theta_3 - \theta_2) \\
& - 2\omega_3\dot{r}_{n3n2}r_{n2}\sin(\theta_3 - \theta_2) - \omega_3^2 r_{n3n2}r_{n2}\cos(\theta_3 - \theta_2)
\end{aligned}
$$

$$
I = r_{n2}r_{n3n4}\sin(\theta_4 - \theta_2) + r_{n2}r_{n3n2}\sin(\theta_3 - \theta_2)
$$

$$
\begin{aligned}
\alpha_2 &= \frac{G}{I} \\
\alpha_3 &= \frac{H}{I}
\end{aligned}
$$

The solution in Eq. (7.56) represents a pair of solutions because of the circuit 1 solution and circuit 2 solution that are discussed in Chap. 5. For example, in the case of the solution in Eq. (7.56), recall the position solution of the inverted slider–crank mechanism from Chap. 5 which is represented by

$$S\tan^2\left(\frac{\theta_4}{2}\right) + T\tan\left(\frac{\theta_4}{2}\right) + U = 0 \tag{7.57}$$

where

$$S = R - Q\quad T = 2P\quad U = Q + R$$
$$P = r_{n2}\sin(\theta_2)\sin(\gamma) + (r_{n2}\cos(\theta_2) - r_{n4})\cos(\gamma)$$
$$Q = -r_{n2}\sin(\theta_2)\cos(\gamma) + (r_{n2}\cos(\theta_2) - r_{n4})\sin(\gamma)$$
$$R = r_{n3n4}\sin(\gamma)$$

The solution in Eq. (7.57) has two values or circuit 1 solution $\theta_{4c1} = \theta_{3c1} - \gamma$ and circuit 2 solution $\theta_{4c2} = \theta_{3c2} - \gamma$. That leads to two angular velocity solutions (ω_{3c1} and ω_{3c2}), two angular acceleration solutions (α_{3c1} and α_{3c2}), two transitional positions (r_{n3n2c1} and r_{n3n2c2}), two translational velocities ($\dot{r}_{n3n2c1}$ and $\dot{r}_{n3n2c2}$), and two translational accelerations$\ddot{r}_{n3n2c1}$ and $\ddot{r}_{n3n2c2}$) as shown in Fig. 7.66.

In addition to Fig. 7.65, there is another configuration called an inverted slider–crank with a fixed crank as shown in Fig. 7.67 that is commonly used in agriculture, construction machines, and other applications. This configuration is a special case from the configuration when $\gamma = 0°$. Thus, when θ_2 increases in counterclockwise, the length r_{n3n2} increases and causes $\theta_3 = \theta_4$ to increase. When $\gamma = 0°$, from Chapt. 5, we have

$$\begin{aligned}(r_{n3n2} + r_{n3n4})^2 &= r_{n2}^2 + r_{n4}^2 - 2r_{n2}r_{n4}\cos(\theta_2 - \delta)\\ r_{n3n2} &= \sqrt{r_{n2}^2 + r_{n4}^2 - 2r_{n2}r_{n4}\cos(\theta_2 - \delta)} - r_{n3n4}\end{aligned} \tag{7.58}$$

where δ is defined as shown in Fig. 7.67. Differentiate Equation (7.58) with respect to the time

$$\dot{r}_{n3n2} = \frac{\omega_2 r_{n2} r_{n4}\sin(\theta_2 - \delta)}{r_{n3n2} + r_{n3n4}} \tag{7.59}$$

Differentiate Equation (7.59) with respect to the time

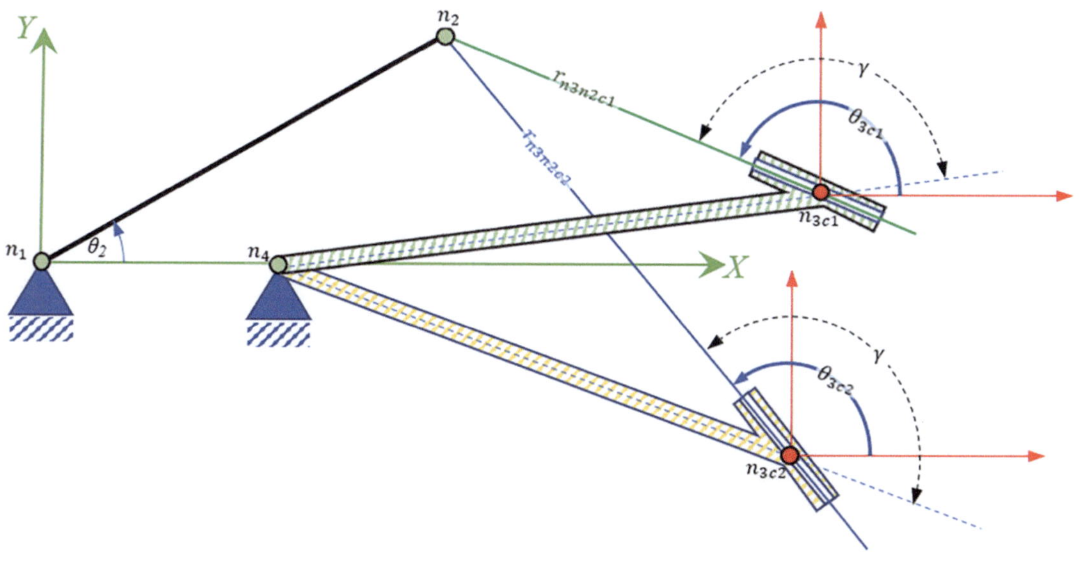

Fig. 7.66 Kinematic diagram of circuit 1 and circuit 2 solutions for inverted slider–crank mechanism

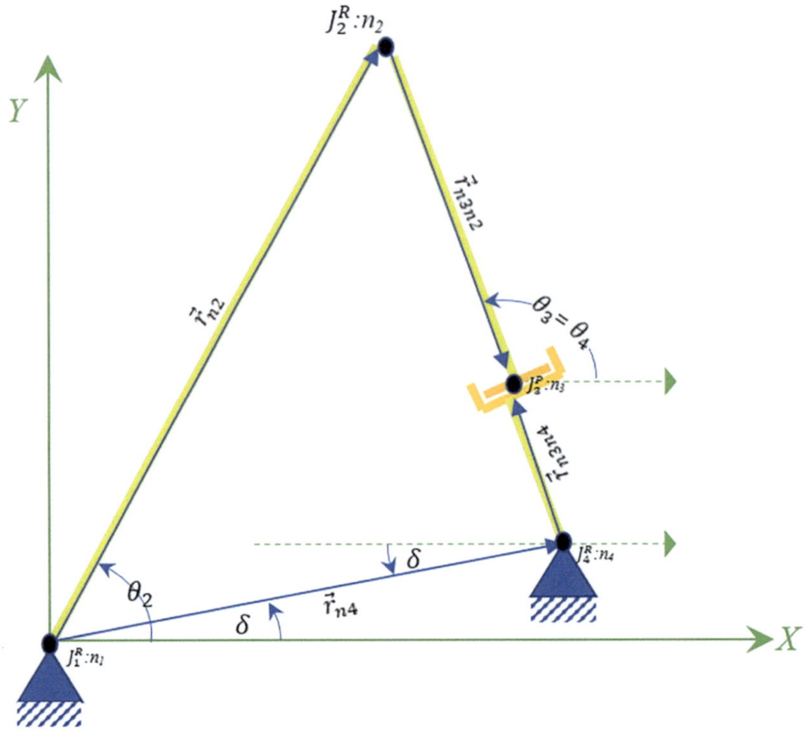

Fig. 7.67 Kinematic diagram inverted Slider–Crank with $\gamma = 0^{\mathrm{o}}$

$$\ddot{r}_{n3n2} = \frac{\alpha_2 r_{n2} r_{n4} \sin(\theta_2 - \delta)}{r_{n3n2} + r_{n3n4}} + \frac{\omega_2^2 r_{n2} r_{n4} \cos(\theta_2 - \delta)}{r_{n3n2} + r_{n3n4}} - \frac{\omega_2 r_{n2} r_{n4} \dot{r}_{n3n2} \sin(\theta_2 - \delta)}{(r_{n3n2} + r_{n3n4})^2} \tag{7.60}$$

or

$$\alpha_2 = \frac{\ddot{r}_{n3n2}(r_{n3n2} + r_{n3n4})}{r_{n2} r_{n4} \sin(\theta_2 - \delta)} - \frac{\omega_2^2 \cos(\theta_2 - \delta)}{\sin(\theta_2 - \delta)} - \omega_2 \dot{r}_{n3n2} \tag{7.61}$$

Using Eq. (7.60), we can calculate the transitional acceleration $\ddot{r}_{n3n2}$ in terms of the angular acceleration α_2 and vice versa using Eq. (7.61). To calculate the angular acceleration α_3, recall the velocity equations of the inverted slider–crank when $\gamma = 0°$ from Chap. 6 as

$$\frac{\omega_3}{\omega_2} = -\frac{r_{n2}\cos(\theta_2 - \theta_3)}{r_{n2n4}} \tag{7.62}$$

Differentiate (7.62) with respect to the time

$$\alpha_3 = -\frac{\alpha_2 r_{n2} \cos(\theta_2 - \theta_3)}{r_{n2n4}} + \frac{\omega_2^2 r_{n2} \sin(\theta_2 - \theta_3)}{r_{n2n4}} - \frac{\omega_2 \omega_3 r_{n2} \sin(\theta_2 - \theta_3)}{r_{n2n4}} + \frac{r_{n2} \dot{r}_{n3n2} \cos(\theta_2 - \theta_3)}{r_{n2n4}^2} \tag{7.63}$$

Example 7.12

For the shown hydraulic gate opener in Fig. 7.17, if the door is open with an angular velocity $\omega_2 = 1$ rad/s from the initial position $\theta_2 = 0°$ to final position $\theta_2 = 50°$, at the position $\theta_2 = 45°$, $\theta_3 = 55°$ and $r_{n3n2} = 405.67$ mm, the cylinder angular velocity is $\omega_4 = \omega_3 = 0.970$ rad/s and the piston transitional velocity is $\dot{r}_{n2n3} = 0.172$ m/s.

- Use the analytical acceleration method to calculate the cylinder angular acceleration and the piston translational acceleration.
- Use the analytical method to evaluate the model-based method and the graphical method.

Solution

Figure 7.68 shows the kinematic diagram with the vector loop. The mechanism has four links: the door frame or the ground L_1, the door L_2, the piston L_3, and the cylinder L_4 that are connected by $J_1^R : n_1, J_2^R : n_2, J_3^P : n_3$, and $J_4^R : n_4$, respectively. The vector loop is shown in Fig. 7.68.

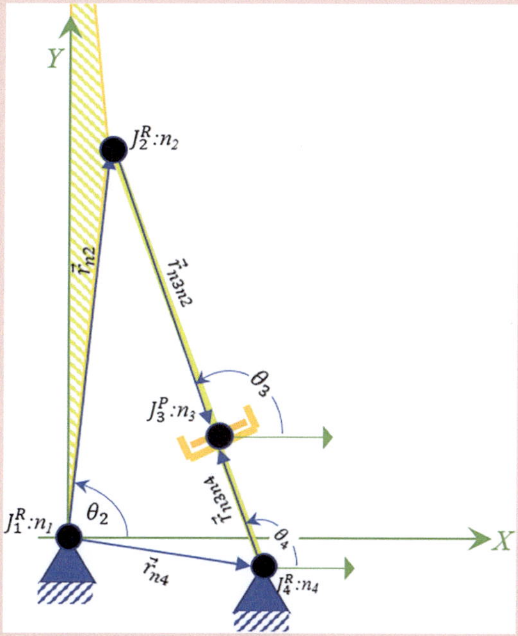

Fig. 7.68 Kinematic diagram with vector loop of hydraulic door opener

First, we need to calculate r_{n4} by

$$r_{n4} = \sqrt{x_{n4}^2 + y_{n4}^2} = \sqrt{125^2 + (-125)^2} = 176.77\,\text{mm}$$

$$\delta = \tan^{-1}(y_{n4}/x_{n4}) = \tan^{-1}(-125/125) = -45°$$

Using Eq. (7.60):

$$\ddot{r}_{n3n2} = \frac{\alpha_2 r_{n2} r_{n4} \sin(\theta_2 - \delta)}{r_{n3n2} + r_{n3n4}} + \frac{\omega_2^2 r_{n2} r_{n4} \cos(\theta_2 - \delta)}{r_{n3n2} + r_{n3n4}} - \frac{\omega_2 r_{n2} r_{n4} \dot{r}_{n3n2} \sin(\theta_2 - \delta)}{(r_{n3n2} + r_{n3n4})^2}$$

$$= \frac{0 \times 0.99 \times 0.177 \times \sin\left(45^\circ - \left(-45^\circ\right)\right)}{0.406 + 0.6} + \frac{1^2 \times 0.99 \times 0.177 \times \cos\left(45^\circ - \left(-45^\circ\right)\right)}{0.406 + 0.6} - \frac{1 \times 0.99 \times 0.177 \times 0.172 \times \sin\left(45^\circ - \left(-45^\circ\right)\right)}{(0.406 + 0.6)^2}$$

$$= 0 + 0 + 0.0298 = 0.0298\ \mathrm{m/s^2}$$

Using Eq. (7.63):

$$\alpha_3 = -\frac{\alpha_2 r_{n2} \cos(\theta_2 - \theta_3)}{r_{n2n4}} + \frac{\omega_2^2 r_{n2} \sin(\theta_2 - \theta_3)}{r_{n2n4}} - \frac{\omega_2 \omega_3 r_{n2} \sin(\theta_2 - \theta_3)}{r_{n2n4}} + \frac{r_{n2} \dot{r}_{n3n2} \cos(\theta_2 - \theta_3)}{r_{n2n4}^2}$$

$$= -\frac{0 \times 0.99 \times \cos\left(45^\circ - 55^\circ\right)}{1.06} + \frac{1^2 \times 0.99 \times \sin\left(45^\circ - 55^\circ\right)}{1.06} - \frac{1 \times 0.970 \times 0.99 \times \sin\left(45^\circ - 55^\circ\right)}{1.06} + \frac{0.99 \times 0.172 \times \sin\left(45^\circ - 55^\circ\right)}{1.06^2} = 0.1444\ \mathrm{rad/s^2}$$

The model-based solution and graphical solution come very close to the analytical solution.

7.7 Jerk

Definition

Jerk:

Jerk is felt as an increasing or decreasing force on a body [1–3].

Students usually study displacement, velocity, and acceleration in dynamics courses, but sometimes they do not study jerk—the time derivative of acceleration with respect to time and mathematically defined as

$$\vec{J} = \frac{\mathrm{d}\vec{a}}{\mathrm{d}t} \tag{7.64}$$

In Eq. (7.64), the derivative is a total derivative and $\vec{J}$ is the absolute jerk rate in m/s^3. We are not going to present the expression of the absolute jerk rate vector $\vec{J}$ which is considered to be high-level work. The reader can find the expression of the absolute jerk rate in Refs. [8].

From Newton's Second Law, we know that the applied forces and/or applied torques are proportional to transitional acceleration and/or rotational acceleration. The definition of jerk is the acceleration rate of change with respect to time. Based on Newton's Second Law, we could say that the jerk rate is considered to be the change in force with respect to time and/or the change in torque with respect to time. The force and/or the torque can be actuated (given by an actuator) or can be transferred through the mechanism.

Jerk has been used in many applications. Humans can experience the influence of jerk rate in many situations. For example, if you are driving a car (or you are a passenger) and suddenly you must brake to avoid hitting something, you (and any passengers) will feel a temporary sudden loss of control or temporary sudden dizzy feeling due to the sudden change in the brake force or the sudden change in acceleration that we refer to as a high jerk magnitude. However, if you gently and smoothly press the brake, the rate of change in the acceleration or jerk rate will be small enough that you (and your passengers) will not notice.

Another example of how humans experience jerk is how comfortable the driver and the passengers feel when a car moves over road irregularities (e.g., bumps or corrugations). If the car is driven on a smooth road, the driver and the passengers feel very comfortable as there is little vertical jerk rate. If the road has some irregularities such as bumps, both the driver and the suspension system are responsible for how much vertical jerk rate the passengers feel. In general, the human reaction to the value of the jerk rate depends on physiological limitations and the attention level of the human brain.

For machine design, it is known that at high jerk magnitude values, there is a very high possibility that the machine's links will vibrate or become flexible with a tendency for noise generation and failure. If we look at what causes the machine to have links vibrate or become flexible, we could find many factors including, but not limited to, the links' geometry (Shape and Dimensions), the links' materials (Density, Young's Modulus, and Poisson's Ratio), the applied loads on the machine and/or the links experience high jerk magnitudes. Machine design includes the selection of the actuator plus the actuator's desired motion that must avoid any high jerk magnitude; otherwise, the machine would vibrate with noise and there would be a possibility of failure. Also, there would be flexibility that can cause inaccuracy in the desired task.

One of the techniques to improve our machine design is to iterate on the desired actuators' motions to make sure that the machine's links are not going to vibrate due to the high jerk magnitude in these desired actuators' motions. The good thing is that we can easily replace the model-based rigid body simulation with the flexible body simulation using MATLAB® Simulink®\ Simscape™ Multibody™. The model-based flexible body is not the topic of this book, but we are going to show one example for the reader's knowledge.

In MATLAB®, Simulink® and Simscape™ Multibody™, there is an option to use the flexible body as shown in Fig. 7.69a. Let's assume that we are going to go with beams as flexible links. More specifically, we are going to select the "Flexible Rectangular Beam" block. If we double-click on the "Flexible Rectangular Beam" block after dragging and dropping it on a new Simulink® model file, we can see the required block parameters as shown in Fig. 7.69b. To simulate the flexible body, we must provide "Geometry" parameters (Solid or Hollow, Width, Height, and Length) and "Stiffness and Inertia" parameters (Density, Young's Modulus, and Poisson's Ratio). We are going to add two flexible body subsystems "Flex Brick with two revolute joints" and "Flex Brick with two revolute joints" to our "Kinematics_Of_Machines.slx" file.

Now, we are going to study the influence of the jerk magnitude of the actuator's desired motion on the machine's flexibility. Recall the door mechanism shown in Fig. 7.21. We are going to use the parameters given in Table 7.4. Note that we are using the same density, Young's modulus, Poisson ratio, link's width, and link's height for all the links.

The comparison between the flexible body simulation and the rigid body simulation using MATLAB®, Simulink® and Simscape™ Multibody™ has the following steps:

1. Write and save "Flex_Rigid_Comp_Chapter7.m" in the below MATLAB® script to declare the parameters for the model.

```
clc
clear all
close all
jerk_flag = 1;%"1" when a high jerk magnitude, "0" when a low jerk
magnitude

%===========================================================================
%Declare the given parameters
%============================

%Dimensions
%=========
unit = 0.1;
r_n2   = 25*unit; %m
r_n5   = 82*unit; %m
r_n3n2 = 33*unit; %m
r_n3n4 = 30*unit; %m
r_n4   = 30*unit; %m
thickness = 0.05*r_n2; %m

%Material
%========
Density = 2700; %kg/m^3
YoungsModulus = 70; %GPa
PoissonsRatio = 0.33;

%Calculation of parameters
%=========================
L2_crosssection = [0 0;r_n2 0;r_n5 0;r_n5 thickness;0 thickness];
L2_dimension    = [r_n2 thickness thickness]; %m
L3_dimension    = [r_n3n2 thickness thickness]; %m
L4_dimension    = [r_n4 thickness thickness]; %m

Frame_L1_n1 = [0 0 0]; %m
Frame_L1_n4 = [r_n4 0 0]; %m
%
Frame_L2_n1 = [0 0 0]; %m
Frame_L2_n2 = [r_n2 0 0]; %m
Frame_L2_n5 = [r_n5 0 0]; %m
Frame_L3_n2 = [-r_n3n2/2 0 0]; %m
Frame_L3_n3 = [r_n3n2/2 0 0]; %m

Frame_L4_n4 = [-r_n3n4/2 0 0]; %m
Frame_L4_n3 = [r_n3n4/2 0 0];%m

%Input Cases
%===========
if jerk_flag == 1
    T = 2; %sec
else
    T = 6;%sec
end
```

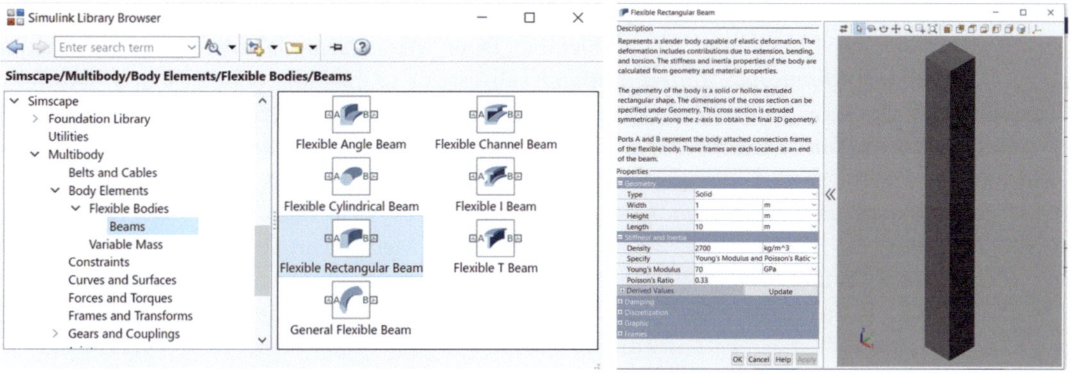

Fig. 7.69 Flexible body in MATLAB®, Simulink® and Simscape™ Multibody™. Designed and implemented by the author. Images used with permission of MathWorks

Table 7.4 Flexible door machine parameters

Geometry and dimensions		Stiffness and inertia	
Type	Solid (m)	Density	2700 kg/m^3
Length of $\lvert r_{n2} \rvert$	2.5	Young's modulus	70 GPa
Length of $\lvert r_{n5} \rvert$	8.2	Poisson ratio	0.33
Length of $\lvert r_{n3n2} \rvert$	3.3		
Length of $\lvert r_{n3n4} \rvert$	3		
Length of $\lvert r_{n4} \rvert$	3		
Width	0.125		
Height	0.125		

2. Open the model file from Example 7.6 and save as "***Rigid_Flexible_Door_Mechanism***
3. ***.xls***" file.
4. Delete all the actuators and sensors.
5. Copy and paste (***Ctrl A, Ctrl C, and Ctrl V***) the entire model. Now, you have two identical models: one on the right-hand side and the second on the left-hand side.
6. In the left-hand side model, replace each "***Brick with two revolute joints***" subsystem with a "***Flex Brick with two revolute joints***" subsystem and replace "***Extrusion with two revolute joints***" with "***Flex Extrusion with two revolute joints***" from our "***Kinematics_Of***
7. ***_Machines.slx***" file.
8. Implement and build the Simulink® subsystem of the equation:

$$\theta_2 = \begin{cases} 0 & 0 \le t \le 0.5\,\text{s} \\ 60\left(\frac{t-0.5}{T} - \frac{1}{2\pi}\sin\left(\frac{2\pi}{T}(t-0.5)\right)\right) & 0.5 < t \le T + 0.5\,\text{s} \\ 60 & t > T + 0.5\,\text{s} \end{cases} \tag{7.65}$$

where $T = 2$ s for the "***High Jerk Magnitude Jerk***" case and $T = 6$ s for the "***Low Jerk Magnitude***" case. At the beginning of "Flex_Rigid_Comp_Chapter7.m" script, we must define the jerk_flag either to be "1" for the high case or to be "0" for the low case. There is an "If" condition at the end of this script to declare the value of T as a function of jerk_flag. The Simulink® implementation of Equation (7.65) is shown in Fig. 7.70.

9. Copy and paste the "***Rotational Motion Actuator***" subsystem twice from our "Kinematics_Of_Machines.slx" file into the "Rigid_Flexible_Door_Mechanism.xls" file. Connect the output ports of the two ***Rotational Motion Actuator*** subsystems to the input ports of the "J_1R_rigid" and "J_1R_flex", respectively. Rename the two ***Rotational Motion Actuator*** subsystems with "***Rotational Motion Actuator Rigid***" and "***Rotational Motion Actuator Flex***" as shown in Fig. 7.71.
10. Copy and paste the "***Transitional Position Sensor***" subsystem twice from our "Kinematics_Of_Machines.slx" file into the "Rigid_Flexible_Door_Mechanism.xls" file. Connect the two "***Transitional Position Sensor***" subsystems to measure the rigid n_5 node and flexible n_5 node. The node n_5 is the "Free Node" in the "L2" subsystem in both cases of rigid and flexible. Change the label names of "***Transitional Position Sensor***" subsystems to "n5 rigid" and "n5 flex" as shown in Fig. 7.71.
11. Adjust the "Stop Time" in the Simulation tab into "T + 1".

Now, the model is ready. First, we adjust the "jerk_flag" to "0" (low jerk magnitude case), run the script, and run the model. During the simulation, you will see the animation on the "Mechanics Explorer" window. We are showing the screenshots of the animation at $t=3$ s and $t=4$ s as shown in Fig. 7.72a, b. After that, we are going to adjust the "switch flag" to "1" (high jerk magnitude case). During the simulation, you will see the animation on the "Mechanics Explorer" window. We are showing the screenshots of the animation at $t=1.35$ s and $t=2.35$ s as shown in Fig. 7.72c, d. We encourage the reader to run the attached model in both cases. Also, the reader can watch the attached animation videos of the low jerk magnitude case and the high jerk magnitude case.

After looking at Fig. 7.72 or watching the animation videos, we can see in the "High Jerk Magnitude" case, the flexible link L_2 oscillates with a noticeable vibration rate that we cannot ignore, and our rigid body assumption is not valid. To understand why our rigid assumption fails at the "High Jerk Magnitude" case, we are going to plot: the angular position, the angular velocity, the angular acceleration, and the angular jerk of both cases plus plotting the measured displacement of node n_5 at each case in Figs. 7.73 and 7.74. The "High Jerk Magnitude" case has a high jerk magnitude with a maximum value of 295.8 deg/s^3 compared to the "Low Jerk Magnitude" case with a maximum value of 11 deg/s^3. The conclusion from our example is that a high jerk magnitude violates our assumptions that the links are rigid bodies, and we cannot use any of our previous and current analyses. We recommend designers or readers to double-check the flexible model to see how much vibration can be observed.

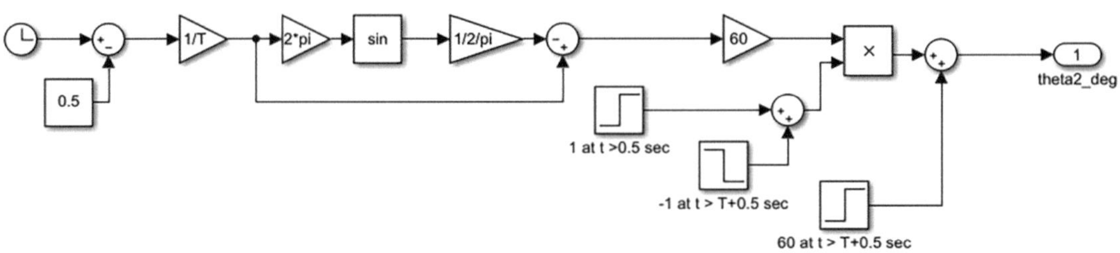

Fig. 7.70 Implementation of input signal of Rigid_Flexible_Door_Mechanism.xls model

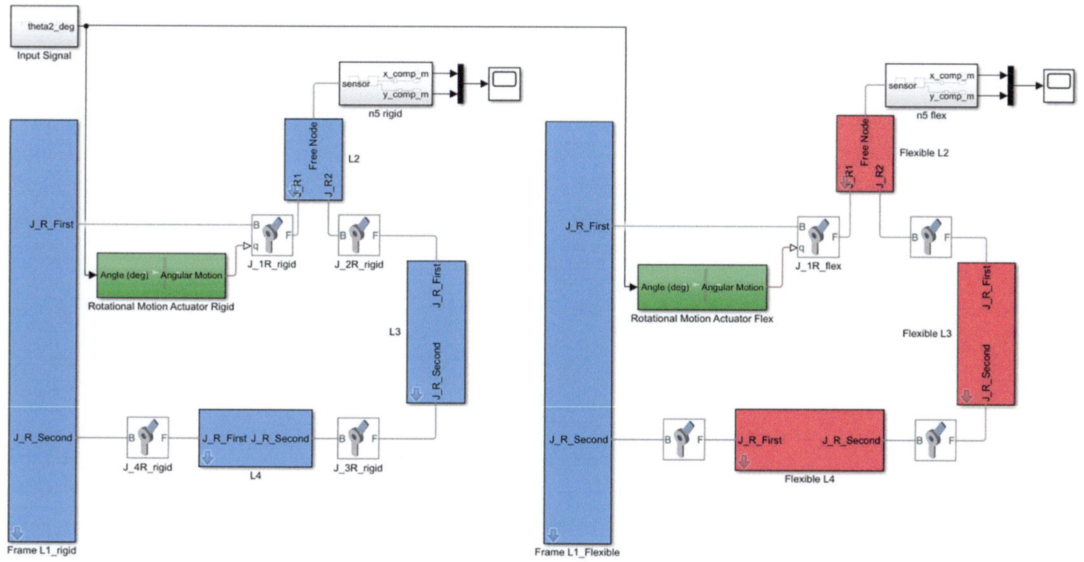

Fig. 7.71 Implementation of rigid and flexible door mechanism

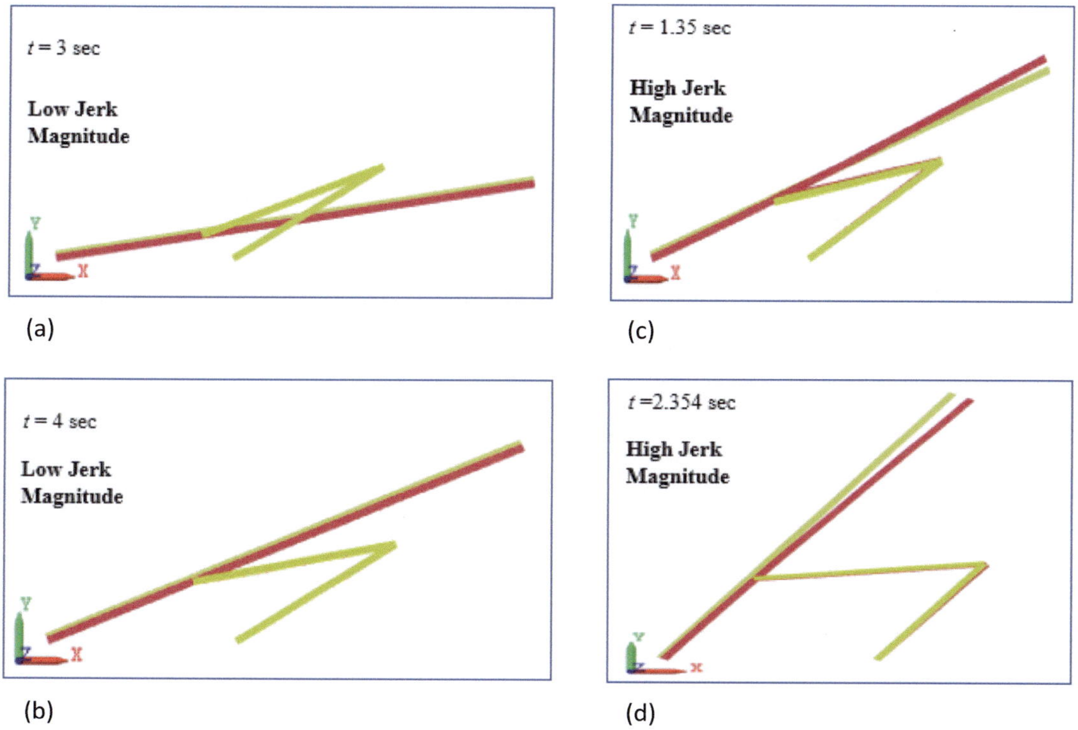

Fig. 7.72 Animations of both rigid (red color) and flexible (yellow color) door mechanisms at low jerk magnitude and high jerk magnitude

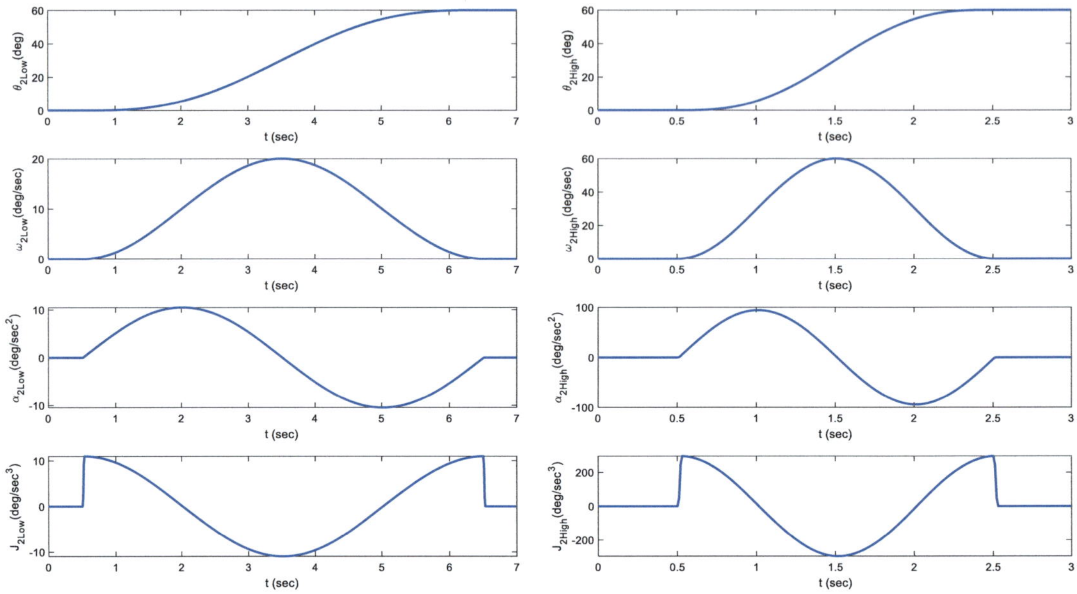

Fig. 7.73 Angular displacement, angular velocity, angular acceleration, and angular jerk of low and high jerk magnitude input

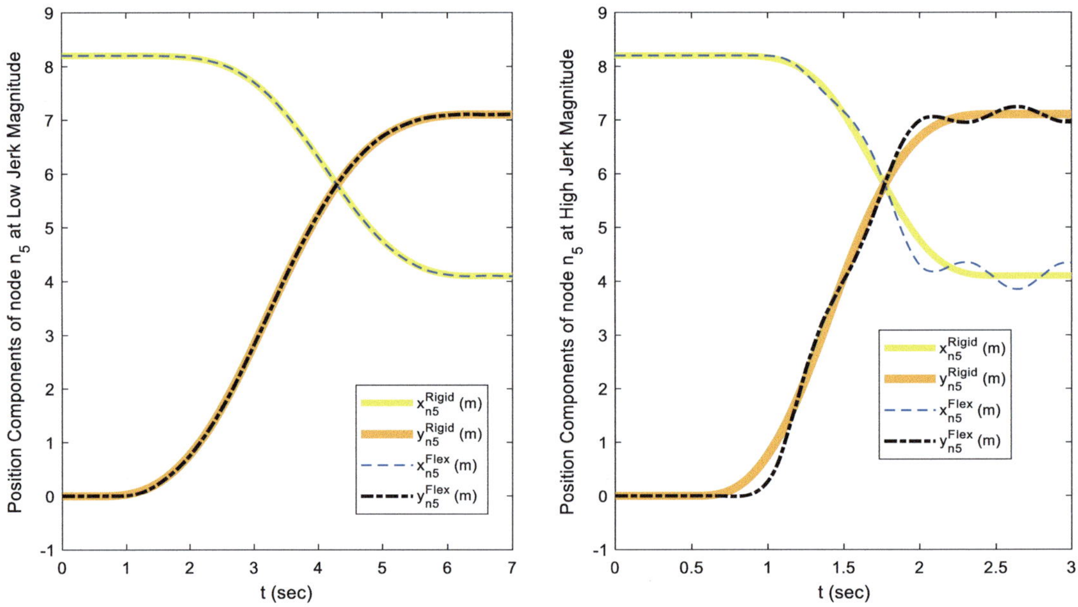

Fig. 7.74 Position of node n_5 of rigid body model and flexible body model at high and low jerk magnitude

Simulink® Tips—Switch Block

Simulink® Switch block is a graphical way to present an "If" condition as shown in Fig. 7.75. The Switch block passes through input 1 when input 2 satisfies the selected criterion; otherwise, it passes through input 3 [11].

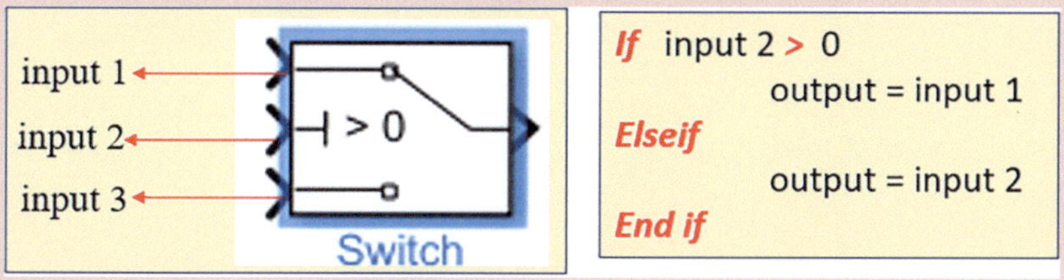

Fig. 7.75 Simulink® switch block as IF condition. Designed and implemented by the author. Images used with permission of MathWorks

7.8 Chapter Summary

Outcome

1. Calculating the translational acceleration vector and the rotational acceleration vector of any mechanism node the using graphical method, analytical method, and model-based Simulink® and Simscape™ Multibody™ for a four-bar mechanism, slider–crank mechanism, and inverted slider–crank mechanism.
2. Understanding the efficiency of the model-based Simulink® and Simscape™ Multibody™ applied to a higher-order bar mechanism.

Problems

Problem 1

For the shown sewing machine in Fig. 7.76, assume the motor moves the needle at node n_5 with the motion below:

$$x_{\text{needle}}(t) = \frac{-A}{2}\left(1 - \cos\left(\frac{\pi t}{T}\right)\right)\text{for}0 \leq t \leq T\ \text{s}$$

$$\dot{x}_{\text{needle}}(t) = \frac{-A\pi}{2T}\sin\left(\frac{\pi t}{T}\right)\text{for}0 \leq t \leq T\ \text{s}$$

Consider the amplitude $A = 32$ mm and $T = 0.05$ s.the time span

1. Calculate the average acceleration over the time span.
2. Calculate the instantaneous acceleration at $t = 0$ s, $t = 0.025$ s, and $t = 0.05$ s.
3. Write MATLAB® script to plot the position displacement $x_{\text{needle}}(t)$ with respect to time, the instantaneous velocity $\dot{x}_{\text{needle}}(t)$ with respect to time, and the instantaneous acceleration $\ddot{x}_{\text{needle}}(t)$ with respect to time.

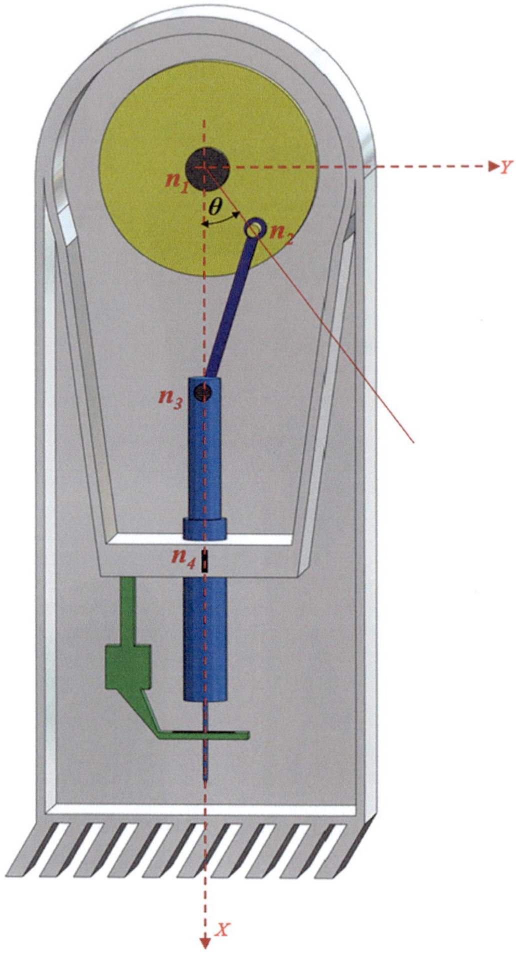

Fig. 7.76 Sewing mechanism. Designed by the author and implemented by both the author and Pranav Dhola. Courtesy of Pranav Dhola

Problem 2

For the shown potato chopper machine in Fig. 7.77, assume a person rotates the handle so that the plate moves following the motion below:

$$x_{\text{plate}}(t) = 50 - 40\left(35\left(\frac{t}{T}\right)^4 - 84\left(\frac{t}{T}\right)^5 + 70\left(\frac{t}{T}\right)^6 - 20\left(\frac{t}{T}\right)^7\right)$$

$$\dot{x}_{\text{plate}}(t) = -10\left(140\left(\frac{t}{T}\right)^3 - 420\left(\frac{t}{T}\right)^4 + 420\left(\frac{t}{T}\right)^5 - 140\left(\frac{t}{T}\right)^6\right)$$

where x in cm, $\mathrm{d}x/\mathrm{d}t$ is in cm/s, the time span $T = 4$ s.

1. Calculate the average acceleration over the time span, T.
2. Calculate the instantaneous acceleration at $t = 0$ s, $t = 2$ s, and $t = 4$ s.
3. Write MATLAB® script to plot the position displacement $x_{\text{plate}}(t)$ with respect to time, the instantaneous velocity $\dot{x}_{\text{plate}}(t)$ with respect to time, and the instantaneous acceleration $\ddot{x}_{\text{plate}}(t)$ with respect to time.

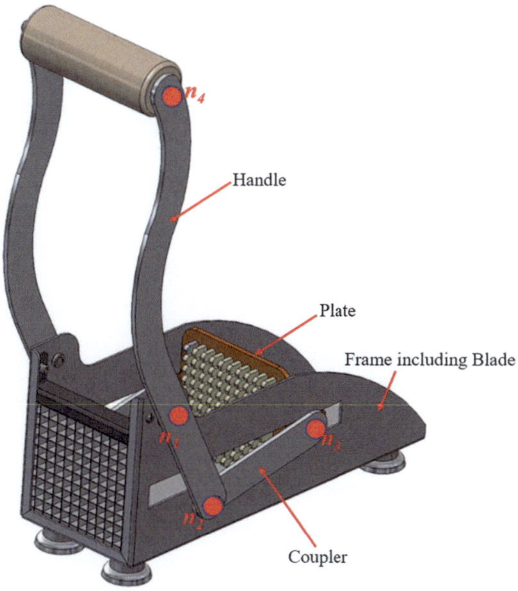

Fig. 7.77 Potato chopper machine. Figure and associated Potato_Chopper_Mechanism.avi video designed by the author and implemented by both the author and Pranav Dhola. Both courtesy of Pranav Dhola

Problem 3

For the pumpjack mechanism shown in Fig. 7.78, the motor is set to rotate the crank with

$$\theta_2(t) = 2\pi\left(\frac{t}{3} - \frac{1}{2\pi}\sin\left(\frac{2\pi t}{3}\right)\right)$$
$$\omega_2(t) = \dot{\theta}_2(t) = \frac{2\pi}{3}\left(1 - \cos\left(\frac{2\pi t}{3}\right)\right)$$

where θ_2 is in rad, ω_2 is in rad/s and t is in seconds. The time span for one cycle is $T=3$ s.

1. Calculate the average acceleration over the time span, T.
2. Calculate the instantaneous acceleration at $t=0$ s, $t=1.5$ s, and $t=3$ s.
3. Write MATLAB® script to plot the position displacement $\theta_2(t)$ with respect to time, the instantaneous velocity $\dot{\theta}_2(t)$ with respect to time, and the instantaneous acceleration $\ddot{\theta}_2(t)$ with respect to time.

Problem 4

For the sewing mechanism shown in Fig. 7.76, the motor is set to rotate at

$$\theta(t) = \frac{250\pi}{9}t^2$$
$$\omega(t) = \dot{\theta}(t) = \frac{500\pi}{9}t$$

where θ is in radians and t is in seconds.

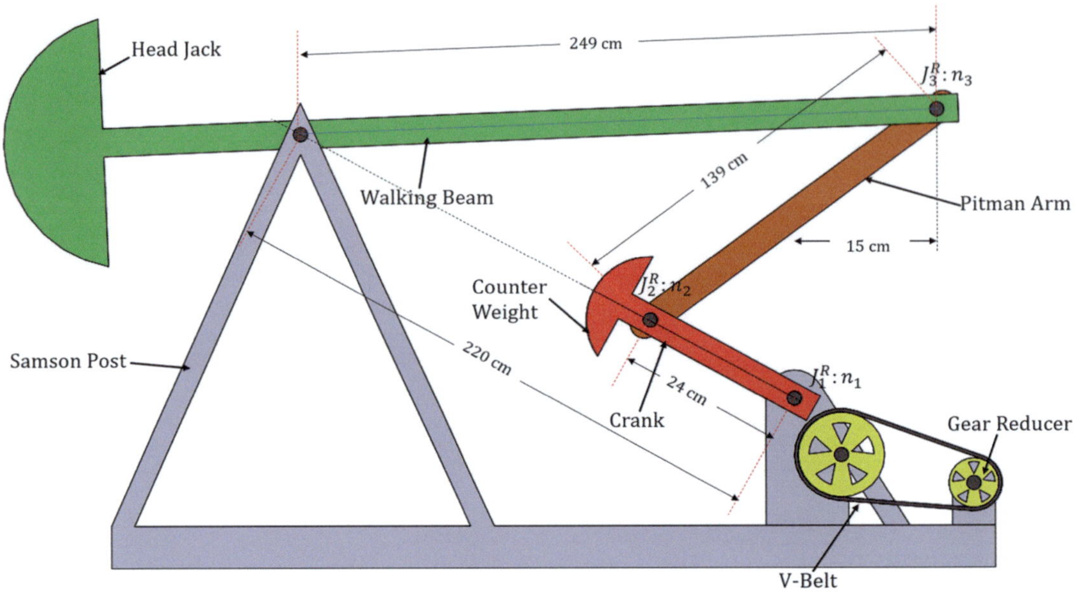

Fig. 7.78 Pumpjack machine. Designed by the author and implemented by both the author and Pranav Dhola. Courtesy of Pranav Dhola

1. Calculate the required time to go over one cycle from $\theta = 0°$ to $\theta = 360°$.
2. Calculate the average acceleration over the time span, T.
3. Calculate the instantaneous acceleration at $t = 0$ s, $t = 0.5T$ s, and $t = T$ s.
4. Write MATLAB® script to plot the position displacement $\theta(t)$ with respect to time, the instantaneous velocity $\omega(t)$ with respect to time, and the instantaneous acceleration $\alpha(t)$ with respect to time.

For *Problem 5, Problem 6,* and *Problem 7,* for the sewing machine in Fig. 7.76, consider that the distance between n_1 and n_2 is 15 cm, the distance between n_2 and n_3 is 30 cm, and the distance between n_1 and n_4 is 70 cm. Assume that the motor rotates the roll with a constant angular velocity $\omega_2 = 333.33$ rpm. At instant time, the roll is making an angle 30° with the positive X-axis, the coupler is making an angle 165.5° with the positive X-axis and rotating with angular velocity 15.61 rad/s and angular acceleration 252.32 rad/s^2, while the needle bar has a sliding velocity of – 3.788 m/s and a sliding acceleration of – 210.24 m/s^2.

Problem 5

At $J_2^R : n_2$,use the needle bar and the coupler kinematic pair to calculate:

- Tangential acceleration component
- Centripetal acceleration component
- Coriolis acceleration component
- Sliding acceleration component

- Total acceleration in polar coordinate
- Total acceleration in Cartesian coordinate.

Problem 6

At $J_2^R : n_2$, use the roller link to calculate:

- Tangential acceleration component
- Centripetal acceleration component
- Total acceleration in polar coordinate
- Total acceleration in Cartesian coordinate

Problem 7

- Compare the calculated total Cartesian acceleration components of node n_2 from Problem 5 with the ones that are calculated from Problem 6 using the calculated values and the acceleration vector diagram.

For *Problem 8, Problem 9,* and *Problem 10,* the Dumpster Truck Mechanism is shown Fig. 7.79. The bed is rotating at a constant angular velocity of 0.5 rpm. At a specific time *t*, the bed is making an angle 30° with the positive *X*-axis, the piston is extracted to the length 1.275 m with a sliding velocity of 0.1975 m/s and a sliding acceleration of -5.059×10^{-3} m/s², the cylinder is at an angle of 139.43° with the positive *X*-axis, rotating with an angular velocity of -0.0227 rad/s, and an angular acceleration of -6.274×10^{-3} rad/s².

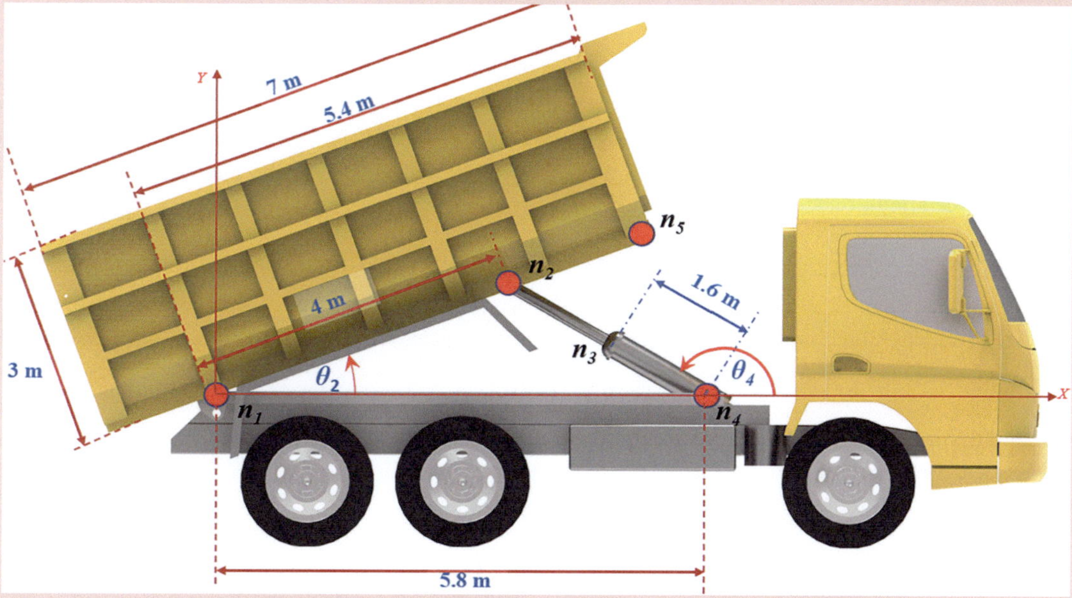

Fig. 7.79 Dumpster truck mechanism. Figure and associated Dumpster Truck Mechanism.mp4 video designed by the author and implemented by both the author and Pranav Dhola. Both courtesy of Pranav Dhola

Problem 8

At $J_2^R : n_2$, use the piston–cylinder kinematic pair to calculate:

- Tangential acceleration component
- Centripetal acceleration component
- Coriolis acceleration component
- Sliding acceleration component
- Total acceleration in polar coordinate
- Total acceleration in Cartesian coordinate.

Problem 9

At $J_2^R : n_2$, use the bed link to calculate:

- Tangential acceleration component
- Centripetal acceleration component
- Total acceleration in polar coordinate
- Total acceleration in Cartesian coordinate.

Problem 10

Compare the calculated total Cartesian acceleration components of node n_2 from Problem 8 with the ones that are calculated from Problem 9.

Problem 11

For the pumpjack machine in Fig. 7.78, when the hydraulic system is actuated to rotate the bed with a constant angular velocity $\omega_2 = 0.5$ rad/s. At the position $\theta_2 = 30°$, $\theta_3 = 91.4°$, and $\theta_3 = 143.7°$, the angular velocities are $\omega_3 = -0.0906$ rad/s and $\omega_4 = -0.0577$ rad/s. Use the analytical acceleration method to the angular accelerations of the Pitman arm and walking beam.

Problem 12

For dumpster truck mechanism in Fig. 7.79, the motor is actuated to rotate the crank with a constant angular velocity $\omega_2 = 0.5$ rad/s.

- Build a Simscape™ Multibody™ model to simulate this scenario for 2 crank cycles.
- Measure the angular accelerations of the Pitman arm.
- Measure the angular accelerations of the walking beam.

Problem 13

For the dumpster truck mechanism in Fig. 7.79, the hydraulic system is actuated to rotate the bed with a constant angular velocity $\omega_2 = 0.5$ rpm. At the position $\theta_2 = 30°$, $\theta_3 = 139.43°$, and $r_{n2n3} = 1.475$ m, the cylinder angular velocity is $\omega_4 = \omega_3 = 0.216$rpm and the piston transitional velocity is $\dot{r}_{n2n3} = 0.172$ m/s. Use the analytical acceleration method to calculate the cylinder angular acceleration α_4 and the piston translational acceleration $\ddot{r}_{n2n3}$.

Problem 14

For dumpster truck mechanism in Fig. 7.79, when the hydraulic system is actuated to rotate the bed with a constant angular velocity $\omega_2 = 0.5$ rpm.

- Build a Simscape™ Multibody™ model to simulate this scenario for 10 s time window.
- Measure the angular accelerations of the cylinder.
- Measure the piston' translational acceleration.

Problem 15

For the sew mechanism in Fig. 7.76, the motor rotates the crank with a constant angular velocity $\omega_2 = 500$ rpm. At the position $\theta_2 = 30°$, $\theta_3 = 165.5°$, and $r_{n3} = 42.038$ cm, the coupler angular velocity is $\omega_3 = -23.416$ rad/s and the transitional velocity of the node n_3 is $\dot{r}_{n3} = -5.683$ m/s. Use the analytical acceleration method to calculate the angular acceleration of the coupler plus the sliding acceleration of node n_4. The distance between n_1 and n_2 is 15 cm, the distance between n_2 and n_3 is 30 cm, and the distance between n_1 and n_4 is 70 cm. The length of the blue needle bar is 55 cm.

Problem 16

Build a Simscape™ Multibody™ model for the mechanism shown in Fig. 7.76. Use the motion actuator subsystem to rotate the crankshaft for two cycles (θ_2 from 0° to 720°) with a constant angular velocity of 500 rpm. Use the acceleration sensors to measure the variation of the angular acceleration of the coupler plus the sliding acceleration of node n_4 with respect to the time. The distance between n_1 and n_2 is 15 cm, the distance between n_2 and n_3 is 30 cm, and the distance between n_1 and n_4 is 70 cm. The length of the blue needle bar is 46 cm.

Problem 17

Build a MATLAB® script that uses the analytical model to calculate θ_3, θ_4, ω_3, ω_4, α_3, and α_4 for a four-bar mechanism. The script receives the mechanism parameters as L_1, L_2, L_3, and L_4. The script receives θ_2 as an input. Try Problem 11 to validate your script.

References

1. "Design of Machinery," Robert L. Norton, McGraw-Hill; 5th Edition, 2011.
2. "Theory of Machines and Mechanisms," John Uicker, Gordon Pennock and Joseph Shigley, Oxford University Press; 5th Edition, 2016.

3. "Machines & Mechanisms: Applied Kinematic Analysis", David Myszka, Pearson; 4th edition, 2011.
4. "Newton Engineering and Physical Science Pocket Book", John Bird, 2nd Edition, 1996.
5. "Simscape™ Multibody™ Getting Started Guide" MathWorks, 2021.
6. "Simscape™ Multibody™ User's Guide", MathWorks, 2021.
7. "Engineering Mechanics: Dynamics", Russel Hibbeler, 14th Edition, 2016.
8. "Kinematic synthesis of linkages," Hartenberg, R.S. & J. Denavit, McGraw-Hill, 1994.
9. "Complete classification of planar four-bar linkage," Baker, C., Journal of Mechanical Design, 1985.
10. "Modelling an inverted slider crank mechanism considering kinematic analysis and Multibody™ aspects, " R.C. Silva, M.A. Nunes, J.P.M. Bento and V.E. Costa, DINAME, Proceedings of the XV international symposium on dynamic problems of mechanics, 2013.
11. "Getting Started with Simulink®", MathWorks, 2021.

Motion Trajectory Design

8

Chapter Objectives

1. Understanding and sketching input–output block diagrams for Machine Motion Trajectory.
2. Understanding the hydraulic actuator and estimating the corresponding motion limits.
3. Understanding the motion limitations coming from using the electrical motor as a motion actuator.
4. Understanding and analyzing different mathematical techniques that are commonly used in trajectory generation.
5. Understanding how to use Simulink® to implement a motion trajectory using the sigmoid function.
6. Understanding how to integrate both the Simscape™ Multibody™ machine model with the Simulink® input motion model for simulation, analysis, and design of a certain machine.

8.1 Introduction

In this chapter, we introduce how to graphically describe the input and output models for displacement, velocity, acceleration, and jerk (SVAJ) of any machine using the block diagram technique. Although in the literature an SVAJ diagram is restricted to CAM motion, in this textbook, we will instead expand it to any machine.

Any mechanism requires a motion source (power source) to work. This motion source could be manual, i.e., using a human or an animal. Many automated motion sources use electrical power, wind power, a falling water source, or combustion power. The motion source can be linear transitional input motion (mainly located at prismatic joints) or rotational input motion (mainly used for revolute joints). A hydraulic actuator system with its corresponding physical motion limits is explained in the chapter as an example of a transitional input motion actuator. Motor systems and their corresponding motion limits are explained with an example of a rotational input motion actuator.

Finally, mathematical techniques to avoid discontinuities in design trajectory as well as lessons on generating trajectories for models in Simscape™ Multibody™ with animation are described.

Supplementary Information The online version contains supplementary material available at https://doi.org/10.1007/978-3-031-72701-6_8.

A. Omran, *Kinematics and Dynamics of Machines*, https://doi.org/10.1007/978-3-031-72701-6_8

8.2 Input–Output Block Diagram of SVAJ

Definition

System Block Diagram

A diagram of a system in which the principal parts or functions are represented by blocks connected by lines that show the relationships of the blocks or the parts [1].

A system block diagram consists of sets of blocks that are connected by a set of arrows to represent the “cause-and-effect” relationships between the inputs and outputs of the system being considered. To sketch a system block diagram, we should know the following:

- ***The system’s purpose***: We must know the purpose of the system. For example, the purpose of an air-conditioner system is to maintain a room’s temperature closer to a certain desired temperature. Another example is that the purpose of a skid steer’s articulated bucket mechanism is to lift or dump a load.
- ***The system’s boundary***: The system’s boundary separates the system that we are studying from everything else. For example, in the air-conditioner system, the boundary is the air-conditioner machine. In another example, the boundary for the skid steer articulated bucket mechanism is the skid steer vehicle.
- ***The system’s inputs***: The system’s inputs can be divided into desired signals (what we wish for) and undesired inputs (disturbance). For the air-conditioner system, the desired input is the desired temperature that we set on the control panel. The disturbances could be the variation in the external temperature or the variation in the room occupancy (how many people are inside the room). In another example, the desired input is to lift or dump the load. The disturbance could be the weight of the load.
- ***The system’s outputs***: The system output is the actual value of the desired signals. For example, the actual room temperature is the air-conditioner system’s output.
- ***The system’s parts or subsystems***: We need to know the major parts that affect the purpose or the process. For example, in the air-conditioner system, the subsystems could be the temperature sensor, controller, and cooling fan motor.
- ***The system’s cause-and-effect connections***: The arrows represent the input–output flow among the subsystems or parts. For example, in the air-conditioner system, there is an arrow coming from the temperature sensor subsystem and ending at the controller.

For a machine’s motion system block diagram, the purpose is to achieve a desired motion. The boundary includes the power source, controller, actuator, mechanism, and sensor which are considered to be the major subsystems as shown in Fig. 8.1.

The power source subsystem can be an electricity source, a combustion source, a wind source, etc. The power source’s functionality is to deliver the needed power to run the mechanism. The power source has no switch to change how much power is to be delivered to the mechanism.

As we defined in Chap. 3, the actuator subsystem is a functional element that can be used to vary the flow of power, mass, or volume from the power source and going to the mechanism with respect to time. Also, the actuator can convert the power from one form (mechanical, electrical, or chemical) to another. For example, a motor connected to an electrical power source or supply to run a mechanism converts electrical power into mechanical rotational power. Another example is the

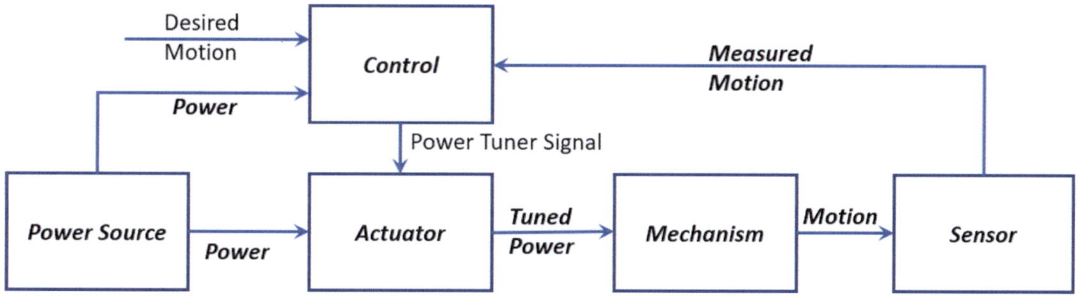

Fig. 8.1 Generic configuration of machine motion system diagram

hydraulic pump plus the control valves that are connected to an engine to run a mechanism. The hydraulic pump converts the mechanical power into hydraulic power.

The mechanism subsystem needs to receive a certain amount of power to perform the required task. For example, the lifting mechanism in a skid steer requires a certain amount of power to move a load or inertia with a specific motion: displacement (S), velocity (V), and acceleration (A) with an acceptable jerk (J) or SVAJ.

The sensor subsystem measures the system's response. When we transfer the power from one subsystem to another, many unknown losses are difficult to estimate. For this reason, we do not measure the power. Instead, we measure the motion (SVAJ), or we measure the applied loads (torques or forces). In most cases, we measure the motion. We recommend that the reader expand his/her knowledge about the motion sensors as listed in Ref. [2].

The control subsystem is to adjust the amount of the power flow from the actuator to the mechanism based on the difference between the user's request and the sensor's measurement. Usually, we do not have a sense of how much power is needed. We usually use the motion as our request or as our measurement.

Example 8.1

Figure 8.2 shows the actual diagram of the motion of the skid steer articulated bucket mechanism. Sketch the block diagram of this motion system showing the kind of power transferred.

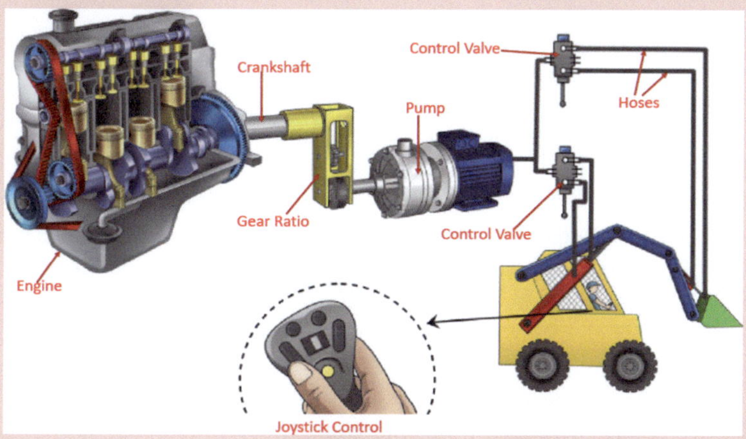

Fig. 8.2 Actual diagram of a *skid steer articulated bucket mechanism* motion system. Designed by the author and implemented by Kenneth Manalansan. Courtesy of Kenneth Manalansan

Solution

As shown in Fig. 8.3, the skid steer motion actuator system includes the following:

- ***The power source subsystem***: The engine is a power supply that provides a source of mechanical rotational motion through the crankshaft.
- ***The actuator subsystem***: The pump and the control valves are the actuator that converts the engine's mechanical power into a tuned power to drive the prismatic joints in the skid steer mechanism.
- ***The mechanism subsystem***: The skid steer articulated bucket mechanism consists of seven links and eight joints. The two hydraulic control valves adjust the flow in order to move the two prismatic joints. The bucket's motion is the output of the skid steer mechanism.
- ***The sensor subsystem***: Human eyes or any digital/analog sensors can be located at the bucket or at the prismatic joints to measure the mechanism's motion.
- ***The control subsystem***: The control process starts from the time the user moves the joystick to request the desired bucket motion. The controller then tunes the valves to achieve the desired motion.

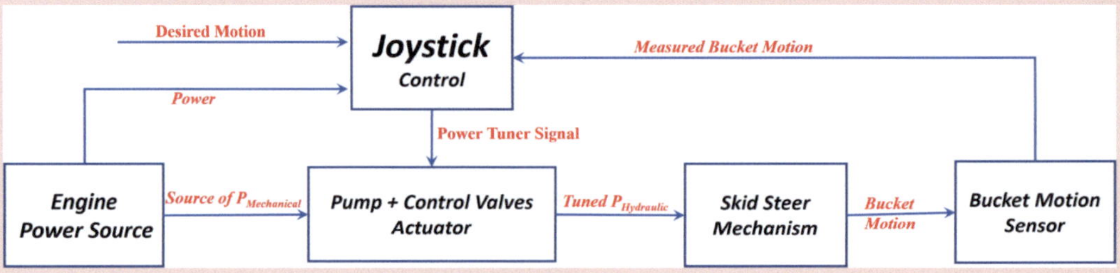

Fig. 8.3 Skid steer mechanism motion system block diagram

8.3 Hydraulic Actuators and Corresponding Motion Limits

A hydraulic actuator subsystem is not restricted to the pump and control valves. There are more hydraulic components in this subsystem. In this section, we are going to mention the major elements of a hydraulic subsystem followed by the motion constraints that are caused by the hydraulic elements.

8.3.1 Hydraulic Circuit

The entire hydraulic subsystem or the hydraulic circuit is usually sketched using a hydraulic circuit diagram in a standard format. Now, we are going to list some of the common and basic elements in our hydraulic subsystem including the definition, functionality, and an example of the shape of each element in Table 8.1. Other elements are not covered in this textbook. Table 8.2 shows the circuit diagram for most of the hydraulic components. Some shorthand is used in describing the valves, such as the "I/J valve", which indicates that the valve is connected to I ports and has J positions. For example, the 4/3 valve shown in Table 8.1 has four connected ports (port A, port B, port T connected to the tank, and port P connected to the pump) and three positions, as shown in Fig. 8.6. We recommend that the reader explore more material about hydraulic subsystems from Ref. [3].

Table 8.1 Basic hydraulic subsystem components

Component	Definition and functionality [3]	Sketch
Reservoir	A hydraulic reservoir is a tank that holds the fluid required to supply the system. A reservoir has many functionalities including, but not limited to, storing fluid, dissipating heat, and separating air, water, or any solid materials. Figure 8.4 shows an example of a hydraulic reservoir	 **Fig. 8.4** Hydraulic reservoir. Designed by the author and implemented by both the author and Pranav Dhola. Courtesy of Pranav Dhola
Pump	A hydraulic pump is a mechanism that converts mechanical power from a source (such as an engine or a motor) into hydraulic power to a sink (such as a prismatic joint). For example, in the skid steer motion system, the hydraulic pump is converting mechanical power or power from an engine into hydraulic power to activate the prismatic joints (pistons and cylinders) When the shaft inside the pump rotates, the fluid is driven out from the pump into the hoses. There are many types of hydraulic pumps We present an example of a gear-type rotary pump in Fig. 8.5, where the primary shaft is connected to either the engine or the motor. The driver gears, upon rotation, transfer motion to the driven gear. Gears remain engaged on the suction side, with fluid being carried between the teeth and the casing. As the gears re-mesh, they expel the liquid. Higher speeds demand increased force to facilitate efficient flow expulsion. We recommend that the reader watch the attached animation video demonstrating the operation of the gear-type rotary pump	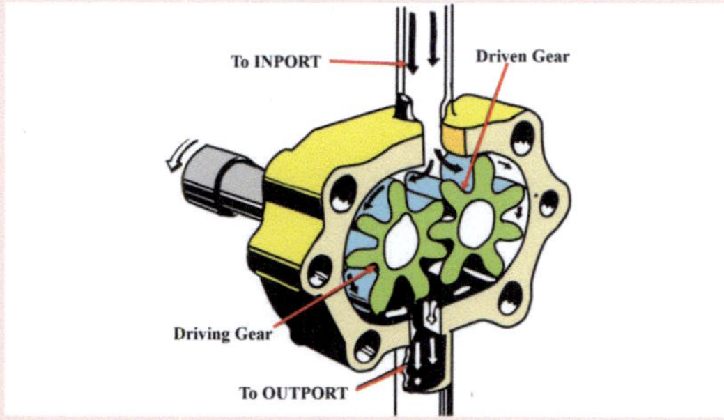 **Fig. 8.5** Gear type rotary pump. Designed by the author and implemented by both the author and Pranav Dhola. Courtesy of Pranav Dhola

(continued)

Table 8.1 (continued)

Component	Definition and functionality [3]	Sketch
Hydraulic directional control valve	A directional control valve is a mechanism that allows hydraulic oil into different paths from one or more sources. This mechanism has a spool inside a cylinder which is mechanically or electrically actuated. The spool's position restricts and controls the amount of fluid flow and/or pressure. Many types of hydraulic directional control valves are given in Refs. [2–5]. One example, a 4/3 valve, is shown in Fig. 8.6. It has three positions (P_1, P_2, and P_3) with four ports (T, P, A, and B). The port "T" is connected to the tank. The port "P" is connected to the pump. The port "A" is connected to the blue chamber inside the cylinder. The port "B" is connected to the red chamber inside the cylinder. The activated position is selected using a joystick • When the joystick is set in the middle, position P_2 is selected. Port A and Port B are blocked, and they do not receive or return any hydraulic fluid. We call this position the "Hold Position" • When the joystick is moved toward the right side, position P_1 is selected. Port A receives hydraulic fluid from the pump and Port B returns hydraulic fluid to the tank. We call this position the "Extraction Position" since the piston slides out from the cylinder • When the joystick is moved toward the left side, position P_3 is selected. Port B receives hydraulic fluid from the pump and Port A returns hydraulic fluid to the tank. We call this position the "Retraction Position" since the piston slides into the cylinder We recommend that the reader watch the attached animation video of the directional control valve	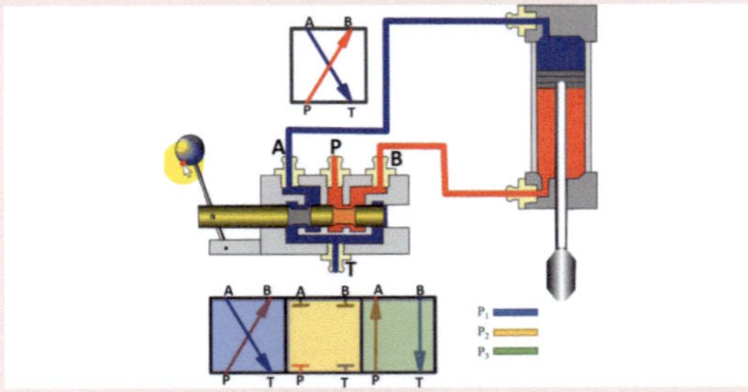 **Fig. 8.6** Mechanism and circuit diagram of a 4/3 hydraulic direction control valve with four ports and three positions. Designed by the author and implemented by both the author and Pranav Dhola. Courtesy of Pranav Dhola
Pressure relief valve	The pressure relief valve is a type of safety valve used to limit the maximum allowed pressure in a hydraulic system. If the pressure exceeds this maximum pressure, the physical hydraulic system can be damaged, or a hose might burst and endanger human lives. The pressure relief valve has a spring acting on a poppet or a ball against the pressure. When the pressure is higher or equal to the maximum allowed pressure, the poppet is pushed letting the fluid go to the reservoir (tank) and keeping the pressure at the allowed maximum value. Figure 8.7 shows an example of a pressure relief valve. We recommend that the reader watch the animation video of the pressure relief valve	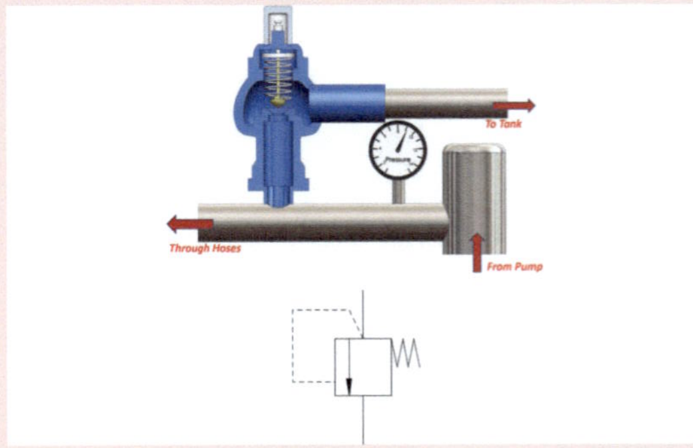 **Fig. 8.7** Example of a pressure relief valve. Designed by the author and implemented by both the author and Pranav Dhola. Courtesy of Pranav Dhola

(continued)

Table 8.1 (continued)

Component	Definition and functionality [3]	Sketch
Filter	Hydraulic filters remove contaminants from hydraulic fluids. The hydraulic filter has a bowl that blocks the contaminants and lets the clean fluid exit through the center. Figure 8.8 shows an example of a hydraulic filter	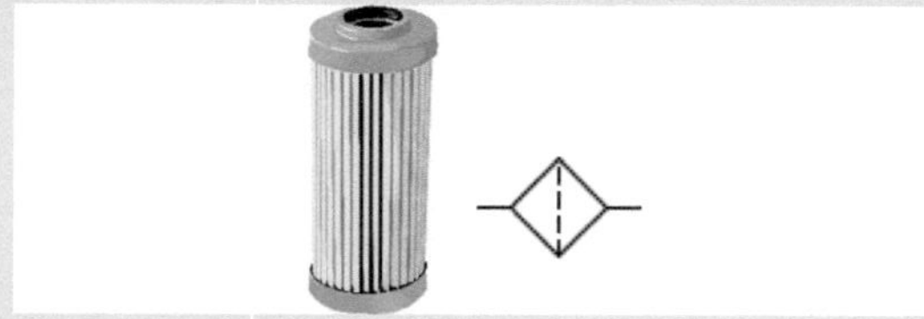 **Fig. 8.8** Example of a hydraulic filter. Designed by the author and implemented by both the author and Pranav Dhola. Courtesy of Pranav Dhola

Table 8.2 Circuit diagrams of hydraulic components

 3/2 directional control valve	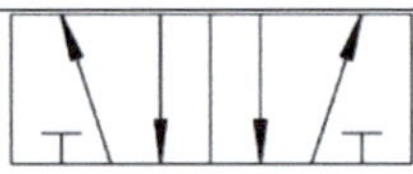 5/2 directional control valve	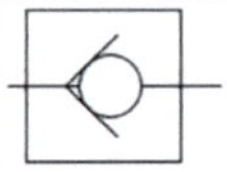 Check valve	 Throttle valve	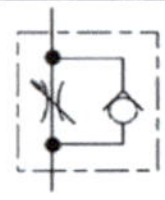 One-way restrictor valve	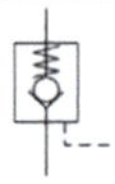 Non-return valve
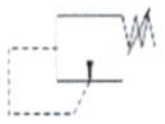 Pressure reducing valve	 Flow control valve	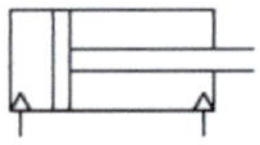 Double-acting cylinder	 Accumulator	 Pressure gauge	 Flow indicator

The hydraulic fluid moves from one element to another in hydraulic hoses or pipes. There are seals, fittings, and connections attaching the hoses to the components. A high-performance hydraulic design should make these connections without or with minimum leaking or losing pressure. Typical hydraulic circuits used in construction or agricultural machines may have thousands of sealed connections.

Figure 8.9 shows a simplified hydraulic circuit that activates a prismatic joint on a skid steer used to lift an articulated bucket mechanism. The engine runs the pump that pushes the hydraulic flow with pressure P_p and flow rate Q_p into the hose. There is a pressure relief valve to saturate the pressure to $P_{\max}$ (making sure that the pressure does not exceed $P_{\max}$). The fluid runs through the hose at flow rate Q_p to reach the P port of the directional control valve. Port T of the directional valve is connected to the tank. The directional valves Port A and Port B are connected to the two sides of a prismatic cylinder. A joystick position is selected by the driver to extend, retract, and hold the position of the piston. There is a simple animation provided to explain this process.

Now we will go over the major motion constraints encountered when a hydraulic actuator is applied, which are as follows:

1. Power losses from the pump, hoses, and cylinder. The power source affects the maximum and minimum motion acceleration value as a function of the motion's speed value.
2. Maximum and minimum speed values of the control valve characteristics.
3. Maximum and minimum displacement values of the hydraulic cylinder.

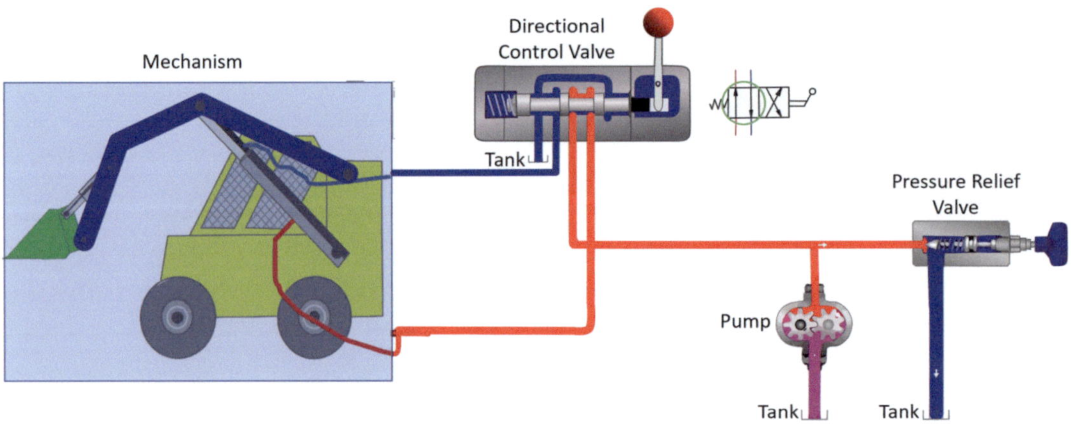

Fig. 8.9 Simplified hydraulic subsystem of a lift cylinder in a skid steer articulated bucket mechanism. Designed by the author and implemented by both the author and Pranav Dhola. Courtesy of Pranav Dhola

8.3.2 Power Losses

The maximum value of the input power to the mechanism is very important in designing the input motion. The hydraulic losses lead to a reduction in the input power from the power source to the machine. The hydraulic losses can be due to leaks or friction. The losses are different from one hydraulic component to another.

For example, Fig. 8.10 shows the power loss distribution from an engine/motor to a mechanism where a hydraulic subsystem is used as an actuator. The values shown in Fig. 8.10 are taken from Ref. [8]. The pump has 15% losses. That means if the engine provides the pump with 100 hp, the pump will deliver 85 hp to the hydraulic plumbing and 15 hp will be lost as friction and leakage. The hydraulic plumbing has around a 10% loss. That means that the hydraulic plumbing would deliver 90% of 85 hp or 0.9×85 hp$=$76.5 hp to the hydraulic cylinder and 10% or $0.1\times85=8.5$ hp will be lost as leakage and friction. The hydraulic cylinder has a 5% loss. It delivers 95% of 76.5 hp or 0.95×76.5 hp$=$72.7 hp to the mechanism with 5% or 0.05×76.5 hp$=$3.8 hp loss. In summary, the example in Fig. 8.10 shows that there are 27.3% total losses when we use this hydraulic system as an actuator.

Studying power limits and losses helps in designing input motion. The actual power delivered (after accounting for power losses from the power source) to the machine is the force multiplied by the transitional speed or the torque and multiplied by the rotational speed that the machine can transfer. For example, if we are going to design an input motion for a skid steer articulated bucket mechanism, we can set two options: one for a light load (low weight or low force) with a faster speed. The second option is for a heavy load (high weight or high force) with a low speed. Unless we have a good estimate for the actual power delivered to the machine, it will be virtually impossible to set up the maximum speed that the machine can deliver. Now, we are going to study the power losses in each hydraulic component.

8.3.2.1 Pump Losses

In any hydraulic circuit, the process starts by running the pump shaft using an engine or a motor. Assume that the engine or the motor is rotating the pump with a power $\mathcal{P}_{\text{engine}}$ or $\mathcal{P}_{\text{motor}}$, torque T_{engine} or T_{motor}, and angular velocity ω_{engine} or ω_{motor}. The input mechanical power to the pump is

$$\begin{aligned}\mathcal{P}_{\text{engine}} &= T_{\text{engine}} \times \omega_{\text{engine}}\\ \mathcal{P}_{\text{motor}} &= T_{\text{motor}} \times \omega_{\text{motor}}\end{aligned} \tag{8.1}$$

Regardless of the type of the hydraulic pump, the output hydraulic power $\mathcal{P}_{\text{pump}}$ from the pump is

$$\mathcal{P}_{\text{pump}} = P_p \times Q_p \tag{8.2}$$

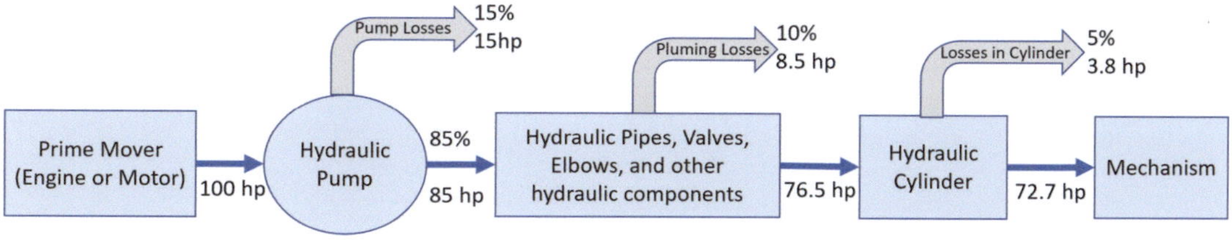

Fig. 8.10 Power losses across hydraulic subsystem

where P_p is the pump pressure and Q_p is the actual delivered pump flow rate. The output hydraulic power $\mathcal{P}_{\text{pump}}$ is less than the input mechanical power $\mathcal{P}_{\text{engine}}$ or $\mathcal{P}_{\text{motor}}$ due to the pump's power losses. The pump dissipated power happens for many reasons such as [6]:

- Cavitation—a lack of sufficient volume of fluid that may cause a pump inlet pressure to drop less than the vapor pressure; liquid–vapor bubbles form and the fluid appears to boil
- Aeration—air entering the oil stream
- Flow leakage and slippage
- Friction between moving parts
- Inertia of moving parts.

The above losses are usually transferred into vibrational losses, loud noise losses, thermal losses, and leaks. We can divide these losses into volumetric losses and mechanical losses.

If the theoretical flow Q_{th} experiences a leak (flow leakage Q_{leak}) and the actual flow rate Q_p delivered by the pump at a given pressure P_p is

$$Q_p = Q_{\text{th}} - Q_{\text{leak}} \tag{8.3}$$

The volumetric efficiency is defined as

$$\eta_{\text{vohumetric}} = \frac{Q_p}{Q_{\text{th}}} \tag{8.4}$$

The torque coming from the engine or the motor T_{ensine} or T_{motor} or is the sum of the actual pump torque T_{pump} and torque losses T_{losses} or

$$\begin{aligned} T_{\text{engine}} &= T_{\text{pump}} + T_{\text{losses}} \\ T_{\text{motor}} &= T_{\text{pump}} + T_{\text{losses}} \end{aligned} \tag{8.5}$$

Torque losses, T_{losses}, are due to the necessity to drive the inertia of the mechanical components and to overcome the friction between the many moving parts. This loss is known as mechanical or torque loss. To express these mechanical losses, we define the mechanical efficiency $\eta_{\text{mechanical}}$ as

$$\begin{aligned} \eta_{\text{mechanical}} &= \frac{T_{\text{pump}}}{T_{\text{ensine}}} = \frac{P_p Q_{\text{th}}}{T_{\text{engine}}\, \omega_{\text{engine}}} \\ \eta_{\text{mechanical}} &= \frac{T_{\text{pump}}}{T_{\text{motor}}} = \frac{P_p Q_{th}}{T_{\text{motor}}\, \omega_{\text{motor}}} \end{aligned} \tag{8.6}$$

To express the total pump losses, we usually define the pump efficiency η_{pump} as the ratio of the output pump power over the input pump power, or

$$\begin{aligned} \eta_{\text{pump}} &= \frac{\mathcal{P}_{\text{pump}}}{\mathcal{P}_{\text{engine}}} = \frac{P_p Q_p}{T_{\text{engine}}\, \omega_{\text{engine}}} = \frac{P_p Q_{\text{th}}}{T_{\text{engine}}\, \omega_{\text{engine}}} \frac{Q_p}{Q_{\text{th}}} \\ &= \eta_{\text{mechanical}} \times \eta_{\text{volumetric}} \\ \eta_{\text{pump}} &= \frac{\mathcal{P}_{\text{pump}}}{\mathcal{P}_{\text{motor}}} = \frac{P_p Q_p}{T_{\text{motor}}\, \omega_{\text{motor}}} = \frac{P_p Q_{\text{th}}}{T_{\text{motor}}\, \omega_{\text{motor}}} \frac{Q_p}{Q_{\text{th}}} \\ &= \eta_{\text{mechanical}} \times \eta_{\text{volumertic}} \end{aligned} \tag{8.7}$$

In most cases, the values of both volumetric efficiency and mechanical efficiency are experimentally or computationally measured and tabulated as a function of angular velocity and pump pressure. The pump losses provide the maximum value of the output pump's power output as a function of the input pump's power input or the maximum of $\mathcal{P}_{\text{pump}} = P_p Q_p$.

Example 8.2

The hydraulic subsystem of a skid steer articulated bucket mechanism has a 0.936 m^3/s theoretical flow rate with a volumetric efficiency of 92% and mechanical efficiency of 90%. The pump is rotated by a motor that provides the pump with a torque $T_{\text{motor}} = 240$ N m and $\omega_{\text{motor}} = 2500$ rpm. If the gear ratio between the angular velocity of the motor and the angular velocity of the pump's input shaft is 1 $(\omega_{\text{engine}} = \omega_{\text{pump}})$ find:

- The actual flow rate of the pump
- The motor losses torque.

Solution

Using Eq. (8.4), we can calculate the actual flow rate as

$$\eta_{\text{vohumetric}} = \frac{Q_p}{Q_{\text{th}}} \text{ or } Q_p = Q_{\text{th}} \eta_{\text{volumetric}}$$

Or

$$= (0.936\,\text{m}^3/\text{s})(0.92) = 0.861\,\text{m}^3/\text{s}$$

Using Eq. (8.6), we can calculate the pump torque as

$$\eta_{\text{mechanical}} = \frac{T_{\text{pump}}}{T_{\text{motor}}}$$

Or

$$T_{\text{pump}} = T_{\text{motor}} \eta_{\text{mechanical}} = (240\,\text{N m})(0.9) = 216\,\text{N m}$$

Using Eq. (8.5), we can calculate the motor losses torque as

$$T_{\text{motor}} = T_{\text{pump}} + T_{\text{losses}}$$

Or

$$T_{\text{losses}} = T_{\text{motor}} - T_{\text{pump}} = (240\,\text{N m}) - (216\,\text{N m}) = 24\,\text{N m}$$

8.3.2.2 Hose and Valve Losses

A pipe can consist of various straight pipes or hoses combined with various types of components (valves, elbows, etc.). There is a power loss (head loss) when the flow moves through the hydraulic pipes. The major power loss of the head loss is the frictional power loss caused by viscosity between the fluid and the hose/pipe wall. The minor power loss of the head loss is due to changing the flow direction (elbows), pipe fittings (pipe connections), and changing the flow area (valves). In this section, we are going to discuss how to estimate these major and minor power losses.

Definition

Laminar Flow:

Laminar flow is defined as that type of flow in which the fluid particles move along well-defined paths or streamlines, and all these streamlines are straight and parallel. Also, laminar flow is defined as flow in which the streamlines are smooth and regular and a fluid element moves smoothly along a streamline [3–8].

Turbulent Flow:

Turbulent flow is a type of flow in which the fluid particles move in a zig-zag way. Also, turbulent flow is defined as flow in which the streamlines break up and a fluid element moves in a random, irregular, and tortuous fashion [3–8].

When the fluid particles follow smooth paths in parallel layers, the flow is called *laminar*. If the fluid particles move irregularly, the flow is called *turbulent*. There are many real-world examples to demonstrate the difference between laminar flow and turbulent flow. The first example is the water flowing from a faucet. When the faucet is slightly opened, the water will come out at low speed down to the sink in a smooth, uniform flow that can be considered laminar (Fig. 8.11a). If the faucet is widely opened, the water will rush out at a higher speed and drop down to the sink with a more irregular and tortuous shape to be considered turbulent (Fig. 8.11b). From this first example, we can conclude that increasing the speed of the fluid will increase the turbulence of the flow. Also, you will notice that the sound becomes louder as the speed of the water increases.

The second example is with two jars—one full of water and the other full of honey. As shown in Fig. 8.12, if we quickly pour from the two jars with exactly the same tilt angle, we notice that the honey drops in a smooth, uniform flow with almost no sound. This can be considered a laminar flow. The water from the other jar drops in a random, irregular shape that can be considered a turbulent flow. There is a possibility that the water will splash with a noticeable sound. From this second example, we notice that while the water flow is turbulent, the honey flow is laminar at the same fluid speed. This is because honey is more viscous than water.

Our focus is to understand losses due to internal fluid flow (hydraulic fluid flow inside a pipe). We are now going to assume a pipe with a length of l, an inner diameter of D_1 at point 1, and an inner diameter D_2 at point 2 with ε being the measure of the roughness of the pipe wall.

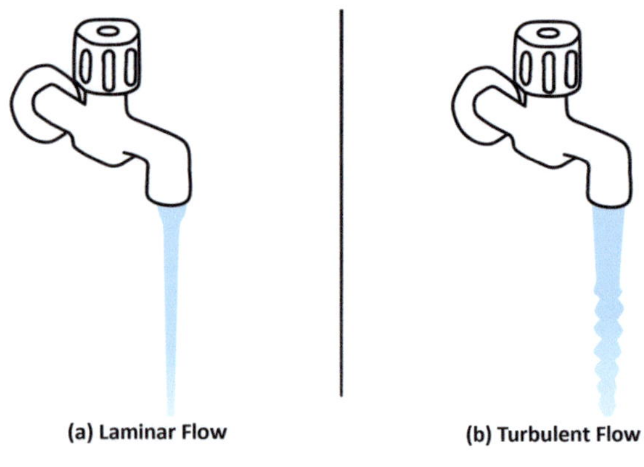

Fig. 8.11 Water laminar flow and turbulent flow from faucet. Designed by the author and implemented by both the author and Pranav Dhola. Courtesy of Pranav Dhola

Fig. 8.12 Honey laminar flow and water turbulent flow. Designed by the author and implemented by both the author and Pranav Dhola. Courtesy of Pranav Dhola

Fig. 8.13 Fluid flow inside pipe

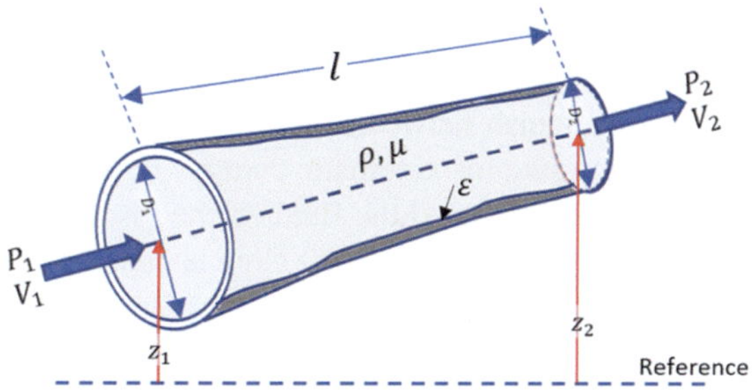

Fluid is passing inside a pipe from point 1 with a mean V_1 to point *2* with a mean velocity V_2, a fluid density ρ, and a fluid viscosityμ as shown in Fig. 8.13. The pipe is tilted so that the altitude of point 1 is z_1 and the altitude of point 2 is z_2. The resistance that the fluid faces from going through a pipe is called the *shear stress* τ. This stress occurs between the fluid and the pipe wall and results in power losses. These power losses are transferred into heat, vibration, and noise and can be observed as the pressure differences $\Delta P = P_1 - P_2$. The material and shape (the pipe's diameter and length) of the pipe and the type of the fluid flow (laminar or turbulent) affect this shear stress and resulting power losses.

Now, we are going to define some terminology used to estimate power losses from the hydraulic fluid inside a pipe.

Definition

Reynolds number:

Reynolds number R_e, as described by Reynolds in 1883, correlates the inertia forces to the viscous forces. Reynolds number R_e is mathematically defined as:

$$\mathrm{Re} = \frac{\rho V D}{\mu} \tag{8.8}$$

where ρ is the fluid density, V is the mean fluid velocity, D is the pipe diameter, and μ is the fluid viscosity [3–9].

When we inject a flow into the middle of a pipe as shown in Fig. 8.14, there are three possibilities:

- The change in the flow rate is small enough that the flow remains in a well-defined line or laminar flow
- The change in the flow rate is intermediate and has increased enough to have an explosion of irregular behavior along the streak
- The change in the flow rate is large enough to have a streak almost immediately become blurred and randomly spread through the entire pipe.

Based on experimental results, Osborne Reynolds proposed in 1882 the so-called Reynolds number, Re, to distinguish between flow that is laminar or turbulent. Reynolds number, Re, is the ratio of the inertial force (or kinematic energy) in part of a fluid flow ρVD and the viscosity μ between the surface of the pipe and the fluid. If the viscosity is excessive, the flow is called *laminar*. If the kinematic energy is excessive, the flow is called *turbulent*. The results from Reynolds' experiments can be summarized as [3–9]

- If $\mathrm{Re} < 2000$, the flow is laminar.
- If $\mathrm{Re} > 4000$, the flow is turbulent.
- If $2000 < \mathrm{Re} < 4000$, the flow is in the transition zone between laminar and turbulent.

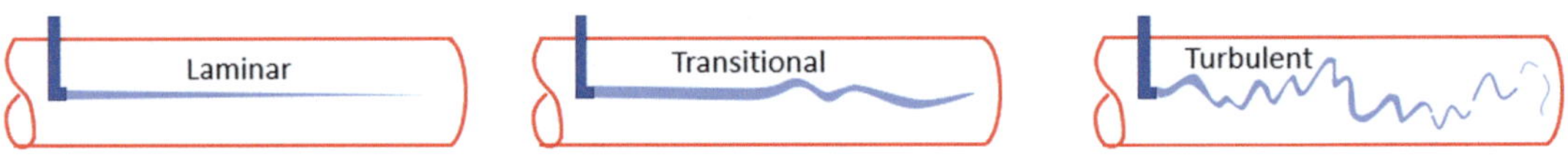

Fig. 8.14 Laminar, transitional, and turbulent flow inside a pipe

Definition

First Law of Thermodynamics:

The first law of thermodynamics for a system is: The time rate of increase of the total stored energy of the systems is equal to the net time rate of energy addition by heat transfer into the system and the net time rate of energy addition by work transfer into the system [3–9].

Based on the first law of thermodynamics, the energy equation of a steady and incompressible fluid flow going from point 1 to point 2 inside an isolated pipe without any work done, is [8, 9]

$$\frac{P_1}{\rho g} + \frac{V_1^2}{2g} + z_1 = \frac{P_2}{\rho g} + \frac{V_2^2}{2g} + z_2 + h_{\text{losses}} \tag{8.9}$$

where h_{losses} is the energy losses that are considered to be the viscous flow losses. The head losses can be broken down into

$$h_{\text{losses}} = h_{L\,\text{major}} + h_{L\,\text{minor}} \tag{8.10}$$

where $h_{L\,\text{major}}$, major losses, are the frictional head losses caused by viscosity between the fluid and the pipe wall and $h_{L\,\text{minor}}$, the minor losses, are the minor head losses that are due to changing the flow direction (elbows), pipe fittings (pipe connections), and changing the flow area (valves). The ***Darcy–Weisbach*** equation expresses the head losses $h_{L\,\text{major}}$ inside a pipe from point 1 to point 2 with [3, 8, 9].

$$h_{L\,major} = f\frac{l}{D}\frac{V^2}{2g} \tag{8.11}$$

where f is the friction factor that has a dependency on the Reynolds number, Re, and the relative roughness, ε/D, where ε is the surface roughness. In 1933, Nikuradse conducted experiments to estimate the friction factor, f, at different flow types (laminar or turbulent) and the pipe's relative roughness. L. F. Moody, along with C. F. Colebrook, created the so-called Moody Chart that shows the functional dependence of f on Re and ε/D as in Fig. 8.15 [8, 9]. The laminar zone in the Moody Chart has the following empirical formula:

$$f = \frac{64}{R_e} \tag{8.12}$$

The mathematical formula to relate the friction coefficient f, Reynolds number Re, and the relative roughness ε/D in the case of turbulent flow is called the Colebrook Formula:

$$\frac{1}{\sqrt{f}} = -2.0 \times \log\left(\frac{\varepsilon/D}{3.7} + \frac{2.51}{R_e\sqrt{f}}\right) \tag{8.13}$$

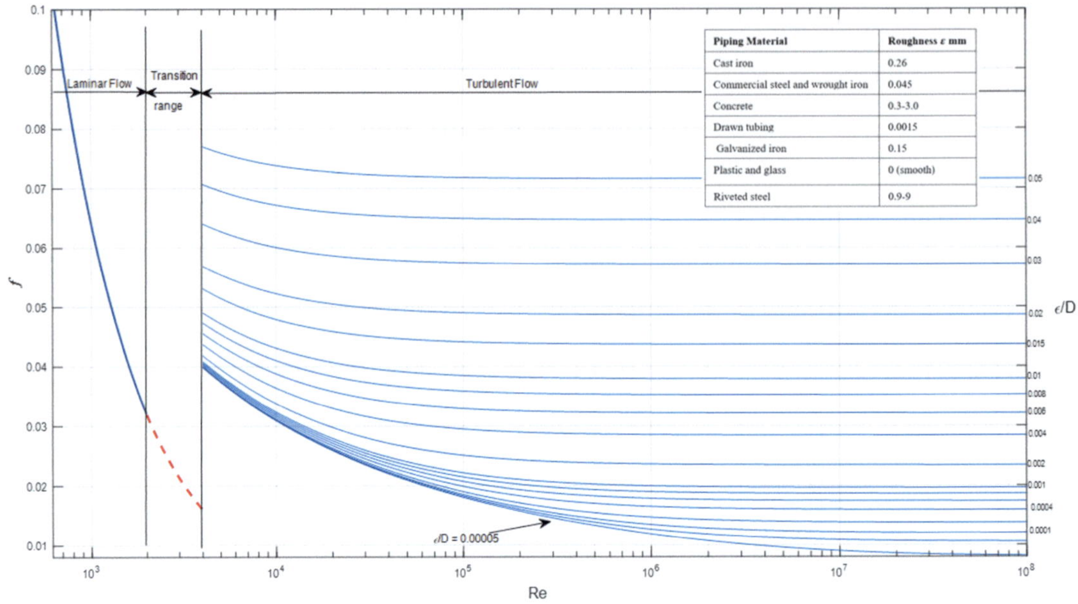

Fig. 8.15 Friction factor as a function of Reynolds number and relative roughness for round pipes—the Moody Chart, Refs. [8, 9]

The head minor losses $h_{L\,\text{minor}}$ associated through the valve, pipe fitting, and/and elbow are evaluated using the formula:

$$h_{L\,\text{minor}} = \frac{K_L V^2}{2g} \tag{8.14}$$

where K_L is the loss coefficient that is a function of geometry and the Reynolds number Re. Table 3.8 lists the empirical K_L values at different elbows (see Fig. 8.16 and Fig. 8.17, see line flow (see Fig. 8.18), and values (see Figs. 8.19–8.22)).

Example 8.3

A hydraulic oil travels with a velocity of 4.6 m/s inside a pipe with a diameter of 1.9 cm and a length of 3.05 m. The pipe has two elbows with regular flanged 90 deg each. The oil has a density of 889.6 kg/m^3 and a viscosity of $\mu = 0.066$ kg/m/s. Calculate:

1. The total head loss through the 3.05 m pipe and two 90 deg elbows
2. The total head loss from the above pipe's specs.

Solution

Using Eq. (8.8) to calculate Reynold's number as

$$\mathrm{Re} = \frac{\rho V D}{\mu} = \frac{\left(889.6\,\mathrm{kg/m^3}\right)(4.6\,\mathrm{m/s})(1.9/100\,\mathrm{m})}{0.066} = 1161.71$$

From Fig. 8.15, $f = 0.0309$, the flow is in the transition range between laminar and turbulent.

Using Eq. (8.11) calculate $h_{L\,\mathrm{major}}$ as

$$h_{L\,\mathrm{major}} = f\frac{l}{D}\frac{V^2}{2g} = 0.0309\frac{(3.05\,\mathrm{m})}{(0.019\,\mathrm{m})}\frac{(4.6\,\mathrm{m/s})^2}{2\left(9.81\,\mathrm{m/s^2}\right)} = 5.35\,\mathrm{m}$$

To calculate the minor losses that are coming from the two elbows, we use Table 8.3 to find $K_L = 0.3$. Using Eq. (8.14) to calculate $h_{L\,\mathrm{minor}}$

$$h_{L\,\mathrm{minor}} = 2\frac{K_L V^2}{2g} = 2\frac{(0.3)(4.6\,\mathrm{m/s})^2}{2\left(9.81\,\mathrm{m/s^2}\right)} = 0.67\mathrm{m}$$

From Eq. (8.10), the total loss of the pipe plus the two elbows is

$$h_L = h_{L\,\mathrm{major}} + h_{L\,\mathrm{minor}} = 5.35\,\mathrm{m} + 0.67\,\mathrm{m} = 5.997\,\mathrm{m}$$

Table 8.3 Loss Coefficient through the valve, pipe fitting, and/or elbow

Component		K_L	Shape
Elbows	Regular 90° flanged	0.3	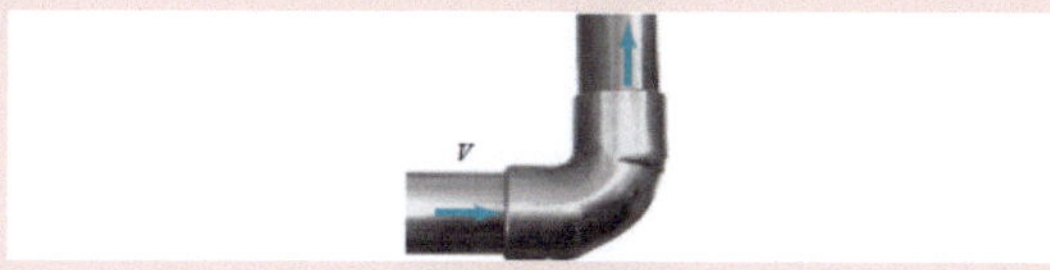 Fig. 8.16 Flanged elbow with 90°. Designed by the author and implemented by both the author and Pranav Dhola. Courtesy of Pranav Dhola 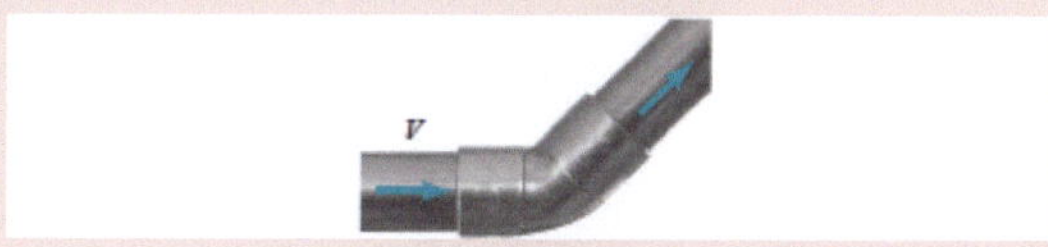 Fig. 8.17 Flanged elbow with 45°. Designed by the author and implemented by both the author and Pranav Dhola. Courtesy of Pranav Dhola
	Regular 90° threaded	1.5	
	Long radius 90° flanged	0.2	
	Long radius 90° threaded	0.7	
	Long radius 45° flanged	0.2	
	Regular 45° threaded	0.4	
Tees	Line flow, flanged	0.2	 Fig. 8.18 Tee line flow. Designed by the author and implemented by both the author and Pranav Dhola. Courtesy of Pranav Dhola
	Line flow, threaded	0.9	
	Branch flow, flanged	1	
	Branch flow, threaded	0.2	
	Line flow, threaded	0.9	
	Branch flow, flanged	1	
	Branch flow, threaded	0.2	

(continued)

Table 8.3 (continued)

Component		K_L	Shape
Valves	Gate fully open	0.15	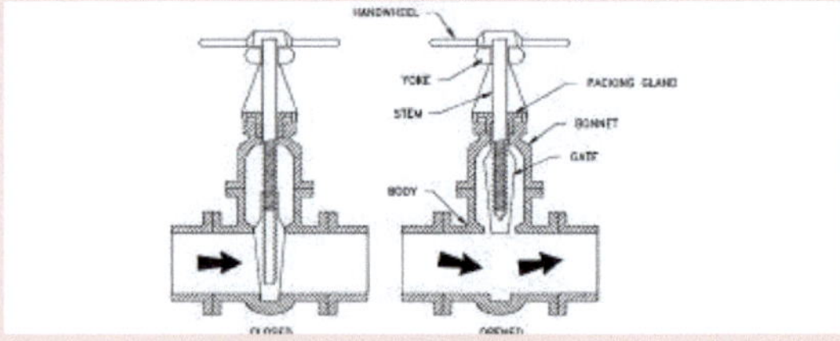 **Fig. 8.19** Gate valve. Designed by the author and implemented by both the author and Pranav Dhola. Courtesy of Pranav Dhola
	Gate, ¼ closed	0.26	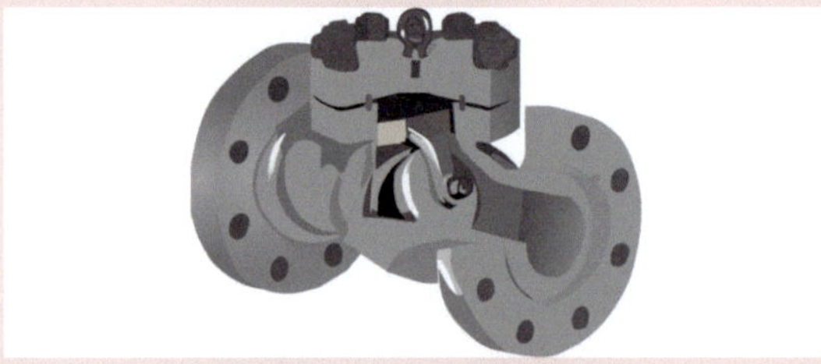 **Fig. 8.20** Swing valve. Designed by the author and implemented by both the author and Pranav Dhola. Courtesy of Pranav Dhola
	Gate, ½ closed	2.1	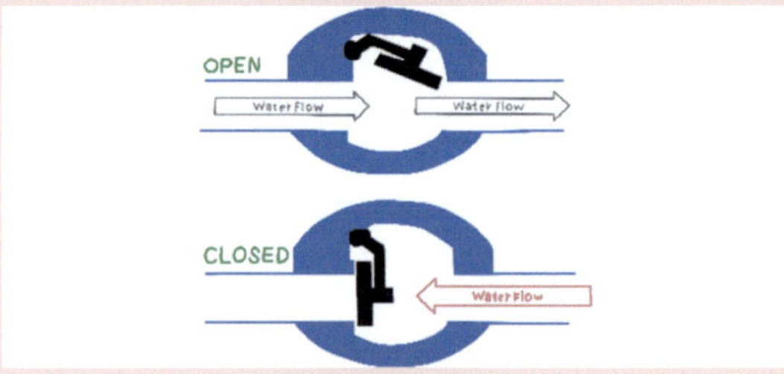 **Fig. 8.21** Check valve. Designed by the author and implemented by both the author and Pranav Dhola. Courtesy of Pranav Dhola
	Gate, ¾ closed	17	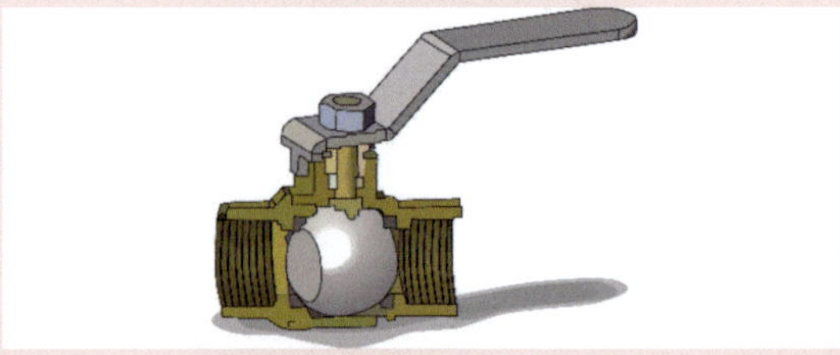 **Fig. 8.22** Ball valve. Designed by the author and implemented by both the author and Pranav Dhola. Courtesy of Pranav Dhola
	Swing check, forward flow	2	
	Swing check, backward flow	Infinity	
	Ball valve, fully open	0.05	
	Ball valve, 1/3 closed	5.5	
	Ball valve, 2/3 closed	210	

8.3.2.3 Cylinder Losses

The main purpose of this subsection is to analyze the power losses that are coming from the hydraulic cylinder. Before going over this losses analysis, we are going to make a brief introduction about different hydraulic cylinder types and the hydraulic cylinder major parts.

A hydraulic cylinder is responsible for converting hydraulic power into mechanical power. Most of the time, we use the hydraulic cylinder as the prismatic kinematic pair, the piston as one link, and the cylinder as the other link. The piston is allowed to slide inside the cylinder with one translational degree of freedom. The piston divides the cylinder into two sides. The piston could have a single rod attached (welded) to only one side, which is why we call it a "single-rod cylinder". Alternatively, the piston could have two rods attached (welded) to each side, in which case we would call it a "double-rod cylinder". Figure 8.23 shows the difference between a single-rod piston and a double-rod piston.

The cylinder could be a single-acting cylinder or a double-acting cylinder. A single-acting cylinder, known as a "plunger", can be a single-rod cylinder and have only one fluid port. This fluid port can let the fluid enter or exit only on one side. Figure 8.24 shows a single-acting cylinder where the port is located on the rod-free side with a return spring on the other side (rod side). The positions of the spring and the port could be flipped. The fluid flow has two cases or two strokes:

1. Extension stroke, where the flow is allowed to enter the cylinder chamber. That causes a higher pressurized force than the load or return spring. In other words, the flow causes an extension force as shown in Fig. 8.24a.
2. Return stroke, where the load's weight or the return spring is higher than the pressure force causing the flow to exit from the pressure port, or a returning force as shown in Fig. 8.24b.

Many applications use the single-acting cylinder such as car jack mechanisms, agricultural machines, and construction machines.

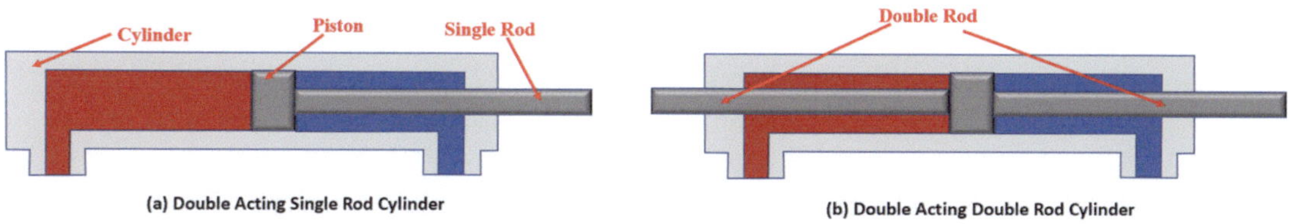

Fig. 8.23 Hydraulic cylinder with single-rod cylinder and double-rod cylinder

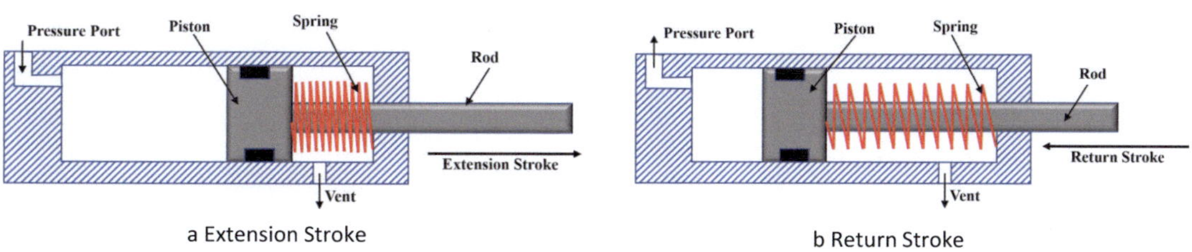

Fig. 8.24 Single-rod and single-acting cylinder, pressurized port at piston side

A double-acting cylinder has two fluid ports. These fluid ports can let the fluid enter and/or exit from each side. The piston in the case of the double-acting cylinder could be a single-rod cylinder or a double-rod cylinder. Figure 8.35 shows a double-acting and single-rod cylinder. The fluid flow has two cases or two strokes:

1. Extension stroke, where the flow is allowed to enter the cylinder chamber from the rod-free side and exit from the rod side causing an extension force as shown in Fig. 8.24a.
2. Return stroke, where the flow is allowed to enter the cylinder chamber from the rod side and exit from the rod-free side causing a returning force as shown in Fig. 8.24b.

A telescopic hydraulic cylinder, shown in Fig. 8.26a, is a compact type of hydraulic cylinder that is designed with a set of nested tubes within one another. The tubes have progressively smaller diameters. The telescopic cylinder could have up to six tubes. A telescopic cylinder has many applications; one of these is the ***Spider Lift*** or the ***Dump Bed*** as shown in Fig. 8.26b. The telescopic cylinder could be a single or double-acting cylinder. The telescopic cylinder provides an exceptionally long output travel from a very compact retracted length. There is an attached video for the telescopic cylinder and another for the dump bed truck with a telescopic cylinder. The reader is encouraged to watch these two videos.

We are now going to introduce some more details about the hydraulic cylinder parts. As an example, Fig. 8.27 shows the double-acting hydraulic cylinder parts. The cylinder assembly consists of a cylinder tube where a piston slides. The piston rod is a hard piece of cold-rolled material that is

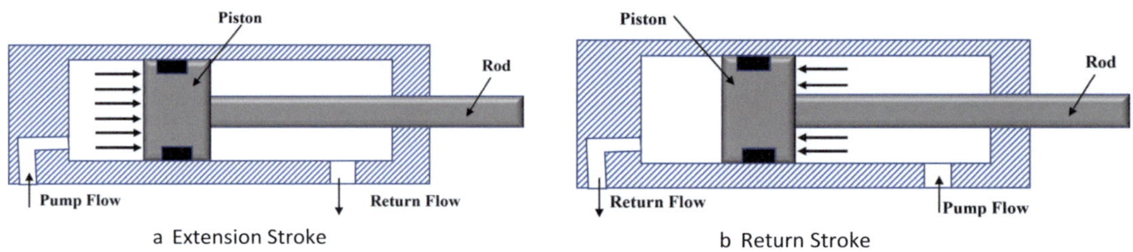

Fig. 8.25 Double-acting cylinder with single-rod piston

Fig. 8.26 Telescopic hydraulic cylinder and application. Figure and associated Dumpster Truck Mechanism.mp4 video designed by the author and implemented by both the author and Pranav Dhola. Both courtesy of Pranav Dhola

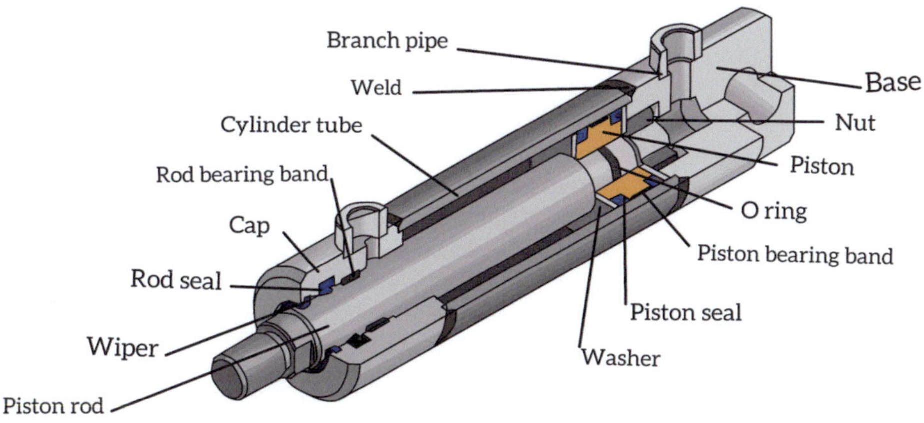

Fig. 8.27 Hydraulic cylinder parts. Designed by the author and implemented by both the author and Pranav Dhola. Courtesy of Pranav Dhola

attached to the piston. There are caps (base and cap) to fence the pressure chamber from the ends. There are wipers to eliminate dust, moisture, and dirt which can cause damage. Rod and piston seals prevent fluid leakage from one chamber to another. A piston-bearing band eliminates metal-to-metal contact. An O-ring creates a barrier between the housing and the casing. A nut locks the piston on the piston rod. Branch pipes allow the fluid to flow in and out.

Hydraulic cylinders have two kinds of power losses—volumetric and mechanical—which can be expressed as volumetric efficiency and mechanical efficiency, respectively. The volumetric efficiency of a hydraulic cylinder comes from leakage losses. In modern sealing technology, the volumetric efficiency is very close to 100% for hydraulic cylinders. The mechanical efficiency of a hydraulic cylinder comes from the forces to overcome both mechanical and fluid friction. In the case of the cylinder, the fluid friction is not as significant as in the case of losses due to the cylinder's aspect ratio (length over diameter ratio). The mechanical friction changes with the velocity and the motion direction (extraction and retraction). Both hydraulic cylinder volumetric efficiency and mechanical efficiency can be calculated with advanced software or experimental tests. Roughly speaking, the average of the hydraulic cylinder's losses is five percent.

The total hydraulic subsystem energy losses are the sum of pump losses, hose losses, valve losses, connection losses, and hydraulic cylinder losses. The conclusion of this losses section is to provide an analysis to help designers determine the losses a mechanism would face from using a hydraulic actuator.

8.3.3 Hydraulic Subsystem SVAJ Constraints

The hydraulic elements control the limits of the actuated motion in terms of position, velocity, and acceleration. For example, the cylinder stroke decides the maximum and minimum position that the hydraulic actuator can deliver. Also, the valve decides the maximum velocity that the hydraulic actuator can deliver to the mechanism. For incompressible flow, Daniel Bernoulli found that the square of the fluid flow rate is directly proportional to the pressure differential across the valve or as written in Eq. (8.15).

$$Q = C_V \sqrt{\Delta P / G} \tag{8.15}$$

where G is the fluid specific gravity and C_v is the valve sizing coefficient that can be measured experimentally or numerically calculated using Computational Fluid Dynamics (CFD) techniques. Although we do not have any limitation on the amount of flow passing across the valve using Eq. (8.15), there is a limitation coming from bubble generation or cavitation due to high-pressure drop. Starting from a specific pressure drop, the bubbles start to be generated and the flow rate stays constant as shown in Fig. 8.28.

The valve manufacturer sets a maximum allowed pressure difference $\Delta P_{\max}$ or a maximum flow rate $Q_{\max}$ to prevent any noise or vibration due to cavitation. Rangeability is defined as the maximum to minimum flow rate that can be controlled by a given valve type and it can be used as a factor for valve sizing and selection [4–8]. The variation of the flow rate versus the maximum pressure is provided by the valve's manufacturer in the so-called valve datasheet or specifications. For example, *Parker Group,* an engineering company for hydraulic valves, provides the $\Delta P \sim Q$ of F types of valves as shown in Fig. 8.29.

As shown in Fig. 8.30 for a double-acting and single-rod cylinder connected to a control valve, the flow that passes across the valve is the same flow going to the cylinder if we assume there is no leakage. The maximum valve flow rate $Q_{\max}$ is the same as the cylinder flow rate $Q_{\max}$. The flow rate of port A, $Q_{A,}$ or the flow rate at port B, Q_B is equal to

$$\begin{aligned} Q_A &= \pm A_A v \\ Q_B &= \mp A_B v \end{aligned} \tag{8.16}$$

where A_A is the piston cross-sectional area and AB is the piston cross-sectional area minus the rod cross-sectional area. From Eq. (8.16) and Fig. 8.30, the maximum velocity limitations when the hydraulic actuator is

$$\begin{aligned} v_{\max}^{\text{Extend}} &= \frac{Q_{A\max}}{A_A} \\ v_{\max}^{\text{Retracted}} &= \frac{Q_{B\max}}{A_B} \end{aligned} \tag{8.17}$$

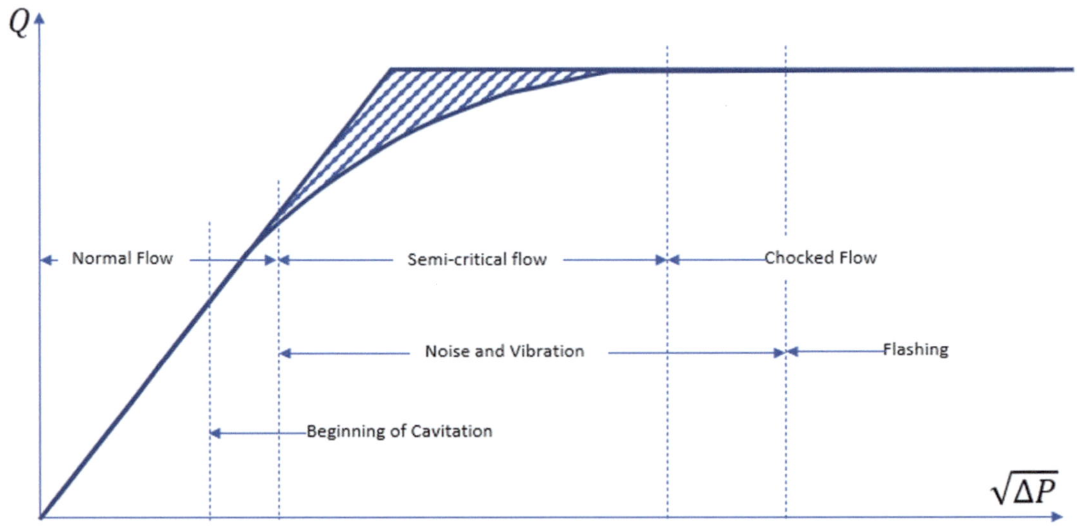

Fig. 8.28 Flow rate diagram of an incompressible fluid flowing through a valve at versus downstream pressure

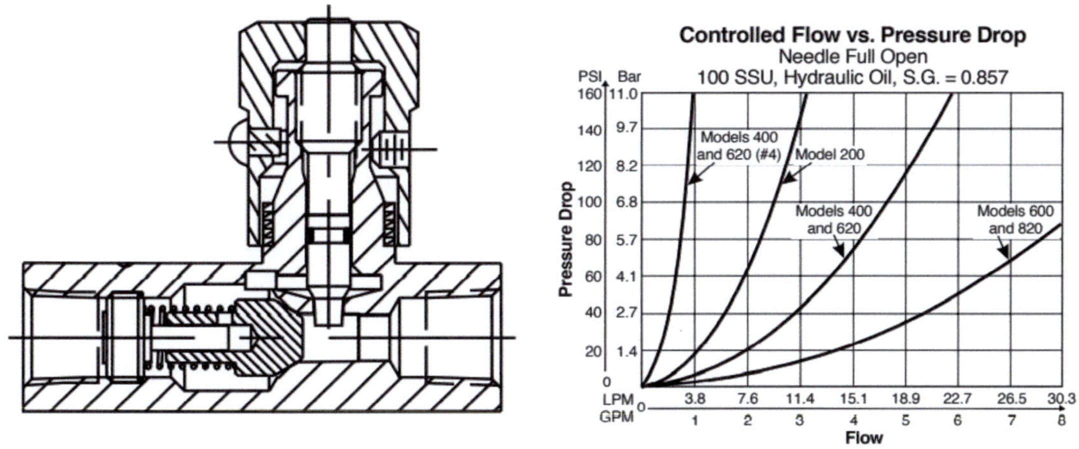

Fig. 8.29 $\Delta P \sim Q$ of Parker Group F-valve type from datasheet

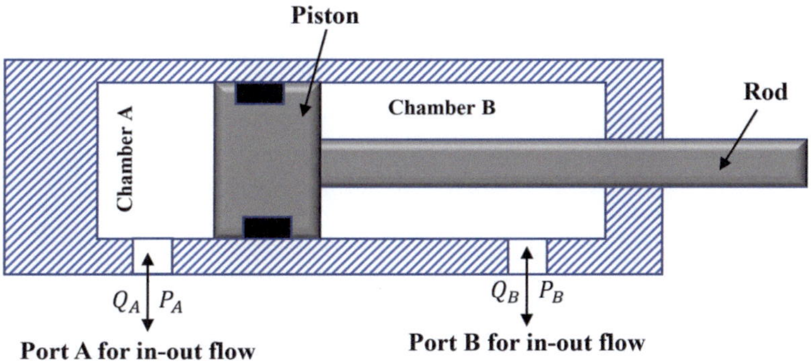

Fig. 8.30 Double-acting and single-rod cylinder

Example 8.4

A skid steer mechanism has a hydraulic single-acting boom cylinder with a control valve that has a maximum flow rate of 22.7 L per minute (lpm). The piston's diameter is 6 cm and the rod's diameter is 4 cm. Find:

- The maximum extended velocity that the boom prismatic joint can move at.
- The maximum retracted velocity that the boom prismatic joint can move at.

Solution

Using Eq. (8.17)

$$v_{\max}^{\text{Extend}} = \frac{Q_{A\max}}{A_A} = \frac{Q_{A\max}}{(\pi/4)d_A^2} = \frac{(22.7\,\text{lpm}) \times 0.001/60}{(\pi/4)(6/100)^2} = 0.134\,\text{m/s}$$

$$v_{\max}^{\text{Retracted}} = \frac{Q_{B\max}}{A_B} = \frac{Q_{B\max}}{(\pi/4)d_B^2} = \frac{(22.7\,\text{lpm}) \times 0.001/60}{(\pi/4)(4/100)^2} = 0.301\,\text{m/s}$$

From the above hydraulic subsystem analysis, usage of the hydraulic actuator has the following limits and restrictions:

- Hydraulic power losses (including but not limited to pump losses, hose losses, cylinder losses, connection losses, and valve losses) reduce the maximum power that the mechanism can be activated by the original power source (engine or motor). Knowing the available actuated power at each speed requires knowing the available force. We have not yet spoken about machine dynamics, or the forces and moments or torques. More details about the machines' dynamics will be explained in Chaps. 11 and 12.
- Cylinder stroke limits the minimum and maximum displacement at which the mechanism can be activated.
- Valve maximum flow rate limits the maximum speed at which the mechanism can be activated.
- Valve time response limits the maximum acceleration and deceleration that the mechanism can be active by.

8.4 Limitations of Electrical Motors

An electrical motor can be powered by a direct current (DC) or an alternating current (AC). The DC motor is powered with a constant current with respect to time that is supplied by a power source such as a generator or battery. The AC motor is powered by alternating current that can be a single-phase or a three-phase AC. In this part, we are not going to talk about motor assembly and construction since it is out of the scope of this textbook. More details about the components and functionality of both DC and AC can be found in Ref. [10]. From the machine point of view, we are going to briefly talk about different types of motors based on the application.

A **gearmotor** is an electro-mechanical device that is powered by AC or DC to produce rotational motion. A gearmotor can be attached to a gearbox to reduce the angular speed, so it works at a slow speed to provide high torque. Gearmotors are known to be cost-effective and have minimum maintenance requirements, so they are used in activating many machines such as conveyor machines and food processing machines [10].

A **stepper motor** is an electro-mechanical device powered by AC to produce rotational motion. A stepper motor rotates through precise angles or positions regardless of the delivered speed and torque as long as the delivered power is less than the stepper motor's available power. A stepper motor achieves positions as a discrete number of steps which makes a stepper motor ideal for packaging machines, medical imaging machines, and printer machines. The stepper motor has some disadvantages such as loss of synchronization when the load changes unexpectedly and sometimes the generation of a high level of vibration and noise [10].

A **servomotor** is an electro-mechanical device powered by AC or DC to produce rotational motion. A servomotor requires a feedback loop to control the output angular position, velocity, and acceleration. Servomotors are used in positioning applications where smooth, controllable motion is required such as flight control surfaces [10].

The motion trajectory is mainly related to the motor torque-speed curve that describes the available torque at each operating speed. In general, the torque-speed curve of the DC motor is a linear inverse relationship as shown in Fig. 8.31 where there are two characteristic points: *no-load speed* and the *stall torque*. The no-load speed point is practically impossible when the motor is running at its maximum speed but applying zero torque. The stall torque point is where the motor is providing the maximum torque but is not spinning.

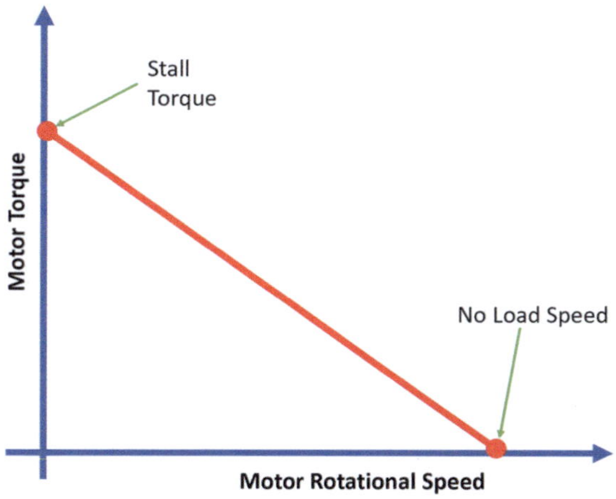

Fig. 8.31 Generic DC motor torque-speed curve [10]

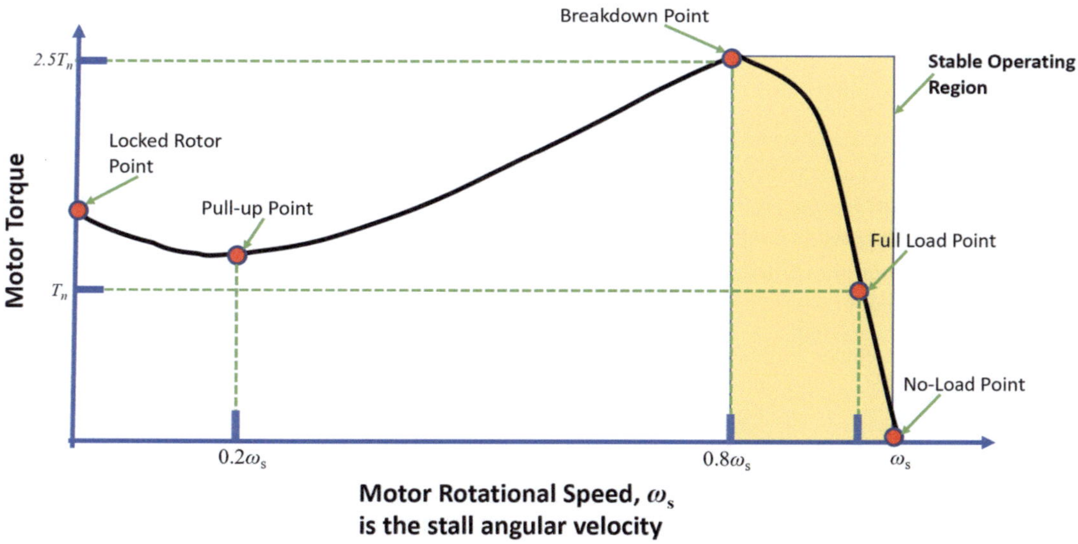

Fig. 8.32 Generic AC motor torque-speed curve

On the other hand, the torque-speed curve of the AC motor has the characteristics shown in Fig. 8.32, where there are five major operating points [10]:

1. The locked rotor point has the torque that the AC motor provides at the start, or where the angular speed is zero. The estimated location of the locked rotor point is at zero angular velocity and torque T_{Locked} to $1.5T_n$, where T_n is the nominal torque.
2. The pull-up point is at the lowest point where the AC motor picks up speed. The estimated location of the pull-up point is at an angular velocity of 20% of the stall angular velocity ω_s.
3. Breakdown point or pull-out torque point is at the maximum torque that an AC motor develops without causing sudden drops in speed. The estimated location of the breakdown point is at an angular velocity of 80% from the stall angular velocity ω_s and at torque $T_{\text{Breakdown}} \approx 2.5T_{\text{n}}$.

4. Full-load point is where the fully loaded motor delivers the maximum torque. The location of the full load is defined to be at the nominal torque T_n.
5. No-load point is where stall happens at the stall angular velocity ω_s.

As shown in Fig. 8.32, the yellow section represents the stable region wherein the motor could operate by increasing the motor angular which reduces the available maximum torque. From the power curve, we can specify the range of the angular velocity and the maximum available torque at each angular velocity [10].

The motor receives the electrical power (from the power source) and converts it to mechanical power (rotational motion). There are many sources of losses in the motor conversion process such as stator resistance loss, rotor resistance loss, hysteresis losses, eddy current losses, friction losses, windage losses, and stray load losses. These losses can be observed as generated heat, vibration, and/or noise that may lead to the motor stopping or a motor failure. The power equation of the motor conversion process is [10]

$$\mathcal{P}_i = \mathcal{P}_o + \underbrace{\mathcal{P}_{\text{ohm}} + \mathcal{P}_h + \mathcal{P}_e + \mathcal{P}_m}_{\text{losses}} \tag{8.18}$$

where

$\mathcal{P}_i$ Input electrical power to the motor
$\mathcal{P}_o$ Output mechanical power from the motor
$\mathcal{P}_{\text{ohm}}$ Ohm power losses
$\mathcal{P}_h$ Hysteresis power losses
$\mathcal{P}_e$ Eddy current power losses
$\mathcal{P}_m$ Mechanical power losses

The motor efficiency η_{motor} is defined as

$$\eta_{\text{motor}} = \frac{\mathcal{P}_o}{\mathcal{P}_i} \tag{8.19}$$

The motor efficiency depends on many factors; one of which is the load torque. The maximum motor efficiency is around 70–80% in the case of the DC and AC motors. The estimating of this efficacy is out of the scope of this textbook.

From the above motor analysis, usage of the motor actuator has the following limits and restrictions:

- Knowing the torque-speed curve provides us with how much available torque to activate the mechanism with at each angular speed.
- Motor power losses reduce the maximum power that can be transferred to the mechanism by the original power source (engine or motor).
- Motor time response limits the maximum acceleration and deceleration that the mechanism can be activated by.

8.5 Trajectory Generation

In the machine input motion, the generated path or mission consists of trajectories. There are three types of trajectories: rise, fall, and dwell.

- ***The rise trajectory*** is defined when the motion of the transitional displacement or the angular displacement increases with time. For example, assume there is a motor attached to a four-bar mechanism at joint J_1^R. When the motor rotates counterclockwise from angle 0° to 45°, the motion is called *rise motion* or *rise trajectory* since the angular displacement θ_2 increases with time.
- ***The fall trajectory*** is defined when the motion of the translational displacement or angular displacement decreases with time. For example, in a four-bar mechanism, when the attached motor to the joint J_1^R rotates θ_2 clockwise from angle 45° to 0°, the motion is called *fall trajectory*.
- ***The dwell trajectory*** is defined when the motion of the translational displacement or angular displacement is at a hold with respect to time or has no change with time.

The machine user wishes to get a task done in no time. That is virtually impossible for many reasons. Actuators have limited minimum and maximum speeds. For example, assume the machine is activated with a hydraulic actuator, as we discussed in Sect. 8.3.3. The hydraulic valve must have a maximum flow rate; otherwise, there is a high possibility that the hydraulic system may explode. In the same way for the electrical motor, the more the speed increases, the less torque that motor can deliver. Every kind of actuator has a power ceiling that cannot be exceeded or there will be an actuator failure. Also, as we discussed in Chap. 7, even if we made an unlimited powered actuator, increasing the jerk magnitude would cause the machine's links to vibrate and generate noise, and machine damage may occur.

Because of the actuator displacement, velocity, plus power limitations, and the mechanism's vibration limitations, engineers must design the input motion request carefully so these limits are not exceeded. We call the input motion path design the *SVAJ diagram design*, where *S* refers to displacement, *V* refers to velocity, *A* refers to acceleration, and *J* refers to jerk. In the literature, the concept of SVAJ is referred to as *CAM motion*. We are going to expand the usage of SVAJ for all kinds of machines' motion.

Most machines have a cyclic job to be repeated after a periodic time *T*. The SVAJ diagram presents the time history of the displacement, velocity, acceleration, and jerk over this period *T*. The first step of drawing or plotting the SVAJ diagram is to plot the displacement function with time *S*(*t*). After that, calculate the first derivative with respect to the time or the velocity *V*(*t*), calculate the second derivative with respect to the time or the acceleration *A*(*t*), and calculate the third derivative with respect to the time or the jerk *J*(*t*). These derivative calculations can be done mathematically, graphically, or by using software. Finally, plot the four SVAJ functions with respect to time in one column as four rows. The SVAJ diagram shows where the displacement of the desired motion function causes a high amplitude in velocity desired motion, in acceleration desired motion, or jerk desired motion that may cause failure for the actuator or the mechanism. Now, we are going to mention some mathematical formulas to design the SVAJ input motion diagram.

Step SVAJ Motion Diagram: In this motion, the requested input displacement goes suddenly or in a step fashion from a certain value to another at a certain time. One example is the dwell-rise-dwell-fail-dwell step SVAJ motion diagram as shown in Fig. 8.33. The displacement motion (translational or rotational) starts at S_1 at $t=0$ s and it stays at S_1 until it goes suddenly to S_2 at time t_1. After that, the displacement stays at S_2 until it goes suddenly back to S_1 at time t_2; assume $S_2>S_1$. The displacement stays at S_2 until $t=T$. The displacement as a function of time is

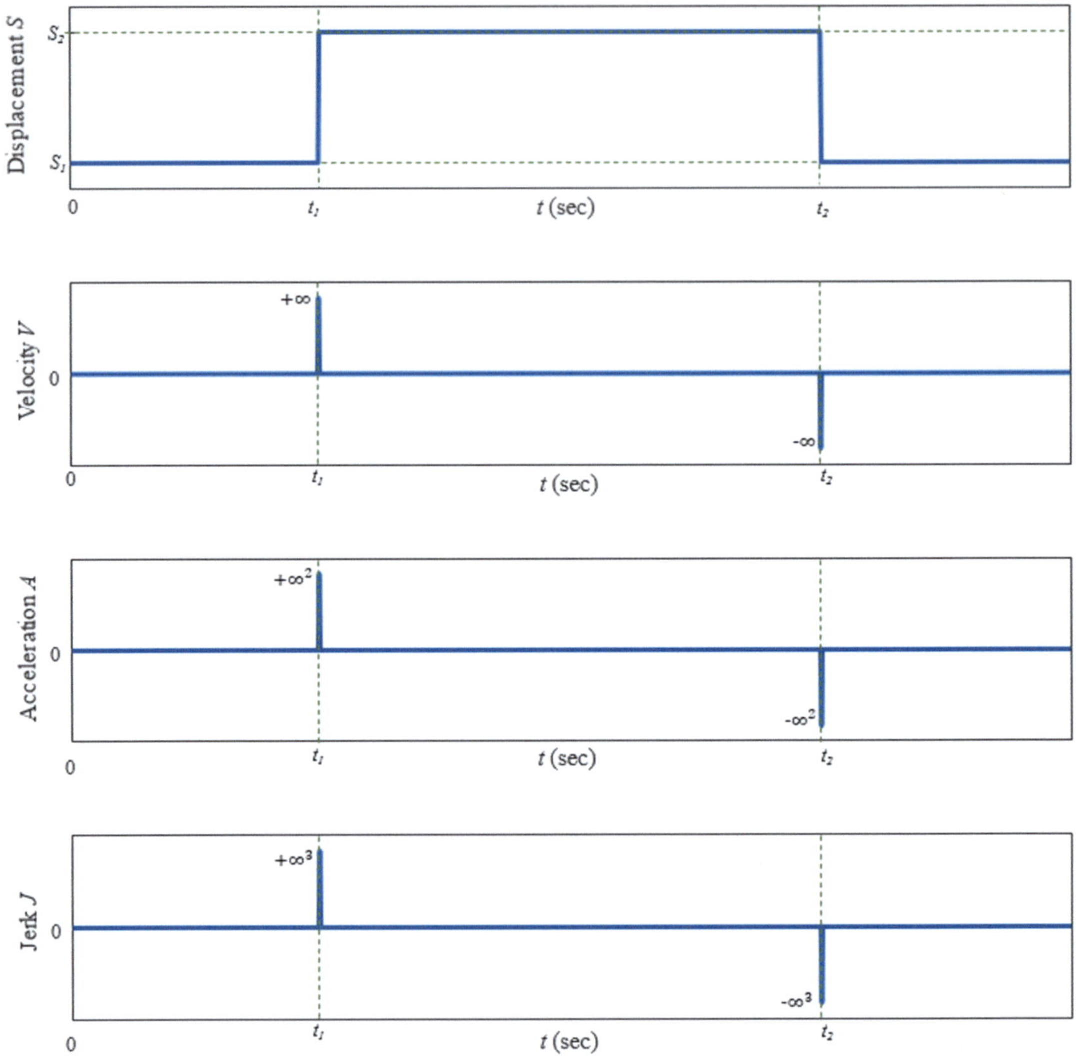

Fig. 8.33 Dwell-rise-dwell-fail-dwell step SVAJ diagram

$$S(t) = \begin{cases} S_1 & 0 \le t \le t_1 \\ S_2 & t_1 \le t \le t_2 \\ S_1 & t_2 \le t \le T \end{cases} \tag{8.20}$$

There are two discontinuities at times t_1 and t_2. The next step to sketch the SVAJ diagram is to calculate the first, second, and third derivatives of $S(t)$ with respect to the time and plot them. The first derivative at t_1 is positive infinity that is mathematically described as:

$$V(t = t_1) = \left.\frac{\mathrm{d}S}{\mathrm{d}t}\right|_{t=t_1} = \lim_{t \to t_1} \frac{\Delta S}{\Delta t} = \lim_{t \to t_1} \frac{S_2 - S_1}{t_1 - t_1} = \frac{+\text{ve}}{0} = +\infty \tag{8.21}$$

The first derivative at t_2 is negative infinity that is mathematically described as:

$$V(t = t_2) = \left.\frac{dS}{dt}\right|_{t=t_2} = \lim_{t\to t_2} \frac{\Delta S}{\Delta t} = \lim_{t\to t_2} \frac{S_1 - S_2}{t_2 - t_2} = \frac{-ve}{0} = -\infty \tag{8.22}$$

The rest of the velocity has a zero value. Simultaneously, we calculate the second derivate with respect to the time or the acceleration $A(t)$ and the third derivative with respect to the time or the jerk $J(t)$. There is a discontinuity in the requested velocity at t_1 and t_2. This leads to the fact that the requested acceleration and the requested jerk have infinite values. Note—when the desired velocity, acceleration, and/or acceleration has an infinite value, the actuator cannot accomplish any of these requests. This may lead to an actuator failure, a sudden stop (stall), and/or total damage. Also, the high jerk leads to a vibration in the mechanism violating the accuracy of the task and may cause damage to the mechanism. For these reasons, the step command is not recommended in machine input mission trajectory design.

Example 8.5

A skid steer mechanism has a single-acting hydraulic bucket cylinder with a bore diameter of 69.9 mm, rod diameter of 34.9 mm, and stroke of 410 mm. From the hydraulic system, the maximum flow speed in the dump process is 0.135 m/s and the maximum flow speed in the roll-back process is 0.18 m/s.

- Use the step SVAJ motion to design the fastest request input motion as dwell-rise-dwell-fall-dwell considering that the first dwell and the third dwell take 1 s each while the second dwell takes 2 s.
- Sketch the resultant step SVAJ diagram and comment on the applicability of the requested motion.

Solution

Using Eq. (8.20), as

$$S(t) = \begin{cases} 0 & 0 \le t \le 1 \\ 410 & 1 \le t \le 3 \\ 0 & 3 \le t \le 4 \end{cases}$$

where S is in mm and t is in seconds.

The sketch of the SVAJ diagram is shown in Fig. 8.34. The velocity has infinite values at $t=1$ s and $t=3$ s. The actuator cannot provide these infinite speeds. The actuator has a maximum rising speed of 0.135 m/s and a maximum falling speed of 0.18 m/s. Also, the infinite jerk values would cause the mechanism to vibrate with a possibility of damaging the mechanism.

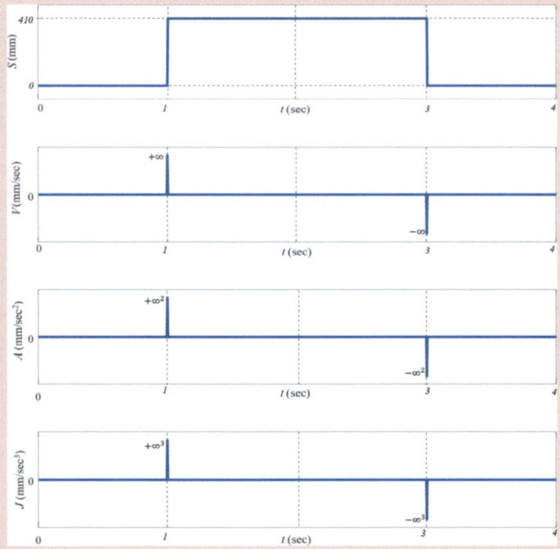

Fig. 8.34 SVAJ diagram of Example 8.5

Ramp SVAJ Motion Diagram: In this motion, the requested input displacement goes linearly from a certain value to another over a certain nonzero time window. One example of the ramp SVAJ is dwell-rise-dwell-fall-dwell as shown in Fig. 8.35. The displacement (translational or rotational) starts at $t=0$ s with a value S_1 and stays the same until t_1. After that, the displacement starts to go linearly from S_1 at t_1 to S_2 at t_2 and stays the same until t_3. The displacement goes down from S_2 at t_3 to S_1 at t_4 and stays at S_1 until T; assume $S_2>S_1$. Equation (8.23) shows the mathematical expression of the dwell-rise-dwell-fall-dwell ramp displacement as:

$$S(t) = \begin{cases} S_1 & t \le t_1 \\ S_1 + m_1(t - t_1) & t_1 < t < t_2 \\ S_2 & t_2 \le t \le t_3 \\ S_2 - m_2(t - t_2) & t_3 < t < t_4 \\ S_1 & t_4 < t < T \end{cases} \tag{8.23}$$

where

$$m_1 = (S_2 - S_1)/(t_2 - t_1)$$
$$m_2 = (S_2 - S_1)/(t_3 - t_2)$$

where T is the periodic time of the machine's trajectory. There is no discontinuity in the displacement curve. The next step to sketch the SVAJ diagram is to calculate the first derivative of $S(t)$ as:

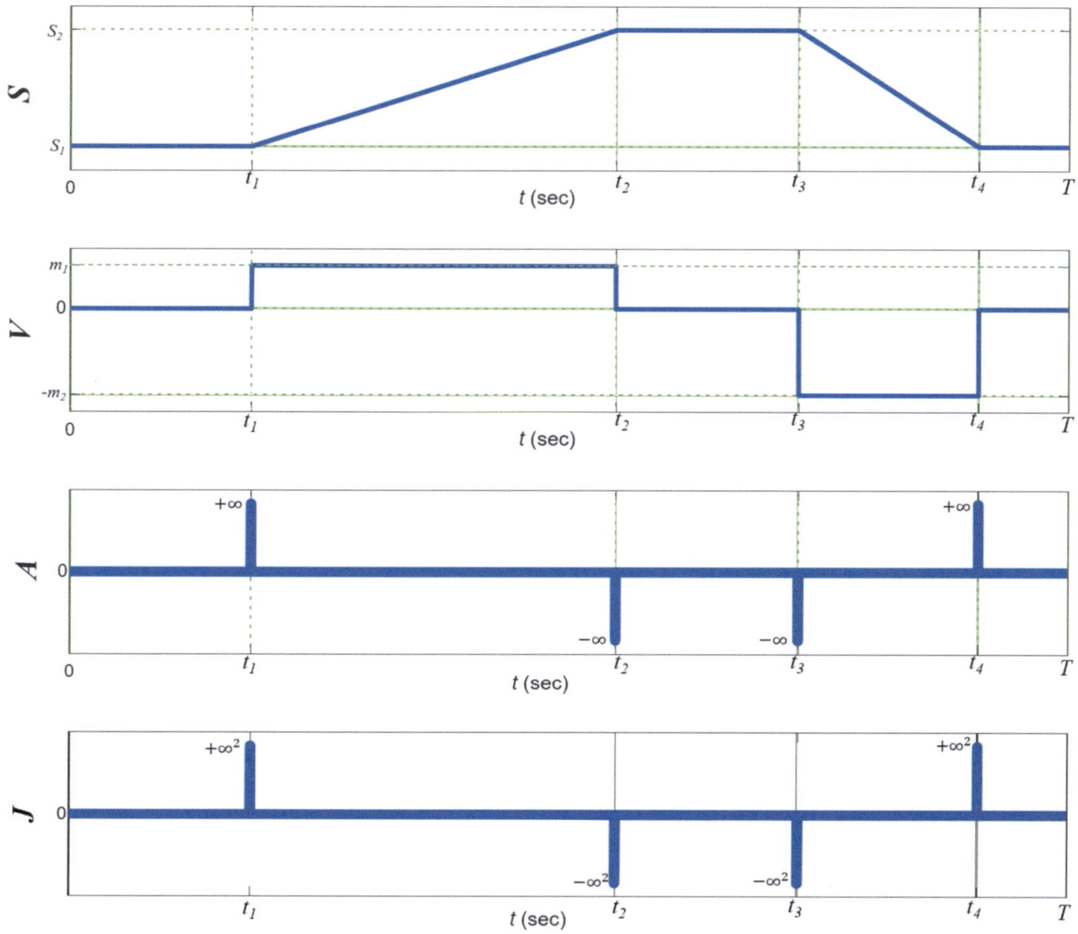

Fig. 8.35 Dwell-rise-dwell-fail-dwell ramp SVAJ diagram

$$V(t) = \begin{cases} o & t \le t_1 \\ m_1 & t_1 < t < t_2 \\ 0 & t_2 \le t \le t_3 \\ -m_2 & t_3 < t < t_4 \\ 0 & t_4 < t < T \end{cases} \tag{8.24}$$

As shown in Fig. 8.35 and from Eq. (8.24), the velocity V has four discontinuities at t_1, t_2, t_3, and t_4 that lead to infinities requested accelerations and infinities requested jerk at times t_1, t_2, t_3, and t_4. The entire ramp SVAJ diagram is virtually impossible since the actuator has limited power and no way to deliver an infinite acceleration. Asking the actuator to deliver such infinite acceleration may lead to stalling the power source. The infinite jerk would lead to mechanism vibration and the possibility of failure.

Example 8.6

The skid steer mechanism has a single-acting hydraulic bucket cylinder with a bore diameter of 69.9 mm, rod diameter of 34.9 mm, and stroke of 410 mm. From the hydraulic system, the maximum flow speed in the dump process is 0.135 m/s and the maximum flow speed in the roll-back process is 0.18 m/s.

- Use the ramp SVAJ motion to design the fastest request input motion of dwell-rise-dwell-fall-dwell considering that the first dwell and the third dwell take 1 s each while the second dwell takes 2 s.
- Sketch the resultant ramp SVAJ diagram and comment on the applicability of the requested motion.

Solution

The givens in Example 8.6 can be summarized as: the rising slope $m_1 = 0.135$ m/s. The falling slope $m_2 = 0.18$ m/s. The low level $S_1 = 0$ m. The up level $S_2 = 410$ mm $= 0.41$ m.

The Ramp Function Times

The difference between $t_o = 0$ and t_1 is given to the first dwell time or 1 s.

$$t_1 = t_0 + 1(\text{first dwell time}) = 0 + 1 = 1\,\text{s}$$

Using the rising ramp $m_1 = 0.135$ m/s, we can calculate t_2 from

$$t_2 = t_1 + \frac{S_2 - S_1}{m_1} = 1 + \frac{0.410}{0.135} = 4.037\,\text{s}$$

The difference between t_3 and t_2 is given to be the second dwell time or 2 s.

$$t_3 = t_2 + 2(\text{second dwell time}) = 4.037 + 2 = 6.037\,\text{s}$$

Using the falling ramp $m_1 = 0.135$ m/s, the time t_4 can be calculated from

$$t_4 = t_3 + \frac{S_2 - S_1}{m_2} = 6.037 + \frac{0.410}{0.18} = 8.315\,\text{s}$$

The entire mission time T is 1 s after t_4 or

$$T = t_4 + 1(\text{third dwell time}) = 8.315 + 1 = 9.315\,\text{s}$$

Using Eq. (8.23), the equation of the dwell-rise-dwell-fall-dwell ramp displacement is

$$S(t) = \begin{cases} 0 & t \le 1 \\ 0.135(t-1) & 1 < t < 4.037 \\ 0.41 & 4.037 \le t \le 6.037 \\ 0.41 - 0.18(t - 4.037) & 6.037 < t < 8.315 \\ 0 & 8.315 < t < 9.315 \end{cases}$$

The sketch of the SVAJ diagram is shown in Fig. 8.34. The requested velocity is within the limit of the actuator's maximum and minimum velocities. There are four infinite acceleration requests at $t_1 = 1$ s, $t_2 = 4.037$ s, $t_3 = 6.037$ s, and $t_4 = 8.315$ s that cause engine stall. Also, the infinite jerk requests, at the same time, would cause the mechanism to vibrate with a possibility of damaging the mechanism.

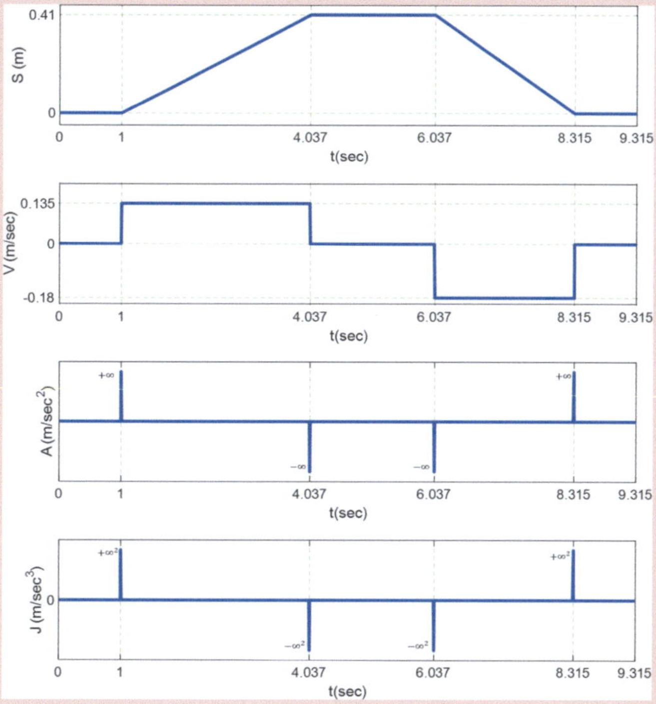

Fig. 8.36 Dwell-rise-dwell-fail-dwell ramp SVAJ diagram of Example 8.6

Harmonic SVAJ Motion Diagram: In this motion, the requested input displacement goes from a certain value $S_1 = S(t_1)$ to another value $S_2 = S(t_2)$ over a certain nonzero time window $\Delta t = t_2 - t_1$ using a sum of trigonometric functions as

$$S(t) = \sum_{i=0}^{n} A_i \sin\left(B_i t + C_i\right) \tag{8.25}$$

where A_i, B_i, and C_i are constants that determine the shape of the SVAJ diagram. These constants can be calculated to satisfy specific conditions or displacement values, velocity values, acceleration values, and/or jerk values. The difficulty of calculating these constants is that the conditions turn into a set of nonlinear equations needing to be solved.

As an example, a simple harmonic, defined as $n = 1$ in Eq. (8.25), rise SVAJ trajectory is

$$\begin{aligned} S(t) &= S_o + \frac{\Delta S}{2}\left(1 - \cos\left(\frac{\pi (t - t_o)}{T_s}\right)\right) \\ V(t) &= \frac{\pi}{T_s}\frac{\Delta S}{2} \sin\left(\frac{\pi (t - t_o)}{T_s}\right) \\ A(t) &= \frac{\pi^2}{T_s^2}\frac{\Delta S}{2} \cos\left(\frac{\pi (t - t_o)}{T_s}\right) \\ J(t) &= -\frac{\pi^3}{T_s^3}\frac{\Delta S}{2} \sin\left(\frac{\pi (t - t_o)}{T_s}\right) \end{aligned} \tag{8.26}$$

where to is the initial time, T_s is the time span, So is the initial displacement, and ΔS is the displacement difference of the trajectory. The velocity $V(t)$, acceleration $A(t)$, and jerk $J(t)$ in Eq. (8.26) are calculated as the first, second, and third derivatives of $S(t)$. Figure 8.37 shows the simple harmonic rise SVAJ diagram. We can design a harmonic SVAJ trajectory that has a difference in the displacement ΔS with maximum velocity Vmax. The trajectory periodic time can be calculated from

$$V_{\max} = V(t = T_s/2 + t_o) = \frac{\pi}{T_s}\frac{\Delta S}{2} \text{ or } T_s = \frac{\pi}{V_{\max}}\frac{\Delta S}{2} \tag{8.27}$$

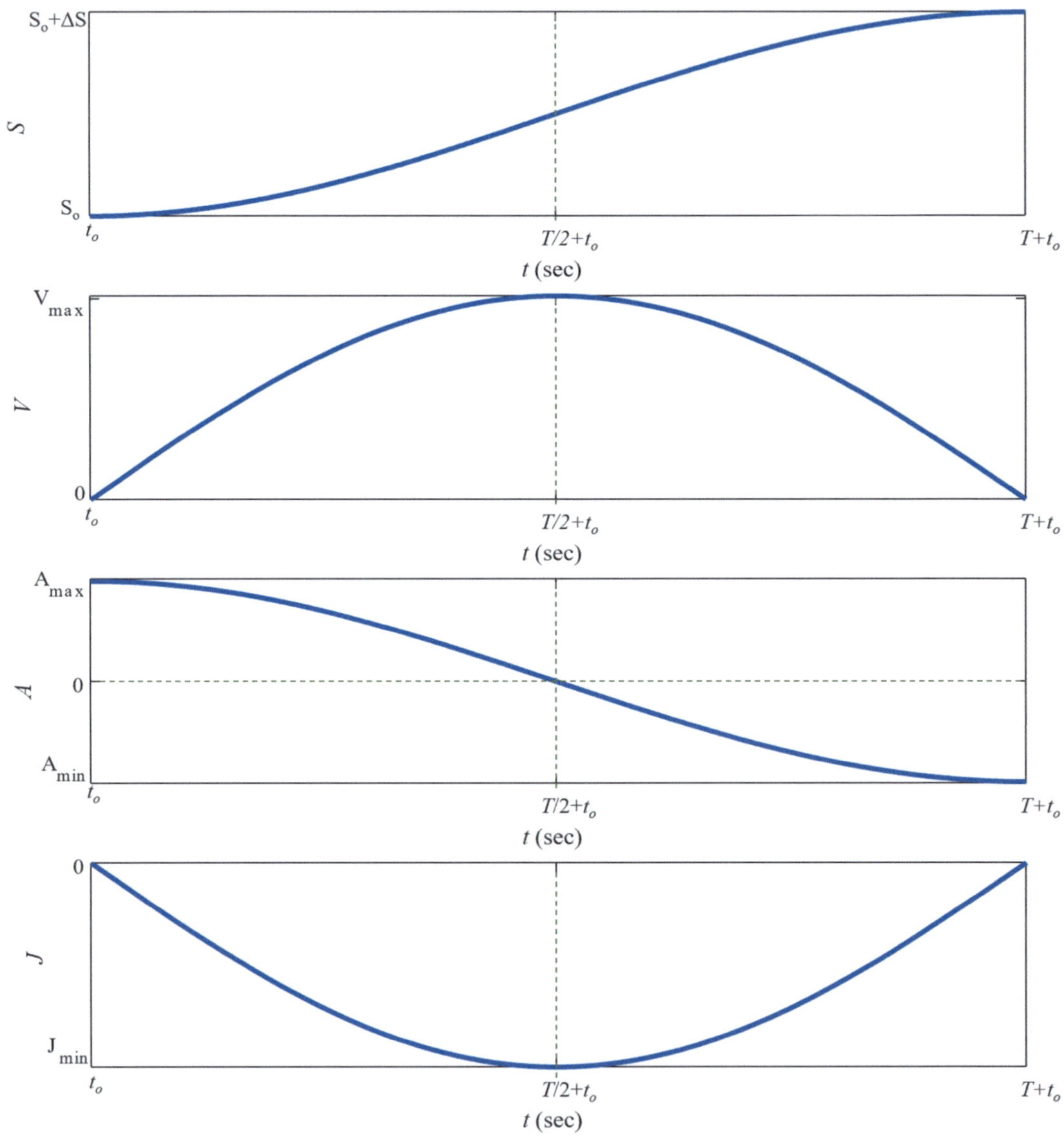

Fig. 8.37 Simple harmonic rise motion

Equation (8.27) helps to keep the requested velocity within the actuator motion limits. The issue with using the simple harmonic is the acceleration of nonzero values at t_o and T that may cause a discontinuity and that may lead to stalling the power source as well as causing an infinite requested jerk.

Example 8.7

The skid steer mechanism has a single-acting hydraulic bucket cylinder with a bore diameter of 69.9 mm, rod diameter of 34.9 mm, and stroke of 410 mm. From the hydraulic system, the maximum flow speed in the dump process is 0.135 m/s and the maximum flow speed in the roll-back process is 0.18 m/s.

- Using the simple harmonic SVAJ motion to design the fastest request input motion as dwell-rise-dwell-fall-dwell considering the motion is going full stroke up and down. Also, the first dwell and the third dwell take 1 s each while the second dwell takes 2 s.
- Sketch the resultant simple harmonic SVAJ diagram and comment on the applicability of the requested motion.

Solution

The requested mission has five trajectories, dwell-rise-dwell-fall-dwell. The first dwell trajectory goes from $t=0$ s to $t=1$ s (1-s duration) with the value 0 m. Using Eq. (8.27), the rise trajectory has

$$t_{o\text{Rise}} = 1\text{s} \ S_{o\text{Rise}} = 0\text{m} \ \Delta S_{\text{Rise}} = 0.41\,\text{m}$$

$$T_{\text{Rise}} = \frac{\pi}{V_{\max}}\frac{\Delta S_{\text{Rise}}}{2} = \frac{\pi}{0.135}\frac{0.41}{2} = 4.77\,\text{s}$$

The second dwell trajectory goes from $t = t_{o\text{Rise}} + T_{\text{Rise}} = 1 + 4.77 = 5.77$ s to $t=7.77$ s (2 s duration) with the value 0.41 m. The fall trajectory has

$$t_{o\text{Fall}} = 7.77\text{s} \ S_{o\text{Fall}} = 0.41\text{m} \ \Delta S_{\text{Fall}} = -0.41\,\text{m}$$

$$T_{\text{Fall}} = \frac{\pi}{V_{\min}}\frac{\Delta S_{\text{Fall}}}{2} = \frac{\pi}{-0.18}\frac{-0.41}{2} = 3.58\,\text{s}$$

The third dwell goes from $t = t_{o\text{Fall}} + T_{\text{Fall}} = 7.77 + 3.58 = 11.35$ s to $t=12.35$ (1 s duration). Using Eq. (8.26), the entire mission is

$$S(t) = \begin{cases} 0 & t \le 1 \\ 0.205\left(1 - \cos\left(\frac{\pi(t-1)}{4.77}\right)\right) & 1 < t < 5.77 \\ 0.41 & 5.77 \le t \le 7.77 \\ 0.41 - 0.205\left(1 - \cos\left(\frac{\pi(t-7.77)}{3.58}\right)\right) & 7.77 < t < 11.35 \\ 0 & 11.35 < t < 12.35 \end{cases}$$

The sketch of the simple SVAJ diagram is shown in Fig. 8.38. The requested velocity is within the actuator motion limits. The acceleration has discontinuities at times: 1 s, 5.77 s, 7.77 s, and 11.35 s that result in an infinite jerk rate that may cause a mechanism vibration.

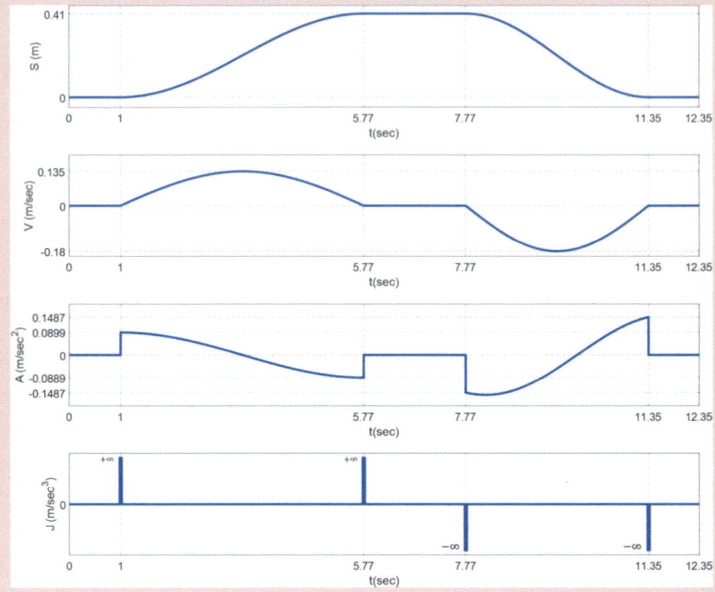

Fig. 8.38 Dwell-rise-dwell-fail-dwell simple harmonic SVAJ diagram of Example 8.7

Polynomial SVAJ Motion Diagram: In this type of motion, we assume that the requested input displacement goes from a certain value $S_1 = S(t_1)$ to another value $S_2 = S(t_2)$ over a certain time window $\Delta t = t_2 - t_1$ using a sum of nth order polynomial function as

$$S(t) = \sum_{i=0}^{n} A_i t^i t_1 \leq t \leq t_2 \tag{8.28}$$

where A_i is a set of constants that determines the shape of the SVAJ diagram. These constants can be calculated to satisfy specific conditions or displacement values, velocity values, acceleration values, and jerk values. The good news is that calculating these constants results in a set of linear equations. Step SVAJ motion design is an example of the polynomial SVAJ motion design where $n = 0$. Ramp SVAJ motion design is another example of the polynomial SVAJ motion design where $n = 1$. More details about polynomial SVAJ motion design will be explain in Chap. 9.

8.6 Recall Simulink®

In this section, we will explain how to use the Simulink® toolbox to build SVAJ input signals to satisfy a specifically required displacement with respect to time taking into consideration the actuator's and the mechanism's constraints as the displacement limits, velocity limits, acceleration limits, and jerk limits. Also, the proposed Simulink® SVAJ avoids the discontinuity in the displacement, the velocity, the acceleration, and the jerk. We are going to review a couple of Simulink® blocks to design an SVAJ that satisfies the actuator and mechanism's constraints without any discontinuity.

Definition[1]

Saturation:

The saturation block produces an output signal that is the value of the input signal bounded to the upper and lower saturation values. The upper and lower limits are specified by the parameters Upper limit and Lower limit [11].

When a user operates a machine, it is sometimes their instinct to push the input all the way up or all the way down. For example, when the driver of a car wants to suddenly accelerate, he or she may push the gas pedal all the way down. Or, when a car driver wants to stop suddenly, he or she may push the brakes all the way down. Suddenly pushing the gas pedal or brake pedal may lead to stalling the car engine in addition to discomfort for the driver and passengers from the car's undesireable vibrations.

A similar scenario is when a tractor driver aggressively pushes the lifting cylinder all the way up. There is a need for logic to prevent pushing the actuator out of its range; otherwise, there will be a possibility of damaging the actuator and/or the mechanism. Saturation is one of our tools to prevent the input signal from exceeding the bounds of the acceptable range. Saturation means bounding the signal between a maximum value and a minimum value. Simulink® offers a block called "saturation" in the discontinuities library as shown in Fig. 8.39, to apply the saturation on any signal. The Saturation block has one input signal and one output signal.

If we select the saturation block for a new model and double-click on the block, a popup menu appears where the user can insert the upper limit (maximum value) and the lower limit (minimum value) as shown in Fig. 8.39. Two test cases about the usage of the saturation block are shown in Fig. 8.40. In the two cases, the saturation block has an upper limit of 0.5 and a lower limit of – 0.5. In Case I, the input is a sinusoidal signal with amplitude 1 that is out of the range. The output cuts out all the out-of-range parts. In Case II, the input is a sinusoidal signal with amplitude 1 that is inside the range and the output is the same as the input.

The saturation block could be used in designing an SVAJ input signal to a hydraulic actuator. The saturation block should consider the hydraulic cylinder stroke to determine the maximum and minimum of the displacement signal. For example, Fig. 8.41 shows when the piston is fully retracted and fully extended. The minimum is defined as zero when the piston is fully retracted. The maximum is defined as the cylinder stroke when the piston is fully extended. Violating these limits may lead to damaging the machine and the hydraulic actuator as previously discussed in Sect. 8.3.

Definition[2]

Rate Limiter:

The rate limiter block limits the first derivative of the signal passing through it. The output changes no faster than the specified limit [11].

[1] Reproduced from [11] by MathWorks with permission from MathWorks.

[2] Reproduced from [11] by MathWorks with permission from MathWorks.

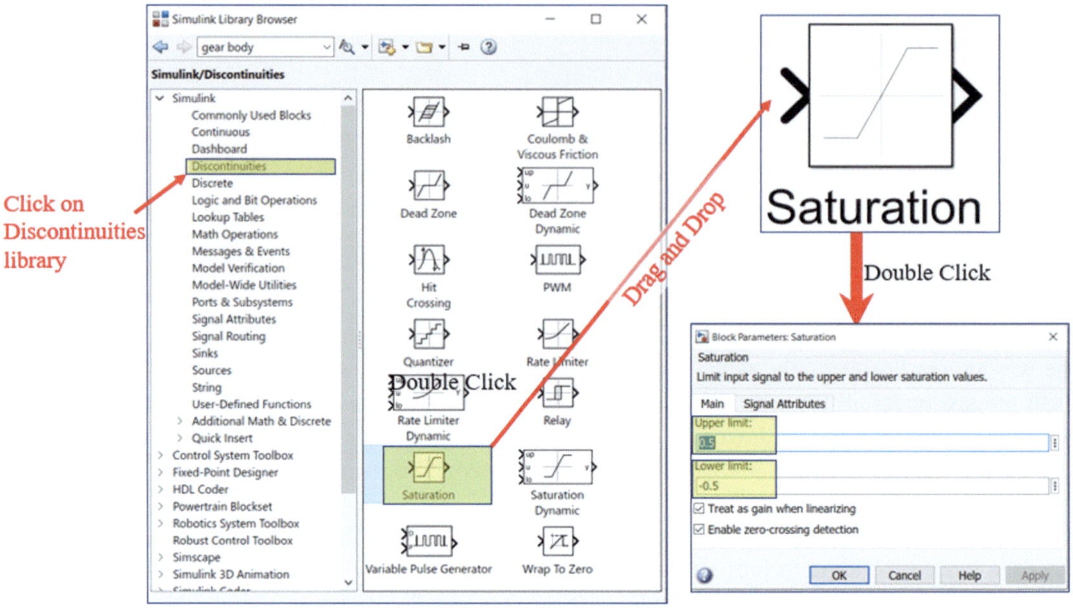

Fig. 8.39 Simulink® saturation block. Designed and implemented by the author. Images used with permission of MathWorks

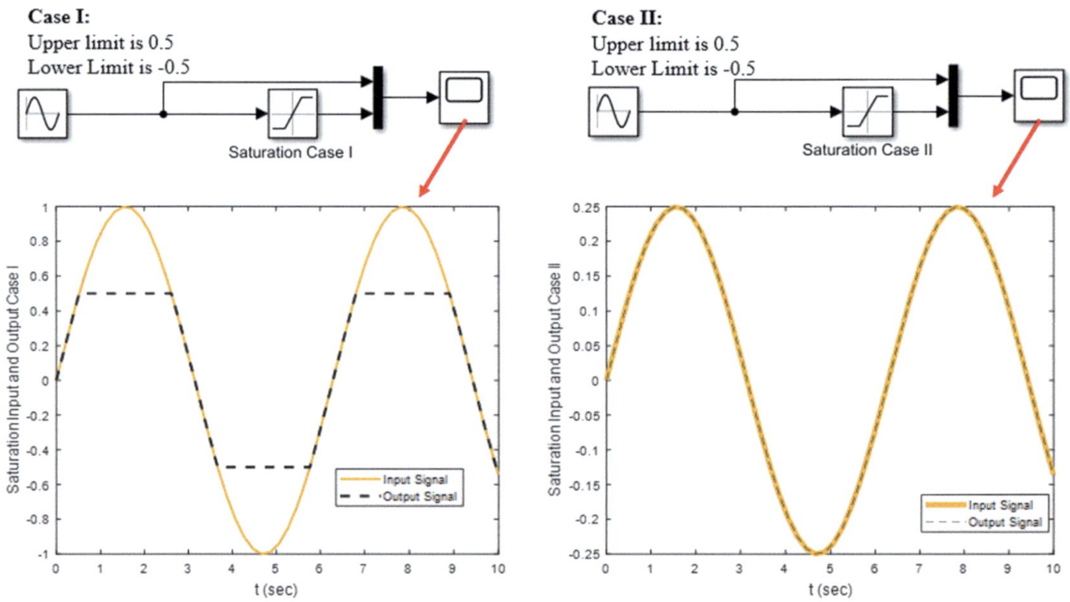

Fig. 8.40 Test cases for Simulink® saturation block

Simulink® offers a block called "rate limiter" in the discontinuities library as shown in Fig. 8.42, to apply the derivative saturation on any signal. The block has one input signal and one output signal. If we select the rate limiter block on a new model and double-click on the block, a popup menu appears where the user can insert the rising slew rate and the falling slew rate as shown in Fig. 8.42.

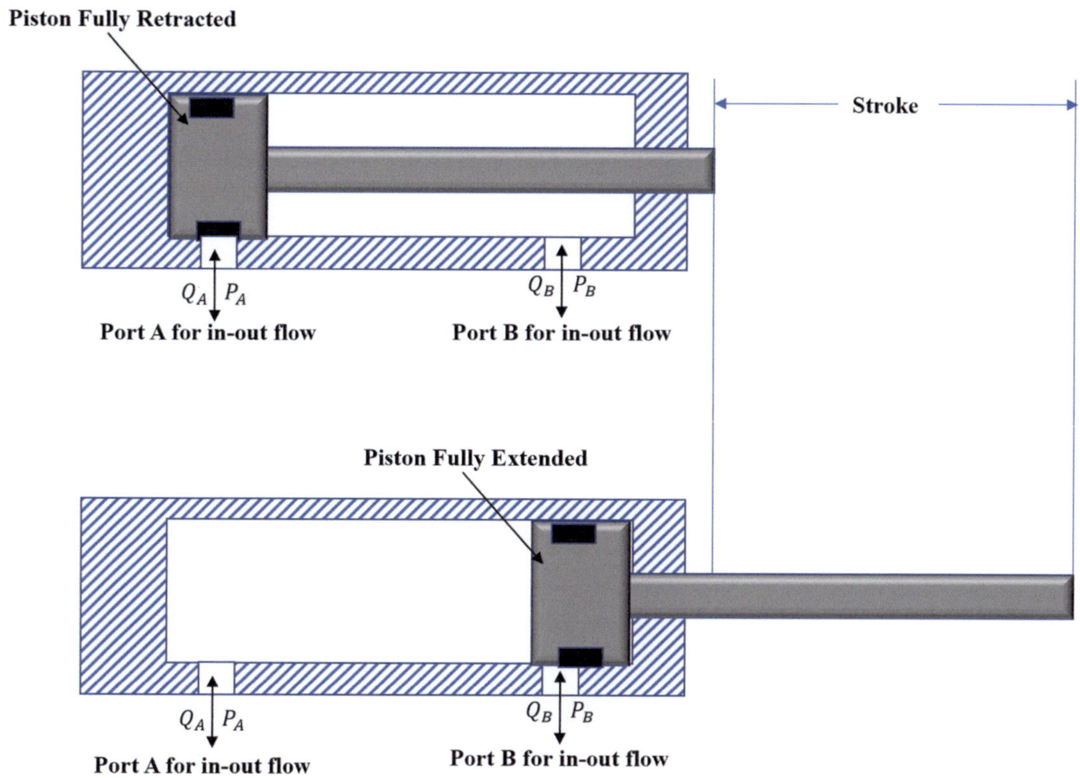

Fig. 8.41 Fully retracted and fully extended piston motion

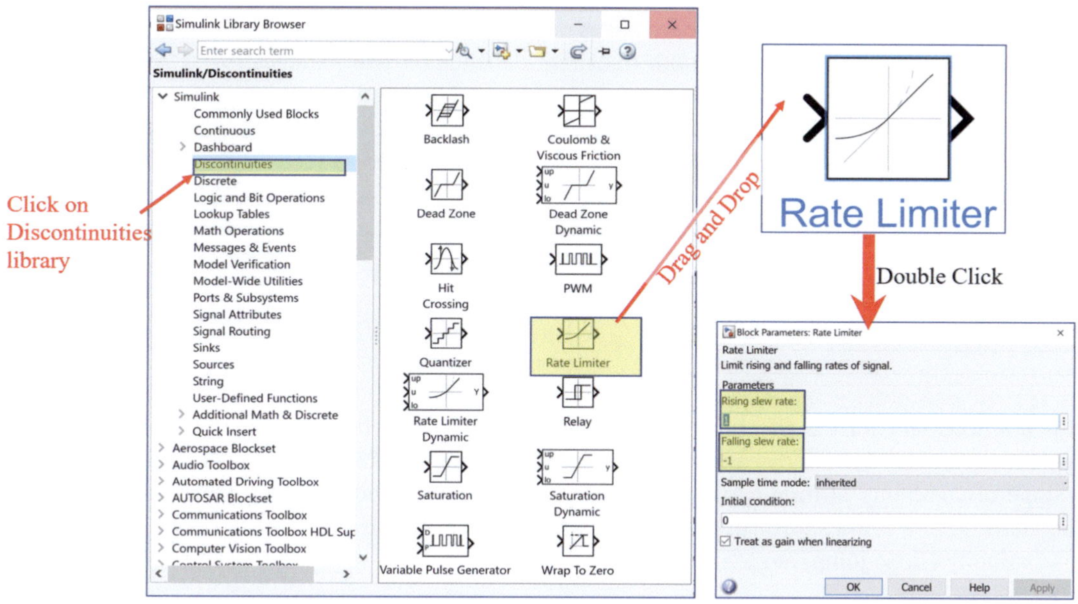

Fig. 8.42 Simulink® rate limiter block. Designed and implemented by the author. Images used with permission of MathWorks

The rate limiter block could be used in designing an SVAJ input signal to a hydraulic actuator. The rate limiter block should consider the hydraulic valve velocity saturation as shown in Fig. 8.29 and Eq. (8.17). Violating these limits may lead to machine hydraulic actuator damage as previously discussed in Sect. 8.3.

Using the saturation and the rate limiter may lead to a slope discontinuity that initiates very high derivatives, which may go to infinite values. For example, let's assume the displacement input is presented in Case I in Fig. 8.40. At the instantaneous time where the displacement signal switches from a sinusoidal wave into a horizontal line, there is a slope or a velocity discontinuity that leads to an infinite second derivative (acceleration) as well as an infinite third derivative (jerk) that are both unacceptable. To eliminate these types of discontinuities, we are going to use "sigmoid function" as explained below.

Definition[3]

Sigmoid Function:

A sigmoid function is a bounded, differentiable, real function that is defined for all real input values and has a non-negative derivative at each point and exactly one inflection point [12].

For an independent variable *t*, the sigmoid function is defined as

$$\text{sigmoid}(t, t_o) = \frac{1}{1 + e^{-k(t-t_o)}} \tag{8.29}$$

where t_o is the inflection time or the central time and *k* is the distribution factor. As shown in Fig. 8.43, the graph of the sigmoid function goes from zero to one in a smooth curve without any discontinuities. Even the high-order derivatives don't have any discontinuities. The curve crosses 0.5 at $t=t_o$. The sigmoid function's smooth transition depends on two parameters—t_o and *k*. The sigmoid function has a time delay t_o that can be located anywhere on the *t*-axis. The distribution parameter *k* presents how much time it would take to go from 0.02 (2%) until it reaches at 0.98 (98%). The sigmoid function is the most commonly known function used in feed-forward neural networks because of its nonlinearity and the computational simplicity of its derivative [12].

Now, we are going to use the sigmoid function to smooth the transition in function *f*(*t*) which is defined by the two functions $f_1(t)$ and $f_2(t)$ as

$$f(t) = \begin{cases} f_1(t) & t \le t_o \\ f_2(t) & t \ge t_o \end{cases} \tag{8.30}$$

It is not necessary for the two functions $f_1(t)$ and $f_2(t)$ to be equal at any time or even time t_o. The two functions may have a discontinuity, or a jump, at time t_o. These two functions may also have a slope discontinuity at time t_o or a slope jump. There are no conditions on these two continuous functions. Figure 8.44a shows an example of a function *f*(*t*) with a discontinuity (jump) and a slope discontinuity (slope jump) at t_o.

[3] Reproduced from [11] by MathWorks with permission from MathWorks.

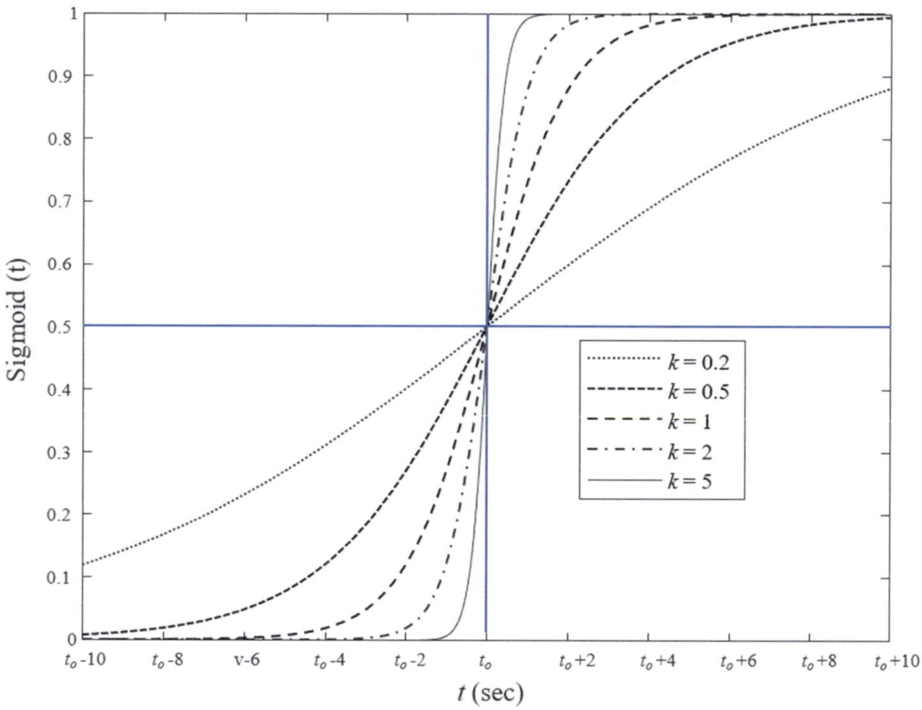

Fig. 8.43 Shape of sigmoid function

Let's define the two continuous functions $g_1(t)$ as a complementary sigmoid function and $g_2(t)$ as a sigmoid function that are defined as

$$\begin{aligned} g_1(t) &= 1 - \text{sigmoid}(t, t_o) = \frac{e^{-k(t-t_o)}}{1 + e^{-k(t-t_o)}} \\ g_2(t) &= \text{sigmoid}(t, t_o) = \frac{1}{1 + e^{-k(t-t_o)}} \end{aligned} \tag{8.31}$$

Figure 8.44b shows the shape of both $g_1(t)$ and $g_2(t)$. To smooth the transition from $f_1(t)$ to $f_2(t)$, we use

$$f_s(t) = g_1(t)f_1(t) + g_2(t)f_2(t) \tag{8.32}$$

Equation (8.32) approximates the function $f(t)$ to a summation of two terms $g_1(t)f_1(t)$ and $g_2(t)f_2(t)$ as shown in Fig. 8.44c, d. The sharpness or smoothness of the sigmoid transition depends on the value of the distribution parameter k. A higher value of k leads to a sharper transition which in turn leads to high derivative values. The sigmoid transition technique is going to be used here to eliminate any discontinuities.

The steps to design the SVAJ input motion using Simulink® based on the sigmoid function are:

1. Define the mission requirements as a set of points and trajectories.
2. If needed, use the saturation blocks and rate limiter blocks to limit all out-of-range points and trajectories.
3. Locate the discontinuity points.

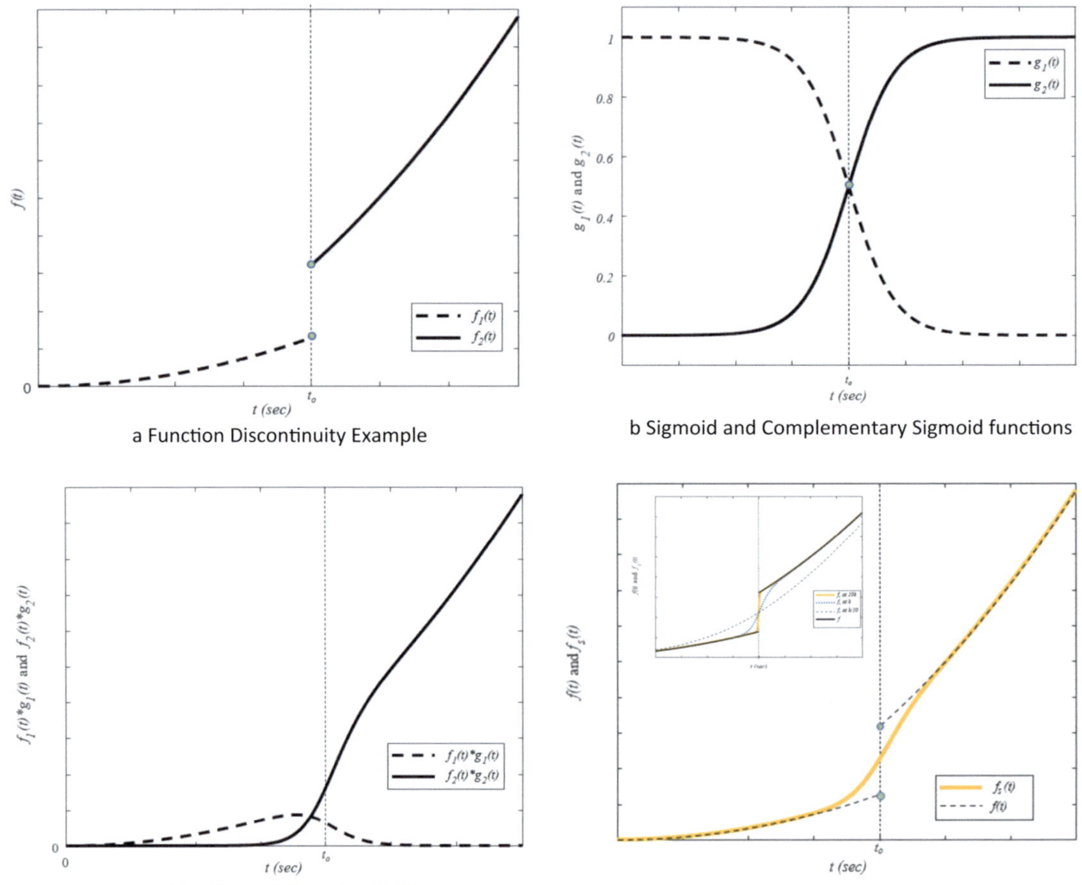

a Function Discontinuity Example

b Sigmoid and Complementary Sigmoid functions

c Two Parts of Equation (8.32)

d Discontinued Function and Smooth Function

Fig. 8.44 Steps to a smooth transition between two functions using sigmoid function

4. Locate the sigmoid function blocks at the times of the discontinuity points. The value of the distribution parameter k is now arbitrary.
5. Connect the output into three derivatives to provide an SVAJ diagram.

In "Kinematics_Of_Machines.slx", the reader can access a developed Simulink® block called "sigmoid function" that has two inputs $f_1(t)$ and $f_2(t)$. When you double-click on the sigmoid function block, the user should enter t_o and k. Figure 8.45 shows an example of the results from running the sigmoid function block over the two displacement functions $S_1(t)$ and $S_2(t)$ that are defined

$$S(t) = \begin{cases} S_1(t) = 3t \;\; t \leq 1 \\ S_2(t) = 3 \;\; t \geq 1 \end{cases} \tag{8.33}$$

The function $S(t)$ has a slope discontinuity at $t_o = 1$ s. In the shown Simulink® example, the two functions S_1 and S_2 are connected to the sigmoid function block with $k = 20$ and $t_o = 1$ s. The output function S_s from the sigmoid function block does not show any significant difference from the actual function S. When we develop the derivatives to calculate V, A, and J, we can see that the V_s, A_s, and

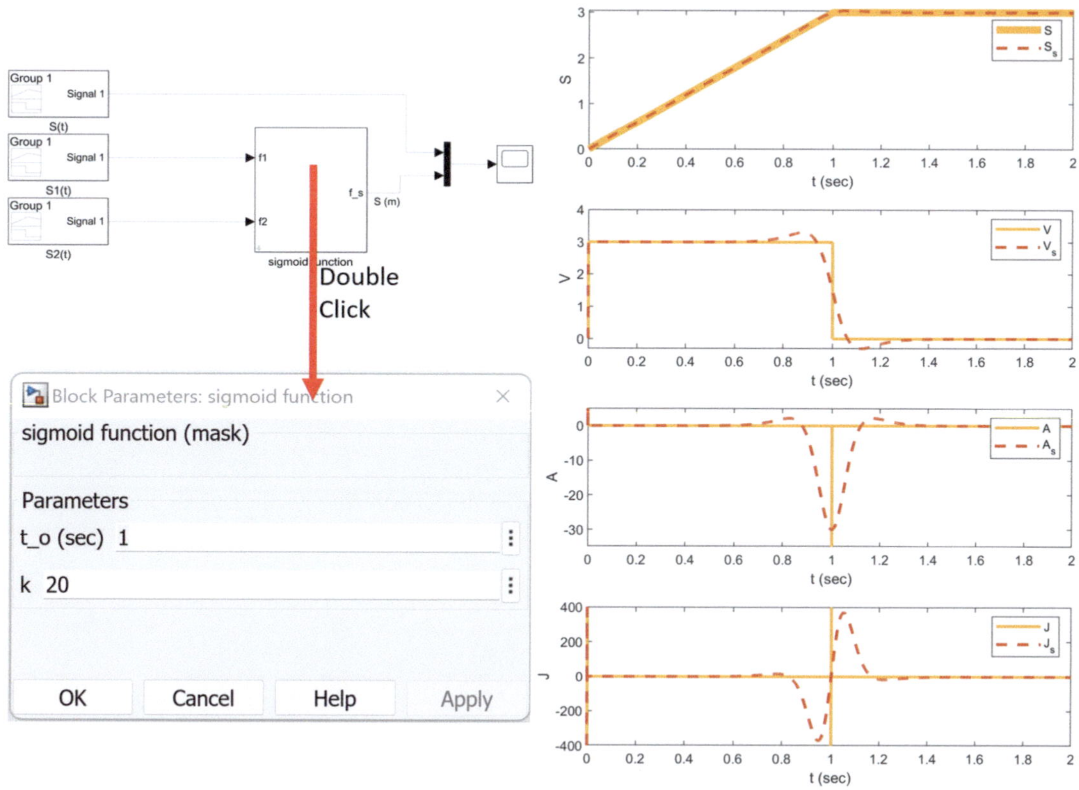

Fig. 8.45 Example of running "sigmoid function" Simulink® block. Designed and implemented by the author. Images used with permission of MathWorks

J_s have finite values while the A and J have infinite value at time $t_o = 1$ s. The values of maximum and minimum V_s, A_s, and J_s can be tuned using the value of k. One of the unique advantages of using the sigmoid function is that there will be no discontinuity for any of the higher-level derivatives.

Example 8.8

The skid steer mechanism has a hydraulic single-acting bucket cylinder with a bore diameter of 69.9 mm, rod diameter of 34.9 mm, and a stroke of 410 mm. From the hydraulic system, the valve maximum flow rate in the dump process is 0.135 m/s and the valve maximum flow rate in the roll-back process is 0.18 m/s.

Use the Simulink® tool to build a lift cylinder SVAJ motion input using the saturation blocks, rate limiter blocks, and sigmoid Transition blocks for dwell-rise-dwell-fall-dwell with the maximum available displacement and the maximum available speed to stay 1 s in the first dwell, 2 s in the second dwell, and 1 s in the third dwell.

Solution

As shown in Fig. 8.46a, the design mission has five trajectories, dwell-rise-dwell-fall-dwell connected by six points Point 1: (t_1, S_1), Point 2: (t_2, S_2), Point 3: (t_3, S_3), Point 4: (t_4, S_4), Point 5: (t_5, S_5), and Point 6: (t_6, S_6).

- The first trajectory, dwell, goes from Point 1: $(t_1, S_1)=(0, 0)$ sec to Point 2: $(t_2, S_2)=(1, 0)$ with a constant $S=0$ m.
- The second trajectory, rise, goes linearly from Point 2: $(t_2, S_2)=(1, 0)$ to Point 3: (t_3, S_3) with $S_3=0.41$ m and rate $m_{23}=V_{max}=0.135$ m/s. The value t_3 can be calculated using the rate by

$$m_{23} = \frac{S_3 - S_2}{t_3 - t_2}$$
$$\text{or } t_3 = \frac{S_3 - S_2}{m_{23}} + t_2 = \frac{0.41 - 0}{0.135} + 1 = 4.037\text{ s}$$

- The third trajectory, dwell, goes from Point 3: $(t_3, S_3)=(4.037, 0.41)$ for 2 s to reach Point 4: $(t_4, S_4)=(6.037, 0.41)$ with a constant $S=0.41$ m.
- The fourth trajectory, fall, goes linearly from Point 4: $(t_4, S_4)=(6.037, 0.41$ to Point 5: (t_5, S_5) with $S_5=0$ m and rate $m_{45}=V_{min}=-0.18$ m/s. The value t_5 can be calculated using the rate by

$$m_{45} = \frac{S_5 - S_4}{t_5 - t_4}$$
$$\text{or } t_5 = \frac{S_5 - S_4}{m_{45}} + t_4 = \frac{0 - 0.41}{-0.18} + 6.037 = 8.315\text{ s}$$

- The fifth trajectory, dwell, goes from Point 5: $(t_5, S_5)=(8.315, 0)$ for 1 s to reach Point 6: $(t_6, S_6)=(9.315, 0)$ with a constant $S=0$ m.

The summary of the design points and each trajectory is listed in Table 8.4. The entire SVAJ input mission can be implemented by:

Step 1: Mathematically implement the five curves in Table 8.4 and as shown in Fig. 8.46b. Collect each function as a subsystem with the names S_{12}, S_{23}, S_{34}, S_{45}, and S_{56}.

Step 2: Add four sigmoid functions from our "Kinematics_Of_Machines.slx" file with the following inflection times: t_2, t_3, t_4, and $t_{5,}$ and set $k=4$. Connect the sigmoid blocks and change their labels as shown in Fig. 8.46c.

Step 3: Drag and drop three derivative blocks and connect them as shown in Fig. 8.46d. Connect four output blocks to the summation block and the three derivative blocks. Rename the blocks as S, V, A, and J. Connect these four outputs to scopes and run the model. The SVAJ diagram is shown in Fig. 8.47.

Table 8.4 Summary of design points and trajectories of Example 8.8

Point	Trajectory
Point 1: $(t_1, S_1)=(0,0)$	$S_{12}(t) = 0$
Point 2: $(t_2, S_2)=(1, 0)$	$S_{23}(t) = 0.135(t - 1)$
Point 3: $(t_3, S_3)=(4.037, 0.41)$	$S_{34}(t) = 0.41$
Point 4: $(t_4, S_4)=(6.037, 0.41)$	$S(t) = -0.18(t - 6.037) + 0.41$
Point 5: $(t_5, S_5)=(8.315, 0)$	$S_{56}(t) = 0$
Point 6: $(t_6, S_6)=(9.315, 0)$	

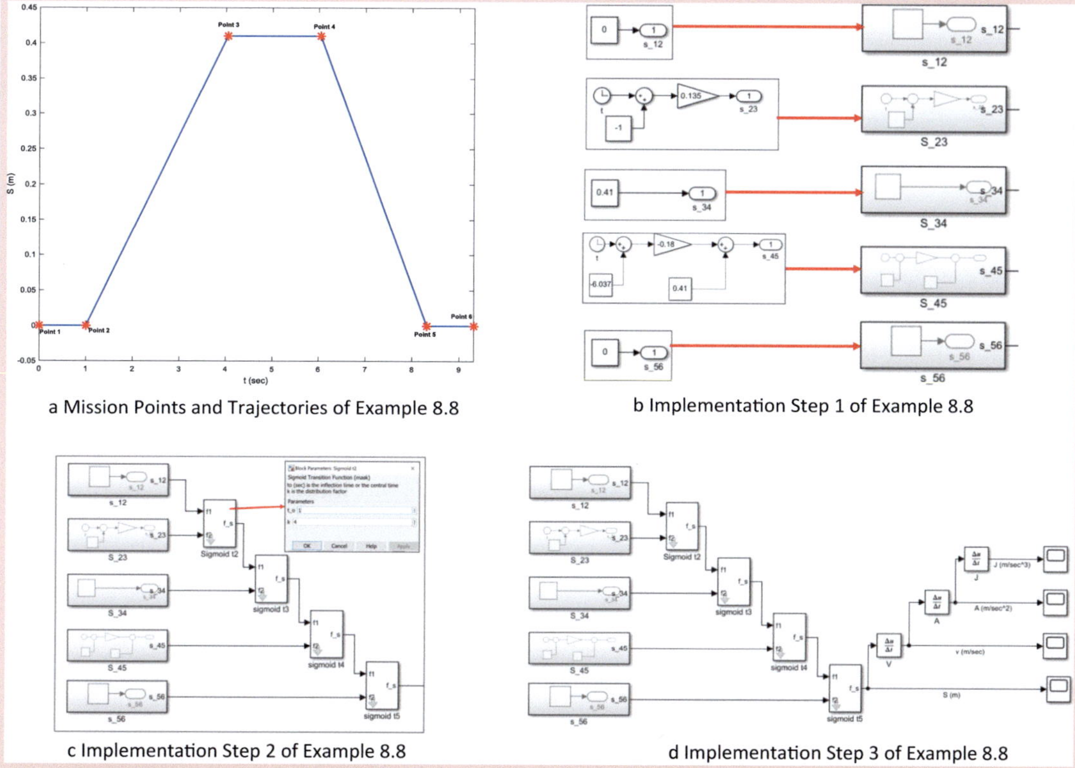

a Mission Points and Trajectories of Example 8.8

b Implementation Step 1 of Example 8.8

c Implementation Step 2 of Example 8.8

d Implementation Step 3 of Example 8.8

Fig. 8.46 Implementation steps of Example 8.8. Designed and implemented by the author. Images used with permission of MathWorks

Fig. 8.47 SVAJ diagram of Example 8.8

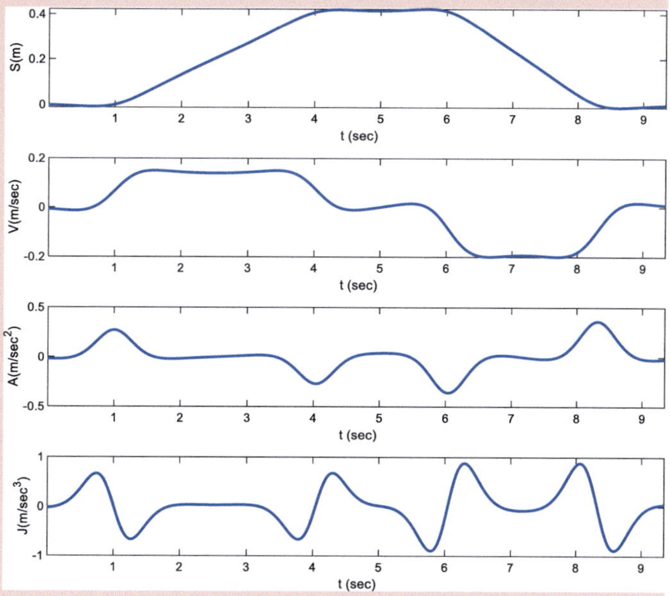

8.7 Animation Using Simulink® SimScape™ Multibody™

In this section, we are going to discuss the above-mentioned SVAJ design techniques on any arbitrary, complete mechanism. We will see the influence of discontinuities in the SVAJ input signal on the output velocity, acceleration, and jerk output signals. We present our comparison with the case study below.

Case Study Skid Steer Machine Modeling Using SimScape™ Multibody™

For the shown skid steer mechanism:

1. Build the Simscape™ Multibody™ model using the dimensions shown in Fig. 8.48.
2. Simulate the lift mechanism to be held at the fully extended position and the bucket mechanism motion using the below techniques. Plot the output SVAJ diagram of the bucket tip.
   - Ramp SVAJ as given in Example 8.6.
   - Harmonic SVAJ as given in Example 8.7.
   - Simulink® using Sigmoid Transition as given in Example 8.8.
3. Compare the results between the above techniques in terms of the SVAJ diagram of the bucket tip or node n_{11}.

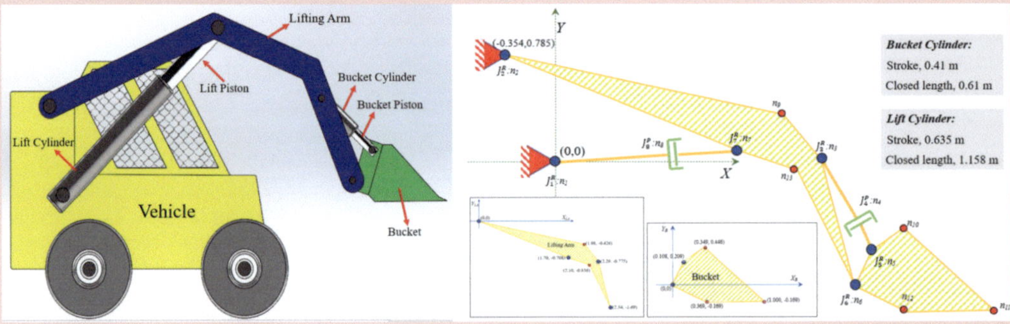

Fig. 8.48 Skid steer bucket mechanism. Figure and associated Skid Steer Articulated Bucket Lift Mechanism.mp4 video designed by the author and implemented by both the author and Pranav Dhola. Both courtesy of Pranav Dhola

Solution and Analysis

Building Simulink® Simscape™ Multibody™ Model

Write and save the MATLAB® script below (SkidSteer_Script_Example8p9.m) that declares the needed model's parameters. The entire model (SkidSteer_Model_Example8p9.slx) is shown in Fig. 8.49. The model is divided into subsystems: Skid steer frame, lift arm, bucket, lift cylinder mechanism, and bucket cylinder mechanism, which are connected to one another through joints. The configurations and the parameters of these subsystems are listed in Table 8.6. There is a scope to measure the tip or n_{11} SVAJ. The input to the lift cylinder mechanism is set as the full stroke with zero velocity and zero acceleration.

```
clc
clear all
%===============================================================================
%Declare the given parameters
%============================
Ground = [ 0          0   0
           -0.3540045 0.7849665  0];

LiftArm_CrossSection =[          0          0
    1.6982    -0.7080
    2.0984    -0.8363
    2.5396    -1.6879
    2.2882    -0.7747
    1.9752    -0.4258];

LiftArm_n2 = [0 0 0];
LiftArm_n7 = [1.6982    -0.7080 0];
LiftArm_n6 = [2.5396    -1.6879 0];
LiftArm_n3 = [2.2882    -0.7747 0];


Bucket_CrossSection = [0          0
    0.3694     -0.1693
    1.0004     -0.1693
    0.3489    0.4464
    0.1077    0.2873];

Bucket_n6 = [0 0 0];
Bucket_n5 = [0.1077    0.2873 0];
Bucket_n11 = [1.0004     -0.1693 0]; %tip

LiftCylinder_Length = 1.158; %m
LiftCylinder_Dimension = [LiftCylinder_Length 0.1 0.1]; %m
LiftCylinder_n1 = [-LiftCylinder_Length/2 0 0]; %m revolute
LiftCylinder_n8 = [LiftCylinder_Length/2 0 0]; %m prismatic

LiftPiston_Length    = 0.635; %m
LiftPiston_Dimension = [LiftPiston_Length 0.05 0.05]; %m
LiftPitson_n7 = [LiftPiston_Length/2 0 0]; %m revolute
LiftPiston_n8 = [-LiftPiston_Length/2 0 0]; %m prismatic

NominalLiftDisplacement = LiftCylinder_Length/2 - LiftPiston_Length/2;

%Bucket Set
%==========
BucketCylinder_Length = 0.65; %m
BucketCylinder_Dimension = [BucketCylinder_Length 0.1 0.1]; %m
BucketCylinder_n3 = [-BucketCylinder_Length/2 0 0]; %m revolute
BucketCylinder_n4 = [BucketCylinder_Length/2 0 0]; %m prismatic

BucketPiston_Length    = 0.41; %m
BucketPiston_Dimension = [BucketPiston_Length 0.05 0.05]; %m
BucketPiston_n5 = [BucketPiston_Length/2 0 0]; %m revolute
BucketPiston_n4 = [-BucketPiston_Length/2 0 0]; %m prismatic

NominaBucketDisplacement = BucketCylinder_Length/2 -
BucketPiston_Length/2;
```

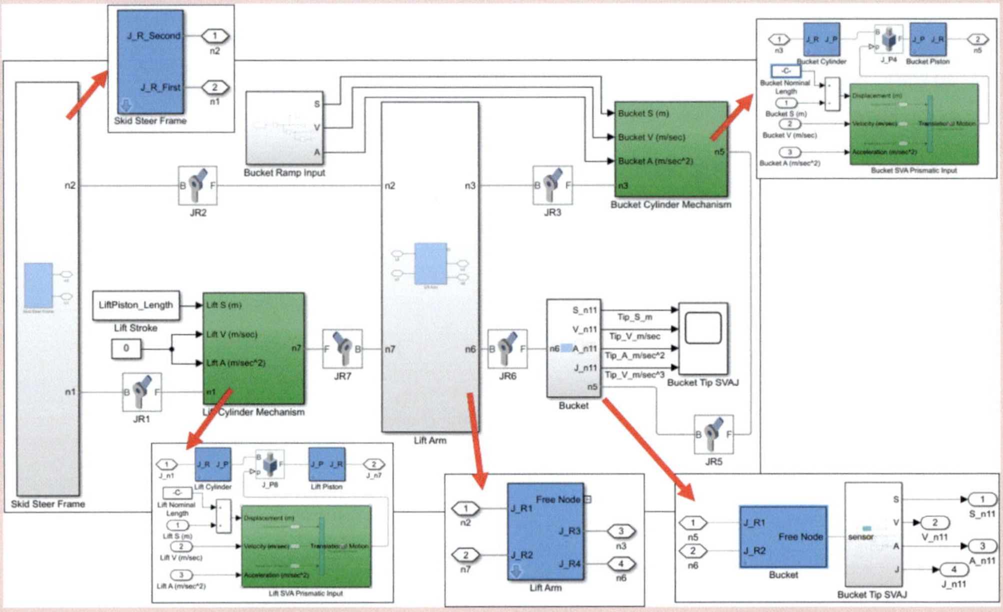

Fig. 8.49 Skid steer model

Table 8.6 Configurations and parameters of skid steer Simulink® SimScape™ Multibody™ model

Layer 1: Subsystem and blocks	Layer 2: Subsystems and blocks	Parameters of subsystems and blocks in layer 2
Skid steer frame subsystem	Skid steer frame subsystem is a "ground with two revolute joints" subsystem	• First Revolute Joint Location = n_1 • Second Revolute Joint Location = n_2
	Two output port blocks labeled as n1 and n2	
Lift arm subsystem	Lift arm subsystem is an "extrusion with four revolute joints" subsystem	• Extrusion cross section = LiftArm_CrossSection • R1 Location = LiftArm_n2 • R2 Location = LiftArm_n7 • R3 Location = LiftArm_n3 • R4 Location = LiftArm_n6
	Four output port blocks labeled as n2, n3, n6, and n7	
Bucket subsystem	Bucket subsystem is an "extrusion with two revolute joints" subsystem	• Extrusion cross section = Bucket_CrossSection • R1 Location = Bucket_n5 • R2 Location = Bucket_n6 • Free Node Location = Bucket_n11
	Four output port blocks labeled as n2, n3, n6, and n7	

(continued)

Table 8.6 (continued)

Layer 1: Subsystem and blocks	Layer 2: Subsystems and blocks	Parameters of subsystems and blocks in layer 2
Lift mechanism subsystem	Lift piston subsystem is a "brick with a revolute joint and a prismatic joint" subsystem	• Brick Dimensions = LiftPiston_Dimension • Revolute Joint Location = LiftPiston_n7 • Prismatic Joint Location = LiftPiston_n8
	Lift cylinder subsystem is a "brick with a revolute joint and a prismatic joint"	• Brick Dimensions = LiftCylinder_Dimension • Revolute Joint Location = LiftCylinder_n1 • Prismatic Joint Location = LiftCylinder_n8
	Lift SVA prismatic input subsystem is a "triple inputs translational motion actuator"	
	Prismatic joint block labeled as JP8	
	Five outputs: Lift S (m), Lift V (m/s), Lift A (m/s^2), n1, and n7	
	Sum block and constant block (labeled as lift nominal length)	• Sum block has a rectangular shape with + + • Constant Value = NominalLiftDisplacement
Bucket mechanism subsystem	Bucket Piston subsystem is a "brick with a revolute joint and a prismatic joint" subsystem	• Brick Dimensions = BucketPiston_Dimension • Revolute Joint Location = BucketPiston_n5 • Prismatic Joint Location = BucketPiston_n4
	Bucket cylinder subsystem is a "brick with a revolute joint and a prismatic joint"	• Brick Dimensions = BucketCylinder_Dimension • Revolute Joint Location = BucketCylinder_n3 • Prismatic Joint Location = BucketCylinder_n4
	Bucket SVA prismatic input subsystem is a "triple inputs translational motion Actuator"	
	Prismatic joint block labeled as JP4	
	Five outputs: Bucket S (m), Bucket V (m/s), Bucket A (m/s^2), n3, and n5	
	Sum block and constant block (labeled as bucket nominal length)	• Sum block has a rectangular shape with + + • Constant Value = NominalBucket Displacement
Six revolute joints: JR1, JR2, JR3, JR5, JR6, and JR7		

Model to Simulate Different SVAJ Bucket Inputs

Now, we are going to recall the Bucket Input SVAJ models from Example 8.6, Example 8.7, and Example 8.8 to be: Bucket Ramp SVAJ input, Bucket Simple Harmonic SVAJ input, and Bucket Sigmoid Transition SVAJ input. We are going to implement Bucket Ramp SVAJ input and Bucket Simple Harmonic SVAJ input as Simulink® subsystems as shown in Fig. 8.50. Bring the sigmoid Simulink® subsystem from Example 8.8. Consequently, connect each Bucket SVAJ input and run the model. Figure 8.51 shows the SVAJ of node n_{11} or the bucket tips.

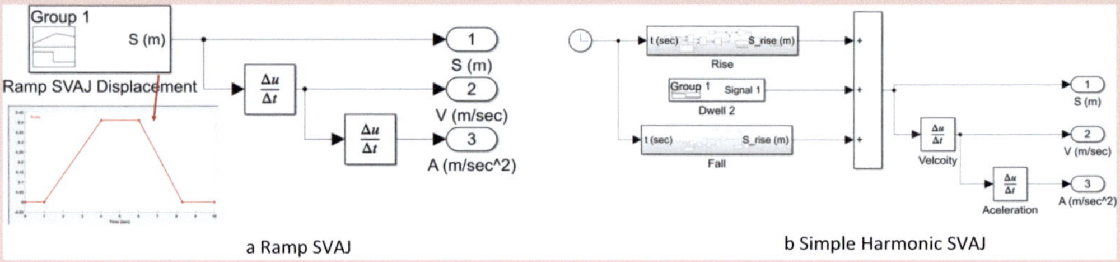

Fig. 8.50 Bucket ramp and simple harmonic SVAJ motion inputs

Comparing Different SVAJ Bucket Inputs

From Fig. 8.51, we conclude that:

1. None of the three inputs violate the hydraulic cylinder displacement limits (stroke) and actuator velocity limits (maximum flow rate and minimum flow rate).
2. The ramp SVAJ input causes a discontinuity in the velocity of the bucket tip node that leads to infinite acceleration and infinite jerk. The infinite acceleration results can cause engine stall. The infinite jerk results can cause vibration and the possibility of damaging the bucket.
3. The simple harmonic SVAJ input causes a discontinuity in the acceleration of the bucket tip node that leads to infinite jerk. The infinite jerk results can cause vibration and the possibility of damaging the bucket.
4. The sigmoid SVAJ input does not cause any discontinuity. The only concern is that the required acceleration is higher than the simple harmonic.

Fig. 8.51 SVAJ results of bucket Tip Node n11 for ramp, simple harmonic, and sigmoid SVAJ inputs

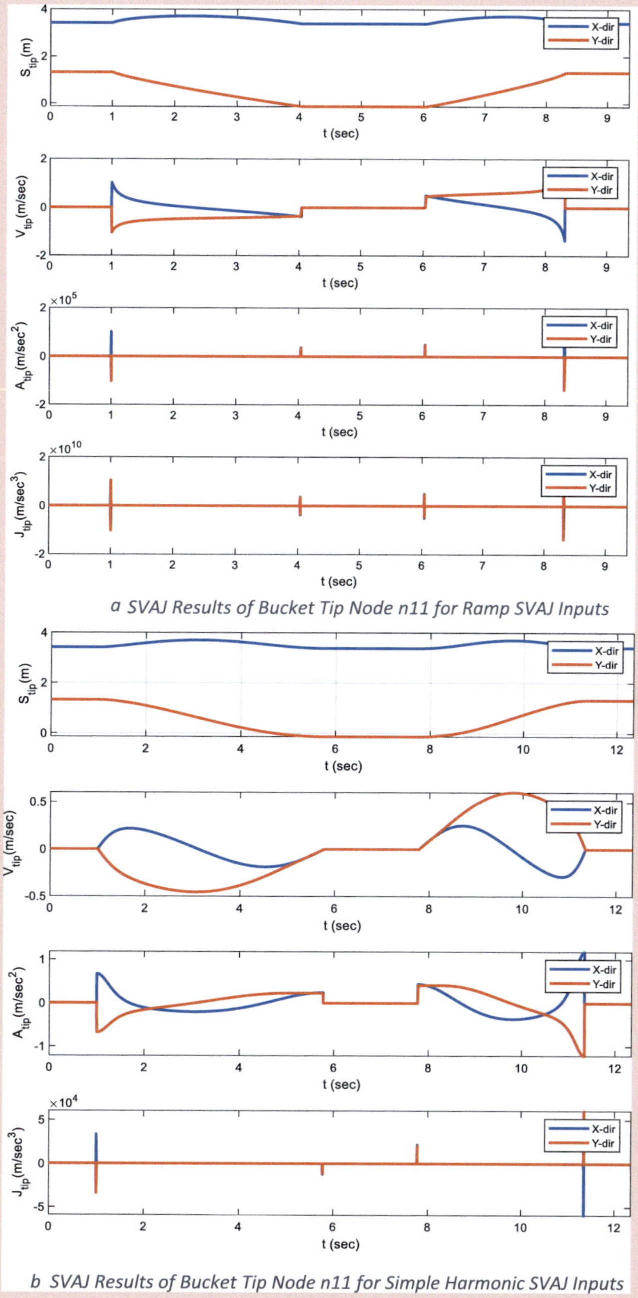

a SVAJ Results of Bucket Tip Node n11 for Ramp SVAJ Inputs

b SVAJ Results of Bucket Tip Node n11 for Simple Harmonic SVAJ Inputs

Fig. 8.51 (continued)

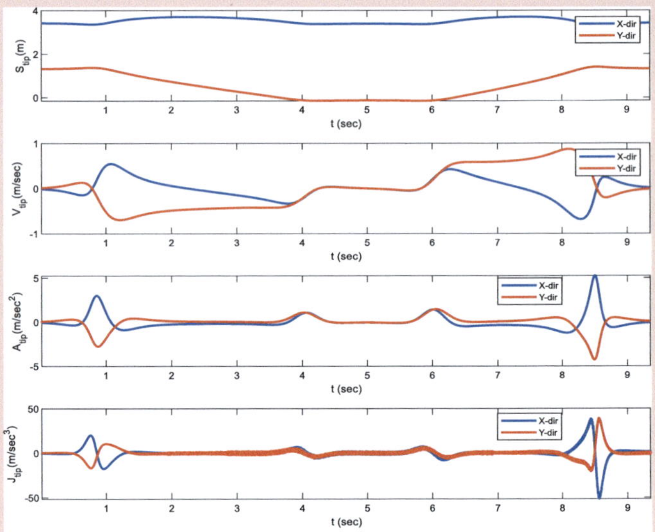

c SVAJ Results of Bucket Tip Node n11 for Sigmoid SVAJ Inputs

8.8 Chapter Summary

This chapter is an introduction to graphically describing input and output models for displacement, velocity, acceleration, and jerk rate (SVAJ) using the block diagram technique. We also discussed the kinematic study of different kinds of actuators found in common machines and their physical limitations. Finally, we used mathematical techniques to avoid discontinuities in design trajectory and lessons on generating trajectories for models in Simscape™ Multibody™ with animation.

Outcomes

1. Understanding the sketching process of a system block diagram of any machine.
2. Understanding the hydraulic circuit and the hydraulic components with power losses analysis and estimation.
3. Understanding the motion limits from using the hydraulic system as an actuator.
4. Understanding different types of electric motors and the limitations and losses of each.
5. Understanding different mathematical techniques to design the input motion trajectory using an SVAJ diagram.
6. Understanding how to implement the input motion trajectory using Simulink®.
7. Understanding how to integrate the designed input motion trajectory with the mechanism simulation to build a 3D animation using Simulink®.

Problems

Problem 1

Figure 8.52 shows the actual diagram of the motion of the lift truck mechanism motion. Sketch the block diagram of this motion system showing the kind of power transferred.

Problem 2

Figure 8.53 shows the actual diagram of the motion of the lift table mechanism motion. Sketch the block diagram of this motion system showing the kind of power transferred.

Problem 3

The hydraulic subsystem of a specific mechanism has a 2×10^{-3} m^3/s theoretical flow rate with a volumetric efficiency of 85% and mechanical efficiency of 95%. The pump is rotated by a motor that provides the pump with a torque $T_{motor}=10$ N m and $\omega_{pump}=1800$ rpm. If the gear ratio between the angular velocity of the motor and the angular velocity of the pump's input shaft is 1 ($\omega_{engine}=\omega_{pump}$) find:

- The actual flow rate of the pump
- The motor losses torque.

Problem 4

The hydraulic subsystem of a specific mechanism has a 67 m^3/s theoretical flow rate with a volumetric efficiency of 82% and mechanical efficiency of 88%. The pump is rotated by a motor

Fig. 8.52 Actual diagram of a lift truck motion system. Designed by the author and implemented by Kenneth Manalansan. Courtesy of Kenneth Manalansan

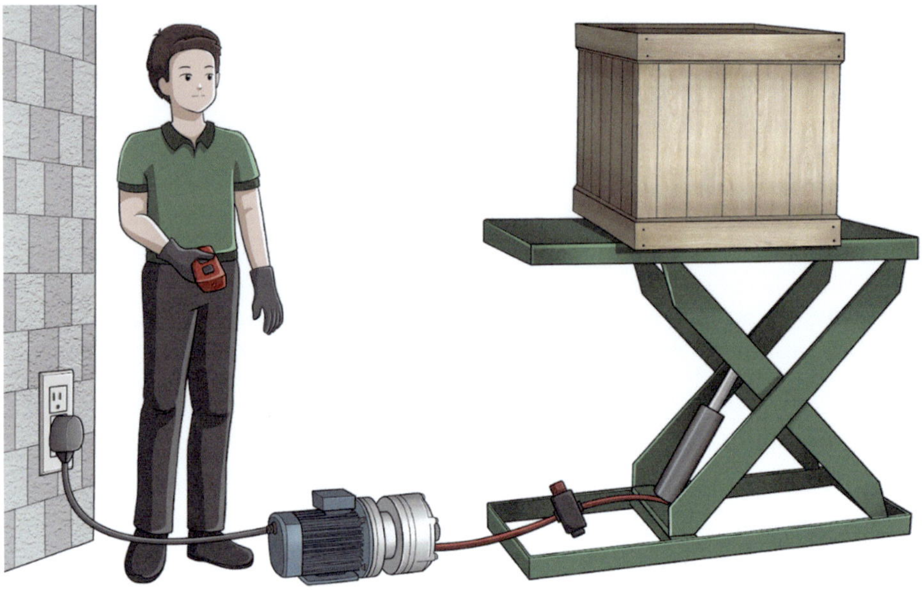

Fig. 8.53 Actual diagram of a lift table motion system. Designed by the author and implemented by Kenneth Manalansan. Courtesy of Kenneth Manalansan

that provides the pump with a torque $T_{\text{motor}} = 46 \times 10^4$ N m and $\omega_{\text{pump}} = 500$ rpm. If the gear ratio between the angular velocity of the motor and the angular velocity of the pump's input shaft is 1 ($\omega_{\text{engine}} = \omega_{\text{pump}}$) find:

- The actual flow rate of the pump
- The motor losses torque.

Problem 5

A hydraulic oil travels with a velocity of 184.62 m/s inside a pipe with a diameter of 10 cm and a length of 1 m. The pipe has one elbow with regular flanged 90 deg. The oil has a density of 889.6 kg/m^3 and viscosity 0.066 kg/m/s. Calculate:

3. The total head loss through the 1 m pipe and the 90 deg elbow.
4. The total head loss from the above pipe's specs.

Problem 6

A hydraulic oil of 2 m/s inside a pipe with a diameter of 6 cm and a length of 6 m. The pipe has two elbows with regular flanged 90 deg each. The oil has a density of 889.6 kg/m^3 and viscosity 0.066 kg/m/s. Calculate:

1. The total head loss through the 6 m pipe and two 90 deg elbows.
2. The total head loss from the above pipe's specs.

Problem 7

A specific mechanism has a hydraulic single-acting cylinder with a control valve that has a maximum flow rate of 2.2 L per min (lpm). The piston's diameter is 5 cm and the rod's diameter is 3 cm. Find:

- The maximum extended velocity that the boom prismatic joint can move at.
- The maximum retracted velocity that the boom prismatic joint can move at.

Problem 8

A specific mechanism has a hydraulic single-acting cylinder with a control valve that has a maximum flow rate of 424 L per min (lpm). The piston's diameter is 14 cm and the rod's diameter is 10 cm. Find:

- The maximum extended velocity that the boom prismatic joint can move at.
- The maximum retracted velocity that the boom prismatic joint can move at.

In Problem 9, Problem 10, and Problem 12, A specific mechanism has a double-acting hydraulic cylinder with a bore diameter of 5 cm, rod diameter of 3 cm, and stroke of 35 cm. From the hydraulic system, the maximum flow speed in pushing the piston to be out of the cylinder is 0.25 m/s and the maximum flow speed in pushing the piston to be inside the cylinder is 0.2 m/s.

Problem 9

- Use the step SVAJ motion to design the fastest request input motion as dwell-rise-dwell-fall-dwell considering that the first and the third dwell take 1 s each while the second dwell takes 2 s.
- Sketch the resultant step SVAJ diagram and comment on the applicability of the requested motion.

Problem 10

- Using the ramp SVAJ motion to design the fastest request input motion of dwell-rise-dwell-fall-dwell considering that the first and the third dwell take 1 s each while the second dwell takes 2 s.
- Sketch the resultant ramp SVAJ diagram and comment on the applicability of the requested motion.

Problem 11

- Using the simple harmonic SVAJ motion to design the fastest request input motion as dwell-rise-dwell-fall-dwell considering the motion is going full stroke up and down. Also, the first and the third dwell take 1 s each while the second dwell takes 2 s.
- Sketch the resultant simple harmonic SVAJ diagram and comment on the applicability of the requested motion.

In Problem 13, Problem 13, and Problem 14, A specific mechanism has a double-acting hydraulic cylinder with a bore diameter of 14 cm, rod diameter of 10 cm, and stroke of 165 cm. From the hydraulic system, the maximum flow speed in pushing the piston out of the cylinder is 2 m/s and the maximum flow speed in pushing the piston inside the cylinder is 1 m/s.

Problem 12

- Use the step SVAJ motion to design the fastest request input motion as dwell-rise-dwell-fall-dwell considering that the first and the third dwell take 1 s each while the second dwell takes 2 s.
- Sketch the resultant step SVAJ diagram and comment on the applicability of the requested motion.

Problem 13

- Using the ramp SVAJ motion to design the fastest request input motion of dwell-rise-dwell-fall-dwell considering that the first and the third dwell take 1 s each while the second dwell takes 2 s.
- Sketch the resultant ramp SVAJ diagram and comment on the applicability of the requested motion.

Problem 14

- Using the simple harmonic SVAJ motion to design the fastest request input motion as dwell-rise-dwell-fall-dwell considering the motion is going full stroke up and down. Also, the first and the third dwell take 1 s each while the second dwell takes 2 s.
- Sketch the resultant simple harmonic SVAJ diagram and comment on the applicability of the requested motion.

References

1. "Systems and Software Engineering Vocabulary," ISO/IEC/IEEE 24765, First Edition, 2010.
2. "Handbook of Modern Sensors", Jacob Fraden, ISBN: 978-0-387-21604-1
3. "Hydraulics Basic Level," Festo, Merkle, D., Schrader, B., and Thomas, M., 2003.
4. "Essential Hydraulics: Fluid Power, Book 1", Winston, M., CreateSpace Independent Publishing Platform, 2014.
5. "Essential Hydraulics: Fluid Power, Book 2", Winston, M., CreateSpace Independent Publishing Platform, 2014.
6. "Hydrodynamics of Pumps", Brennen, C., Concepts ETI, Inc. and Oxford University Press, 1994.
7. "Fluid Mechanics and Hydraulic Machines," Bansal, K., Laxmi Publications, 2005.
8. "Hydraulic System Volume 1: Introduction to Hydraulics for Industry Professionals", Khalil, M., ISBN: 978-0-692-62236-0, 2019.
9. "Fundamentals of Fluid Mechanics", Munson, Young, and Okiishi, 8th Edition, Wiley, 2016.
10. "Electric Motors and Drives, Fundamentals, Types and Applications", Hughes, A., 4th Edition, Elsevier, 2014.
11. "Getting Started with Simulink®", MathWorks, 2023.
12. "The influence of the sigmoid function parameters on the speed of backpropagation learning," Han, J. and Moraga, C., Natural to Artificial Neural Computation, 1995.

Cam Analysis 9

Chapter Objectives

1. Understanding different types of cams, followers, and cam joints.
2. Understanding how to use *SVAJ* and *svaj* diagrams to design and select the follower motion profile.
3. Understanding the usage of different mathematical formulas to design the follower motion.
4. Understanding the configuration parameters of the cam mechanism.
5. Understanding how to design the cam profile based on the follower motion profile and other selected design parameters.
6. Learning different applications for the cam mechanism.
7. Understanding how to use Simscape™ Multibody™ in MATLAB® and Simulink® to calculate the kinematics of the cam mechanism based on a model simulation.

9.1 Introduction

In this chapter, different types of cams and cam followers are introduced followed by a definition of the configuration parameters and the performance parameters of the cam mechanism. Displacement, velocity, acceleration, and jerk rate diagrams (also called SVAJ cam diagrams) are introduced, leading to an algorithm for the cam design. Finally, some practical examples of the cam mechanism are introduced.

9.2 Different Types of Cams and Cam Followers

The cam mechanism consists of the following parts:

- Cam link.
- Follower link
- Cam joint that connects the follower link with the cam link.

Supplementary Information The online version contains supplementary material available at https://doi.org/10.1007/978-3-031-72701-6_9.

A. Omran, *Kinematics and Dynamics of Machines*, https://doi.org/10.1007/978-3-031-72701-6_9

The cam joint keeps the follower sliding on a specific curved outline on the cam link or within a curved groove in the cam link. The cam mechanism has a 2D motion if the follower link moves inside a 2D plane. The sliding 2D plane is called the *follower plane*. The axis on which the cam link has a translational or a rotational motion is called the *cam axis*. Figures 9.1 and 9.2 show the cam mechanism undergoing 2D motion. The cam mechanism aims to convert an oscillating rotational (Fig. 9.1) or an oscillating translational cam link motion (Fig. 9.2) into an oscillating sliding follower motion on the follower plane. Both the cam link and the follower link produce oscillating motions. The reader is encouraged to watch the attached animation videos of the cam mechanisms for both Figs. 9.1 and 9.2.

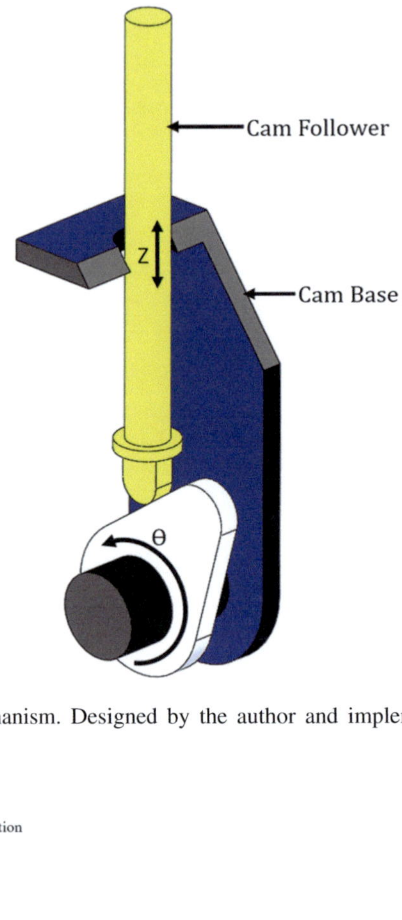

Fig. 9.1 2D rotational cam mechanism. Designed by the author and implemented by both the author and Pranav Dhola. Courtesy of Pranav Dhola

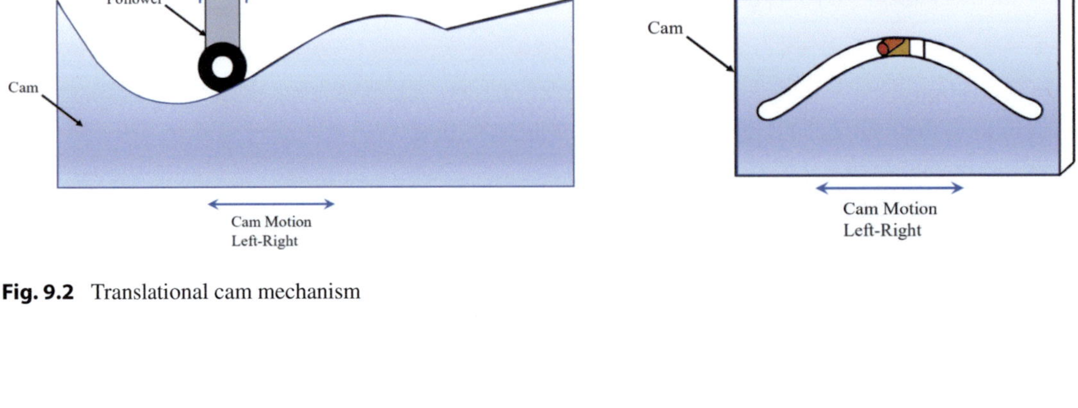

Fig. 9.2 Translational cam mechanism

The cam link is called the *camshaft* when it rotates as shown in Fig. 9.1. The camshaft rotates around a perpendicular axis on the follower plane, where the follower moves up and down over the curved outline. The cam link may also undergo translatory motion as shown in Fig. 9.2. As the cam link moves left and right and vice versa, the follower moves up and down on the follower plane. In both the 2D rotational cam mechanism and 2D translational cam mechanism, the aim is to get the follower to move on a specific trajectory within a specific timeframe.

The same concepts of the 2D rotational and translational cam mechanisms can be applied to 3D. Figure 9.3 shows a cam cylindrical mechanism, where the follower rides on the surface of a cylinder, or in a groove cut into the surface of a cylinder. In the case shown in Fig. 9.3, the cam cylinder mechanism is used to convert rotational motion (rotation of the cam cylinder) into linear parallel motion (follower motion) in the direction of the cylinder's rotational axis. The reader is encouraged to watch the attached video of the cam cylinder mechanism animation.

Now, this section will explain how the usage of cams is important in machine design with the famous example of the cam mechanism in an engine, whereby cam mechanisms are used to open and close both the intake valve and the exhaust valve. Figure 9.4 shows an engine mechanism that consists of four pistons, four cylinders, four connecting rods, one crankshaft, one camshaft, a crankshaft sprocket, a camshaft sprocket, a timing belt, four intake valves, four exhaust valves, four intake cams, and four exhaust cams. The reader is encouraged to watch the attached video of the four-stroke engine animation.

This engine mechanism has four strokes: Intake, Compression, Combustion, and Exhaust. This four-stroke cycle is shown in Fig. 9.5. Figure 9.5a shows the piston position where crankshaft is at 0° during the intake stroke. During the intake stroke, the crankshaft rotates clockwise from a 90° angle where the piston is initially at the top dead center (T.D.C.) until the crankshaft reaches an angle of 270°, with the piston oriented at the bottom dead center (B.D.C.). During the piston motion from T.D.C. to B.D.C., the intake valve must be in the open position to pull the air or the air–fuel mixture into the cylinder. In other words, during the intake stroke, the intake valve must be in the open position.

Figure 9.5b shows the piston position during the compression stroke that starts after the intake stroke. During the compression stroke, the crankshaft keeps rotating clockwise from angle 270°

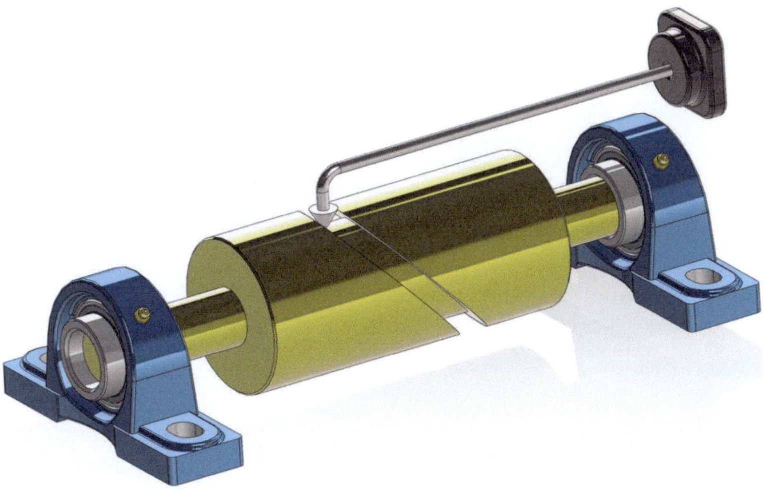

Fig. 9.3 Cylindrical cam mechanism. Designed by the author and implemented by both the author and Pranav Dhola. Courtesy of Pranav Dhola

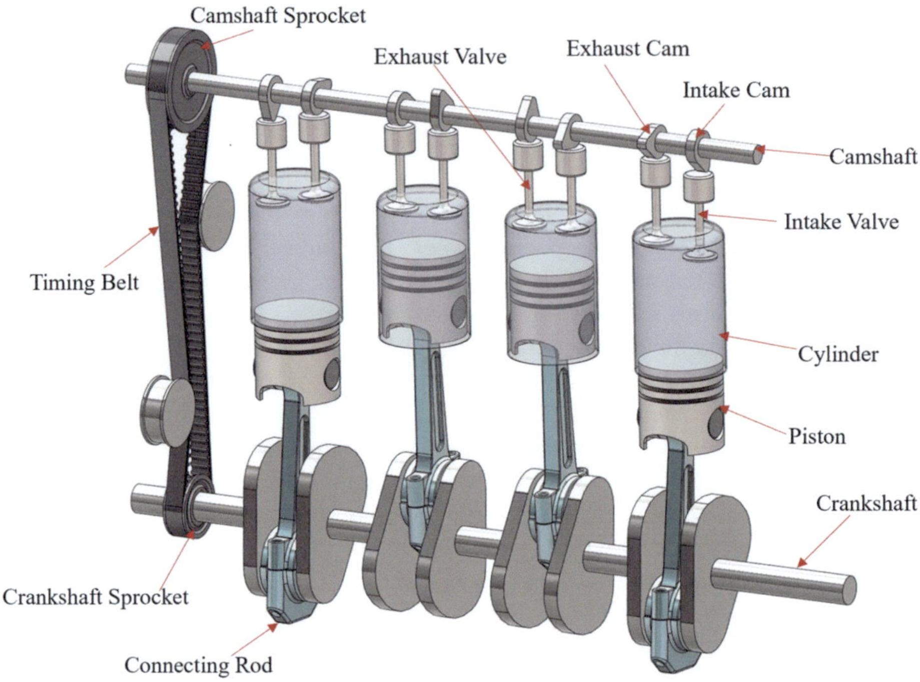

Fig. 9.4 Four-stroke engine mechanism. Figure and associated Single Piston–Cylinder Mechanism.mp4 video designed by the author and implemented by both the author and Pranav Dhola. Both courtesy of Pranav Dhola

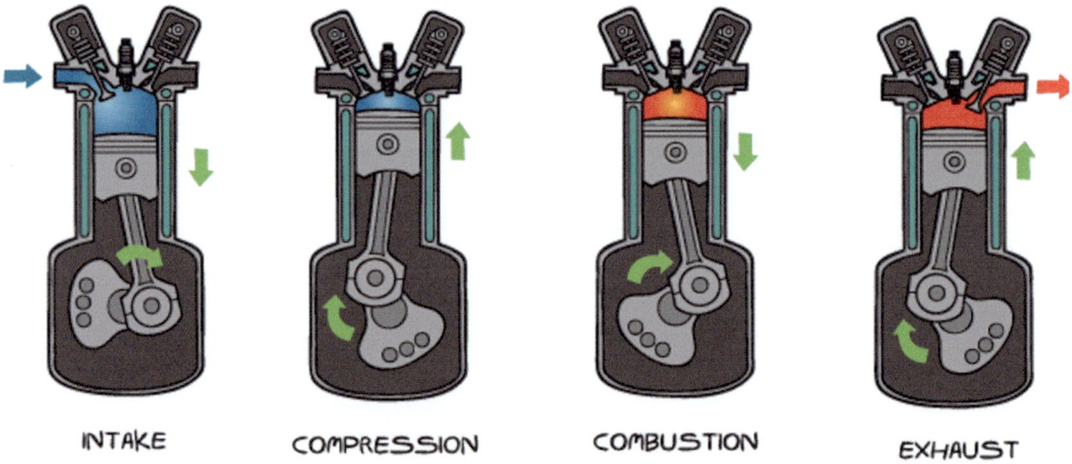

a Intake Stroke *b Compression Stroke* *c Combustion Stroke* *d Exhaust Stroke*

Fig. 9.5 Four-stroke piston–cylinder. Designed by the author and implemented by both the author and Pranav Dhola. Courtesy of Pranav Dhola

(piston is at B.D.C.) to 90° (piston is at T.D.C.), causing the piston to compress the air or the air–fuel mixture. The intake valve as well as the exhaust valve should be in the closed position for the entirety of the stroke. Note that the crankshaft makes a complete cycle of 360° throughout the intake stroke and the compression stroke.

Figure 9.5c shows the piston position during the combustion stroke that starts after the compression stroke, while the piston is at T.D.C. The air–fuel mixture is ignited either by a spark plug (gasoline engines) or simply from high compression (diesel engines). The increased pressure caused by the combustion of the fuel/air mix forces the piston back downwards into B.D.C. In other words, in the combustion stroke, the crankshaft rotates clockwise from 90° (T.D.C.) to 270° (B.D.C.), and both the intake and exhaust valves must be in the closed position.

Figure 9.5d shows the piston position during the exhaust stroke that starts after the combustion stroke while the piston is at B.D.C. The crankshaft rotates from 270° (B.D.C.) to 90° (T.D.C.). The exhaust valve should be in the open position to force the combusted air out of the cylinder. Note that the crankshaft makes a complete 360° cycle throughout the combustion stroke and the exhaust stroke.

From the above explanation of the four strokes during two crankshaft cycles, one can come up with two mechanism goals for the intake and exhaust valves:

- Over two back-to-back crankshaft cycles, the intake valve must be in the open position only during the intake stroke. Elsewhere, the intake valve must be in the closed position.
- Over two back-to-back crankshaft cycles, the exhaust valve is in the open position only during the exhaust stroke. Elsewhere, the intake valve must be in the closed position.

To accomplish the two goals above, one first adds a camshaft that rotates with twice the rotational speed of the crankshaft. As such, the two cycles from the crankshaft will lead to one cycle from the camshaft. This rotational ratio of the camshaft relative to the crankshaft can be accomplished by many mechanisms, such as the timing belt that is shown in Fig. 9.4. The crankshaft rotates a crankshaft sprocket and the camshaft sprocket in turn rotates the camshaft. The timing belt translates the rotation from the crankshaft to the camshaft through these two sprockets. The sizes of the camshaft sprocket and the crankshaft sprocket are such that the camshaft rotates at twice the speed of the crankshaft. The intake valve and exhaust valve motion goals can be rewritten concerning the camshaft as:

- During one camshaft cycle, the intake valve must be in the open position during the intake stroke. Otherwise, the intake valve must be in the closed position.
- During one camshaft cycle, the exhaust valve must be in the open position during the exhaust stroke. Otherwise, the intake valve must be in the closed position.

To accomplish these goals, one can make the camshaft drive two cams that function to open and close the two valves. Figure 9.6 shows the intake valve cam mechanism. The cam link profile, which is simply called the cam profile, causes the intake valve (follower) to slide down and up to open and close the valve. Later, this chapter will explain the cam profile in more detail. The same intake valve cam mechanism is used for the exhaust valve. The design of these two cam mechanisms is crucial for accomplishing the motion goals.

More examples of engineering applications of the cam mechanism are introduced later on, in Sect. 9.6. For now, design aspects of the cam mechanism are introduced, as well as the different cam mechanism types that can be specified based on the cam link, follower link, or the cam joint.

9.2.1 Cam Mechanism Types Based on Cam Link

The cam link may have either a rotational motion, as shown in Fig. 9.1, or a translational motion, as shown in Fig. 9.2. In the case of the rotational cam link motion, the cam link is called a *camshaft*. There are two types of rotational camshafts: the radial camshaft and the axial camshaft. The

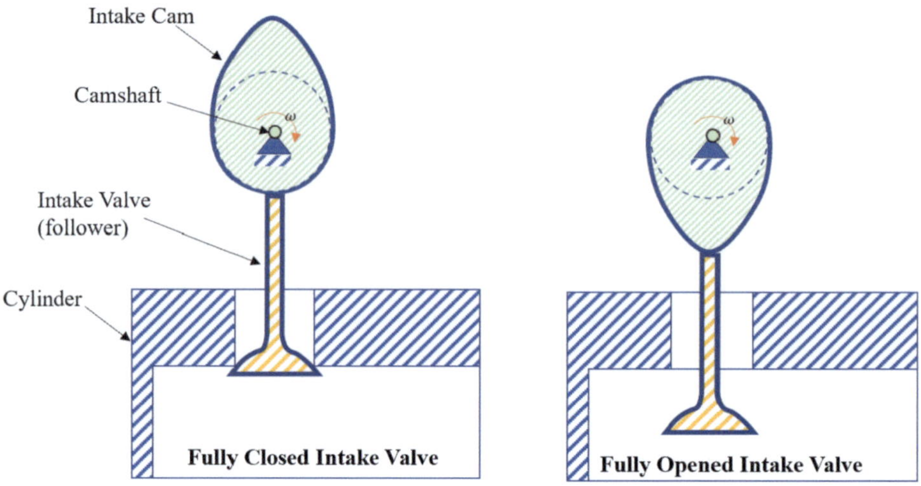

Fig. 9.6 Intake valve cam mechanism kinematic diagram

direction of the follower's motion relative to the axis of camshaft rotation determines whether it is radial or axial.

The ***radial rotational camshaft*** is where the follower moves in a plane perpendicular to the camshaft's axis of rotation. One example of the radial camshaft is shown in Fig. 9.1, where a disk plate (camshaft) rotates around a local z-axis. There is a camshaft outlined curve located on the x–y plane, perpendicular to the z-axis. The follower keeps sliding on this outlined curve inside the x–y plane.

The ***axial rotational camshaft*** is where the follower moves parallel to the axis of camshaft rotation. One example of the axial camshaft is shown in Fig. 9.3 where a cylinder (camshaft) rotates around a local z-axis. There is a camshaft outlined curve that is located on the cylinder's surface. The follower keeps sliding in the same direction of rotation on the z-axis.

The cam mechanism types can be itemized based on the cam shape. Table 9.1 shows a list of ten different cam link shapes. There are more cam link shapes that are not included in this textbook. The reader is encouraged to watch the attached video for each cam mechanism in Table 9.1.

9.2.2 Cam Mechanism Types Based on Follower Link

The follower link's edge, the part of the follower that is physically sliding on the cam link, has different shapes or types, as listed in Table 9.2. Every shape has advantages and disadvantages related to the generated friction between the follower link and the cam link and to the precision of the expected motion. For example, a knife edge produces a more precise expected motion but generates more friction than a rolling edge. Each shape is selected based on the design, production capabilities, and requirements. The reader is encouraged to watch the attached video for each cam mechanism in Table 9.2.

The follower offset distance (follower eccentricity as described later in this chapter) is defined as the distance between the follower's line of movement and the camshaft's center of rotation. The follower offset distance is used to classify two types of followers: the in-line follower type and the offset follower type. The in-line follower type is defined when the follower offset distance is zero.

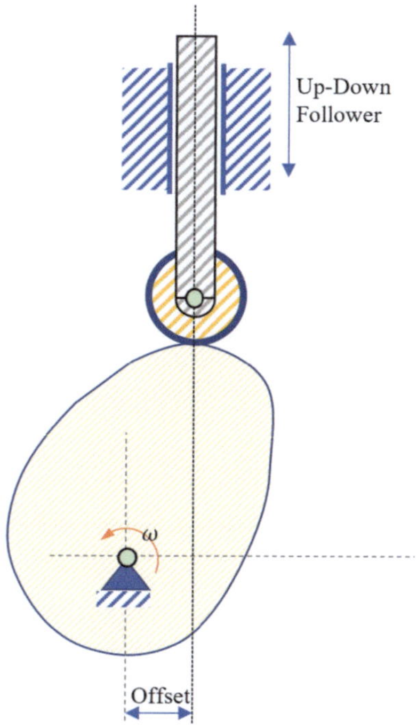

Fig. 9.7 Offset distance between line of follower movement and camshaft's center of rotation

In other words, the in-line follower type is defined when the follower's axis passes through the camshaft's center of rotation in the plane of rotation. The offset follower type is defined when the follower offset distance is not zero, as shown in Fig. 9.7. In other words, the radial follower type is defined when the movement of the follower's axis is not collinear with the camshaft rotational axis.

9.2.3 Cam Mechanism Types Based on Cam Joint

The cam joint (the connection between the follower link and the cam link) can be constructed as either a force closure type or a form closure type. The force closure type requires an external force to be applied to the joint to keep the follower link attached to the cam link, as shown in Fig. 9.8 where the spring can only push the follower toward the cam to maintain contact between the two of them. In general, the cam external force is defined as positive only in the direction that closes the cam joint. The external force could be gravity, springs, air cylinders, or hydraulic cylinders.

The form-closed type is defined when the joint between the follower link and the cam link closes by geometry without any external force. The curved groove consists of two cam surfaces, one surface on either side of the follower. These two surfaces drive the follower to move inside the curved groove. Figure 9.9 shows two examples of the form closure cam mechanism in both follower rotational motion and follower translational motion. Also, the conjugate cam in Table 9.1 is an example of the form closure cam mechanism.

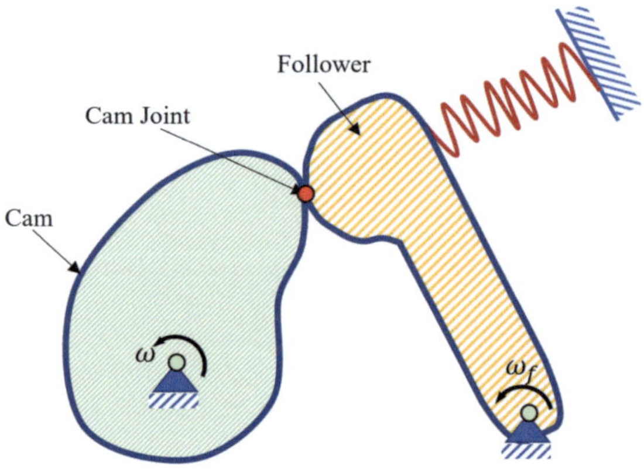

Fig. 9.8 Force closure cam mechanism type with spring external force

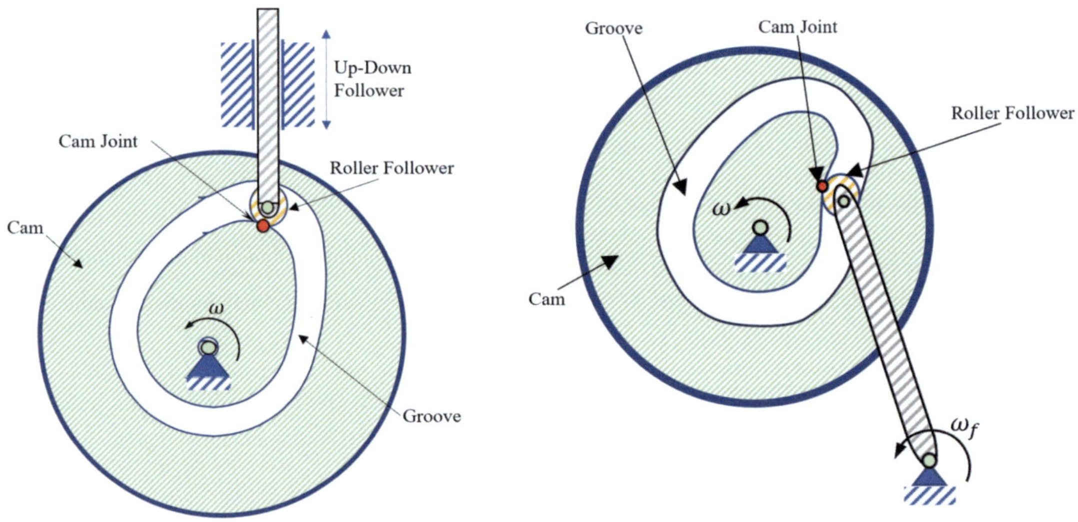

Fig. 9.9 Form closure cam mechanism type

9.3 Follower Motion Design

9.3.1 Follower Motion Segments

In general, the follower motion could have one or more motion segments: rise, fall, or dwell. The definitions from Chap. 8 are listed below.

- ***The rise motion*** is when the motion of the translational displacement or the angular displacement increases with time.
- ***The fall motion*** is defined when the motion of the translational displacement or angular displacement decreases with time.
- ***The dwell motion*** is defined when the motion of the translational displacement or angular displacement is held constant with time or has no change with time.

Table 9.1 Different cam link shapes; associated videos: Disc or Plate with Knife Edge Follower Cam Mechanism.mov, Heart-Shaped Cam Mechanism.mov, Convex Globoidal Cam Mechanism.mov, Concave Globoidal Cam Mechanism.mov and Spherical Cam Mechanism.mov designed by the author and implemented by both the author and Pranav Dhola. Courtesy of Pranav Dhola

Cam link type	Type of motion	Shape
Wedge or flat cam:The wedge-shaped cam link has a straight shape that the follower moves over	Translational oscillatory motion	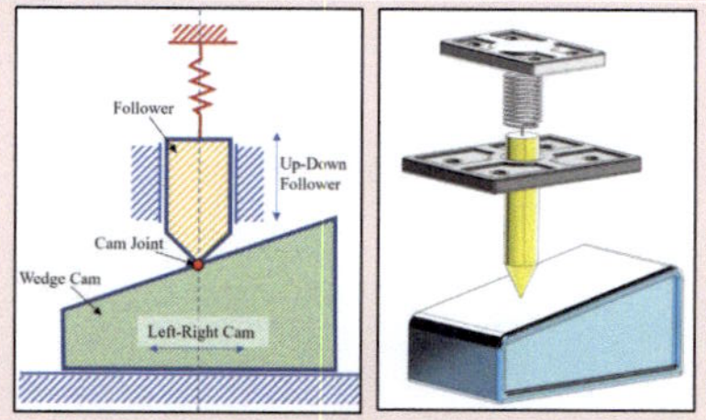
Translating cam: The translating cam link has a specific groove that the follower slides over	Translational oscillatory motion	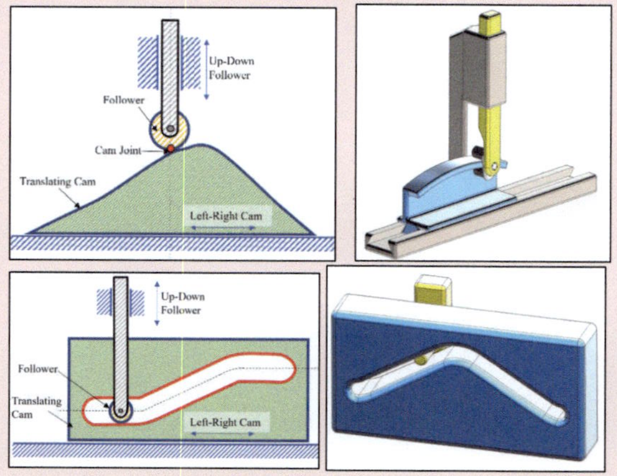
Disk or plate cam:The follower moves over a specific curved groove in the disk-shaped cam link	Radial rotational motion	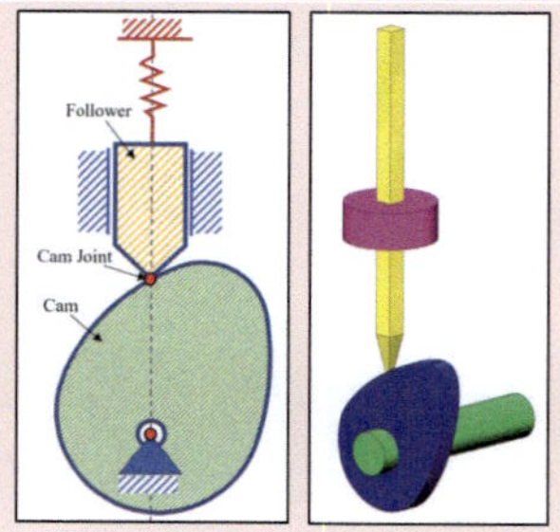

(continued)

Table 9.1 (continued)

Cam link type	Type of motion	Shape
Spiral cam: The camshaft has a spiral-shaped groove contour that the follower moves on	Radial rotational motion	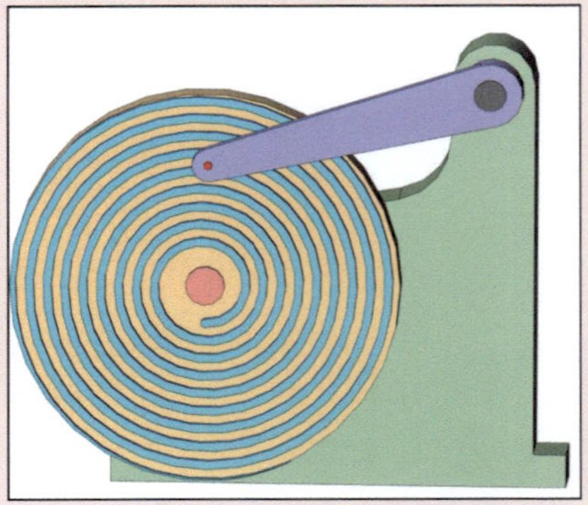
Heart-shaped cam: The follower roll moves over the heart-curved groove on the heart-shaped cam link	Radial rotational motion	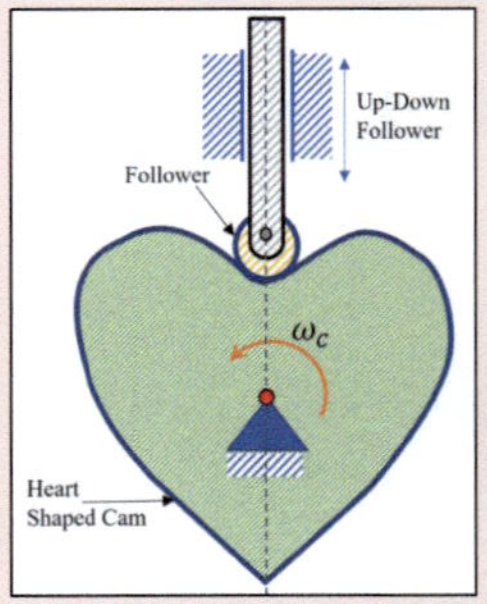 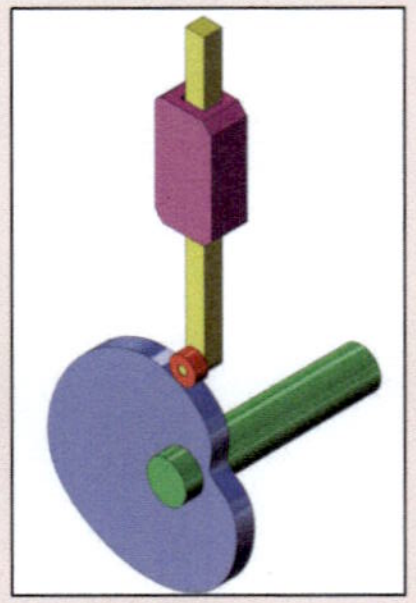
Snail drop cam: The cam link is like the disk or plate-shaped cam, with a sudden drop in the curved groove	Radial rotational motion	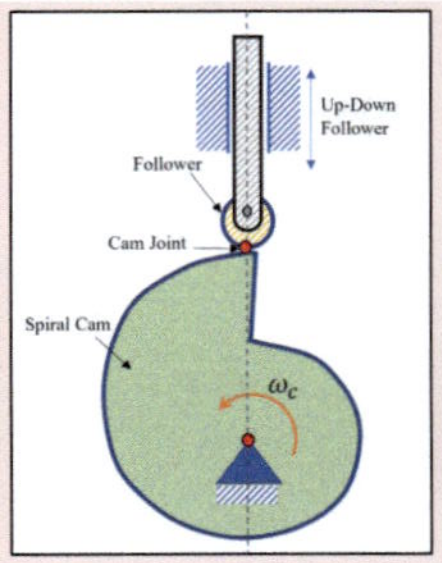 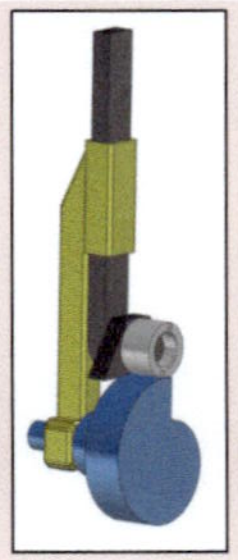

(continued)

Table 9.1 (continued)

Cam link type	Type of motion	Shape
Conjugate cam: The cam link consists of bolted double disks. The follower moves over the two double disks' grooves	Radial rotational motion	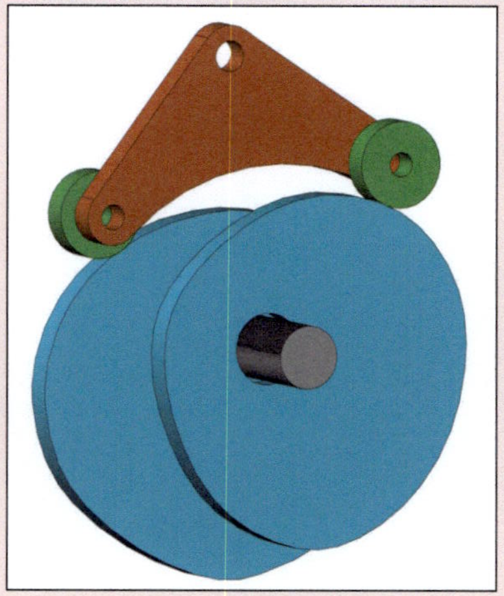
Cylindrical cam: The follower moves on a groove that is located on the cylinder's surface (cam link)	Axial rotational motion	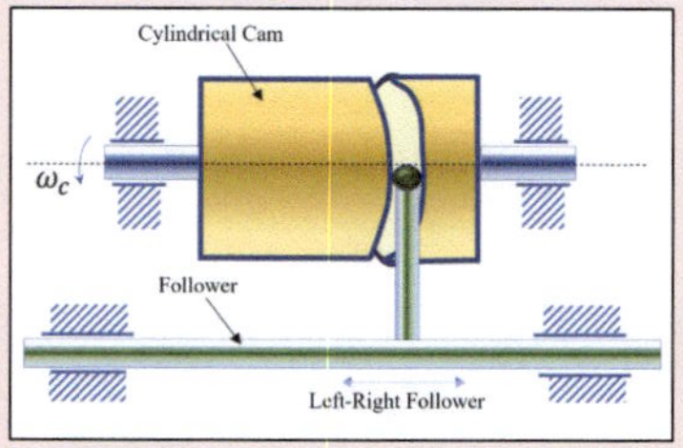 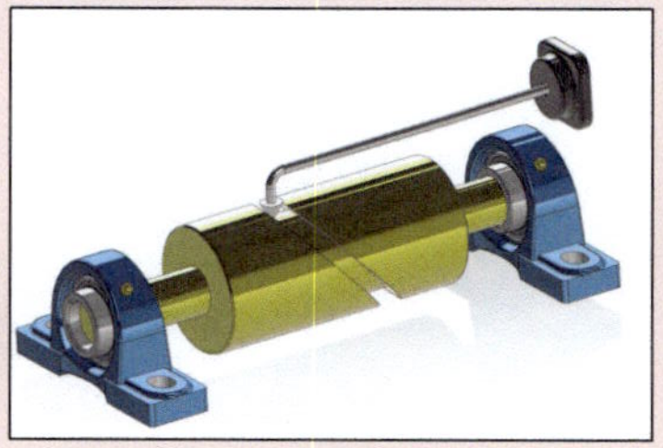

(continued)

Table 9.1 (continued)

Cam link type	Type of motion	Shape
Globoidal cam: The globoidal cam is similar to the cylindrical cam but uses a concave or convex surface instead of a uniform one	Axial rotational motion	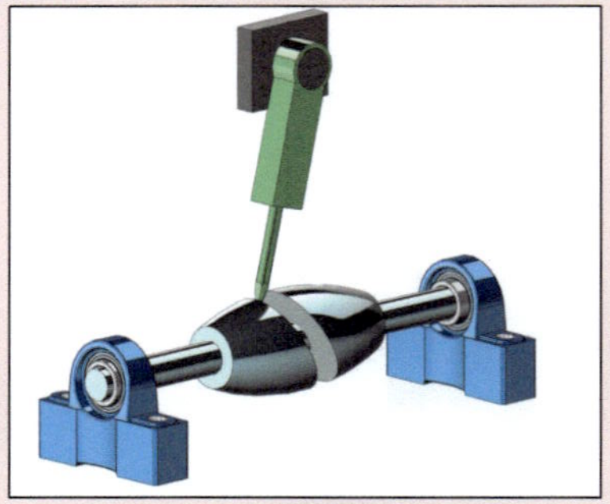 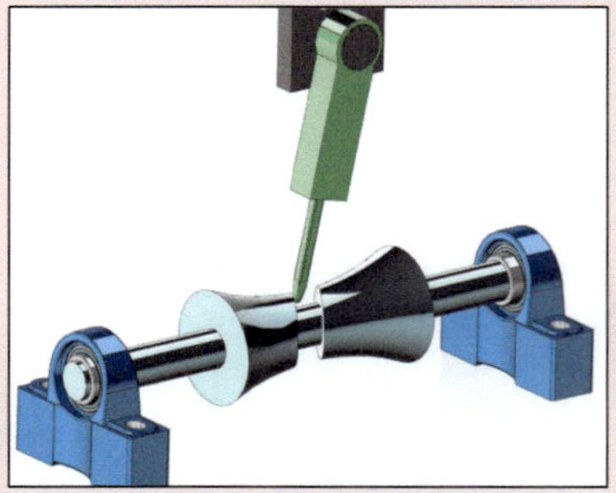
Spherical cam: The spherical cam is similar to the cylindrical cam but uses a spherical surface instead	Axial rotational motion	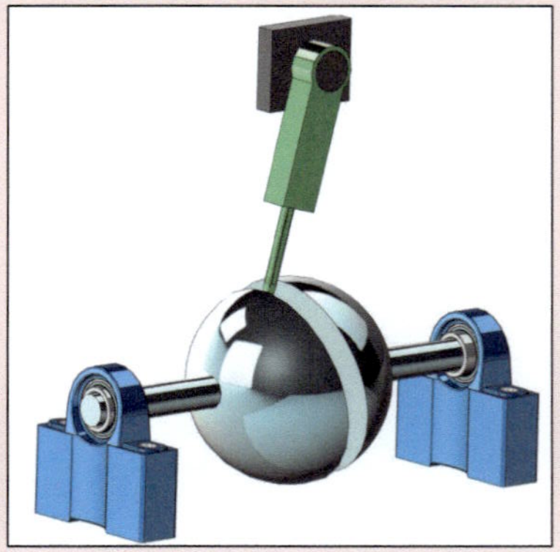

Table 9.2 Different cam link shapes; associated videos: Knife-Edge Follower.mov, Roller Follower.mov and Spherical or Mushroom Follower.mov designed by the author and implemented by both the author and Pranav Dhola. Courtesy of Pranav Dhola

Follower link's edge type	Shape
Roller follower	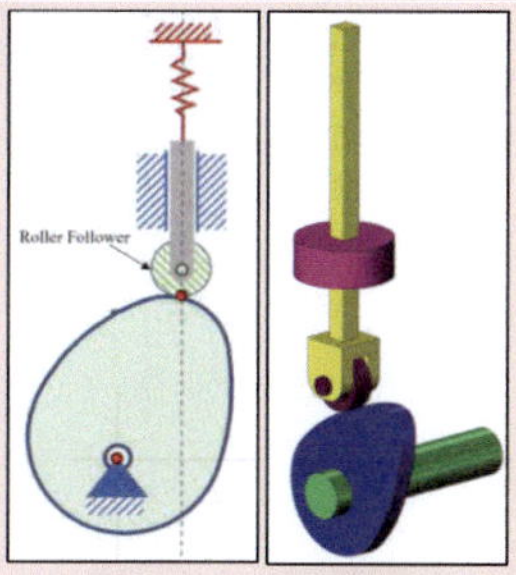
Knife-edge follower	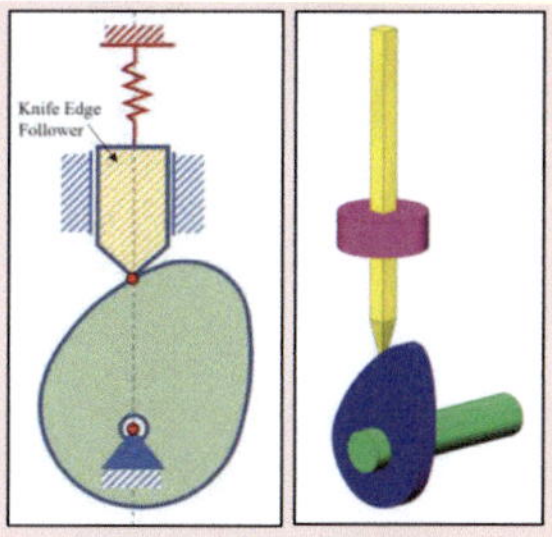
Flat-faced follower	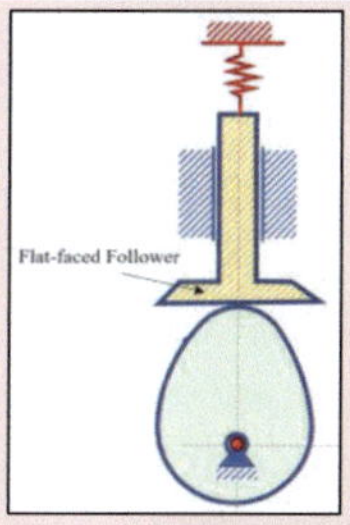
Spherical or mushroom follower	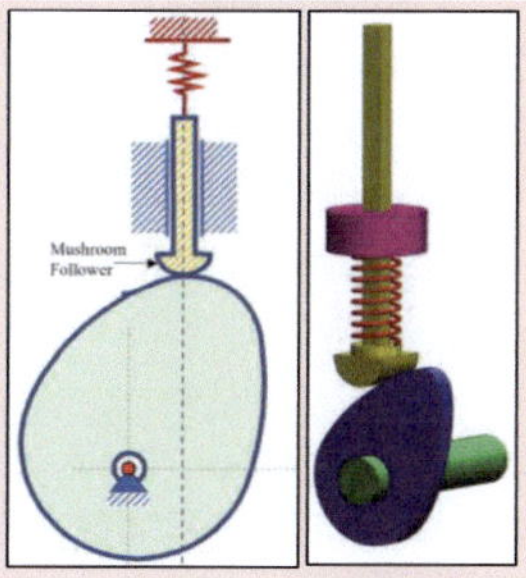

One good example to explain the fall and rise motion segments is given in Fig. 9.7 where fall means moving in the "Down-follower" direction and rise means moving in the "Up-follower" direction. Another example is the four-stroke engine in Sect. 9.2. Figure 9.10 shows the status of the exhaust valve versus the camshaft rotation angle. The yellow zone (in Fig. 9.10) represents the status of the exhaust valve. The theoretical process has the exhaust valve suddenly open and suddenly close at the beginning of the exhaust stroke and the end of the exhaust stroke respectively. This is virtually impossible. Instead, the real process is required and can be accomplished by the cam mechanism. The red line

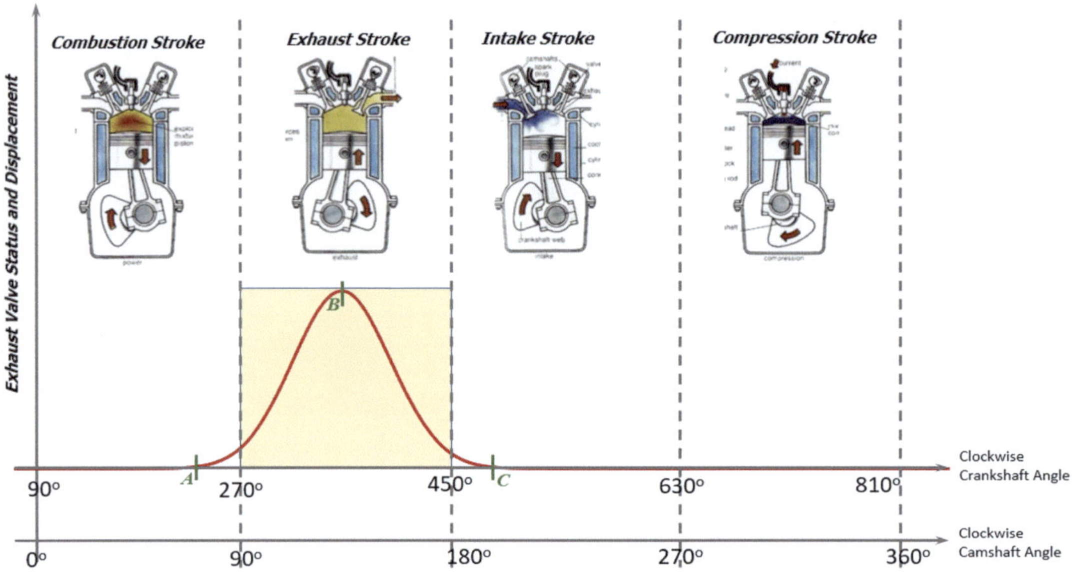

Fig. 9.10 Exhaust valve status and displacement w.r.t camshaft and crankshaft rotations. Designed by the author and implemented by both the author and Pranav Dhola. Courtesy of Pranav Dhola

in Fig. 9.10 represents an example of the real motion that the cam mechanism should open the exhaust valve with. The valve stays in the open position from point ***A*** to point ***C***. From point ***A*** to point ***B*** the valve is moving in the rising direction (downward). From point B to point C, the valve is moving in the falling direction (upward). Otherwise, the valve is in a state of dwell motion.

In Fig. 9.10, the cam profile is designed to have three subcurves that cause the dwell follower motion, the rise follower motion (from point A to point B), and the fall follower motion (point B to point C). To design such a cam profile, it is first assumed that the design is a radial rotational motion type, a force closure type, a flat-faced follower type, and a disk cam type. Also, the camshaft is rotating clockwise. The cam mechanism is shown in Fig. 9.11 for the intake valve. The camshaft's curved groove can be divided into three subcurves (dwell, rise, and fall). The three points A, B, and C are located between the three curves based on the angle distribution in Fig. 9.10. More details on how to come up with the cam profile will be explained later in this chapter. The valve opens and closes based on the scenario below:

- The valve is fully closed when the follower setting is at point A.
- When the camshaft rotates clockwise, the follower moves along the blue curve, leading to the opening of the valve. The valve is fully open when the follower reaches point B. The process that takes place from point A to point B is called the ***rise motion***.
- The camshaft keeps rotating clockwise. The follower now moves from the blue curve to the red curve; see Fig. 9.11. The valve begins to close and continues to do so until it becomes fully closed at point C. The process that takes place from point B to point C is called the ***fall motion***.
- The camshaft keeps rotating clockwise. The follower now moves from the red curve to the green curve, see Fig. 9.11. The valve stays fully closed because the green curve is circular as shown in Fig. 9.11. The process that takes place from point C to point B is called the ***dwell motion***.

The reader is encouraged to watch the attached video of the exhaust valve's cam mechanism. From this example, the conclusion can be made that the shape of the camshaft or the cam curved groove determines the follower motion profile.

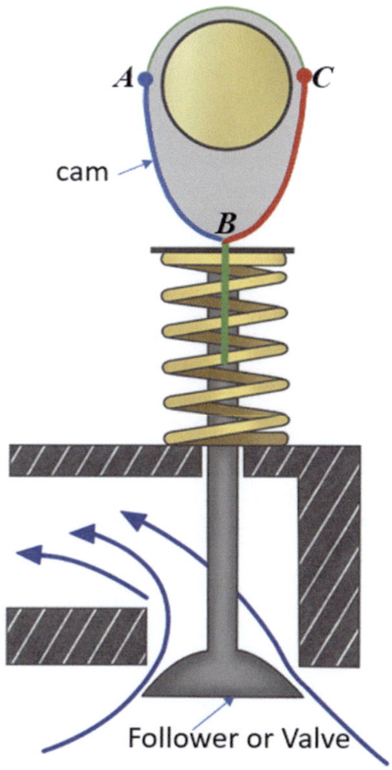

Fig. 9.11 Engine intake valve cam mechanism Designed by the author and implemented by both the author and Pranav Dhola. Courtesy of Pranav Dhola

After using the engine valve mechanism to understand the rise, fall, and dwell of the follower motion, the mathematical foundation to design such motions is explained. The cam machine has a cyclic job to be repeated after a periodic time T or a camshaft rotational cycle 2π. The SVAJ diagram presents the displacement S, velocity V, acceleration A, and jerk J of the follower versus time t or camshaft angle θ over this periodic time T or the camshaft period 2π. If the mathematical function of S is defined, then calculate the first derivative (the velocity) V, the second derivative (the acceleration) A, and the third derivative (the jerk) J. These derivative calculations can be done graphically, using software, or mathematically as

$$V(t) = \dot{S}(t) = \frac{\mathrm{d}S(t)}{\mathrm{d}t} \tag{9.1}$$

$$A(t) = \dot{V}(t) = \frac{\mathrm{d}V(t)}{\mathrm{d}t} = \ddot{S}(t) = \frac{\mathrm{d}^2S(t)}{\mathrm{d}t^2} \tag{9.2}$$

$$J(t) = \dot{A}(t) = \frac{\mathrm{d}A(t)}{\mathrm{d}t} = \ddot{V}(t) = \frac{\mathrm{d}^2V(t)}{\mathrm{d}t^2} = \dddot{S}(t) = \frac{\mathrm{d}^3S(t)}{\mathrm{d}t^3} \tag{9.3}$$

In Eqs. (9.1), (9.2), and (9.3), S, V, A, and J are functions of time. The time dependency of SVAJ can be replaced by another variable such as $s(\theta)$, $v(\theta)$, $a(\theta)$, or $j(\theta)$. In this case, we rewrite Eqs. (9.1), (9.2), and (9.3) as

$$v(\theta) = \frac{\mathrm{d}s(\theta)}{\mathrm{d}\theta} \tag{9.4}$$

$$a(\theta) = \frac{\mathrm{d}v(\theta)}{\mathrm{d}\theta} = \frac{\mathrm{d}^2 s(\theta)}{\mathrm{d}\theta^2} \tag{9.5}$$

$$j(\theta) = \frac{\mathrm{d}a(\theta)}{\mathrm{d}\theta} = \frac{\mathrm{d}^2 v(\theta)}{\mathrm{d}\theta^2} = \frac{\mathrm{d}^3 s(\theta)}{\mathrm{d}\theta^3} \tag{9.6}$$

The upper-case letters *SVAJ* are used when the dependent variable is time, t. The lower-case letters *svaj* are used when the dependent variable is the camshaft angle, θ. Over the periodic time T, or one cycle angle $\theta_{cycle} = 2\pi = 360°$, there are n motion segments. Each segment could be rise, fall, or dwell that occurs in the time interval ΔT_i or in the angle interval $\Delta\theta_i$, where $i = \{1,2, 3, \ldots, n\}$. The periodic time is the summation of the time intervals as

$$T = \sum_{i=1}^{n} \Delta T_i \tag{9.7}$$

The summation of the total angle intervals should be equal to 2π in radians or 360°in degrees as

$$\theta_{\text{cycle}} = \sum_{i=1}^{n} \Delta\theta_i = 2\pi \tag{9.8}$$

Most camshafts rotate with a constant angular velocity ω that can be calculated from

$$\omega = \frac{2\pi}{T} \tag{9.9}$$

In the case of constant angular velocity, each angle interval is related to its own time interval by

$$\Delta\theta_i = \omega \Delta T_i \tag{9.10}$$

Plot the four *SVAJ* or *svaj* functions in one column to obtain the *SVAJ* or *svaj* diagrams that show where the displacement desired function causes a high amplitude in the desired velocity, acceleration, or jerk. It is of interest to identify this because a high amplitude of velocity, acceleration, or jerk may cause noise, vibration, and/or failure for the actuator or the mechanism.

To use the *SVAJ* or *svaj* diagram in designing the follower motion, there are two types of motion requirements for the camshaft's curved groove: critical extreme position (CEP), endpoint specification, and critical path motion (CPM). The CEP represents the design specifications as a table or a sketch of the desired positions S_d of the follower (i.e., extreme positions) with respect to time t or camshaft angle θ but does not specify any constraints on the path motion between these desired points. Figure 9.12 shows a CEP specification consisting of nine position points with specific times and/or camshaft angles. The green line proposes a stepped trajectory that passes by the desired nine points. The green line is called the *constant displacement trajectory*. The red line proposes a ramp trajectory that passes through the desired nine points. The red line is called the *constant velocity trajectory*. The blue line proposes a smooth trajectory that passes through the nine points.

In the follower motion's design process, one first splits the trajectory over the periodic time $\underline{T}$ or the entire camshaft complete rotation 2π into motion segments that could be rise, fall, or dwell. Each motion segment has a total displacement ΔS that occurs within a time window ΔT or an angle window $\Delta\beta$ where β is angular variable. The motion can be designed separately using ramp, harmonic function, or polynomial methods. These methods will be explained in the next subsections.

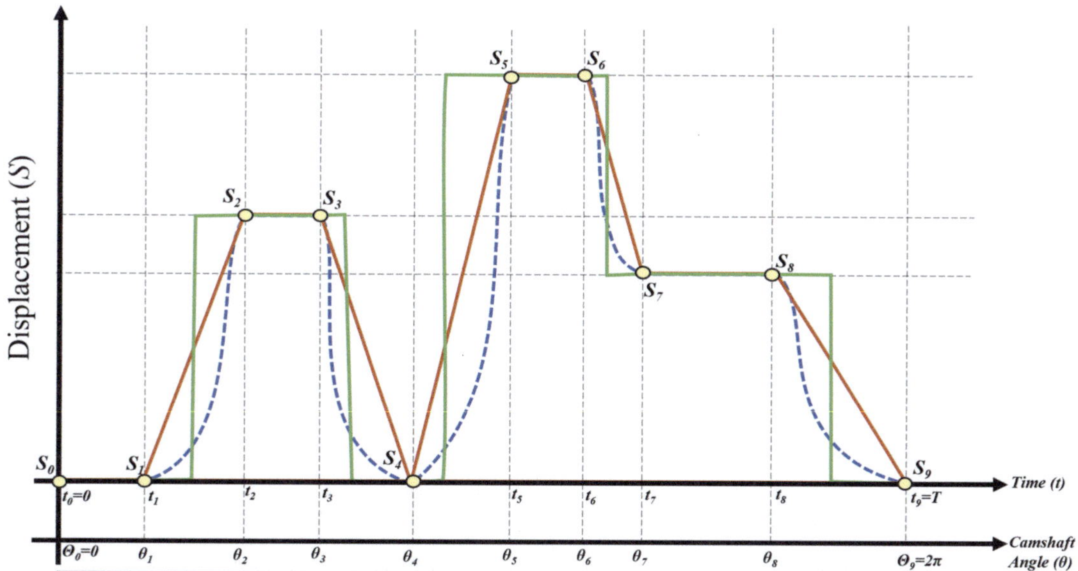

time	$t_0=0$	t_1	t_2	t_3	t_4	t_5	t_6	t_7	t_8	$t_9=T$
Camshaft Angle	$\Theta_0=0$	θ_1	θ_2	θ_3	θ_4	θ_5	θ_6	θ_7	θ_8	$\Theta_9=2\pi$
Displacement	S_0	S_1	S_2	S_3	S_4	S_5	S_6	S_7	S_8	S_9

Fig. 9.12 CEP desired points and proposed trajectories

Mathematical Note: Slope Discontinuity

The cam trajectory consists of motion segments such as rise, fall, and dwell. The trajectory motion segments are set in order with a guaranteed continuity in the value between every two back-to-back segments. However, there is no guarantee of continuity in the first slope or the high-order slope. It is assumed that a function $f(x)$ is continuous at x_o but has a slope discontinuity at that same point x_o as

$$f\left(x_o^{-\text{ve}}\right) = f\left(x_o^{+\text{ve}}\right) \text{ and } \frac{\mathrm{d}f}{\mathrm{d}x}\left(x_o^{-\text{ve}}\right) \neq \frac{\mathrm{d}f}{\mathrm{d}x}\left(x_o^{+\text{ve}}\right)$$

where $f\left(x_o^{-\text{ve}}\right)$ and $\frac{\mathrm{d}f}{\mathrm{d}x}\left(x_o^{-\text{ve}}\right)$ are the values of $f(x_o - \Delta x)$ and $\frac{\mathrm{d}f}{\mathrm{d}x}(x_o - \Delta x)$, respectively, when Δx tends to zero and $f\left(x_o^{+\text{ve}}\right)$ and $\frac{\mathrm{d}f}{\mathrm{d}x}\left(x_o^{+\text{ve}}\right)$ are the values of $f(x_o + \Delta x)$ and $\frac{\mathrm{d}f}{\mathrm{d}x}(x_o + \Delta x)$, respectively, when Δx tends to zero. In other words, the function $f(x)$ has a slope discontinuity if the slope on the right side of point x_o is different from the slope on the left side.

When the function $f(x)$ has slope discontinuity at x_o and $\frac{\mathrm{d}f}{\mathrm{d}x}\left(x_o^{-\text{ve}}\right) < \frac{\mathrm{d}f}{\mathrm{d}x}\left(x_o^{+\text{ve}}\right)$, the function $f(x)$ has positive undefined high (second and up) order slopes. The higher-order slopes of $f(x)$ are

$$\begin{aligned}
\frac{\mathrm{d}^2 f}{\mathrm{d}x^2}(x_o) &= +\infty \\
\frac{\mathrm{d}^3 f}{\mathrm{d}x^3}(x_o) &= +\infty^2 \\
&\vdots \\
\frac{\mathrm{d}^n f}{\mathrm{d}x^n}(x_o) &= +\infty^{n-1}
\end{aligned}$$

When the function $f(x)$ has slope discontinuity at x_o and $\frac{df}{dx}\left(x_o^{-ve}\right) > \frac{df}{dx}\left(x_o^{+ve}\right)$, the function $f(x)$ has negative undefined high (second and up) order slopes. The higher-order slopes of $f(x)$ are

$$\frac{d^2f}{dx^2}(x_o) = -\infty$$
$$\frac{d^3f}{dx^3}(x_o) = -\infty^2$$
$$\vdots$$
$$\frac{d^nf}{dx^n}(x_o) = -\infty^{n-1}$$

This slope discontinuity note is used when the *SVAJ* or *svaj* diagram of the follower trajectory that consists of back-to-back motion segments is sketched. The mathematical derivation of this note was delivered in Chap. 8.

For example, Fig. 9.13 shows two linear displacement functions $S_1(x)$ and $S_2(x)$ with a velocity (first slope) discontinuity at point x_o. In the case of the first function $S_1(x)$ in Fig. 9.13a, the slope on the right side of x_o is greater than the slope on the left side which

Fig. 9.13 Slope discontinuity examples

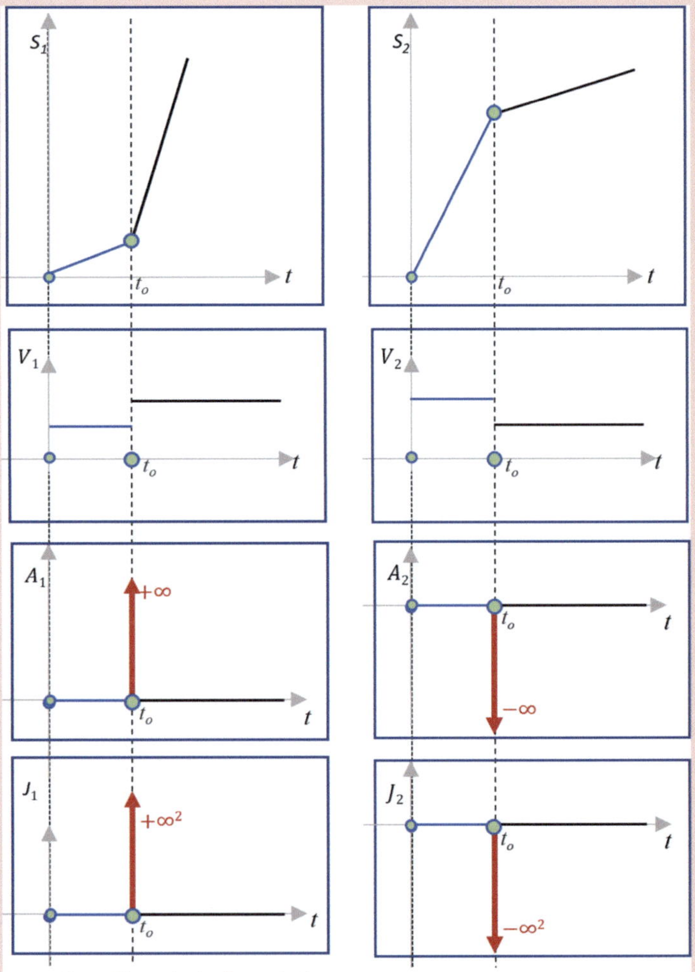

leads to the velocity function $V\left(x_o^{-\mathrm{ve}}\right) < V\left(x_o^{+\mathrm{ve}}\right)$. In this case of Fig. 9.13a, the second slope (A_1) and the third slope (J_1) are

$$A_1(x_o) = +\infty$$
$$J_1(x_o) = +\infty^2$$

In the case of the second function $S_2(x)$ in Fig. 9.13b, the slope on the right side of x_o is less than the slope on the left side which leads to the velocity function $V_2\left(x_o^{-\mathrm{ve}}\right) > V_2\left(x_o^{+\mathrm{ve}}\right)$. In the case of Fig. 9.13b, the second slope (A_2) and the third slope (J_2) are

$$A_2(x_o) = -\infty$$
$$J_2(x_o) = -\infty^2$$

Keep in mind that the infinite acceleration and infinite jerk can cause vibration, noise, and a possibility of damaging the mechanism as discussed in Chap. 7.

Example 9.1

A camshaft rotates through an angle θ and with the follower displacement trajectory

$$s(\theta) = \begin{cases} 0 & 0 \le \theta < 90^\circ \\ \frac{20}{45}(\theta - 90^\circ) & 90^\circ \le \theta < 135^\circ \\ -\frac{20}{45}(\theta - 180^\circ) & 135^\circ \le \theta < 180^\circ \\ 0 & 180^\circ \le \theta \le 360^\circ \end{cases}$$

where θ is in degrees and *s* is in mm.

1. Calculate the velocity *v*(θ), the acceleration *a*(θ), and the jerk *j*(θ).
2. Sketch the *svaj* diagram of the follower motion.

Solution

Using Eq. (9.4) to calculate *v*(θ):

$$v(\theta) = \frac{\mathrm{d}s(\theta)}{\mathrm{d}\theta} = \begin{cases} 0 & 0 \le \theta \le 90^\circ \\ \frac{20}{45} & 90^\circ \le \theta \le 135^\circ \\ -\frac{20}{45} & 135^\circ \le \theta \le 180^\circ \\ 0 & 180^\circ \le \theta \le 360^\circ \end{cases}$$

Using Eq. (9.5) to calculate *a*(θ):

$$a(\theta) = \frac{\mathrm{d}v(\theta)}{\mathrm{d}\theta} = \begin{cases} +\infty & \theta = 90^\circ \\ -\infty & \theta = 135^\circ \\ +\infty & \theta = 180^\circ \\ 0 & \text{otherwise} \end{cases}$$

Using Eq. (9.6) to calculate $j(\theta)$:

$$j(\theta) = \frac{da(\theta)}{d\theta} = \begin{cases} +\infty^2 & \theta = 90^\circ \\ -\infty^2 & \theta = 135^\circ \\ +\infty^2 & \theta = 180^\circ \\ 0 & \text{otherwise} \end{cases}$$

The sketch of the *svaj* diagram is shown in Fig. 9.14.

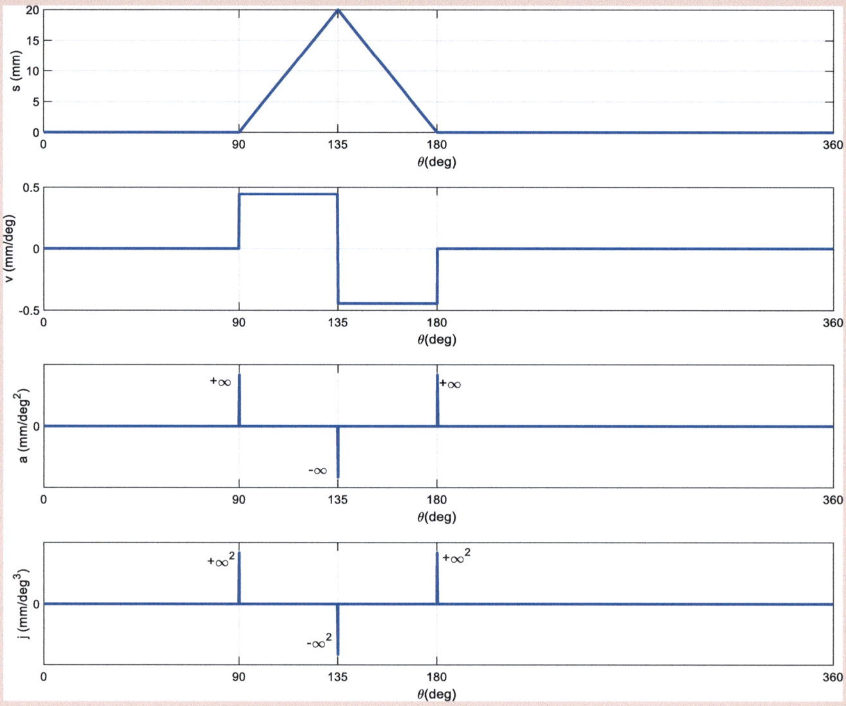

Fig. 9.14 svaj diagram of Example 9.1

9.3.2 Polynomial CEP Motion

A polynomial CEP motion segment goes from a certain value $S_1 = S(t_1)$ to another value $S_2 = S(t_2)$ over a certain time window ΔT using a sum of nth-order polynomial functions as

$$S(t) = \sum_{i=0}^{n} C_i t^i \quad t_1 \le t \le t_2 \tag{9.11}$$

where C_i is a set of constants that determines the shape of the SVAJ diagram. These constants can be calculated to satisfy specific conditions or displacement values, velocity values, acceleration values, and jerk values. These constants turn out to be a set of linear equations. Step SVAJ motion design is an example of polynomial *SVAJ* motion design where $n = 0$. Ramp SVAJ motion design is another example of polynomial *SVAJ* motion design where $n = 1$.

Table 9.3 Mathematical equations and step of ramp CEP trajectory

	Position, S	Velocity, V	Acceleration, A	Jerk, J
Rise CEP Trajectory	$S(t) = S_o + \Delta S$ 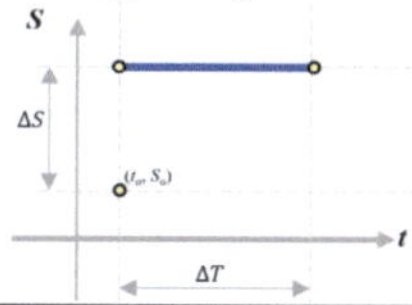	$V(t) = 0$ 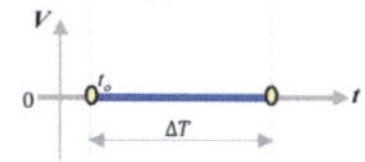	$A(t) = 0$ 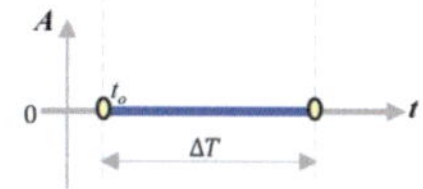	$J(t) = 0$ 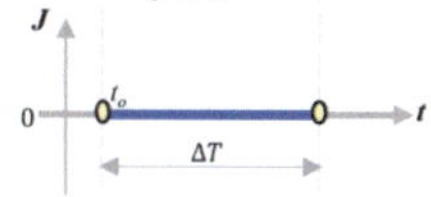
Fall CEP Trajectory	$S(t) = S_o - \Delta S$ 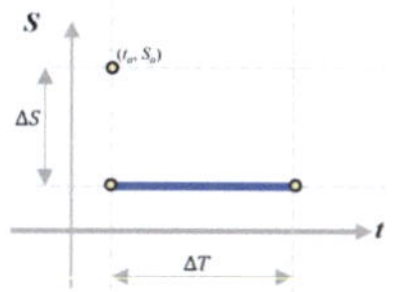	$V(t) = 0$ 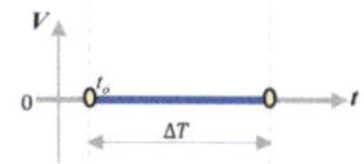	$A(t) = 0$ 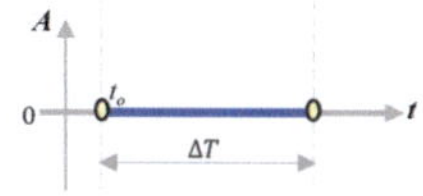	$J(t) = 0$ 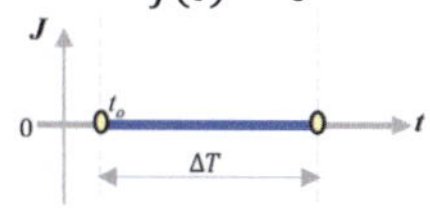

9.3.2.1 Step CEP Motion or Constant Displacement CEP Motion

Step CEP motion design is a sudden change in the displacement to rise, fall, or dwell at time t_o. The displacement ΔS is positive in the case of rise, negative in the case of fall, and zero in the case of dwell. More details about step motion are provided in Chap. 8. Table 9.3 lists the mathematical equations and plots of both the rise and fall Step *CEP* trajectory motion segments. The step CEP motion causes discontinuities in velocity, acceleration, and jerk. Step CEP motion is not often utilized but is still needed for certain applications, such as for the case of a *Snail Drop Cam* as shown in Table 9.1.

9.3.2.2 Ramp CEP Motion or Constant Velocity CEP Motion

Ramp CEP motion design goes from time t_o to time $t_o + \Delta T$ with a constant velocity or with a straight line to rise, fall, or dwell with a total displacement ΔS and within a time window ΔT. The displacement ΔS is positive in the case of rise, negative in the case of fall, and zero in the case of dwell. More details about ramp motion are provided in Chap. 8. Table 9.4 lists the mathematical equations and plots of both the rise and fall ramp *CEP* trajectory motion segments.

Example 9.2

A disk cam mechanism with a knife-edge follower has the follower trajectory motion segments sequentially as

- ***Motion segment 1:*** Rise 10 mm in 1 s.
- ***Motion segment 2:*** Dwell for 0.5 s.
- ***Motion segment 3:*** Fall 5 mm for 0.5 s.
- ***Motion segment 4:*** Dwell for 0.5 s.
- ***Motion segment 5:*** Fall 5 mm for 0.5 s.
- ***Motion segment 6:*** Dwell for 0.5 s.

The ramp CEP motion (constant velocity CEP motion) is used to design the follower trajectory motion segments.

1. Find the cam constant angular speed.
2. Find the initial angle and final angle of each motion segment (rise, fall, and dwell).
3. Sketch the *svaj* diagram and *SVAJ* diagram.

Table 9.4 Mathematical equations and plot of ramp CEP trajectory

	Position, S	Velocity, V	Acceleration, A	Jerk, J
Rise CEP Trajectory	$S(t) = S_o + \frac{\Delta S}{\Delta T}(t - t_o)$ 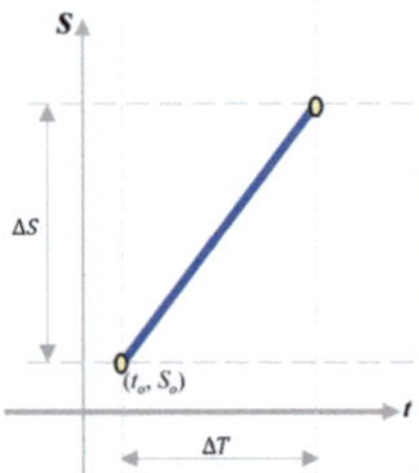	$V(t) = +\frac{\Delta S}{\Delta T}$ 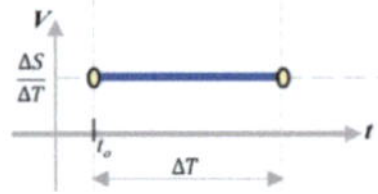	$A(t) = 0$ 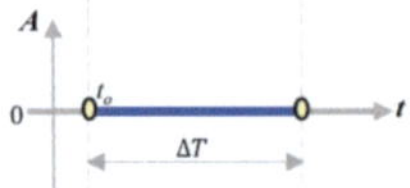	$J(t) = 0$ 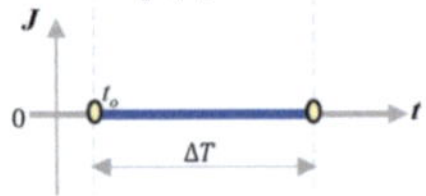
Fall CEP Trajectory	$S(t) = S_o - \frac{\Delta S}{\Delta T}(t - t_o)$ 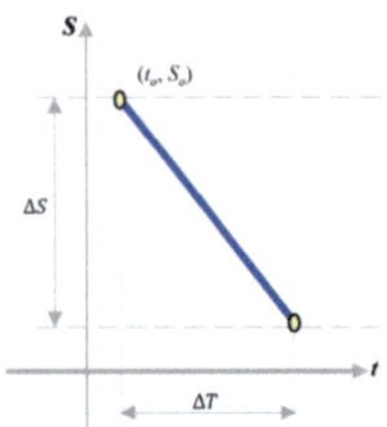	$V(t) = -\frac{\Delta S}{\Delta T}$ 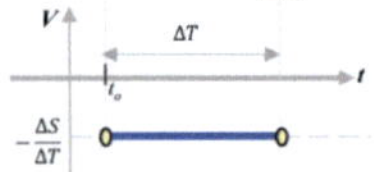	$A(t) = 0$ 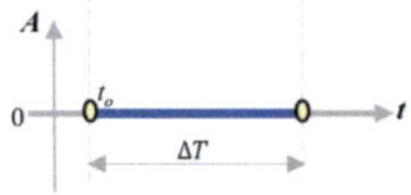	$J(t) = 0$ 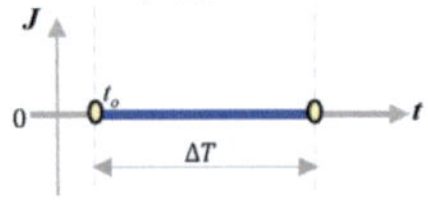

Solution

First, calculate the periodic time using Eq. (9.7)

$$T = \Delta T_1 + \Delta T_2 + \Delta T_3 + \Delta T_4 + \Delta T_5 + \Delta T_6$$
$$= 1\,\text{s} + 0.5\,\text{s} + 0.5\,\text{s} + 0.5\,\text{s} + 0.5\,\text{s} + 0.5\,\text{s} = 3.5\,\text{s}$$

From Eq. (9.9), the constant angular velocity is calculated as

$$\omega = \frac{2\pi}{T} = \frac{2\pi}{3.5\,\text{s}} = 1.795\,\text{rad/s}$$

Table 9.5 shows the time interval ΔT, the angle interval $\Delta\theta$, the initial angle θ_i, and the final angle θ_f of each trajectory motion segment. The angle interval from Eq. (9.10) is calculated as

Table 9.5 Example 9.2 Trajectory motion segments

Segment number	Time interval, ΔT (s)	Angle interval, $\Delta\theta$ (deg)	Initial angle, θ_i (deg)	Final angle, θ_f (deg)
1	1	102.86°	0°	102.86°
2	0.5	51.43°	102.86°	154.29°
3	0.5	51.43°	154.29°	205.73°
4	0.5	51.43°	205.73°	257.15°
5	0.5	51.43°	257.15°	308.58°
6	0.5	51.43°	308.58°	360°

$$\Delta\theta_i = \omega\Delta T_i = (1.795\,\text{rad/s}) \times \Delta T_i \times \left(\frac{180}{\pi}\frac{\text{deg}}{\text{rad}}\right)$$

$$= 102.86\Delta T_i\,\text{deg} \quad i = \{1, 2, 3, 4, 5, 6\}$$

Start the first segment from 0°. The final angle is calculated from

$$\theta_f = \theta_i + \Delta\theta_i \quad i = \{1, 2, 3, 4, 5, 6\}$$

Both the *SVAJ* and *svaj* diagrams are shown in Fig. 9.15. A linear line between the trajectory points is used since the design is a ramp CEP motion (constant velocity).

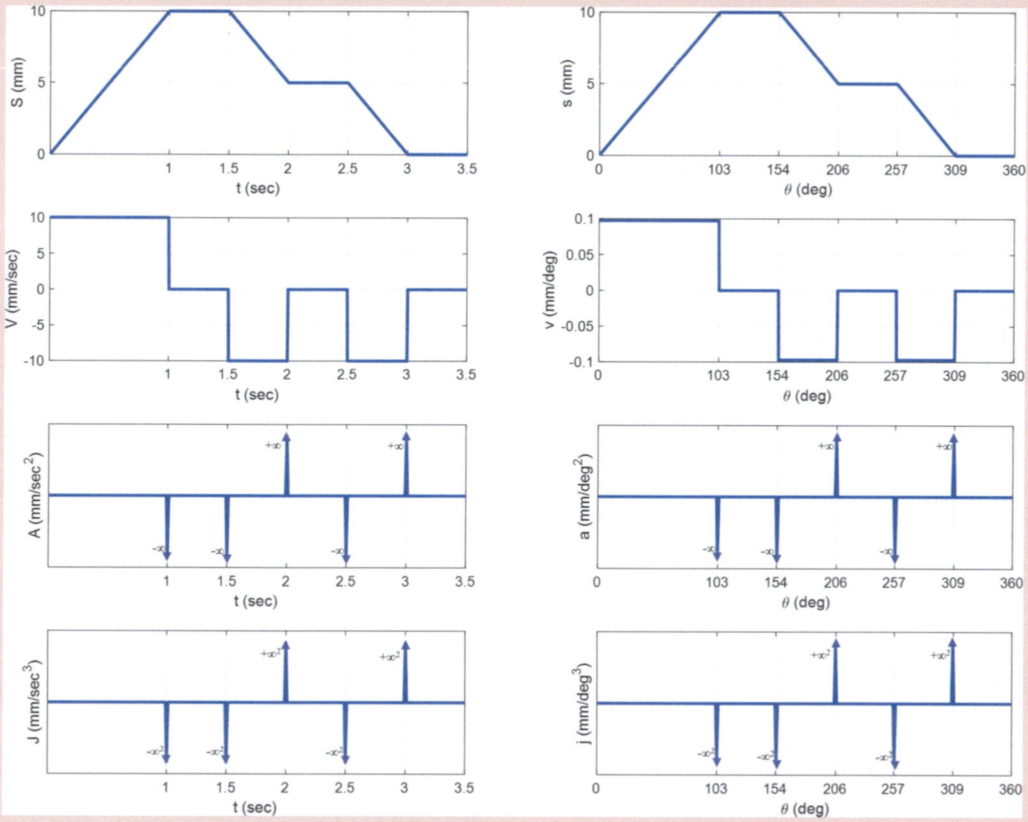

Fig. 9.15 Example 9.2 SVAJ diagram and svaj diagram

9.3.2.3 Higher-Order Polynomial Motion

Let's assume that a polynomial CEP motion is used to design a rise motion or a fall motion that starts at time $t = t_o$ or crankshaft angle $\theta = \theta_o$ with a time window ΔT or a camshaft angle window $\Delta\beta$ to achieve a total displacement ΔS. It is desired to achieve initial and final conditions of zero for the velocity, acceleration, and jerk. First, calculate this polynomial in terms of θ. The variables are introduced below.

$$\overline{\theta} = \frac{\theta - \theta_o}{\Delta\beta} \text{ and } 0 \leq \overline{\theta} \leq 1 \text{ when } \theta_o \leq \theta \leq \theta_o + \Delta\beta$$
$$\overline{s} = \frac{s - s_o}{\Delta s} \text{ and } 0 \leq \overline{S} \leq 1 \text{ when } s_o \leq s \leq s_o + \Delta s \tag{9.12}$$

Select the designed motion with constraints below:

1. The displacement shall be s_o at time $t = t_o$ and $\theta = \theta_o$ or $\overline{s}(\overline{t} = 0) = \overline{s}(\overline{\theta} = 0) = 0$.
2. The velocity shall be zero at time $t = t_o$ or $\theta = \theta_o$ or $v(\overline{t} = 0) = v(\overline{\theta} = 0) = 0$.
3. The acceleration shall be zero at time $t = t_o$ or $\theta = \theta_o$ or $a(\overline{t} = 0) = a(\overline{\theta} = 0) = 0$.
4. The jerk shall be zero at time $t = t_o$ or $\theta = \theta_o$ or $j(\overline{t} = 0) = j(\overline{\theta} = 0) = 0$.
5. The displacement shall be $s_o + \Delta s$ at time $t = t_o + \Delta T$ and $\theta = \theta_o + \beta$ or $\overline{s}(\overline{t} = 1) = s(\overline{\theta} = 1) = 1$.
6. The velocity shall be zero at time $t = t_o + \Delta T$ and $\theta = \theta_o + \beta$ or $v(\overline{t} = 1) = v(\overline{\theta} = 1) = 0$.
7. The acceleration shall be zero at time $t = t_o + \Delta T$ and $\theta = \theta_o + \beta$ or $a(\overline{t} = 1) = a(\overline{\theta} = 1) = 0$.
8. The jerk shall be zero at time $t = t_o + \Delta T$ and $\theta = \theta_o + \beta$ or $j(\overline{t} = 1) = j(\overline{\theta} = 1) = 0$.

The polynomial order is 7th to satisfy the above eight conditions or

$$s(\overline{\theta}) = A_o + A_1\overline{\theta} + A_2\overline{\theta}^2 + A_3\overline{\theta}^3 + A_4\overline{\theta}^4 + A_5\overline{\theta}^5 + A_6\overline{\theta}^6 + A_7\overline{\theta}^7 \tag{9.13}$$

$$v(\overline{\theta}) = A_1 + 2A_2\overline{\theta} + 3A_3\overline{\theta}^2 + 4A_4\overline{\theta}^3 + 5A_5\overline{\theta}^4 + 6A_6\overline{\theta}^5 + 7A_7\overline{\theta}^6 \tag{9.14}$$

$$a(\overline{\theta}) = 2A_2 + 6A_3\overline{\theta} + 12A_4\overline{\theta}^2 + 20A_5\overline{\theta}^3 + 30A_6\overline{\theta}^4 + 42A_7\overline{\theta}^5 \tag{9.15}$$

$$j(\overline{\theta}) = 6A_3 + 24A_4\overline{\theta} + 60A_5\overline{\theta}^2 + 120A_6\overline{\theta}^3 + 210A_7\overline{\theta}^4 \tag{9.16}$$

Applying the first four conditions on the polynomials in Eqs. (9.13)–(9.16), leads to

$$s(\overline{\theta} = 0) = A_o = 0 \tag{9.17}$$

$$v(\overline{\theta} = 0) = A_1 = 0 \tag{9.18}$$

$$a(\overline{\theta} = 0) = 2A_2 = 0 \tag{9.19}$$

$$j(\overline{\theta} = 0) = 6A_3 = 0 \tag{9.20}$$

From Eqs. (9.17) to (9.20), the polynomial is 4–5–6–7th order, respectively. Applying the second four conditions leads to four equations with four unknowns (A_4, A_5, A_6, and A_7), or

$$s(\overline{\theta} = 1) = A_4 + A_5 + A_6 + A_7 = h \tag{9.21}$$

$$v(\overline{\theta} = 1) = 4A_4 + 5A_5 + 6A_6 + 7A_7 = 0 \tag{9.22}$$

$$a(\overline{\theta} = 1) = 12A_4 + 20A_5 + 30A_6 + 42A_7 = 0 \tag{9.23}$$

$$j(\overline{\theta} = 1) = 6A_3 + 24A_4 + 60A_5 + 120A_6 + 210A_7 = 0 \tag{9.24}$$

Solving Eqs. (9.21)–(9.24), the *svaj* functions are then calculated to be

$$
\begin{aligned}
s(\theta) &= s_o + (\Delta s)\left(35\left(\frac{\theta-\theta_o}{\Delta\beta}\right)^4 - 84\left(\frac{\theta-\theta_o}{\Delta\beta}\right)^5 + 70\left(\frac{\theta-\theta_o}{\Delta\beta}\right)^6 - 20\left(\frac{\theta-\theta_o}{\Delta\beta}\right)^7\right)\\
v(\theta) &= \left(\frac{\Delta s}{\Delta\beta}\right)\left(140\left(\frac{\theta-\theta_o}{\Delta\beta}\right)^3 - 420\left(\frac{\theta-\theta_o}{\Delta\beta}\right)^4 + 420\left(\frac{\theta-\theta_o}{\Delta\beta}\right)^5 - 140\left(\frac{\theta-\theta_o}{\Delta\beta}\right)^6\right)\\
a(\theta) &= \left(\frac{\Delta s}{\Delta\beta^2}\right)\left(420\left(\frac{\theta-\theta_o}{\Delta\beta}\right)^2 - 1680\left(\frac{\theta-\theta_o}{\Delta\beta}\right)^3 + 2100\left(\frac{\theta-\theta_o}{\Delta\beta}\right)^4 - 840\left(\frac{\theta-\theta_o}{\Delta\beta}\right)^5\right)\\
j(\theta) &= \left(\frac{\Delta s}{\Delta\beta^3}\right)\left(840\left(\frac{\theta-\theta_o}{\Delta\beta}\right) - 5040\left(\frac{\theta-\theta_o}{\Delta\beta}\right)^2 + 8400\left(\frac{\theta-\theta_o}{\Delta\beta}\right)^3 - 4200\left(\frac{\theta-\theta_o}{\Delta\beta}\right)^4\right)
\end{aligned}
\tag{9.25}
$$

Equation (9.25) provides a fall motion or a rise motion with a total displacement Δs that starts and ends with zero velocity, zero acceleration, and zero jerk. These zeros guarantee that there is no discontinuity in all of the higher-order derivatives up to the third-order derivative (the jerk). However, if any further derivatives are developed, there are nonzero high-order derivatives that may lead to some discontinuities in the fourth, fifth or any higher-order derivatives. Other polynomial motions are driven using different motion conditions. The *svaj* diagram of Eq. (9.25) can be used in both the fall motion and rise motion. Note that the previous derivation was done for *svaj*, but a similar derivation can be done for *SVAJ*. The results would be the same if θ_o was replaced by t_o and $\Delta\beta$ by ΔT.

Example 9.3

A disk cam mechanism with a knife-edge follower has the follower trajectory motion segments sequentially as

- ***Motion segment 1***: Rise 10 mm in 1 s.
- ***Motion segment 2***: Dwell for 0.5 s.
- ***Motion segment 3***: Fall 5 mm for 0.5 s.
- ***Motion segment 4***: Dwell for 0.5 s.
- ***Motion segment 5***: Fall 5 mm for 0.5 s.
- ***Motion segment 6***: Dwell for 0.5 s.

The 7th-order polynomial is used to design the follower trajectory motion segments.

1. Find the cam constant angular speed.
2. Find the initial angle and the final angle of each motion segment (rise, fall, and dwell).
3. Use MATLAB® script to plot the *svaj* diagram.

Solution

First, start calculating the periodic time using Eq. (9.7)

$$T = \Delta T_1 + \Delta T_2 + \Delta T_3 + \Delta T_4 + \Delta T_5 + \Delta T_6$$
$$= 1\,\text{s} + 0.5\,\text{s} + 0.5\,\text{s} + 0.5\,\text{s} + 0.5\,\text{s} + 0.5\,\text{s} = 3.5\,\text{s}$$

From Eq. (9.9), calculate the constant angular velocity as

$$\omega = \frac{2\pi}{T} = \frac{2\pi}{3.5\,\text{s}} = 1.795\,\text{rad/s}$$

Table 9.6 shows the time interval ΔT, the angle interval $\Delta\theta$, the initial angle θ_i, and the final angle θ_f of each trajectory motion segment. Calculate the angle interval from Eq. (9.10) as

$$\Delta\theta_i = \omega\Delta T_i = (1.795\,\text{rad/s})\times\Delta T_i\times\left(\frac{180\,\text{deg}}{\pi\ \ \text{rad}}\right) = 102.86\Delta T_i\,\text{deg} \quad i = \{1,2,3,4,5,6\}$$

Start the first segment from 0°. The final angle is calculated from

$$\theta_f = \theta_i + \Delta\theta_i \quad i = \{1,2,3,4,5,6\}$$

Now, calculate the *svaj* diagram using Eq. (9.25) as implemented in the MATLAB® script using the steps below:

1. Clean the variable history by opening a new script and saving it as "Poly7th_Cam_script.m"
2. Add all the given parameters.
3. Calculate the needed parameters of each segment.

As mentioned previously, the reader is encouraged to attend the "MATLAB® Onramp" and "Simulink® Onramp" free online courses (Fig. 9.16).

Table 9.6 Example 9.3 Trajectory motion segments

Segment number	Time interval, ΔT (s)	Angle interval, $\Delta\beta$ (deg)	Initial angle, θ_i (deg)	Final angle, θ_f (deg)
1	1	102.86°	0°	102.86°
2	0.5	51.43°	102.86°	154.29°
3	0.5	51.43°	154.29°	205.73°
4	0.5	51.43°	205.73°	257.15°
5	0.5	51.43°	257.15°	308.58°
6	0.5	51.43°	308.58°	360°

```
%This file to declare the parameters of Example 9.3
%======================================
%Clear all the variables and close all the figures
%==================================
clc
clear all
close all
%==========================================================================
%Declare the given parameters
%=============================
dT = [ 1 0.5 0.5 0.5 0.5 0.5]; % time steps of each  motion segment (sec)
dS = [10 0 -5 0 -5 0]; % displacement steps for each motion segment (mm)
s_intial = [0 10 10 5 5 0]; %initial displacement at each segment (mm)
omega = 2*pi/sum(dT); % cam angular velocity (rad/sec)
dtheta = dT*omega*180/pi; %anglar difference (deg)
theta_step(1) = 0;
for i =1:length(dT)
    theta_step(i+1) = theta_step(i)+ dtheta(i); % angle of each  motion segment (deg)
end

%Calculating svaj
%==============
theta = linspace(0,360,1000); % cam angle (deg)

%Intilization of svaj
s = 0;
v = 0;
a = 0;
j = 0;

for i =1:length(theta)
    for k = 1:length(dT)
        if theta(i)>theta_step(k) && theta(i)<=theta_step(k+1)
            ds = dS(k);  dB = theta_step(k+1)-theta_step(k);
            dtheta = theta(i)-theta_step(k);
            s(i) = s_intial(k)+ds*(35*(dtheta/dB)^4-
                   84*(dtheta/dB)^5+70*(dtheta/dB)^6-20*(dtheta/dB)^7);
            v(i) = ds/dB*(140*(dtheta/dB)^3-
                  420*(dtheta/dB)^4+420*(dtheta/dB)^5-140*(dtheta/dB)^6);
            a(i) = ds/(dB^2)*(420*(dtheta/dB)^2-
                1680*(dtheta/dB)^3+2100*(dtheta/dB)^4-840*(dtheta/dB)^5);
            j(i) = ds/(dB^3)*(840*(dtheta/dB)^1-
               5040*(dtheta/dB)^2+8400*(dtheta/dB)^3-4200*(dtheta/dB)^4);
        end
    end
end
%===========================
%Plotting svaj
%==============
subplot(4,1,1);plot(theta,s);xlabel('\theta (deg)');ylabel('s (mm)');
xlim([0 360])
subplot(4,1,2);plot(theta,v);xlabel('\theta (deg)');ylabel('v (mm/deg)');
xlim([0 360])
subplot(4,1,3);plot(theta,a);xlabel('\theta (deg)');ylabel('a (mm/deg^2)');
xlim([0 360])
subplot(4,1,4);plot(theta,j);xlabel('\theta (deg)');ylabel('j (mm/deg^3)');
xlim([0 360])
```

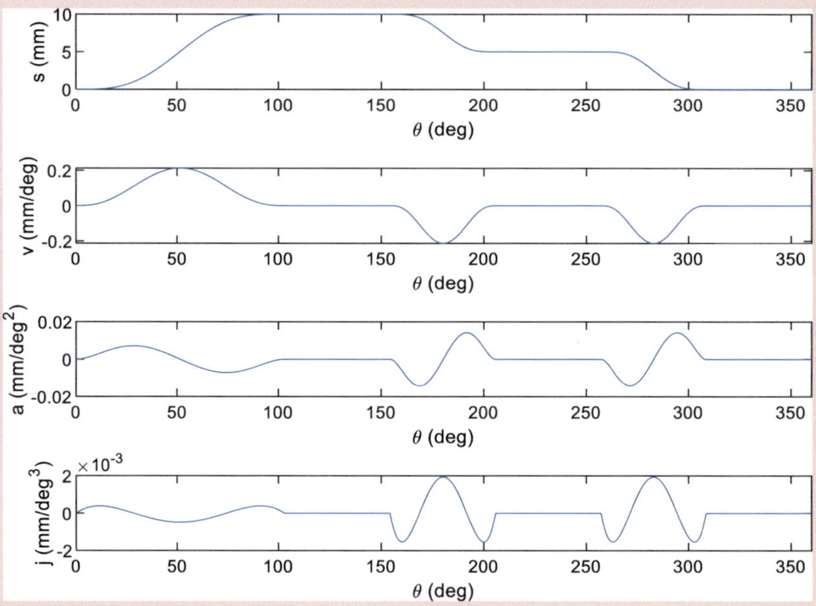

Fig. 9.16 Example 9.3 svaj diagram

9.3.3 Harmonic CEP Motion

A harmonic CEP trajectory segment design that goes from one point *i* to a second point *j*, is derived from trigonometric functions. The harmonic function provides a smooth transition between the two points. More details about harmonic motion are provided in Chap. 8. As an example, the Simple Harmonic CEP trajectory segment for the rise and fall with a total displacement ΔS and a time window ΔT is listed in Table 9.7. The harmonic motion segment provides zero initial values at $t=t_o$ and

Table 9.7 Mathematical equations and plot of simple harmonic CEP trajectory

	Displacement S	Velocity	Acceleration	Jerk
Harmonic Rise CEP Trajectory	$S(t) = S_o + \frac{\Delta S}{2}\left(1 - \cos\left(\frac{\pi(t-t_o)}{\Delta T}\right)\right)$ 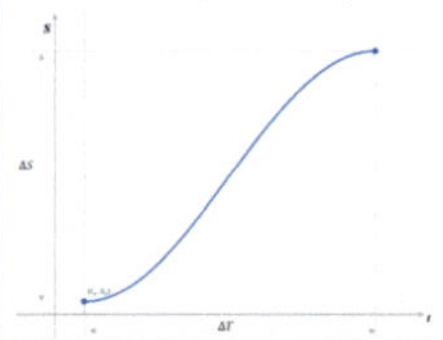	$V(t) = \frac{\pi}{\Delta T}\frac{\Delta S}{2}\sin\left(\frac{\pi(t-t_o)}{\Delta T}\right)$ 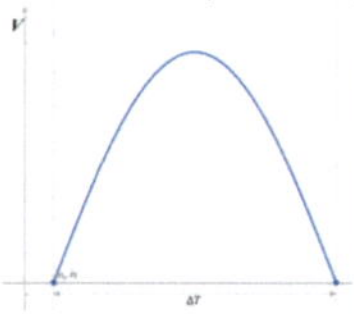	$A(t) = \frac{\pi^2}{\Delta T^2}\frac{\Delta S}{2}\cos\left(\frac{\pi(t-t_o)}{\Delta T}\right)$ 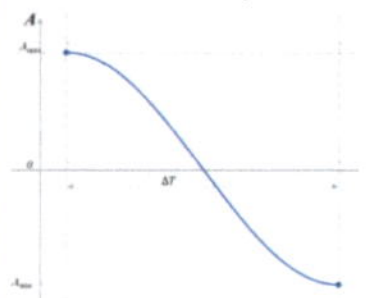	$J(t) = -\frac{\pi^3}{T_s^3}\frac{\Delta S}{2}\sin\left(\frac{\pi(t-t_o)}{T_s}\right)$ 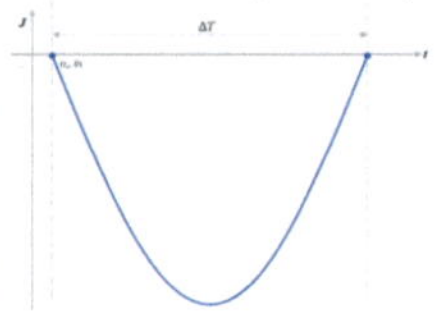
Harmonic Fall CEP Trajectory	$S(t) = S_o - \frac{\Delta S}{2}\left(1 - \cos\left(\frac{\pi(t-t_o)}{\Delta T}\right)\right)$ 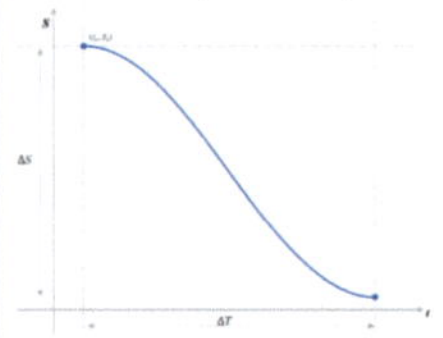	$V(t) = \frac{-\pi}{\Delta T}\frac{\Delta S}{2}\sin\left(\frac{\pi(t-t_o)}{\Delta T}\right)$ 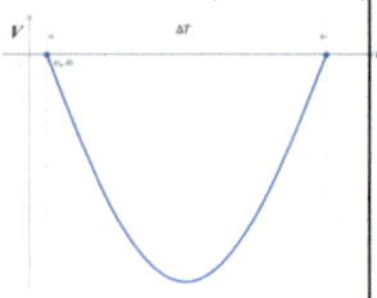	$A(t) = \frac{-\pi^2}{\Delta T^2}\frac{\Delta S}{2}\cos\left(\frac{\pi(t-t_o)}{\Delta T}\right)$ 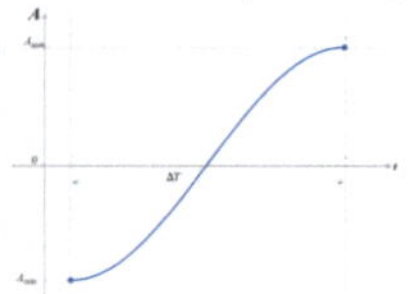	$J(t) = \frac{\pi^3}{T_s^3}\frac{\Delta S}{2}\sin\left(\frac{\pi(t-t_o)}{T_s}\right)$ 

final values at $t = t_o + T$ for both velocity and jerk, but there are nonzero acceleration values at both $t = t_o$ and $t = t_o + T$. This may lead to a discontinuity in acceleration when there is a switch from one segment to another.

9.3.4 Cycloidal CEP Motion

A cycloidal CEP trajectory motion segment design that goes from one point i to a second point j, is derived from a combination of linear and sinusoidal functions. The cycloidal function provides a smooth transition between the two points. Table 9.8 lists the mathematical equations and the plots of both the rise and fall of the cycloidal *CEP* trajectory motion segments. One of the major advantages of cycloidal motion compared to simple harmonic motion is that the velocity and acceleration both start and end with values of zero. This avoids any discontinuities when the cycloidal is used as one of the trajectory motion segments. This advantage reduces vibration and noise that could be generated from the possibility of infinite jerk as in the cases of step motion, ramp motion, or even simple harmonic motion. There are nonzero jerk values at both $t = t_o$ and $t = t_o + T$ that may cause a discontinuity in the jerk signal.

There are more motion segment types than what is presented above. The reader can explore more types in Refs. [1–3]. When designing a cam profile, the design engineer can use a single motion segment type or can mix different motion segment types. For example, the design engineer can design the cam profile as a set of ramp motion segments only. Another design engineer may design the same cam profile as a combination of cycloidal motion segments and ramp motion segments. Some comparisons between the maximum velocity, maximum acceleration, and maximum jerk are listed in Refs. [1–3].

9.3.5 Constant Velocity CPM Motion

Constant velocity motion segment means that V is held to be constant from point $t = t_i$ to point $t = t_j$ or from angle $\theta = \theta_i$ to angle $\theta = \theta_j$. Since the displacement (S or s) is defined as the integration of the velocity, the displacement is a straight line. Both acceleration and jerk, as the first and second derivatives of

Table 9.8 Mathematical equations and plot of cycloidal CEP trajectory

	Displacement S	Velocity	Acceleration	Jerk
Cycloidal Rise CEP Trajectory	$S(t) = S_o + \Delta S\left(\frac{t-t_o}{\Delta T} - \frac{1}{2\pi}\sin\left(\frac{2\pi(t-t_o)}{\Delta T}\right)\right)$ 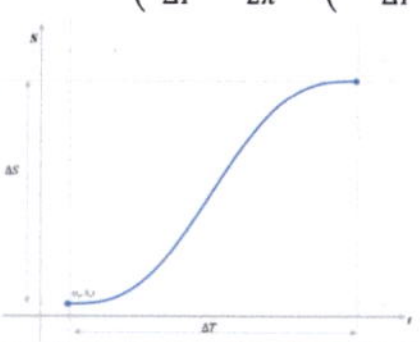	$V(t) = \frac{\Delta S}{\Delta T}\left(1-\cos\left(\frac{2\pi(t-t_o)}{\Delta T}\right)\right)$ 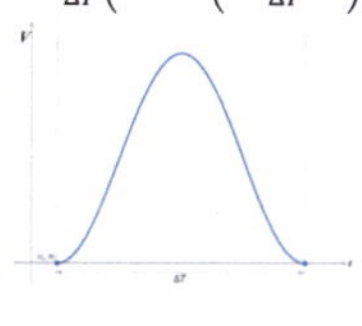	$A(t) = \frac{2\pi\Delta S}{\Delta T^2}\sin\left(\frac{2\pi(t-t_o)}{\Delta T}\right)$ 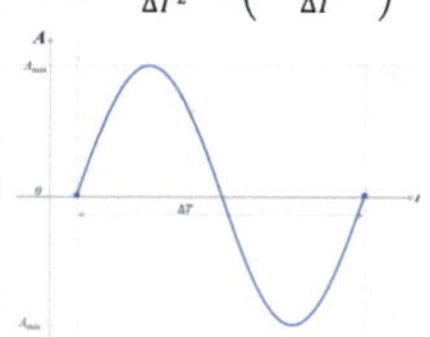	$J(t) = \frac{4\pi^2\Delta S}{\Delta T^3}\cos\left(\frac{2\pi(t-t_o)}{\Delta T}\right)$ 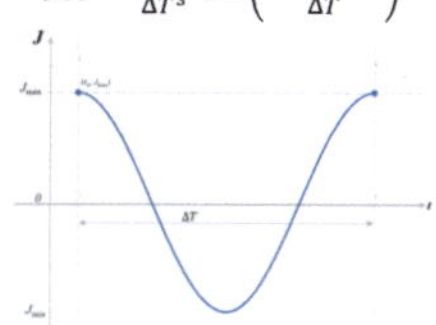
Cycloidal Fall CEP Trajectory	$S(t) = S_o - \Delta S\left(\frac{t-t_o}{\Delta T} - \frac{1}{2\pi}\sin\left(\frac{2\pi(t-t_o)}{\Delta T}\right)\right)$ 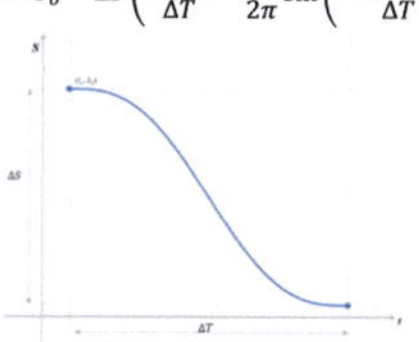	$V(t) = \frac{-\Delta S}{\Delta T}\left(1-\cos\left(\frac{2\pi(t-t_o)}{\Delta T}\right)\right)$ 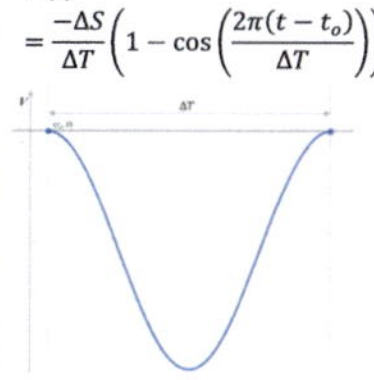	$A(t) = \frac{-2\pi\Delta S}{\Delta T^2}\sin\left(\frac{2\pi(t-t_o)}{\Delta T}\right)$ 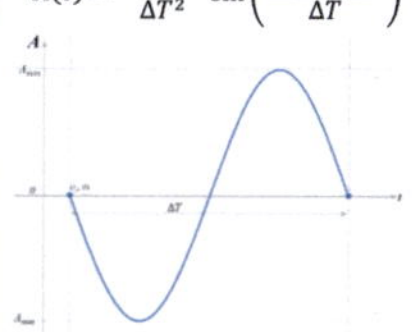	$J(t) = -\frac{4\pi^2\Delta S}{\Delta T^3}\cos\left(\frac{2\pi(t-t_o)}{\Delta T}\right)$ 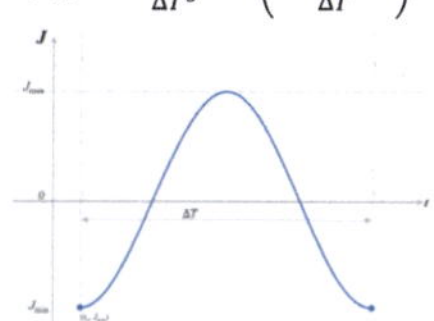

Table 9.9 Mathematical equations and plot of acceleration CEP trajectory

	Displacement S	Velocity	Acceleration	Jerk
Constant Acceleration Rise CPM Motion Segment	$S(t)=\frac{A_o}{2}(t-t_o)^2+V_o(t-t_o)+S_o$ 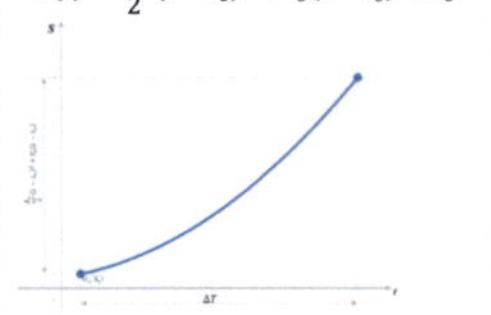	$V(t)=A_o(t-t_o)+V_o$ 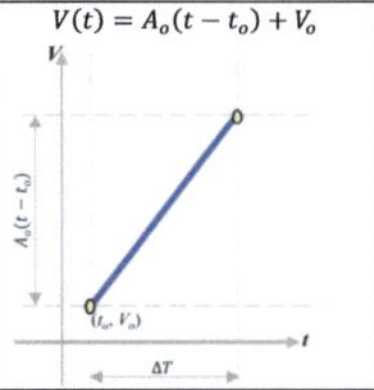	$A(t)=A_o$ 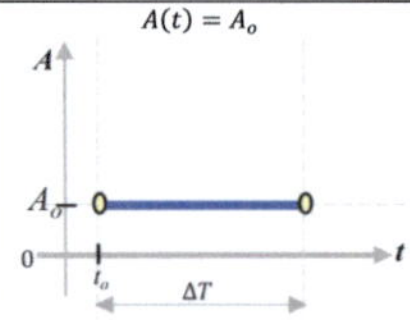	$J(t)=0$ 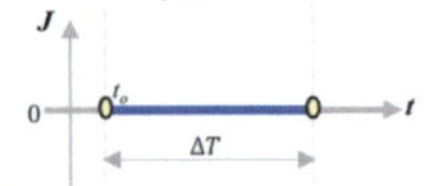
Constant Acceleration Fall CPM Motion Segment	$S(t)=-\frac{A_o}{2}(t-t_o)^2-V_o(t-t_o)+S_o$ 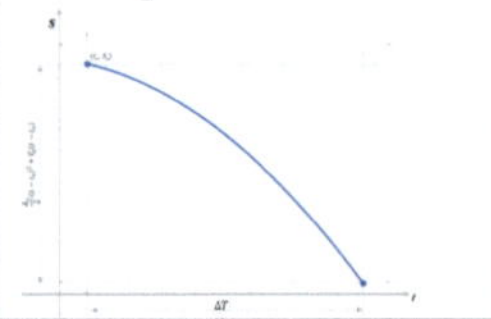	$V(t)=-A_o(t-t_o)-V_o$ 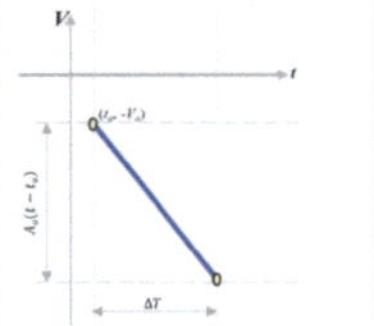	$A(t)=-A_o$ 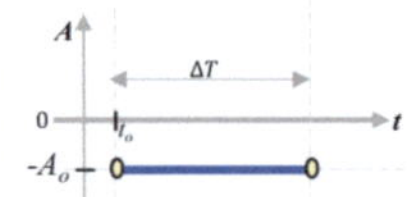	$J(t)=0$ 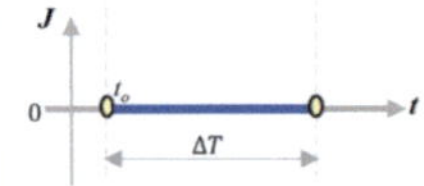

velocity, respectively, are zero. All the higher derivatives are also zero. The characteristics of the constant velocity motion segment are exactly the same as the ramp CEP motion segment. In other words, the constant velocity motion segment is the ramp CEP motion segment, as listed in Table 9.4.

Making two back-to-back ramp motion (constant velocity) segments leads to a jump in the velocity at the point of switch between the two segments. In Example 9.2, it is shown how an infinite acceleration and jerk arise due to the jump in velocity caused by using ramp motion segments. A ramp motion segment is recommended at low velocities to avoid any higher jump in the velocity. Sometimes, the design engineer is required to use a constant velocity motion ramp for some applications wherein a constant velocity is a request.

9.3.6 Constant Acceleration CPM Motion

Constant acceleration motion segment means that $A=A_o$ is held to be constant from point $t=t_o$ to point $t=t_o+T$, or $a=a_o$ is held to be constant from point $\theta=\theta_o$ to point $\theta=\theta_o+\Delta\beta$. Since the velocity ($V$ or v) is defined as the integration of the acceleration, the velocity (V or v) is a straight line. Displacement is the integration of the velocity, which results in a parabolic displacement (S or s). For such a reason, the constant acceleration motion segment is called "***Parabolic or Gravity Motion Segment***." The jerk, along with all other higher derivatives, is zero.

Table 9.9 lists the mathematical equations and the plots of the rise and the fall constant acceleration *CMP* motion segments. Note that the negative velocity "$-V_0$" in the fall was assumed to ensure that the mathematical model guarantees fall displacement. It is clear that the constant acceleration CPM is the 2nd polynomial CEP motion segment. The two back-to-back constant acceleration CPM motion segments cause a discontinuity in the acceleration curve, resulting in an infinite jerk at the point of switch. A constant acceleration motion segment is recommended at low acceleration when the design engineer is required to have a constant acceleration for a specific motion segment.

9.4 Cam Sizing

Cam sizing is a process for initially calculating the cam's dimensions and configurations after designing the follower motion *SVAJ* or *svaj*. In other words, the x_c and y_c coordinates of the cam groove or cam curvature (also called the cam profile) will be determined while the cam center's location is known. Before the cam sizing steps are explored, the next section will go over some cam geometry and configuration terminologies such as the base circle, prime circle, follower circle, trace point, pitch curve, cam profile (cam curve), eccentricity (offset), and pressure angle.

9.4.1 Cam Geometric Parameters

Definitions

Cam Base Radius and Base Circle:

The base circle, with the base radius R_b, is defined as the smallest circle that can be drawn tangent to the physical cam surface. The base circle's center is located at the cam's center of rotation [1–3].

Prime Radius or Prime Circle:

The prime circle, with the prime radius R_p, is defined as the smallest circle that can be drawn tangent to the locus of the follower's centerline [1–3].

Follower Radius of Follower Circle:

The radius of the roller follower [1–3].

Trace Point:

A point on the follower that is used to generate the pitch curve [1–3].

Pitch Curve:

The pitch curve is the locus of the follower's centerline. [1–3].

Cam Profile or Cam Curve:

The cam curve is a curved outline or groove which, by its oscillation or rotation motion, gives a predetermined and specified motion to the follower [1–3].

Eccentricity or Offset:

The eccentricity is defined as the perpendicular distance between the follower's axis of motion and the cam's center [1–3].

Pressure Angle:

The pressure angle is the angle between the follower's direction of motion (velocity) and the direction of the axis of transmission. Also, the pressure angle is the angle between the follower movement and the direction in which the cam is pushing it [1–3].

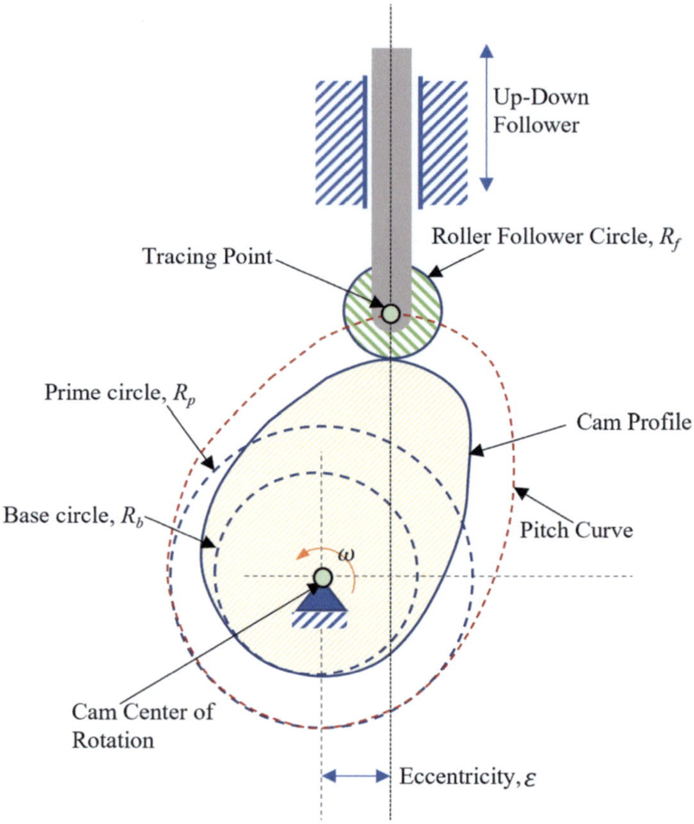

Fig. 9.17 Cam configuration and geometry nomenclature

Figure 9.17 shows the configurations of the plate cam mechanism in the roller follower case, the knife-edge case, and the flat-face case. The spherical follower, or the mushroom follower case, has a similar configuration to the roller follower case. The cam profile, or the cam surface, is defined as a working contour at which the cam link meets the follower link to operate.

Base Circle R_b: The base circle is the smallest circle that one can draw tangent to the cam surface. The center of the base circle is the center of the cam rotation. The base circle is represented as a blue dotted line in Fig. 9.17.

Trace Point: The follower is initially defined by the trace point, or the reference point, on the follower by which the follower's *svaj* motion is measured. As shown in Fig. 9.17, the trace point is defined as the center of the roller in the case of the roller follower or mushroom follower. In the case of a knife-edge follower, the trace point is defined as the knife-edge tip or the point at which the cam and follower come into contact. The trace point is located on the contact surface of the follower in the case of a flat-faced follower.

Pitch Curve: If one were to take the follower and move it around the cam profile, one could trace its path along the cam surface. This path, called the *pitch curve*, is represented by a dotted red line in Fig. 9.17. In other words, the trace point generates the pitch curve. The pitch curve is the profile that goes about the cam. In the case of both the knife-edge follower and flat-face follower, the pitch curve is the same as the cam profile.

Prime Circle $\boldsymbol{R_p}$: The prime circle, with the prime radius R_p, is the smallest circle that can be drawn tangent to the locus of the centerline of the follower. In the case of the roller, or mushroom follower, the prime circle is similar to the base circle, except for the fact that it is extended by the additional radius of the follower. If the follower were to roll over the base circle, the follower center would generate the prime curve. One of the blue dotted lines in Fig. 9.17 shows the prime circle. For a knife-edge and a flat-faced follower, the prime circle and the base circle are the same as shown in Fig. 9.17.

Eccentricity or Offset $\boldsymbol{\varepsilon}$: The eccentricity is known as the offset between the follower's line of action and the parallel line that passes through the rotational center of the cam. In other words, it is the perpendicular distance between the follower's axis of motion and the cam's rotational axis as shown in Fig. 9.17.

Pressure Angle $\boldsymbol{\delta}$: The cam link's surface and the follower link's surface meet at a point that is called the *trace point* for a 2D case. Passing through this trace point exists a tangential plane that effectively bisects the cam link's surface and the follower link's surface. This tangential plane is also called the *common tangent*, or the *axis of slip*. The cam joint prohibits motion in the perpendicular direction plane (common normal or axis of transmission) on the tangential plane with a contact force F_c. The pressure angle, δ, is located between the path of the follower motion and the axis of transmission as shown in Fig. 9.18. This pressure angle changes with the cam rotational angle θ. In other words, the pressure angle changes at each point on the cam surface (cam profile). The transmission angle, $\mu = 90° - \delta$, is the angle between the follower motion plane and the tangential plane.

It is recommended that the pressure angle δ change from 0° to 30° to avoid an excessive side force on the sliding follower. To calculate the pressure angle as a function of cam geometry and follower translational motion s or S, start from

$$V = \dot{S} = \frac{dS}{dt}$$
$$V = \frac{dS}{dt} = \underbrace{\frac{dS}{d\theta}}_{v}\underbrace{\frac{d\theta}{dt}}_{\omega} = v\omega = b\omega \tag{9.26}$$

or

$$b = v$$

where b is the ***Instantaneous Center of Velocity*** of both the follower link and the cam link. From the trigonometry shapes in Figs. 9.19 and 9.20, one can come up with:

$$\tan(\delta) = \frac{b - \varepsilon}{\sqrt{R_P^2 - \varepsilon^2} + s} = \frac{v - \varepsilon}{\sqrt{R_b^2 - \varepsilon^2 + s}} \tag{9.27}$$

Using Eq. (9.27), the pressure angle δ can be calculated from the cam geometry, R_p and ε, and the follower's *svaj* motion.

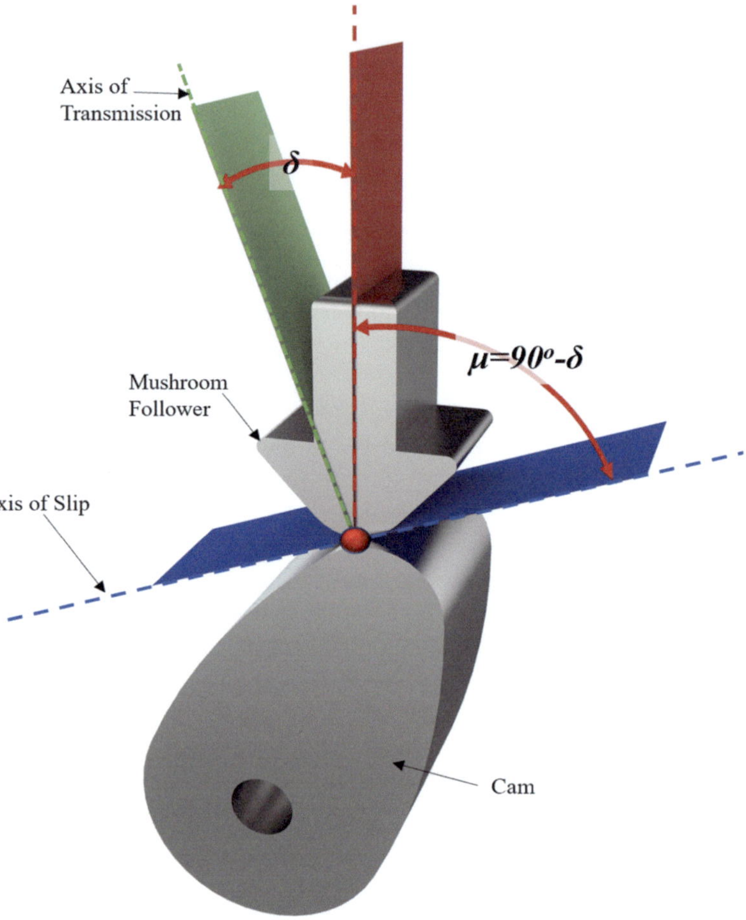

Fig. 9.18 Pressure angle in cam mechanism with Mushroom follower. Designed by the author and implemented by both the author and Pranav Dhola. Courtesy of Pranav Dhola

Technical Reminder: Kinem-atic Pair Instantaneous Center of Velocity

Recall the "Kinematic Pair Instantaneous Center of Velocity" from Chap. 6, Sect. 6.6.2. There it is stated, "***For each kinematic pair, there is an instantaneous center of velocity or velocity pole at which the two links have the same velocity***." The cam mechanism can be seen as a kinematic pair (two links and a joint). The two links are the follower link and the cam link. Assume an instantaneous, arbitrary node n_4 belongs to both the cam link and the follower link simultaneously. The velocity of node n_4 from the cam link is calculated, $V_{n4} = r_{n4n1}\omega = b\omega$. The velocity of node n_4 from the follower link is calculated, $V = V_{n4}$. Equating the values of V_{n4}

$$V = b\omega$$

Using the first part of Eq. (9.26),

$$V = \frac{dS}{dt} = \underbrace{\frac{dS}{d\theta}}_{v}\underbrace{\frac{d\theta}{dt}}_{\omega} = v\omega = b\omega \text{ or } v = b$$

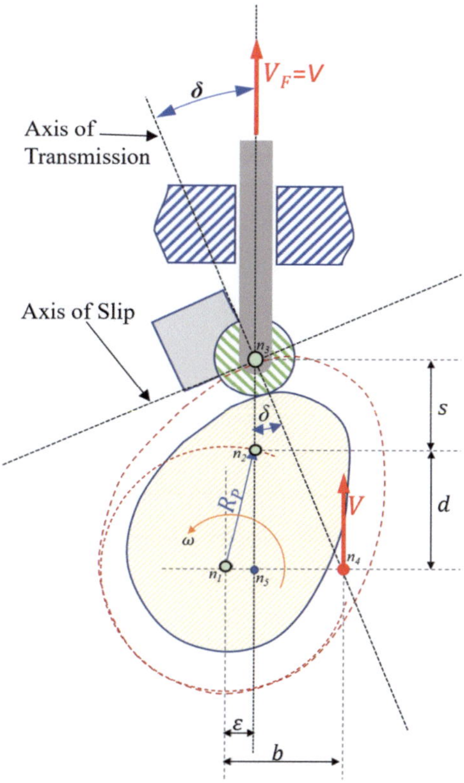

Fig. 9.19 Instantaneous center of velocity of cam mechanism

Derivation of Eq. (9.23)

In Fig. 9.20, the right triangle n_1–n_5–n_2 and the right triangle n_5–n_4–n_3 are drawn out from the cam mechanism.

From the trigonometry of the right triangle n_1–n_5–n_2

$$R_b^2 = d^2 + \varepsilon^2$$

or

$$d = \sqrt{R_b^2 - \varepsilon^2}$$

From the trigonometry of the right triangle n_5–n_4–n_3 and substituting from the previous equation

$$\tan(\delta) = \frac{b-\varepsilon}{d+s} = \frac{b-\varepsilon}{\sqrt{R_b^2 - \varepsilon^2} + s}$$

From Eq. (9.26), $b = v$, By substituting in the previous equation

$$\tan(\delta) = \frac{v-\varepsilon}{\sqrt{R_b^2 - \varepsilon^2} + s}$$

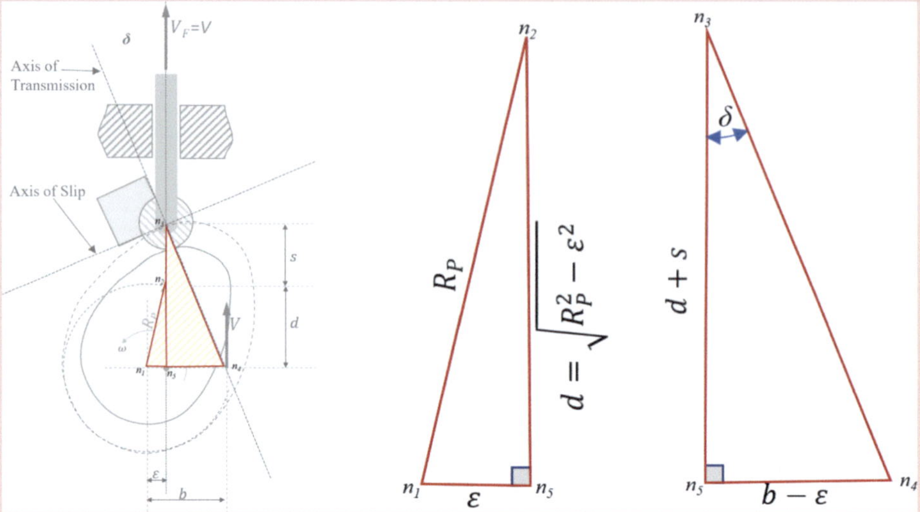

Fig. 9.20 Cam mechanism trigonometry for pressure angle

9.4.2 Cam Profile Design

After the *svaj* or *SVAJ* follower motion has been designed, this section will show how to continue the process and design the cam profile. The cam profile design could be done using either a graphical or an analytical method. For each follower type, there are different steps to develop the cam profile from a known *svaj* or *SVAJ* diagram and a known base radius R_b.

For the graphical technique, assume that the cam is stationary and the follower is rotated in the opposite direction of the original cam rotation. This assumption is based on the relative motion concept. For example, if the cam rotates clockwise, the follower should be rotated counterclockwise while the cam is stationary. The cam profile is drawn using n number of points. These points are measured at specific angles of rotation θ_k where $k=1, 2, \ldots, n$. At each angle θ_k, s_k from the *svaj* diagram is measured. s_k is called ΔR_{θ_k}. Take for example the s diagram shown in Fig. 9.21. The angle θ_k changes from zero to 360° as located on the x-axis. For the selected θ_k where $k=1, 2, \ldots, n$, $\Delta R_{\theta_k} = s_k$ is measured, as shown in Fig. 9.21. Note, the more points used in drawing the cam profile, the more accurate shape and dimension the diagram will have.

For the analytical technique, the cam profile is developed as a closed form of the known formula for s as well as the base radius R_b. The cam profile closed form is developed for each follower profile as a function of geometry. In graphical and analytical methods, a tabulated cam profile is developed to be manufactured by using the cam's profile points (x_c, y_c). As an example, the upcoming sections will now develop the cam profile for the knife-edge follower type.

9.4.2.1 Knife-Edge Follower Using Graphical Method

The steps below are used to design the cam profile in the case of the knife-edge follower:

1. Draw a local non-rotating x_c–y_c frame (gray) at the center of the cam.
2. Draw the base circle (red) from the origin with a radius of R_b, as shown in Fig. 9.22.
3. The home position is defined as the y-axis. The angle θ starts from zero at the home position and increases in the counterclockwise direction to reach a maximum of 360°.

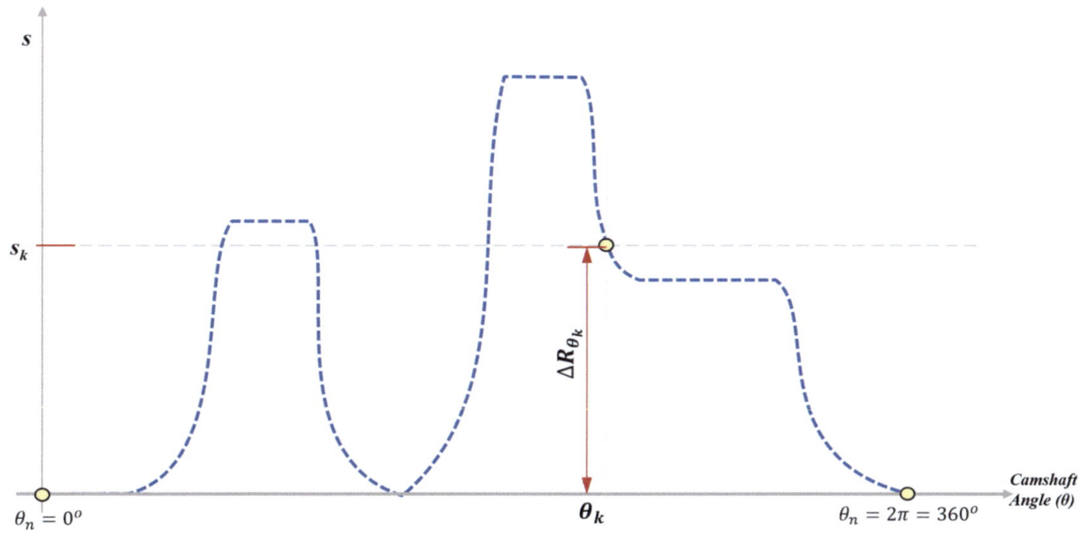

Fig. 9.21 Generic follower displacement diagram

4. Draw n radial lines (blue) from the center of the cam. The angle between two back-to-back radial lines can be constant (not a must) and equals to

$$\Delta\theta = \frac{360^\circ}{n} \tag{9.28}$$

5. For example, if 10 radial lines are drawn, the constant difference angle between every two back-to-back lines will be $\Delta\theta = 360^\circ/10 = 36^\circ$. For each radial line, calculate the angle θ_i, measured from the home position, as

$$\theta_i = \frac{360^\circ(i-1)}{n},\ i = 1, 2, \ldots, n \tag{9.29}$$

From the *svaj* diagram, find $\Delta R_{\theta_i} = s(\theta_i)$. Each radial line intersects with the base circle at point P_{θ_i}.

6. Draw the follower (green) at the home position, as shown in Fig. 9.22. The tip of the follower's knife edge should be oriented on the base circle if the *svaj* diagram has a zero s value when $\theta = 0$, which is the most common case.
7. Draw point Q_{θ_i} on each radial line away from point P_{θ_i} by a distance ΔR_{θ_i} in the outward direction.
8. Draw a smooth curve through the point Q_{θ_i} that represents the cam profile (blue with yellow shading).

Example 9.4

The camshaft rotates through an angle θ. The follower displacement is

$$s(\theta) = \begin{cases} 0 & 0 \le \theta < 90^\circ \\ \frac{20}{45}(\theta - 90^\circ) & 90^\circ \le \theta < 135^\circ \\ -\frac{20}{45}(\theta - 180^\circ) & 135^\circ \le t < 180^\circ \\ 0 & 180^\circ \le t \le 360^\circ \end{cases}$$

where θ is in degree and s is in mm. If R_b=50 mm, design the cam profile graphically for a knife-edge follower.

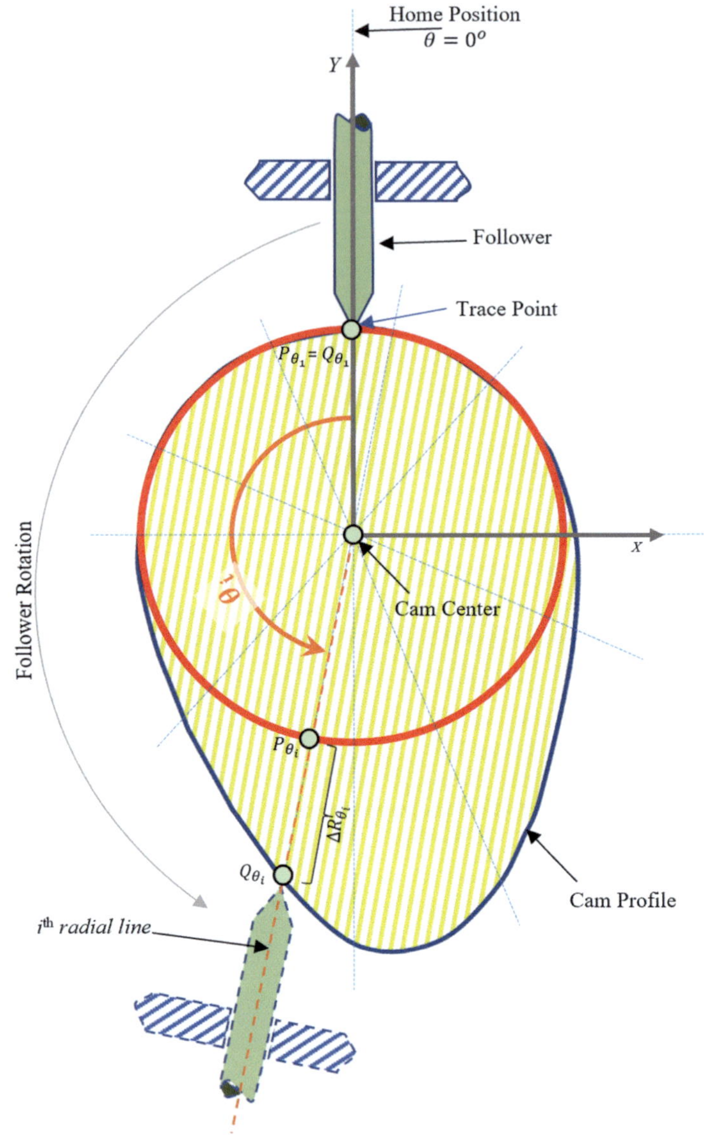

Fig. 9.22 Cam profile graphical design, knife-edge follower

Solution

This example recalls the s diagram from Example 9.1 and a more detailed version of this diagram is shown in Fig. 9.23. $\Delta R_{\theta_k} = s_k$ at $\theta = \{0°, 90°, 110°, 135°, 150°, 180°, 360°\}$ is measured to be $\Delta R = \{0, 0, 8.89, 20, 13.33, 0, 0\}$ mm.

Following the same steps of the graphical method for knife-edge as:

1. Draw the x_c–y_c frame and draw the base circle (R_b = 50 mm) centered at the origin.
2. Draw 6 radial lines from the origins with angles $\theta = \{0°, 90°, 110°, 135°, 150°, 180°\}$. The intersection between these lines and the base circle is defined as $P_k, k = \{1, 2, 3, 4, 5, 6\}$ as shown in Fig. 9.24.

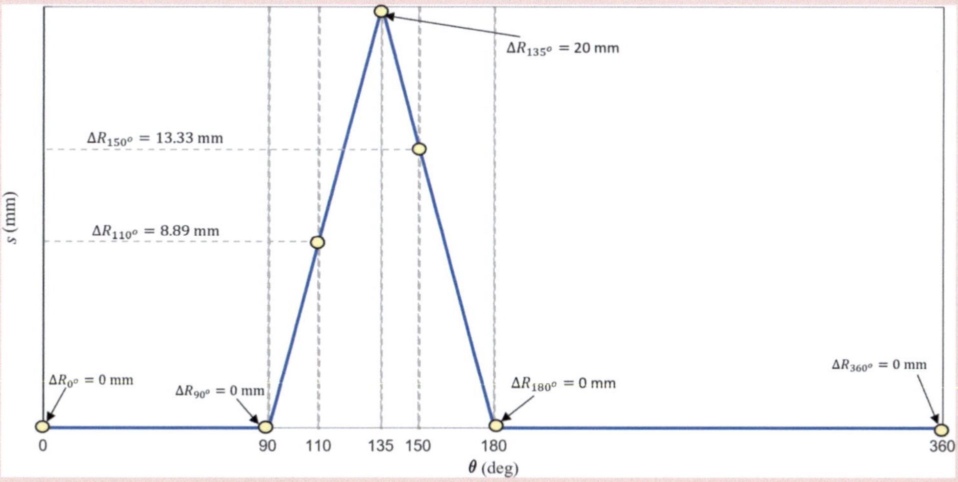

Fig. 9.23 s-Diagram of Example 9.1

Fig. 9.24 Cam profile of Example 9.4

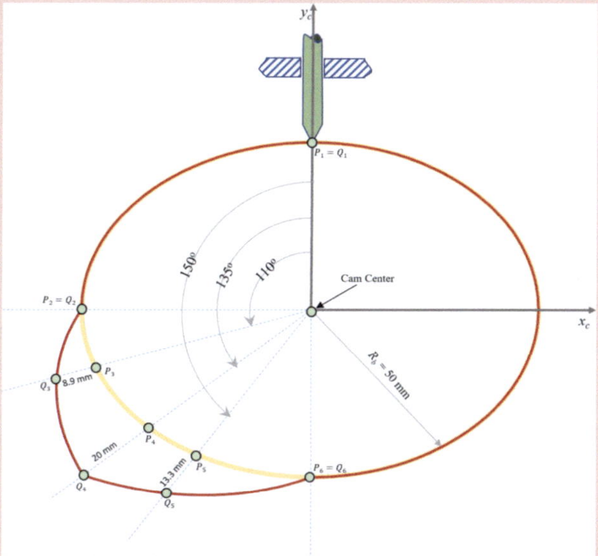

3. On each radial line, draw the point Q_k far from the point P_k by a distance ΔR_{θ_k} in the outward direction as shown in Fig. 9.24. For example, on the radial line with an angle $\theta_3 = 110°$, the point P_3 acts as the intersection of this line with the base circle. The point Q_3 is drawn a distance $\Delta R_{110^\circ} = s_3 = 8.89$ mm in the outward direction as shown in Fig. 9.24.
4. Connect the points Q_k, $k = \{1, 2, 3, 4, 5, 6, 1\}$ with a smooth curve to get the cam profile.

9.4.2.2 Knife-Edge Follower Using Analytical Method

From the geometry in Fig. 9.25, the cam profile can be calculated as

$$\begin{aligned} R_c &= R_b + \Delta R = R_b + s \\ x_c(\theta) &= R_c \sin(\theta) \\ y_c(\theta) &= R_c \cos(\theta) \end{aligned} \tag{9.30}$$

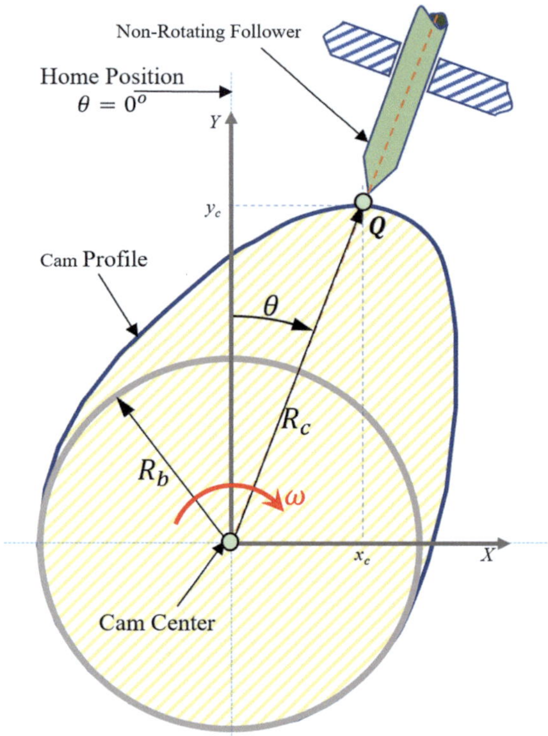

Fig. 9.25 Cam with knife-edge follower geometry

where R_c is the distance from the cam center to the interaction point, surface point Q, which has an x_c and y_c component. The rotational angle θ changes from zero to 360° starting from the home position and heading clockwise. At each angle θ, the R_c, x_c, and y_c variables can be calculated from the known R_b and s as a function of θ. The advantage of this analytical method is that a script can be developed that is capable of calculating many more points than the graphical method. Also, for manufacturing, the expression in Eq. (9.30) is commonly used because the analytical method has more precision than the graphical method, especially when computer numerical control (CNC) machines are used to manufacture the cam profile.

Example 9.5

An engineer designs the cam profile:

$$s_1(\theta) = \begin{cases} 0 & 0 \leq \theta < 90^\circ \\ \frac{20}{45}(\theta - 90^\circ) & 90^\circ \leq \theta < 135^\circ \\ -\frac{20}{45}(\theta - 180^\circ) & 135^\circ \leq t < 180^\circ \\ 0 & 180^\circ \leq t \leq 360^\circ \end{cases}$$

where θ is in degrees and s is in mm. If $R_b = 50$ mm, use a MATLAB® script to plot the cam profile.

Solution

The outline of the MATLAB® script is listed below:

1. Declare the cam profile parameter R_b.
2. Declare the independent variable $0 \le \theta \le 360°$ with an incremental step of 1°.
3. Calculate the dependent variables s from the given formulas in the problem statement.
4. Calculate x_c, y_c from Eq. (9.30).
5. Plot the cam profile (Fig. 9.26).

```
%Clear all the variables and close all the figures
%=================================
clc
clear all
close all
%=========================================================================
%Declare the given parameters
%============================
Rb = 50; %mm
%=========================================================================
%Calculating s as a function of theta
%=============
theta = [0:360]; % cam angle (deg)

for i =1:length(theta)

    if theta(i)>=0 && theta(i)<90
        s(i) =0; %mm
    end

    if theta(i)>=90 && theta(i)<135
        s(i) =20/45*(theta(i)-90); %mm
    end

    if theta(i)>=135 && theta(i)<180
        s(i) =-20/45*(theta(i)-180); %mm
    end

    if theta(i)>=180 && theta(i)<=360
        s(i) =0; %mm
    end
end
%=========================================================================
%Calculating the cam profile
%=============
x_c = (Rb+s).*sind(theta); %mm
y_c = (Rb+s).*cosd(theta); %mm
%===========================
%Plotting cam profile
%=============
plot(x_c,y_c);xlabel('x_c (mm)');ylabel('y_c (mm)');
```

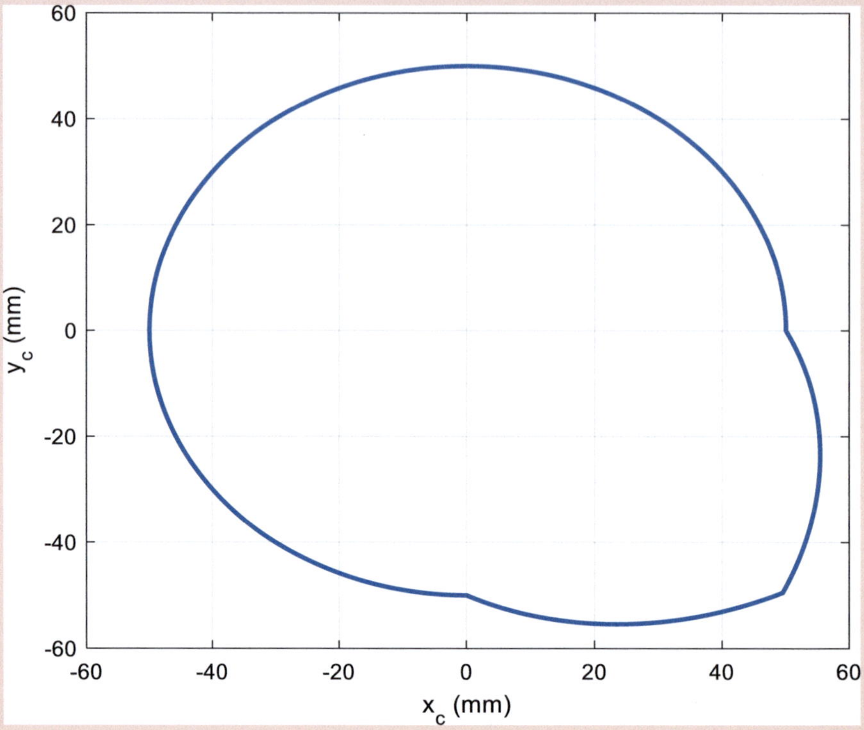

Fig. 9.26 Cam profile of Example 9.5

Note: In this textbook, it is discussed how to plot the cam profile for the knife-shaped follower only. The rest of the follower shapes follow a similar approach but require more complicated mathematics that is out of the scope of this textbook.

9.4.2.3 Cam Mechanism Design and Limitations

From the follower motion design in Sect. 9.3 and cam sizing in Sect. 9.4, the steps of cam design can be summarized as follows:

Step 1: The design engineer should have a specific application to be accomplished such as the cam valve mechanism in the piston engine or any similar example.

Step 2: The design engineer should select the cam shape and follower shape from the different types as listed in Sect. 9.2. In the selection process, consideration should be made of

- Whether the follower motion is translational motion or rotational motion
- Whether the cam mechanism is a force or a form-closed.
- Whether the follower shape is to be a knife, roller, or another shape

Step 3: The design engineer should select the base radius R_b and the offset ε from the available dimensions. In other words, select R_b as $R_{b\min} \le R_b \le R_{b\max}$. Also, the design engineer should select the follower dimensions so that the follower radius R_f is $R_{f\min} \le R_b \le R_{f\max}$.

Step 4: The design engineer should select one of the mathematical models to design the follower motion as mentioned in Sect. 9.3.

Step 5: The design engineer should design the cam profile using methods described in Sect. 9.4.

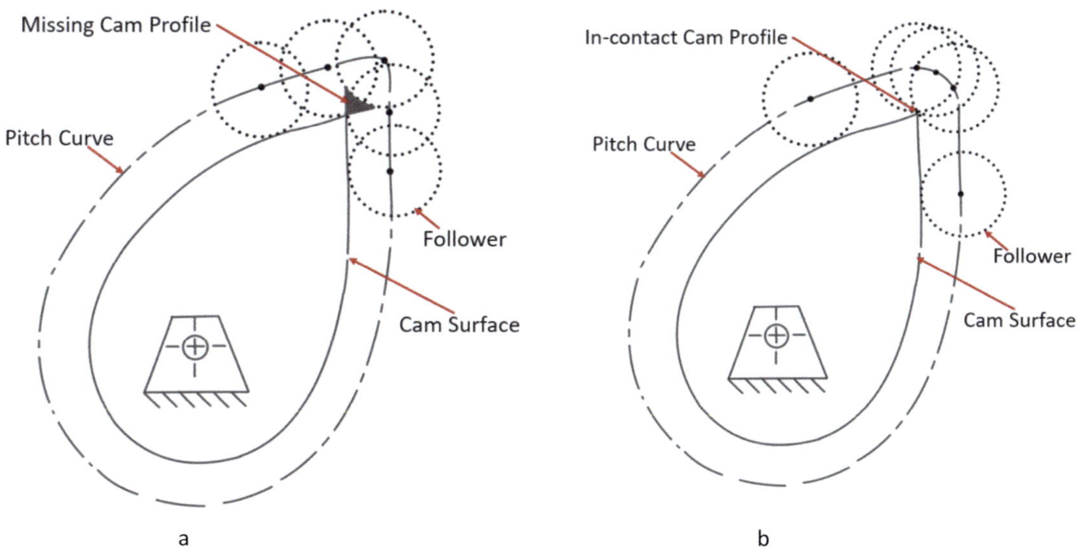

Fig. 9.27 Influence of base circle radius on follower missing cam profile zone

The above steps are iterative on the selected design parameters and are based on budget, manufacturing facilities, and some other optimization criteria as listed below.

The selection of both R_b and R_f would normally come to using the minimum value for R_b (saving space) and maximum value for R_f (more strength in the follower); the first choice being important to save space and money. However, it is occasionally necessary for the design engineer to increase the base radius R_b and simultaneously reduce R_f to make sure that the cam is always in contact with the follower at all the profile zones. For example, Fig. 9.27a shows the case when the rolling follower misses the cam profile due to undercutting. In Fig. 9.27b, the same follower motion profile as Fig. 9.27a is used, but with a larger base radius R_b to ensure that the cam profile is in contact with the follower at all times.

The selection of the offset ε starts with a value of zero as a simple manufacturing and design choice. However, a zero offset value sometimes cannot balance the pressure angle with a smoother running cam. Equation (9.27) shows that increasing the offset reduces the pressure angle. The design engineer may need to increment the offset to reduce the pressure angle with the same follower motion and selected R_b.

Another aspect of the design process is the follower motion profile selection. The design engineer should avoid any discontinuity in the motion profile; otherwise, the cam design will experience noise, vibration, the possibility of damaging the machine, and/or stalling of the power source such as an engine or motor. Sometimes, the design engineer may insert dwells between rise and fall. Having a dwell segment helps to let the vibration settle down after the fall or rise motion.

Cam Design Aspects:

Reference 4 provides full details about how to design the cam. The reader is recommended to read this textbook if any further details about cam design are needed. One of the most important cam design aspects in Ref. [4] is listed below.

- The first step in designing a cam system is to establish the proposed design speed of the complete machine. This decision is a critical one and should be based on the best experience and judgment available. The customer should be helpful in this decision. After the design speed has been determined, establish a time chart to synchronize the cam system with other actions of the machine.

- Positive drive, closed-track radial cams, or conjugate dual cams with roller followers are the most popular choices.
- The cam contour should be smooth with no abrupt changes in its curvature. Note that curvature at any point on a cam is directly related to the follower's acceleration.
- The minimum curvature or sharpness of a convex cam contour is dependent on the value of the follower's maximum negative acceleration. That is, the larger the negative acceleration, the sharper the cam surface will be.
- The cam size should be as small as possible to minimize the cam–follower sliding velocity, surface wear, torque on the camshaft, and cost and space requirements.
- The pressure angle should be kept to a minimum; 30 degrees is a general arbitrary limit for all followers.
- Proper dynamic cam–follower design necessitates the study of the cam–follower acceleration curve.
- The maximum follower acceleration should be as low as possible to keep the inertia forces and stresses small.
- The noise, surface wear, and vibration of a cam–follower system are dependent on the shape of the follower acceleration curve; hence, smoothness and continuity of the acceleration curve are essential.
- The moving parts of the cam–follower mechanism should be lightweight and as rigid as possible to keep the inertia forces, noise, and wear to a minimum, especially at high speeds.
- The torque curve should be investigated in addition to the system's force distribution.
- Manufacturing methods and accuracy of cam cutting and inspection are of paramount importance in ensuring a system's anticipated performance. Small surface errors that are imperceptible to the eye may produce high stress and vibrations in the follower linkages.

9.5 Implementing Cams Using Simscape™ Multibody™

The cam mechanism consists of the cam link, follower link, and cam joint. The cam link and follower link are recommended to be modeled using the "Extruded Solid" block or by importing them from any CAD software such as SolidWorks or AutoCAD. The CAD file could be imported with the "*.stl" extension. The cam joint is modeled using the "***Point on Curve Constraint***" block and "***Spline***" block.

Definition[1]

Spline Block:

The Spline block creates a continuous spline curve based on cubic interpolation between the specified points. The curve can be two-dimensional, such as a planar cam profile, or three-dimensional, such as a roller coaster track. The dimension of the spline depends on the

[1] Reproduced from [5–8] by MathWorks with permission from MathWorks.

dimension of the coordinate matrix. An N-by-2 matrix specifies a 2-D curve in the *xy* plane. An N-by-3 matrix specifies a 3-D curve. All the coordinates are resolved in the local reference frame of the block. Moreover, according to the specified end conditions, the curve can be either open or closed [5–8].

Point on Curve Constraint Block

The Point on Curve Constraint block confines a point to a curve. The frame attached to the point is free to rotate depending on other constraints in the model. Use the Point on Curve Constraint block to model point-on-curve constraints, such as that between a roller coaster and a track or between a cam follower and a cam [5–8].

The idea of the cam mechanism is to slide two surfaces on top of each other. When the cam rotates, the follower should slide on the cam contour (cam profile in Sect. 9.4.2). In Simscape™ Multibody™, the cam profile can be modeled using the Spline block. The results from the cam profile design are a vector defined by $\{x_c, y_c\}$ or $\{x_c, y_c, z_c\}$. To use the Spline block, follow the steps below:

1. Open the Simulink® Library Browser and scroll down to reach the Spline block, located underneath "Simscape™\Multibody™\Curves and Surface" as shown in Fig. 9.28a. Drag and drop the Spline block into the Simulink® model.
2. Another way to bring a Spline block into a Simulink® model is to go to a blank area inside the Simulink® model and double-click. Type *Spline*. You will see a drop menu with *Spline* as one of the options as shown in Fig. 9.28b. Select the *Spline* option. The Spline block will show up in the

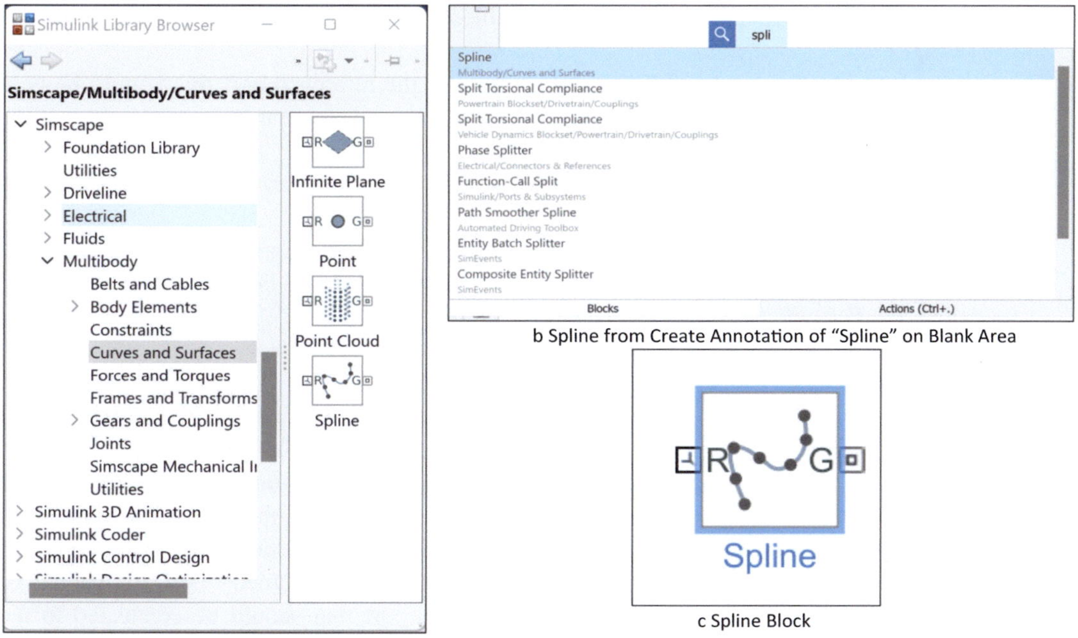

a Spline from Simulink Library Browser

b Spline from Create Annotation of "Spline" on Blank Area

c Spline Block

Fig. 9.28 How to bring spline block into Simulink® model. Designed and implemented by the author. Images used with permission of MathWorks

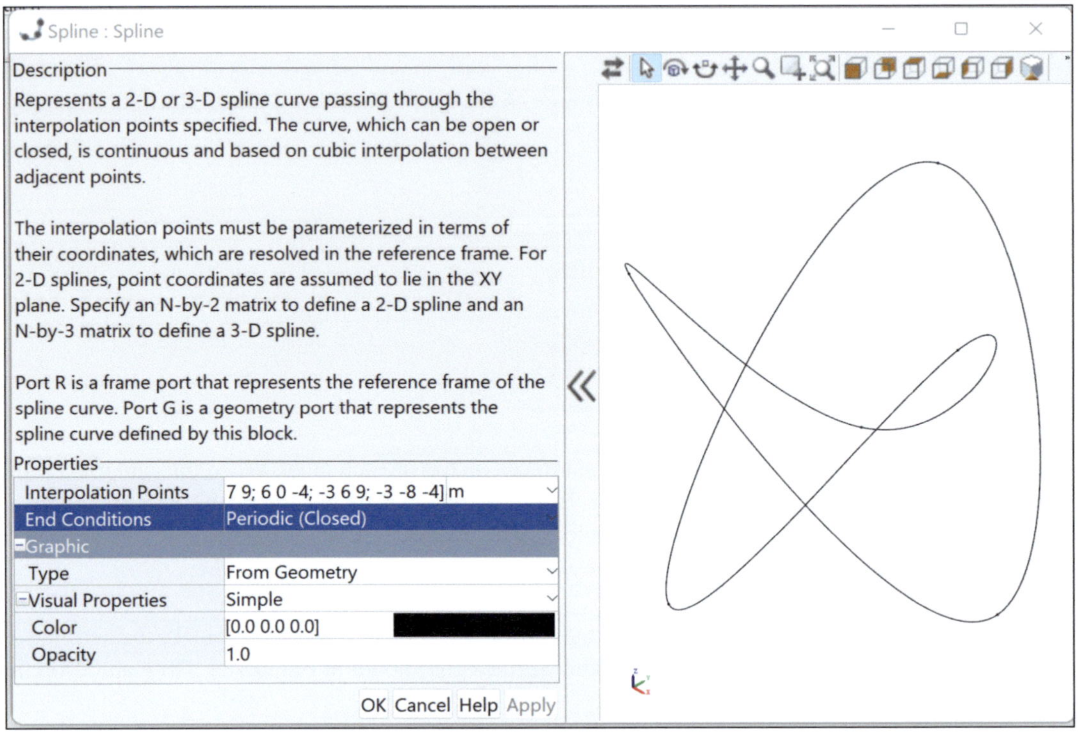

Fig. 9.29 Spline's editing menu. Designed and implemented by the author. Images used with permission of MathWorks

Simulink® model. Using Step 1 or Step 2, a Spline block will be created in a Simulink® model as shown in Fig. 9.28c.

3. The Spline block has two ports: "R" and "G." Port R is the reference port that should be connected to the cam. Port G is a geometry port that should be connected to a moving frame or the "Point on Curve Constraint" block.
4. If the Spline block is double-clicked, the editing menu will be revealed (as shown in Fig. 9.29) that has the properties below:
   - ***Interpolation Point***: The cam profile vector $[x_c, y_c, z_c]$ should be plugged into this section.
   - ***End Condition***: There are two options–Periodic (closed) or Open. For this discussion's cam modeling, the *Periodic* option should be selected.
   - ***Graphic***: This property is to pick the color of the cam profile.

More details will come in the solved examples. Also, the reader is recommended to read more about the Spline block in Refs. [5–8].

The "*Point on Curve Constraint*" block can be obtained from the Simulink® Library Browser (Simscape™ Multibody™ Constraints as shown in Fig. 9.30) or from typing the words, "Point on Curve Constraint" on a blank area inside a Simulink® model. Figure 9.30 shows the "Point on Curve Constraint" block that has two ports—port B and port F. Port B is a geometry port that should be connected to port G of the Spline block. Port F is a frame port that should be connected to the follower. The editing menu (which appears when the "*Point on Curve Constraint*" block is double-clicked) is shown in Fig. 9.30. There are two Direction options: Follower on Base or Base on Follower. In this discussion's cam mechanism model, use Follower on Base since the follower is moving on the base (cam). More details will be provided in the solved examples section. Also, the reader is recommended to read more about the Spline block in Refs. [5–8].

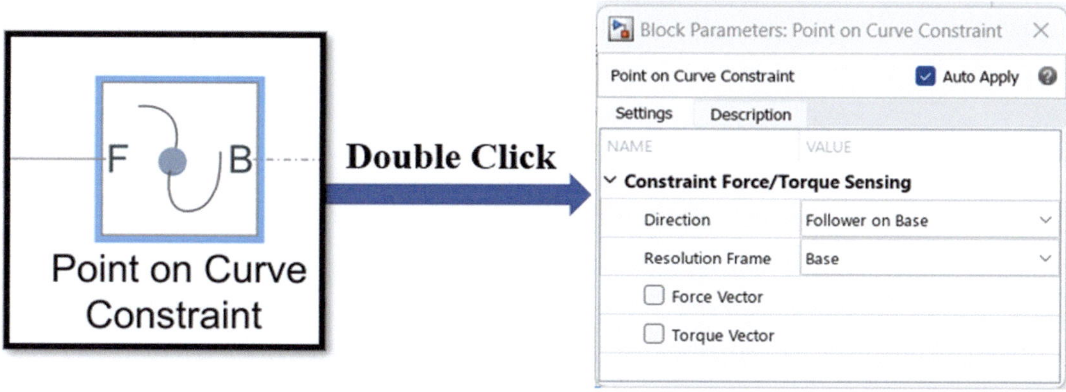

Fig. 9.30 Point on curve constraint block. Designed and implemented by the author. Images used with permission of MathWorks

Example 9.6

The camshaft rotates through an angle θ. The follower displacement is

$$s(\theta) = \begin{cases} 0 & 0 \le \theta < 90^\circ \\ \frac{20}{45}(\theta - 90^\circ) & 90^\circ \le \theta < 135^\circ \\ -\frac{20}{45}(\theta - 180^\circ) & 135^\circ \le t < 180^\circ \\ 0 & 180^\circ \le t \le 360^\circ \end{cases}$$

where θ is in degree and s is in mm and R_b=50 mm. The follower is a knife edge with the given dimension in Fig. 9.31. The thickness of both the cam and follower is 70 mm. Use Simscape™ Multibody™ to simulate the cam mechanism where $\omega_c = 180\,\text{deg/s}$ over one period. Plot the follower SVAJ diagram from the simulation. Comment on your results.

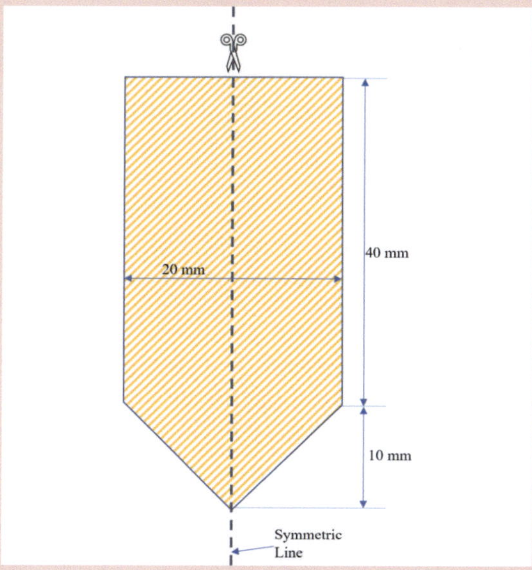

Fig. 9.31 Follower cross-section; Example 9.6

Solution

From Example 9.5, we have

$$x_c(\theta) = \begin{cases} 50\sin(\theta) & 0 \leq \theta < 90^\circ \\ \left(\frac{20}{45}(\theta - 90^\circ) + 50\right)\sin(\theta) & 90^\circ \leq \theta < 135^\circ \\ \left(50 - \frac{20}{45}(\theta - 180^\circ)\right)\sin(\theta) & 135^\circ \leq t < 180^\circ \\ 50\sin(\theta) & 180^\circ \leq t \leq 360^\circ \end{cases}$$

$$y_c(\theta) = \begin{cases} 50\cos(\theta) & 0 \leq \theta < 90^\circ \\ \left(\frac{20}{45}(\theta - 90^\circ) + 50\right)\cos(\theta) & 90^\circ \leq \theta < 135^\circ \\ \left(50 - \frac{20}{45}(\theta - 180^\circ)\right)\cos(\theta) & 135^\circ \leq t < 180^\circ \\ 50\cos(\theta) & 180^\circ \leq t \leq 360^\circ \end{cases}$$

Write a MATLAB® script to declare the simulation parameters as listed below. In the MATLAB® script, shift θ_c (theta) 90° since the home position point is on the line of 90°, as shown in Fig. 9.22.

```
clc
clear all
theta = [0:360];
for i = 1:length(theta)
    if theta(i)>= 90 && theta(i)<180
        xc(i) = 50*sind(theta(i));
        yc(i) = 50*cosd(theta(i));
    end

    if theta(i)>=180 && theta(i)<225
        xc(i) = (20/45*(theta(i)-180)+50)*sind(theta(i));
        yc(i) = (20/45*(theta(i)-180)+50)*cosd(theta(i));
    end

    if theta(i)>=225 && theta(i)<=270
        xc(i) = (50-20/45*(theta(i)-270))*sind(theta(i));
        yc(i) = (50-20/45*(theta(i)-270))*cosd(theta(i));
    end

    if theta(i)>270 && theta(i)<=360
        xc(i) = 50*sind(theta(i));
        yc(i) = 50*cosd(theta(i));
    end
        if theta(i)>=0 && theta(i)<90
        xc(i) = 50*sind(theta(i));
        yc(i) = 50*cosd(theta(i));
    end
end

% locations of nodes
%====================
Frame_L1_n1 = [0 0 0]; %revolute joint location (m)
Frame_L1_n2 = [0 50 0]/1000; %pristamtic joint location (m)

%Links
%=====
Follower_CorssSection =[0 0; 10 10; 10 40; -10 40;-10 10]/1000; %m
Follower_Thickness = 70/1000; %m
```

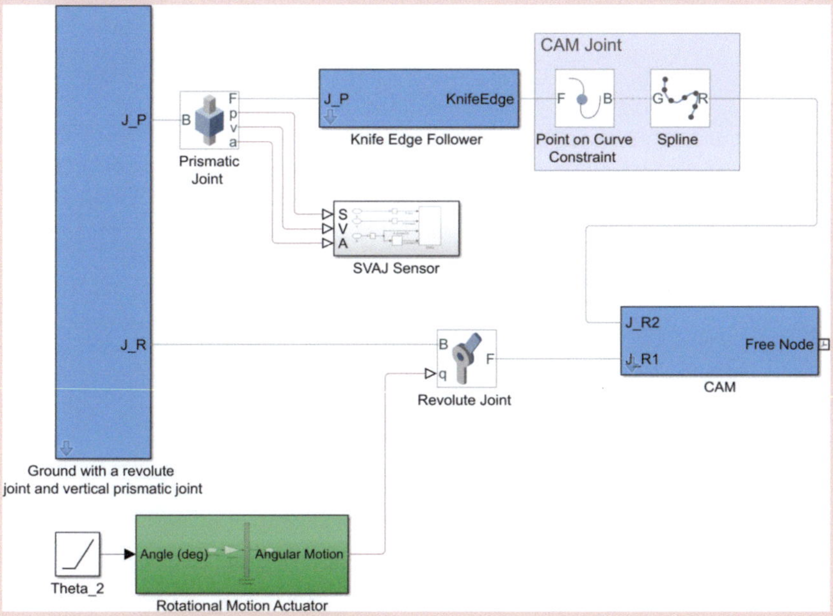

Fig. 9.32 CAM knife follower Simscape™ Multibody™ model; Example 9.6

To build the Simulink® model, follow the steps below:

Step 1: Open a new model and save it as "CAMKnifeEdge_model.slx." Build the model, as shown in Fig. 9.32, with the listed blocks and subsystems in Table 9.10. The blocks "Ground with a revolute joint and vertical prismatic joint," "Knife-Edge Follower," "Extrusion with two revolute joints," and "Rotational Motion Actuator" are from Kinematics_Of_Machines.xls. The rest of the blocks are from the Simulink® Library Browser.

Step 2: Run the model for 2 s (1 cycle) to get the results from the scopes as shown in Fig. 9.33. During the simulation, an animation will be shown in the Mechanics Explorer Window. Figure 9.34 shows a screenshot from the simulation animation. There is a copy of this model for the reader to explore included with this textbook.

Comment: The SVAJ diagram shows how the discontinuity in the velocity leads to a high acceleration and jerk rate.

Table 9.10 CAM Knife Follower Blocks and Subsystem; third column with row 2, row 3, row 4, row 6, row 9 and row 11. Designed and implemented by the author. Images used with permission of MathWorks

Block or subsystem	Name	Plugged in parameters
Ground with a revolute joint and vertical prismatic joint	Ground with a revolute Joint and vertical prismatic joint	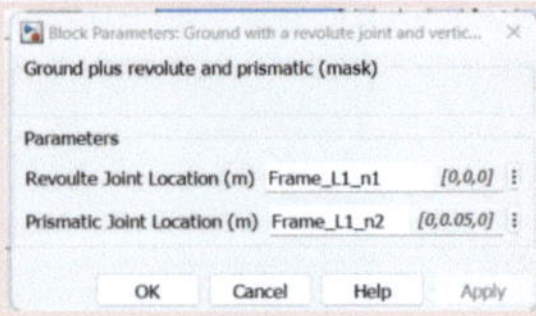
Knife-edge follower	Knife-edge follower	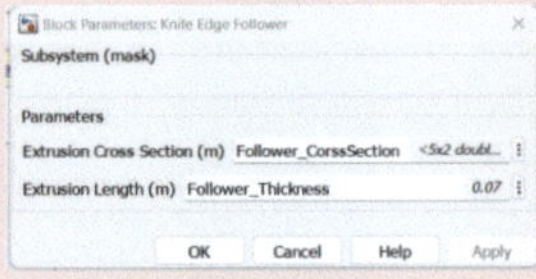
Extrusion with two revolute joints	Cam	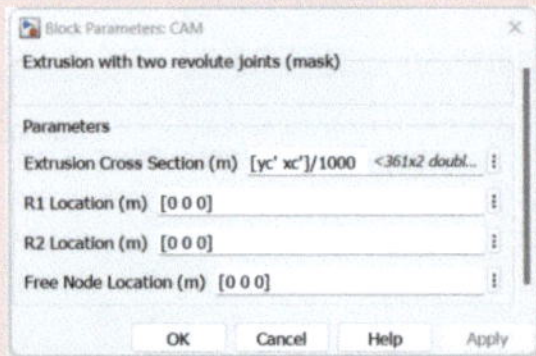
Point on curve constraint	Point on curve constraint	
Spline	Spline	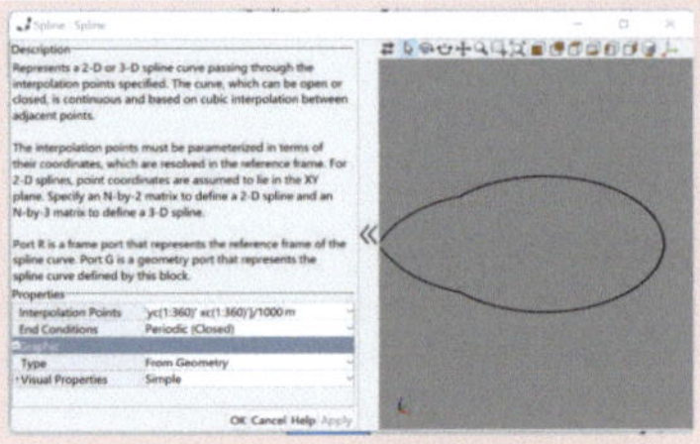
Revolute joint	CAM center	Select the "Motion" to be "Provided by Input" and the "Torque" to be "Automatically Computed" in the actuation properties from the revolute joint editing menu
Prismatic joint	Prismatic joint	Select the "Position", "Velocity", and "Acceleration" in the sensing properties from the prismatic joint editing menu
SVAJ sensor		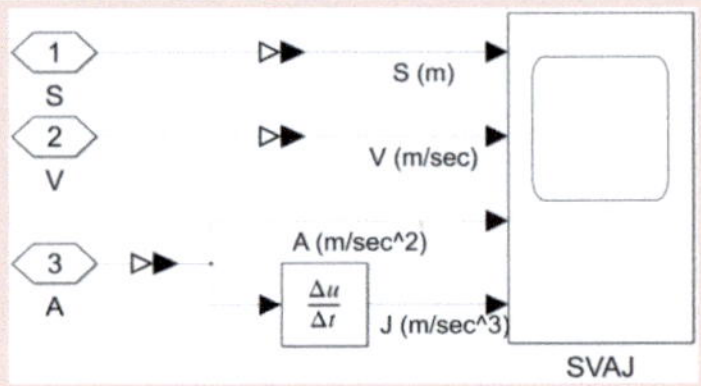
Rotational motion actuator	CAM motor	
Ramp	Theta_2	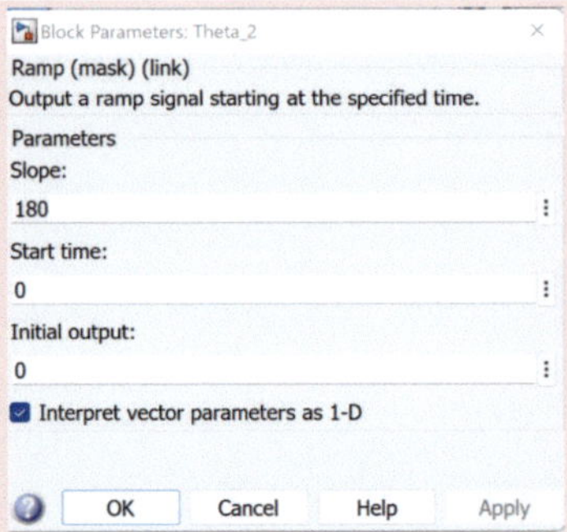

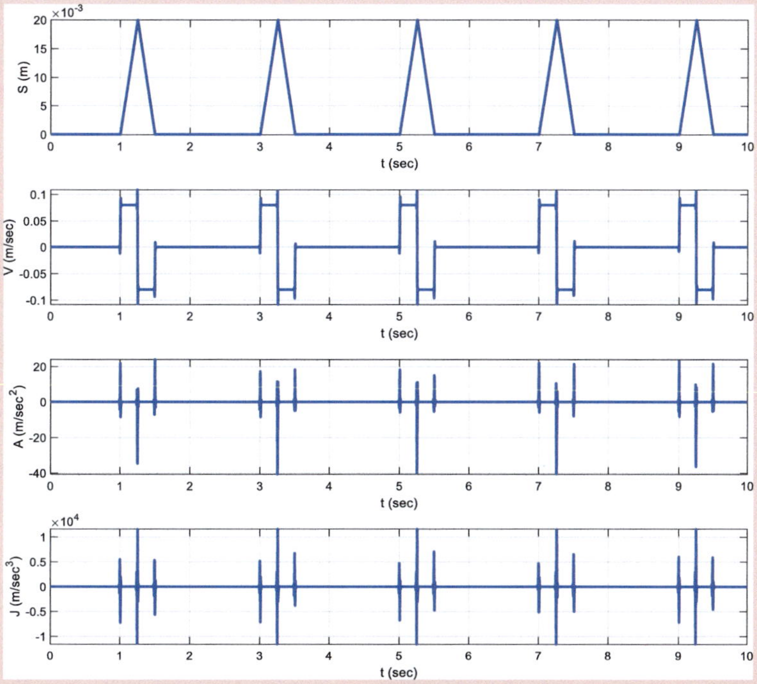

Fig. 9.33 CAM knife-edge follower SVAJ; Example 9.6

Fig. 9.34 Cam with knife-edge follower mechanism animation snapshot; Example 9.6

Definition[2]

File Solid Block:

The File Solid block models a solid element with geometry, inertia, color, and reference frame derived from an external file. The file must be of a part model, which is to say that it contains at least solid geometry data. Some formats may provide color and inertia data, though such properties can be specified manually if necessary [5–8].

One of the more unique blocks in Simscape™ Multibody™ is the "***File Solid***" block which allows the user to import a CAD file for a rigid body that includes the shape and the dimensions of a specific link. Engineers and customers are more likely to watch a 3D animation with CAD models than a 3D animation with a generic shape. Figure 9.35 shows the File Solid Block. After double-clicking on the File Solid block, the editing menu will pop up, as shown in Fig. 9.35. Click on the "File Name" to locate the CAD file that needs to be imported. For example, in Fig. 9.35, the CAD file "Cam_Shaft.SLDPRT" is selected. The file will show up in the display window. Select the "Graphic" editing option to change the color. More details about editing options can be found in Refs. [5–8].

Engineering Note: Imported CAD File Extension

Among the supported formats are those native to CATIA (V4, V5, and V6), Creo, Inventor, Unigraphics NX, Solid Edge, SolidWorks, and Parasolid (all CAD applications common in industry and academia). These include CATPART, PRT, IPT, SLDPRT, and X_T (and its binary version, X_B). Other valid formats not associated with a specific application but common in 3-D modeling include SAT (often referred to as ACIS), JT, STL, and STEP [5–8].

Note: CAD drawing and assembly files, which do not contain the necessary data for a solid element, cannot be imported to the block [5–8].

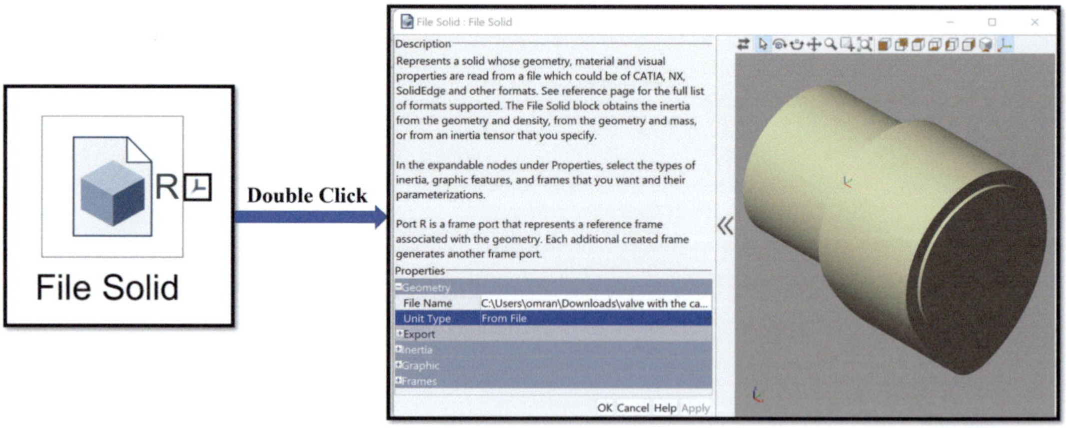

Fig. 9.35 File solid block. Designed and implemented by the author. Images used with permission of MathWorks

[2] Reproduced from [5–8] by MathWorks with permission from MathWorks.

Case Study

This case study shows how to build this chapter's model from CAD files. The model of this example is called "CamFromFiles.xls" and it is available with this textbook. This is not a solved example. The reader is not required to build this model and can explore this case simply for knowledge. The steps for building this model are:

1. Using SolidWorks or any CAD software, draw the cam link and save the model as "Cam_Shaft.SLDPRT" Also, draw the follower link and save the model as "Rod.SLDPRT".
2. Open a new Simulink® file, and bring "Ground with a revolute joint and prismatic joint" block and "Rotational Motion Actuator" block from the library file "Kinematics_Of_Machines."
3. Drag and drop Prismatic Joint block, Revolute Joint block, two File Solid block, Spline block, Point on Curve Constraint block, and Ramp block from the Simulink® libraries.
4. Connect all these blocks, as shown in Fig. 9.36a. The two Solid File blocks imported the two CAD files, "Cam_Shaft.SLDPRT" and "Rod.SLDPRT." Note, the first File Solid block is

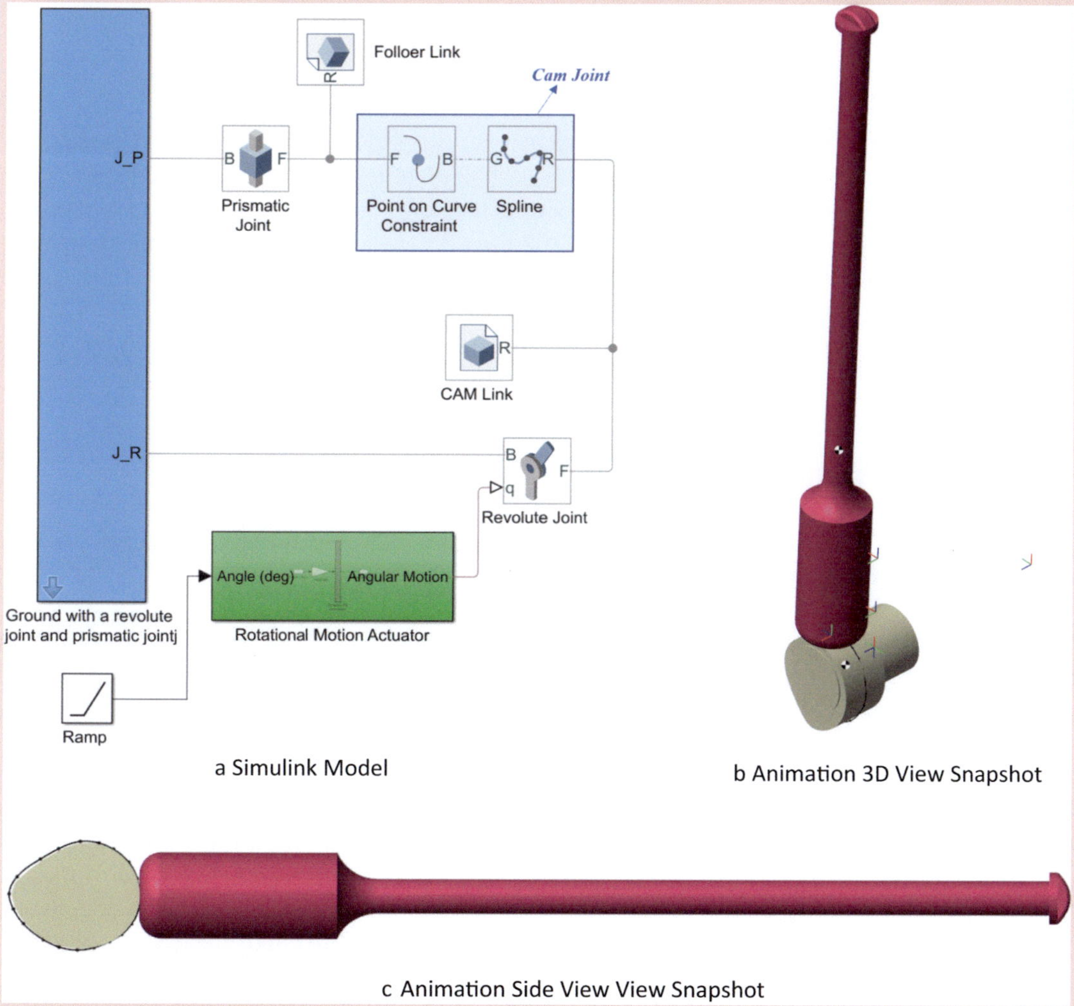

Fig. 9.36 Simulink® model of CAM mechanism using solid file blocks and animation snapshot

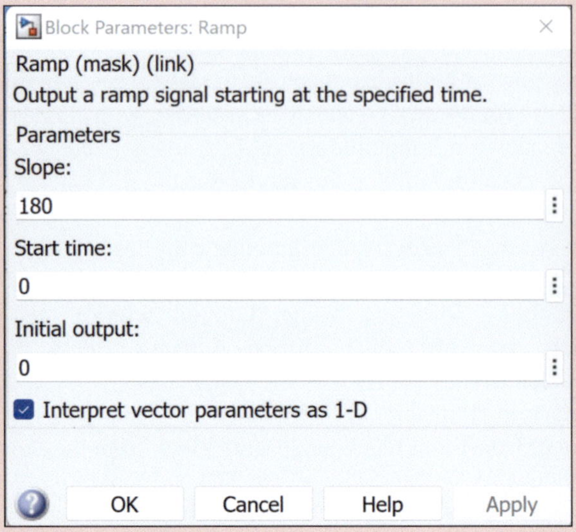

Fig. 9.37 Ramp in Simulink® model of cam mechanism using solid file blocks. Designed and implemented by the author. Images used with permission of MathWorks

called the CAM link and the second File Solid block is called the Follower Link. Figure 9.36b, c show the 3D view and the side view from the "Mechanics Explorer Window."

5. For the Ramp block, tune the editing parameters to provide an angular speed of 180 deg/s, as shown in Fig. 9.37.
6. The plugged-in values of the "Ground with a revolute joint and prismatic joint" block and Spline block are provided from the CAD files. After running the model, an animation including the two CAD files will appear. The reader is recommended to run this file and watch the animation. The reader can use the same techniques to build any machine from CAD files.

9.6 Examples of Machines Utilizing Cam Mechanism

In this subsection, some applications that the cam mechanism is used for are introduced. Most of these applications are mechanically related ones.

Packing Machines: The cam mechanism is used in designing packing machines as shown in Fig. 9.38 whereby two cam mechanisms are attached to a single motor. Each cam mechanism provides a different periodic motion that the packing machine is operating at. The reader is recommended to watch the attached video about this packing machine.

Mechanical Paper Punch: As shown in Fig. 9.39, there is a lever that, when pushed down, activates a gear system (called a rack and pinion) that moves the flat plate profile to the left. This pushes down the followers, which in turn punch two holes in a piece of paper/card. The edge of the flat plate cam can be shaped to give different vertical movements of the cam follower. Flat plate/linear cams are used frequently in machines that carry out the same repetitive movements. The reader is recommended to watch the attached videos about the paper punch machine.

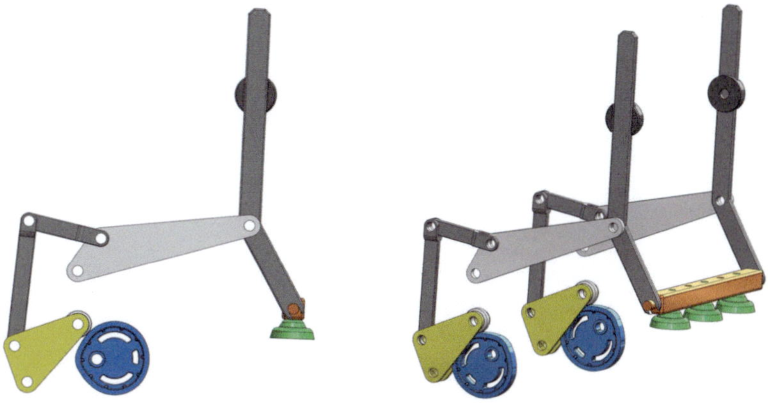

Fig. 9.38 Cam mechanism in packing machine. Designed by the author and implemented by both the author and Pranav Dhola. Courtesy of Pranav Dhola

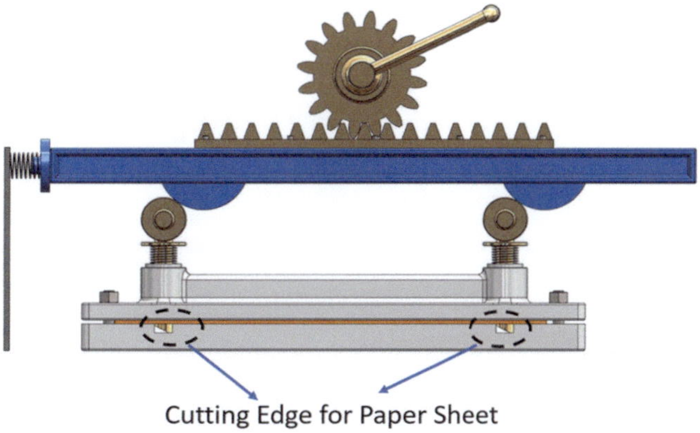

Fig. 9.39 Cam mechanism in paper punch machine. Figure and associated Paper Punch Machine.mp4 video designed by the author and implemented by both the author and Pranav Dhola. Both courtesy of Pranav Dhola

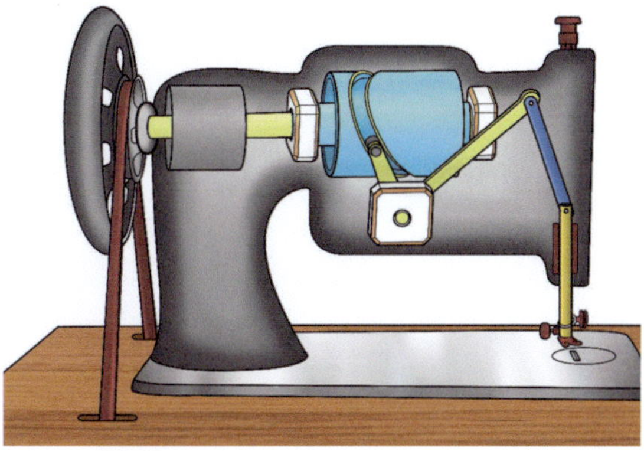

Fig. 9.40 Cylindrical cam mechanism in sewing machine. Figure and associated Sewing Maching.mp4 video designed by the author and implemented by both the author and Pranav Dhola. Both courtesy of Pranav Dhola

Sewing Head Mechanism: The sewing head mechanism is one of the most famous applications of the cam cylindrical mechanism and is shown in Fig. 9.40. The reader is recommended to watch the attached videos about the cylindrical cam mechanisms and the sewing head mechanism.

Toy Mechanism: The cam mechanism is considered one of the easiest mechanisms with which to build toys. A parent can use the hand-powered crank to rotate the shaft, whereby the cam is centered. The follower is attached to a cute shape, such as a man (Fig. 9.41), bird (Fig. 9.42), or any other shapes a child may be attracted to. The follower motion causes the shape to periodically

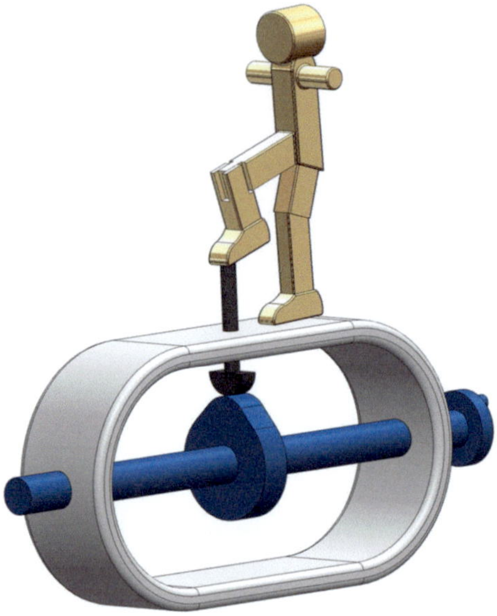

Fig. 9.41 Cam mechanism in man-toy. Designed by the author and implemented by both the author and Pranav Dhola. Courtesy of Pranav Dhola

Fig. 9.42 Cam mechanism in bird-toy. Figure and associated Bird Toy Machine.mp4 video designed by the author and implemented by both the author and Pranav Dhola. Both courtesy of Pranav Dhola

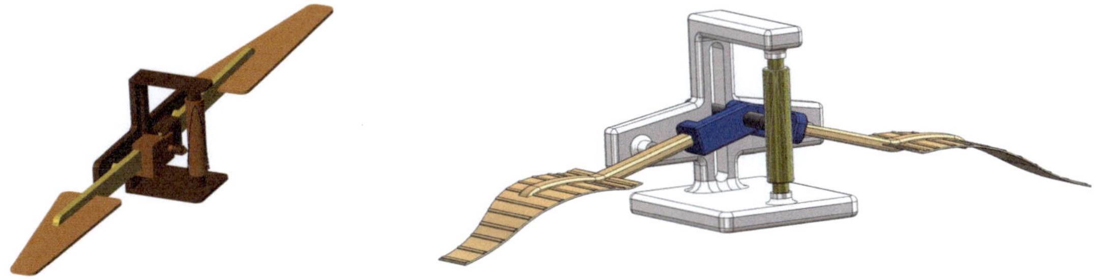

a Simscape Multibody™ Model Graphics *b Model Graphics*

Fig. 9.43 Cam mechanism in flapping wing. Figure and associated Flapping Wing Mechanism.mp4 video designed by the author and implemented by both the author and Pranav Dhola. Both courtesy of Pranav Dhola

move up and down, amazing the child. The cam types can be any of the types that are discussed in Sect. 9.2.1. The flat plate follower is commonly used for simple toy applications, although another follower type could be used. The reader is recommended to watch the attached videos about the cam-based toy mechanisms.

MathWorks provides an interesting flapping wing mechanism as one of the Simscape™ Multibody™ examples, shown in Fig. 9.43. The mechanism consists of a cylindrical cam that rotates by a motor. The two wings flap in sync with each other when the follower moves over the cam groove. The reader is recommended to watch this flapping wing mechanism on the MathWorks website.

9.7 Chapter Summary

Outcomes

1. Defining different types of cams, cam followers, and cam joints.
2. *SVAJ* and *svaj* diagrams are very useful tools to design the follower motion profile and avoid any discontinuity that may cause noise, vibration, damage to the machine, and/or stalling of the power source.
3. Polynomial motion, harmonic motion, and cycloidal motion are three common ways to develop any fall, rise, or dwell segment in the follower motion profile.
4. Cam base radius, prime radius, follower radius, trace point, pitch curve, cam profile, eccentricity, and pressure angle are the configuration parameters to describe the cam mechanism.
5. The follower motion profile, plus the selected base radius and offset, can be used to design the cam profile.
6. Using Simscape™ Multibody™ to simulate the gear train kinematic models.

Problems

Problem 1

A camshaft rotates one cycle from 0° to 360° to move the follower displacement trajectory as listed below.

$$s(\theta) = \begin{cases} \frac{50}{45^\circ}\theta & 0 \le \theta < 45^\circ \\ 50 & 45^\circ \le \theta < 135^\circ \\ 50 - \frac{50}{60^\circ}(\theta - 135^\circ) & 135^\circ \le \theta < 195^\circ \\ 0 & 195^\circ \le \theta \le 360^\circ \end{cases}$$

where θ is in degrees and *s* is in mm.

1. Calculate the velocity $v(\theta)$, the acceleration $a(\theta)$, and the jerk $j(\theta)$.
2. Sketch the *svaj* diagram of the follower motion.

Problem 2

A camshaft rotates one cycle from 0° to 360° to move the follower displacement trajectory as listed below.

$$s(\theta) = \begin{cases} \frac{50}{40}\theta & 0 \le \theta < 40^\circ \\ 50 & 40^\circ \le \theta < 140^\circ \\ 50 - \frac{30}{90}(\theta - 140^\circ) & 140^\circ \le \theta < 230^\circ \\ 20 & 230^\circ \le \theta < 250^\circ \\ 50 - \frac{20}{50}(\theta - 250^\circ) & 250^\circ \le \theta < 300^\circ \\ 0 & 300^\circ \le \theta < 360^\circ \end{cases}$$

where θ is in degrees and *s* is in cm.

1. Calculate the velocity $v(\theta)$, the acceleration $a(\theta)$, and the jerk $j(\theta)$.
2. Sketch the *svaj* diagram of the follower motion.

Problem 3

A camshaft rotates one cycle from 0° to 360° to move the follower displacement trajectory as listed below.

$$s(\theta) = \begin{cases} 0 & 0 \le \theta < 20^\circ \\ \frac{5}{30}(\theta - 20^\circ) & 20^\circ \le \theta < 50^\circ \\ 5 & 50^\circ \le \theta < 90^\circ \\ 5 + \frac{5}{30}(\theta - 90^\circ) & 90^\circ \le \theta < 120^\circ \\ 10 & 120^\circ \le \theta < 150^\circ \\ 10 - \frac{10}{150}(\theta - 150^\circ) & 150^\circ \le \theta < 300^\circ \\ 0 & 300^\circ \le \theta < 360^\circ \end{cases}$$

where θ is in degrees and *s* is in cm.

1. Calculate the velocity $v(\theta)$, the acceleration $a(\theta)$, and the jerk $j(\theta)$.
2. Sketch the *svaj* diagram of the follower motion.

Problem 4

A disk cam mechanism with a knife-edge follower has the follower trajectory motion segments sequentially as

- ***Motion segment 1***: Rise 50 mm in 0.393 s.
- ***Motion segment 2***: Dwell for 0.785 s.
- ***Motion segment 3***: Fall 50 mm for 0.524 s.
- ***Motion segment 4***: Dwell for 1.440 s.

The ramp CEP motion (constant velocity CEP motion) is used to design the follower trajectory motion segments.

1. Find the cam constant angular speed.
2. Find the initial angle and the final angle of each motion segment (rise, fall, and dwell).
3. Sketch the *svaj* diagram and *SVAJ* diagram.

Problem 5

A disk cam mechanism with a knife-edge follower has the follower trajectory motion segments sequentially as

- ***Motion segment 1***: Rise 50 mm in 6.981×10^{-2} s.
- ***Motion segment 2***: Dwell for 0.175 s.
- ***Motion segment 3***: Fall 30 mm for 0.157 s.
- ***Motion segment 4***: Dwell for 3.491×10^{-2} s.
- ***Motion segment 5***: Fall 20 mm for 8.727×10^{-2} s.
- ***Motion segment 6***: Dwell for 0.105 s.

The ramp CEP motion (constant velocity CEP motion) is used to design the follower trajectory motion segments.

1. Find the cam constant angular speed.
2. Find the initial angle and the final angle of each motion segment (rise, fall, and dwell).
3. Sketch the *svaj* diagram and *SVAJ* diagram.

Problem 6

A disk cam mechanism with a knife-edge follower has the follower trajectory motion segments sequentially as

- ***Motion segment 1***: Rise 50 mm in 0.393 s.
- ***Motion segment 2***: Dwell for 0.785 s.
- ***Motion segment 3***: Fall 50 mm for 0.524 s.
- ***Motion segment 4***: Dwell for 1.440 s.

The 7th-order polynomial is used to design the follower trajectory motion segments.

4. Find the cam constant angular speed.
5. Find the initial angle and the final angle of each motion segment (rise, fall, and dwell).

6. Sketch the *svaj* diagram.
7. Use MATLAB® script to plot the *svaj* diagram.

Problem 7

A disk cam mechanism with a knife-edge follower has the follower trajectory motion segments sequentially as

- ***Motion segment 1***: Rise 50 mm in 0.393 s.
- ***Motion segment 2***: Dwell for 0.785 s.
- ***Motion segment 3***: Fall 50 mm for 0.524 s.
- ***Motion segment 4***: Dwell for 1.440 s.

The harmonic CEP motion is used to design the follower trajectory motion segments.

1. Find the cam constant angular speed.
2. Find the initial angle and the final angle of each motion segment (rise, fall, and dwell).
3. Sketch the *svaj* diagram.
4. Use MATLAB® script to plot the *svaj* diagram.

Problem 8

A disk cam mechanism with a knife-edge follower has the follower trajectory motion segments sequentially as

- ***Motion segment 1***: Rise 50 mm in 0.393 s.
- ***Motion segment 2***: Dwell for 0.785 s.
- ***Motion segment 3***: Fall 50 mm for 0.524 s.
- ***Motion segment 4***: Dwell for 1.440 s.

The cycloidal CEP motion is used to design the follower trajectory motion segments.

1. Find the cam constant angular speed.
2. Find the initial angle and the final angle of each motion segment (rise, fall, and dwell).
3. Sketch the *svaj* diagram.
4. Use MATLAB® script to plot the *svaj* diagram.

Problem 9

The camshaft rotates through an angle θ. The follower displacement is

$$s(\theta) = \begin{cases} \frac{50}{45^\circ}\theta & 0 \le \theta < 45^\circ \\ 50 & 45^\circ \le \theta < 135^\circ \\ 50 - \frac{50}{60^\circ}(\theta - 135^\circ) & 135^\circ \le \theta < 195^\circ \\ 0 & 195^\circ \le \theta \le 360^\circ \end{cases}$$

where θ is in degree and s is in mm. If $R_b = 100$ mm, design the cam profile graphically for a knife-edge follower.

Problem 10

The camshaft rotates through an angle θ. The follower displacement is

$$s(\theta) = \begin{cases} \frac{50}{40}\theta & 0 \le \theta < 40^\circ \\ 50 & 40^\circ \le \theta < 140^\circ \\ 50 - \frac{30}{90}(\theta - 140^\circ) & 140^\circ \le \theta < 230^\circ \\ 20 & 230^\circ \le \theta < 250^\circ \\ 50 - \frac{20}{50}(\theta - 250^\circ) & 250^\circ \le \theta < 300^\circ \\ 0 & 300^\circ \le \theta < 360^\circ \end{cases}$$

where θ is in degree and s is in mm. If $R_b = 100$ mm, design the cam profile graphically for a knife-edge follower.

Problem 11

An engineer designs the cam profile:

$$s(\theta) = \begin{cases} \frac{50}{45^\circ}\theta & 0 \le \theta < 45^\circ \\ 50 & 45^\circ \le \theta < 135^\circ \\ 50 - \frac{50}{60^\circ}(\theta - 135^\circ) & 135^\circ \le \theta < 195^\circ \\ 0 & 195^\circ \le \theta \le 360^\circ \end{cases}$$

where θ is in degrees and s is in mm. If $R_b = 50$ mm, use a MATLAB® script to plot the cam profile.

Problem 12

An engineer designs the cam profile:

$$s(\theta) = \begin{cases} \frac{50}{40}\theta & 0 \le \theta < 40^\circ \\ 50 & 40^\circ \le \theta < 140^\circ \\ 50 - \frac{30}{90}(\theta - 140^\circ) & 140^\circ \le \theta < 230^\circ \\ 20 & 230^\circ \le \theta < 250^\circ \\ 50 - \frac{20}{50}(\theta - 250^\circ) & 250^\circ \le \theta < 300^\circ \\ 0 & 300^\circ \le \theta < 360^\circ \end{cases}$$

where θ is in degrees and s is in mm. If $R_b = 100$ mm, use a MATLAB® script to plot the cam profile.

In Problem 13 and Problem 14, there are two options for the cam motion profile. In preliminary design stage, if you are hired to pick one of these two options, use Simscape™ Multibody™ to simulate the cam mechanism for each option where $\omega_c = 180\text{deg/sec}$ over one period. Plot the follower SVAJ diagram from the simulation for each case. Comment on your results.

In your simulation, the follower is a knife edge with the given dimension in Fig. 9.44. The thickness of both the cam and follower is 70 mm and $R_b = 100$ mm.

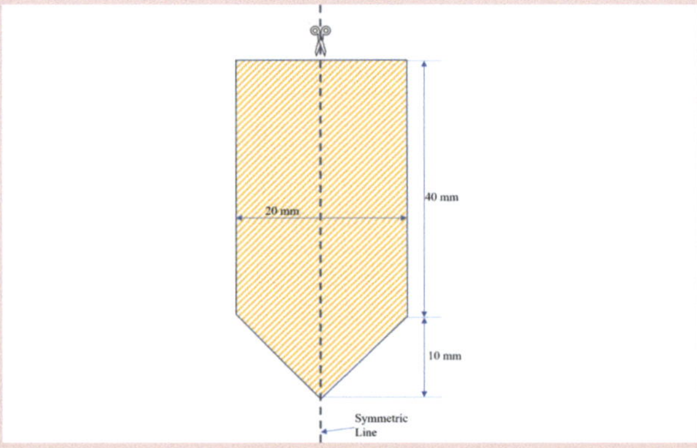

Fig. 9.44 Follower cross-section

Problem 13

Option 1: Ramp CEP Motion or Constant Velocity CEP Motion

$$s(\theta) = \begin{cases} \frac{50}{45^\circ}\theta & 0 \le \theta < 45^\circ \\ 50 & 45^\circ \le \theta < 135^\circ \\ 50 - \frac{50}{60^\circ}(\theta - 135^\circ) & 135^\circ \le \theta < 195^\circ \\ 0 & 195^\circ \le \theta \le 360^\circ \end{cases}$$

where θ is in degrees and *s* is in mm.

Problem 14

Option 2: 7th-order polynomial

$$s(\theta) = \begin{cases} 50\left(35\left(\frac{\theta}{45}\right)^4 - 84\left(\frac{\theta}{45}\right)^5 + 70\left(\frac{\theta}{45}\right)^6 - 20\left(\frac{\theta}{45}\right)^7\right) & 0 \le \theta < 45^\circ \\ 50 & 45^\circ \le \theta < 135^\circ \\ 50 - 50\left(35\left(\frac{\theta-135}{60}\right)^4 - 84\left(\frac{\theta-135}{60}\right)^5 + 70\left(\frac{\theta-135}{60}\right)^6 - 20\left(\frac{\theta-135}{60}\right)^7\right) & 135^\circ \le \theta < 195^\circ \\ 0 & 195^\circ \le \theta \le 360^\circ \end{cases}$$

where θ is in degrees and *s* is in mm.

References

1. "Design of Machinery," Robert L. Norton, McGraw-Hill; 5th Edition, 2011.
2. "Theory of Machines and Mechanisms," John Uicker, Gordon Pennock and Joseph Shigley, Oxford University Press; 5th Edition, 2016.
3. "Machines & Mechanisms: Applied Kinematic Analysis", David Myszka, Pearson; 4 edition, 2011.
4. "Cam Design Handbook," Harold A. Rothbart, McGraw-Hill, 1st Edition, 2003.
5. "Simscape™ Multibody™ Getting Started Guide," MathWorks, 2023.
6. "Simscape™ Multibody™ User's Guide," MathWorks, 2023.
7. "Simscape™ Multibody™ Reference," MathWorks, 2023.
8. "Simscape™ Multibody™ Release Notes," MathWorks, 2023.

Gear Analysis 10

Chapter Objectives

1. Understanding different types of gears.
2. Understanding the configuration parameters of the spur gear.
3. Understanding the configuration parameters of the meshed gears mechanism.
4. Understanding how to design gears based on standardized gear formulas.
5. Understanding the kinematics of two meshed gears.
6. Understanding the kinematics of a simple gear train.
7. Understanding the kinematics of a planetary gear train.
8. Understanding the kinematics of differentials.
9. Understanding how to use Simscape™ Multibody™ in MATLAB® and Simulink® to calculate the angular velocity and angular acceleration of gears in any mechanism.

10.1 Introduction

In this chapter, we introduce several types of gear that are used in common machines. After that, we provide a more detailed analysis of the spur gear in terms of the definition of the configuration parameters of a single gear. This is followed by explaining the two meshed gears and how to design a tooth profile. Standardized gear manufacturing and design are introduced to set up a procedure for a straightforward gear design. The kinematics of two meshed gears is explained using analytical methods as well as Simscape™ Multibody™ simulation. Finally, we explain the kinematic model of a simple gear train, compound gear train, planetary gear train, and differentials.

10.2 Gear Types

Rolling cylinders are a very simple way to transfer rotary motion from a power source to a power sink. For example, a car needs to transfer the rotational motion from the engine shaft (power source) with an average speed of 3000 rpm into the wheel shaft (power sink) with an average speed of

Supplementary Information The online version contains supplementary material available at https://doi.org/10.1007/978-3-031-72701-6_10.

A. Omran, *Kinematics and Dynamics of Machines*, https://doi.org/10.1007/978-3-031-72701-6_10

50 rpm. The rolling cylinders could be arranged to provide such a car rotational motion transfer. Unfortunately, cylinders arranged in this manner are not usually capable of applying consistent friction between each other resulting in slipping.

One very successful solution to this problem is the addition of meshing teeth between the rolling cylinders. These toothed cylinders are known as *gears*. When two or more of these toothed cylinders are meshed, we have a ***gear pair***. The gear pair node connects two meshed gears at the contact point where there is no relative motion along the line of action. The meshed gears guarantee a speed ratio between the shafts with almost no slipping. So, for the car example, adding a set of gear trains (transmission) will transfer the rotational motion from the engine shaft to the gear shaft with almost no slipping.

Many different types of gears and gear sets have been and continue to be developed for specific performance criteria. Below is a list of some of the most common with their definitions to set them apart.

10.2.1 Spur Type

A spur gear, the most common gear type, is a cylinder or disk with teeth parallel to the axis of rotation as shown in Fig. 10.1a. A spur gear is also known as a straight-cut gear. A pair of spur gears, as shown in Fig. 10.1b, can be used to transfer the power from a power source shaft to a power sink shaft. The spur gear has many applications, including transmissions, conveyor systems, and driving pumps. It is common for spur gears to be driven by a motor. It is recommended that the reader watches the animation video of the pair of spur gears.

10.2.2 Rack Type

The rack is a limited-length flat bar with a stack of parallel teeth alongside one another as shown in Fig. 10.2a. Usually, the rack is coupled or assembled with different types of gears. For example, a spur gear can be coupled with the rack. This set is called a *rack–pinion* and is shown in Fig. 10.2b. The pinion refers to the spur gear. The rack can be used to convert rotational motion into transitional

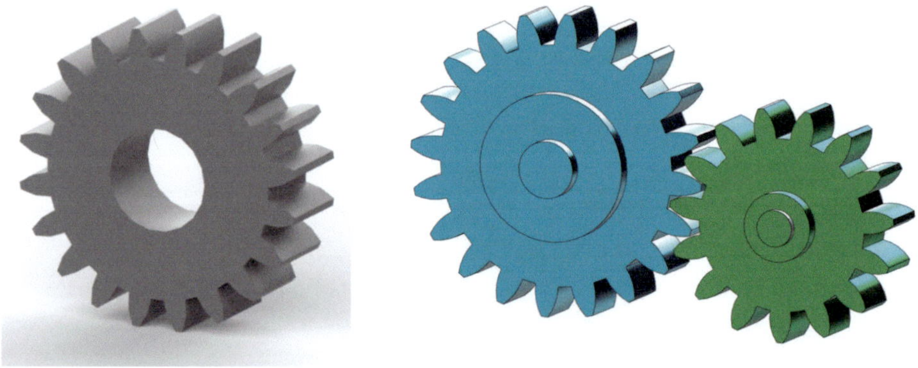

a Single Spur Gear

b Spur Gear Pair

Fig. 10.1 Spur type. Figure courtesy of Pranav Dhola; associated Meshing Spur Gear.mp4 and Meshing Spur Gear_V2.mp4 videos designed by the author and implemented by both the author and Pranav Dhola. Both courtesy of Pranav Dhola

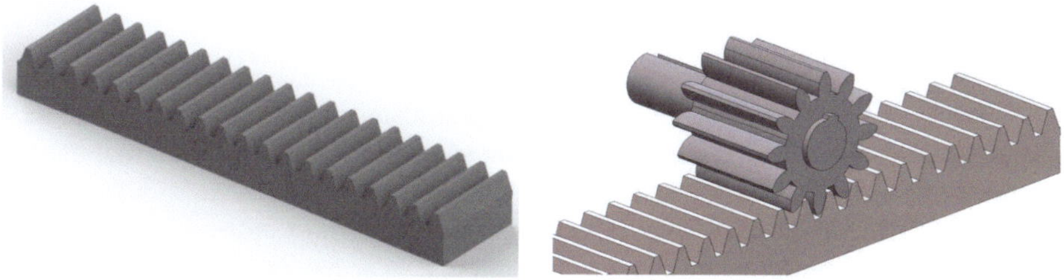

Fig. 10.2 Rack type. Figure courtesy of Pranav Dhola; associated Pinion-Rack Mechanism.mp4 video designed by the author and implemented by both the author and Pranav Dhola; also courtesy of Pranav Dhola

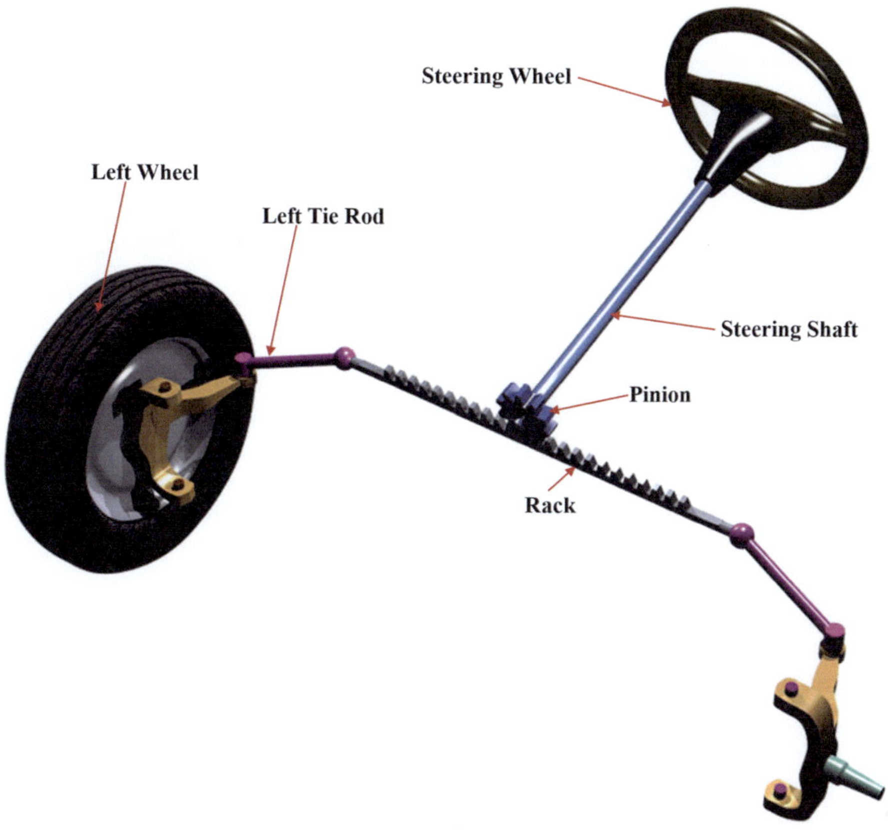

Fig. 10.3 Car steering system. Figure courtesy of Pranav Dhola; associated Car_Steering_Mechanism.mp4 video designed by the author and implemented by both the author and Pranav Dhola; also courtesy of Pranav Dhola

motion or vice versa. Rack gears have many applications, such as in the steering system of a car as shown in Fig. 10.3. More details about the car steering system are provided in Chap. 1. It is recommended that the reader watches the animation video of the rack–pinion set.

10.2.3 Internal Type

An internal gear has its teeth stacked inside an inner circular surface as shown in Fig. 10.4a. The internal gear is one of the great ideas used to reduce the distance between coupled gears. Imagine the amount of space that can be reduced by using one spur gear and another internal gear—see Fig. 10.4b—compared to simply using two spur gears.

The internal gear can be coupled with several gear types; in most cases, a spur gear and a helical gear are used as an internal gear coupler. The internal gear has many applications such as the internal gear pump in Fig. 10.5. Later in Sect. 10.6, we will introduce more details about using the internal gear in the development of the so-called planetary gear train and its usage in transmissions. It is recommended that the reader watches the animation video of the coupled internal gear with a spur gear.

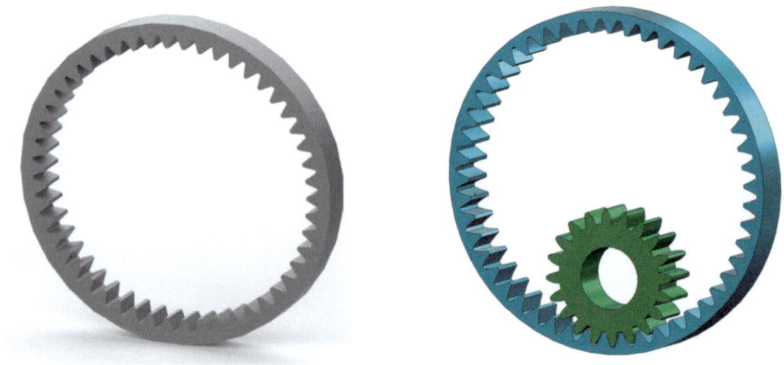

a Single Internal Gear b Internal-Spur Set

Fig. 10.4 Internal gear type. Figure and associated Internal Gear.mp4 video designed by the author and implemented by both the author and Pranav Dhola. Both courtesy of Pranav Dhola

Fig. 10.5 Internal gear pump. Designed by the author and implemented by both the author and Pranav Dhola. Courtesy of Pranav Dhola

10.2.4 Helical Type

The helical gear is similar to a spur gear as a cylinder or disk with a stack of teeth on a rounded surface but the teeth are not parallel to the axis of rotation. The helical gear teeth are oriented with the helix angle, ψ, that can be up to 45° with a left- or right-handed orientation, as shown in Fig. 10.6. The spur gear is a special case of helical gear when the helix angle is zero.

When two helical gears are in contact (meshed), the orientation plus the helical angle determine the angle between the two shafts. For example, as in Fig. 10.7a, when the two gears have the same helical angle with opposite orientation (right-/left-hand pair), the two shafts are parallel. Another example is shown in Fig. 10.7b when the two gears are set as right/right-hand pair for non-parallel

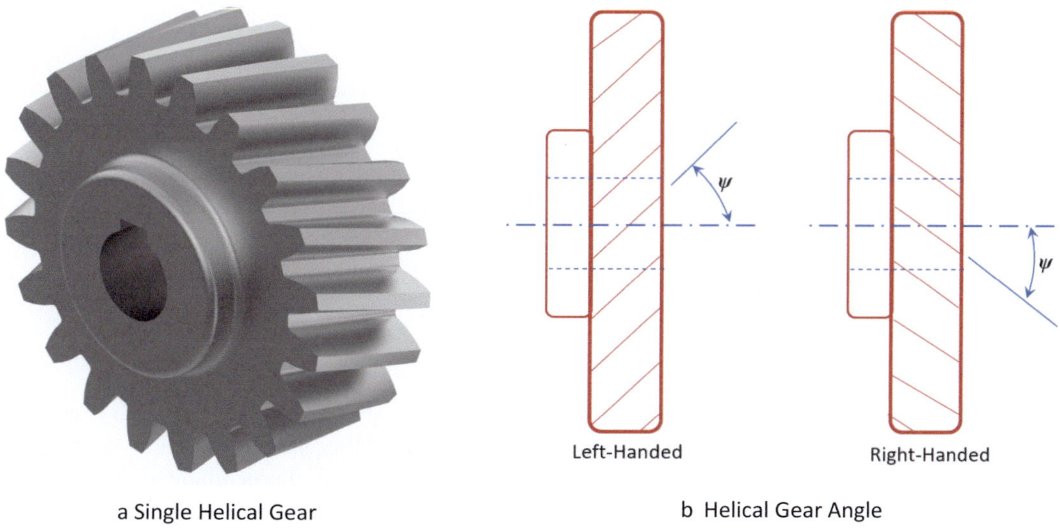

Fig. 10.6 Helical type. Designed by the author and implemented by both the author and Pranav Dhola. Courtesy of Pranav Dhola

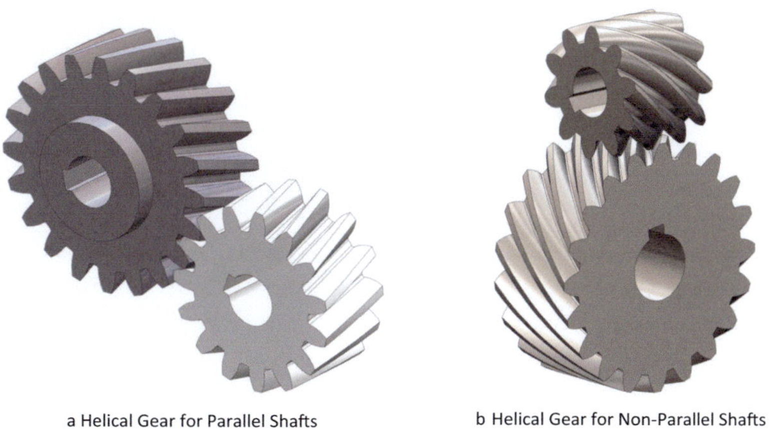

Fig. 10.7 Different shaft orientations for helical gear. Figure and associated Helical Gear.mp4 and Helical Gear_V2.mp4 videos designed by the author and implemented by both the author and Pranav Dhola. All courtesy of Pranav Dhola

shafts. The capability to transfer power from two shafts that are not parallel is considered an advantage of the helical gear pair over the spur gear pair.

The nonzero helical angle causes the helical gear to have more tooth surfaces in contact. (This is commonly quantified as the contact ratio.) This makes the helical gear have less noise, a higher load-carrying capacity, and a greater reduction of wear than the spur gear. In other words, the helical angle provides a gradual engagement between the two teeth from the coupled gears resulting in smoother and quieter operation than the spur gear [1–3].

The helical gear is a bit of a challenge for a design engineer to come up with the helix profile since the helical gear could have straight teeth or spiral teeth. For high speed, the helical gear is preferred over the spur gear especially because of its noise reduction and high capability of carrying higher stress than the spur gear. The more complicated the teeth profile, the more challenging for the design engineer in the production phase. It is recommended that the reader watches the animation video of the pair of helical gears.

10.2.5 Worm Type

The worm gear type, as shown in Fig. 10.8, consists of a cylindrical "worm" meshing with a gear. The worm gear type transmits the power between non-parallel shafts. Worm gear types are capable of providing a very high gear ratio in a small package but with lower efficiency.

10.2.6 Bevel Type

The bevel-type gear, as shown in Fig. 10.9 has a semi-cone shape whereby the teeth are located on the rounded surface area. Bevel gears connect two non-parallel shafts. For most applications, the angle between the shafts is 90°. In other words, the two shafts are perpendicular. When the two

Fig. 10.8 Worm gear type. Figure and associated Worm Gear.mp4 video designed by the author and implemented by both the author and Pranav Dhola. Both courtesy of Pranav Dhola

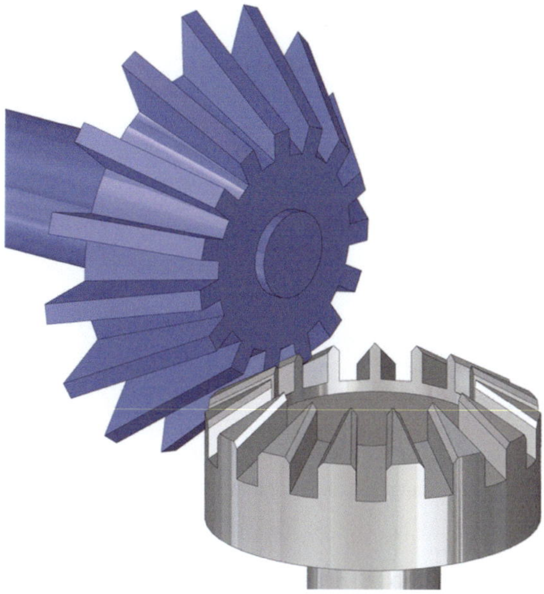

Fig. 10.9 Bevel gear type. Designed by the author and implemented by both the author and Pranav Dhola. Courtesy of Pranav Dhola

mated bevel gears are the same size and connect two perpendicular shafts, the bevel gears are called *miter gears*. The car differential, later explained in Sect. 10.7, is a very good example of a bevel gear application.

10.3 Spur Gear Design and Analysis

10.3.1 Configurations and Parameters of Spur Gear Teeth

In this section, we are going to talk about the spur-type gear teeth configuration and profile. Most of the other types have similar configurations and profiles with more advances that are out of the scope of this textbook. The spur gear has a cylindrical shape with two base surfaces forming the gear's circular shapes. These are perpendicular to the axis of rotation. Figure 10.10 shows the spur gear's profile that has the characteristics described below.

As shown in Fig. 10.10, the gear has an imaginary circle called a ***pitch circle*** that is the size of the gear's pure roller circle. There is also an ***addendum circle*** with an addendum diameter d_a, and a ***dedendum circle*** with a dedendum diameter d_d at the bottom of the gear tooth. The radial distance from the pitch circle to the top of the gear tooth is called ***addendum*** a, while ***dedendum*** b is the radial distance from the pitch circle to the bottom of a gear tooth. The summation of both addendum and dedendum is the tooth's ***whole depth***, h_T. The difference between a dedendum and an addendum is called the ***clearance***, c. The diameter of the pitch circle is called the ***pitch diameter***, d, while the imaginary curved surface of the pitch circle is called the ***pitch surface***.

Each tooth has a curved surface above the pitch surface and in an outward direction from the center of rotation called the ***face surface*** and it has a ***face width***, the tooth length in the parallel direction with the rotational axis. The tooth surface below the pitch surface and in the inward direction from the center of rotation is called the ***flank surface.*** Each tooth has a top flat surface, namely, ***top land***. The ***bottom land*** is the bottom curved surface between the two back-to-back teeth.

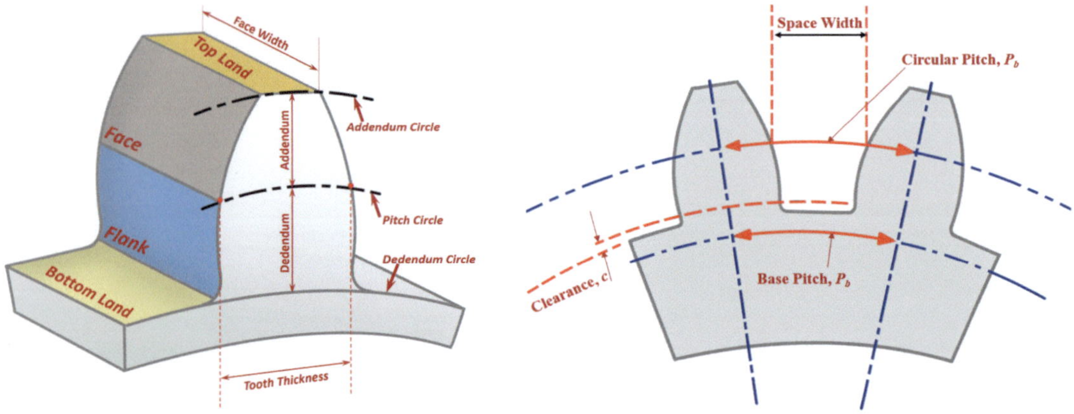

a Spur Gear Tooth 3D View b Two Back-to-Back Spur Gear Teeth

Fig. 10.10 Spur gear teeth profile

Sometimes, to avoid any sharp edges in the gear profile, a ***corner round*** is added between the ***top land*** and the ***face***.

As shown in Fig. 10.10, the circular distance between two back-to-back teeth is called the ***circular pitch***, P_c, which can be calculated using the pitch diameter d and number of the teeth N from

$$P_c = \frac{\pi d}{N} \tag{10.1}$$

The American Gear Manufacturers Association (AGMA) defined the ***diametral pitch***, P_d, as a terminology to characterize the gear. It is known to be

$$P_d = \frac{N}{d} \tag{10.2}$$

The ***module***, m, is the ratio of the pitch diameter to the number of teeth in a gear.

$$m = \frac{d}{N} \tag{10.3}$$

The module is defined in the SI unit system where d is measured in millimeters. The two definitions of P_d and m in Eqs. (10.2) and (10.3) lead to measuring the size of a gear tooth or tooth size. The diametral pitch, P_d, and module, m, seem to be reciprocal of each other but that is not numerically true. If we look at the source of each definition, the diameter pitch, P_d, is in US customary units using inches for measuring the length of the pitch diameter d in Eq. (10.2). On the other hand, module, m, is in the SI unit system using millimeters for measuring the unit length or the pitch diameter d in Eq. (10.3). To consider the unit difference, we use

$$m = \frac{25.4}{P_d} \tag{10.4}$$

The value 25.4 comes from the unit conversion between inches and millimeters.

Example 10.1

A spur gear is designed to have 16 teeth with a diametral pitch of 10. Determine the following:

1. The diameter of the pitch circle.
2. The circular pitch.
3. The module.

Solution

To calculate the diameter of the pitch circle, d, we use Eq. (10.2).

$$d = \frac{N}{P_d} = \frac{16\ \text{teeth}}{10\ \text{in}^{-1}} = 1.6\ \text{in}$$

To calculate the circular pitch, P_c, we use Eq. (10.1).

$$P_c = \frac{\pi d}{N} = \frac{\pi(1.6\ \text{in})}{16\ \text{teeth}} = 0.314\ \text{in}$$

To calculate the module, m, we use Eq. (10.4).

$$m = \frac{25.4}{P_d} = \frac{25.4}{10\ \text{in}^{-1}} = 2.54\ \text{mm}$$

Engineering Note: What is AGMA?

The American Gear Manufacturers Association (AGMA), established in 1916, is a voluntary association of companies, consultants, and academicians. Up until 2022, the AGMA had 495 of the world's top gearing companies as members across more than 30 nations. The AGMA's main interest is in the design, manufacture, and application of gears, couplings, and related power transmission components and equipment. The AGMA holds annual meetings to present and discuss manufacturing technologies. There is an AGMA magazine and website that the reader is recommended to explore.

10.3.2 Spur Gear Tooth Profile

The tooth profile could be designed in many shapes. The most common shape is the involute curve shape that is based on the geometry of two meshed gears as shown in Fig. 10.11. To understand the involute curve, we are going to first introduce some geometric parameters between two meshed gears.

- The ***pitch point*** is the common point at which the meshing gear pitch circles are in contact.
- The ***base circle,*** also called the ***root circle,*** is a circle from which the involute teeth profile is generated.
- The ***pressure line*** is a straight line passing through the pitch point and tangent to both base circles.
- The ***path of contact*** is the distance between the two tangent points, N and M.

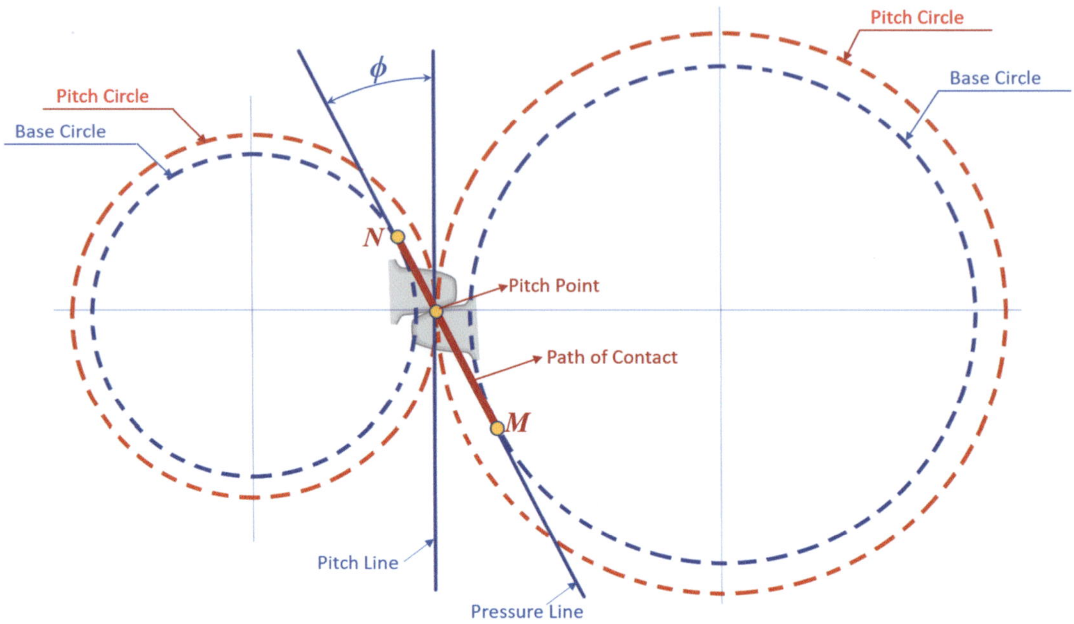

Fig. 10.11 Base circle and pressure angle, spur gear

- The ***pressure angle***, ϕ, is located at the pitch point and between the pitch line and the pressure line. From the geometry in Fig. 10.11, the relationship between the pressure angle, base diameter d_b, and pitch diameter d is

$$d_b = d\cos(\phi) \tag{10.5}$$

- The ***center distance***, C, is the center-to-center distance between two meshed gears.

$$C = r_1 + r_2 = \frac{d_1 + d_2}{2} \tag{10.6}$$

Using

$$P_d = \frac{N}{d}$$

Then

$$C = \frac{N_1 + N_2}{2P_d}$$

- The ***contact ratio***, m_p, is the average number of teeth in contact with two meshed gears.

Engineering Note: Gear Schematic Diagram

A gear schematic diagram is a simple sketch used to present a geartrain—a set of connected gears and shafts. The schematic diagram of some of the gear types is listed in Table 10.1. Sometimes, this textbook will use schematic diagrams. The reader is encouraged to practice using the gear schematic diagram. For example, the schematic diagram of the supergear is sketched as a vertical line with two horizontal lines at the end as a simple side view of the gear.

Table 10.1 Spur gear and internal gear schematic diagram

Geartrain	Schematic diagram	Geartrain	Schematic diagram
Spur gear with a shaft; similar to Fig. 10.1a	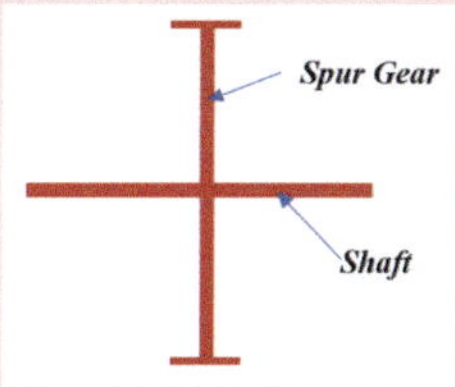	Two meshed spur gears with two separate shafts; similar to Fig. 10.1b	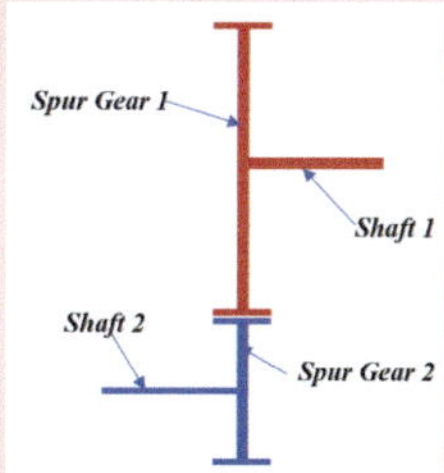
Internal gear with a shaft; similar to Fig. 10.4a	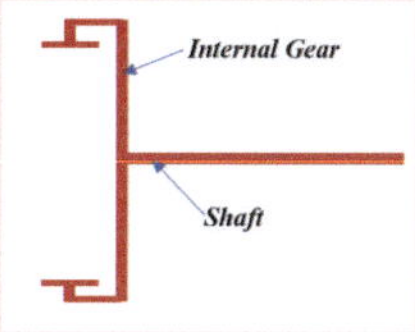	Two meshed spur gears and internal gear with two separate shafts; similar to Fig. 10.4b	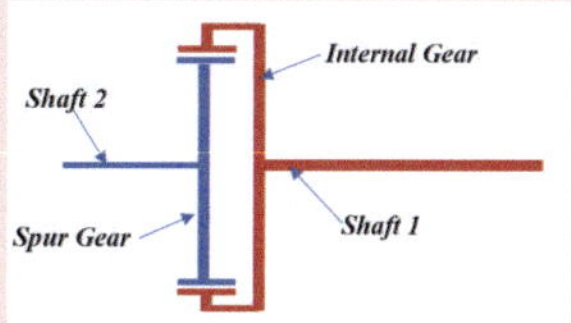

Example 10.2

The water lift mechanism, as shown in Fig. 10.12, consists of a motor that rotates a pinion with 15 teeth and a bull with 30 teeth. Both pinion and bull are meshed. The designer selected 5-pitch and 20° as the pressure angle. Determine:

1. The pitch diameter of each gear.
2. The base diameter of each gear.
3. The center distance.

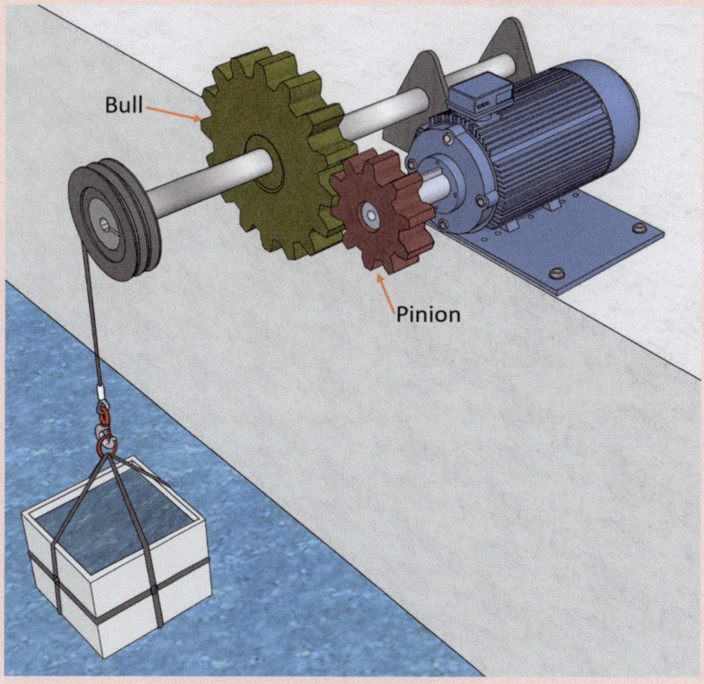

Fig. 10.12 Water lift mechanism. Designed by the author and implemented by Kenneth Manalansan. Courtesy of Kenneth Manalansan

Solution

Since the two gears are meshed, the diametral pitch is the same and the pitch diameter of both gears can be calculated by using Eq. (10.2).

$$d_1 = \frac{N_1}{P_d} = \frac{15}{5} = 3 \text{ in}$$

$$d_2 = \frac{N_2}{P_d} = \frac{30}{5} = 6 \text{ in}$$

The base diameter of each gear can be calculated by using Eq. (10.5).

$$d_{b1} = d_1 \cos(\phi) = (3\text{in}) \cos(20°) = 2.82 \text{ in}$$

$$d_{b2} = d_2 \cos(\phi) = (6\text{in}) \cos(20°) = 5.64 \text{ in}$$

The center distance can be calculated by using Eq. (10.6).

$$C = \frac{d_1 + d_2}{2} = \frac{3 \text{ in} + 6 \text{ in}}{2} = 4.5 \text{ in}$$

For a single tooth, the base circle divides the shape into an outward and an inward section—see Fig. 10.13. The outward section is cornered from the addendum circle, base circle, B involute curve, and C involute curve. The inward section is cornered from the dedendum circle, base circle, B trochoid, and C trochoid curve. The tooth profile is defined by the involute curves and the trochoid curves along with the addendum circle. The counter $i = \{1, 2, 3, \ldots, N\}$ refers to the ith tooth from a gear with N number of teeth. The goal of the base circle is to divide the tooth profile into a desired contact curve (involute curve) and undesired contact curve (trochoid curve).

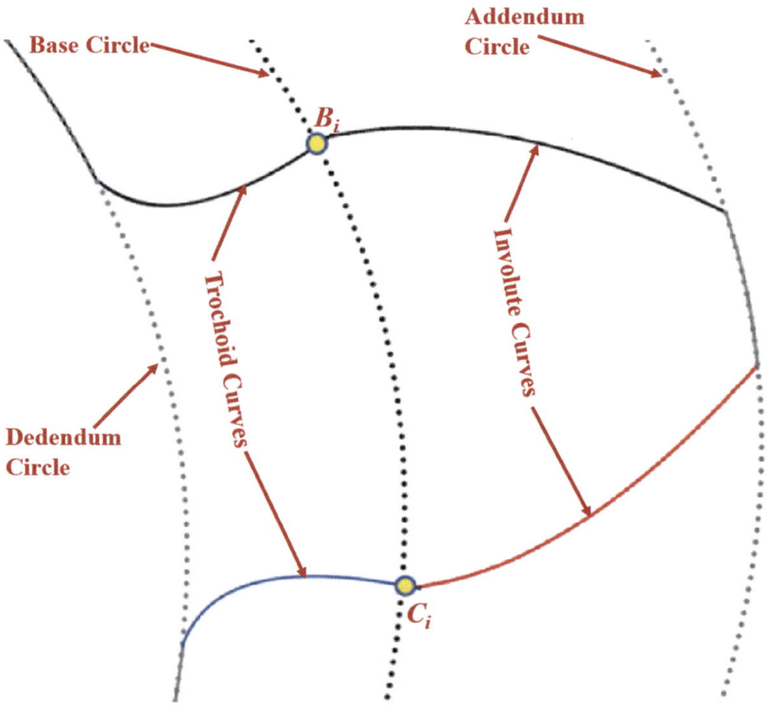

Fig. 10.13 Tooth profile, spur gear

A good gear design should not allow the gear tip to be in contact with the trochoid curve of the other meshed gear, otherwise, interference phenomena happen causing noise and vibration that may damage the meshed gears. The intersections between the two curves are points B_i and C_i; located on the base circle.

The B involute curve of tooth number i is generated by unwrapping an infinite number of taut strings from the base circle to be a tangent to. Each unwrapped taut string starts as a tangent at an arbitrary point j located on the base circle and ends as a free end with the same circular length between point j and point B_i; see Fig. 10.14. The free ends of the taut strings generate the B involute curve up to the addendum circle.

To clarify the process of generating the involute curve, let's assume that we have the pitch diameter d, the number of the teeth N, pressure angleϕ, and addendum diameter d_a as knowns. We could use hand drawing or any CAD software to draw the involute curve and outward profile, using the steps below.

1. Plot the pitch circle and the dedendum circle with the two diameters d and d_b, respectively.
2. Calculate the base circle as $d_b = d\cos(\phi)$.
3. Plot the base circle.
4. Plot N radial lines to present each tooth. The angle of each radial line is $\theta_i = 2\pi(i-1)/N$, where $i=\{1, 2, \ldots, N\}$.
5. For each tooth i:
   - Go clockwise, count each intersection point between the radial lines and base circle with B_j, where $j=\{1, 2, \ldots, N\}$. For example, you can see B_1, B_2, B_3, and B_4 in Fig. 10.14.
   - From each point B_j, plot a tangent to the base circle heading toward $j=1$ with a length $L_j = 2\pi d_d(j-1)/N$.
   - Connect the free ends of each tangent line to plot the ith tooth B involute curve up to the addendum circle.
   - Use the tooth symmetry around the radial line to plot the ith tooth C involute curve.
6. Repeat the same substeps in step 5 for all the other gear teeth.

In the above procedure, we could use more than N points by changing $L_j = 2\pi d_d(j-1)/M$ instead of $L_j = 2\pi d_d(j-1)/N$, where M is the number of points that are used to plot the involute curve.

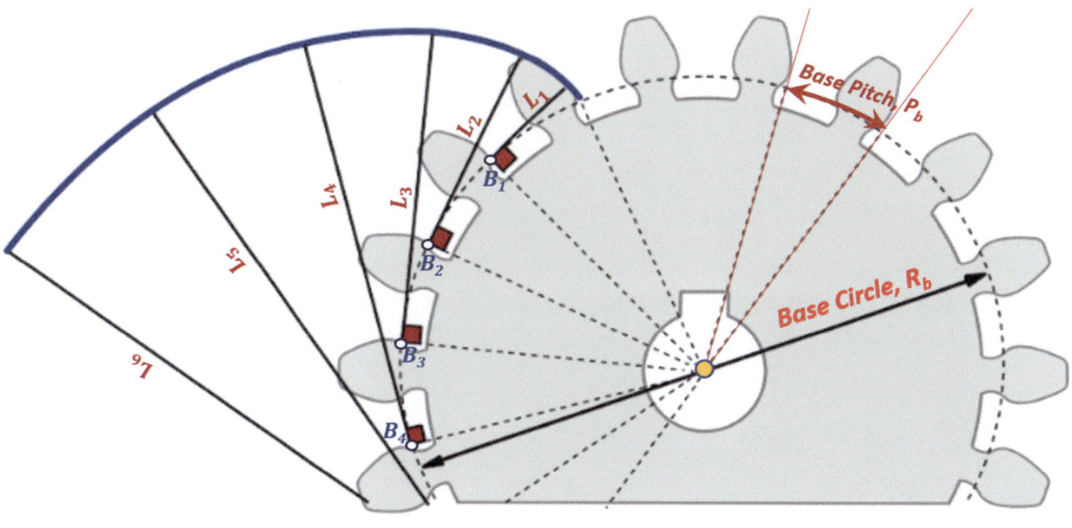

Fig. 10.14 Spur tooth involute circle

The inward section is defined by the trochoid curves, the path motion of the rack cutter forming the area under the involute curve to the dedendum circle. To generate the trochoid curves, we simply plot two radial lines from B_i and C_i that end at the dedendum circle with an undercut filet. This filet should be carefully designed to avoid the so-called interference phenomena.

The interference phenomenon happens when the tooth tip comes into contact with the trochoid curve of the other gear. Figure 10.15 shows a pinion driving a gear. The gear tooth experiences interference with the pinion's trichoid curve. The pinion's tooth does not experience any interference with the gear's trichoid curve.

There are three common ways to avoid this interference—increase the pressure angle, increase the radius of the undercutting filet, and/or decrease the path of the contact.

We could use Eq. (10.5) to show how the pressure angle affects the entire profile shape. In Fig. 10.16, we see that increasing the pitch angle increases the radial distance between the base circle and the pitch circle. In other words, increasing the pitch angle will decrease the base circle which leads to an increase in the involute curve with less possibility of interference.

The undercutting filet can also be designed to avoid interference. For example, in Fig. 10.17, we can see the possibility of interference when the filet radius is not big enough and how increasing the filet radius would avoid it. The only challenge in increasing the filet radius is that it leads to a weaker gear structure and could cause teeth failure.

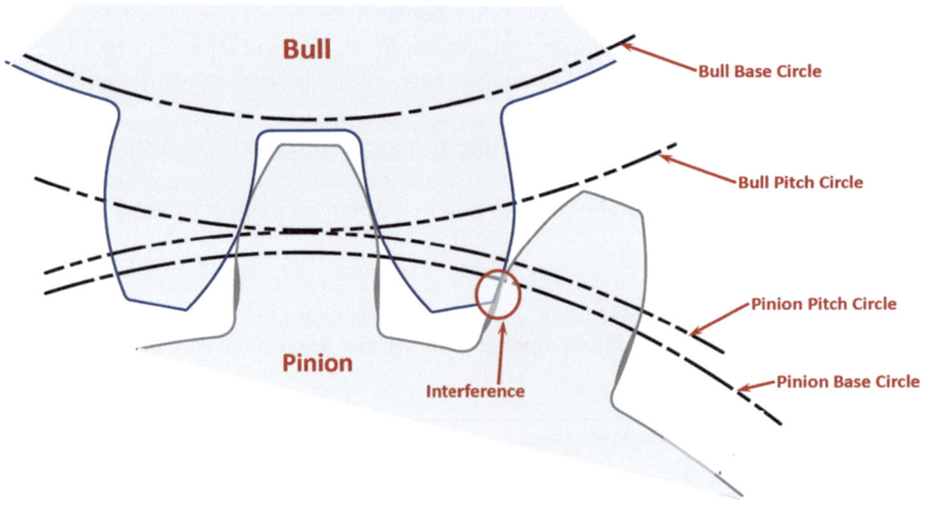

Fig. 10.15 Meshed gear interference

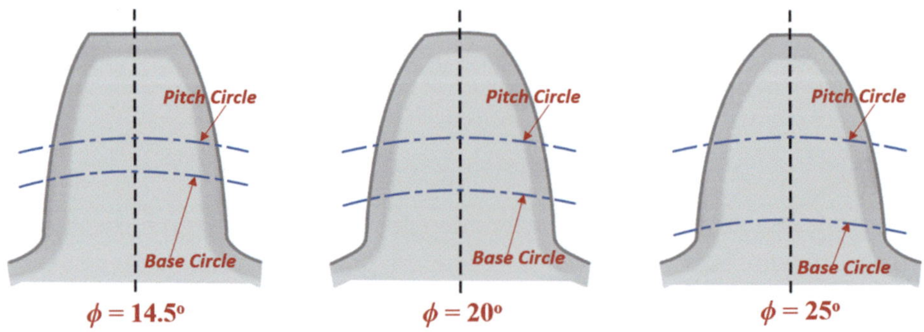

Fig. 10.16 Pressure angle influence on tooth profile

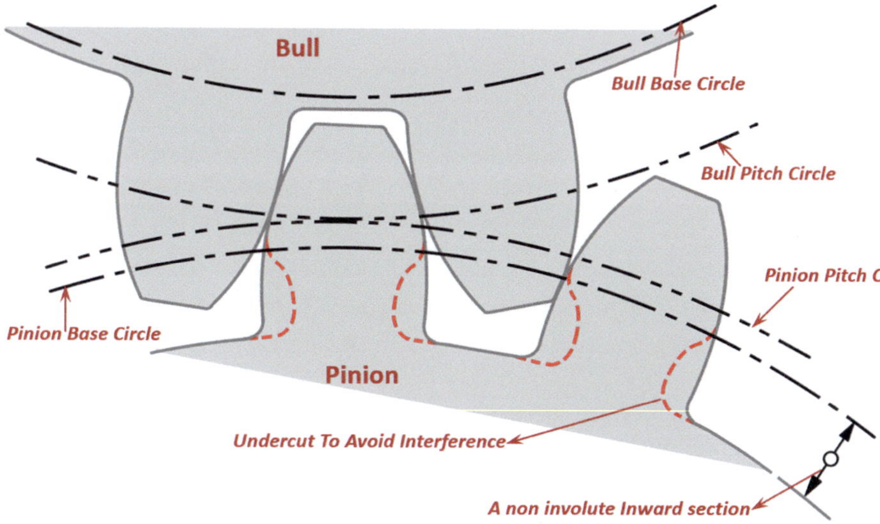

Fig. 10.17 Undercut to avoid interference

The gear ratio can also play an important role in avoiding interference. The higher the gear ratio, the more possibility of having interference. In other words, meshing a small gear with a much larger gear could cause interference. To avoid this, Ref. [3] develops a relationship of the number of teeth for the meshing gears. Let's assume two gears with N_1 and N_2 numbers of teeth. To avoid interference, the number of teeth N_2 should satisfy:

$$N_2 < \frac{N_1^2 \sin^2(\phi) - 4k^2}{4k - 2N_1 \sin^2(\phi)} \tag{10.7}$$

where k is related to the addendum a through

$$a = \frac{k}{P_d}$$

Example 10.3

For the pinion and bull in the water lift mechanism, as shown in Fig. 10.12, check if the selected pressure angle and number of teeth of each gear will cause interference or not. Assume the addendum $a = 1/P_d$.

Solution

We are using Eq. (10.7) and calculating the left-hand side (LHS).

$$k = aP_d = \left(\frac{1}{P_d}\right)P_d = 1$$

$$\text{LHS} = \frac{N_1^2 \sin^2(\phi) - 4k^2}{4k - 2N_1 \sin^2(\phi)} = \frac{15^2 \sin^2(20°) - 4(1^2)}{4(1) - 2(15)\sin^2(20°)} = 45.49$$

Since $N_2 = 30 < 45.49$, the selected pressure angle and the selected number of teeth are not going to cause any interference. However, the interference could happen from other sources.

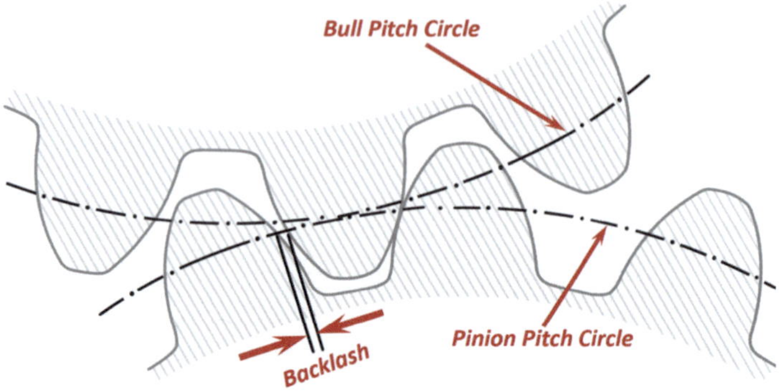

Fig. 10.18 Contact geometry of two meshed gears

The *backlash* is the difference between the tooth spacing of a gear and the tooth width of the mating gear, measured along the pitch circle arc of the first gear. Note, from Fig. 10.18, we could see another definition for the clearance, *c*, as the space between the top of a gear tooth and the bottom of the mating gear tooth. Many formulas and standards have proposed a range of the backlash as a function of the diametral pitch P_d as listed in [1].

10.3.3 Spur Gear Design

In the previous two subsections, we introduce many parameters to identify the shape of the gear tooth. Designing a gear could be done in many ways and by many standards. Both the AGMA and the American National Standards Institute (ANSI) offer standardized gears that can mesh with other gears because they have standardized diametral pitch and pressure angle. The AGMA publishes newly updated standards, certified by the ANSI, every year. Table 10.2, Ref. [4] provides the teeth specification based on AGMA requirements. Once both the pressure angle and diametral pitch are known, the formulas in Table can be used to calculate the tooth configuration parameters.

To know the pressure angle, the big standard gear manufacturers such as Boston Gear, Browning Gears, or Martin Sprocket & Gear are using a standard pressure angle at values 14.5°, 20°, or 25°. As design engineers, we should pick one of these pressure angles. Earlier, engineers used to use 14.5° as the pressure angle as a common value because it provides more power transmission and less pressure on the bearing. Later, from the strain analysis, it was discovered that gears with small pressure angles are more easily damaged. For this reason, the most common standard pressure angle is now 20°. A pressure angle of 25° could be used if necessary.

Table 10.2 AGMA full-depth gear tooth specifications

Tooth feature	Spur
Addendum, a	$1/P_d$
Dedendum, b	$0.002+1.2/P_d$
Working depth, h_k	$2/P_d$
Whole depth, h_t	$2.2/P_d+0.002$
Clearance (ground tooth), c	$0.2/P_d+0.002$
Maximum filet radius, r_{fmax}	$c/(1-\sin(\varphi))$
Circular tooth thickness, t	$\pi/(2P_d)$

Example 10.4

The pinion in Example 10.2 has been made as 15 teeth mid-steel cut full depth with a pressure angle of 20°. The average workload on the motor is 15 hp and it runs at 50 rpm that brings the diametral pitch to be 2 1/in. Use AGMA standard to determine:

1. The base circle diameter.
2. The thickness of the teeth.
3. The addendum and addendum radius.
4. The dedendum and dedendum radius.

Solution

To calculate the diameter of the pitch circle, *d*, we use Eq. (10.2).

$$\mathrm{d} = \frac{\mathrm{N}}{\mathrm{P_d}} = \frac{15\ \text{teeth}}{2\ \text{in}^{-1}} = 7.5\ \text{in}$$

To calculate the base circle diameter d_b, we use Eq. (10.5).

$$d_b = d\cos(\phi) = (7.5\ \text{in})\cos(20^o) = 7.05\ \text{in}$$

Using the formulas in Table 10.2, we calculate:

- the thickness of the teeth,

$$t = \frac{\pi}{2P_d} = \frac{1.571}{2\ \text{in}^{-1}} = 0.786\ \text{in}$$

- the addendum,

$$a = \frac{1}{P_d} = \frac{1}{2\ \text{in}^{-1}} = 0.5\ \text{in}$$

- and the dedendum,

$$b = \frac{1.2}{P_d} + 0.002 = \frac{1.2}{2\ \text{in}^{-1}} + 0.002 = 0.602\ \text{in}$$

For the geometry in Fig. 10.10, we can calculate the addendum radius.

$$r_a = \frac{d}{2} + a = \frac{7.5}{2} + 0.5 = 4.25\ \text{in}$$

and the dedendum radius

$$r_b = \frac{d}{2} - b = \frac{7.5}{2} - 0.602 = 3.148\ \text{in}$$

10.3.4 Spur Gear Kinematics

To get two gears to couple, pair, or roll over each other, “meshing” is the terminology that we use to describe these two gears. We may call them “meshing gears”, a “two-gear mesh”, or a “gear pair” as

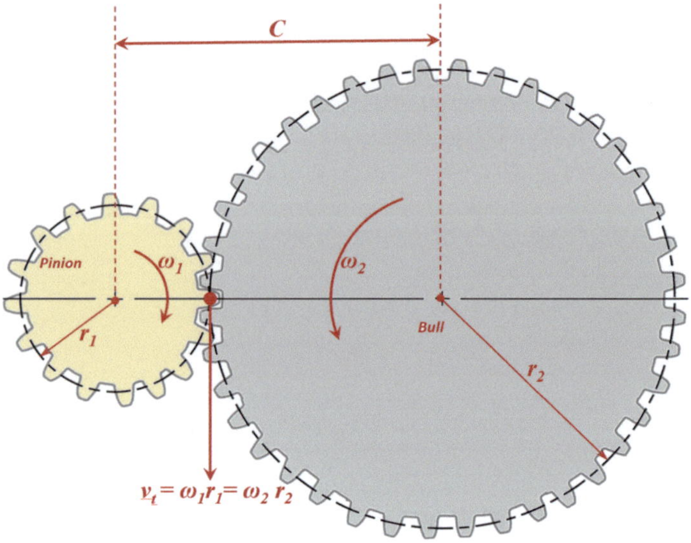

Fig. 10.19 Two meshed gears. Designed by the author and implemented by both the author and Pranav Dhola. Courtesy of Pranav Dhola

shown in Fig. 10.19. The smaller gear is called a ***pinion***. The bigger gear is called a ***bull***. From now on, we are going to call the two coupled gears "meshed gears".

The two gears are designed to provide a constant velocity ratio m_V, between their shafts. This gear ratio is defined as the driver gear angular velocity ω_1 divided by the driven angular velocity ω_2 as

$$m_V = -\frac{\omega_{\text{driven}}}{\omega_{\text{driver}}} = -\frac{\omega_2}{\omega_1} \tag{10.8}$$

Note the negative value coming from the fact that the two gears rotate against each other; see Fig. 10.19.

The tangential velocity at the pitch point, also known as the pitch line velocity, v_t, can be calculated by using the driver angular velocity ω_1 multiplied by the driver radius r_1 or it can be calculated by using the driven angular velocity ω_2 multiplied by the driven radius $r_{2.}$

$$v_t = \omega_1 r_1 = \omega_2 r_2 \tag{10.9}$$

or

$$\frac{\omega_2}{\omega_1} = \frac{r_1}{r_2}$$

Using both Eq. (10.8) and Eq. (10.9), we end up with

$$m_V = -\frac{\omega_2}{\omega_2} = -\frac{r_1}{r_2} = -\frac{d_1}{d_2} \tag{10.10}$$

For the two meshed gears, the diametral pitch P_d and the pitch angle φ must be the same which leads to

$$P_{d1} = \frac{N_1}{d_1} = P_{d2} = \frac{N_2}{d_2} \tag{10.11}$$

or

$$\frac{d_1}{d_2} = \frac{N_1}{N_2}$$

Therefore

$$m_V = -\frac{\omega_2}{\omega_1} = -\frac{r_1}{r_2} = -\frac{d_1}{d_2} = -\frac{N_1}{N_2}$$

For the spur internal gear type, the equations are the same except we use a "+" sign instead of "-" since the two gears rotate in the same direction. In other words, for the internal gear type

$$m_V = +\frac{\omega_2}{\omega_1} = +\frac{r_1}{r_2} = +\frac{d_1}{d_2} = +\frac{N_1}{N_2} \tag{10.12}$$

From both Eq. (10.11) and Eq. (10.12), we conclude that the smaller gear (pinion) runs with a higher angular velocity than the bigger gear (bull).

The two meshed gears can be described by three ratios: velocity ratio m_V, torque ratio m_T, or gear ratio m_G. The velocity ratio m_V is equal to the driven angular velocity divided by the driver angular velocity as described in Eq. (10.11) and Eq. (10.12) for both spur external gear type and spur internal gear type. The torque ratio m_T is the driven torque T_2 divided by the driver torque T_1.

$$m_T = \mp\frac{T_2}{T_1} \tag{10.13}$$

In Eq. (10.13), the negative sign is for the external spur gear type, and the positive sign is for the internal spur gear type. The definition of the driver power P_1 and the driven power P_2 is

$$P_1 = T_1\omega_1 \text{ and } P_2 = T_2\omega_2 \tag{10.14}$$

If we assume there are no power losses between the two gears' connections, the two powers should be the same. In other words, the power will be transferred from the driver to the driven with no losses.

$$P_1 = T_1\omega_1 = P_2 = T_2\omega_2 \tag{10.15}$$

Or

$$\frac{T_1}{T_2} = \frac{\omega_2}{\omega_1}$$

Using Eq. (10.13) and Eq. (10.15), the torque ratio is

$$m_T = \mp\frac{T_2}{T_1} = \mp\frac{\omega_1}{\omega_2} = \mp\frac{r_2}{r_1} = \mp\frac{d_2}{d_1} = \mp\frac{N_2}{N_1} = \frac{1}{m_V} \tag{10.16}$$

Both the velocity ratio and torque ratio could be more or less than one. For example, if the driver has a smaller number of teeth than the driven, the speed ratio would be less than 1 and the torque ratio will be the reciprocal of the speed ratio (more than 1) and the other way around. The gear ratio m_G is defined as always being more than 1. In other words, the gear ratio m_G can be expressed in terms of either the velocity ratio or the gear ratio.

$$m_G = \begin{cases} |m_T| & |m_T| > |m_V| \\ |m_V| & |m_T| < |m_V| \\ 1 & |m_T| = |m_V| \end{cases} \tag{10.17}$$

Example 10.5

Figure 10.12 shows a water lift mechanism that consists of a motor that rotates a pinion with a pitch diameter of 75 mm with 15 teeth and a bull with a pitch diameter of 300 mm. The pinion and bull are meshed. The motor rotates at 130 rpm. Determine:

1. The mechanism's velocity ratio.
2. The mechanism's torque ratio.
3. The mechanism's gear ratio.
4. The mechanism's circular pitch.
5. The bull's number of teeth.
6. The bull gear's angular velocity.
7. The translational velocity that the container is lifted by if the bull radius is 100 mm.

Solution

To calculate the velocity ratio m_V, we use Eq. (10.11).

$$\mathrm{m_V} = -\frac{\mathrm{d_1}}{\mathrm{d_2}} = -\frac{75\ \mathrm{mm}}{300\ \mathrm{mm}} = -\frac{1}{4} = -0.25$$

To calculate the torque ratio m_T, we use Eq. (10.16).

$$m_T = -\frac{d_2}{d_1} = -\frac{300\ \mathrm{mm}}{75\ \mathrm{mm}} = -4$$

To calculate the gear ratio m_G, we use Eq. (10.17).

$$m_G = |m_T| = |-4| = 4$$

To calculate the circular pitch P_c, we use Eq. (10.1).

$$P_c = \frac{\pi d_1}{N_1} = \frac{\pi(75\ \mathrm{mm})}{15} = 15.708\ \mathrm{mm}$$

To calculate the bull's number of teeth N_2, we use Eq. (10.11).

$$\frac{d_1}{d_2} = \frac{N_1}{N_2}$$

$$\text{or} \quad N_2 = \frac{d_2}{d_1}N_1 = \left(\frac{300\ \mathrm{mm}}{75\ \mathrm{mm}}\right)(15) = 60\ \text{teeth}$$

To calculate the bull's angular velocity ω_2, we use Eq. (10.11).

$$\frac{\omega_2}{\omega_1} = -\frac{d_1}{d_2}$$

$$\text{or} \quad \omega_2 = -\frac{d_1}{d_2}\omega_1 = \left(\frac{75\ \mathrm{mm}}{300\ \mathrm{mm}}\right)(130\mathrm{rpm}) = 32.5\mathrm{rpm}$$

The container's translational velocity is the same as the bull's tangential velocity v_t which can be calculated by using Eq. (10.9) using the bull's radius r_3 and the angular speed of the bull which is the same as the bull gear, ω_2.

$$v_t = \omega_2 r_3 = (32.5\ \text{rpm})\left(\frac{\pi}{30}(\text{rad/s})/(\text{rpm})\right)(100\ \text{mm})\left(\frac{1}{1000}\ \text{m/mm}\right) = 0.34\ \text{m/s}$$

Note, that π/30 is to convert the rpm into rad/s. The bull's radius, r_3, has a conversion of 1/100 from mm into m.

Simscape™ Multibody™ offers a sublibrary called "Gears and Couplings" that has four standard blocks as listed in Table 10.3 which includes the name of each block, the shape of each block, and the editing menu for each block. There are two connections for each block: Base and Follower. Each connection should be plugged into a shaft. For example, if we have two shafts that rotate at different speeds, we can use one of the "Gears and Couplings" blocks to connect them. We connect the first shaft to the Base connection, and we connect the second shaft to the Follower connection. The speed difference between the two shafts is the job of the "Gears and Couplings" block.

Definition[1]

Simscape™ Multibody™ Common Gear Constraint:

Represents a Common Gear Constraint between two frames. The block constrains the base and follower frames to corotate as meshed gears. The base and follower gears rotate about the base and follower z-axes, respectively.

Either the center distance and gear ratio (follower teeth/base teeth) or the pitch circle radii of the base and follower gears can be specified to define the geometry of the gears. During simulation, the gears must be held in alignment by the rest of the mechanism.

Ports B and F are frame ports that represent the base and follower frames, respectively [5–8].

In this section, we are going to focus on the Common Gear Constraint block to model the external spur gears connection or internal spur gears connection. The editing menu of the Common Gear Constraint asks for "Type" to be an external type or internal type. Also, the editing menu asks for "Specification Method" which could be "Center Distance and Ratio" or "Pitch Circle Radii".

Now, we are going to present the steps for how to build a model with two spur gears. Figure 10.20 shows a model that has two shafts. The centers of rotation of the two shafts are separated from each other by a distance C that is equal to the sum of the radii of the two spur gears r_1 and r_2.

The base shaft has a base spur gear. The follower shaft has a follower spur gear. Both base and follower gears are connected by a spur gear constraint. To build this model in Simscape™ Multibody™, we take the following steps:

[1] Reproduced from [5–8] by MathWorks with permission from MathWorks.

Table 10.3 Simscape™ Multibody™ gears and couplings blocks. Figures in the second and third columns are designed and implemented by the author. Images used with permission of MathWorks

<table>
<tr><th>Block Name</th><th>Block Shape</th><th>Editing Menu</th></tr>
<tr><td>Common Gear Constraint</td><td>
</td><td>External Type
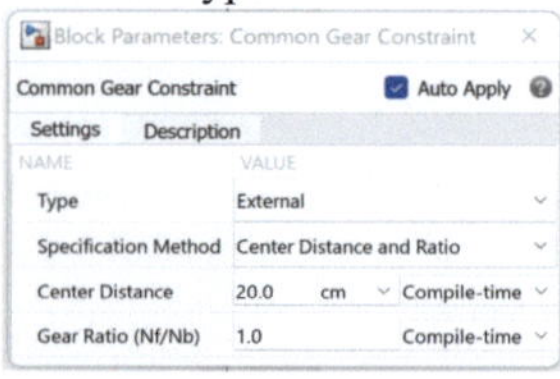

Internal Type
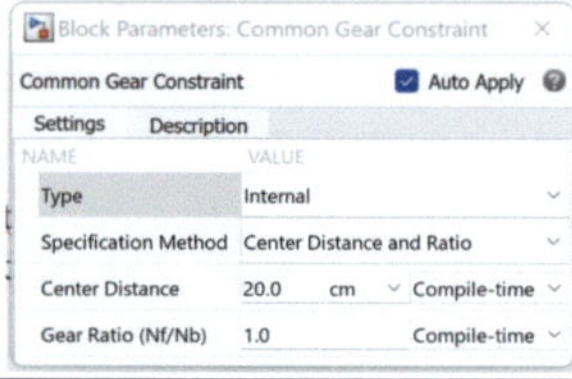
</td></tr>
<tr><td>Bevel Gear Constraint</td><td>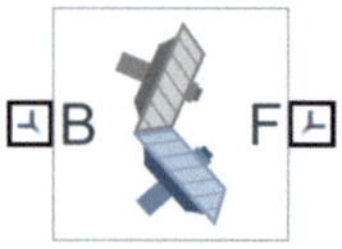
</td><td>Perpendicular Shafts
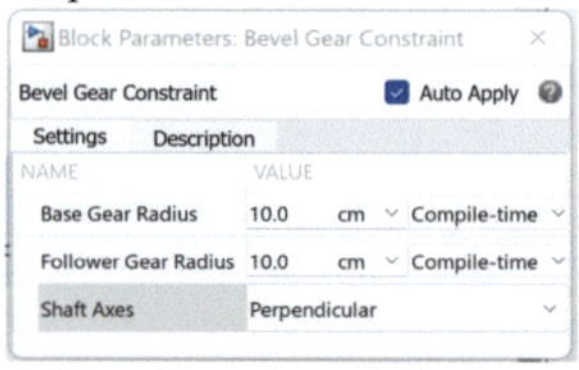

Arbitrarily Oriented Shafts

</td></tr>
<tr><td>Worm and Gear Constraint</td><td>
</td><td>Right-Hand Worm Direction

</td></tr>
<tr><td>Rack and Pinion Constraint</td><td>
</td><td>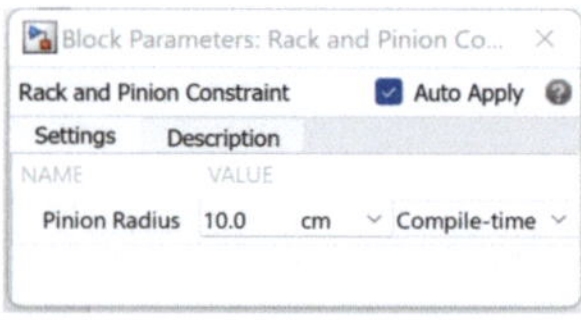
</td></tr>
</table>

Step 1: Build a ground link that has two nodes far from each other by a distance of C. The first node is the base, and the second node is the follower.

Step 2: Connect two revolute joint blocks with the follower node and the base node separately. The ground should be the base for any of the two revolute joint blocks.

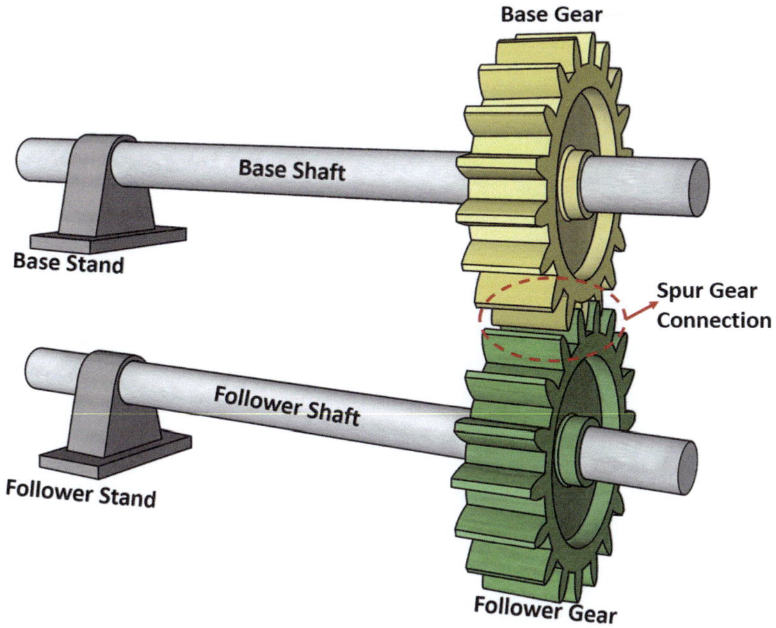

Fig. 10.20 Base–follower gears connection. Designed by the author and implemented by both the author and Pranav Dhola. Courtesy of Pranav Dhola

Step 3: Connect the follower connections of the two revolute joint blocks to two cylindrical shafts: follower shaft and base shaft.

Step 4: Connect the follower gear with the follower shaft and connect the base gear with the base shaft.

Step 5: Connect the two follower gears and the follower shaft with a “gear connection” block.

Example 10.6

Build a Simscape™ Multibody™ model to simulate Example 10.5. Measure the rotational velocity of the bull gear and compare it with the results from the solution of Example 10.5. For the model, use 200 mm as the length of the two shafts from the hubs to the gears. Both the radii of the two shafts are 10 mm. Both gears have a face width of 10 mm.

Solution

First, we make the gear train schematic diagram as shown in Fig. 10.21. The *xyz* coordinates are located at C_1; whereby the motor is attached to the pinion shaft. The pinion has a center C_4. The bull shaft has a center, C_2, at the hub. The bull has a center C_3. All these nodes are defined in *xyz* frame as

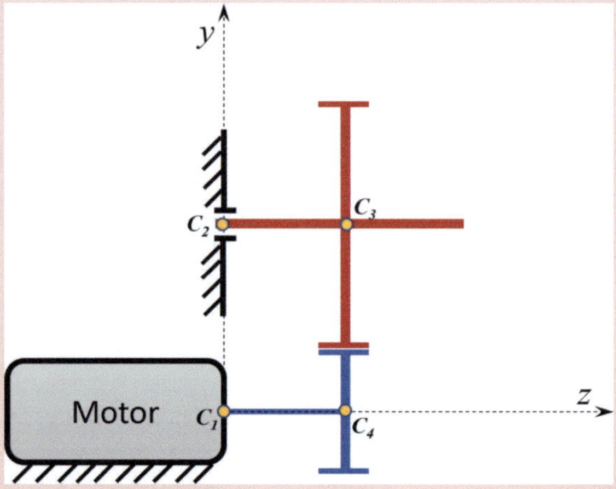

Fig. 10.21 Mechanism schematic diagram; Example 10.6

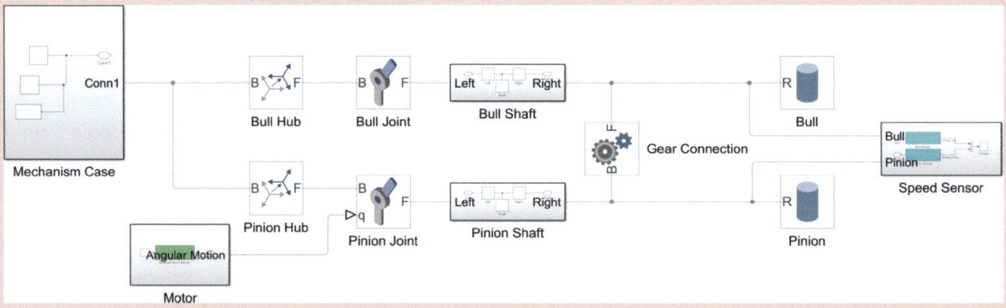

Fig. 10.22 Water lift mechanism Simscape™ Multibody™; Example 10.6

$$C_1 = \begin{bmatrix} 0 & 0 & 0 \end{bmatrix}$$

$$C_2 = \begin{bmatrix} 0 & (d_1 + d_2)/2 & 0 \end{bmatrix} = \begin{bmatrix} 0 & (75\text{ mm} + 300\text{ mm})/2 & 0 \end{bmatrix} = \begin{bmatrix} 0 & 187.5 & 0 \end{bmatrix}\text{ mm}$$

$$C_3 = \begin{bmatrix} 0 & (d_1 + d_2)/2 & L_2 \end{bmatrix} = \begin{bmatrix} 0 & (75\text{ mm} + 300\text{ mm})/2 & 1000 \end{bmatrix} = \begin{bmatrix} 0 & 187.5 & 1000 \end{bmatrix}\text{ mm}$$

$$C_4 = \begin{bmatrix} 0 & 0 & L_1 \end{bmatrix} = \begin{bmatrix} 0 & 0 & 1000 \end{bmatrix}\text{ mm}$$

To build the Simulink® model, we follow the steps below:

Step 1: Open a new model and save it as "Example_10p2_model". Build the model as shown in Fig. 10.22 with the listed blocks and in Table 10.4. All the blocks are from the Simulink® Library Browser and our library file; "Kinematics_Of_Machines".

Step 2: Run the model for 10 s to get the results from the scope as shown in Fig. 10.23. During the simulation, there is an animation shown in the Mechanics Explorer Window. Figure 10.24 shows a screenshot from the simulation animation. There is a copy of this model for the reader to explore included with this textbook.

Table 10.4 Example 10.6 blocks and subsystems. Designed and implemented by the author. Images used with permission of MathWorks

Block or subsystem	Name	Plugged in parameters
Ground subsystem includes • Word frame block • Mechanism configuration block • Solver configuration	Mechanism case	Not available
Rigid transformation	Pinion hub	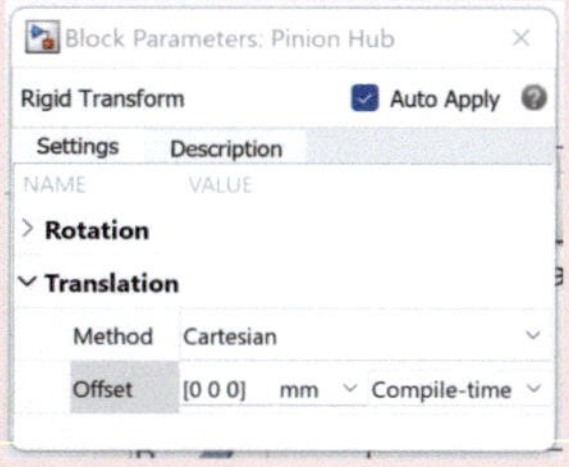
Rigid transformation	Bull hub	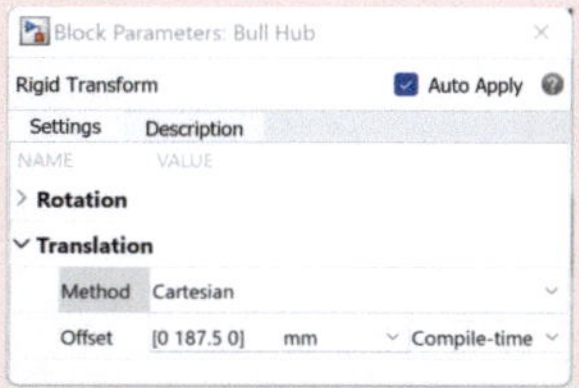
Revolute joint	Pinion revolute	Select the "Motion" to be "Provided by Input" and the "Torque" to be "Automatically Computed" in the actuation properties from the revolute joint editing menu
Revolute joint	Bull revolute	Not available
Two shaft subsystems	Pinion shaft and Bull shaft	With length of 200 mm and radius of 10 mm
Cylindrical solid	Pinion	Change properties/geometry as listed below 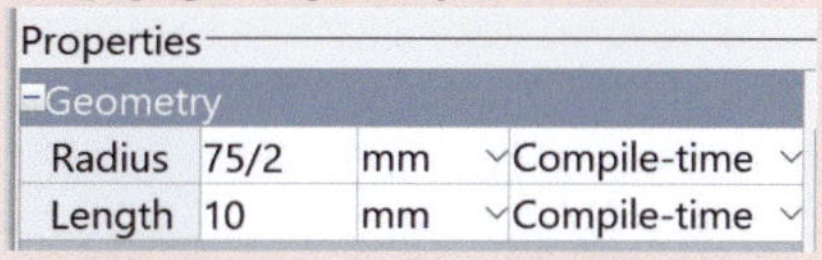
Cylindrical solid	Bull	Change properties/geometry as listed below 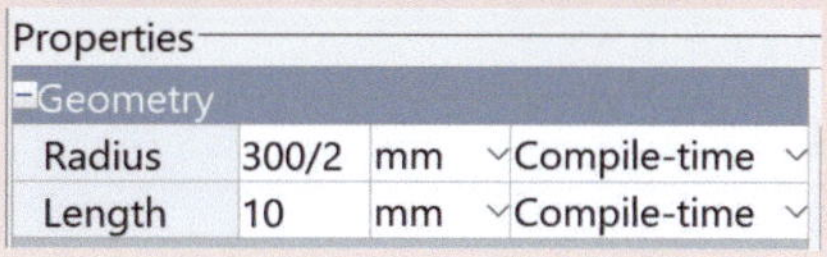
Common gear constraint to connect gear 1 as a base and gear 2 as a follower	Gear connection	
Measuring subsystem that consists of two angular velocity sensor subsystems, Mux, and two Scope blocks	Sensors	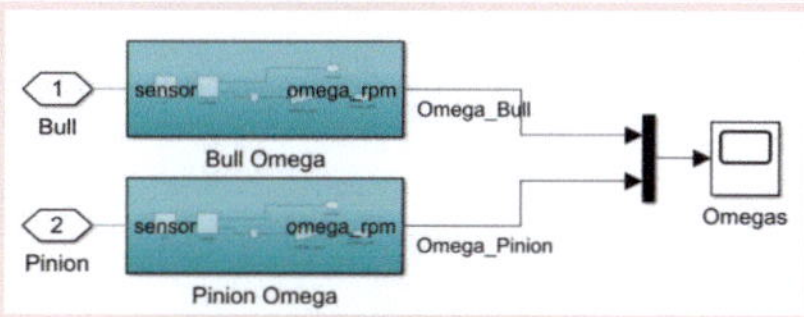
Actuator subsystem that consists of a rotational motion actuator subsystem and a ramp with 360 as a gain	Actuator	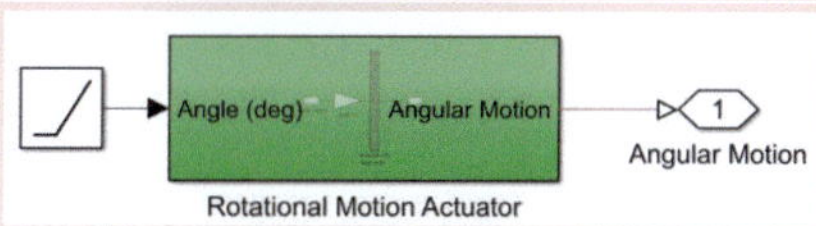

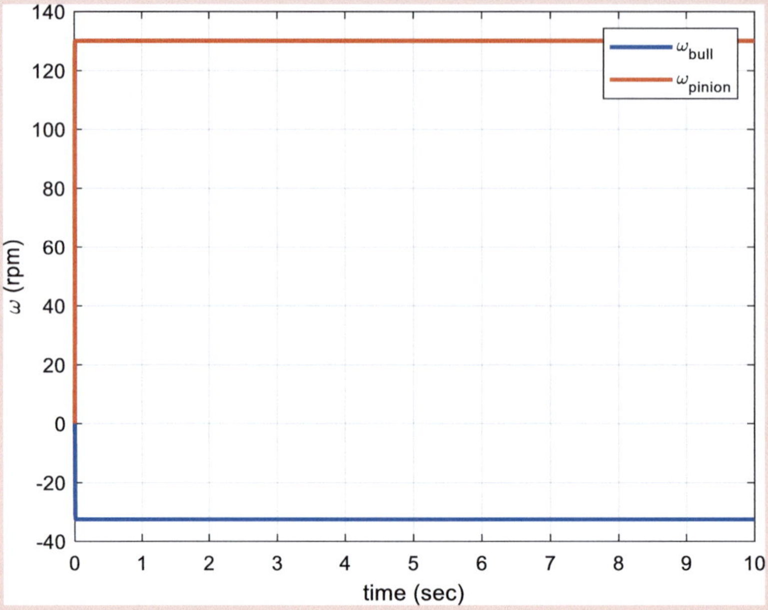

Fig. 10.23 Measured angular velocities; Example 10.6

Model Validation

As shown in Fig. 10.24, the measured bull's angular velocity is 13.5 rpm which is consistent with the answer in Example 10.5.

The hand-calculated results are consistent with what the simulation provided in Fig. 10.27.

Fig. 10.24 Lift water mechanism animation snapshot; Example 10.6

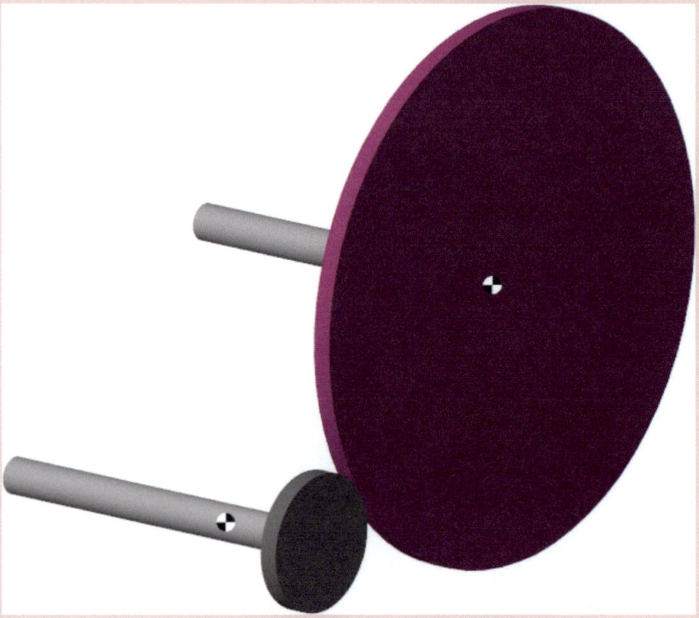

10.4 Simple Gear Trains

A gear train consists of two or more meshed gears. The gear train aims to transfer the power from the driving shaft to the driven shaft at a specific speed ratio. The easiest way to accomplish this is to select two gears with the appropriate speed ratio. At a very high gear ratio, there is a high possibility of gear tooth bending which may lead to fatigue or premature failure. For this reason, a gear train consisting of only two gears is not the best choice at high gear ratios.

A gear train is defined as "simple" when there is only one gear located on each shaft. In other words, the gear train happens in a serial action. For example, if we have M number of gears, each gear should have a separate shaft to be in a simple gear train. Figure 10.25 shows a simple gear train that consists of four gears with N_1, N_2, N_3, and N_4 as the number of teeth for each gear. The total velocity ratio of this simple four-gear train is

$$i_{4/1} = \left(i_{2/1}\right)\left(i_{3/2}\right)\left(i_{4/3}\right) = \left(-\frac{N_1}{N_2}\right)\left(-\frac{N_2}{N_3}\right)\left(-\frac{N_3}{N_4}\right) = -\frac{N_1}{N_4} \tag{10.18}$$

Note that the negative sign shows that the two gears rotate in opposite directions.

In general, the total velocity ratio of M number of gears arranged as a simple gear train is

$$i_{k/1} = (-1)^{k-1} \prod_{j=1}^{j=k-1} \left(\frac{N_{j+1}}{N_j}\right) \tag{10.19}$$

From Eq. (10.19), in a simple gear train, if the number of gears is even, the input and output gears rotate in opposite directions. If the number of gears is odd, the input and output gears rotate in the same direction.

Fig. 10.25 Four gears simple gear train. Figure and associated Simple 4-Gear Train.mp4 video designed by the author and implemented by both the author and Pranav Dhola. Both courtesy of Pranav Dhola

To implement the simple gear train in Simscape™ Multibody™, first, we start from the given parameters which are the circular pitch, $P_{c,}$ and a list of the number of gears in order from the input to the output with the number of teeth for each. After that, we follow the steps below.

1. Bring a ground block with a configuration block.
2. Adjust the position of each rigid transformation block to match the coordinates of the center of each shaft.
3. Connect each transformation block with the base of a revolute joint.
4. Connect each revolute joint block's follower with a cylindrical solid block and add the gear constraint block between the cylinders.

Example 10.7

A simple gear train consists of three gears with three different shafts. A motor is connected to the first shaft with a constant angular speed of 60 rpm. Table 10.5 lists the parameters of both gears and shafts. There are three sensors to measure the angular velocities of the three gears.

1. Implement this mechanism using Simscape™ Multibody™.
2. Measure the sensors' speeds and compare them with the hand calculation from Eq. (10.19).

Table 10.5 Mechanism parameters; Example 10.7

Element	Parameters
Gear 1	$r_1=1$ and $L_1=0.05$ m
Gear 2	$r_2=0.5$ and $L_2=0.05$ m
Gear 3	$r_3=0.25$ and $L_3=0.05$ m
Shaft 1	$r_{s1}=0.05$ and $L_{s1}=1$ m
Shaft 2	$r_{s2}=0.05$ and $L_{s2}=1$ m
Shaft 3	$r_{s3}=0.05$ and $L_{s3}=1$ m

Solution

The distance C_{12} between the center of Shaft 1 and the center of Shaft 2 is

$$C_{12} = r_1 + r_2 = 1\text{ m} + 0.5\text{ m} = 1.5\text{ m}$$

The distance C_{13} between the center of Shaft 1 and the center of Shaft 3 is

$$C_{13} = r_1 + 2r_2 + r_3 = 1\text{ m} + 2 \times 0.5\text{ m} + 0.25 = 2.25\text{ m}$$

The location of each shaft center is

$$\text{First Shaft Hub } C_1 = \begin{bmatrix} 0 & 0 & 0 \end{bmatrix}$$

$$\text{Second Shaft Hub } C_2 = \begin{bmatrix} 0 & C_{12} & 0 \end{bmatrix} = \begin{bmatrix} 0 & 1.5 & 0 \end{bmatrix}\text{m}$$

$$\text{Third Shaft Hub } C_3 = \begin{bmatrix} 0 & C_{13} & 0 \end{bmatrix} = \begin{bmatrix} 0 & 2.25 & 0 \end{bmatrix}\text{m}$$

To build the Simulink® model, we follow the steps below.

Step 1: Open a new model and save it as "Example_10p2_model". Build the model as shown in Fig. 10.26 with the listed blocks and in Table 10.6. All the blocks are from the Simulink® Library Browser and our library file, "Kinematics_Of_Machines".

Step 2: Run the model for 10 s to obtain the results from the scope as shown in Fig. 10.27 During the simulation, there is an animation shown in the Mechanics Explorer Window. Figure 10.28 shows a screenshot from the simulation animation. There is a copy of this model for the reader to explore included with this textbook.

Model Validation

To validate the model, we are going to use Eq. (10.9) to calculate the angular speed of gear 2 from gear 1 as.

$$\omega_2 = -\frac{r_1}{r_2}\omega_1 = -\frac{(1\ \mathrm{m})}{(0.5\ \mathrm{m})}(60\ \mathrm{rpm}) = -120\ \mathrm{rpm}$$

The angular velocity of gear 3 as a function of the aspect ratio of gear 2 is also calculated from Eq. (10.9)

$$\omega_3 = -\frac{r_2}{r_3}\omega_2 = -\frac{(0.5\ \mathrm{m})}{(0.25\ \mathrm{m})}(-120\mathrm{rpm}) = +240\ \mathrm{rpm}$$

The hand-calculated results are consistent with what the simulation provides in Fig. 10.27.

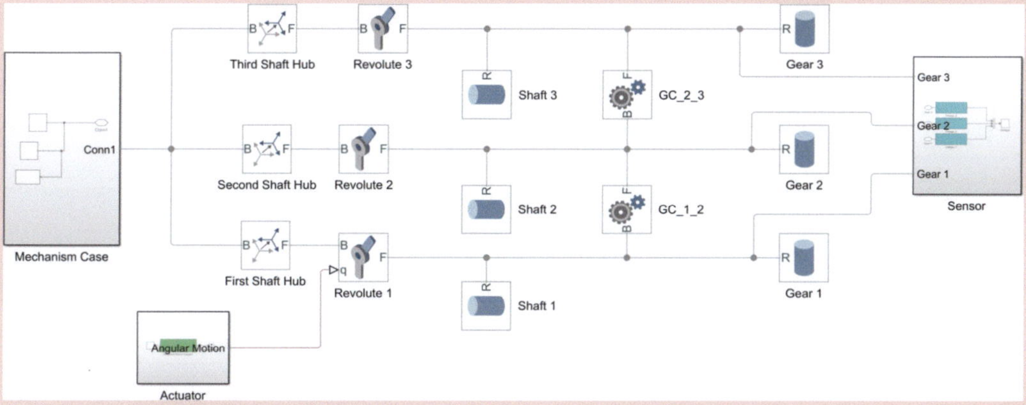

Fig. 10.26 Simple gear train Simscape™ Multibody™ model; Example 10.7

Table 10.6 Example 10.7 blocks and subsystems. Designed and implemented by the author. Images used with permission of MathWorks

Block or subsystem	Name	Plugged in parameters
Ground subsystem includes • Word frame block • Mechanism configuration block • Solver configuration	Mechanism case	Not available
Rigid transformation	First shaft hub	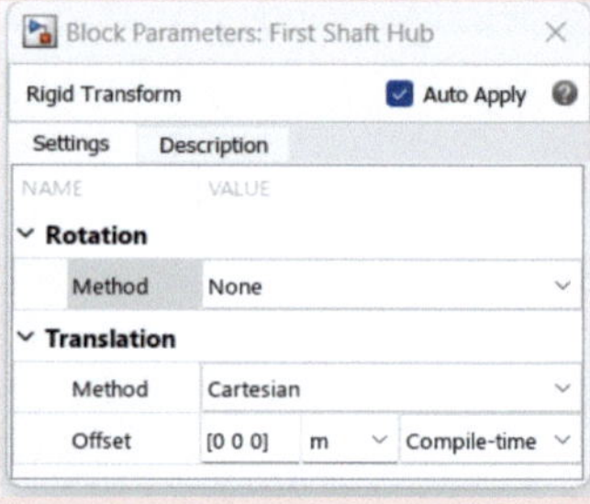
Rigid transformation	Second shaft hub	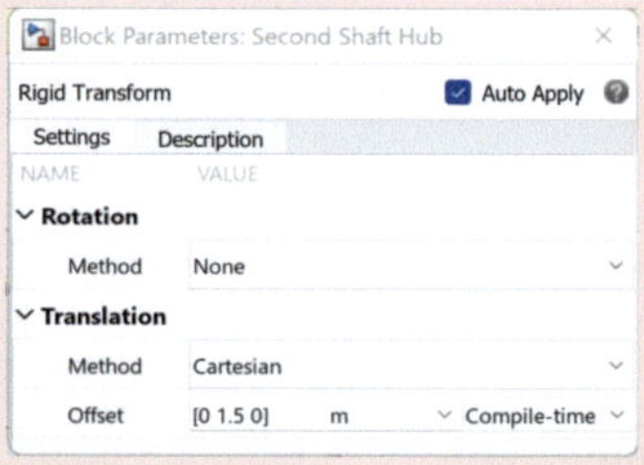
Rigid transformation	Third shaft hub	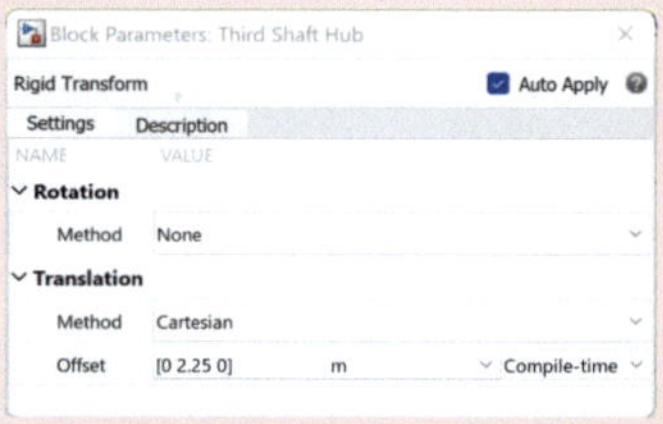
Revolute joint	Revolute 1	Select the "Motion" to be "Provided by Input" and the "Torque" to be "Automatically Computed" in the actuation properties from the revolute joint editing menu
Two revolute Joint	Revolute 2 and Revolute 3	N/A
Three cylindrical solid	Shaft 1, Shaft 2, and Shaft 3	Change properties/geometry as listed below 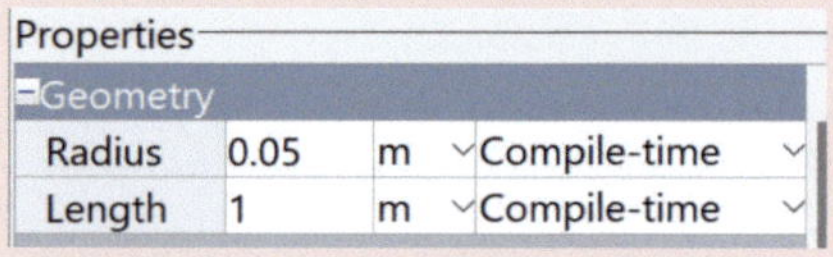
Cylindrical solid	Gear 1	Change properties/geometry as listed below 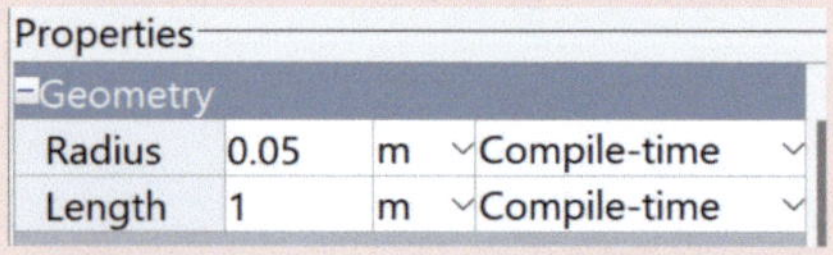
Cylindrical solid	Gear 2	Change properties/geometry as listed below 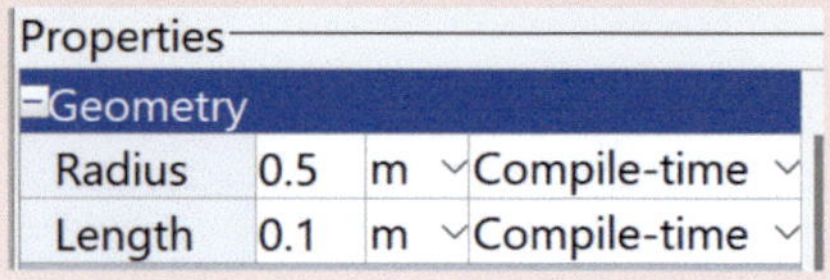

(continued)

Table 10.6 (continued)

Block or subsystem	Name	Plugged in parameters
Cylindrical solid	Gear 3	Change properties/geometry as listed below Properties Geometry Radius 0.25 m Compile-time Length 0.1 m Compile-time
Common gear constraint to connect gear 1 as a base and gear 2 as a follower	GC_1_2	Block Parameters: GC_1_2 Common Gear Constraint — Auto Apply Settings — Description NAME — VALUE Type: External Specification Method: Pitch Circle Radii Base Gear Radius: 1 m Compile-time Follower Gear Radius: 0.5 m Compile-time
Common gear constraint to connect gear 2 as a base and gear 3 as a follower	GC_2_3	Block Parameters: GC_2_3 Common Gear Constraint — Auto Apply Settings — Description NAME — VALUE Type: External Specification Method: Pitch Circle Radii Base Gear Radius: 0.5 m Compile-time Follower Gear Radius: 0.25 m Compile-time
Measuring subsystem that consists of three angular velocity sensor subsystems, Mux block, and three Scope blocks	Sensors	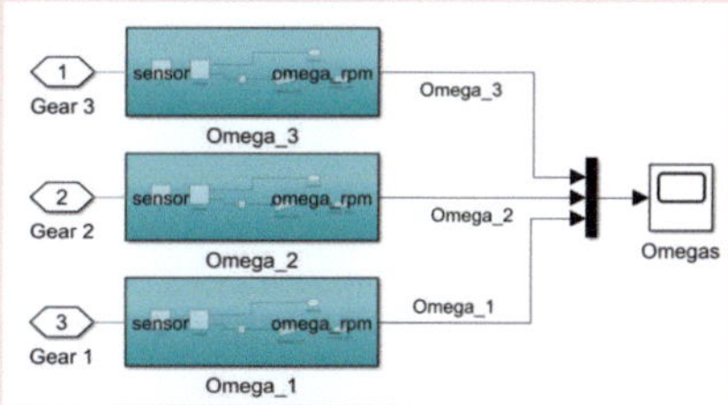
Actuator subsystem that consists of a rotational motion actuator subsystem and a ramp with 360° as a gain	Actuator	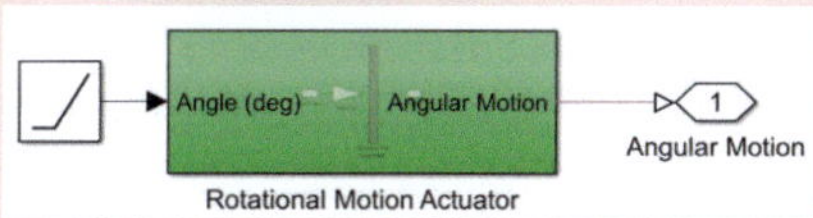

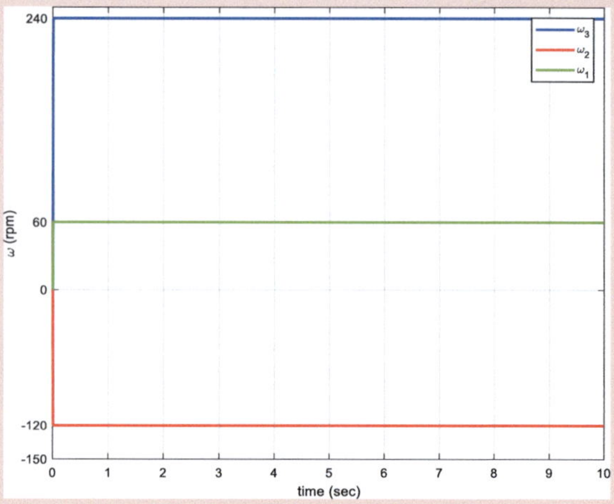

Fig. 10.27 Measured angular velocities; Example 10.7

Fig. 10.28 Simple gear train mechanism animation snapshot; Example 10.7

10.5 Compound Gear Trains

Designing a simple gear train causes design challenges such as assembling a different shaft for each gear and assembling these shafts. Also, the more gears, the more required space to assemble the entire gear train. All these simple gear train design issues make it a good choice for a total small gear ratio of less than 10.

For a higher gear ratio, a compound gear train is one of the best solutions. A compound gear train is defined as when at least one shaft carries more than one gear. There is also another type called a "reverted compound gear train" that is defined as being when the shafts are assembled in the reverted way. For example, we can convert the four-gear simple gear train in Fig. 10.25 into a compound train by having both of the two middle gears acting on the same shaft as shown in Fig. 10.29 where Fig. 10.29a shows the simple gear train, Fig. 10.29b shows the equivalent compound gear train, and Fig. 10.29c shows the equivalent reverted compound gear train. Note that the values of N_1, N_2, N_3, and N_4 of the simple gear train are not the same as in any of the compound gear train or the reverted gear train. We can still use the same rule of the spur gear kinematics to calculate the angular velocities across the compound gear train or the reverted compounded gear train.

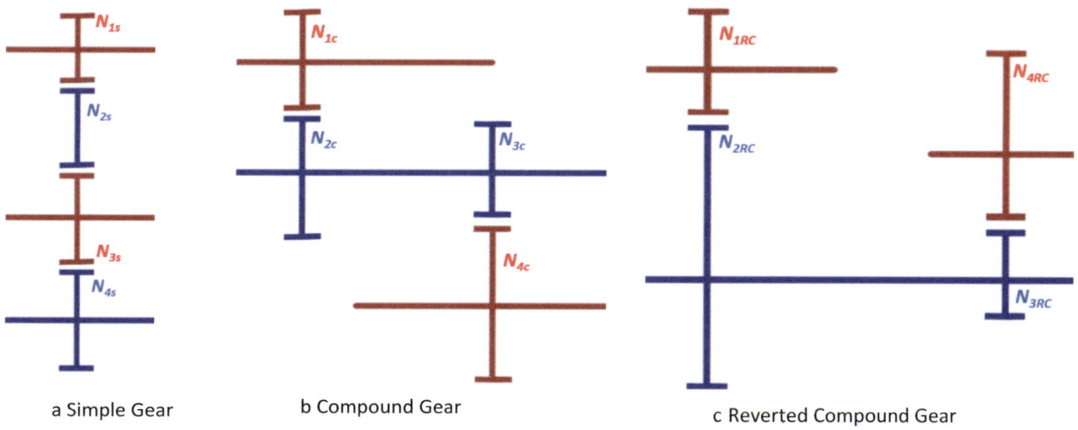

Fig. 10.29 Schematic diagrams for simple gear train compared with compound gear train and reverted compound gear train

10.6 Planetary Gear Trains

Planetary gear trains, also known as epicyclic gear trains, have links that hold the center of the gears. Planetary gear trains' links can rotate. Note that in both simple and compound gear trains, the shafts about which the gears rotate are in fixed locations. On the other hand, the shafts about which the gears of a planetary set rotate can themselves be rotating about a different axis.

The name "planetary" comes from the assembly of the train resembling a set of planets rotating around a sun. Following a sun-planet orbit, Fig. 10.30 shows a 3D view of the planetary gear train plus both the front and side views. The center gear is called the "sun". The gears that orbit around the sun are called "planets" that are held by the carrier. The ring gear meshes with the planets as an internal gear set. Planetary trains are common in automotive transmission because of the capability to alter the train's value by holding one of the gears or carriers using a brake or clutch mechanism.

The kinematics of a planetary train could be presented by superposition tabulated data using the steps below.

Step 1: Build up as shown in Table 10.7. Fill the gear up with gear ratios from the number of teeth or the radius of each gear. Fill the table up with the given arm speed and the other given gear speed (sun, ring, or planet).

Step 2: Use the row whereby the arm speed is known, and the gear speed is known to calculate the relative speed $\omega_{gear/arm}$.

Step 3: Use the gear ratios from the fifth column to calculate the relative gear speeds in the fourth column.

Step 4: Calculate the other unknown gear speeds from $\omega_{gear} = \omega_{arm} + \omega_{gear/arm}$.

Example 10.8

A planetary gear train with the carrier is driven at 1200 rpm while the sun is stationary. The number of teeth on each gear is $N_{sun} = 30$, $N_{planet} = 20$, $N_{ring} = 120$. Determine the rotational velocity of the planet gear and the ring gear using the superposition.

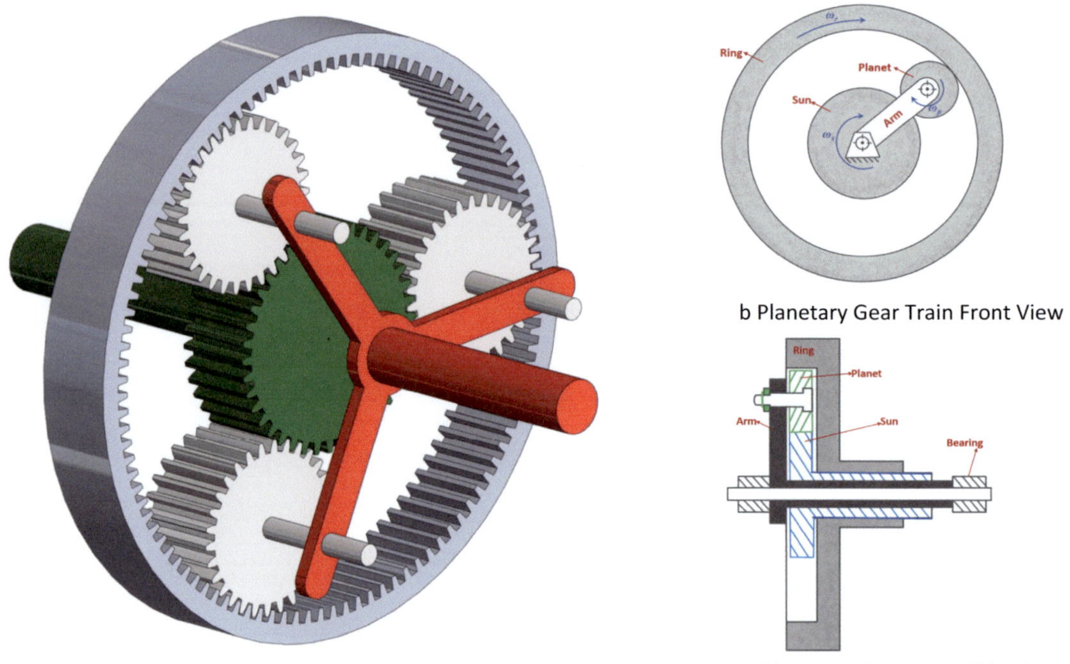

a Planetary Gear Train 3D View

b Planetary Gear Train Front View

c Planetary Gear Train Side View

Fig. 10.30 Planetary gear train. Figure and associated Planetary Gear.mp4 video designed by the author and implemented by both the author and Pranav Dhola. Both courtesy of Pranav Dhola

Table 10.7 Solution for superposition of planetary gear

Gear	$\omega_{gear}=$	$\omega_{arm}+$	$\omega_{gear/arm}$	Gear Ratio
Sun				
Planet				$\omega_{plant}/\omega_{sun}$
Ring				$\omega_{ring}/\omega_{planet}$

Solution

We follow the steps in the superposition. In Step 1, build up Table 10.8 with the givens, $\omega_{sun}=0$ rpm and $\omega_{arm}=1200$ rpm. The gear ratios are calculated from the number of teeth of each gear as listed in Table 10.8.

Table 10.8 Planetary speeds; Step 1 in Example 10.8

Gear	$\omega_{gear}=$	$\omega_{arm}+$ (rpm)	$\omega_{gear/arm}$	Gear ratio
Sun	0 rpm	1200		
Planet		1200		$\omega_{plant}/\omega_{sun} = -N_{sun}/N_{planet} = -30/20 = -1.5$
Ring		1200		$\omega_{ring}/\omega_{planet} = N_{planet}/N_{ring} = 20/120 = 0.1667$

In Step 2, fill the fourth cell in the second row with $\omega_{sun/arm} = \omega_{sun} - \omega_{arm} = 0–1200$ rpm = −1200 rpm as shown in Table 10.9.

In Step 3, use the gear ratio to calculate the relative angular velocities of the planet gear and the ring gear as listed in Table 10.10. For example, $\omega_{planet/arm} = \omega_{sun/arm} \mathrm{x} (\omega_{plant}/\omega_{sun}) = (-1200 \text{ rpm})\mathrm{x}(-1.5) = 1800$ rpm.

In Step 4, calculate the planet angular velocity and the ring angular velocity from $\omega_{gear} = \omega_{arm} + \omega_{gear/arm}$ as shown in Table 10.11.

Table 10.9 Planetary speeds; Step 2 in Example 10.8

Gear	$\omega_{gear} =$	$\omega_{arm}+$(rpm)	$\omega_{gear/arm}$	Gear ratio
Sun	0 rpm	1200	−1200 rpm	
Planet		1200		−1.5
Ring		1200		0.1667

Table 10.10 Planetary speeds; Step 3 in Example 10.8

Gear	$\omega_{gear} =$	$\omega_{arm}+$(rpm)	$\omega_{gear/arm}$ (rpm)	Gear ratio
Sun	0 rpm	1200	−1200	
Planet		1200	1800	−1.5
Ring		1200	300	0.1667

Table 10.11 Planetary speeds; Step 4 in Example 10.8

Gear	$\omega_{gear} =$(rpm)	$\omega_{arm}+$(rpm)	$\omega_{gear/arm}$ (rpm)	Gear ratio
Sun	0	1200	−1200	
Planet	3000	1200	1800	−1.5
Ring	1500	1200	300	0.1667

Note: These detailed steps with different tables are for explaining the steps. The student could make the algorithm on a single table

Another way to calculate the angular velocities of a planetary gear train is to develop a mathematical model from the relative motion. In the planetary gear train, there is the driver gear (sun, arm, or ring) and there is a driven gear (sun, arm, or ring) plus the carrier. The relative speed of both driver and driven relative to the carrier is

$$\begin{aligned} \omega_{driver}/\omega_{carrier} &= \omega_{driver} - \omega_{carrier} \\ \omega_{driven}/\omega_{carrier} &= \omega_{driven} - \omega_{carrier} \end{aligned} \tag{10.20}$$

Dividing the two relative velocities from Eq. (10.21) leads to

$$\frac{\omega_{driver}/\omega_{carrier}}{\omega_{driven}/\omega_{carrier}} = \frac{\omega_{driver} - \omega_{carrier}}{\omega_{driven} - \omega_{carrier}} = \pm\frac{N_{driven}}{N_{driver}} = \pm\frac{r_{driven}}{r_{driver}} \tag{10.21}$$

Equation (10.21) provides a solution to find the speed of a planetary gear train.

Example 10.9

A planetary gear train with the carrier is driven at 1200 rpm while the sun is stationary. The number of teeth of each gear is $N_{sun}=30$, $N_{planet}=20$, $N_{ring}=120$. Determine the rotational velocity of the planet gear and the ring gear using the mathematical planetary gear train model.

Solution

We start with the driver as the sun gear and the driven as the planet gear. Writing Eq. (10.21) as

$$\frac{\omega_{\text{sun}} - \omega_{\text{carrier}}}{\omega_{\text{planet}} - \omega_{\text{carrier}}} = -\frac{N_{\text{planet}}}{N_{\text{sun}}}$$

or

$$\frac{0\text{ rpm} - 1200\text{ rpm}}{\omega_{\text{planet}} - 1200\text{ rpm}} = -\frac{20}{30}$$

$$\omega_{\text{planet}} = -\frac{30}{20}(0\text{ rpm} - 1200\text{ rpm}) + 1200\text{ rpm} = 3000\text{ rpm}$$

Now, we are going to apply the rule to the sun and the ring. Writing Eq. (10.21) as

$$\frac{\omega_{\text{sun}} - \omega_{\text{carrier}}}{\omega_{\text{ring}} - \omega_{\text{carrier}}} = -\frac{N_{\text{ring}}}{N_{\text{sun}}}$$

or

$$\frac{0\text{ rpm} - 1200\text{ rpm}}{\omega_{\text{ring}} - 1200\text{ rpm}} = -\frac{120}{30}$$

$$\omega_{ring} = -\frac{30}{120}(0\text{rpm} - 1200\text{ rpm}) + 1200\text{ rpm} = 1500\text{ rpm}$$

Simscape™ Multibody™ does not provide a single block to simulate the planetary gear train. Instead, a planetary gear train could be built as an assembly of the sun, planets, ring, and carrier. MathWorks provides this planetary gear train model assembly in "`smdoc_planetary_gear_e.slx`" as shown in Fig. 10.31. The model has four planets–gear, sun gear, carrier, and ring gear. The model requires input speeds for the ring gear and sun gear. The reader can explore more about this model from MathWorks.

There is another toolbox, called Simscape™ Driveline™, that provides component libraries for modeling and simulating rotational and translational mechanical systems. It includes models of worm gears, lead screws, and vehicle components such as engines, tires, transmissions, and torque converters. Simscape™ Driveline™ is another tool to model gears more like Simscape™ Multibody™. Simscape™ Driveline™ models the rotational motion only as 1 DoF whereas Simscape™ Multibody™ models the 3D motion with 6 DoF. Also, Simscape™ Driveline™ provides a model for clutches and synchronizers that are needed in modeling transmissions and gearboxes.

Besides the planetary train in Fig. 10.30, various other planetary trains have been proposed over the years, especially for the automotive and transmission industry. For example, Levali (Ref. 9) proposes 12 different types of planetary trains as shown in Fig. 10.32. Sometimes, design engineers come up with new planetary trains by combining the two regular planetary trains from Figure 10.32 to make a single planetary train.

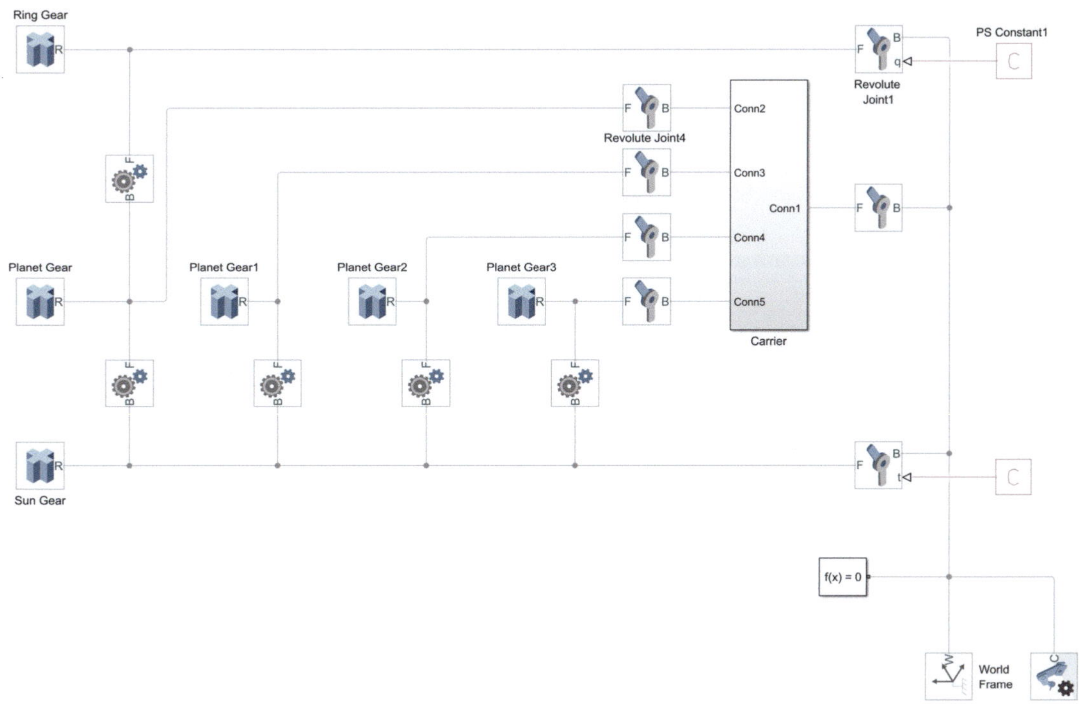

Fig. 10.31 Simscape™ Multibody™ Planetary Model, "smdoc_planetary_gear_e.slx"

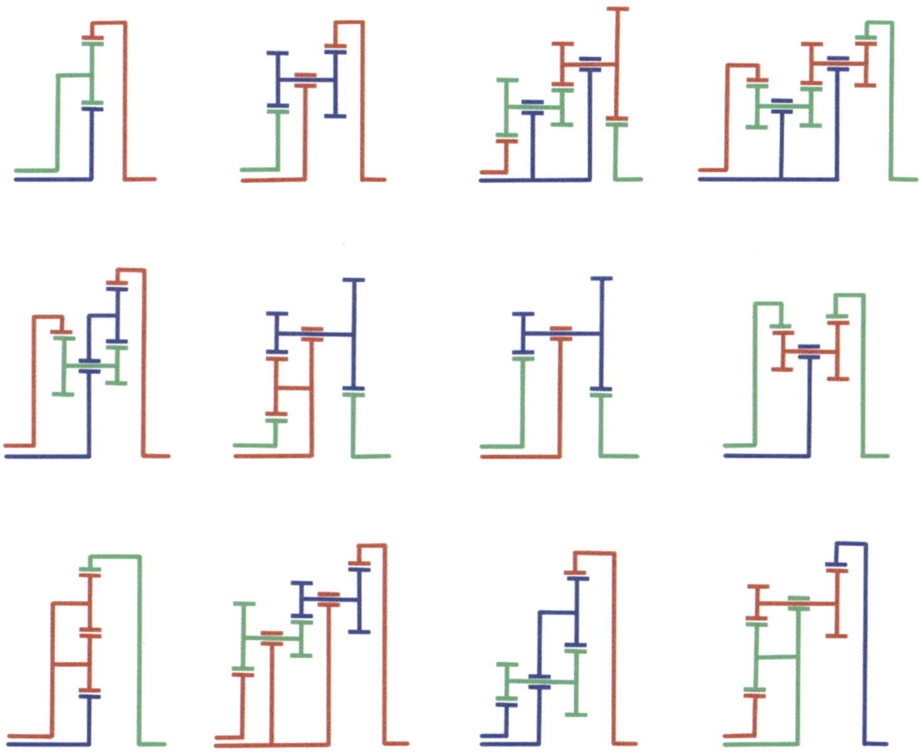

Fig. 10.32 Levali's 12 proposed planetary trains

10.7 Differential

Differential gear arrangement is a mechanism that allows a difference in velocity between two elements. For example, the car's differential allows the inner and outer vehicle wheels to travel different distances (different in velocity) in a turn as explained later in detail in this section. As shown in Fig. 10.33, a differential assembly is a gear train assembly that consists of

- Differential side gear or sun gears.
- Pinion shaft or cross pin.
- Axle shafts or half shafts.
- Ring gear or crown wheel.
- Drive pinion or bevel pinion.
- Differential pinions, planet gears, or spider gears.
- Differential case or housing.

Note that

Figure 10.33 shows a difference assembly which has the differential gear arrangement plus ring/pinion. In other words, Fig. 10.33 is not just the differential gear arrangement, but also ring/pinion (crown/pinion) gears. The differential gear arrangement is a self-contained assembly which is independent of the ring/pinion gears.

Figure 10.34 shows the differential assembly in a car. The ring/pinion gears, as a part of the differential assembly, work as gear reduction (around 3–4 as a gear ratio) in the vehicle by slowing the rotational speed of the transmission just before moving to the wheels. Also, the ring/pinion gears turn the rotational power flow by 90°. Thus, in the car, the power source is the engine, and the power sinks are the wheels with the transmission in between. The wheels rotate 90° from the engine assembly with the transmission. The bevel gear connection between the pinion gear and ring gear is doing this job by turning the rotational power by 90°.

The engine rotates the transmission that rotates the drive shaft. The drive pinion, located at the end of the drive shaft, rotates the ring gear that rotates the two differential side gears through the differential pinion. The left and right wheels rotate at the same speed when the car is driven on a straight smooth road with equal resistance on both wheels and the two differential side gears are rotating at the same speed. In other words, there is no slipping between the left and right wheels or difference in the speed between the left and right differential side gears.

If the car is driven over a left-curved road, the left wheel gear should rotate more slowly than the right. In other words, the left differential side gear should rotate slower than the right differential side gear. The differential pinion rotates as a spin that allows the left differential gear to slow down from the right differential side. With no differential allowance for the two wheels to rotate at different speeds, there would be undue wear on the tires.

Simscape™ Multibody™ does not provide a single block to simulate the differential assembly. Similar to the planetary gear train, the differential could be built as an assembly of the drive pinion, ring, spider gears, and differential pinion gears. If the reader is interested in developing gearbox and transmission models, it is recommended to read the Simscape™ Driveline™ manual. Many companies are using the Simscape™ Driveline™ as a modeling tool, especially for developing the transmission controller.

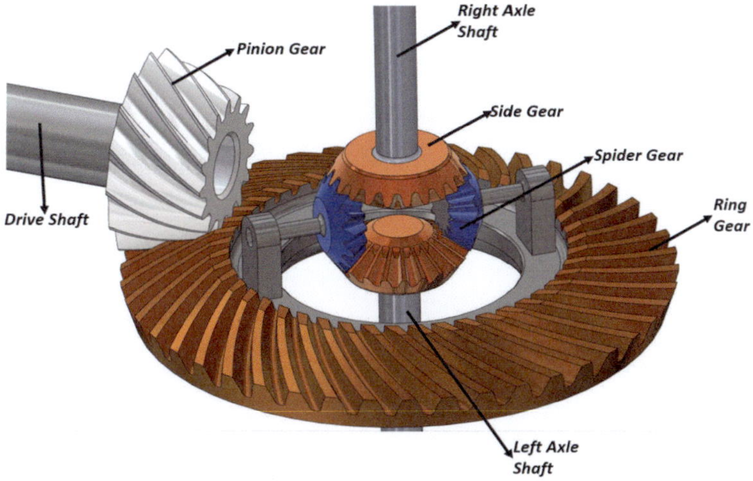

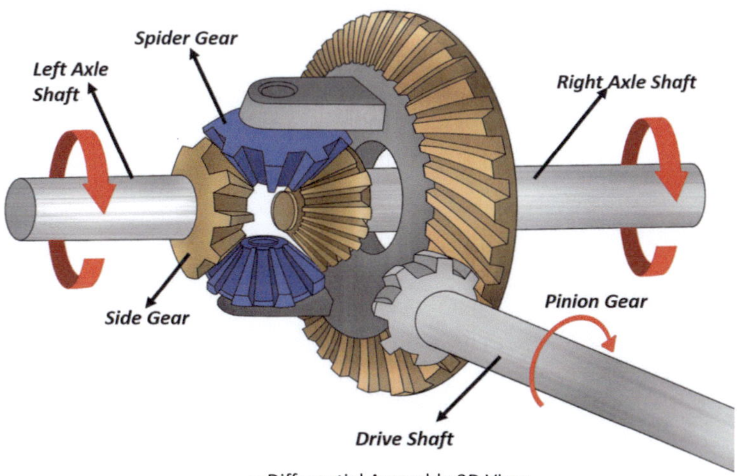

a Differential Assembly 3D View

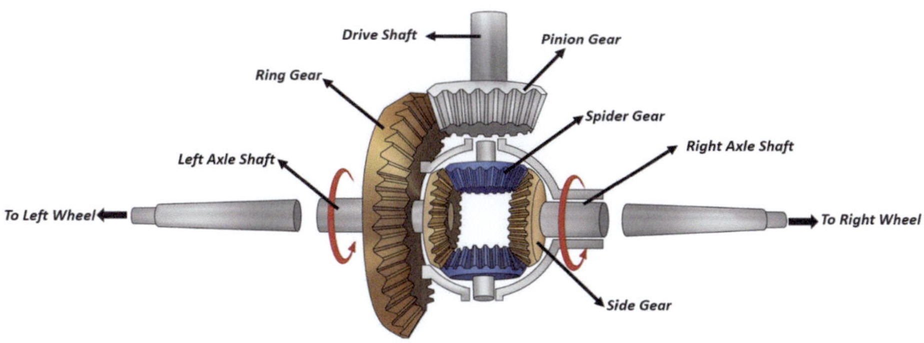

b Differential Assembly 2D View

Fig. 10.33 Differential assembly elements. Figure and associated Differential Gear Assembly.mp4 video designed by the author and implemented by both the author and Pranav Dhola. Both courtesy of Pranav Dhola

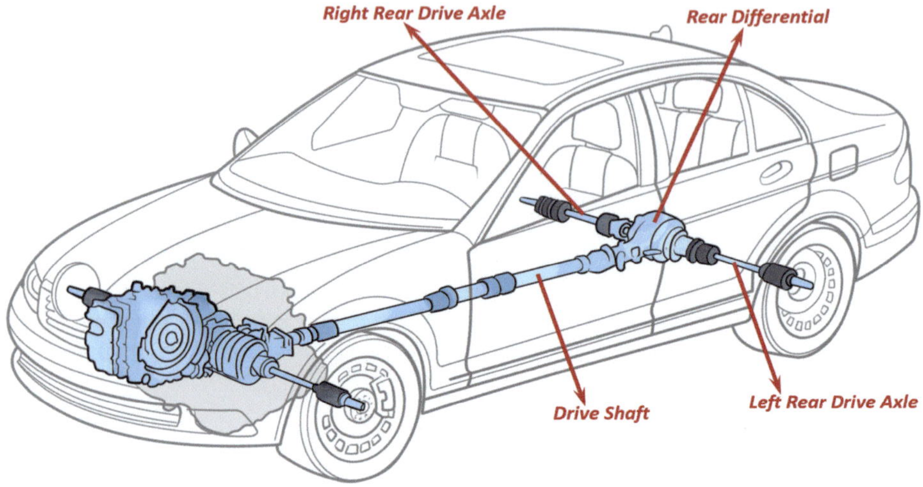

Fig. 10.34 Car differential gear assembly. Designed by the author and implemented by Kenneth Manalansan. Courtesy of Kenneth Manalansan

10.8 Chapter Summary

Outcomes

1. Defining different types of gears and their applications.
2. Identifying the spur gear's configuration parameters.
3. The spur gear's tooth profile is defined from the meshed gear mechanism.
4. Pressure angle, number of teeth, increase in the radius of the undercutting filet, and/or decrease in the path of the contact can be used to avoid the two gears' interference.
5. The published AGMA standardized formulas can be used to design or select a gear for a specific application.
6. Velocity ratio, torque ratio, and gear ratio are used to calculate the angular velocity between two meshed gears or in the angular velocities all across simple/compound gear trains.
7. Use the superposition tabulated method or the analytical equations method to calculate the angular velocities across the planetary trains.
8. Differentials play a very important part in vehicle transmissions.

Problems

For Problem 1, Problem 2, and Problem 3 determine the following:

1. the diameter of the pitch circle,
2. the circular pitch, and
3. the module.

Problem 1

A spur gear is designed to have 20 teeth with a diametral pitch of 4.

Problem 2

A spur gear is designed to have 30 teeth with a diametral pitch of 12.

Problem 3

A spur gear is designed to have 40 teeth with a diametral pitch of 8.

For Problem 4, Problem 5, and Problem 6, a motor rotates a pinion that rotates a bull. If both pinion and bull are meshed, determine the following:

1. the pitch diameter of each gear,
2. the base diameter of each gear, and
3. the center distance.

Problem 4

A pinion with 10 teeth and a bull with 30 teeth. The designer selected 10-pitch and 20° as the pressure angle.

Problem 5

A pinion with 9 teeth and a bull with 25 teeth. The designer selected 6-pitch and 25° as the pressure angle.

Problem 6

A pinion with 15 teeth and a bull with 60 teeth. The designer selected 20-pitch and 25° as the pressure angle.

For Problem 7, Problem 8, and Problem 9, a pinion rotates a bull with the assumption that the addendum a = 1/Pd. Check if the selected pressure angle and number of teeth of each gear will cause interference or not.

Problem 7

A pinion has 10 teeth, and a bull has 30 teeth plus the pressure angle is 20°.

Problem 8

A pinion has 9 teeth, and a bull has 25 teeth plus the pressure angle is 25°.

Problem 9

A pinion has 16 teeth, and a bull has 60 teeth plus the pressure angle is 14°.

For Problem 10, Problem 11, and Problem 12, a pinion is made as a mid-steel cut full depth. Use AGMA standard to determine:

1. the base circle diameter,
2. the thickness of the teeth,
3. the addendum and addendum radius, and.
4. the dedendum and dedendum radius.

Problem 10

A pinion has 10 teeth, the pressure angle is 20°, and 10-pitch.

Problem 11

A pinion has 15 teeth, the pressure angle is 14.5°, and 2-pitch.

Problem 12

A pinion has 5 teeth, the pressure angle is 25°, and 1-pitch.

For Problem 13, Problem 14, and Problem 15, if a mechanism with meshed inion and bull, determine:

1. the mechanism's velocity ratio,
2. the mechanism's torque ratio,
3. the mechanism's gear ratio,
4. the mechanism's circular pitch,
5. the bull's number of teeth, and
6. the bull gear's angular velocity.

Problem 13

A pinion with a pitch diameter of 1.8 in with 18 teeth and a bull with a pitch diameter 7.2 in. The pinion rotates at 1200 rpm.

Problem 14

A pinion with a pitch diameter 25 mm with 16 teeth and a bull with a pitch diameter 200 mm. The pinion rotates at 600 rpm.

Problem 15

A pinion with a pitch diameter 1 cm with 15 teeth and a bull with a pitch diameter 3 cm. The pinion rotates at 100 rpm.

Problem 16

Build a Simscape™ Multibody™ model to simulate Problem 13. Measure the rotational velocity of the bull gear and compare it with the results from your solution of Problem 13-. For the model, use 8 inches as the length of the two shafts from the hubs to the gears. Both the radii of the two shafts are 0.4 inches. Both gears have a face width of 0.4 inches.

Problem 17

Build a Simscape™ Multibody™ model to simulate Problem 14. Measure the rotational velocity of the bull gear and compare it with the results from your solution of Problem 14. For the model, use 100 mm as the length of the two shafts from the hubs to the gears. Both the radii of the two shafts are 5 mm. Both gears have a face width of 5 mm.

Problem 18

Build a Simscape™ Multibody™ model to simulate Problem 15. Measure the rotational velocity of the bull gear and compare it with the results from your solution of Problem 15. For the model, use 8 cm as the length of the two shafts from the hubs to the gears. Both the radii of the two shafts are 0.15 cm. Both gears have a face width of 0.3 cm

For Problem 19 to Problem 24, we are going to use Fig. 10.35 whereby there are four gears (Gear 1 with N_1 number of teeth, Gear 2 with N_2 number of teeth, Gear 3 with N_3 number of teeth, and Gear 4 with N_4 number of teeth) connected through four shafts. Calculate:

- the angular velocity of each gear, and
- the diameter for Gear 2, Gear 3, and Gear 4.

Problem 19

The number of teeth of each gear is $N_1 = 6$, $N_2 = 12$, $N_3 = 24$, and $N_4 = 48$. The diameter of the first gear is $d_1 = 10$ cm. The shaft of the first gear rotates at 100 rpm.

Problem 20

The number of teeth of each gear is $N_1 = 4$, $N_2 = 10$, $N_3 = 20$, and $N_4 = 40$. The diameter of the first gear is $d_1 = 0.2$ m. The shaft of the first gear rotates at 500 rpm.

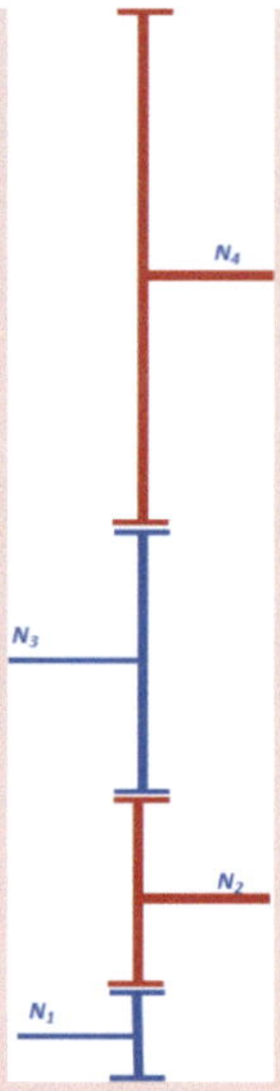

Fig. 10.35 Gear train for Problem 19 to Problem 24

Problem 21

The number of teeth of each gear is $N_1 = 5$, $N_2 = 15$, $N_3 = 30$, and $N_4 = 40$. The diameter of the first gear is $d_1 = 15$ mm. The shaft of the first gear rotates at 3000 rpm.

Problem 22

Build a Simscape™ Multibody™ model to simulate Problem 19. Measure the rotational velocity of each gear and compare it with the results from your solution of Problem 19. For the model, use 20 cm as the length of each shaft from the hub to the gear. The radius of every shaft is 2 cm. All gears have a face width of 3 cm.

Problem 23

Build a Simscape™ Multibody™ model to simulate Problem 20. Measure the rotational velocity of each gear and compare it with the results from your solution of Problem 20. For the model, use 0.4 m as the length of each shaft from the hub to the gear. The radius of every shaft is 0.05 m. All gears have a face width of 0.03 m.

Problem 24

Build a Simscape™ Multibody™ model to simulate Problem 21. Measure the rotational velocity of each gear and compare it with the results from your solution of Problem 21. For the model, use 30 mm as the length of each shaft from the hub to the gear. The radius of every shaft is 3 mm. All gears have a face width of 3 mm.

For Problem 25 to Problem 28, we are going to use Fig. 10.36 whereby there are four gears (Gear 1 with N_1 number of teeth, Gear 2 with N_2 number of teeth, Gear 3 with N_3 number of teeth, and Gear 4 with N_4 number of teeth) connected through three shafts. Calculate:

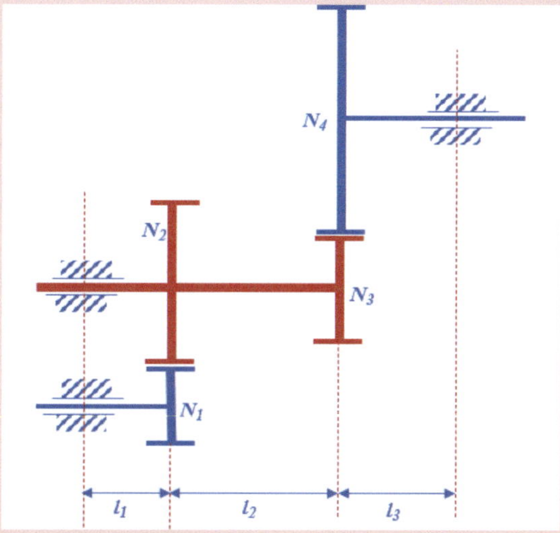

Fig. 10.36 Gear train for Problem 25 to Problem 28

- the angular velocity of each gear, and
- the diameter for Gear 2 and Gear 4.

Problem 25

The number of teeth of each gear is $N_1=6$, $N_2=24$, $N_3=6$, and $N_4=12$. The diameter of the first gear and the diameter of the third gear are $d_1=d_3=10$ cm. The shaft of the first gear rotates at 100 rpm.

Problem 26

The number of teeth of each gear is $N_1=4$, $N_2=10$, $N_3=4$, and $N_4=16$. The diameter of the first gear and the diameter of the third gear are $d_1=d_1=0.2$ m. The shaft of the first gear rotates at 500 rpm.

Problem 27

Build a Simscape™ Multibody™ model to simulate Problem 25. Measure the rotational velocity of each gear and compare it with the results from your solution of Problem 25. For the model, use $l_1=l_2=l_3=20$ cm. The radius of every shaft is 2 cm. All gears have a face width of 3 cm.

Problem 28

Build a Simscape™ Multibody™ model to simulate Problem 26. Measure the rotational velocity of each gear and compare it with the results from your solution of Problem 26. For the model, use $l_1=l_2=l_3=0.4$ m. The radius of every shaft is 0.02 m. All gears have a face width of 0.03 m.

For Problem 29 to Problem 32, we are going to use Fig. 10.37 whereby there are four gears (Gear 1 with N_1 number of teeth, Gear 2 with N_2 number of teeth, Gear 3 with N_3 number of teeth, and Gear 4 with N_4 number of teeth) connected through three shafts. Calculate:

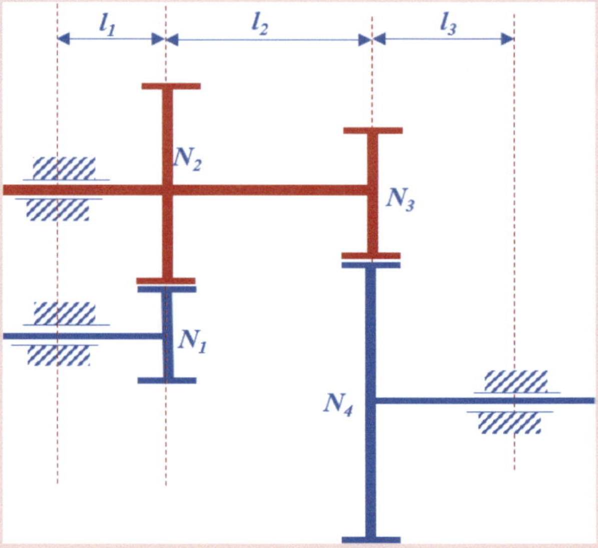

Fig. 10.37 Gear train for Problem 29 to Problem 32

- The angular velocity of each gear, and
- The diameter for Gear 2 and Gear 4.

Problem 29

The number of teeth of each gear is $N_1 = 6$, $N_2 = 24$, $N_3 = 6$, and $N_4 = 12$. The diameter of the first gear and the diameter of the third gear are $d_1 = d_3 = 10$ cm. The shaft of the first gear rotates at 100 rpm.

Problem 30

The number of teeth of each gear is $N_1 = 4$, $N_2 = 10$, $N_3 = 4$, and $N_4 = 16$. The diameter of the first gear and the diameter of the third gear are $d_1 = d_1 = 0.2$ m. The shaft of the first gear rotates at 500 rpm.

Problem 31

Build a Simscape™ Multibody™ model to simulate Problem 29. Measure the rotational velocity of each gear and compare it with the results from your solution of Problem 29. For the model, use $l_1 = l_2 = l_3 = 20$ cm. The radius of every shaft is 2 cm. All gears have a face width of 3 cm.

Problem 32

Build a Simscape™ Multibody™ model to simulate Problem 30. Measure the rotational velocity of each gear and compare it with the results from your solution of Problem 30. For the model, use $l_1 = l_2 = l_3 = 0.4$ m. The radius of every shaft is 0.02 m. All gears have a face width of 0.03 m.

For Problem 33 to Problem 38, we are going to use Fig. 10.38.

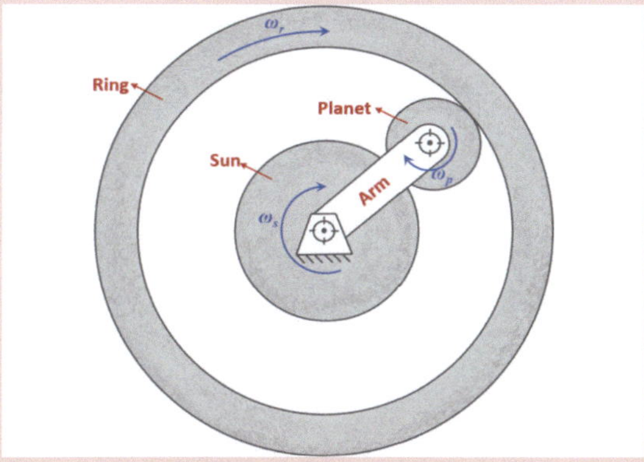

Fig. 10.38 Gear train for Problem 33 to Problem 38

Problem 33

A planetary gear train with the carrier is driven at 600 rpm while the sun is stationary. The number of teeth on each gear is $N_{sun}=8$, $N_{planet}=16$, $N_{ring}=60$. Determine the rotational velocity of the planet gear and the ring gear using the superposition.

Problem 34

A planetary gear train with the planet is driven at 800 rpm while the sun is rotating at 2800 rpm. The number of teeth on each gear is $N_{sun}=12$, $N_{planet}=28$, $N_{ring}=84$. Determine the rotational velocity of the carrier gear and the ring gear using the superposition.

Problem 35

A planetary gear train with the ring is driven at 1200 rpm while the sun is rotating at 2000 rpm. The number of teeth on each gear is $N_{sun}=12$, $N_{planet}=9$, $N_{ring}=27$. Determine the rotational velocity of the planet gear and the carrier gear using the superposition.

Problem 36

A planetary gear train with the carrier is driven at 600 rpm while the sun is stationary. The number of teeth on each gear is $N_{sun}=8$, $N_{planet}=16$, $N_{ring}=60$. Determine the rotational velocity of the planet gear and the ring gear using the mathematical planetary gear train model.

Problem 37

A planetary gear train with the planet is driven at 800 rpm while the sun is rotating at 2800 rpm. The number of teeth on each gear is $N_{sun}=12$, $N_{planet}=28$, $N_{ring}=84$. Determine the rotational velocity of the carrier gear and the ring gear using the mathematical planetary gear train model.

Problem 38

A planetary gear train with the ring is driven at 1200 rpm while the sun is rotating at 2000 rpm. The number of teeth on each gear is $N_{sun} = 12$, $N_{planet} = 9$, $N_{ring} = 27$. Determine the rotational velocity of the planet gear and the carrier gear using the mathematical planetary gear train model.

References

1. "Machines & Mechanisms: Applied Kinematic Analysis", David Myszka, Pearson; 4th edition, 2011.
2. "Engineering Mechanics: Statics", James L. Meriam, and L. G. Kraige, 7th edition, ISBN-13 : 978–0470917879
3. "Mechanisms and Dynamics of Machinery", Mabie Hamilton and Reinholtz Charles, 4th ed., John Wiley and Sons Inc., New York, 1987.
4. "AMERICAN NATIONAL STANDARD: Tooth Proportions for Fine-Pitch Spur and Helical Gearing," ANSI/AGMA 1003- H07, February 6, 2020.
5. "Simscape™ Multibody™ Getting Started Guide," MathWorks, 2023.
6. "Simscape™ Multibody™ User's Guide," MathWorks, 2023.
7. "Simscape™ Multibody™ Reference," MathWorks, 2023.
8. "Simscape™ Multibody™ Release Notes," MathWorks, 2023.
9. "Structure and Analysis of Planetary Gear Trains", Zoltan Levai, Jnl. Mechanisms Volume 3, pp.131–14 8, 1968.

Statics of Machines

11

Chapter Objectives

1. To understand the definition of external forces and external moments plus contact/reaction forces and contact/reaction moments.
2. To understand how to sketch a free body diagram of an entire linkage or a linkage's part.
3. To understand how to use Newton's first law to calculate acting forces and acting moments when a machine is in a steady state.
4. To understand how to use Simscape™ Multibody™ in MATLAB® Simulink® to calculate forces and moments when a machine is in a steady state.

11.1 Introduction

In this chapter, we introduce the static analysis of machines including the definitions of external forces and external moments plus their source as actuators or external loads. After that, we will introduce contact forces and contact moments of any kinematic pair (two links and one joint) when the kinematic pair is disassembled. The free body diagram (FBD) is introduced to sketch acting forces and moments on any linkage's part. Newton's first law is introduced to come up with and develop 2D equilibrium equations. We will use equilibrium equations to calculate the required actuators' forces and moments as well as the reaction or contact forces and moments for any determined linkage's part. Lastly, we present how to use a Simscape™ Multibody™ model to calculate the required actuator forces and moments, and reaction forces and moments acting on a machine in a steady position.

11.2 External Forces

Definition

Force

A force is a vector quantity that represents a pushing or pulling action on a part [1].

Supplementary Information The online version contains supplementary material available at https://doi.org/10.1007/978-3-031-72701-6_11.

A. Omran, *Kinematics and Dynamics of Machines*, https://doi.org/10.1007/978-3-031-72701-6_11

External Force

An external force occurs as a result of interaction between the machine and the surroundings [1, 2].

In general, an external force, $\overrightarrow{F}_i$, is a push or pull action that is externally applied at a specific node n_i. This could be a free node or a node where joints could be located. Figure 11.1 shows how a person applies a pushing force to close a door and a pulling force to open the door. The person is applying the pushing or the pulling forces at a node n_i located on the handle.

An external force has a vector form that can be expressed in any coordinate system. In this textbook, we will express any force in the Cartesian coordinate system and our analysis will be restricted to be in 2D. Sometimes, we can express the force as a unit vector and amplitude. In general, the mathematical expression of the external force, $\vec{F}_i$, is

$$\vec{F}_i = F_{xi}\hat{i} + F_{yi}\hat{j} \tag{11.1}$$

where F_{xi} is the external force component in the x-direction and F_{yi} is the external force component in the y-direction that is applied at node n_i. This external force follows the same conventional vector operations; i.e., we can add or subtract two forces as adding and subtracting two vectors. We sketch the force as a straight arrow pointing out of the node. The force is measured in kg m/s^2 or "N" in the International System of Units (SI) and "lb" in the US customary system.

For any machine, the external force $\vec{F}_i$ could be an actuator's force that's applied at node n_i providing a transitional input motion SVAJ. Also, the external force $\vec{F}_i$ could be a load force such as a weight that is acting at a specific node n_i.

When we speak about the statics and dynamics of any machine, we should sketch the external forces as arrows that are located at specific nodes. For example, let's recall the slide crank mechanism shown in Fig. 11.2 Assume a hydraulic actuator is used to push and pull node n_3 on the prismatic joint J_3^P. In this case, the external force $\vec{F}_3$ should be applied in the same direction that the

Fig. 11.1 Applied pushing and pulling faces on door. Designed by the author and implemented by Kenneth Manalansan. Courtesy of Kenneth Manalansan

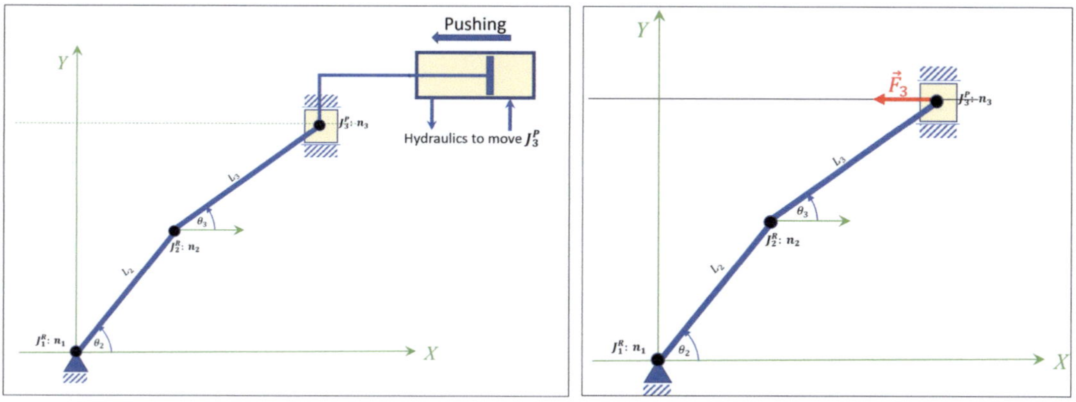

Fig. 11.2 Hydraulic external force

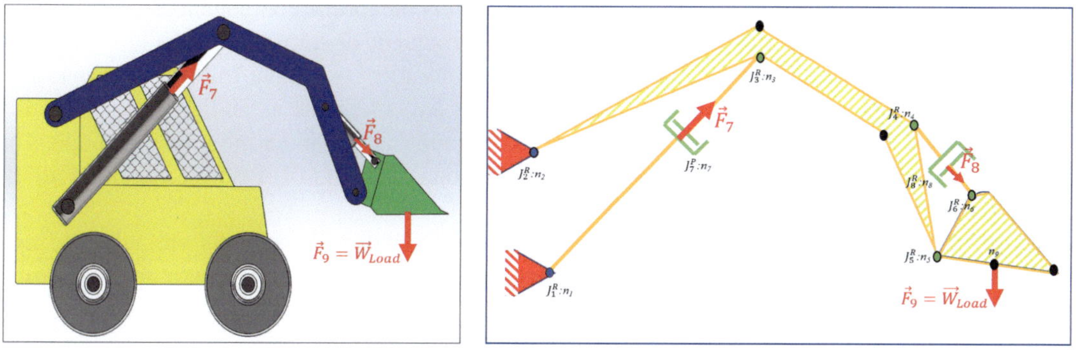

Fig. 11.3 Skid steer articulated bucket mechanism. Figure and associated Skid Steer Articulated Bucket Lift Mechanism.mp4 video designed by the author and implemented by both the author and Pranav Dhola. Both courtesy of Pranav Dhola

prismatic joint allows the motion to occur in—the x-direction. The external force $\vec{F}_3$ is sketched as an arrow acting at node n_3 as shown in Fig. 11.2.

Another example, the skid steer articulated bucket machine, shown in Fig. 11.3, has two actuators' forces $\vec{F}_7$ and $\vec{F}_8$ that are sketched as arrows applied at nodes n_7 and n_8 in the directions wherein the prismatic joints J_7^P and J_8^P allow the motion to be. Also, the weight of load $\vec{W}_{\text{load}}$ or $\vec{F}_9$ is assumed to be concentrated at node n_9 and pointing downward due to gravity as shown in Fig. 11.3.

11.3 External Moment

Definition

Moment and Torque:

A moment is the twisting action produced by a force [1].

External Moment and Torque:

An external moment occurs as a result of interaction between the machine and the surroundings [1, 2].

The moment, $\vec{M}_i$, is a twisting action around a reference node n_i. The torque $\vec{T}_i$ is any moment where the reference node n_i is located at the center of a twisting shaft. An external moment or torque is the result of any external load or actuator. For example, when we use a motor to rotate one of the revolute joints, a moment is generated at the shaft. In this case, we call this moment a torque. The moment or torque has a vector form that can be expressed in any coordinate system. In this textbook, we will express a moment in the Cartesian coordinate system and our analysis will be restricted to be in 2D (x–y plane). In other words, the moment will be around the z-axis only. Following the right-hand rule, mentioned in Chap. 2, the direction of a moment could be clockwise (negative −ve) or counterclockwise (positive + ve).

In 2D, the mathematical expression of the moment $\vec{M}_i$ or torque $\vec{T}_i$ is

$$\begin{aligned} \vec{M}_i &= M_i \hat{z} \\ \vec{T}_i &= T_i \hat{z} \end{aligned} \tag{11.2}$$

where M_i is the moment component about the z-axis and T_i is the torque component about the z-axis. For simplicity, since in this textbook we deal with the moment or torque around the z-axis only (2D), we replace $\vec{M}_i$ with M_i and $\vec{T}_i$ with T_i. The external moment or torque follows the same conventional vector operations; i.e., we can add or subtract two moments as adding and subtracting two vectors. We sketch the moment or the torque as a circular arrow about the reference node. The moment is measured in kg.m^2/s^2 or "N.m" in the International System of Units (SI) and "lb.ft" in the US customary system.

When we study machines, we define that the source of the external moments around the z-axis located at the reference node n_i could be caused by a force $\vec{F}_j$ that is acting at node n_j as shown in Fig. 11.4. The resultant moment is calculated as

For

$$\vec{r}_{ninj} = x_{ninj}\hat{i} + y_{ninj}\hat{j} \quad \text{and} \quad \vec{F}_j = F_{xj}\hat{i} + F_{yj}\hat{j} \tag{11.3}$$

The moment is

$$M_i = \vec{r}_{ninj} \times \vec{F}_j$$

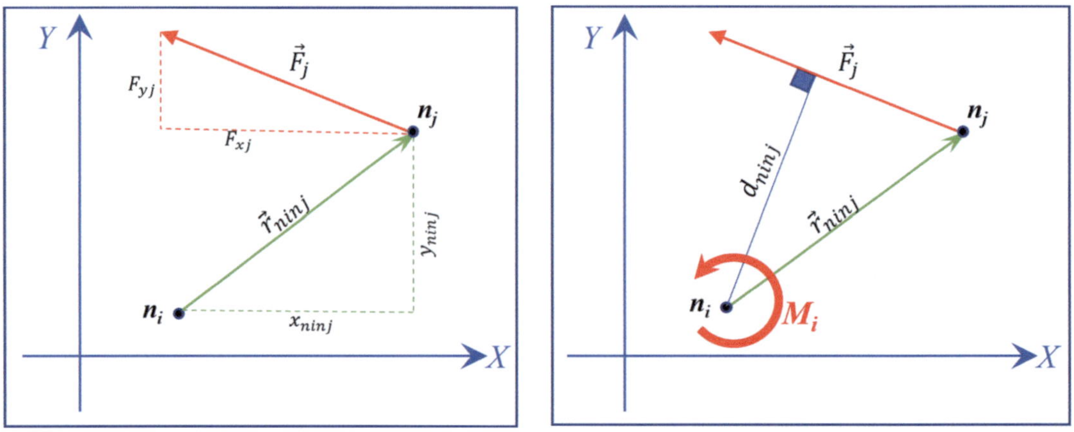

Fig. 11.4 Moment as force multiplied by distance

where "×" is the cross product that provides

$$M_i = \left(x_{ninj}F_{yj} - y_{ninj}F_{xj}\right) = d_{ninj}F_j$$

where

$$F_j = \left|\vec{F}_j\right|$$

Note that the moment M_i is around the z-axis that is located at the reference node n_i. The value d_{ninj} is the shortest distance from the reference node n_i to the vector $\vec{F}_j$. Or, d_{ninj} is the perpendicular distance from n_i to vector $\vec{F}_j$.

Note that when the extension of the force vector $\vec{F}_j$ passes through the reference node n_i, this force does not cause any moment around this reference node since $\vec{r}_{ninj} \times \vec{F}_j = 0$. Figure 11.5 shows a force $\vec{F}_j$ acting at node n_j. The extended line of the force $\vec{F}_j$ passes through the node n_i. This causes $\vec{r}_{ninj} \times \vec{F}_j = 0$ or the moment is zero because $d_{ninj}=0$.

The external moment M_i could be from an actuator such as an engine or motor. For example, we can add a motor at joint J_1^R to a four-bar mechanism as shown in Fig. 11.6. The motor shaft is located by a gear set to rotate the revolute joint J_1^R around the z-axis. The motor provides a torque T_1 that develops an input rotational motion SVAJ at n_1. We sketch the torque as a circular arrow about node n_1 as shown in Fig. 11.6.

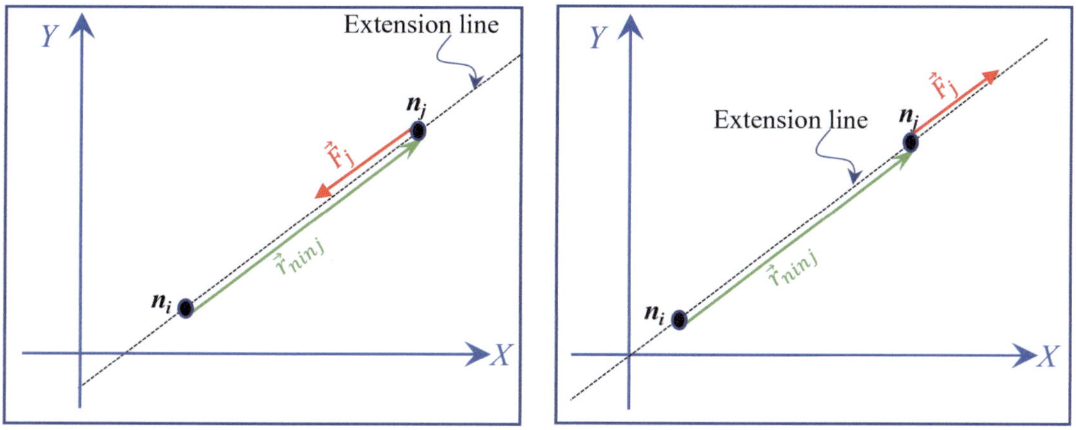

Fig. 11.5 Zero moment

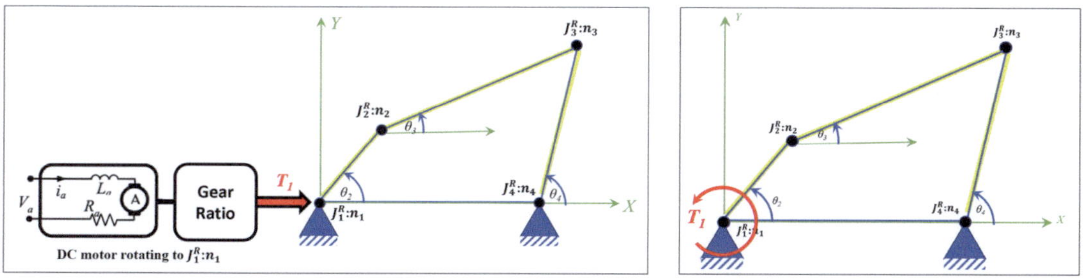

Fig. 11.6 Motor torque in four-bar mechanism

11.4 Contact Forces and Moments

A machine is a linkage that consists of multiple solid bodies. In addition to external forces and moments, there are contact forces and moments between every two links that are connected through a joint (kinematic pair). These contact forces and moments between these two connected links ensure that motion is restricted to the direction that the joint is designed to go in.

For example, if we have a kinematic pair with a revolute joint or two connected links L_i and L_j through a revolute joint J_k^R, that restricts any relative motion in the x *or* y-direction (i.e., $\Delta x = 0$ and $\Delta y = 0$ between the two links at node n_k). The joint J_k^R allows only a rotational motion θ_k between the two links L_i and L_j at node n_k. When we disassemble these two links, there are two contact forces F_{xk} and F_{yk} that prevent relative motion in the x-direction and the y-direction as shown in Fig. 11.7. These two contact forces are applied to the link L_i in an opposite direction than at L_j.

Table 11.1 shows the contact forces and moments present when we disassemble different types of kinematic pairs. In each kinematic pair type, the two links L_i and L_j are connected with a specific joint (prismatic, gear, and cam) at node n_k. In the case of the prismatic joint, the allowable motion is in the tangential direction (t-direction) and the prevented motion is in the normal direction (n-direction). The prismatic joint exerts a contact force F_{nk} on the two links to prevent relative motion in the n-direction plus a contact torque to prevent the rotational motion.

For the three joints listed in Table 11.1 two surfaces slide relative to each other; in other words, there is an allowable motion in the tangential direction to the two surfaces at the point of contact. Motion is restricted in the normal direction on the two surfaces at each joint node. For example, in a prismatic joint, let's assume that a piston is sliding inside a cylinder. The piston (first surface) slides on the cylinder (second surface). Motion is restricted in the normal direction. To block the normal motion, there is a contact/reaction force. The same scenario occurs in the gear joint and the cam joint.

In the case of a gear joint, the contact force is acting in the direction normal to the meshed teeth at the contact point as shown in Fig. 11.8. The direction of the gear contact force makes an angle φ known as the pressure angle which is on a line tangent to both pitch circles of the mating gears. More details about the pressure angle were given previously in Chap. 10. Let's assume that two gears are centered at node n_i and n_j with two radii $\vec{r}_{nink}$ and $\vec{r}_{nink}$, respectively. Node n_i is at a distance of $r_{ninj} = r_{nink} + r_{njnk}$ from the node n_j. The two gears contact at node n_k with gear joint J_k^G. If we disassemble the two gears as shown in Fig. 11.8, we can calculate the two torques T_i and T_j at node n_i and at node n_j as:

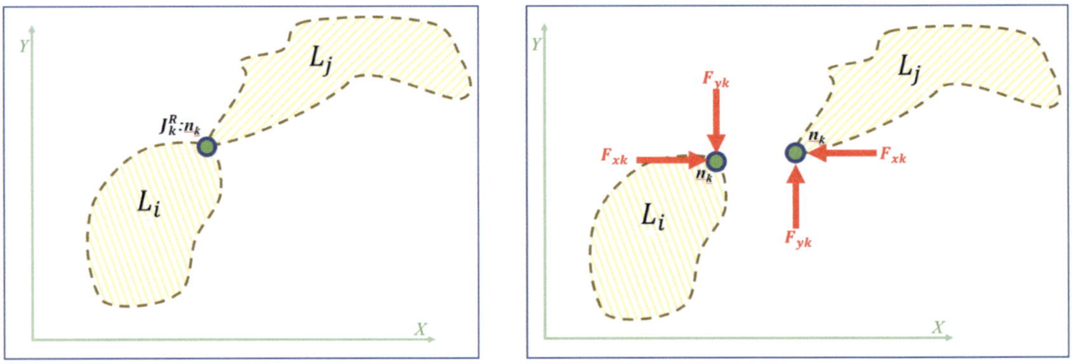

Fig. 11.7 Assembled and disassembled two links with revolute joint

Table 11.1 Contact forces and moments between two links connected by joint

Joint Name	Assembled	Disassembled
Prismatic, sliding, or piston joint between two links.	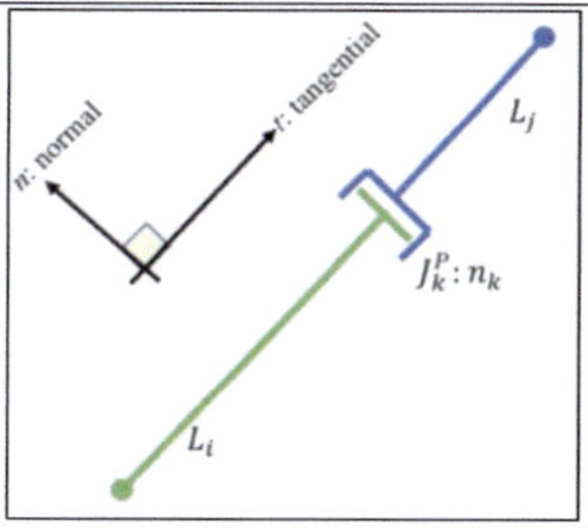	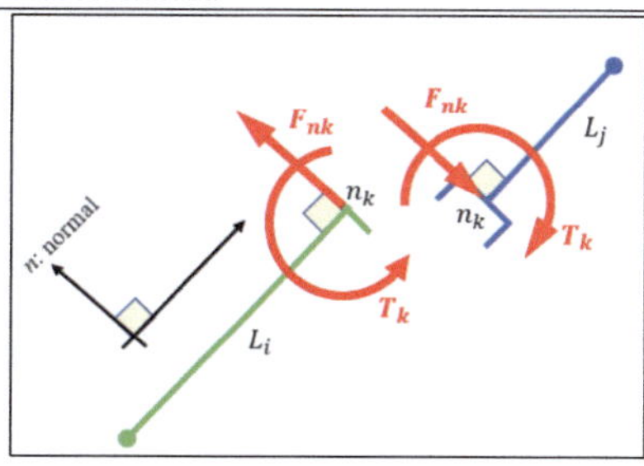
Gear front view joint. The angle φ is the pressure angle that's coming from the gears' teeth contact.	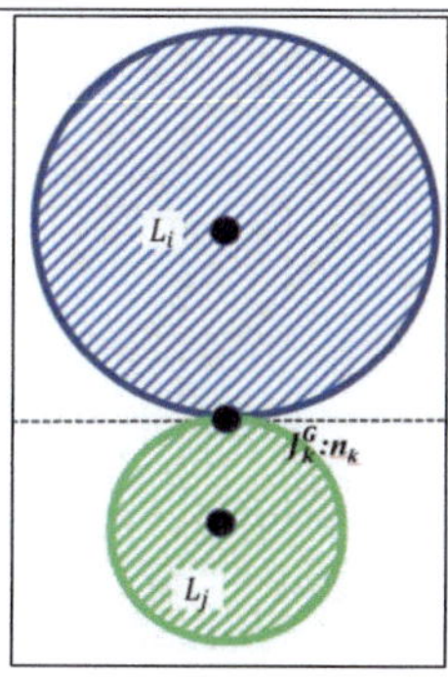	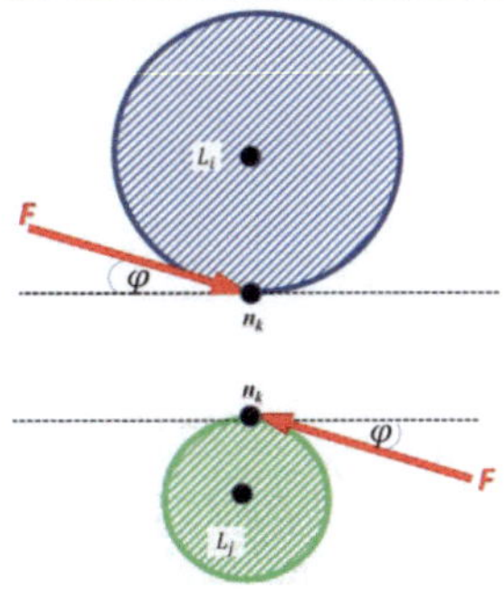
CAM joint	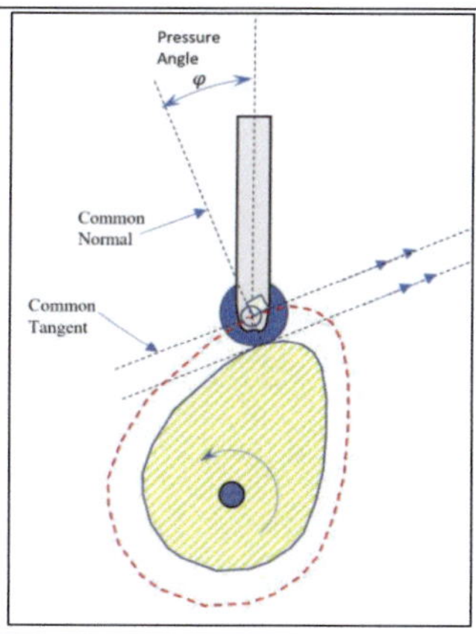	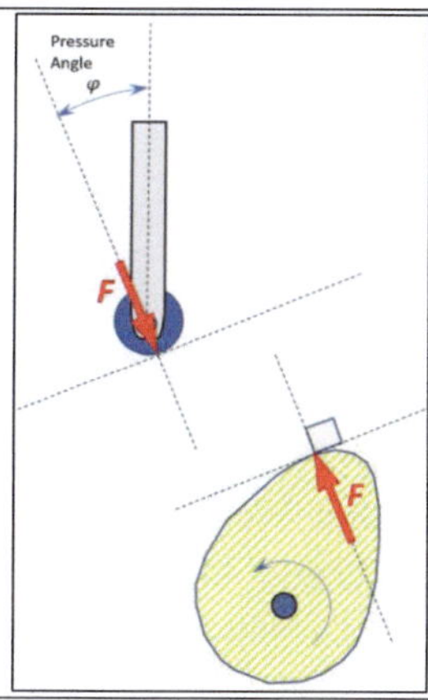

$$T_i = r_{nink} \times F\cos(\varphi) \quad \text{and} \quad T_j = r_{njnk} \times F\cos(\varphi) \tag{11.4}$$

or

$$\frac{T_i}{T_j} = \frac{r_{nink}}{r_{njnk}}$$

In the ideal cases, we assume that there is no force in the tangential direction when two surfaces slide relative to each other. This assumes that the two surfaces are perfectly smooth and that there is no possibility of any obstacles or resisting forces when the two surfaces slide relative to each other.

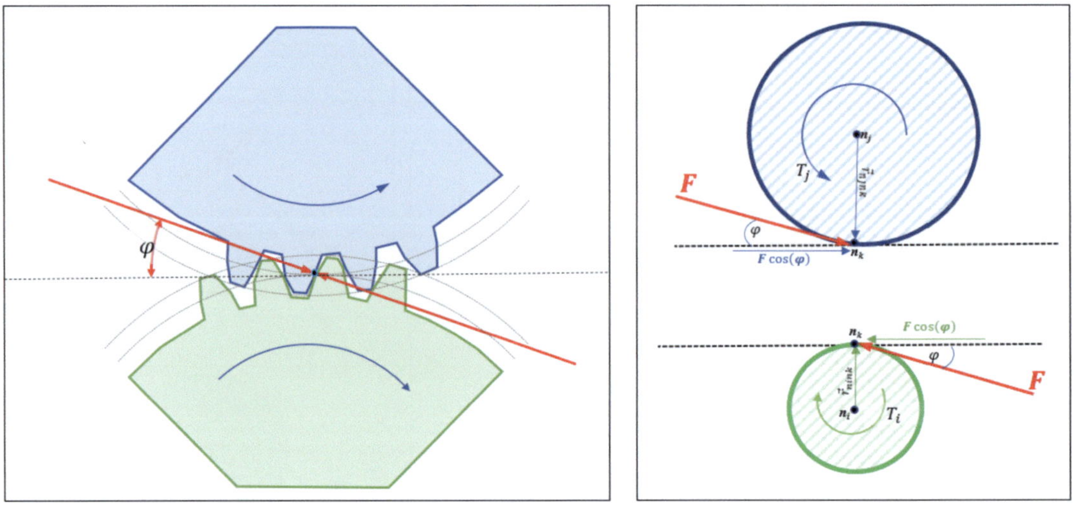

Fig. 11.8 Direction of gear joint's contact force

> **Definition**
>
> **Friction Forces:**
>
> Friction is a force that resists the movement of contacting surfaces that slide relative to each other. This force always acts tangent to the surface at the points of contact and is directed so as to oppose the possible or existing motion between the surfaces [2, 3].

Sometimes the frictionless assumption is not valid and there is an additional force—a friction force that acts in the tangential direction to the two surfaces at the contact points. This friction force exists because the two sliding surfaces are not perfectly smooth. There are surface irregularities that tend to interlock with one another as shown in Fig. 11.9. When one surface tries to slide on the other, the interlocking of these irregularities opposes the motion. This opposing force is called the *friction force* and is a function of the surface roughness.

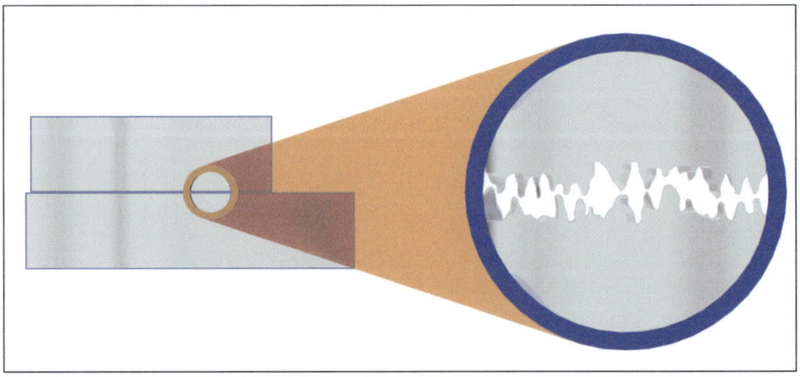

Fig. 11.9 Two sliding surfaces relative to each other with irregularities. Designed by the author and implemented by Kenneth Manalansan. Courtesy of Kenneth Manalansan

Table 11.2 Static friction coefficients

		Dry	Lubricated
Hard steel	On hard steel	0.45	0.08
	On Babbitt	0.35	0.15
Mild steel	On mild steel	0.6	0.12
	On bronze	0.34	0.16
	On brass	0.44	Not available
	On copper lead	0.36	0.15
	On cast iron	0.23	0.13
	On lead	0.95	0.3
	On aluminum	0.5	Not available
	On laminated plastic	0.35	0.05
	On Teflon	Not available	0.04
Cast iron	On cast iron	0.15	0.07
	On bronze	0.22	Not available
	On brass	0.3	Not available
	On copper	0.29	Not available
	On zinc	0.21	Not available
Aluminum	On aluminum	1.40	Not available

In a static analysis, the frictional force holds back a stationary link just before it starts moving which is why we may call it a *friction force*. The friction force F_f is defined as [2, 3]

$$F_f = \mu F_n \tag{11.5}$$

where F_n is the normal force on the surface or else we may call it the *contact/reaction force*. The friction coefficient μ depends on many factors that include but are not limited to the materials and the surface conditions of the contacting surfaces. The friction coefficient is measured and recorded experimentally. Table 11.2 lists the average values of friction coefficients for common materials [1]. As you can see, there are two common causes for the friction coefficient—dry or lubricated. The lubricated case has a lower friction coefficient because using a lubricant smooths down the friction between the two surfaces. Most machines are designed to work using lubrication to reduce this friction. Less friction means less actuator force and less money to run the machine. Also, there is a friction torque that occurs in revolute joints. In this textbook, we are not going to consider frictional forces between any sliding surfaces. More details about friction analysis are available in Refs. [2, 3].

11.5 Free Body Diagram

Definition

Free Body Diagram:

A free body diagram (FBD) is a sketch of the outlined shape of the body, which represents it as being isolated or "free" from its surroundings, i.e., a "free body" [2].

For initial design and analysis, an FBD is a great tool for understanding the influence of each external load (force or moment) and each of the reaction or contact loads (force or moment) on any specific part of the linkage. This could be a single link, a single joint, a collection of links and joints (connected or separate), or the entire linkage model. The idea of sketching the FBD of any linkage's part is to isolate and make this part free and stand-alone. After isolating or cutting the part of interest from the whole, the reaction or contact loads should be applied to this part and sketched as described in Sect. 11.5. Also, if any of the external loads are directly applied to this part, they should be sketched as shown in Sects. 11.1 and 11.3.

To sketch the FBD of any linkage's part, we could follow the steps below:

1. Sketch this part alone by cutting the joints attaching this part to the rest of the linkage.
2. Sketch the external loads that are already present on this part.
3. Sketch the contact or reaction loads due to the cuts.

Example 11.1

For the hydraulic door opener shown in Fig. 11.10, the hydraulic cylinder pushes to open the door with an external force F. Sketch the FBD of the linkage without the ground and the FBD of the door alone.

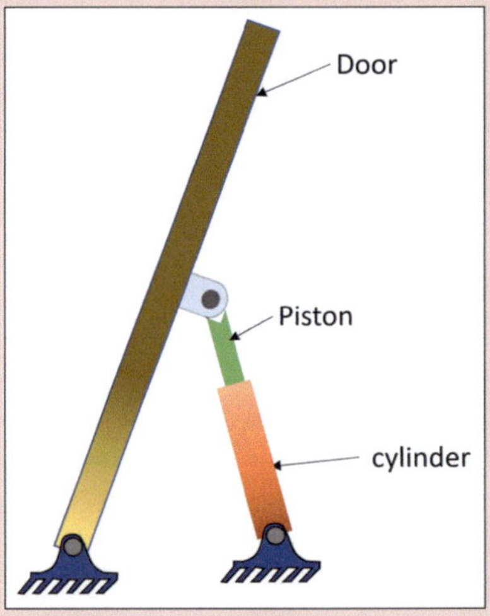

Fig. 11.10 Hydraulic door opener

Solution

The steps to sketch the linkage's FBD without the ground:

- Label the nodes as shown in Fig. 11.11.
- Locate the x-axis and y-axis at node n_1.
- Sketch the external force (blue arrow) acting at the prismatic joints in the allowable piston motion direction.
- Cut the linkage from the ground at the two revolute joints.
- Sketch the reaction forces (red arrows) acting on the linkage due to the two revolute joints.

Follow similar steps to sketch the door FBD as shown in Fig. 11.12.

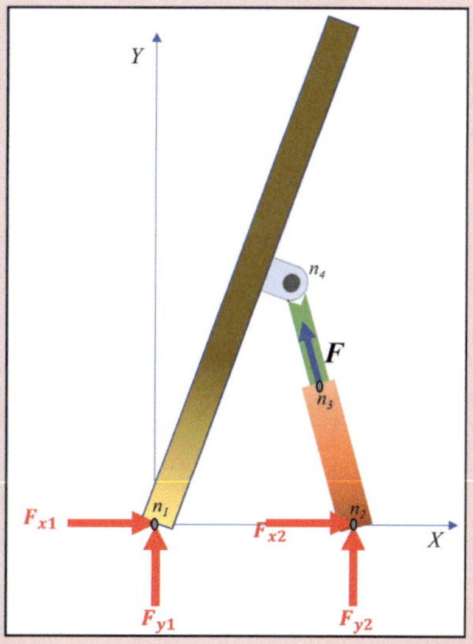

Fig. 11.11 FBD of hydraulic door opener

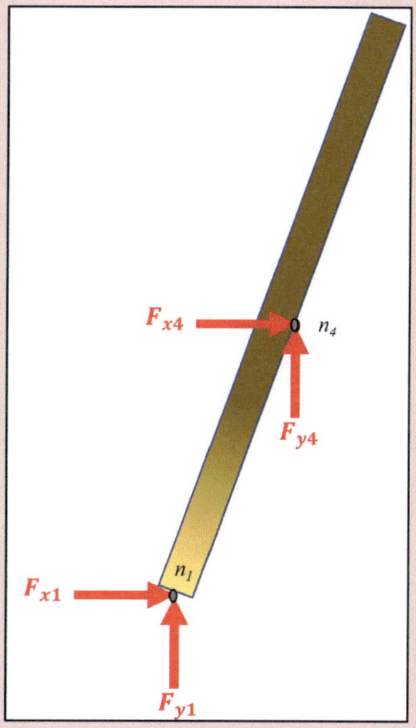

Fig. 11.12 FBD of door

11.6 Static Equilibrium

Definition

Newton's First Law:

A particle originally at rest or moving in a straight line with constant velocity tends to remain in this state provided the particle is not subjected to an unbalanced force [2].

A particle remains at rest or continues to move with uniform velocity (in a straight line with a constant speed) if there is no unbalanced force acting on it [3].

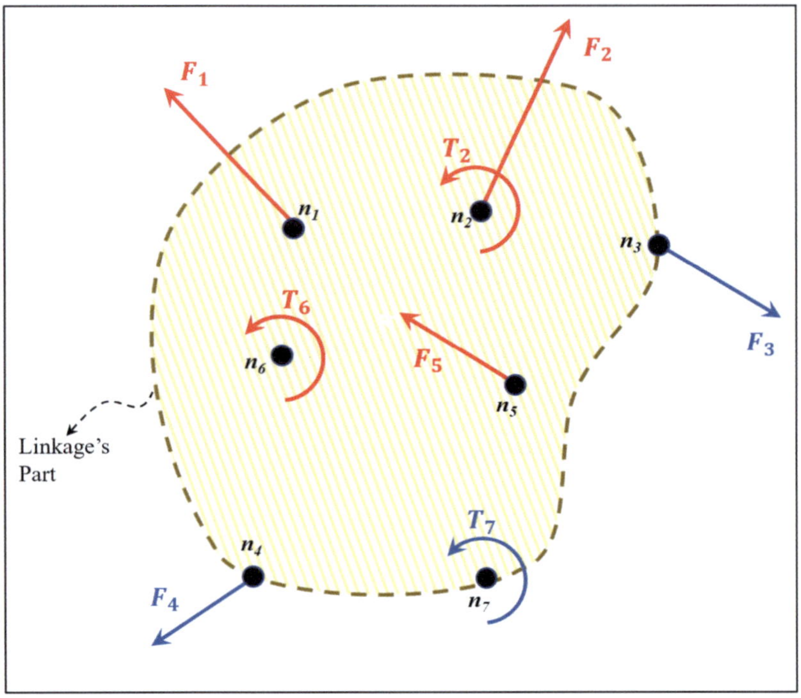

Fig. 11.13 Forces and moments acting on linkage's part

Let's assume a linkage's part with a set of acting forces and acting torques as shown in Fig. 11.13 The part consists of a set of links and a set of joints that are all connected. In our analysis, the linkage's part could be a single link, single joint, or any set of arbitrary links and joints that are all connected. Sometimes, the linkage's part could be a linkage cut from the ground. The sources of the acting forces could be from external loads (red arrows) or contact forces (blue arrows). The sources of the moments could be from external torques (red circular arrows) or from reaction moments (blue circular arrows). The contact forces and contact moments come from isolating this linkage's part from the entire linkage.

If the linkage's part (shown in Fig. 11.13) is at rest or moving in a straight line with constant velocity, then we can apply Newton's first law in 2D by [2, 3]

$$\begin{aligned} \sum F_x &= 0 \\ \sum F_y &= 0 \\ \sum M_A &= 0 \end{aligned} \tag{11.6}$$

Equation (11.6) indicates that the summation of the forces acting on the linkage's part in both x-direction $\sum F_x$ and y-direction $\sum F_y$ is zero and the summation of the moments acting about the z-axis around a reference node n_A is zero. The summation of the moments is the summation of the actual acting torques plus the summation of the cross products of the radii and the acting forces. The radii are from node "A" to the node whereby the acting forces are located.

For example, if we apply Eq. (11.6) to the linkage's part in Fig. 11.13, considering the reference node n_A to be n_1, the summations are

$$\begin{aligned}\sum F_x &= F_{x1} + F_{x2} + F_{x3} + F_{x4} + F_{x5} = 0\\ \sum F_y &= F_{y1} + F_{y2} + F_{y3} + F_{y4} + F_{y5} = 0\\ \sum M_1 &= T_2 + T_6 + T_7 + \vec{r}_{n1n2} \times \vec{F}_2 + \vec{r}_{n1n3} \times \vec{F}_3 + \vec{r}_{n1n4} \times \vec{F}_4 + \vec{r}_{n1n5} \times \vec{F}_5 = 0\end{aligned} \quad (11.7)$$

where $\vec{r}_{n1ni}$, $i=\{2, 3, 5, 4\}$ is the radius from the reference node n_1 to the node n_i whereby F_i is located. In 2D, the number of static equilibrium equations is three. We can use these equilibrium equations to find three unknown reaction forces, reaction moments, actuator forces, or/and actuator torques. If we have more than three unknowns, we call the static problem *indeterminate*.

We can use Eq. (11.6) to prove that the two-force link ("two-force member" in other textbooks) such as a coupler and a connecting rod has only axial force (tension or compression). The two-force link has two forces that are acting along the same line with the same value. The two-force link could be a link between two aligned revolute joints (along the same line) with no applied external force on the link away from this joint and no moment acting on the joints or the link as shown in Fig. 11.14.

Let's assume the x-axis to be in the normal direction and the y-axis to be in a tangential direction as shown in Fig. 11.14. The origin is assumed to be located at node n_j. If we apply the moment part of Eq. (11.6) considering $n_A = n_j$

$$\sum M_j = d_{njni} F_{ti} = 0 \quad \text{or} \quad F_{ti} = 0 \quad (11.8)$$

Applying the y-direction force component in Eq. (11.6)

$$\sum F_x = F_{tj} - \underbrace{F_{ti}}_{0} = 0 \quad \text{or} \quad F_{tj} = 0 \quad (11.9)$$

From Eqs. (11.8) and (11.9), it turns out that the only acting forces are F_{ni} and F_{nj}. Another example is the hydraulic door opener mechanism, shown in Fig. 11.10, where we can replace F_{x2} and F_{y2} by one single force F_2 acting in the same direction of $\vec{r}_{n2n3}$ as shown in Fig. 11.15.

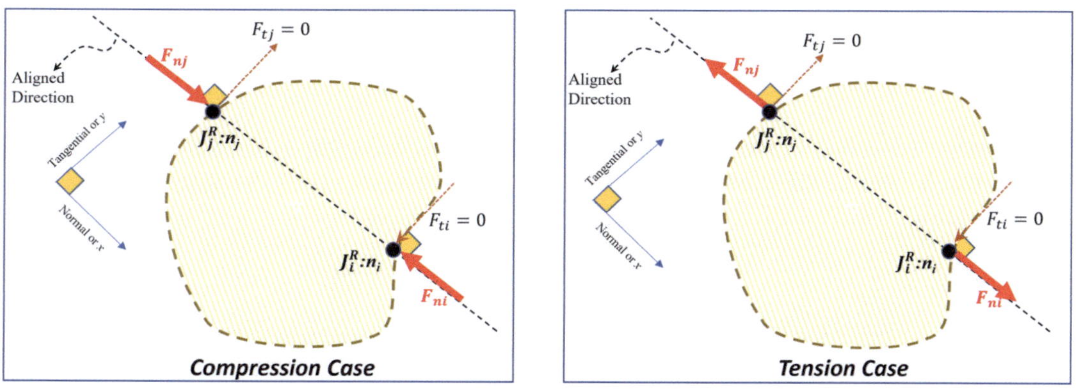

Fig. 11.14 Two-force member

Fig. 11.15 Hydraulic door opener FBD using two-force link

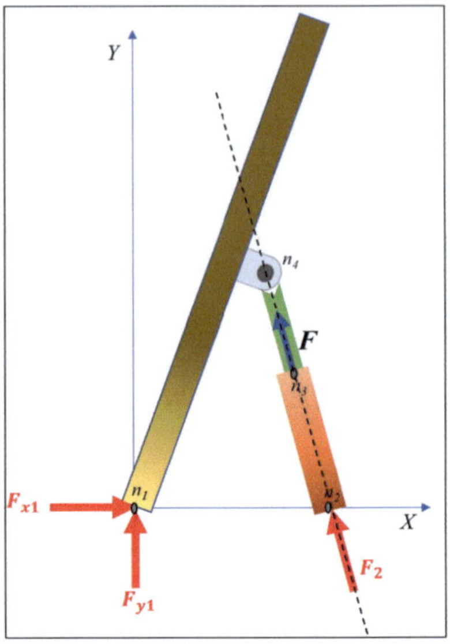

Example 11.2

Figure 11.16 shows the piston link as a linkage's part with acting external and contact forces applied. Can you tell if this link is at rest or not?

Fig. 11.16 Piston link from piston engine mechanism

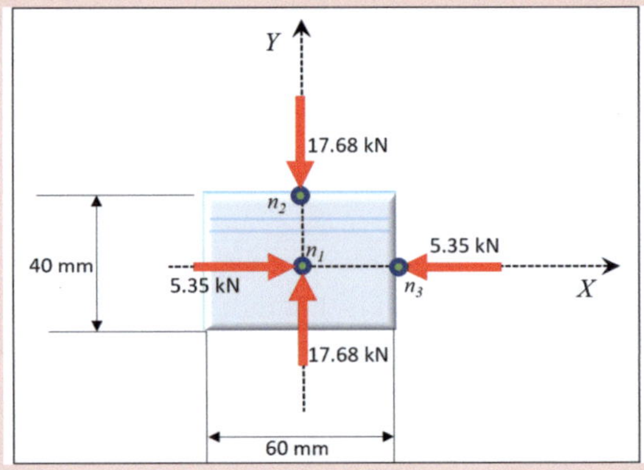

Solution

To check if the piston in Fig. 11.16 is at rest or not, we calculate the summation of the forces and moments as listed in Eq. (11.16) by

$$\sum F_x = 5.35\ \text{kN} - 5.35\ \text{kN} = 0$$

$$\sum F_y = 17.68\ \text{kN} - 17.68\ \text{kN} = 0$$

$$\sum M_1 = 0$$

The three mathematical conditions are satisfied. This means that the piston is at rest. Note the moment around the reference node n_1 is zero since all forces' vectors are going through this node.

Example 11.3

Figure 11.17 shows the dumpster truck mechanism ready to dump a load of sand with a weight of 5000 lb. in a steady position. The weight is located at node n_5. The mechanism consists of four links: L_1 (Vehicle Frame), L_2 (Bed), L_3 (Piston), and L_4 (Cylinder). The mechanism has four joints: three revolute joints, J_1^R at node n_1, J_2^R at node n_2, and J_4^R at node n_4, plus one prismatic joint J_3^P at node n_3. Calculate the hydraulic force needed to support the shown position in Fig. 11.17. Also, calculate the contact forces at revolute joint J_1^R.

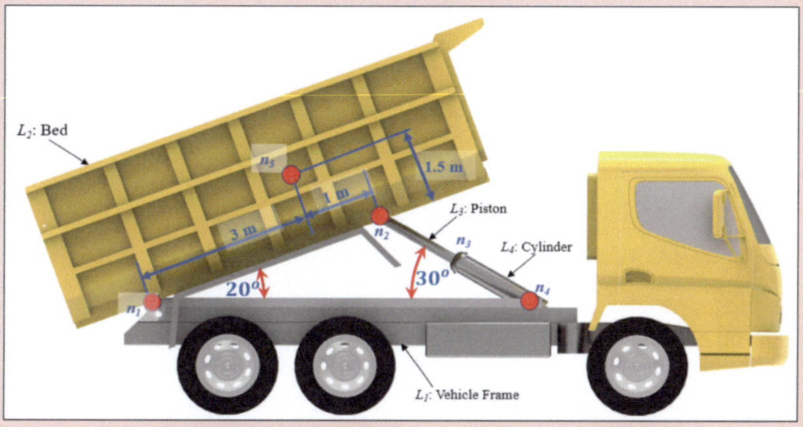

Fig. 11.17 Dumpster truck mechanism. Figure and associated Dumpster Truck Mechanism.mp4 video designed by the author and implemented by both the author and Pranav Dhola. Both courtesy of Pranav Dhola

Solution

The steps of the solution are:

1. Sketch the kinematic diagram of the entire mechanism as shown in Fig. 11.18a. Locate the origin of the X-axis and y-axis at node n_1.
2. Sketch the FBD of the bed link L_2 with two cuts, one from the ground link L_1 (vehicle frame) and the second from the piston link L_3 as shown in Fig. 11.18b.
3. As shown in Fig. 11.18b, the forces acting on the bed are:
   - F_3, the reaction force between the piston link L_3 and bed link L_2 due to the revolute joint J_2^R. Note, the piston link L_3 is a two-force link so the reaction force is located in the same direction $\vec{r}_{n2n3}$.
   - F_5 at node n_5 as the weight of the sand, that's known to be 5000 lb. in the negative y-direction for force due to gravity.
   - The two forces F_{x1} and F_{y1} at node n_1 as the reaction forces between the ground link L_1 (vehicle frame) and the bed link L_2.
4. Decompose the force F_3 in x-direction as "$-F_3 \cos(30°)$" and in y-direction as "$+F_3 \sin(30°)$".
5. Calculate the length of $|\vec{r}_{n1n5}| = r_{n1n5}$ as

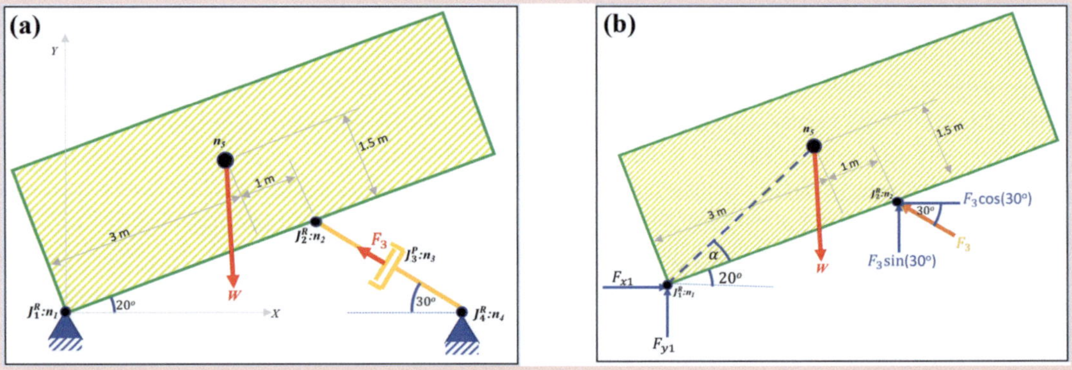

Fig. 11.18 Dumpster truck kinematic diagram and bed FBD. **a** Dumpster truck kinematic diagram**b** FBD of dumpster truck bed link L_2

$$r_{n1n5} = \sqrt{(3\ \mathrm{m})^2 + (1.5\ \mathrm{m})^2} = 3.354\ \mathrm{mD}$$

6. Define the angle α as shown in Fig. 11.18b. Calculate the angle α as

$$\alpha = \tan^{-1}\left(\frac{1.5}{3}\right) = 26.56^\circ$$

Now, we are going to apply the moment part of Eq. (11.6), considering $n_A = n_1$ as

$$\sum M_1 = (4\,\mathrm{m})(\cos(20^\circ)) \times (F_3 \sin(30^\circ) + (4\,\mathrm{m})(\sin(20^\circ)) \times (F_3 \cos(30^\circ) - (5000\,\mathrm{lb})(3.354\,\mathrm{m}) \times \cos(\alpha + 20^\circ) = 0$$

Solving the above moment summation equation in terms of F_3 to have

$$F_3 = 3763.13\,\mathrm{lb}$$

Applying the forces part of Eq. (11.6) as

$$\sum F_x = F_{x1} - F_3 \cos(30^\circ) = F_{x1} - (3763.13\,\mathrm{lb}) \times \cos(30^\circ) = 0$$

Solving the above x-direction force summation equation in terms of F_{x1} to have

$$F_{x1} = 3259\,\mathrm{lb}$$

$$\sum F_y = F_{y1} + F_3 \sin(30^\circ) - 5000 = F_{y1} + (3763.13\,\mathrm{lb}) \times \sin(30^\circ) - 5000\,\mathrm{lb} = 0$$

Solving the above y-direction force summation equation in terms of F_{y1} to have

$$F_{y1} = 3118.4\,\mathrm{lb}$$

The summary of the solution is the hydraulic force F_3=3763.13 lb. and the contact force at revolute joint J_1^R is $\vec{F}_2 = 3259\hat{i} + 3118.4\hat{j}$ lb. There is an attached video to show how the "Dumpster Truck Mechanism" works.

Example 11.4

Figure 11.19 shows a pumpjack mechanism that consists of four links: A Samson post and the ground frame link L_1, the crank and counterweight link L_2, the pitman arm link L_3, and the walking beam and horse head link L_4 plus four revolute joints at nodes n_1, n_2, n_3, and n_4. The load to be lifted is 250 lb. and it's located at node n_7. The walking beam weighs 130 lb. and is located at node n_5. The horse head weighs 60 lb. and is located at node n_6. The counterweight is 200 lbs. and centered at node n_8. There is a motor that drives the revolute joint at node n_1 through a gear ratio of *1:4*. The smaller gear is attached to the motor shaft. Find the motor torque to set the mechanism at the stationary position shown in Fig. 11.19 where the walking beam is horizontal.

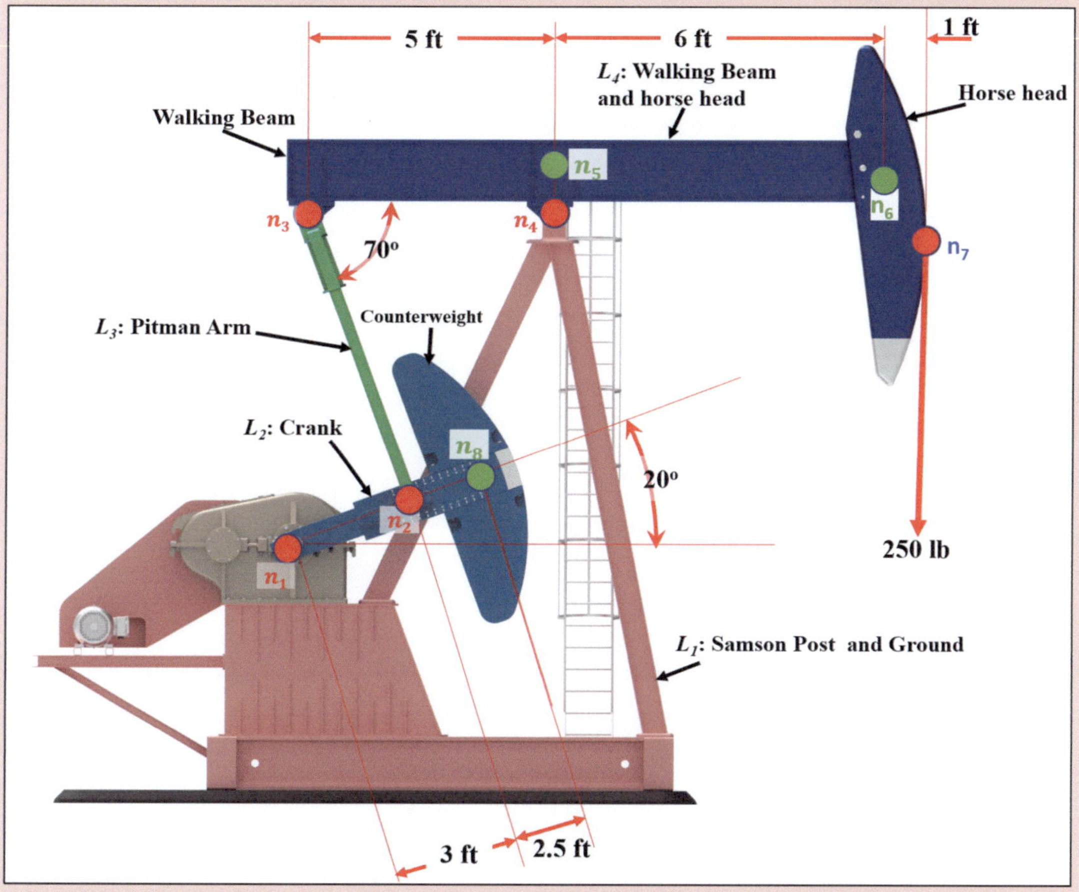

Fig. 11.19 Pumpjack mechanism. Designed by the author and implemented by both the author and Pranav Dhola. Courtesy of Pranav Dhola

Solution

The first step to thinking about sketching the FBD of the pumpjack mechanism is when we cut the ground link as shown in Fig. 11.20. We would have two reaction forces (F_{x1} and F_{y1}) at node n_1 because of the revolute joint J_1^R and two reaction forces (F_{x4} and F_{y4}) at node n_4 because of the revolute joint J_4^R. The problem is considered indeterminate because the number of the reaction forces is higher than three reaction forces and we have an unknown torque T_1.

The next is to sketch the pumpjack mechanism's FBD that is cut from the Samson post and ground link L_1 as shown in Fig. 11.20. We add two reaction forces (F_{x1} and F_{y1}) at node n_1 because of the revolute joint J_1^R and two reaction forces (F_{x4} and F_{y4}) at node n_4 because of the revolute joint J_4^R. Also, we have an unknown torque T_1. The number of unknowns is 5 (F_{x1}, F_{y1}, F_{x4}, F_{y4}, and T_1).

Now, we can use the fact that link L_3 is a two-force link and sketch the FBD of link L_4 as shown in Fig. 11.21. We assume that link L_3 has an axial force in tension. Applying the moment equilibrium equation (from Eq. (11.6)) on L_4 leads to

$$\sum M_4 = (5\,\text{ft}) \times (F_3 \sin(70°) - (6\,\text{ft}) \times (60\,\text{lb}) - (7\,\text{ft}) \times (250\,\text{lb}) = 0$$

Solving the above moment summation equation in terms of F_3 to have

$$F_3 = 449.083\ \text{lb}$$

Now, we are going to sketch the FBD of link L_2 as shown in Fig. 11.22. Applying the moment equilibrium equation (from Eq. (11.6)) on L_2 leads to

$$\sum M_1 = T_1 + (3\,\text{ft}) \times (449.083\ \text{lb}) - (5.5\,\text{ft}) \cos(20°) \times (200\,\text{lb}) = 0$$

Solving the above moment summation equation in terms of T_1 to have

$$T_1 = -313.58\ \text{lb.ft}$$

Now we can use Eq. (11.4) to calculate the motor torque as

$$T_m = \frac{T_1}{4} = 78.39\ \text{lb.ft}$$

There is an attached video to show how the "pumpjack mechanism" works.

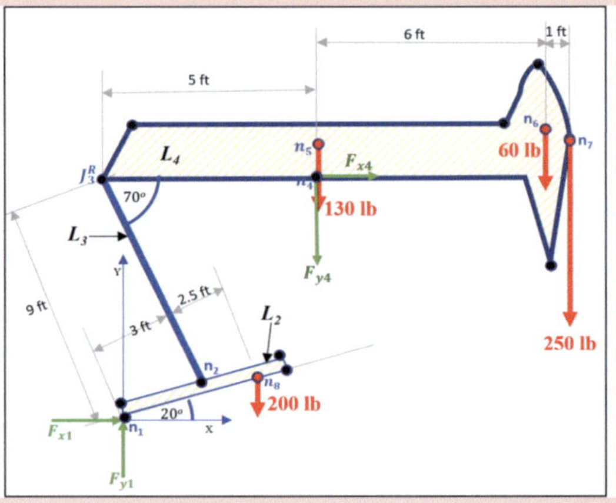

Fig. 11.20 FBD of pumpjack mechanism

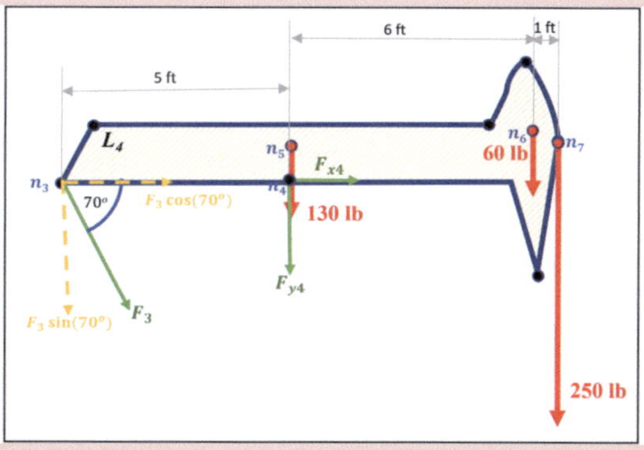

Fig. 11.21 FBD of L_4 in pumpjack mechanism

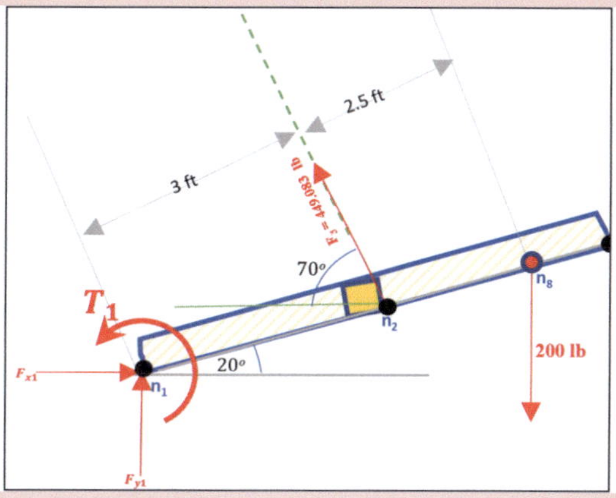

Fig. 11.22 FBD of L_2 in pumpjack mechanism

11.7 Calculating Equilibrium Forces and Moments Using Simscape™ Multibody™

For static analysis, we are interested in knowing the values of forces and torques at nodes or joints with specific external loads for a specific linkage's stationary position. For example, consider the articulated skid steer bucket mechanism in Fig. 11.3. Let's assume that we have a specific load weight W. We would like to know the required actuator forces $\overrightarrow{F}_7$ and $\overrightarrow{F}_8$ to hold the mechanism at a specific position for the displacements at joints J_7^P and J_8^P. To do this, we will need to know how to measure the forces and torques at specific nodes or joints and how to apply the forces and torques at specific nodes.

Definition[1]

Force and Torque Sensor Block:

Blocks with force and torque sensing appear in two Simscape™ Multibody™ libraries [4–6]:

Forces and Torques—Sense the magnitude of certain forces not explicitly provided by input. Blocks with force sensing include Inverse Square Law Force and Spring and Damper Force. Each block can sense only the magnitude of its own force.

Joints—Sense various forces and torques acting directly at a joint. All joint blocks provide force and torque sensing. However, the specific force and torque types that you can sense vary from joint to joint. Force and torque sensing are available strictly between the bodies the joint connects.

In Simscape™ Multibody™, we can measure the forces and the torques at any joint with an optional output from the joint block. For example, in Fig. 11.23, when we double-click on the Revolute Joint block, an editor block opens. We click on the Sensing "+" sign to open the sensor's options. We pick "Actuator Torque" followed by selecting "Auto Apply". The editor block closes. An additional output port "t" is added to the joint block. We use this "t" to measure the "Actuator Torque". We can also measure the reaction forces and torques at the joint. For example, in Fig. 11.24, when we double-click on the Revolute Joint block, an editor block opens. We then click on the Composite Force/

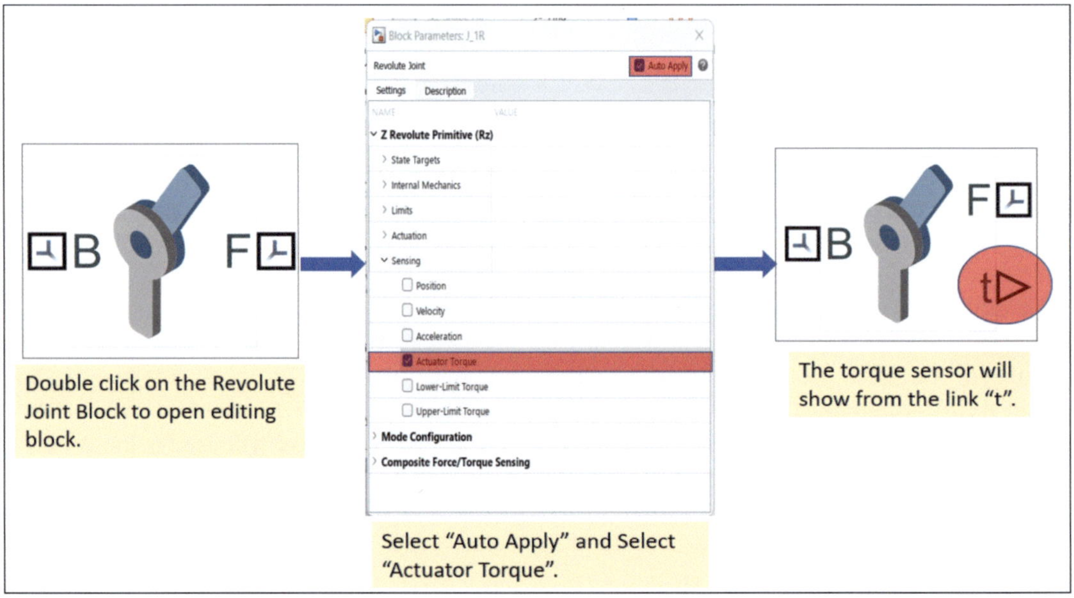

Fig. 11.23 Revolute joint actuator torque sensor. Designed and implemented by the author. Images used with permission of MathWorks

[1] Reproduced from [4–7] by MathWorks with permission from MathWorks.

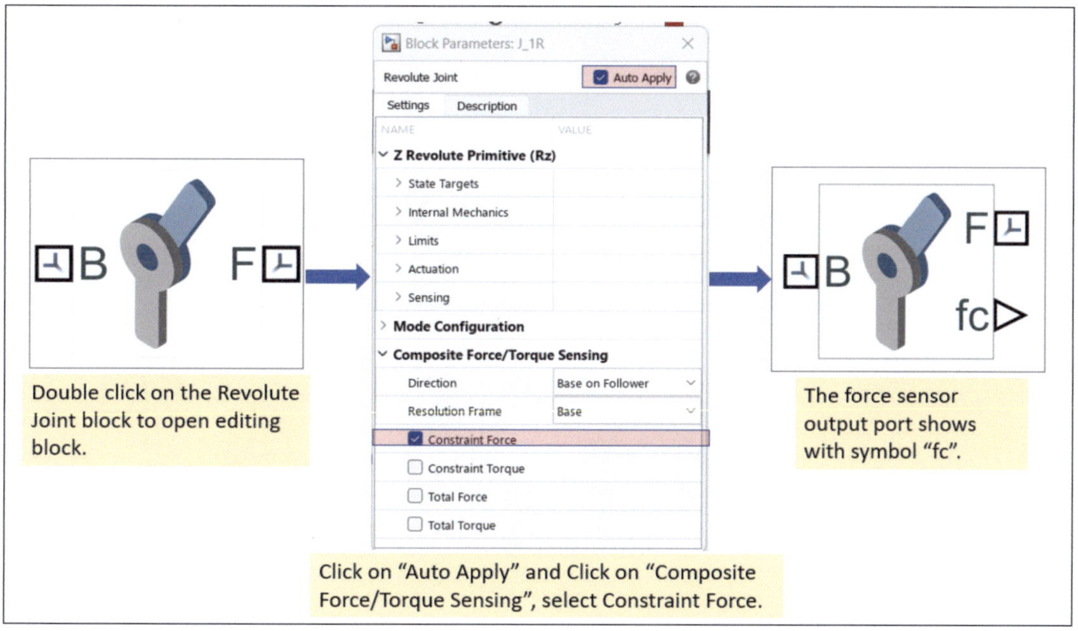

Fig. 11.24 Revolute joint reaction or contact force sensor. Designed and implemented by the author. Images used with permission of MathWorks

Torque Sensing "+" sign to access the sensor's options. We pick "Constraint Force" followed by Selecting "Auto Apply". The editor block closes. An additional output port "fc" is added to the joint block. We use this "fc" to measure the reaction forces' components. There is also an option to select the Direction and the Resolution Frame. More details about the editing of actuator forces/torques and the reaction forces/torques can be found in Refs. [4–7].

Similar force and torque sensors can be added for the rest of the 2D joints as listed in Table 11.3. Note that we do not have the force between two gears because the gear is actuated by the torque applied at the revolute joint in the center of the driver gear. The same scenario is applied in the case of a cam-and-follower scenario.

Definition[2]

External Force and Torque Block:

External Force and Torque blocks appear in two Simscape™ Multibody™ libraries [4–7].

External Force and Torque Block represent a general force and torque that an external agency applies on a rigid body frame. The force and torque input can be constant or it can vary with time. The block provides a set of physical signal ports that you use to specify this input. The ports are hidden by default. Selecting an actuation mode exposes the corresponding physical signal port.

[2] Reproduced from [4–7] by MathWorks with permission from MathWorks.

Table 11.3 Actuator force and torque joint sensors. Designed and implemented by the author. Images used with permission of MathWorks

Joint name	Actuator force or torque	Block with the actuator force or torque sensor
Prismatic joint Actuator force	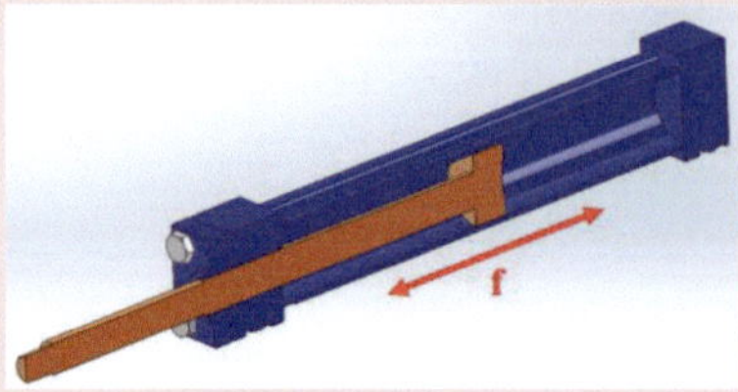	
Pin slot joint Actuator force in the local *X*-direction and actuator moment around Z-axis	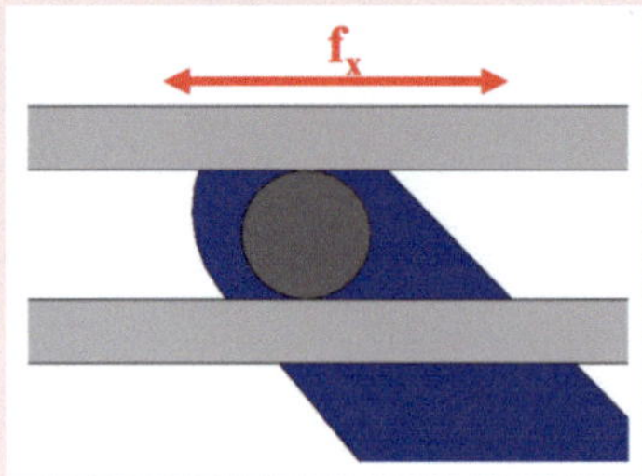	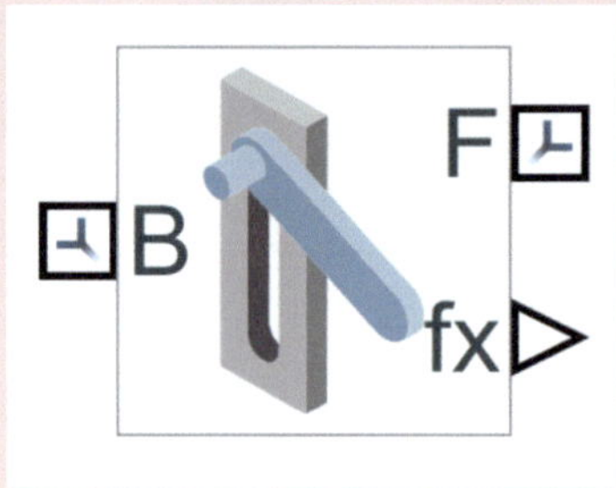

Each force and torque input acts on the origin of the follower frame in a direction that depends on the chosen force type and resolution frame. A force/torque vector component acts along/about the associated Cartesian axis, resolved in the chosen resolution frame. For example, the Force (*X*) input acts along the *X*-axis of the resolution frame. A complete force/torque vector acts in the general direction that its components specify, resolved in the chosen resolution frame.

A force input with a positive value acts along the positive direction of the associated axis. A torque input with a positive value acts about the associated axis according to the right-hand rule.

To implement any external load in a Simscape™ Multibody™ linkage model, we will need a force and/or torque source block to attach at a specific node. Such a block is called "***External Force and Torque Block***". If we scroll in the Simulink® library to reach Simscape\Multibody\Forces and Torques as shown in Fig. 11.25, we can see a few blocks. We will drag and drop the "***External Force and Torque Block"*** into a Simulink® model. After that, we double-click on the "***External Force and Torque Block***" to get the editing menu block as shown in Fig. 11.25. We can pick one or more of the forces (Force (*X*), Force (*Y*), Force (*Z*), or Force), and/or the torques (Torque (*X*), Torque (*Y*), Torque (*Z*), or Torque) and press "Auto Apply". An input port (or input ports) will be added to the "External Force and Torque Block" as shown in Fig. 11.25. We can feed the input port with the desired forces' and/or torques' signals. There are two options to select the Force Resolution Frame to be in World or Attached Frame and to select the Torque Resolution Frame to be in World or Attached Frame. More details about the editing of External Force and Torque blocks can be found in Refs. [4–7].

After knowing how to implement force and torque sensors and force and torque actuators, we can summarize static analysis using Simscape™ Multibody™ as:

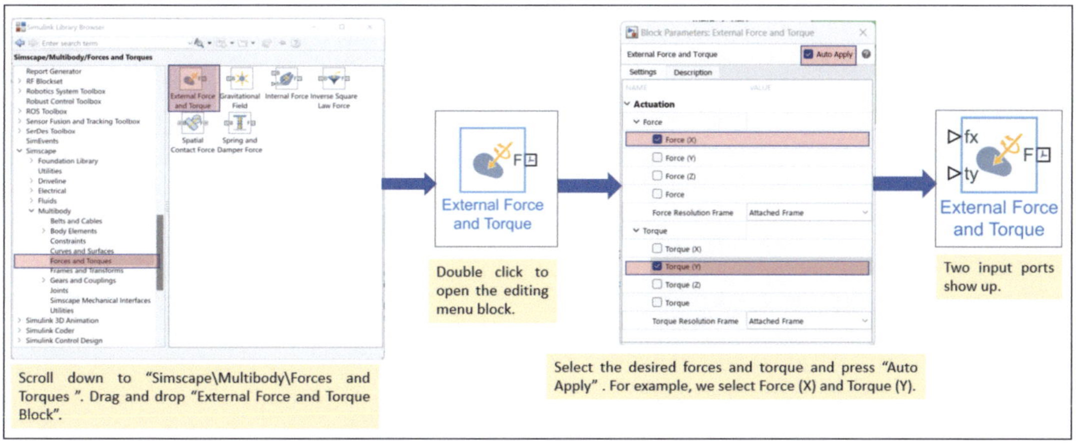

Fig. 11.25 Implementation of "external force and torque" block. Designed and implemented by the author. Images used with permission of MathWorks

1. Build the Simscape™ Multibody™ model.
2. Add the needed "Triple Inputs Rotational Motion Actuator" and/or "Triple Inputs Transitional Motion Actuator" to set the linkage at the desired position and orientation.
3. Add "External Force and Torque" Blocks at the known acting nodes.
4. Enable the forces and/or torques at the joints where we are interested in measuring the actuating force.

Example 11.5

For the stationary dumpster truck mechanism shown in Fig. 11.17 , use the Simscape™ Multibody™ model to determine the force required by the hydraulic cylinder to maintain the bed's position.

Solution

The steps to build the model for static analysis are:

1. Sketch the kinematic diagram and label the links, joints, and nodes as shown in Fig. 11.17.
2. Implement the 4 links as listed in Table 11.4 Also, with the dimensions provided in Fig. 11.17, we can use the geometric shape shown in Fig. 11.26 to calculate the rest of the dimensions as

$$r_{n1n7} = r_{n1n2}\cos(20°) = (4\,\text{m}) \times \cos(20°) = 3.76\,\text{m}$$
$$r_{n2n7} = r_{n1n2}\sin(20°) = (4\,\text{m}) \times \sin(20°) = 1.37\,\text{m}$$
$$r_{n2n4} = \frac{r_{n1n2}}{\sin(30°)} = \frac{(1.3681\,\text{m})}{\sin(30°)} = 2.73\,\text{m}$$
$$r_{n7n4} = \frac{r_{n1n2}}{\tan(30°)} = \frac{(1.3681\,\text{m})}{\tan(30°)} = 2.37\,\text{m}$$
$$r_{n1n4} = r_{n1n7} + r_{n7n4} = 3.76\,\text{m} + 2.37\,\text{m} = 6.13\,\text{m}$$

Table 11.4 Vehicle bucket mechanism four links' parameters. Designed and implemented by the author. Images used with permission of MathWorks

Link	Parameters
Vehicle Frame (L1): Ground with two revolute joints	Block Parameters: Vehicle Frame L1 Ground plus two revolute (mask) Parameters First Revoulte Joint Location (m) [0 0 0] Second Revoulte Joint Location (m) [6.1284 0 0] OK Cancel Help Apply
Arm (L4): Extrusion with two revolute joints	Block Parameters: Bed L2 Extrusion with two revolute joints (mask) Parameters Extrusion Cross Section (m) [0 0;5 0;5 3;0 3] R1 Location (m) [0 0 0] R2 Location (m) [4 0 0] Free Node Location (m) [3 1.5 0] OK Cancel Help Apply
Piston (L3):	Block Parameters: Piston L3 Brick plus revolute and prismatic (mask) Parameters Brick Dimensions [1.4 0.05 0.05] Revolute Joint Location [0.7 0 0] Prismatic Joint Location [-0.7 0 0] OK Cancel Help Apply
Cylinder (L4):	Block Parameters: Cylinder L4 Brick plus revolute and prismatic (mask) Parameters Brick Dimensions [1.4 0.1 0.1] Revolute Joint Location [-0.7 0 0] Prismatic Joint Location [0.7 0 0] OK Cancel Help Apply

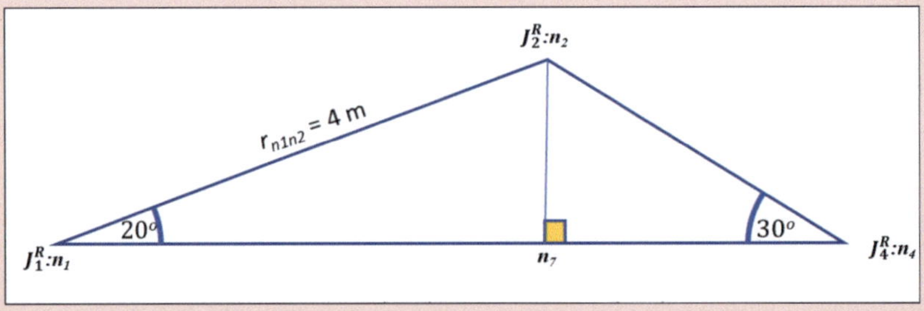

Fig. 11.26 Geometric shape of dump truck mechanism

3. The length $r_{n2n4}=2.7362$ m is the length of the cylinder plus the extended length of the piston. If we assume the length of the cylinder $r_{n3n4}=1.4$ m, then the extended length of the piston $r_{n2n3}=r_{n2n4}-r_{n3n4}=(2.73\text{ m})-(1.4\text{ m})=1.33$ m.
4. Connect the four links with two revolute joints and one prismatic joint.
5. Enable motion actuation and the actuator force sensing in the prismatic joint JP_3.

6. Enable the Constraint Force in Composite Force/Torque Sensing to the revolute joint JR_1.
7. Connect the JP_3 motion actuation input port to "Triple Inputs Translational Motion Actuator" with a displacement value $D = r_{n2n3} = 1.33$ m and $V = A = 0$.
8. Connect the "External Force and Torque Block" to the free node in the Bed L2 link. Select the force as "fy". Feed the input port "fy" with -5000 lb using "Simulink®-PS Converter block" and a Constant block.
9. Connect the JP_3 force-sensing port with a PS-Simulink® Converter block followed by a Display block.
10. Connect the JR_1 force-sensing port with a PS-Simulink® Converter block followed by a Display block.
11. Run the simulation to get the value of the force at joint JP_3 from the Display block to be 3763.1 N. That's almost the same as we get from Example 11.3. Also, the reaction force values at joint JR_1 from the Display block are shown as: $F_{x1} = 3259$ lb and $F_{y1} = 3119$ lb. These values are almost the same as what we get from Example 11.3. The entire model is shown in Fig. 11.27. Note that the model is attached for the reader to explore.

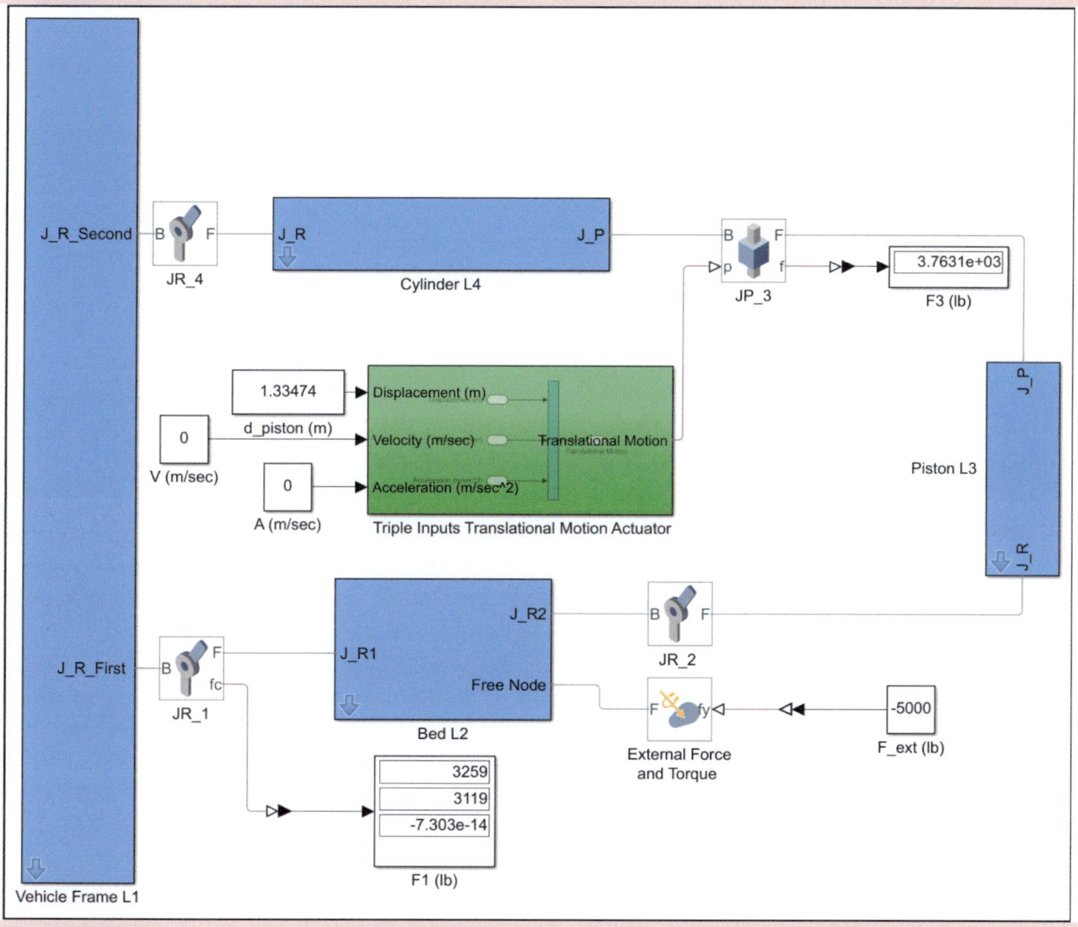

Fig. 11.27 Dump truck mechanism Simscape™ Multibody™ model

11.8 Chapter Summary

An introduction to equilibrium has been given in terms of forces and moments that support a machine's weight and inertial load plus the external forces and torques that can be from actuators or loads. An explanation of the FBD and internal acting forces and moments has been given. It has been shown how to measure equilibrium forces and moments using Simscape™ Multibody™.

Outcomes

1 Understanding the external forces and moments and their sources to be actuators or/and external loads.
2 Understanding the contact forces and moments in any 2D kinematic pair with a specific type of joint.
3 Understanding and sketching the FBD for any 2D linkage's part.
4 Understanding and using the 2D equilibrium equations (derived from Newton's first law) in calculating the determinate acting forces and the moments on any 2D machine at steady state.
5 Understanding how to add force/torque actuators and sensors to a machine Simscape™ Multibody™ model.
6 Understanding the steps to set a machine Simscape™ Multibody™ model for a specific steady state and measuring the applied forces and moments.

Problems

Problem 1

For the sewing machine shown in Fig. 11.28, the v needle bar tip against an external force F. Sketch the FBD of the linkage without the ground and the FBD of each link.

Problem 2

For the sewing machine shown in Fig. 11.28, if the external force is measured to be 10 N upward at the position where $\theta_2 = 30°$, find the required motor torque and the reaction forces at nodes n_4 and n_1. The distance between n_1 and n_2 is 15 mm. The distance between n_2 and n_3 is 30 mm. The distance between n_1 and n_4 is 70 mm. Ignore the weight of each link and each joint in your calculation.

Problem 3

For the sewing machine shown in Fig. 11.28, if the external force is measured to be 10 N upward at the position where $\theta_2 = 30°$, use the Simscape™ Multibody™ model to find the required motor torque and the reaction forces at nodes n_4 and n_1. The distance between n_1 and n_2 is 15 mm. The distance between n_2 and n_3 is 30 mm. The distance between n_1 and n_4 is 70 mm. Build your model that ignores the weight of each link and each joint.

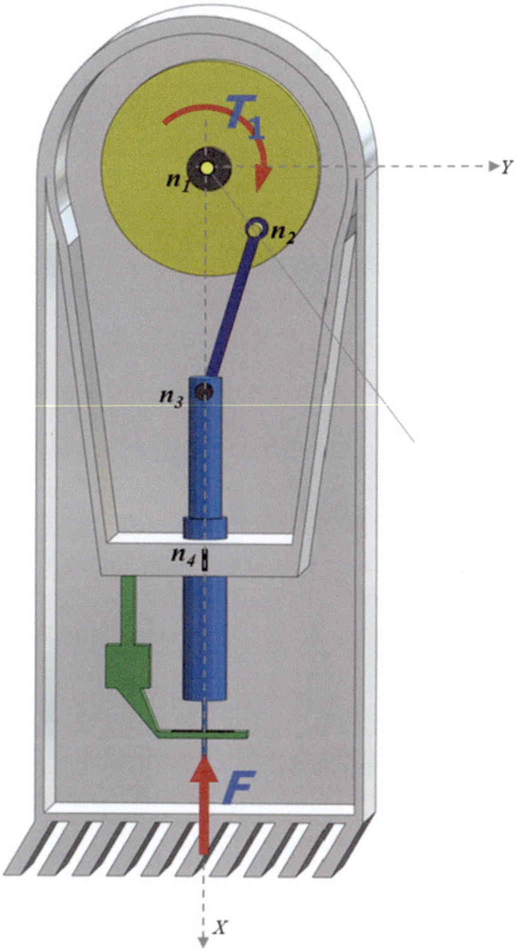

Fig. 11.28 Sewing mechanism. Designed by the author and implemented by both the author and Pranav Dhola. Courtesy of Pranav Dhola

Problem 4

For the scissor lift table mechanism shown in Fig. 11.29, the lifted mass is concentrated at the center with 250 kg. The hydraulic actuator applies a force $\boldsymbol{F}$ to hold the table in the position below. Sketch the FBD of the linkage without the ground and the FBD of each link.

Problem 5

For the scissor lift table machine shown in Fig. 11.29, find the required hydraulic actuator force F and the reaction forces at the two ground nodes. The distance between n_1 and n_2 is 15 mm. The distance between n_2 and n_3 is 30 mm. The distance between n_1 and n_4 is 70 mm. Ignore the weight of each link and each joint in your calculation.

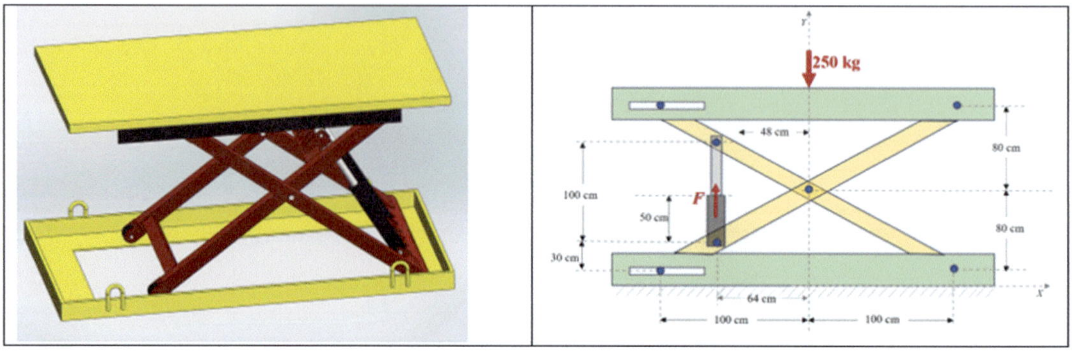

Fig. 11.29 Scissor lift table. Figure and associated Scissor Lift Table Mechanism.mp4 video designed by the author and implemented by both the author and Pranav Dhola. Both courtesy of Pranav Dhola

Problem 6

For the scissor lift table machine shown in Fig. 11.29, build and use the Simscape™ Multibody™ model to find the required hydraulic actuator force F and the reaction forces at the two ground nodes. Build your model that ignores the weight of each link and each joint.

Problem 7

For the lift vehicle mechanism shown in Fig. 11.30, a hydraulic force F is applied to lift the weight of 1000 N. Sketch the FBD of the linkage without the ground and the FBD of each link.

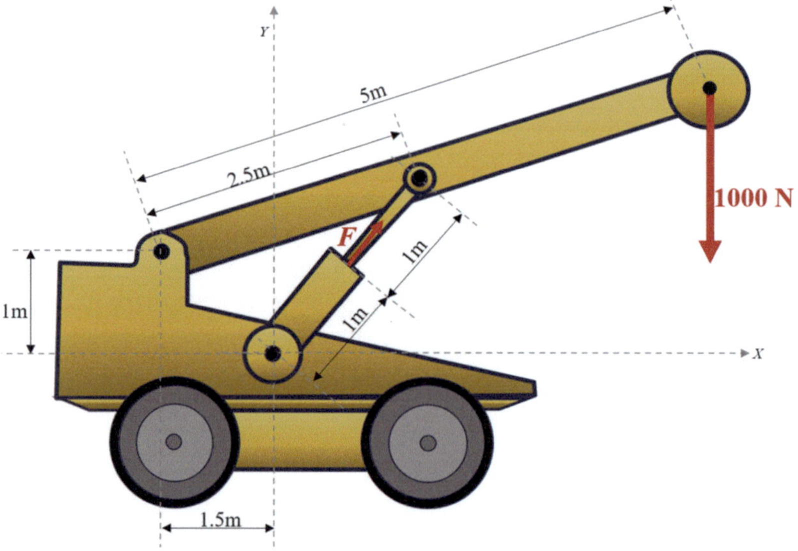

Fig. 11.30 Lift vehicle mechanism

Problem 8

For the lift vehicle mechanism shown in Fig. 11.30, find the required hydraulic force F and the reaction forces at the ground nodes. Ignore the weight of each link and each joint in your calculation.

Problem 9

For the lift vehicle mechanism shown in Fig. 11.30, build and use the Simscape™ Multibody™ model to find the required hydraulic force F and the reaction forces at the ground nodes. Build your model that ignores the weight of each link and each joint.

Problem 10

For the conveyor machine shown in Fig. 11.31 a motor torque T_1 is applied to move the box with a constant speed facing a resistance force of 30 N at node n_5. Sketch the FBD of the linkage without the ground and the FBD of each link.

Problem 11

For the conveyor machine shown in Fig. 11.31, find the required motor torque T_1 and the reaction forces at the ground nodes when r_{n2n1} has an angle 30° with the positive X-axis. Ignore the weight of each link and joint in your calculation.

Problem 12

For the conveyor machine shown in Fig. 11.31, build and use the Simscape™ Multibody™ model to find the required motor torque T_1 and the reaction forces at the ground nodes when r_{n2n1} has an angle of 30° with the positive X-axis. Build your model that ignores the weight of each link and joint.

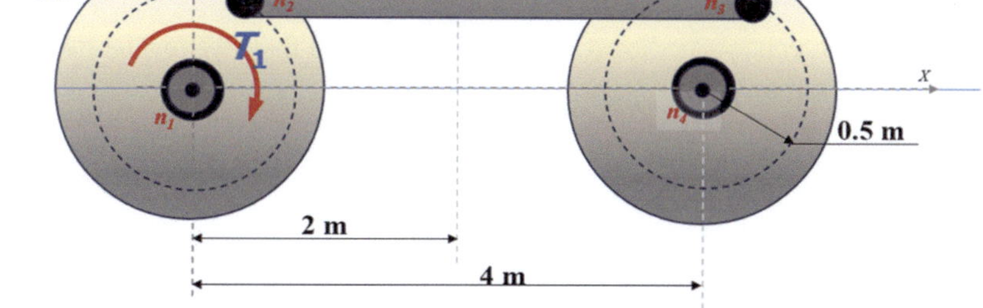

Fig. 11.31 Conveyor machine

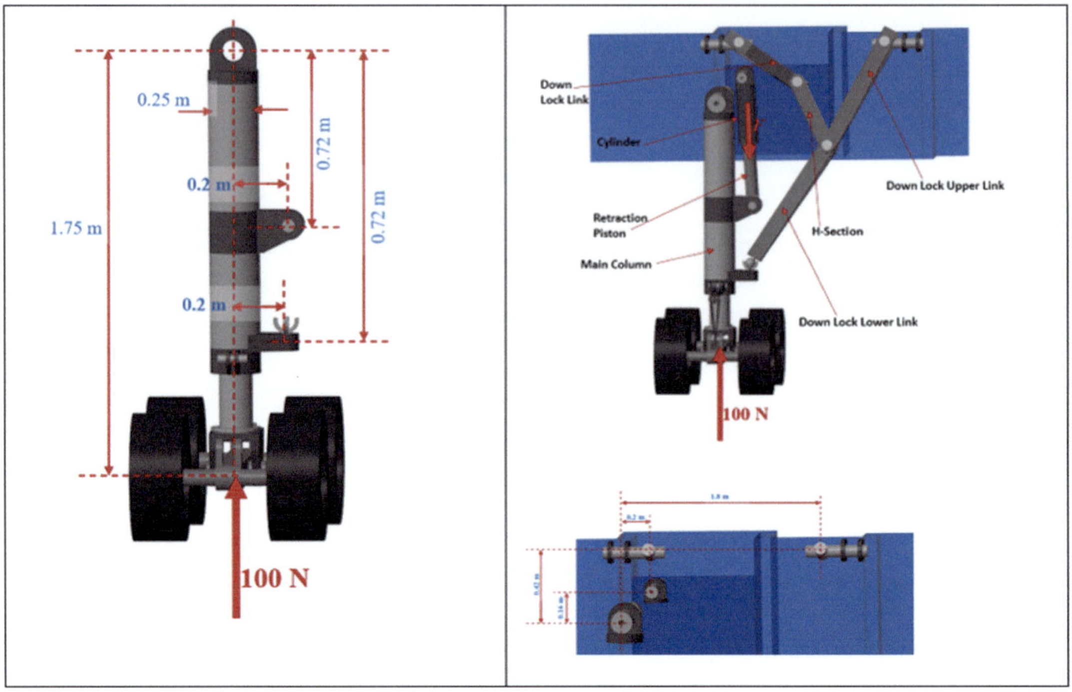

Fig. 11.32 Landing gear mechanism. Designed by the author and implemented by both the author and Pranav Dhola. Courtesy of Pranav Dhola

Problem 13

For the landing gear mechanism shown in Fig. 11.32, sketch the FBD of the linkage without the ground and the FBD of each link.

Problem 14

For the landing gear mechanism shown in Fig. 11.32, build and use the Simscape™ Multibody™ model to find the required hydraulic force and the reaction forces at the ground nodes when the piston is extended by 0.41 m to make the main column almost vertically to touch the runway and receives a reaction force of 100 N. Build your model that ignores the weight of each link and joint.

Problem 15

For the skid steer bucket mechanism shown in Fig. 11.33, sketch the FBD of the linkage without the ground and the FBD of each link.

Problem 16

For the skid steer bucket mechanism shown in Fig. 11.33 use the Simscape™ Multibody™ model to find the required hydraulic forces and the reaction forces at the ground nodes when the bucket stroke is 0.2 m, the lift stroke is 0.5 m, and the load $W_{Load} = F_{12} = 1000$ N. Build your model that ignores the weight of each link and joint.

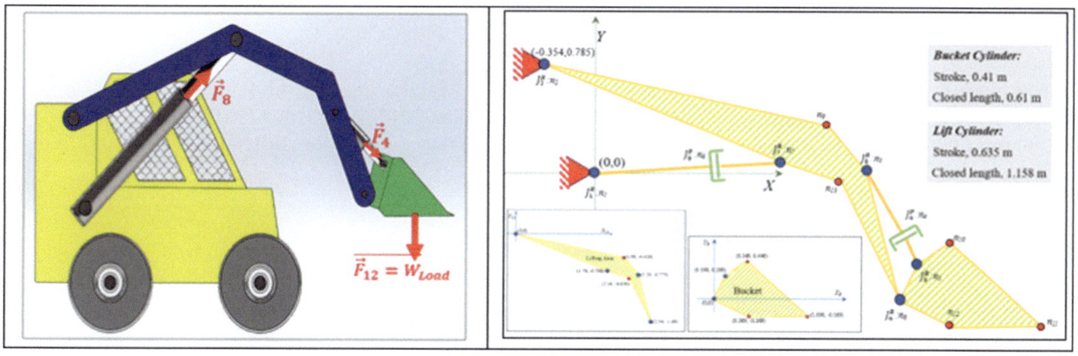

Fig. 11.33 Skid steer bucket mechanism. Figure and associated Skid Steer Bucket Mechanism.mp4 video designed by the author and implemented by both the author and Pranav Dhola. Both courtesy of Pranav Dhola

Problem 17

For the car wiper mechanism shown in Fig. 11.34, a motor torque T_1 is applied to overcome the resistance forces that applied at nodes n_6 and n_8 with a value of 5 N and 5 N. Sketch the FBD of the linkage without the ground.

Problem 18

For the car wiper mechanism shown in Fig. 11.34, build and use the Simscape™ Multibody™ model to find the required motor torque T_1 and the reaction forces at the ground nodes when the crank has an angle 30° with the positive X-axis and the wiper has angle – 30° with the positive X-axis. Build your model that ignores the weight of each link and joint.

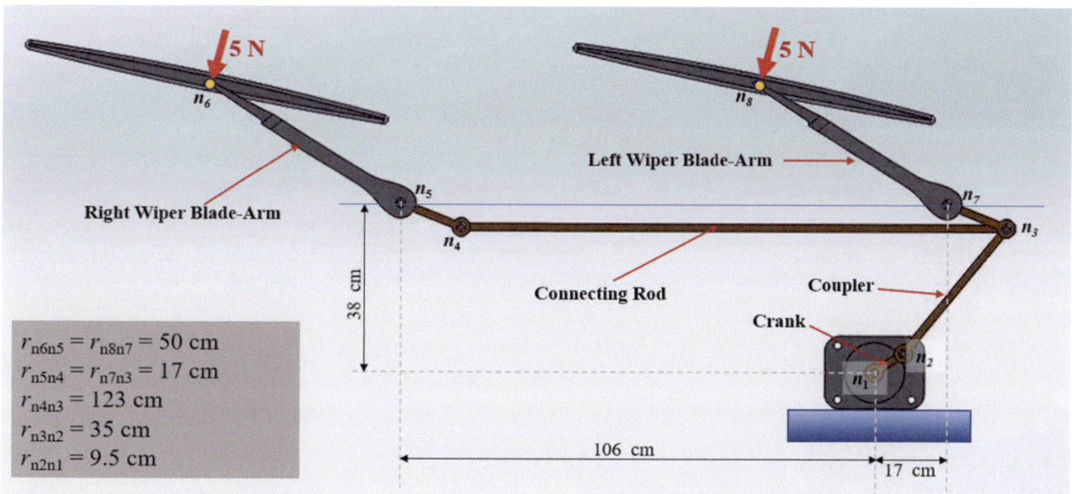

Fig. 11.34 Car wiper mechanism. Figure and associated Car_Wiper_Mechanism.mp4 video designed by the author and implemented by both the author and Pranav Dhola. Both courtesy of Pranav Dhola

References

1. “Machines & Mechanisms: Applied Kinematic Analysis”, David Myszka, Pearson; 4th Edition, 2011.
2. “Engineering Mechanics: Statics”, Russell C. Hibbeler, 14th edition, 2016, ISBN-13: 9780133918922.
3. “Engineering Mechanics: Statics”, James L. Meriam, and L. G. Kraige, 7th edition, ISBN-13: 978–0470917879
4. “Simscape™ Multibody™ Getting Started Guide,” MathWorks, 2023.
5. “Simscape™ Multibody™ User’s Guide,” MathWorks, 2023.
6. “Simscape™ Multibody™ Reference,” MathWorks, 2023.
7. “Simscape™ Multibody™ Release Notes,” MathWorks, 2023.

12 Dynamics of Machines

Chapter Objectives

1. Understanding how to calculate the mass, the center of gravity, and the mass moment of inertia of a single link.
2. Understanding how to apply Newton's second law to a single link with unbalanced forces and/or torques.
3. Understanding how to use Newton's second law to calculate the acting forces and the acting moments as an inverse dynamic model solution or to calculate the acceleration, velocity, and position as a forward dynamic model solution.
4. Understanding how to use Simscape™ Multibody™ in MATLAB® and Simulink® to calculate the forces and the moments when the machine is performing a specific motion trajectory.

12.1 Introduction

In this chapter, we start with the definitions of a link's mass, a link's center of gravity, and a link's mass moment of inertia as well as the way of calculating them from the link's shape, dimensions, and material. After that, we introduce how we use Newton's second law in developing a dynamic model. The solution of this model could be developed as a forward dynamic model solution (calculating the forces and/or the moments from a known motion) or as an inverse dynamic model solution (calculating the motion from known forces and/or moments). Two full dynamic models are developed: the four-bar mechanism dynamic model and the slider–crank mechanism dynamic model. Lastly, we present how to use Simscape™ Multibody™ to build a dynamic model-based simulation of any machine and to calculate the required actuator forces and torques and reaction forces and moments acting on a machine during a specific motion.

Supplementary Information The online version contains supplementary material available at https://doi.org/10.1007/978-3-031-72701-6_12.

A. Omran, *Kinematics and Dynamics of Machines*, https://doi.org/10.1007/978-3-031-72701-6_12

12.2 Mass

Definition

Mass:

Mass is the measure of a body's resistance to acceleration [1–3].

Engineers size and shape a set of rigid bodies (links and joints). After that, they assemble these rigid bodies to make a mechanism and/or a machine. When we, as engineers, would like to design, analyze, and simulate our machine/mechanism, we should be able to characterize the material, shape, and size of each rigid body in terms of a set of physical and mathematical parameters. One of these parameters is the mass, which we give the symbol m. The mass, m, is measured by the unit of kg in the International System of Units (SI) and "lbm" or "slug" in the U.S. Customary system. The mass, m, is a scalar value which means we can apply the scalar operations such as addition, subtraction, multiplication, and division to it.

To know how to estimate the mass of each machine link, we recall the rigid body definition in Chap. 2: "*Physically speaking, a rigid body is an idealization of a solid body in which deformation is neglected. Mathematically speaking, the distance between any two points on a rigid body remains constant regardless of any force being applied* [4]."

Alternatively, the rigid body (link) is a collection of point masses such that the distance between any two of these point masses stays constant under any force or torque. From this definition, we can calculate the mass of a rigid body as the summation or the integration of the point masses that belong to the link or

$$m = \int \mathrm{d}m \quad \text{or} \quad m = \sum_{k=1}^{n} m_k \tag{12.1}$$

where dm and m_k are point masses. The integration formula in Eq. (12.1) is used for an infinite number of point masses or a continuous model. This integration results in a scalar value, not a function, because it is done over the link. All the mass, area, and volume integrations are limited in this chapter since these integrations are all over the link. The summation formula in Eq. (12.2) is for a finite number, n, of point masses. The integration or the summation methods are used based on how we are going to analyze the shape.

The mass distribution of a link (rigid body) can be measured in many ways. One common way to measure the mass distribution is the unit mass over the unit volume or density. We give this measurement the symbol "ρ" in kg/m^3 for the International System of Units (SI) and "lbm/ft^3" or "slug/ft^3" in the U.S. Customary system. Every point mass belonging to a rigid body has a density ρ and a finite volume dV. The entire mass of the rigid body can be calculated as

$$m = \int \rho \mathrm{d}V \quad \text{or} \quad m = \sum_{k=1}^{n} \rho V_k \tag{12.2}$$

where

$$\mathrm{d}m = \rho \mathrm{d}V \quad \text{or} \quad m_k = \rho V_k$$

Density is one of a material's physical properties. Appendix A lists different density values for common machine materials.

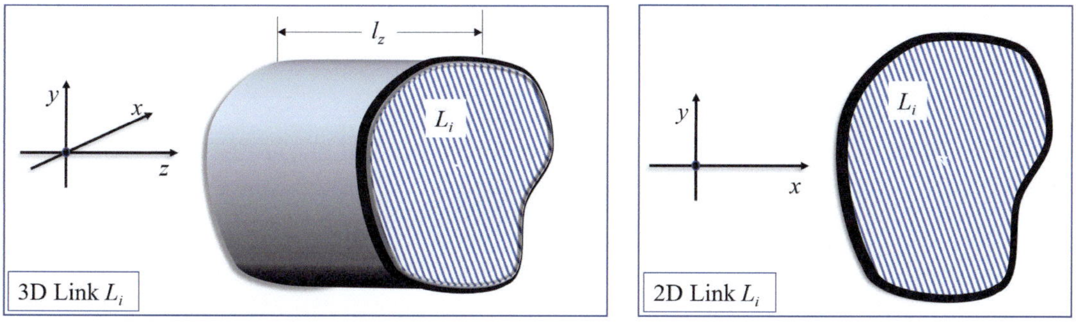

Fig. 12.1 3D and 2D link shape

In this textbook, our interest is restricted to 2D analysis. That means we assume a uniform distribution in the Z-direction. This can also be thought of as a length of extrusion or thickness. Figure 12.1 shows how we assume the 3D shape of any link in this textbook. The volume could then be condensed to

$$dV = l_z dA \tag{12.3}$$

or

$$m = l_z \int \rho(A) dA \quad \text{or} \quad m = l_z \sum_{k=1}^{n} \rho A_k$$

where l_z is the length of the element in the Z-direction as shown in Fig. 12.1. Note that the density could be constant or variable. When the link is made from a single material, it has a constant density and is called a homogenous link. When the link is made from multiple different materials, then it does not have a constant density and we call it a non-homogenous link or composite material link. Composite materials are used in many machines. For example, sometimes a single link could be designed from a wooden part and a steel part that are bonded together. Our focus here is the single material or the homogenous link, so that:

$$m = \rho l_z \int dA = \rho l_z A \quad \text{or} \quad m = \rho l_z \sum_{k=1}^{n} A_k = \rho l_z A \tag{12.4}$$

The integrations and the summations described in Eqs. (12.2) and (12.4) can be done by hand calculation or with any CAD software. Appendix B lists the volumes of some common shapes such as a cylinder, rectangular prism, triangular prism, etc. In our analysis, we are going to focus primarily on summation (thinking of a more complicated shape as the sum of its individual simple parts or discrete elements). First, we divide the link's geometry into a set of parts (i.e., discretization). Each part is selected to have one of the shapes as listed in Appendix B. We calculate the volume of each part from the given formulas in Appendix B. We use Eqs. (12.2) and (12.4) to calculate the center of gravity (C.G) of the link.

Example 12.1

Figure 12.2 shows a link shape that's made from steel. The link is brick-shaped with two circular holes. Determine the mass of this link.

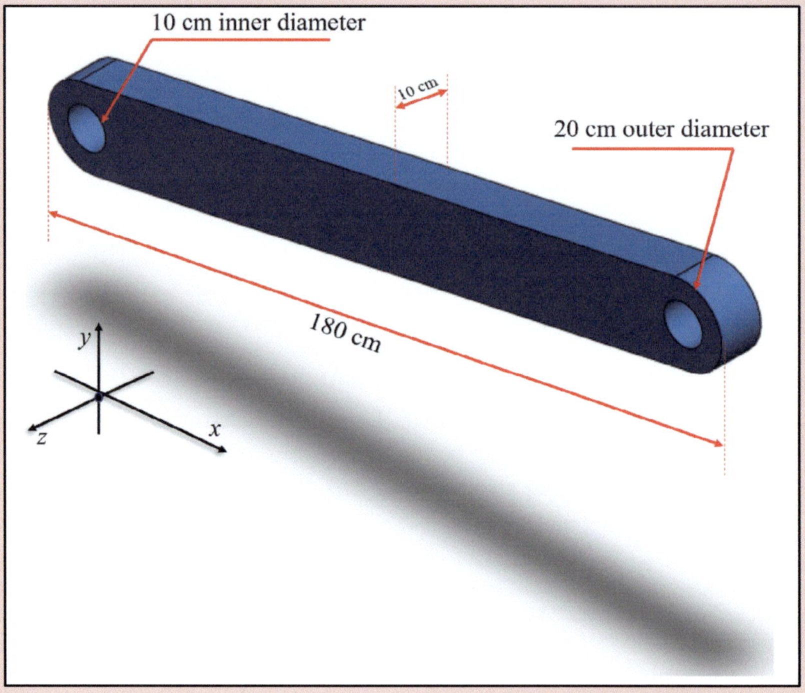

Fig. 12.2 Link shape. Designed by the author and implemented by both the author and Pranav Dhola. Courtesy of Pranav Dhola

Solution

First, we look at the steel from Appendix A to get $\rho = 7850$ kg/m^3. The *Z*-direction length, l_z = 10 cm = 0.1 m. As shown in Fig. 12.3, we break down the link into five common parts:

1. Rectangular shape, Part 1, with dimension 160 × 20 cm
2. Left outer half-circle, Part 2, with radius 10 cm
3. Left inner circle, Part 3, with radius 5 cm
4. Right outer half-circle, Part 4, with radius 10 cm
5. Right inner circle, Part 5, with radius 5 cm

We calculate the area of each part as listed in Table 12.1.

The total mass is then calculated from Eq. (12.4) as

$$\begin{aligned} m_{total} &= \rho l_z \sum_{k=1}^{5} A_k = \rho l_z (A_{Part1} + A_{Part2} - A_{Part3} + A_{Part4} - A_{Part5}) \\ &= \left(7850 \frac{\text{kg}}{\text{m}^3}\right)(0.1\ \text{m})\left(0.32\ \text{m}^2 + 0.0157\ \text{m}^2 - 0.0079\ \text{m}^2 + 0.0157\ \text{m}^2 - 0.0079\ \text{m}^2\right) \\ &= 263.446\ \text{kg} \end{aligned}$$

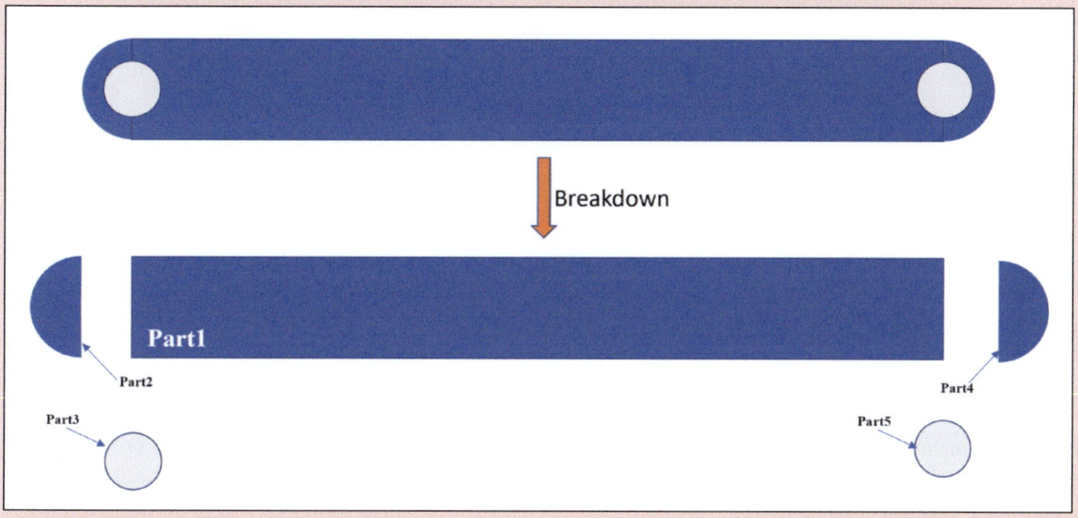

Fig. 12.3 Link breakdown (discretization) to parts

Table 12.1 Areas of parts for Example 12.1

Part	Area, from appendix B
1: Rectangular	$A_{Part1} = (1.6\text{ m}) \times (0.2\text{ m}) = 0.32\text{ m}^2$
2: Half Circle	$A_{Part2} = \frac{\pi(0.1\text{ m})^2}{2} = 0.0157\text{ m}^2$
3: Circle	$A_{Part3} = \pi(0.05\text{ m})^2 = 0.0079\text{ m}^2$
4: Half circle	$A_{Part4} = \frac{\pi(0.1\text{ m})^2}{2} = 0.0157\text{ m}^2$
5: Circle	$A_{Part5} = \pi(0.05\text{ m})^2 = 0.0079\text{ m}^2$

12.3 Center of Gravity

Definition

Center of Gravity:

The center of gravity (C.G.) of an object is the balance point of that object. [1–3]

The single body (link) total weight is a force that is equal to the total mass, m, multiplied by the gravity constant, g. As we know from Chap. 11, the force should be applied at a specific node (point). The total weight of a link is effectively applied at a node that we call the center of gravity (C.G.). Sometimes, we refer to this point as the center of mass. The C.G. is a virtual point whereby we could approximately replace the link with a point mass, m. Figure 12.4 shows how we could approximate the entire link (as a rigid body) with a collection of point masses (m_k at node n_k, where $k = \{1,2, \ldots, n\}$) to a single point mass, m, that's located at the C.G.

Knowing the location of the C.G. simplifies the solution of dynamic problems. Instead of dealing with several point masses belonging to a single link, we deal with a single point mass that's located at the C.G. of this link.

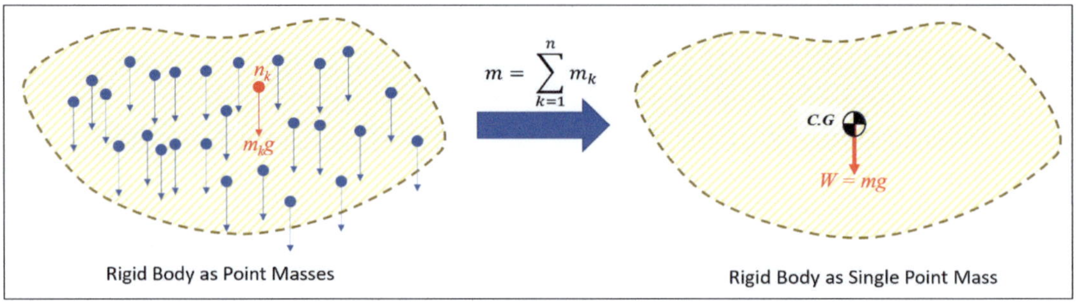

Fig. 12.4 Single body as point masses and as a single point mass

To develop the formula for calculating the C.G.'s location, let's assume a link that consists of infinite point masses. Each point mass has the mass dm and is located at a position (x,y). The C.G.'s node is defined as the average location of these point masses or

$$\begin{aligned} x_{C.G} &= \frac{\int x\mathrm{d}m}{\int \mathrm{d}m} = \frac{\int x\rho \mathrm{dV}}{\int \rho \mathrm{dV}} = \frac{\int x\rho \mathrm{dA}}{\int \rho \mathrm{dA}} \\ y_{C.G} &= \frac{\int y\mathrm{d}m}{\int \mathrm{d}m} = \frac{\int y\rho \mathrm{dV}}{\int \rho \mathrm{dV}} = \frac{\int y\rho \mathrm{dA}}{\int \rho \mathrm{dA}} \\ z_{C.G} &= \frac{\int z\mathrm{d}m}{\int \mathrm{d}m} = \frac{\int z\rho \mathrm{dV}}{\int \rho \mathrm{dV}} = \frac{\int z\rho \mathrm{dA}}{\int \rho \mathrm{dA}} \end{aligned} \tag{12.5}$$

where $x_{C.G}$,$y_{C.G}$, and $z_{C.G}$ are the components of the C.G.'s position. When we calculate the C.G.'s location for a certain link, the density ρ could be constant (homogenous link) or the density ρ could be variable (non-homogenous link or composite material link).

In the case of a discreet model, we would have a finite number of point masses. Each point mass has a mass m_k and a location (x_k, y_k) where $k = \{1, 2, \ldots, n\}$. The C.G. node is defined as the average location of these point masses or

$$\begin{aligned} x_{C.G} &= \frac{\sum_{k=1}^{n} x_k m_k}{\sum_{k=1}^{n} m_k} = \frac{\sum_{k=1}^{n} x_k \rho_k V_k}{\sum_{k=1}^{n} \rho_k V_k} = \frac{\sum_{k=1}^{n} x_k \rho_k A_k}{\sum_{k=1}^{n} \rho_k A_k} \\ y_{C.G} &= \frac{\sum_{k=1}^{n} y_k m_k}{\sum_{k=1}^{n} m_k} = \frac{\sum_{k=1}^{n} y_k \rho_k V_k}{\sum_{k=1}^{n} \rho_k V_k} = \frac{\sum_{k=1}^{n} y_k \rho_k A_k}{\sum_{k=1}^{n} \rho_k A_k} \\ z_{C.G} &= \frac{\sum_{k=1}^{n} z_k m_k}{\sum_{k=1}^{n} m_k} = \frac{\sum_{k=1}^{n} z_k \rho_k V_k}{\sum_{k=1}^{n} \rho_k V_k} = \frac{\sum_{k=1}^{n} z_k \rho_k A_k}{\sum_{k=1}^{n} \rho_k A_k} \end{aligned} \tag{12.6}$$

For the homogenous links, the ρ_k is constant and our analysis is for 2D only. Thus, the location of the C.G. can be then calculated as

$$
\begin{aligned}
x_{C.G} &= \frac{\int x\mathrm{dV}}{\int \mathrm{dV}} \quad \text{and} \quad y_{C.G} = \frac{\int y\mathrm{dV}}{\int \mathrm{dV}} \\
x_{C.G} &= \frac{\int x\mathrm{dA}}{\int \mathrm{dA}} \quad \text{and} \quad y_{C.G} = \frac{\int y\mathrm{dA}}{\int \mathrm{dA}} \\
&\text{or} \\
x_{C.G} &= \frac{\sum_{k=1}^{n} x_k V_k}{\sum_{k=1}^{n} V_k} \quad \text{and} \quad y_{C.G} = \frac{\sum_{k=1}^{n} y_k V_k}{\sum_{k=1}^{n} V_k} \\
x_{C.G} &= \frac{\sum_{k=1}^{n} x_k A_k}{\sum_{k=1}^{n} A_k} \quad \text{and} \quad y_{C.G} = \frac{\sum_{k=1}^{n} y_k A_k}{\sum_{k=1}^{n} A_k}
\end{aligned} \tag{12.7}
$$

The integrations and the summations listed in Eq. (12.7) can be done by hand calculation or by any CAD software. Appendix B lists the areas, volumes, and the C.G. locations of some common shapes such as a cylinder, a rectangular prism, a triangular prism, etc. In our analysis, we are going to focus more on the summation (discrete) technique. First, we divide the link's geometry into a set of parts (discretization). Each part is selected to have one of the shapes in Appendix B. We calculate the volume or the area plus the local C.G. of each part from the given formulas in Appendix B. We use Eq. (12.7) to calculate the C.G. of the link as a whole.

Engineering Tips—C.G. of a Symmetrical Link

If a homogenous link has a plane (line) of symmetry, the C.G. should lie on this plane (line). For example, for the rectangular shape, there are two symmetrical lines. The C.G. should lie on these two symmetrical lines. In other words, the C.G. is the intersection of these two symmetrical lines as shown in Fig. 12.5.

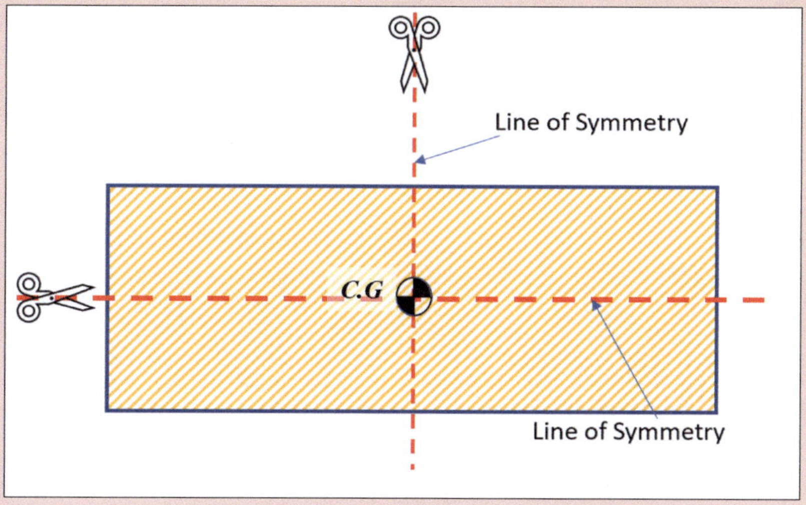

Fig. 12.5 Rectangle's C.G

Example 12.2

Figure 12.6 shows a skid steer articulated bucket mechanism that has seven links. Figure 12.7 shows a simplified shape of the bucket link L_7 with dimensions. Determine the mass of this bucket link L_7 and its C.G. location. This bucket link L_7 consists of four welded parts: Part 1 is a rectangular prism shape (5 × 600 × 1540 mm), Part 2 is a right triangular prism shape (760 × 600 × 5 mm), Part 3 is a right triangular prism shape (760 × 600 × 5 mm), and Part 4 is a rectangular prism shape (760 × 5 × 1550 mm). All the parts are made of steel.

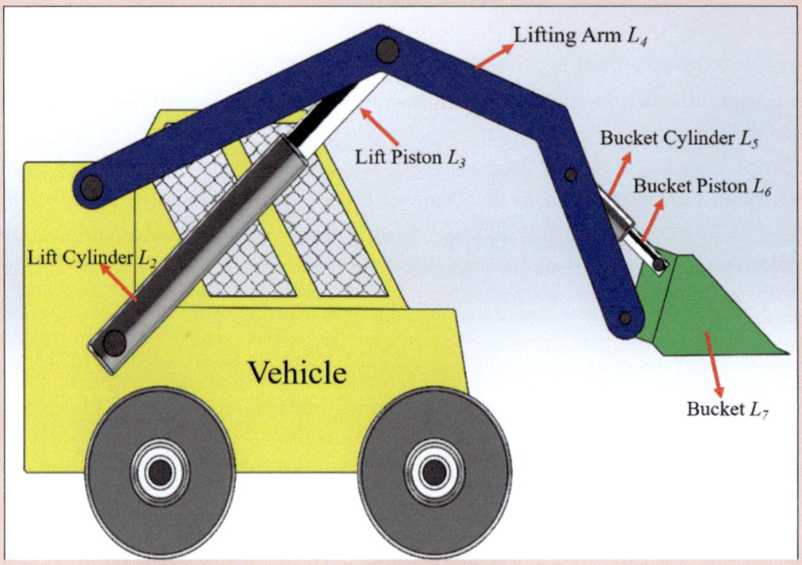

Fig. 12.6 Skid steer articulated bucket mechanism. Figure and associated Skid Steer Articulated Bucket Lift Mechanism.mp4 video designed by the author and implemented by both the author and Pranav Dhola. Both courtesy of Pranav Dhola

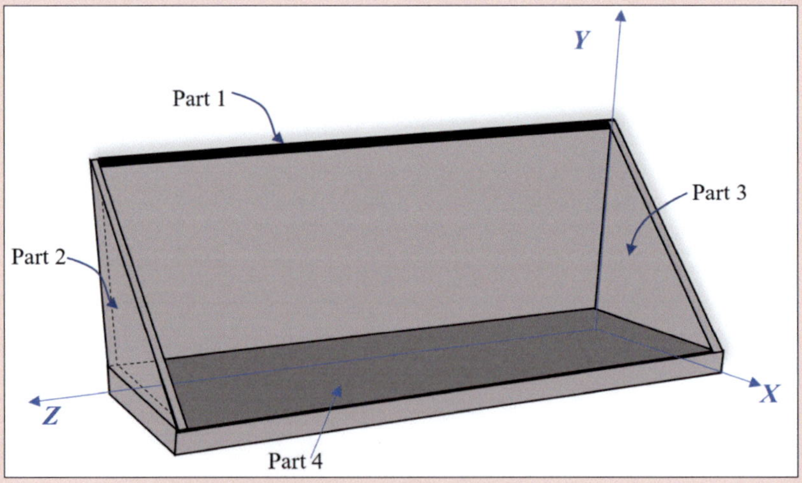

Fig. 12.7 Bucket link from skid steer articulated bucket mechanism

Solution

From Appendix A, the density of all four parts is 7850 kg/m^3. Figure 12.8 shows the bucket link in the *X-Y* plane. Using Appendix B, we list the volume and the C.G. location of each in Table 12.2.

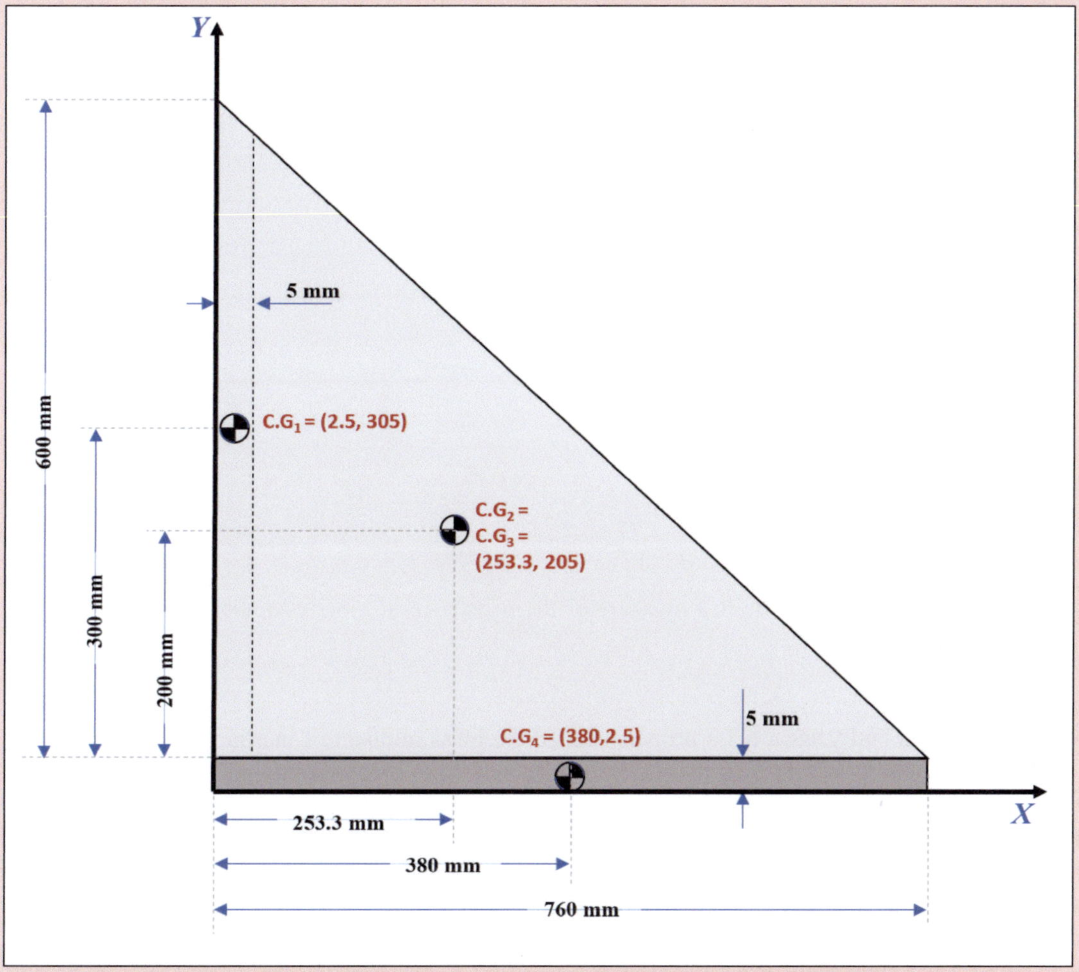

Fig. 12.8 Skid steer bucket in X-Y plane

Table 12.2 Volume and C.G. parameters of skid steer parts, Example 12.2

Part number	V (mm^3)	$x_{C.G}$ (mm)	$y_{C.G}$ (mm)	$x_{C.G}V$ (mm^4)	$y_{C.G}V$ (mm^4)
1 Rectangular Prism	$V_1 = (5)(600)(1540) = 4{,}620{,}000$	2.5	305	11,550,000	1,409,100,000
2 Triangular Prism	$V_2 = 1/2(760)(600)(5) = 1{,}140{,}000$	253.3	205	288,800,000	233,700,000
3 Triangular Prism	$V_3 = 1/2(760)(600)(5) = 1{,}140{,}000$	253.3	205	288,800,000	233,700,000
4 Rectangular Prism	$V_4 = (760)(5)(1550) = 5{,}890{,}000$	380	2.5	2,238,200,000	14,725,000
Summation	12,790,000			2,827,350,000	1,891,225,000

To calculate the total mass, we apply Eq. (12.2)

$$m_7 = \sum_{k=1}^{4} \rho V_k = \rho \sum_{k=1}^{4} V_k = \left(7850 \frac{\text{kg}}{\text{m}^3}\right)\left(12,790,000 \times 10^{-9}\ \text{m}^3\right) = 100.4\ \text{kg}$$

To calculate the C.G., we apply Eq. (12.7)

$$y_{C.G7} = \frac{\sum_{k=1}^{4} y_k V_k}{\sum_{k=1}^{4} V_k} = \frac{1,891,225,000}{12,790,000} = 147.9\ \text{mm}$$

$$x_{C.G7} = \frac{\sum_{k=1}^{4} x_k V_k}{\sum_{k=1}^{4} V_k} = \frac{2,827,350,000}{12,790,000} = 221.06\ \text{mm}$$

Note the location of the bucket L_7's C.G. (221.06, 147.9) is located at the average of the four parts' C.G.s.

12.4 Mass Moment of Inertia

Definition

Mass Moment of Inertia:

The mass moment of inertia is a measure of the resistance of a body to angular acceleration. [1–3]

A link, as a rigid body, has a mass property called the mass moment of inertia I_ξ around the ξ-axis. The mass moment of inertia I_ξ measures the resistance of the link to rotational acceleration around the ξ-axis. In other words, the higher the moment of inertia I_ξ, the more resistance there is for the link to speed the rotation up or to slow the rotation down around the ξ-axis. For example, we could have two links with the same mass and the same C.G. location but with different shapes. The link which has a higher moment of inertia I_ξ around the ξ-axis will be more difficult to speed up or slow down the rotation around the ξ-axis.

The link, as a rigid body, is a collection of point masses. We consider one point mass dm belonging to the link that rotates around the ξ-axis as shown in Fig. 12.9. The moment of inertia of this point mass dm is calculated as

$$I_\xi = \int r^2 \mathrm{d}m \tag{12.8}$$

where r is the shortest distance from the location of this point mass dm to the ξ-axis. In other words, r is the perpendicular distance from the ξ axis to the point mass dm.

In general, the mass moment of inertia, I, is measured by kg.m^2 in the International System of Units (SI) and "lbm.ft^2" or "slug.ft^2" in the U.S. Customary system. The mass moment of inertia, I, is a scalar value which means we can apply the scalar operations such as addition, subtraction, multiplication, and division.

Let's consider the point mass dm which has a density ρ and a volume dV. Equation (12.8) can be written as

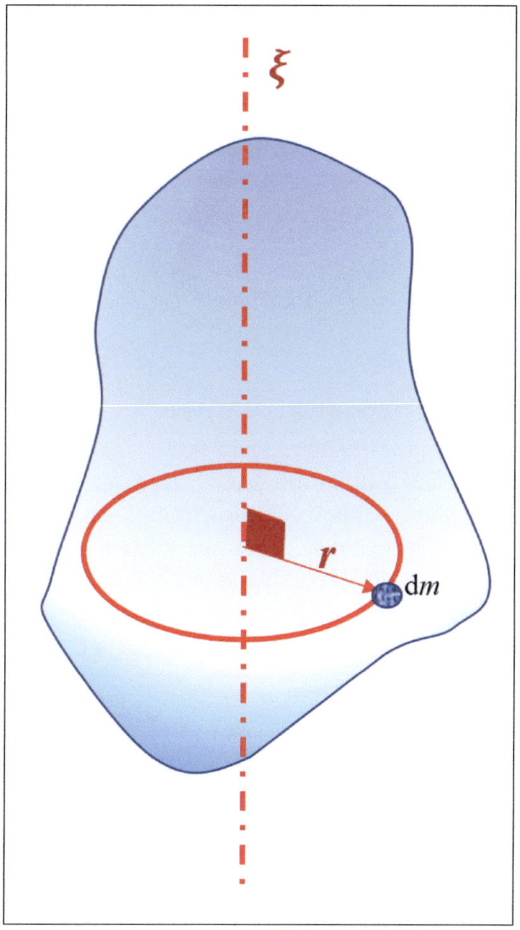

Fig. 12.9 General link shape

$$I_{\xi} = \int r^2 \mathrm{d}m = \int r^2 \rho \mathrm{d}V \tag{12.9}$$

where

$$\mathrm{d}m = \rho \mathrm{d}V$$

In the case of uniform distribution in the Z-direction, which is our interest in this textbook, Eq. (12.8) could be written as

$$I_{\xi} = \int r^2 \mathrm{d}m = \int r^2 \rho \mathrm{d}V = \int r^2 \rho l_z \mathrm{d}A \tag{12.10}$$

where

$$\mathrm{d}V = l_z \mathrm{d}A$$

where l_z is the thickness or the length of the element in the Z-direction as shown in Fig. 12.1. The integration in Eq. (12.10) can be hand calculated or computer calculated using CAD software. Appendix B lists the mass moments of inertia around a Z-axis that is located at the C.G of some common shapes.

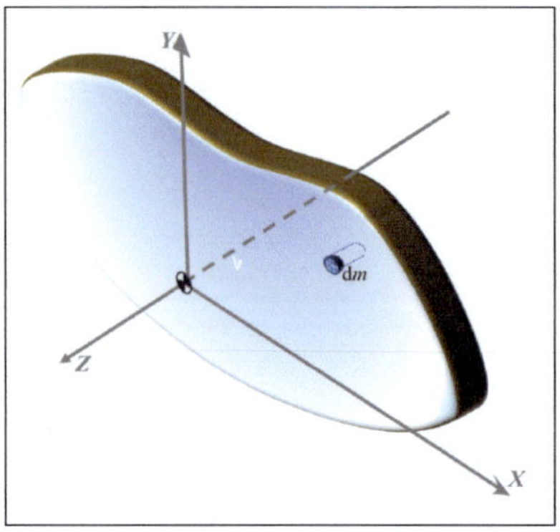

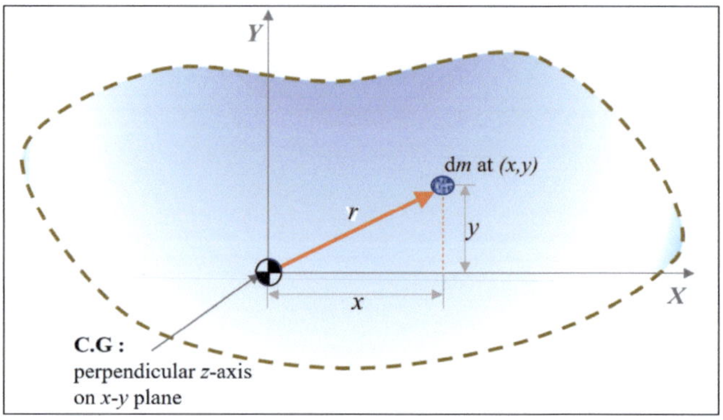

Fig. 12.10 General link in X-Y plane

In this textbook, we restrict our study to 2D motion. In our case, this is rotation around the *Z*-axis. For example, when we study a four-bar mechanism, the rotational motions ($\theta_1, \theta_2, \theta_3$, and θ_4) are about the *Z*-axis. We don't consider any rotational motion around the other two axes (*X*-axis or *Y*-axis). Figure 12.10 shows an *X-Y* plane view of a link rotating around the *Z*-axis, which is perpendicular to the *X-Y* plane at the link's C.G. The moment of inertia around the *Z*-axis can be written as

$$I_z = \int r^2 \mathrm{d}m = \int r^2 \rho \mathrm{d}V = \int r^2 \rho l_z \mathrm{d}A \tag{12.11}$$

or

$$I_z = \int (x^2 + y^2) \mathrm{d}m = \int (x^2 + y^2) \rho \mathrm{d}V = \int (x^2 + y^2) \rho l_z \mathrm{d}A$$

where

$$r^2 = x^2 + y^2$$

In Eq. (12.11), we show that the moment of inertia I_z can be calculated in Cartesian coordinates by replacing *r* with *x* and *y* from the geometry as shown in Fig. 12.10 where $r^2 = x^2 + y^2$.

Note, Fig. 12.10 is an example where the location of the *Z*-axis is not necessarily located at the link's C.G. The *Z*-axis could be located at any n_j. We use the symbol I_i for the moment of inertia of a link L_i around the Z-axis that's located at the link's C.G. We use the symbol I_{inj} for the moment of inertia of a link L_i around the *Z*-axis that's located at a node n_j. Now, we are going to introduce the parallel axis theorem to calculate I_{inj} as a function of I_i (around the C.G).

Definition

Parallel Axis Theorem:

The parallel axis theorem states that the moment of inertia of a body about an axis parallel to another axis passing through the center of a mass is equal to the sum of the moment of inertia of the body about an axis passing through the center of the mass and the product of the mass and the square of the distance between the two axes. [1–3]

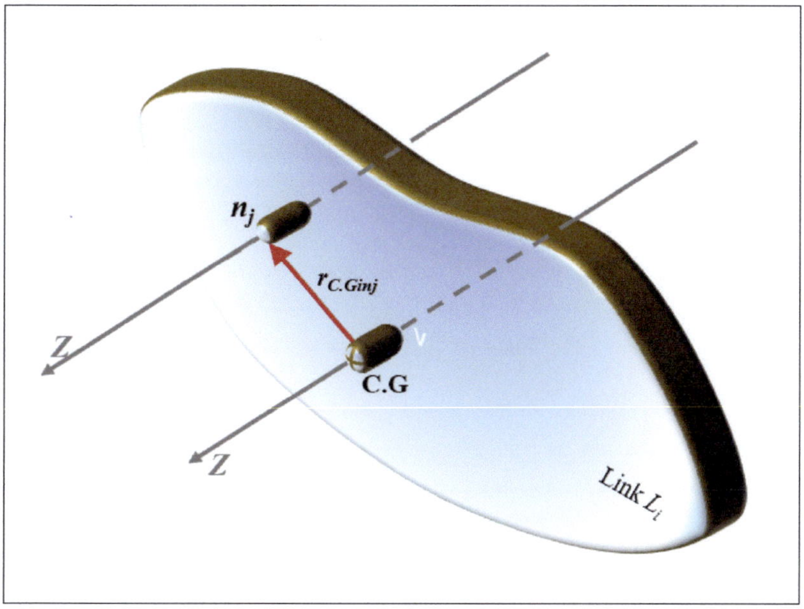

Fig. 12.11 Link L_i with two parallel Z-axes

Figure 12.11 shows a link L_i and two parallel Z-axes that are located at node n_j and the C.G. node. Using the parallel axis theorem, the relationship between I_{inj} and I_i is given as

$$I_{inj} = I_i + mr_{C.Ginj}^2 \tag{12.12}$$

where $r_{C.Ginj}$ is the radius from the link L_i's C.G. node to the node n_j.

In this textbook, we use this parallel axis theorem to calculate the moment of inertia for a link L_i using the steps below.

Calculating Link's Moment of Inertia

- Divide the link's geometry into a set of n parts, $k = \{1,2, 3, \ldots, n\}$. Each part should have the same shape as one of the shapes in Appendix B.
- Calculate each part's volume using the formulas in Appendix B.
- Calculate each part's C.G. ($x_{C.Gk}$ and $y_{C.Gk}$) using the formulas in Appendix B.
- Calculate each part's moment of inertia I_k using the formulas in Appendix B.
- Calculate each part's mass using Eq. (12.2) as

$$m_k = \rho V_k$$

- Calculate the link's C.G. ($x_{C.Gi}$ and $y_{C.Gi}$) using Eq. (12.6) as

$$x_{C.Gi} = \frac{\sum_{k=1}^{n} x_k m_k}{\sum_{k=1}^{n} m_k}$$

$$y_{C.Gi} = \frac{\sum_{k=1}^{n} y_k m_k}{\sum_{k=1}^{n} m_k}$$

- Calculate the summation of the moments of inertia of all the link's parts using Eq. (12.12) as

$$I_{inj} = \sum_{k=1}^{4} \left(I_k + m_k r^2_{C.GiC.Gk}\right)$$

where

$$r^2_{C.GiC.Gk} = x^2_{C.GiC.Gk} + y^2_{C.GiC.Gk}$$

$$x_{C.GiC.Gk} = x_{C.Gk} - x_{C.Gi} \quad \text{and} \quad y_{C.GiC.Gk} = y_{C.Gk} - y_{C.Gi}$$

The radius $r_{C.GiC.Gk}$ is from the link L_i's C.G. location to the kth part's C.G. location.

Example 12.3

Previously in Example 12.2 we calculated the mass and the C.G. of the bucket link L_7. Continue the calculation and find the mass moment of inertia I_7 around the Z-axis that's passing through the link L_7's C.G.

Solution

The bucket link has four parts, $k = \{1, 2, 3, 4\}$, that are individually shown in Fig. 12.12 with the calculated mass m_k, C.G. location ($x_{C.Gk}$ and $y_{C.Gk}$), and moment of inertia I_k in Table 12.3. These calculated masses, C.G.s, and moments of inertia are based on Appendix A and Appendix B.

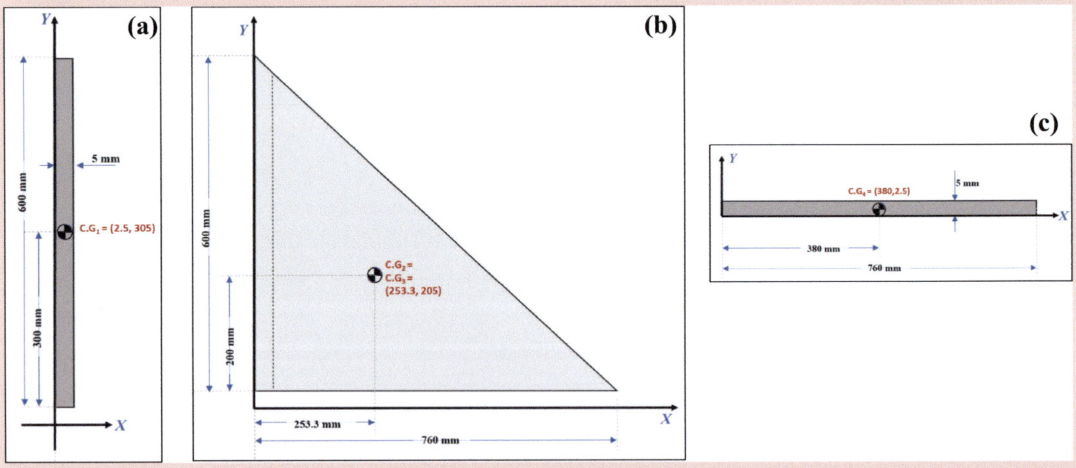

Fig. 12.12 Skid steer bucket's parts

Table 12.3 Skid steer bucket's parts: mass, moment of inertia, and C.G.

Part	Dimension	V	$m = \rho V$ ($\rho = 7850$ kg/m^3)	I	C.G. location
Part 1 Rectangular Prism	$w = 5$ mm $h = 600$ mm $d = 1540$ mm	$V = w \times h \times d$ $= 4.62\times 10^6$ mm^3	$m = 36.27$ kg	$I = \frac{m}{12}\left(w^2 + h^2\right)$ $= 1.088$ kg m^2	$x_{c.g} = 2.5$ mm $y_{c.g} = 305$ mm
Part 2 and Part 3 Triangular Prism	$w = 760$ mm $h = 600$ mm $d = 5$ mm	$V = 1/2 \times w \times h \times d$ $= 1.14\times 10^6$ mm^3	$m = 8.95$ kg	$I = \frac{m}{36}\left(w^2 + h^2\right)$ $= 0.233$ kg m^2	$x_{c.g} = 253.3$ mm $y_{c.g} = 205$ mm
Part 4 *Rectangular Prism*	$w = 760$ mm $h = 5$ mm $d = 1550$ mm	$V = w \times h \times d$ $= 5.89\times 10^6$ mm^3	$m = 46.24$ kg	$I = \frac{m}{12}\left(w^2 + h^2\right)$ $= 2.226$ kg m^2	$x_{c.g} = 380$ mm $y_{c.g} = 2.5$ mm

From Example 12.2, we have $x_{C.G7} = 221.06$ mm and $y_{C.G7}$ =147.9 mm Now, we apply Eq. (12.12) as

$$
\begin{aligned}
I_7 = \sum_{k=1}^{4} \left(I_k + m_k r^2_{C.GiC.Gk}\right) &= \left(1.088 \text{ kg m}^2\right) + \left(0.233 \text{ kg m}^2\right) + \left(0.233 \text{ kg m}^2\right) + \left(2.226 \text{ kg m}^2\right) \\
&+ (36.27 \text{ kg}) \times \left((2.5 \text{ mm} - 221.06 \text{ mm})^2 + (305 \text{ mm} - 147.9 \text{ mm})^2\right) \times 10^{-6} + 2 \\
&\times (8.95 \text{ kg}) \times \left((253.3 \text{ mm} - 221.06 \text{ mm})^2 + (205 \text{ mm} - 147.9 \text{ mm})^2\right) \times 10^{-6} \\
&+ (46.24 \text{ kg}) \times \left((380 \text{ mm} - 221.06 \text{ mm})^2 + (2.5 \text{ mm} - 147.9 \text{ mm})^2\right) \times 10^{-6} \\
&= 8.6253 \text{ kg.m}^2
\end{aligned}
$$

The above substitution can be done in the form of the equation form as listed above or it could be done using a table.

12.5 Dynamic Model

Dynamic modeling is very important in the design and analysis of machines. There are many techniques to develop a dynamic model such as the lumped parameter dynamic method, the principle of D'Almbert, and the energy method—virtual work. More details about these different techniques are given in Ref. [4]. In this textbook, we will use Newton's second law to develop the dynamic model for a machine.

12.5.1 Newton's Second Law

Definition

Newton's Second Law:

A body that has an unbalanced force has an acceleration that is proportional to the force, in the direction of the force, and inversely proportional to the mass of the object [1–4].

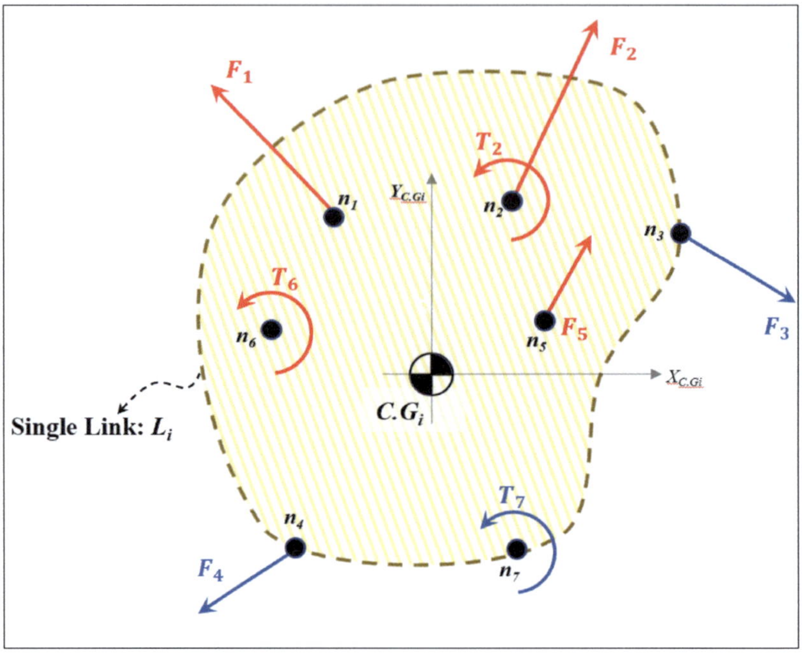

Fig. 12.13 Forces and moments acting on linkage's part

Newton's second law states that an unbalanced force and/or moment on a body causes an acceleration. Thus, if a single link (shown in Fig. 12.13) is exposed to unbalanced forces and/or unbalanced moments, then we can locate a local non-rotating frame at a $C.G_i$ and we can apply Newton's second law in 2D as [2, 3].

$$\begin{aligned}\sum F_x &= m_i a_{xC.Gi}\\ \sum F_y &= m_i a_{yC.Gi}\\ \sum M_A &= I_i \alpha_i\end{aligned} \tag{12.13}$$

The sources of the acting forces on L_i can be external loads (red arrows) or contact forces (blue arrows). The sources of the moments acting on L_i can be from external torques (red circular arrows) or from reaction moments (blue circular arrows). Equation (12.13) indicates that the summation of the forces acting on the link in both the X-direction $\sum F_x$ and the Y-direction $\sum F_y$ are equal to the link's mass, m_i, multiplied by transitional acceleration components ($a_{xC.Gi}$and$a_{yC.Gi}$) that are measured at the link's C.G. in both the X-direction and the Y-direction. Also, the summation of the moments acting around the Z-axis located at the link's C.G. is equal to the moment of inertia, I_i, multiplied by the link's angular acceleration α_i. The summation of the moments is the summation of the actual acting torques plus the summation of the cross products of the radii and the acting forces. The radii are from the link's C.G. to the node whereby the acting forces are located. For example, if we have a force F_j located at node n_j, this force will produce a moment around the Z-axis (located at the C.G. and perpendicular to *the xy* plane) that's equal to the cross product of r_{CGnj} and F_j.

For example, let's apply Eq. (12.13) to the single link L_i (in Fig. 12.13) as

$$\begin{aligned}&\sum F_x = F_{x1} + F_{x2} + F_{x3} + F_{x4} + F_{x5} = m_i a_{xC.Gi}\\ &\sum F_y = F_{y1} + F_{y2} + F_{y3} + F_{y4} + F_{y5} = m_i a_{yC.Gi}\\ &\sum M_1 = T_2 + T_6 + T_7 + \vec{r}_{C.Gin2} \times \vec{F}_2 + \vec{r}_{C.Gin3} \times \vec{F}_3 + \vec{r}_{C.Gin4} \times \vec{F}_4 + \vec{r}_{C.Gin5} \times \vec{F}_5 = I_i \alpha_i\end{aligned} \tag{12.14}$$

where $\vec{r}_{C.Ginj}$, $j = \{2, 3, 5, 4\}$ is the radius from the link's C.G. to the node n_j at where F_j is located. In 2D, the number of dynamic equations is three. We can use these equations to find three unknown reaction forces, reaction moments, actuator forces, and/or actuator torques.

In machines, a linkage consists of a set of assembled links and joints. Each link is assumed to be a rigid body that has its own acceleration (both transitional and rotational), mass, mass moment of inertia, and center of gravity location. Equation (12.13) can be applied to a single body or a single link. We can't apply Eq. (12.13) to the entire machine or to any linkage part that has more than a single link. To determine the dynamic model of the entire machine using Eq. (12.13), we should first sketch the linkage FBDs as disassembled links (sketch the FBD for each individual link), then write the Eq. (12.13) for each link, and finally solve the resultant equations.

For example, let's discuss how to develop the dynamic model of a four-bar mechanism. There are four links, one ground link L_1 (we don't care about the dynamics of the ground since the ground has no motion), and three non-ground links or three rigid bodies (L_2, L_3, L_4) with three different masses (m_2, m_3, m_4), three different moments of inertia (I_2, I_3, I_4), and three different C.G.s ($C.G_2$, $C.G_3$, $C.G_4$). Applying Eq. (12.13) for each of the non-ground links, we will end up with three equations for each link for a total of nine equations. After that, we solve these nine equations. If we have a machine with more links, we end up with more equations. In the next subsection, we will show the dynamic model of the four-bar mechanism and of the slider–crank mechanism.

A machine's dynamic equations or Eq. (12.13) can represent two general models: a direct (forward) dynamic model and an inverse dynamic model. In the direct dynamic model, the actuators' forces and/or torques are known while the links' motions (position, velocity, and acceleration) are unknown. In other words, the input is the actuators' forces and/or the actuators' torques. The outputs are the accelerations described by a kinematic model used to calculate the links' velocities and positions. For the inverse dynamic model, these motions are known and the actuators' forces and/or torques are unknown. In other words, the input signals are the motions. The output signals are the actuators' forces and/or the actuators' torques. In both of these two models, the reaction forces and moments are middle unknown variables that need to be calculated.

Example 12.4

For the gate opener shown in Fig. 12.14, consider a motor that opens the door from 0° to 85° using the SVAJ below

$$\theta_2(t) = \begin{cases} 0 & t < 0.5 \\ 42.5(t-0.5) & 0.5 \le t \le 2.5 \\ 85 & 2.5 < t \le 3 \end{cases}$$

$$\omega_2(t) = \begin{cases} 0 & t < 0.5 \\ 42.5 & 0.5 \le t \le 2.5 \\ 0 & 0.5 \le t \le 3 \end{cases}$$

$$\alpha_2(t) = \begin{cases} 0 & t < 0.5 \\ 0 & 0.5 \le t \le 2.5 \\ 0 & 2.5 < t \le 3 \end{cases}$$

where $\theta_2(t)$ is measured in deg, $\omega_2(t)$ is measured in deg/s, and $\alpha_2(t)$ is measured in deg/s^2. The inertia parameters of the door, follower, and coupler are listed in Table 12.4.

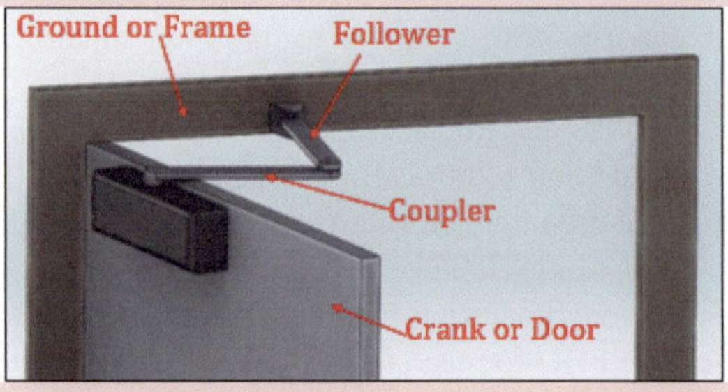

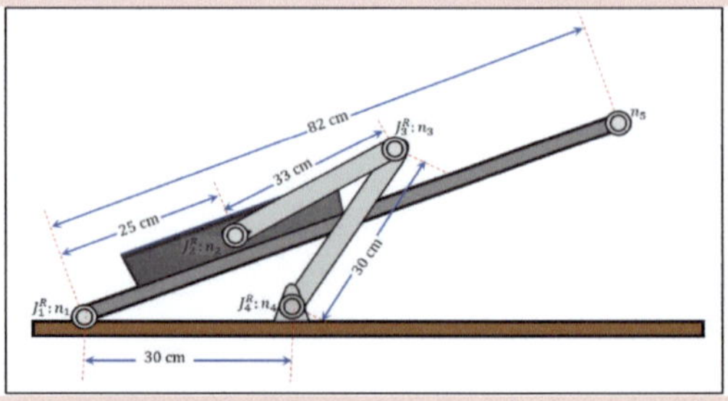

Fig. 12.14 Door mechanism. Figure and associated Door_4Bar_Mechanism.mp4 video designed by the author and implemented by both the author and Pranav Dhola. Both courtesy of Pranav Dhola

Table 12.4 Inertia parameters of door mechanism

Link	Mass (kg)	Moment of Inertia (kg m^2)	C.G location
Wooden door (2.1 × 0.82 × 0.06 m)	60	303	In the middle between nodes n_1 and n_5
Follower aluminum (30 × 5 × 5 cm)	2.025	0.187	In the middle between nodes n_3 and n_4
Follower aluminum (33 × 5 × 5 cm)	2.228	0.248	In the middle between nodes n_2 and n_3

At $t = 1.5$ s, the FBD of the door is shown in Fig. 12.15. The reaction forces at node n_2 are measured to be $F_{x2} = -0.2$ N and $F_{y2} = 0.067$ N. The transitional acceleration components of C.G$_2$ are $a_{xC.G2} = -0.155$ m/s^2 and $a_{yC.G2} = -0.165$ m/s^2. Calculate the reaction forces at node n_1 (F_{x1} and F_{y1}). Calculate the actuator torque at node n_1, T_1.

Fig. 12.15 FBD of door

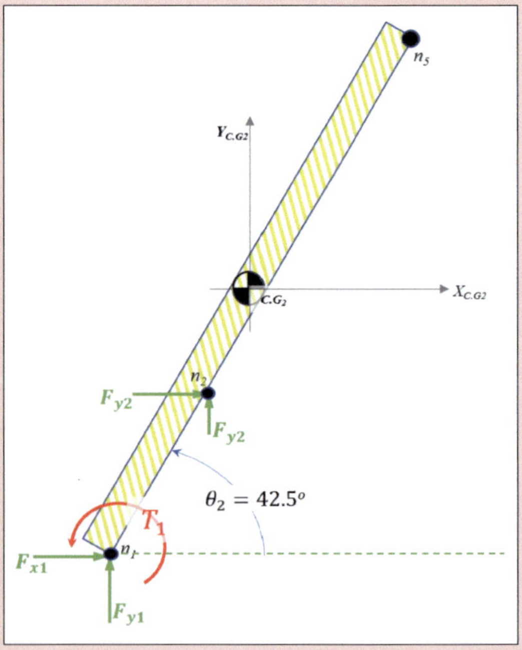

Solution

We start by applying Eq. (12.13) in the X-direction as

$$F_{x1} + F_{x2} = m_2 a_{xC.Gi}$$

or

$$F_{x1} + 0.2\text{ N} = (60\ \text{kg}) \times \left(-0.155\ \text{m/s}^2 \right)$$

Then

$$F_{x1} = -9.5N$$

Next, we apply Eq. (12.13) in the Y-direction as

$$F_{y1} + F_{y2} = m_2 a_{yC.Gi}$$

or

$$F_{y1} - 0.067\text{ N} = (60\ \text{kg}) \times \left(-0.165\ \text{m/s}^2 \right)$$

Then

$$F_{y1} = -9.95N$$

At time $t = 1.5$ sec

$$\theta_2(t = 1.5\text{s}) = 42.5((1.5\ \text{s}) - 0.5) = 42.5^\circ$$

Now, we apply Eq. (12.13) around the Z-axis as

$$F_{x1} y_{C.G2} - F_{y1} x_{C.G2} + F_{x2} y_{C.G2n2} - F_{y2} x_{C.G2n2} + T_1 = I_i \alpha_i$$

or

$$(-9.5N) \times \left(\left(\frac{82}{2}\ \text{m} \right) \times \sin(42.5^\circ) \right) - (-9.95N) \times \left(\left(\frac{82}{2}\ \text{m} \right) \times \cos\left(42.5^0\right) \right) + (-0.2\text{ N}) \times \left(\left(\frac{82}{2} - 25 \right) \text{m} \right)$$
$$\times \sin\left(42.5^0\right)) - (0.067\text{ N}) \times \left(\left(\frac{82}{2} - 25 \right) \text{m} \right) \times \cos\left(42.5^0\right)) + T_1 = \left(303\ \text{kg m}^2\right) \times \left(0 \frac{\text{rad}}{\text{s}^2} \right)$$

$$T_1 = 0.0168\text{N. m}$$

Example 12.5

Figure 12.16 shows a pendulum mechanism that consists of a ground link L_1 and a brick link L_2 (1 × 0.1 × 0.1 m) with a mass of 1 kg. The two links L_1 and L_2 are connected by a revolute joint J_1^R. The mechanism is activated by a torque T_1 using an electric motor. The motor provides the rotational motion:

$$\theta_2(t) = \frac{\pi}{2}\left(1 - \cos\left(\frac{\pi t}{2}\right)\right)$$
$$\omega_2(t) = \frac{\pi^2}{4} \sin\left(\frac{\pi t}{2}\right)$$
$$\alpha_2(t) = \frac{\pi^3}{8} \cos\left(\frac{\pi t}{2}\right)$$

where $\theta_2(t)$ is measured in rad, $\omega_2(t)$ is measured in rad/s, and $\alpha_2(t)$ is measured in rad/s^2. Calculate the motor torque T_1 and the reaction forces between the two links L_1 and L_2 at $t = 1$ sec.

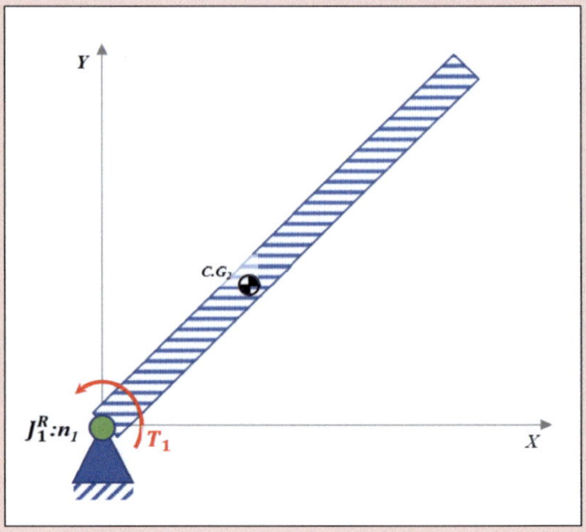

Fig. 12.16 Pendulum mechanism

Solution

Figure 12.17 shows the kinematic pendulum mechanism's kinematic diagram and FBD of the link L_2. The link L_2 is a brick or a rectangular prism with $w = 1$ m, $h = 0.1$ m, and $d = 0.1$ m plus $m_2 = 1$ kg. From Appendix B, the mass moment of inertia is

$$I_2 = \frac{m}{12}\left(w^2 + h^2\right) = \frac{(1\ \text{kg})}{12}\left((1\ \text{m})^2 + (0.1\ \text{m})^2\right) = 0.0842\ \text{kg m}^2$$

Using Appendix B, the link L_2's C.G. is located at

$$r_{C.G2} = \frac{h}{2} = 0.5\ \text{m}$$

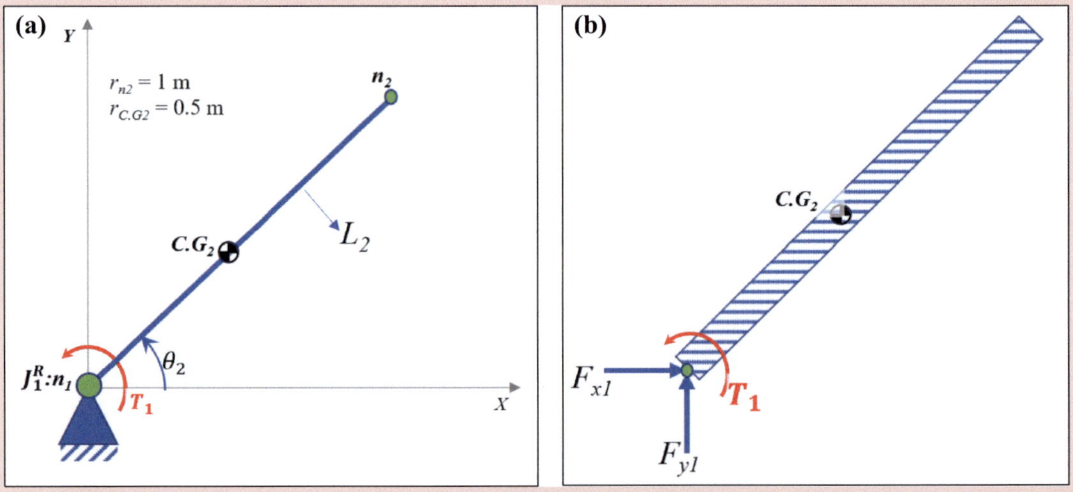

Fig. 12.17 Pendulum mechanism kinematic diagram and FBD

At $t = 1$ s, the angular position θ_2, the angular velocity ω_2, and the angular acceleration α_2 are

$$\theta_2(t = 1\,\mathrm{s}) = \frac{\pi}{2}\left(1 - \cos\left(\frac{\pi(1\,\mathrm{s})}{2}\right)\right) = \frac{\pi}{2}\mathrm{rad}$$
$$\omega_2(t = 1\,\mathrm{s}) = \frac{\pi^2}{4}\sin\left(\frac{\pi(1\,\mathrm{s})}{2}\right) = \frac{\pi^2}{4}\frac{\mathrm{rad}}{\mathrm{s}} = 2.467\frac{\mathrm{rad}}{\mathrm{s}}$$
$$\alpha_2(t = 1\,\mathrm{s}) = \frac{\pi^3}{8}\cos\left(\frac{\pi(1\,\mathrm{s})}{2}\right) = 0\,\mathrm{rad/s^2}$$

Recall the acceleration expression from Chap. 7. The acceleration components of link L_2's C.G. are

$$\vec{a}_{C.G2} = \alpha_2 r_{C.G2}\hat{\theta} - \omega_2^2 r_{C.G2}\hat{r} = (0)(0.5\,\mathrm{m})\hat{\theta} - \left(\frac{\pi^2}{4}\frac{\mathrm{rad}}{\mathrm{s}}\right)^2(0.5\,\mathrm{m})\hat{r} = \frac{\pi^3}{32}\hat{r}\frac{\mathrm{m}}{\mathrm{s}^2} = 0.9689\hat{r}\frac{\mathrm{m}}{\mathrm{s}^2}$$

Or

$$a_x = a_r\cos(\theta) = \left(0.9689\frac{\mathrm{m}}{\mathrm{s}^2}\right)\cos\left(\frac{\pi}{2}\right) = 0\frac{\mathrm{m}}{\mathrm{s}^2}$$
$$a_y = a_r\sin(\theta) = \left(0.9689\frac{\mathrm{m}}{\mathrm{s}^2}\right)\sin\left(\frac{\pi}{2}\right) = 0.9689\frac{\mathrm{m}}{\mathrm{s}^2}$$

Using the FBD of link L_2 and applying Eq. (12.13):

$$F_{x1} = m_2 a_{xC.G2} = (1\,\mathrm{kg}) \times \left(0\frac{\mathrm{m}}{\mathrm{s}^2}\right) = 0\mathrm{N}$$
$$F_{y1} = m_2 a_{yC.G2} = (1\,\mathrm{kg}) \times \left(0.9689\frac{\mathrm{m}}{\mathrm{s}^2}\right) = 0.9689\,\mathrm{N}$$
$$T_1 = I_2\alpha_2 = \left(0.0842\mathrm{kg\,m^2}\right) \times \left(0\frac{\mathrm{rad}}{\mathrm{s}^2}\right) = 0\,\mathrm{N\,m}$$

12.5.2 Four-Bar Mechanism Dynamic Model

In this section, we are going to develop an inverse dynamic model for a four-bar mechanism. Figure 12.18 shows a 4-bar mechanism with an external actuator torque T_1. This mechanism is in the *X-Y* plane. In other words, the weight has no influence since it is in the negative *Z*-direction. The mechanism is exposed to external forces and torques:

- $\vec{F}_5$ and T_5 at node n_5 which belongs to link L_2. Note that T_5 is not necessarily located at node n_5 and T_5 could be located at any node belonging to link L_2.
- $\vec{F}_6$ and T_6 at node n_6 which belongs to link L_3. Note that T_6 is not necessarily located at node n_6 and T_6 could be located at any node belonging to link L_3.
- $\vec{F}_7$ and T_7 at node n_7 which belongs to link L_4. Note that T_7 is not necessarily located at node n_7 and T_7 could be located at any node belonging to link L_4.

The first step is to break the mechanism down into a set of free body diagrams for each link as shown in Fig. 12.19. The actuator torque T_1 is sketched in red and the external loads ($\vec{F}_l$ and T_l where $l = \{5, 6, 7\}$) are sketched in blue. There are reaction forces (F_{xk} and F_{yk} where $k = \{1,2,3,4\}$) shown in green at nodes n_1, n_2, n_3, and n_4 coming from the joints J_1^R, J_2^R, J_3^R, and J_4^R. For each

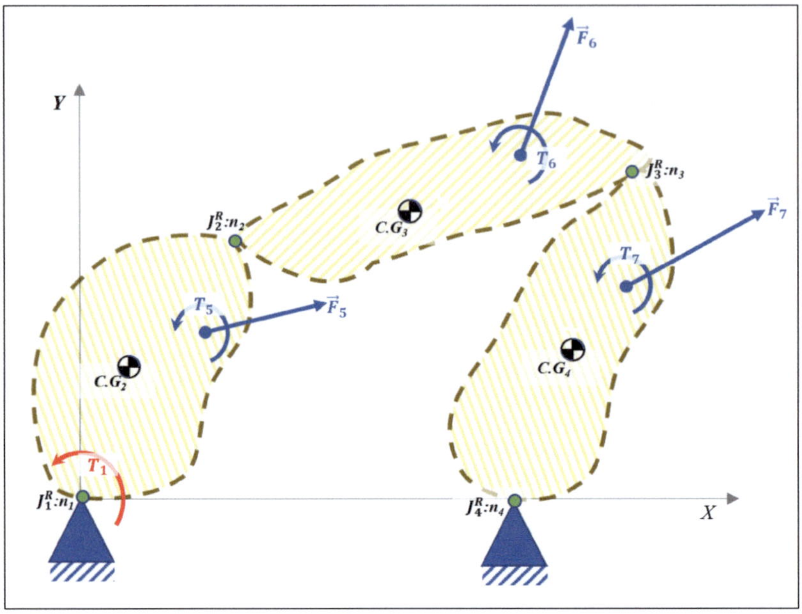

Fig. 12.18 Kinematic diagram of four-bar mechanism

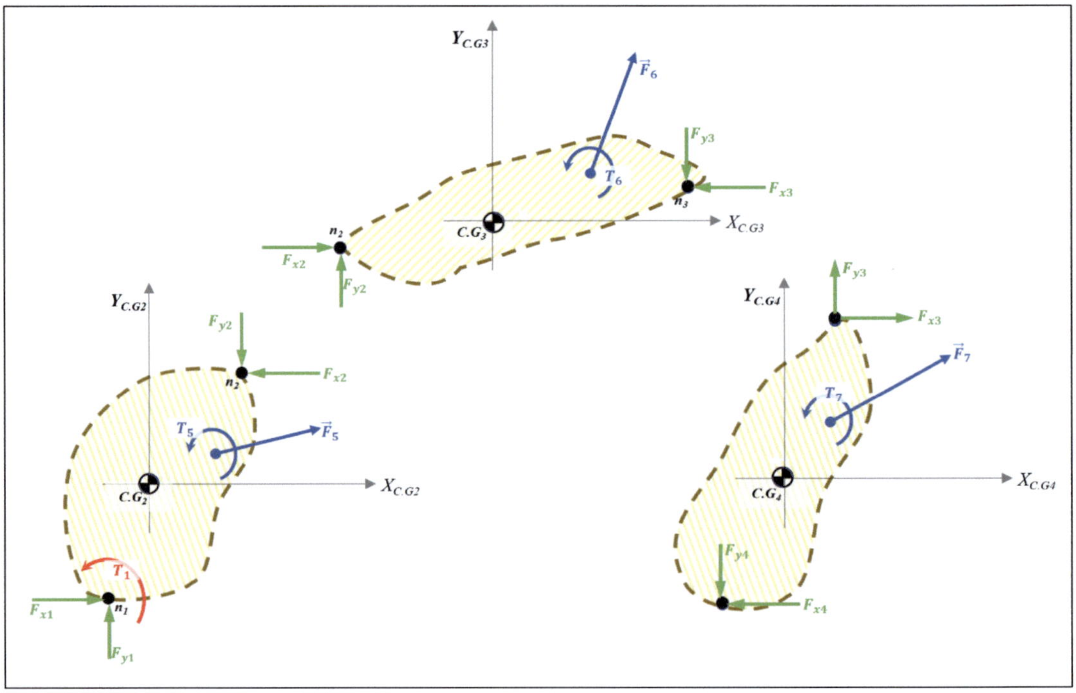

Fig. 12.19 Free body diagrams of four-bar mechanism

kinematic pair (two links L_i and L_j that are connected by a revolute joint J_k^R at node n_k), there are two contact (reaction) forces, F_{xk} and F_{yk}, that are applied to the link L_i in an opposite direction than at link L_j as shown in Fig. 12.19.

Now, we are going to apply Newton's second law as listed in Eq. (12.13) for each link as

For link L_2

$$\begin{aligned} F_{x1} - F_{x2} + F_{x5} &= m_2 a_{xC.G2} \\ F_{y1} - F_{y2} + F_{y5} &= m_2 a_{yC.G2} \\ T_1 + T_5 + \vec{r}_{C.G2n1} \times \vec{F}_1 + \vec{r}_{C.G2n2} \times \vec{F}_2 + \vec{r}_{C.G2n5} \times \vec{F}_5 &= I_2 \alpha_2 \end{aligned} \tag{12.15}$$

where

$$\begin{aligned} \vec{r}_{C.G2n1} &= -x_{C.G2n1} \hat{i}_{C.G2} - y_{C.G2n1} \hat{j}_{C.G2} \\ \vec{r}_{C.G2n2} &= x_{C.G2n2} \hat{i}_{C.G2} + y_{C.G2n2} \hat{j}_{C.G2} \\ \vec{r}_{C.G2n5} &= x_{C.G2n5} \hat{i}_{C.G2} + y_{C.G2n5} \hat{j}_{C.G2} \end{aligned}$$

$$\begin{aligned} \vec{F}_1 &= F_{x1} \hat{i}_{C.G2} + F_{y1} \hat{j}_{C.G2} \\ \vec{F}_2 &= -F_{x2} \hat{i}_{C.G2} - F_{y2} \hat{j}_{C.G2} \\ \vec{F}_5 &= F_{x5} \hat{i}_{C.G2} + F_{y5} \hat{j}_{C.G2} \end{aligned}$$

For link L_3

$$\begin{aligned} F_{x2} - F_{x3} + F_{x6} &= m_3 a_{xC.G3} \\ F_{y2} - F_{y3} + F_{y6} &= m_3 a_{yC.G3} \\ T_6 + \vec{r}_{C.G3n2} \times \vec{F}_2 + \vec{r}_{C.G3n3} \times \vec{F}_3 + \vec{r}_{C.G3n6} \times \vec{F}_6 &= I_3 \alpha_3 \end{aligned} \tag{12.16}$$

where

$$\begin{aligned} \vec{r}_{C.G3n2} &= -x_{C.G3n2} \hat{i}_{C.G3} - y_{C.G3n2} \hat{j}_{C.G3} \\ \vec{r}_{C.G3n3} &= x_{C.G3n3} \hat{i}_{C.G3} + y_{C.G3n3} \hat{j}_{C.G3} \\ \vec{r}_{C.G3n6} &= x_{C.G3n6} \hat{i}_{C.G3} + y_{C.G3n6} \hat{j}_{C.G3} \end{aligned}$$

$$\begin{aligned} \vec{F}_2 &= F_{x2} \hat{i}_{C.G3} + F_{y2} \hat{j}_{C.G3} \\ \vec{F}_3 &= -F_{x3} \hat{i}_{C.G3} - F_{y3} \hat{j}_{C.G3} \\ \vec{F}_6 &= F_{x6} \hat{i}_{C.G3} + F_{y6} \hat{j}_{C.G3} \end{aligned}$$

For link L_4

$$\begin{aligned} F_{x3} - F_{x4} + F_{x7} &= m_4 a_{xC.G4} \\ F_{y3} - F_{y4} + F_{y7} &= m_4 a_{yC.G4} \\ T_7 + \vec{r}_{C.G4n3} \times \vec{F}_3 + \vec{r}_{C.G4n4} \times \vec{F}_4 + \vec{r}_{C.G4n7} \times \vec{F}_7 &= I_4 \alpha_4 \end{aligned} \tag{12.17}$$

where

$$\begin{aligned} \vec{r}_{C.G3n2} &= -x_{C.G3n2} \hat{i}_{C.G3} - y_{C.G3n2} \hat{j}_{C.G3} \\ \vec{r}_{C.G3n3} &= x_{C.G3n3} \hat{i}_{C.G3} + y_{C.G3n3} \hat{j}_{C.G3} \\ \vec{r}_{C.G3n6} &= x_{C.G3n6} \hat{i}_{C.G3} + y_{C.G3n6} \hat{j}_{C.G3} \end{aligned}$$

$$\begin{aligned} \vec{F}_2 &= F_{x2} \hat{i}_{C.G3} + F_{y2} \hat{j}_{C.G3} \\ \vec{F}_3 &= -F_{x3} \hat{i}_{C.G3} - F_{y3} \hat{j}_{C.G3} \\ \vec{F}_6 &= F_{x6} \hat{i}_{C.G3} + F_{y6} \hat{j}_{C.G3} \end{aligned}$$

Note, that the positive "+ve" signs and negative "−ve" signs in Eqs. (12.15), (12.16), and (12.17) are based on the sketches shown in Figs. 12.18 and 12.19. These signs can change based on the location of each node relative to the C.G.s. For Eqs. (12.15), (12.16), and (12.17), the motor torque T_1 is unknown. The reaction forces (F_{xk} and F_{yk} where $k = \{1,2,3,4\}$) are 8 unknowns as well. The total number of unknowns is nine. If we know the assigned trajectory, we should be able to calculate the translational accelerations and the angular accelerations. Now, we have nine equations derived from Eqs. (12.15), (12.16), and (12.17) for the nine unknowns, the actuation torque, and the eight reaction forces. Equation (12.18) shows these nine equations in matrix form.

$$\begin{bmatrix} 1 & 0 & -1 & 0 & 0 & 0 & 0 & 0 & 0 \\ 0 & 1 & 0 & -1 & 0 & 0 & 0 & 0 & 0 \\ 0 & 0 & 1 & 0 & -1 & 0 & 0 & 0 & 0 \\ 0 & 0 & 0 & 1 & 0 & -1 & 0 & 0 & 0 \\ 0 & 0 & 0 & 0 & 1 & 0 & -1 & 0 & 0 \\ 0 & 0 & 0 & 0 & 0 & 1 & 0 & -1 & 0 \\ C_1 & C_2 & C_3 & C_4 & 0 & 0 & 0 & 0 & 1 \\ 0 & 0 & C_5 & C_6 & C_7 & C_8 & 0 & 0 & 0 \\ 0 & 0 & 0 & 0 & C_9 & C_{10} & C_{11} & C_{12} & 0 \end{bmatrix} \begin{bmatrix} F_{x1} \\ F_{y1} \\ F_{x2} \\ F_{y2} \\ F_{x3} \\ F_{y3} \\ F_{x4} \\ F_{y4} \\ T_1 \end{bmatrix} = \begin{bmatrix} m_2 a_{xC.G2} \\ m_2 a_{yC.G2} \\ m_3 a_{xC.G3} \\ m_3 a_{yC.G3} \\ m_4 a_{xC.G4} \\ m_4 a_{xC.G4} \\ I_2 \alpha_2 \\ I_3 \alpha_3 \\ I_4 \alpha_4 \end{bmatrix} + \begin{bmatrix} F_{x5} \\ F_{y5} \\ F_{x6} \\ F_{y6} \\ F_{x7} \\ F_{y7} \\ T_5 - y_{C.G2n5} F_{x5} + x_{C.G2n5} F_{y5} \\ T_6 - y_{C.G3n6} F_{x6} + x_{C.G3n6} F_{y6} \\ T_7 - y_{C.G4n7} F_{x7} + x_{C.G4n7} F_{y7} \end{bmatrix}$$

where

$$C_1 = y_{C.G2n1} \quad C_2 = -x_{C.G2n1} \quad C_3 = y_{C.G2n2} \quad C_4 = -x_{C.G2n2}$$
$$C_5 = y_{C.G3n2} \quad C_6 = -x_{C.G3n2} \quad C_7 = y_{C.G3n3} \quad C_8 = -x_{C.G3n3}$$
$$C_9 = y_{C.G4n3} \quad C_{10} = -x_{C.G4n3} \quad C_{11} = y_{C.G4n4} \quad C_{12} = -x_{C.G4n4}$$

(12.18)

If the givens at each instance in time are the position θ_2,angular velocity ω_2, and angular acceleration α_2 of the revolute joint J_1^R, we first use the equations introduced in Chaps. 5, 6, and 7 to find the positions { θ_3 and θ_4}, velocities { ω_3 and ω_4}, and accelerations {α_3 and α_4} as well as the accelerations at the C.Gs $\{a_{xC.G2}, a_{yC.G2}, a_{xC.G3}, a_{yC.G3}, a_{xC.G4}, \text{and} a_{yC.G4}\}$. After that, we solve Eq. (12.18) to find the actuator torque T_1 and the reaction forces (F_{xk} and F_{yk} where $k = \{1,2,3,4\}$).

Figure 12.20 shows a fully developed block diagram that integrates both the kinematic model and the dynamic model of the four-bar mechanism we have been discussing.

The matrix form shown in Eq. (12.18) can be solved using a numerical calculator or MATLAB®. Many calculations are needed to fill up this matrix. The calculations required to solve the four-bar mechanism's dynamic model are overwhelming and very time-consuming. Imagine making these calculations at each time step! For this reason, it is recommended to write a MATLAB® script to develop a Simscape™ Multibody™ model which we will explain later in this chapter. A mathematical algorithm used to calculate the actuator's torque and the resulting reaction forces is listed below.

Dynamic Calculation Algorithm of Four-Bar Mechanism

Inputs:

- The input signals are $\theta_2(t)$, $\theta_2(t)$, and $\theta_2(t)$.
- The input parameters are mass m_i, mass moment of inertia I_i , locations of the revolute joints' nodes r_{ninj} and C.G as $x_{C.Ginj}$and$y_{C.Ginj}$ where ($j = \{i{-}1$ and $i\}$).

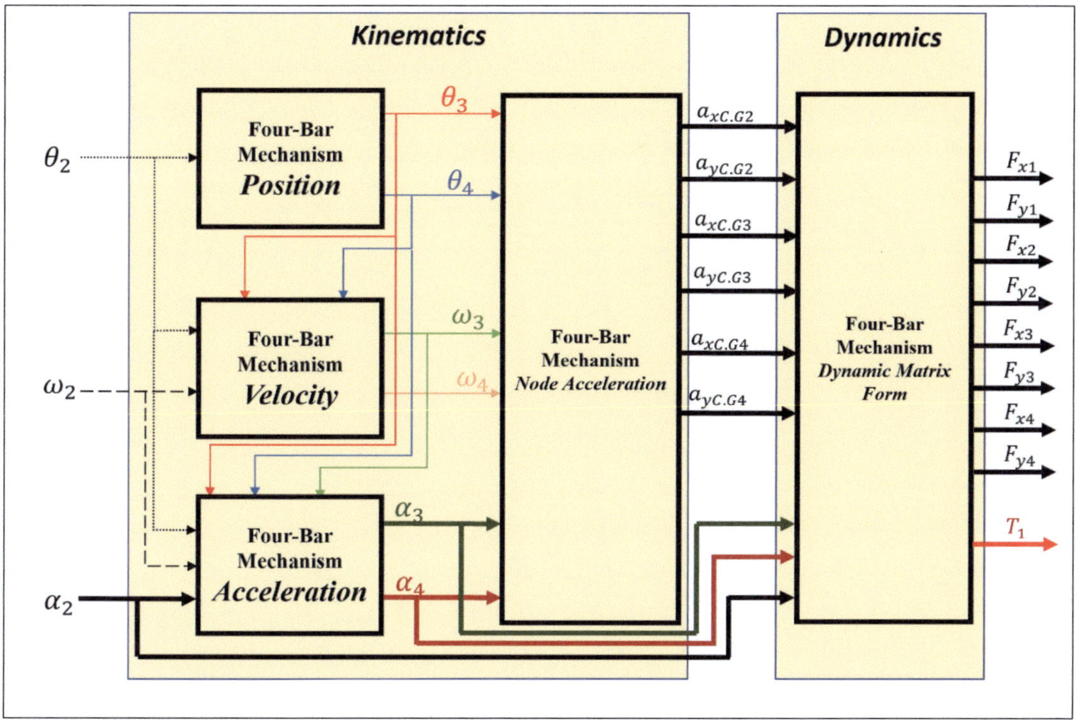

Fig. 12.20 Four-bar mechanism kinematics and dynamics diagram

Calculation steps at each time *t*:

1. Calculate the angular positions θ_3 and θ_4 from our analysis in Chap. 5 as

$$K_1 = \frac{r_{n4}}{r_{n2}} \quad K_2 = \frac{r_{n4}}{r_{n3n4}} \quad K_3 = \frac{r_{n2}^2 - r_{n3n2}^2 + r_{n3n4}^2 + r_{n4}^2}{2r_{n2}r_{n3n4}}$$

$$A_1 = (1 - K_2)\cos(\theta_2) - K_1 + K_3$$

$$B_1 = -2\sin(\theta_2)$$

$$C_1 = K_1 - (K_2 + 1)\cos(\theta_2) + K_3$$

$$A_1 \tan^2\left(\frac{\theta_4}{2}\right) + B_1 \tan\left(\frac{\theta_4}{2}\right) + C_1 = 0$$

$$\theta_3 = \cos^{-1}\frac{r_{n3n4}\cos(\theta_4) + r_{n4} - r_{n2}\cos(\theta_2)}{r_{n3n2}}$$

2. Calculate the angular velocities, ω_3 and ω_4 from our analysis in Chap. 6 as

$$\frac{\omega_3}{\omega_2} = \frac{r_{n2}}{r_{n2n3}} \times \frac{si\overleftarrow{n}\sin(\theta_4 - \theta_2)}{\sin\sin(\theta_4 - \theta_3)}$$

$$\frac{\omega_4}{\omega_2} = \frac{r_{n2}}{r_{n3n4}} \times \frac{\sin\sin(\theta_3 - \theta_2)}{\sin\sin(\theta_4 - \theta_3)}$$

3. Calculate the angular accelerations, α_3 and α_4 from our analysis in Chap. 7 as

$$A = -r_{n3n2}\sin(\theta_3) \quad B = r_{n3n4}\sin(\theta_4) \quad C = r_{n3n2}\cos(\theta_3) \quad D = -r_{n3n4}\cos(\theta_4)$$
$$E = -r_{n2}\alpha_2\sin(\theta_2) + r_{n2}\omega_2^2\cos(\theta_2) + r_{n3n2}\omega_3^2\cos(\theta_3) - r_{n3n4}\omega_4^2\cos(\theta_4)$$
$$F = -r_{n2}\alpha_2\cos(\theta_2) - r_{n2}\omega_2^2\sin(\theta_2) - r_{n3n2}\omega_3^2\sin(\theta_3) - r_{n3n4}\omega_4^2\sin(\theta_4)$$
$$\alpha_3 = \frac{B \times F - D \times E}{B \times C - A \times D} \qquad \alpha_4 = \frac{C \times E - A \times F}{B \times C - A \times D}$$

4. Calculate the translational accelerations $a_{xC.G2}$, $a_{yC.G2}$, $a_{xC.G3}$, $a_{yC.G3}$, $a_{xC.G4}$, and $a_{yC.G4}$ from our analysis in Chap. 7 as

$$a_{xC.G2} = -\omega_2^2 x_{C.G2} - \alpha_2 y_{C.G2}$$
$$a_{yC.G2} = -\omega_2^2 y_{C.G2} + \alpha_2 x_{C.G2}$$
$$a_{xC.G3} = -\omega_2^2 x_{n2} - \alpha_2 y_{n2} - \omega_3^2 x_{C.G3n2} - \alpha_3 y_{C.G3n2}$$
$$a_{yC.G3} = -\omega_2^2 y_{n2} + \alpha_2 x_{n2} - \omega_3^2 y_{C.G3n2} + \alpha_3 x_{C.G3n2}$$
$$a_{xC.G4} = -\omega_4^2 x_{C.G4n4} - \alpha_4 y_{C.G4n4}$$
$$a_{yC.G4} = -\omega_4^2 y_{C.G4} + \alpha_4 x_{C.G4n4}$$

5. Fill up the matrix in Eq. (12.18) and solve to calculate the actuation torque T_1 and the reaction forces F_{xi} and F_{yi} where $i = \{1, 2, 3, 4\}$.

Results and Plotting

We can write a MATLAB® script to repeat the above calculation steps at each time t from $t_i = 0$ sec to $t_f = T$ sec. After finishing the calculation, the script plots the actuator's torque T_1 and the reaction forces F_{xi} and F_{yi} (where $i = \{1, 2, 3, 4\}$) versus time. Writing this script is a bit of a challenge for many students because of editing and debugging. For this reason, we recommend the use of Simscape™ Multibody™. This modeling technique will be further explained at the end of this chapter.

12.5.3 Slider–Crank Mechanism Dynamic Model

Figure 12.21 shows the slider–crank mechanism with an external actuator torque T_1. This mechanism is in the X-Y plane. The weight has no influence since the weight is in the negative Z-direction. The mechanism is exposed to external forces and torques:

- $\vec{F}_5$ and T_5 at node n_5 which belongs to link L_2. Note that T_5 is not necessarily located at node n_5 and T_5 could be located at any node belonging to link L_2.
- $\vec{F}_6$ and T_6 at node n_6 which belongs to link L_3. Note that T_6 is not necessarily located at node n_6 and T_6 could be located at any node belonging to link L_3.
- $\vec{F}_7$ at node n_7 which belongs to link L_4.

The first step is to break the mechanism down into a set of free body diagrams for each link as shown in Fig. 12.22. The actuator torque T_1 is sketched in red and the external loads ($\vec{F}_l$ where $l = \{5, 6, 7\}$ and T_n where $n = \{5, 6,\}$) are sketched in blue. There are reaction forces (F_{xk} and F_{yk} where $k = \{1, 2, 3\}$ and F_{y4}) shown in green at nodes n_1, n_2, n_3, and n_4 coming from the joints J_1^R, J_2^R, J_3^R, and J_4^P.

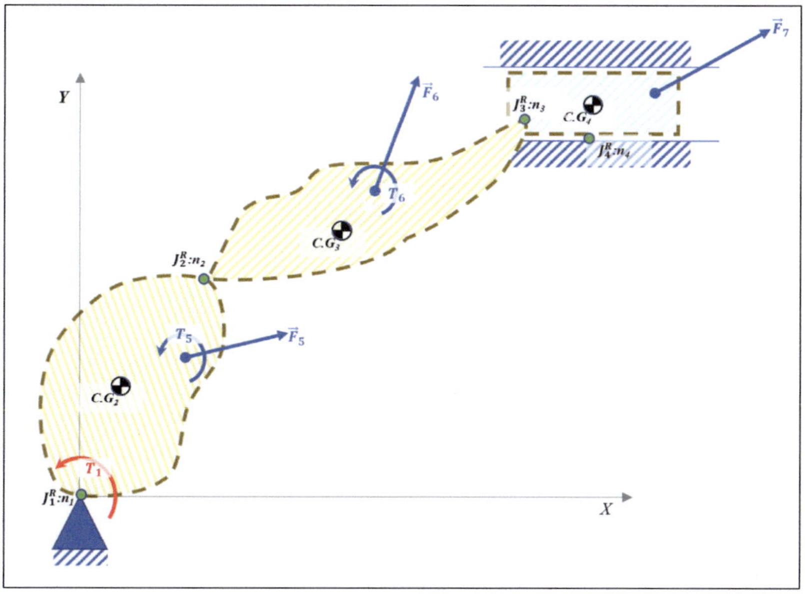

Fig. 12.21 Kinematic diagram of slider–crank mechanism

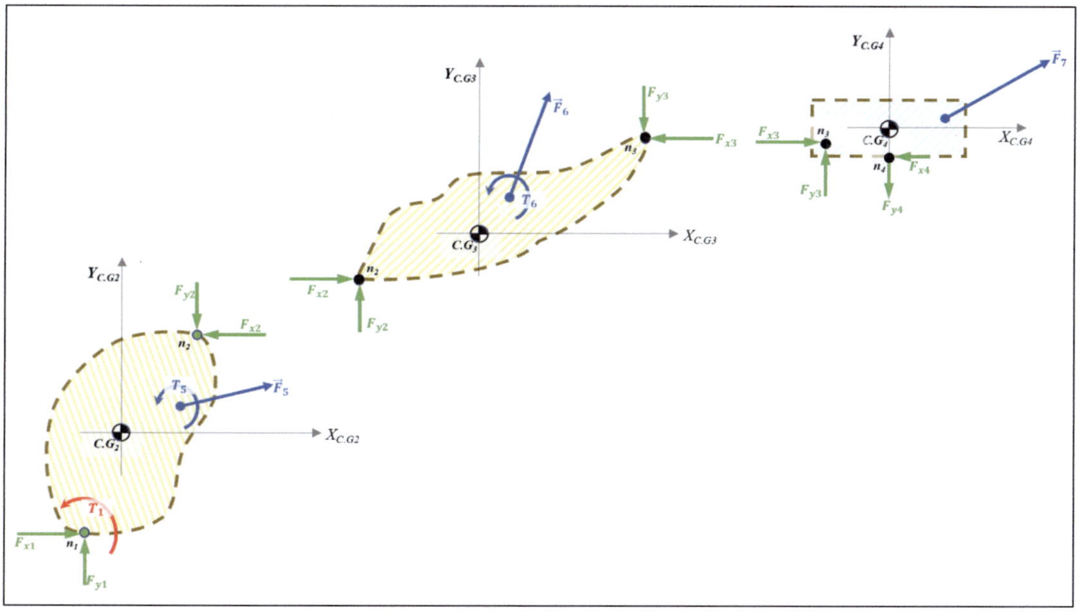

Fig. 12.22 FBD of slider–crank mechanism

Now, we are going to apply Newton's second law as listed in Eq. (12.13) for each link as

For link L_2

$$F_{x1} - F_{x2} + F_{x5} = m_2 a_{xC.G2}$$
$$F_{y1} - F_{y2} + F_{y5} = m_2 a_{yC.G2}$$
$$T_1 + T_5 + \vec{r}_{C.G2n1} \times \vec{F}_1 + \vec{r}_{C.G2n2} \times \vec{F}_2 + \vec{r}_{C.G2n5} \times \vec{F}_5 = I_2\alpha_2 \tag{12.19}$$

For link L_3

$$
\begin{aligned}
&F_{x2} - F_{x3} + F_{x6} = m_3 a_{xC.G3} \\
&F_{y2} - F_{y3} + F_{y6} = m_3 a_{yC.G3} \\
&T_6 + \vec{r}_{C.G3n2} \times \vec{F}_2 + \vec{r}_{C.G3n3} \times \vec{F}_3 + \vec{r}_{C.G3n6} \times \vec{F}_6 = I_3 \alpha_3
\end{aligned}
\tag{12.20}
$$

For link L_4

$$
\begin{aligned}
&F_{x3} + F_{x7} = m_4 a_{xC.G4} \\
&F_{y3} - F_{y4} + F_{y7} = m_4 a_{yC.G4} = 0 \\
&\vec{r}_{C.G4n3} \times \vec{F}_3 + \vec{r}_{C.G4n4} \times \vec{F}_3 + \vec{r}_{C.G4n7} \times \vec{F}_7 = I_4 \alpha_4
\end{aligned}
\tag{12.21}
$$

The acceleration term $a_{yC.G4} = 0$ since the link L_4 has no transitional motion in the Y-direction. Note, that the positive "+ve" signs and negative "−ve" signs in Eqs. (12.19), (12.20), and (12.21) are based on the shown sketches shown in Figs. 12.21 and 12.22. These signs can change based on the location of each node relative to the C.G.s. Equations (12.19), (12.20), and (12.21) have the motor torque T_1 as unknown. The reaction forces (F_{xk} and F_{yk} where $k = \{1,2,3\}$ and F_{y4}) are 7 unknowns as well. The total number of unknowns is eight. If we know the trajectory, we should be able to calculate the transitional accelerations and the angular accelerations. Now, we have eight equations derived from Eqs. (12.19), (12.20), and (12.21) for the eight unknowns, the actuation torque, and the seven reaction forces. We disregard the static equation describing the link L_4's force in the Y-direction. Equation (12.22) shows these eight equations in matrix form.

$$
\begin{bmatrix}
1 & 0 & -1 & 0 & 0 & 0 & 0 & 0 \\
0 & 1 & 0 & -1 & 0 & 0 & 0 & 0 \\
0 & 0 & 1 & 0 & -1 & 0 & 0 & 0 \\
0 & 0 & 0 & 1 & 0 & -1 & 0 & 0 \\
0 & 0 & 0 & 0 & 1 & 0 & 0 & 0 \\
0 & 0 & 0 & 0 & 0 & 1 & -1 & 0 \\
C_1 & C_2 & C_3 & C_4 & 0 & 0 & 0 & 1 \\
0 & 0 & C_5 & C_6 & C_7 & C_8 & 0 & 0
\end{bmatrix}
\begin{bmatrix}
F_{x1} \\ F_{y1} \\ F_{x2} \\ F_{y2} \\ F_{x3} \\ F_{y3} \\ F_{y4} \\ T_1
\end{bmatrix}
=
\begin{bmatrix}
m_2 a_{xC.G2} \\ m_2 a_{yC.G2} \\ m_3 a_{xC.G3} \\ m_3 a_{yC.G3} \\ m_4 a_{xC.G4} \\ 0 \\ I_2 \alpha_2 \\ I_3 \alpha_3
\end{bmatrix}
+
\begin{bmatrix}
F_{x5} \\ F_{y5} \\ F_{x6} \\ F_{y6} \\ F_{x7} \\ F_{y7} \\ T_5 - y_{C.G2n5} F_{x5} + x_{C.G2n5} F_{y5} \\ T_6 - y_{C.G3n6} F_{x6} + x_{C.G3n6} F_{y6}
\end{bmatrix}
\tag{12.22}
$$

where

$$
\begin{aligned}
&C_1 = y_{C.G2n1} \quad C_2 = -x_{C.G2n1} \quad C_3 = y_{C.G2n2} \quad C_4 = -x_{C.G2n2} \\
&C_5 = y_{C.G3n2} \quad C_6 = -x_{C.G3n2} \quad C_7 = y_{C.G3n3} \quad C_8 = -x_{C.G3n3}
\end{aligned}
$$

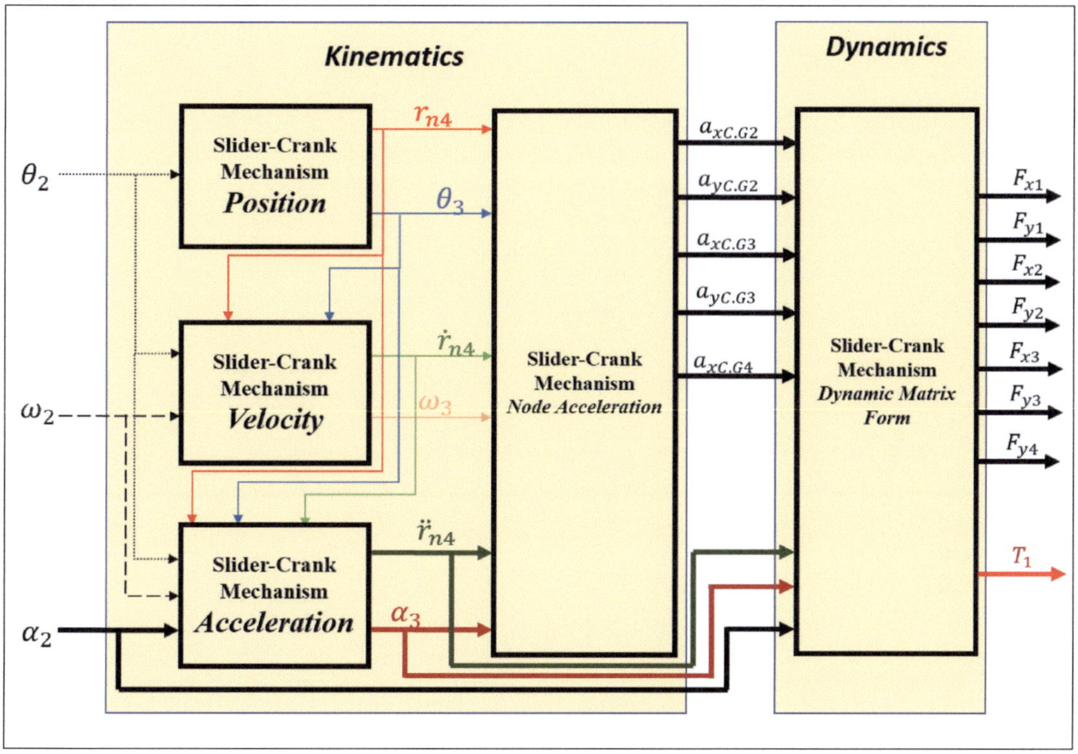

Fig. 12.23 Slider–crank mechanism kinematics and dynamics diagram

If the givens at each instance time are the position θ_2, velocity ω_2, and acceleration α_2 of the revolute joint J_1^R, we first use the equations introduced in Chaps. 5, 6, and 7 to find the positions $\{\theta_3$ and $r_{n4}\}$, velocities $\{\omega_3$ and $\dot{r}_{n4}\}$, and accelerations $\{\alpha_3$ and $\ddot{r}_{n4}\}$ as well as the accelerations at the C.G.s $\{a_{xC.G2}, a_{yC.G2}, a_{xC.G3}, a_{yC.G3}, \text{and} a_{xC.G4}\}$. After that, we solve Eq. (12.22) to find the actuator torque T_1 and the reaction forces (F_{xk} and F_{yk} where $k = \{1, 2, 3\}$ and F_{y4}). Figure 12.23 shows a fully developed block diagram that integrates both the kinematic model and the dynamic model of the slider–crank mechanism. A mathematical algorithm used to calculate the actuator's torque and the resulting reaction forces is listed below.

Dynamic Calculation Algorithm of Slider–Crank Mechanism

Inputs:

- The input signals are $\theta_2(t)$, $\theta_2(t)$, and $\theta_2(t)$.
- The input parameters are mass m_i, mass moment of inertia I_i, locations of the revolute joints' nodes r_{ninj} and C.G. as $x_{C.Ginj}$and$y_{C.Ginj}$ where i = {2, 3, 4} and $j = \{i-1, i\}$

Calculation steps at each time t:

1. Calculate the angular positions θ_3 and r_{n4} from our analysis in Chap. 5 as

$$\sin(\theta_3) = \frac{r_{n3n4} + r_{n2}\sin(\theta_2)}{r_{n3n2}}$$
$$r_{n4} = r_{n2}\cos(\theta_2) - r_{n3n2}\cos(\theta_3)$$

2. Calculate the angular velocities, ω_3 and $\dot{r}_{n4}$ from our analysis in Chap. 6 as

$$\frac{\omega_3}{\omega_2} = \frac{r_{n2}}{r_{n2n3}} \times \frac{\cos(\theta_2)}{\cos(\theta_3)}$$
$$\dot{r}_{n4} = -r_{n2}\omega_2 \sin(\theta_2) + r_{n2}\omega_2 \cos(\theta_2) \tan(\theta_3)$$

3. Calculate the angular accelerations, α_3 and $\ddot{r}_{n4}$ from our analysis in Chap. 7 as

$$\ddot{r}_{n4} = -r_{n2}\alpha_2 \sin(\theta_2) - r_{n2}\omega_2^2 \cos(\theta_2) + r_{n3n2}\alpha_3 \sin(\theta_3) + r_{n3n2}\omega_3^2 \cos(\theta_3)$$
$$\alpha_3 = \frac{\left(r_{n2}\omega_2^2 - r_{n3n2}\omega_3^2 \cos(\theta_3 - \theta_2) + \ddot{r}_{n4} \cos(\theta_2)\right)}{r_{n2}\sin(\theta_3 - \theta_2)}$$

4. Calculate the translational accelerations $a_{xC.G2}$, $a_{yC.G2}$, $a_{xC.G3}$, $a_{yC.G3}$, $a_{xC.G4}$, and $a_{yC.G4}$ from our analysis in Chap. 7 as

$$a_{xC.G2} = -\omega_2^2 x_{C.G2} - \alpha_2 y_{C.G2}$$
$$a_{yC.G2} = -\omega_2^2 y_{C.G2} + \alpha_2 x_{C.G2}$$
$$a_{xC.G3} = -\omega_2^2 x_{n2} - \alpha_2 y_{n2} - \omega_3^2 x_{C.G3n2} - \alpha_3 y_{C.G3n2}$$
$$a_{yC.G3} = -\omega_2^2 y_{n2} + \alpha_2 x_{n2} - \omega_3^2 y_{C.G3n2} + \alpha_3 x_{C.G3n2}$$
$$a_{xC.G4} = \ddot{r}_{n4}$$

5. Fill up the matrix in Eq. (12.22) and solve to calculate the actuation torque T_1 and the reaction forces F_{xi} and F_{yi} where $i = \{1, 2, 3, 4\}$.

Results and Plotting

We can write a MATLAB® script to repeat the above calculation steps at each time t from $t_i = 0$ sec to $t_f = T$ sec. After finishing the calculation, the script plots the actuator's torque T_1 and the reaction forces F_{xi} and F_{yi} where $i = \{1, 2, 3, 4\}$ versus time. Writing this script is a bit of a challenge for many students because of the editing and debugging. For this reason, we recommend the use of Simscape™ Multibody™. This modeling technique will be further explained at the end of this chapter.

12.6 Calculating Dynamic Forces and Dynamic Moments using Simscape™ Multibody™

From our previous dynamical model analysis (in Sect. 12.5.2 for the four-bar mechanism and Sect. 12.5.3 for the slider–crank mechanism), we end up with a lengthy algorithm and a 9 × 9 or 8 × 8 matrix to be inverted at each time instant. Even using MATLAB® scripting to complete this task is a bit of a challenge because of the editing and debugging required. Imagine a machine with more links and more joints. We would end up with an increasingly complicated algorithm and a very high-order matrix to be inverted. For this reason, we will introduce a Simscape™ Multibody™ model to be used. It is a straightforward modeling environment with animation and all kinds of visualizations. Such a model helps us to analyze and optimize our initial design efficiently.

The only difference between dynamic analysis and static analysis is to include the mass and the density of each link inside the model. To accomplish this goal, we are going to make a new version of the links with an option to enter the mass properties. We add the word "Dynamic" at the beginning of each link's type name. The simulation of the model will calculate the centers of gravity and the moments of inertia for each link. We can use the same Simscape™ Multibody™ model from the static analysis in Chap. 11 to perform the dynamic analysis in this chapter. We summarize the dynamic analysis using Simscape™ Multibody™ as:

1. Build the Simscape™ Multibody™ model.
2. Add the needed "Triple Inputs Rotational Motion Actuator" and/or "Triple Inputs Transitional Motion Actuator" to set the linkage at the desired position and orientation.
3. Add "Extremal Force and Torque" blocks at the known acting nodes.
4. Enable the forces or the torques at the joints that we are interested in measuring the actuating forces and/or torques at.

Example 12.6

For the shown gate opener in Fig. 12.14, if a motor opens the door from 0° to 85° using the SVA below

$$\theta_2(t) = \begin{cases} 0 & t < 0.5 \\ 42.5(t-0.5) & 0.5 \le t \le 2.5 \\ 85 & 2.5 < t \le 3 \end{cases}$$

$$\omega_2(t) = \begin{cases} 0 & t < 0.5 \\ 42.5 & 0.5 \le t \le 2.5 \\ 0 & 0.5 \le t \le 3 \end{cases}$$

$$\alpha_2(t) = \begin{cases} 0 & t < 0.5 \\ 0 & 0.5 \le t \le 2.5 \\ 0 & 2.5 < t \le 3 \end{cases}$$

where $\theta_2(t)$ is measured in deg, $\omega_2(t)$ is measured in deg/s, and $\alpha_2(t)$ is measured in deg/s^2.

1. Use a Simscape™ Multibody™ model to determine the time history of the actuator torque T_l and the reaction forces between links.
2. Compare the results of Example 12.4 to the simulation results of your developed Simscape™ Multibody™ model.

Solution

The steps to implement the Simscape™ Multibody™ model are:

1. Open a new script (*.m file) and declare the model parameters as shown below. Save the file under the name DoorOpener_script.m. This file includes the dimensional parameters and the dynamic parameters (masses). A copy of this m file is included with this textbook. Run the file.

```
clc
clear all
close all
%=========================================================================
%Declare the given parameters
%============================
r_n2   = 25/100; %m
r_n5   = 82/100; %m
r_n3n2 = 33/100; %m
r_n3n4 = 30/100; %m
r_n4   = 30/100; %m
T_cycle = 2; %sec
theta_2_max = 85; %deg
omega_2 = theta_2_max/2; %deg/sec
alpha_2 = 0 %deg/sec^2

%Calculation of parameters
%=========================
L2_crosssection = [0 0;r_n2 0;r_n5 0;r_n5 r_n2/10;0 r_n2/10];
L2_dimension    = [1 0.1 0.1]*r_n2; %m
L3_dimension    = [r_n3n2 0.1*r_n2 0.1*r_n2]; %m
L4_dimension    = [r_n4 0.1*r_n2 0.1*r_n2]; %m
Frame_L1_n1 = [0 0 0]; %m
Frame_L1_n4 = [r_n4 0 0]; %m
Frame_L2_n1 = [0 0 0]; %m
Frame_L2_n2 = [r_n2 0 0]; %m
Frame_L2_n5 = [r_n5 0 0]; %m
Frame_L3_n2 = [-r_n3n2/2 0 0]; %m
Frame_L3_n3 = [r_n3n2/2 0 0]; %m
Frame_L4_n4 = [-r_n3n4/2 0 0]; %m
Frame_L4_n3 = [r_n3n4/2 0 0]; %m

%Dynamic parameters
%=========================
m2 = 60; %kg
m3 = 2.025; %kg
m4 = 2.228; %kg
```

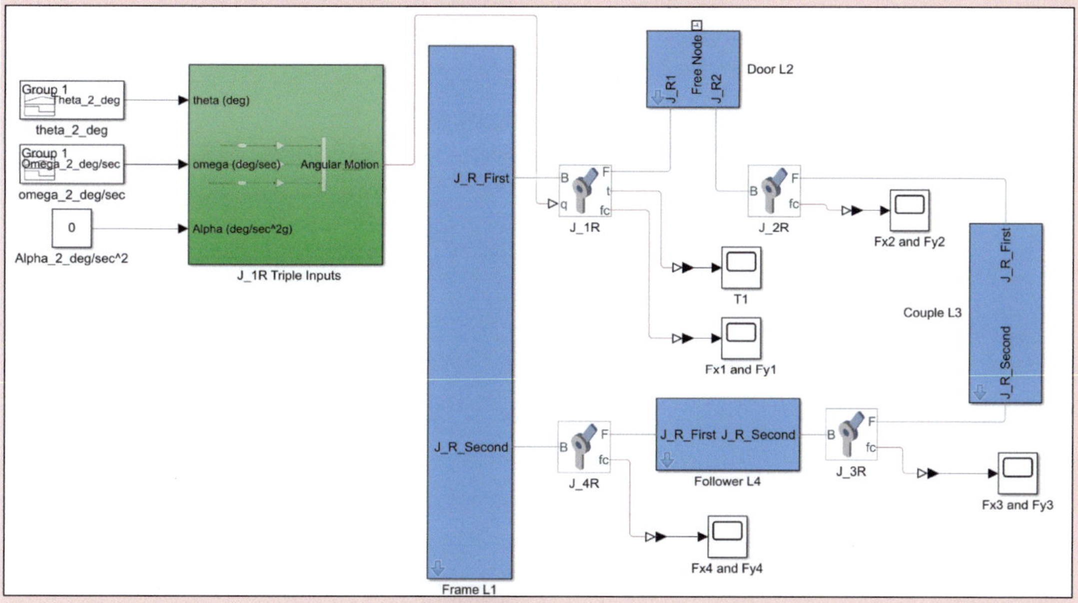

Fig. 12.24 Door opener Simscape™ Multibody™ model

2. Open a new model and save it as "DoorOpener_model". Build the model as shown in Fig. 12.24 with the listed blocks and subsystems in Table 12.5.
3. Use the given equations of θ_2, ω_2, and α_2 in the problem statement to build the input signals using the two Signal Builder blocks and the single Constant block. The first Signal block is for θ_2(Theta_2_deg in Simulink®), which has 4 points {(0,0), (0.5,0), (2.5,85), (3,8.5))} as shown in Fig. 12.25a. The second Signal block is for ω_2 (Omega_2_deg/sec in Simulink®), which has 6 points {(0.0), (0.5,0), (0.5,42.5), (2.5,4.5), (2.5,0), (3, 0)}as shown in Fig. 12.25b.

Run the model for 3 seconds to get the results from the scopes. During the simulation, there is an animation shown in the Mechanics Explorer Window. Figure 12.26 shows a screenshot from the simulation animation. There is a copy of this model for the reader to explore included with this textbook. Figure 12.27 shows the reaction forces on the four revolute joints. Figure 12.28 shows the actuator torque T_1. The results of this simulation confirm the results in Example 12.4.

Table 12.5 Door opener model's blocks and subsystems. Designed and implemented by the author. Images used with permission of MathWorks

Block or subsystem	Name	Plugged in parameters
Dynamic brick with two revolute joints	Coupler L3	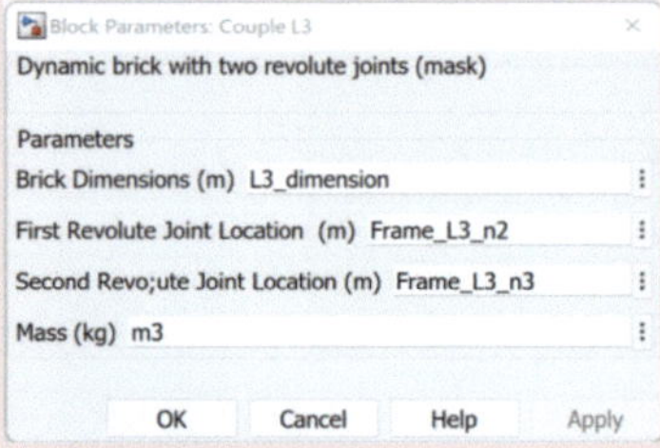
Dynamic brick with two revolute joints	Follower L4	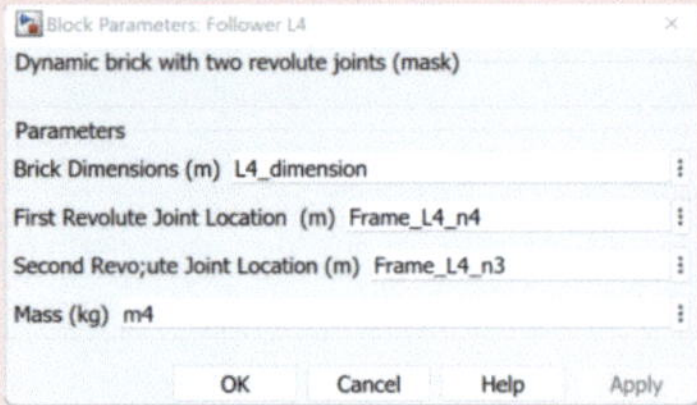
Extrusion with two revolute joints	Door L2	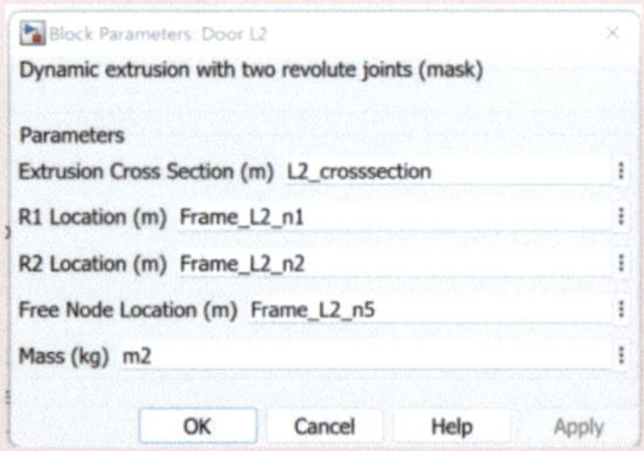
Four revolute joints	J_1R, J_2R, J_3R, and J_$R	In each joint, select "Composite Force/Torque Sensing/ Constraint Force" to "fc" output port. Connect each fc with "PS Simulink® Converter" and Scope For joint J_1R, select "Sensing/ Actuator Torque" to "t" output port. Connect "t" output port with "PS Simulink® Converter" and Scope.
Triple inputs rotational motion actuator	J_1R Triple Inputs	
Two signal builder blocks	theta_2_deg and omega_2_deg/sec	These two blocks implement the angular displacement input and the angular velocity input. These two blocks are connected to the first two inputs of the "J_1R Triple Inputs"
Constant block	Alpha_2_deg/s^2	This block implements the angular acceleration input with a zero value
Scope blocks	Names as listed in Fig. 12.24	These Scope blocks display the results

12.7 Chapter Summary

An introduction to methods of calculating a link's mass, center of gravity, and mass moment of inertia has been given. Some explanation of how to apply Newton's second law to a single link with unbalanced acting forces and/or movements has also been given. Techniques used to develop dynamic models for the four-bar link mechanism and the slider–crank mechanism have been described. An explanation of how to measure the reaction forces and/or moments as well as the actuator forces and/or torques using Simscape™ Multibody™ simulations are included.

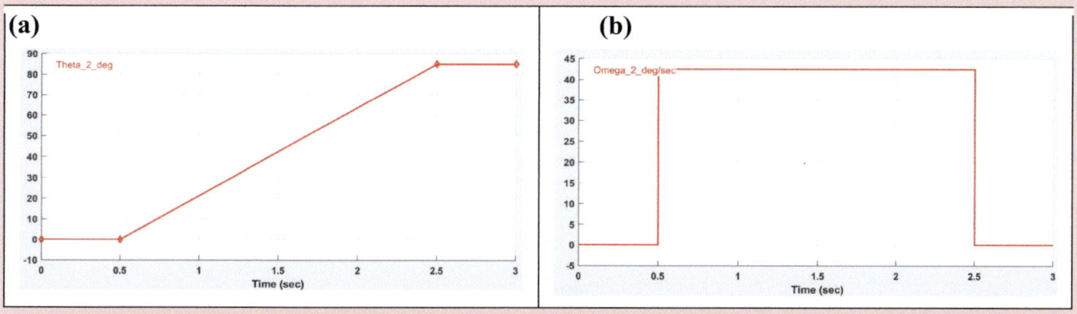

Fig. 12.25 Input signal two signal builder block for door opener

Fig. 12.26 Door opener animation snapshot

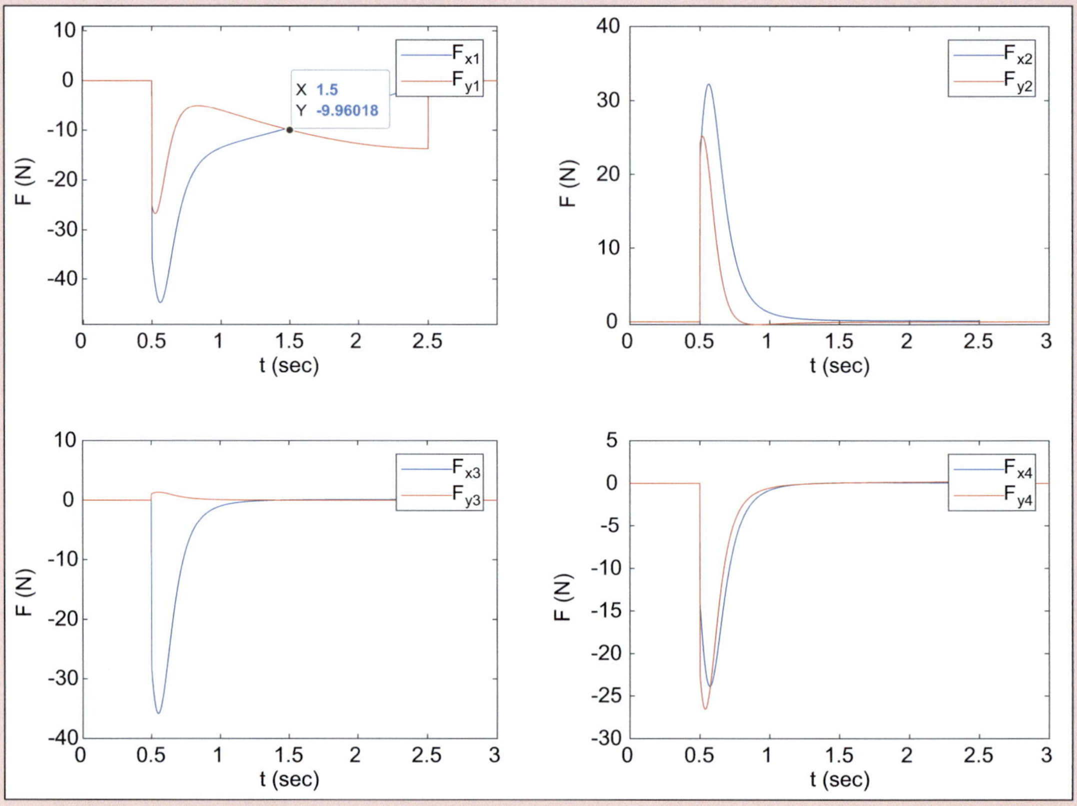

Fig. 12.27 Door opener reaction forces on the four revolute joints

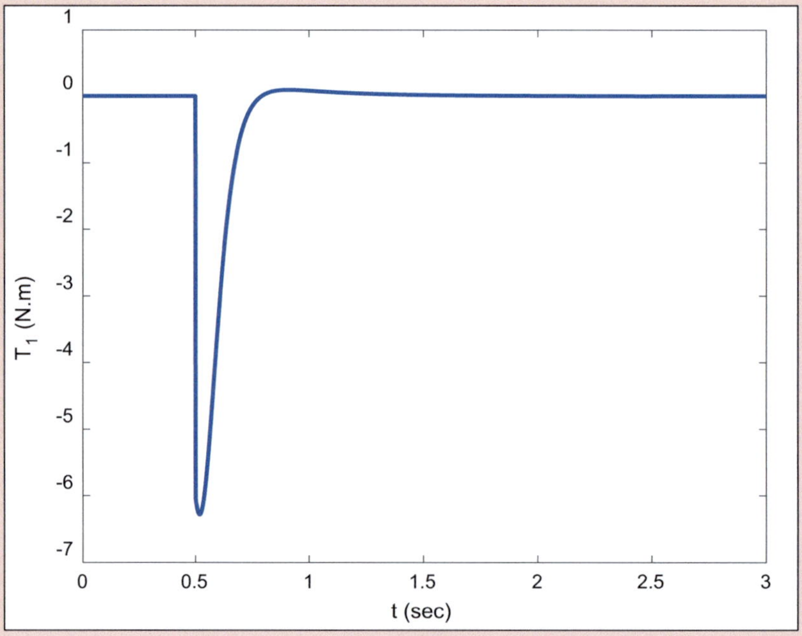

Fig. 12.28 Door opener actuator torque at first revolute joint

Outcomes

1. Understanding how to calculate any link's mass, center of gravity, and mass moment of inertia.
2. Understanding how to apply Newton's second law to a link with unbalanced forces and/or torques.
3. Understanding how to develop a complete dynamic model with two examples: the four-bar mechanism and the slider–crank mechanism.
4. Understanding the steps needed to set up and measure the reaction forces and/or moments as well as the actuators' forces and/or moments using Simscape™ Multibody™ models.

Problems

Problem 1

Figure 12.29a shows the dumpster truck mechanism. Figure 12.29b shows the simplified shape of the dumpster truck mechanism's bed that is manufactured from steel, $\rho = 7850$ kg/m^3. Find:

1. the bed's mass,
2. the bed's C.G. location, and
3. the bed's mass moment of inertia around the *Z*-axis that's passing through the bed's C.G.

Problem 2

Figure 12.30a shows the pumpjack mechanism. Figure 12.30b shows the simplified shape of the pumpjack mechanism's crank link as a brick plus a half cylinder that is done from steel, $\rho = 7850$ kg/m^3. Find:

1. the crank's mass,
2. the crank's C.G. location, and
3. the crank's mass moment of inertia around the *Z*-axis that's passing through the crank's C.G.

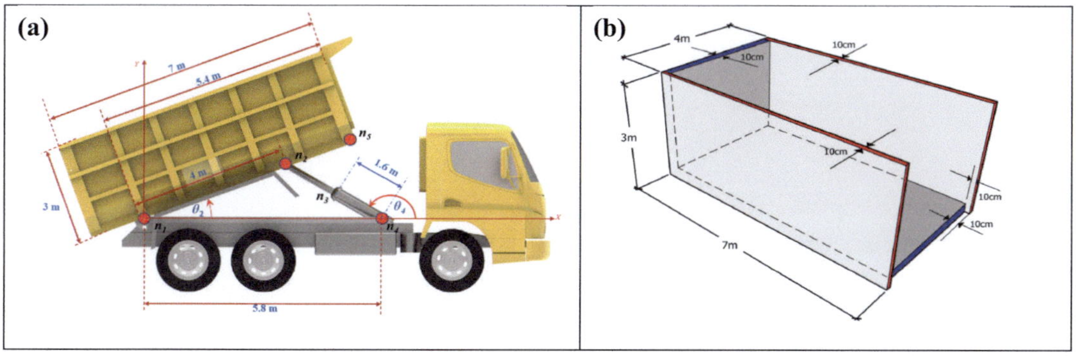

Fig. 12.29 Dumpster truck mechanism and bed link. Figure and associated Truck Mechanism.mp4 video designed by the author and implemented by both the author and Pranav Dhola. Both courtesy of Pranav Dhola

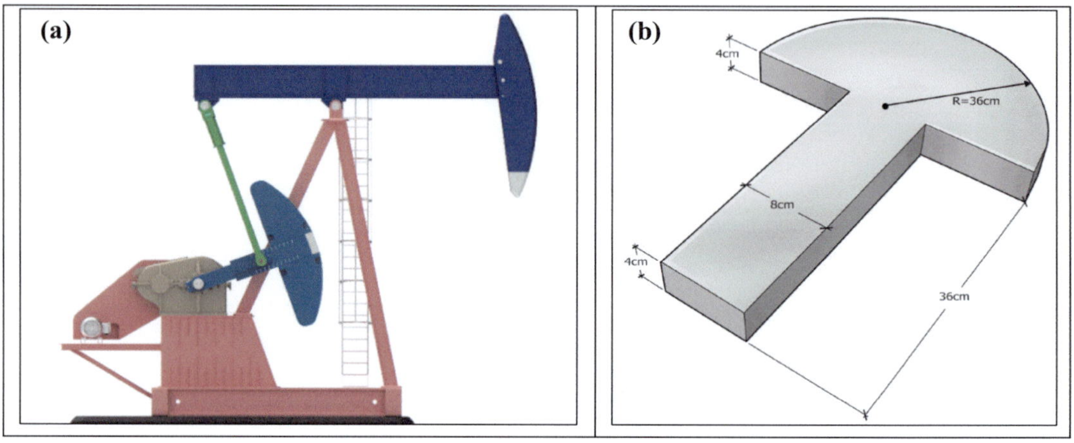

Fig. 12.30 Pumpjack mechanism and crank link. Designed by the author and implemented by both the author and Pranav Dhola. Courtesy of Pranav Dhola

Problem 3

Figure 12.31a shows the lift table mechanism. Figure 12.31b shows the simplified shape of the upper table (steel $\rho = 7850$ kg/m^3) carrying a box (load $\rho = 2735$ kg/m^3). If we assume that upper table along with the box are treated as a one link, find:

1. the table-box's mass,
2. the table-box's C.G. location, and
3. the table-box's mass moment of inertia around the Z-axis that's passing through the table-box's C.G.

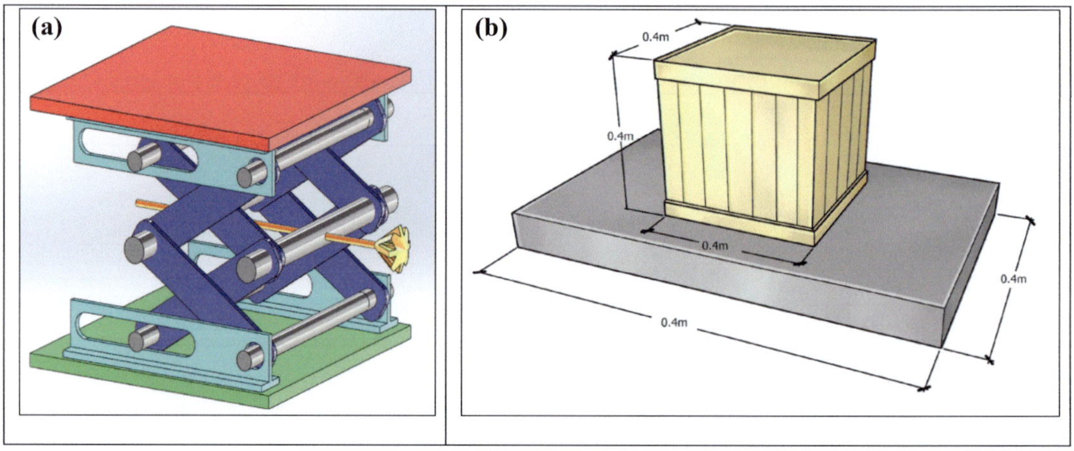

Fig. 12.31 Lift table mechanism and upper table with load box link. a. Designed by the author and implemented by both the author and Pranav Dhola. Courtesy of Pranav Dhola b. Designed by the author and implemented by Kenneth Manalansan. Courtesy of Kenneth Manalansan

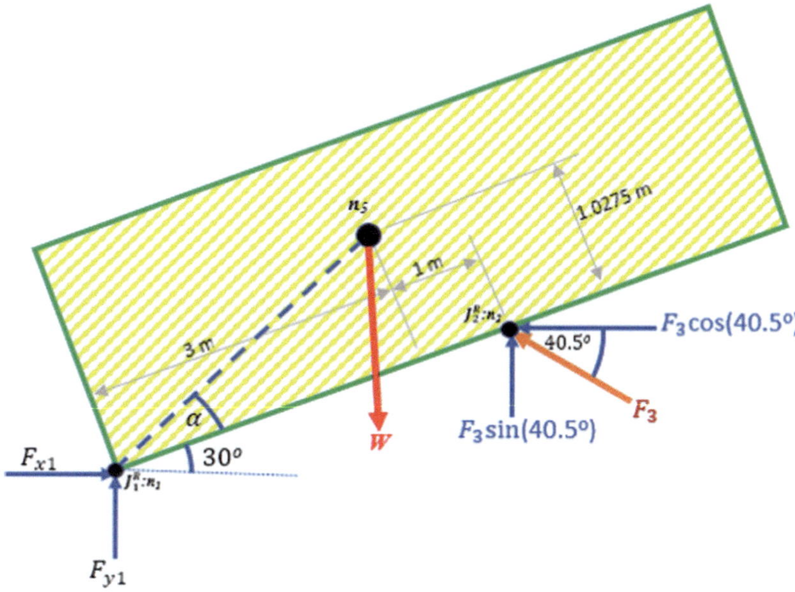

Fig. 12.32 FBD of dumpster truck mechanism bed

Problem 4

At a specific time t, the FBD of the bed link inside the dumpster truck mechanism is shown in Fig. 12.32 that has a translational acceleration $a_{xC.G2} = 4.05 \times 10{-}^{4}$ m/s^2 and $a_{yC.G2} = -5.372 \times 10{-}^{3}$ m/s^2 plus the angular acceleration $\alpha_2 = 0$ rad/s^2 whereby $I_{zbed} = 3.642 \times 10^5$ kg m^2 and $m_{\text{bed}} = 62203.4$ kg. Find the reaction force and the actuator force at this specific time t.

Figure 12.33 shows FDBs of each link in the piston–cylinder mechanism. Problems 5, 6, and 7 will be using these FBDs. At a certain time t:

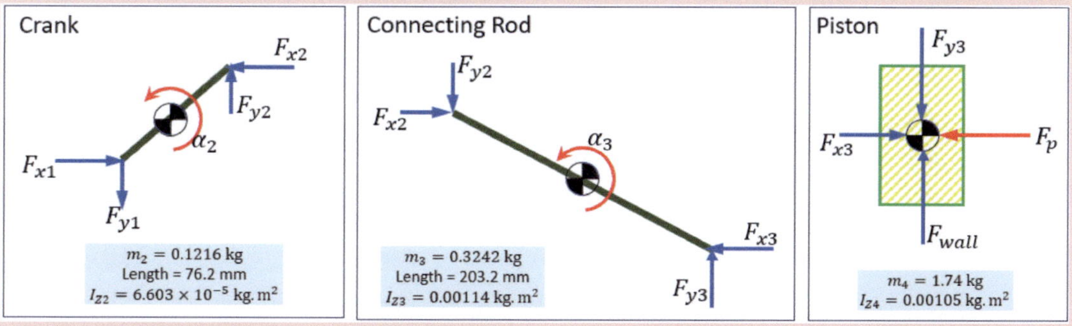

Fig. 12.33 FBD of dumpster truck mechanism bed

- the applied pressure force is $F_p = -33.07$ N,
- the piston moves with an acceleration $a_{3x} = +27.89$ m/s^2,
- the connecting rod has an angular acceleration $\alpha_3 = 9994$ rad/s^2 and a translational acceleration $a_{CG3x} = -333.6$ m/sec^2 and $a_{CG3y} = -948.9$ m/s^2, plus
- The crank has an angular an angular acceleration $\alpha_2 = 4486$ rad/s^2 and a translational acceleration $a_{CG3x} = -347.5$ m/s^2 and $a_{CG2y} = -948.9$ m/s^2.

Problem 5

Find the value of the connecting force F_{3x} using the piston FBD.

Problem 6

Find the connecting forces F_{3y}, F_{2x}, and F_{2y} using the connecting rod FBD.

Problem 7

Find the reaction forces F_{x1} and F_{y1} using the crank FBD.

References

1. "Machines & Mechanisms: Applied Kinematic Analysis", David Myszka, Pearson; 4 edition, 2011.
2. "Engineering Mechanics: Dynamics", Russell C. Hibbeler, 14th edition, 2016, ISBN-13: 9780134228266.
3. "Engineering Mechanics: Dynamics", James L. Meriam, and L. G. Kraige, 9th edition, ISBN: 978-1-119-39098-5
4. "Design of Machinery," Robert L. Norton, McGraw-Hill; 5th Edition, 2011.
5. "Simscape™ Multibody™ Getting Started Guide," MathWorks, 2020.
6. "Simscape™ Multibody™ User's Guide," MathWorks, 2020.
7. "Simscape™ Multibody™ Reference," MathWorks, 2020.
8. "Simscape™ Multibody™ Release Notes," MathWorks, 2020.

Appendix A: Metal Material Density

Metal	Density (g/cm^3)	Density (kg/m^3)	Density (lb/in^3)	Density (lb/ft^3)
Aluminum	2.70	2700	0.098	169
Antimony	6.70	6697	0.242	418
Arsenic	5.78	5778	0.209	361
Barium	3.5	3500	0.126	218.5
Beryllium	1.85	1848	0.067	115
Bismuth	9.81	9808	0.354	612
Boron	2.45	2450	0.089	153
Brass	8.50–8.80	8500–8800	0.307–0.318	531–499
Cadmium	8.64	8642	0.312	540
Calcium	1.55	1550	0.056	97
Carbon (graphite)	2.25	2250	0.081	140.5
Cerium	8.16	8160	0.295	509
Cesium	1.90	1903	0.069	119
Chromium	7.19	7190	0.260	449
Cobalt	8.83	8832	0.319	551
Copper	8.93	8930	0.323	557.5
Dysprosium	8.55	8551	0.309	534
Erbium	9.07	9066	0.328	566
Europium	5.24	5244	0.189	327
Gadolinium	7.90	7901	0.285	493
Gallium	5.91	5907	0.213	369
Germanium	5.32	5323	0.192	332
Gold	19.30	19,302	0.697	1205
Hafnium	13.31	13,310	0.481	831
High carbon steel	7.81	7810	0.282	488
Holmium	8.795	8795	0.318	549
Indium	7.3	7300	0.264	456
Iridium	22.562	22,562	0.815	1409
Iron	7.87	7870	0.284	491

A. Omran, *Kinematics and Dynamics of Machines*, https://doi.org/10.1007/978-3-031-72701-6

Metal	Density (g/cm^3)	Density (kg/m^3)	Density (lb/in^3)	Density (lb/ft^3)
Lanthanum	6.15	6146	0.222	384
Lead	11.34	11,340	0.410	708
Lithium	0.533	533	0.019	33.3
Lutetium	9.84	9841	0.356	614
Magnesium	1.74	1738	0.063	108.5
Manganese	7.43	7430	0.268	464
Medium carbon steel	7.83	7830	0.283	489
Mercury	13.55	13,546	0.489	846
Mild Steel	7.85	7850	0.284	490
Molybdenum	10.22	10,220	0.369	638
Neodymium	7.01	7008	0.253	438
Nickel	8.90	8902	0.322	556
Niobium	8.57	8570	0.310	535
Osmium	22.587	22,587	0.816	1410
Palladium	12.02	12,020	0.434	750
Phosphorus (white)	1.83	1830	0.066	114
Platinum	21.45	21,450	0.775	1339
Plutonium	19.86	19,860	0.717	1240
Potassium	0.855	855	0.031	53
Praseodymium	6.773	6773	0.245	423
Promethium	7.264	7264	0.262	453
Protactinium	15.43	15,430	0.557	963
Rhenium	21.02	21,020	0.759	1312
Rhodium	12.41	12,410	0.448	775
Rubidium	1.532	1532	0.055	96
Ruthenium	12.45	12,450	0.450	777
Samarium	7.52	7520	0.272	469.5
Scandium	2.99	2989	0.108	187
Selenium	4.81	4809	0.174	300
Silicon	2.33	2329	0.084	145
Silver	10.49	10,490	0.379	655
Sodium	0.97	967	0.035	60
Stainless steel	7.7–8.0	7700–8000	0.278–0.289	481–499
Strontium	2.6	2600	0.094	162
Tantalum	16.6	16,600	0.600	1036
Technetium	11.5	11,500	0.415	718
Tellurium	6.24	6237	0.225	389
Terbium	8.23	8230	0.297	514
Thallium	11.87	11,872	0.429	741

Metal	Density (g/cm^3)	Density (kg/m^3)	Density (lb/in^3)	Density (lb/ft^3)
Thorium	11.8	11,800	0.426	737
Thulium	9.32	9321	0.337	582
Tin	5.765	5765	0.208	360
Titanium	4.51	4507	0.163	281
Tungsten	19.25	19,254	0.696	1202
Uranium	19.05	19,050	0.688	1189
Vanadium	6.16	6160	0.223	385
Ytterbium	6.90	6903	0.249	431
Yttrium	4.47	4469	0.161	279
Zinc	7.13	7133	0.258	445
Zirconium	6.51	6505	0.235	406

Appendix B: Geometric Properties

A. Omran, *Kinematics and Dynamics of Machines*, https://doi.org/10.1007/978-3-031-72701-6

Shape	Volume	Center of gravity	Mass moment of inertia
Figures are designed by the author and implemented by Kenneth Manalansan. Courtesy of Kenneth Manalansan 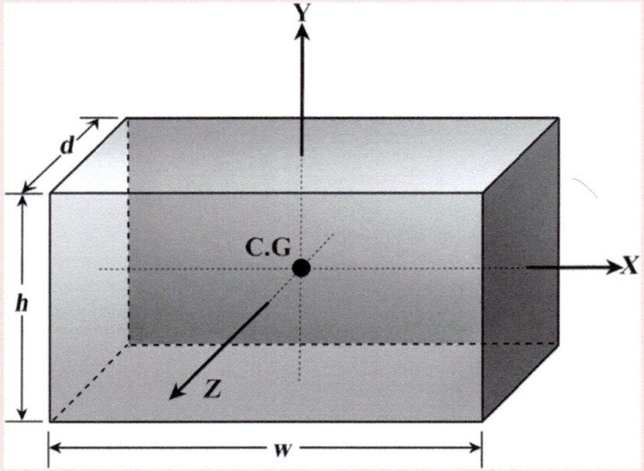	$V = d \times h \times w$	$x_{C.G} = 0$ $y_{C.G} = 0$ $z_{C.G} = 0$	$I_{xx} = \frac{m}{12}(h^2 + d^2)$ $I_{yy} = \frac{m}{12}(w^2 + d^2)$ $I_{zz} = \frac{m}{12}(h^2 + w^2)$
Figures are designed by the author and implemented by Kenneth Manalansan. Courtesy of Kenneth Manalansan 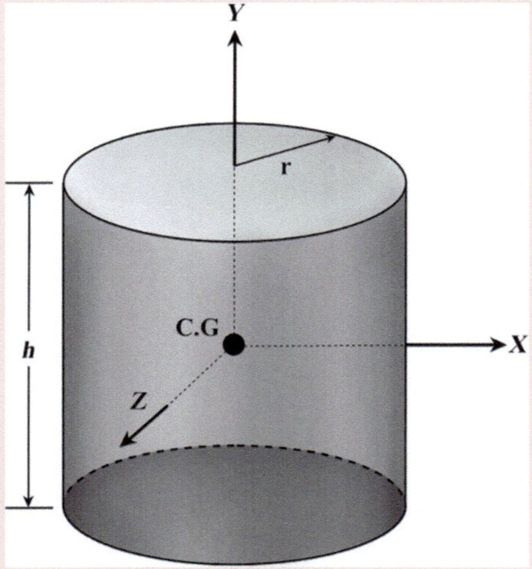	$V = \pi r^2 h$	$x_{C.G} = 0$ $y_{C.G} = 0$ $z_{C.G} = 0$	$I_{xx} = I_{xx} = \frac{m}{12}(3r^2 + h^2)$ $I_{yy} = \frac{m}{2} r^2$

Shape	Volume	Center of gravity	Mass moment of inertia
Figures are designed by the author and implemented by Kenneth Manalansan. Courtesy of Kenneth Manalansan 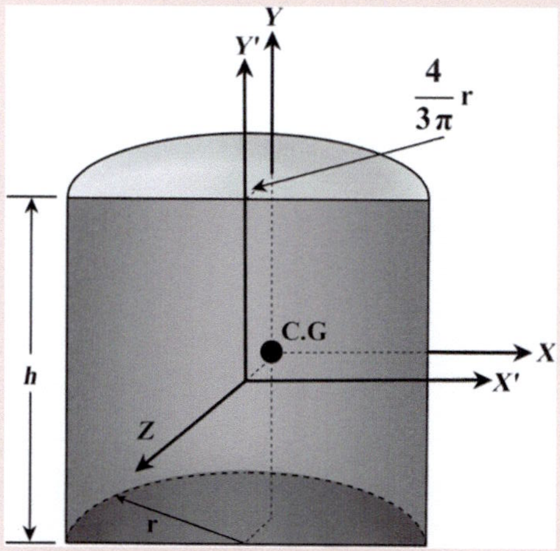	$V = \frac{1}{2}\pi r^2 h$	$x_{C.G} = 0$ $y_{C.G} = 0$ $z_{C.G} = -\frac{4r}{3\pi}$	$I_{xx} = I_{xx} = \left(\frac{1}{4} - \frac{16}{9\pi^2}\right)mr^2 + \frac{mh^2}{12}$ $I_{yy} = \left(\frac{1}{2} - \frac{16}{9\pi^2}\right)mr^2$
Figures are designed by the author and implemented by Kenneth Manalansan. Courtesy of Kenneth Manalansan 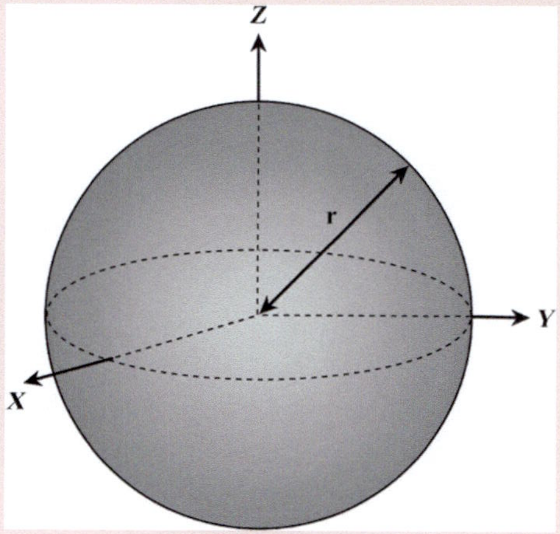	$V = \frac{4}{3}\pi r^3$	$x_{C.G} = 0$ $y_{C.G} = 0$ $z_{C.G} = 0$	$I_{xx} = I_{xx} = I_{yy} = \frac{2}{5}mr^2$

Shape	Volume	Center of gravity	Mass moment of inertia
Figures are designed by the author and implemented by Kenneth Manalansan. Courtesy of Kenneth Manalansan 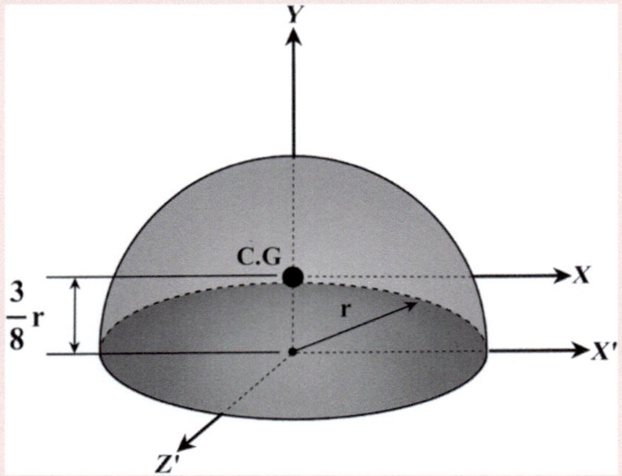	$V = \frac{2}{3}\pi r^3$	$x_{C.G} = 0$ $y_{C.G} = \frac{3r}{8}$ $z_{C.G} = 0$	$I_{xx} = I_{xx} = \frac{83}{320}mr^2$ $I_{yy} = \frac{2}{5}mr^2$
Figures are designed by the author and implemented by Kenneth Manalansan. Courtesy of Kenneth Manalansan 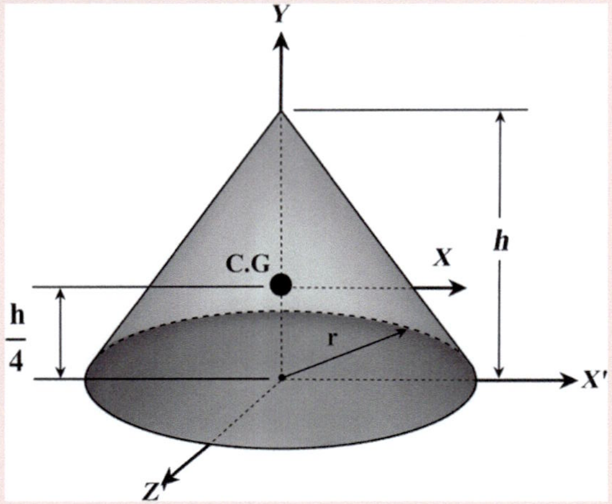	$V = \frac{1}{3}\pi r^2 h$	$x_{C.G} = 0$ $y_{C.G} = \frac{h}{4}$ $z_{C.G} = 0$	$I_{xx} = I_{xx} = \frac{3m}{80}(4r^2 + h^2)$ $I_{yy} = \frac{3}{10}mr^2$

Index

A

absolute acceleration, 321, 332, 345
absolute acceleration vectors, 345
absolute jerk rate vector, 383
absolute polar acceleration components, 319
AC (alternating current), 420
acceleration and jerk, 481, 482, 501
acceleration and jerk analysis, 309
acceleration Cartesian components, 320
acceleration component in r-direction, 321
acceleration component in θ-direction, 321
acceleration components
  average angular, 316
  average translational, 310
  centripetal, 324, 328, 331, 332, 337, 338, 346, 392–394
  instantaneous angular, 316
  normal relative, 339
  polar, 319, 321, 322, 337, 341
  relative centripetal, 339, 346
  relative Coriolis, 346
  relative normal, 348, 353, 361
  relative tangential, 346, 361
  total relative, 346
  transitional, 319, 610, 612
acceleration
  graphic, 42, 43, 309, 345–348, 350, 352, 353, 355, 361, 362, 364, 367, 368, 381, 389, 397, 423, 448, 467
  instantaneous, 65, 310, 311, 313, 316, 317, 319, 324, 325, 328, 332, 337, 371, 376, 389–392, 436
  maximum, 404, 417, 420, 422, 428, 446, 481, 496
  requested, 425, 427, 431
  required, 346, 352, 361, 392, 399, 420, 432, 446, 482
  tangential, 42, 309, 321, 324, 326, 328, 329, 331, 332, 337–339, 341, 346, 348, 349, 353, 354, 361, 362, 392, 394
  transitional, 319, 374, 379, 380, 383, 395, 610–612, 622
  zero, 66, 310, 314, 316, 324, 425, 431, 442, 475–477, 481, 482, 628
acceleration curve, 482, 496
acceleration rate, 383
acceleration sensor, 340, 341, 345, 356, 365, 395
acceleration vector
  absolute transitional, 319
  absolute translational, 323
  angular, 309, 310, 316, 323, 331, 346, 359, 362, 364, 368, 371
  average angular, 316
  average translational, 310
  instantaneous, 313, 316
  relative, 118, 309, 323, 324, 327, 345, 346, 349, 353, 359
  total, 310, 348, 353, 359, 362, 393
acceleration vector diagram, 309, 315, 319, 320, 327, 328, 330, 332, 335, 336, 348, 349, 351, 353, 355, 358, 359, 364, 366, 368, 393
accumulator, 404
accuracy, 220, 225, 425, 496
AC electricity, 128
AC motor, 420–422
acting forces, 42, 563, 574–576, 588, 595, 610, 628
acting moments, 563, 595
acting torques, 574, 575, 610
actuation, 127, 128, 130, 213, 215, 216, 274, 341, 539, 544, 583, 586, 587, 618, 620, 622, 624
actuator and mechanism constraints, 432
actuator displacement, 423
actuator failure, 423, 425
actuator forces, 563, 575, 581, 583, 595, 611, 628
actuator forces/torques, 575, 583, 584, 595, 611, 628
actuators
  angular, 211, 212, 217, 304, 306, 395
  desired, 383, 384, 398, 423, 425, 468, 585, 625
  direct, 75, 158, 263, 264, 611
  linear, 129, 130, 161, 208, 209, 211
  manual, 70, 127
  pneumatic, 75, 127, 130
  required, 127, 130, 209, 563, 581, 589, 590, 595, 618
  unlimited powered, 423
actuators and sensors, 163, 385, 588
actuator torque, 575, 582, 612, 615, 618, 620, 623, 625, 627, 628, 630
actuator velocity limits, 446
actuator's location, 354, 364

A. Omran, *Kinematics and Dynamics of Machines,* https://doi.org/10.1007/978-3-031-72701-6

addendum, 521, 527, 529–531, 555, 556
addendum and addendum radius, 531, 556
addendum and dedendum, 521
addendum circle, 521
addendum diameter, 521, 527
addendum radius, 531, 556
AGMA (American Gear Manufacturers Association), 522, 523
AGMA magazine, 523
AGMA requirements, 530
agriculture, 233, 300, 379
aircraft, 24, 25, 34–36, 85, 94, 160
aircraft frame, 36
air cylinders, 459
algorithm
  graphical, 284, 287, 291, 295, 346, 347, 352, 361, 362, 367
  graphical acceleration, 347, 352, 367
  mathematical, 216, 618, 623
aligned axes, 178, 179
aligned axes rotational, 178
aligned axes rotational method, 178
alignment
  initial, 188
alternating current, *see* AC
analysis
  graphic, 1, 98, 141, 221–223, 277, 284, 287, 288, 291, 294, 295, 345, 352, 359, 362, 364, 367, 368
  static, 3, 4, 563, 571, 581, 584, 585, 625
  strain, 530
Amercian Gear Manufacturers Association, *see* AGMA
American National Standards Institute, *see* ANSI
analytical acceleration analysis, 368, 376
analytical method, 41, 201, 248, 250, 252, 255, 298, 301, 302, 306, 309, 371, 381, 389, 488, 492, 515
analytical position models, 201
analytical solution, 41, 225, 239, 242, 245, 253, 275, 295, 298
analytical solution and plots, 41
analytical solution and velocity, 373, 377
analytical velocity analysis, 295
analytical velocity models, 255, 309
angle
  helix, 519
  pitch, 528, 532
  polar, 321, 322, 341
  revolution, 167
  tilt, 408
  toggle, 245, 253, 307
  transmission, 483, 485, 530
angle acceleration, 324
angle interval, 468, 474, 478
angle motion sensor, 215
angles of rotation, 488
angle velocity, 261, 262, 303, 304, 325
angular, 30, 34, 42, 120, 121, 162, 201, 208–210, 212, 213, 215–217, 219, 248, 255, 256, 261–267, 269–272, 274, 281, 285, 286, 288, 291, 293, 295, 296, 298, 301, 303–306, 309, 310, 315–318, 323, 324, 326, 328, 330, 331, 333, 337, 339, 343, 345–347, 350–353, 355, 359, 362, 364, 367, 368, 371, 374, 376, 379–381, 386, 388, 392–395, 405, 407, 420–423, 449, 460, 468, 473, 474, 477, 506, 511, 512, 515, 532–534, 540, 542, 543, 545, 546, 549, 554, 556, 557, 559, 560, 604, 610, 615, 618–620, 622, 623, 628, 633
angular acceleration, 213, 214, 272, 310, 315–318, 323, 324, 328, 330, 337, 339, 343, 348, 350, 355, 359, 364, 374, 376, 379, 380, 386, 388, 392–395, 515, 615, 624, 628, 634
angular acceleration of motor, 318
angular actuation, 216
angular displacement, 162, 208–210, 212, 248, 256, 388, 423, 460, 628
angular displacement input, 628
angular jerk, 386, 388
angular speed, 34, 315, 324, 420–422, 473, 477, 506, 511, 512, 535, 542, 543
angular speed of gear, 543
angular velocity
  driver, 532, 533
  driver gear, 532
  measured, 261, 316, 337, 376, 386, 407, 540, 546
  relative, 261, 270, 271, 325, 339, 346
angular velocity displacements, 316
angular velocity input, 628
angular velocity motion, 120
angular velocity of gear, 543
angular velocity ratio, 285
angular velocity sensor, 289, 539, 545
animation, 1, 4, 20, 32, 33, 41–43, 85, 102, 103, 105, 106, 110, 112–114, 116–118, 120, 122, 124, 131, 160, 171, 195, 206, 224, 260, 263, 264, 269, 275, 289, 290, 294, 314, 318, 345, 365, 386, 387, 397, 401, 402, 404, 448, 454, 455, 501, 503–506, 516, 518, 520, 538, 540, 543, 546, 625, 627, 629
ANSI/AGMA, 530
ANSI (American National Standards Institute), 530
applications, 2, 41, 43, 48, 72, 75, 76, 97, 116, 119, 124, 126, 130, 247, 248, 255, 284, 379, 383, 415, 416, 420, 453, 457, 473, 482, 494, 504, 506, 507, 509, 516–518, 520, 554
Arbitrary Axis, 178
archimedes, 3
areas, 98, 141, 145, 407, 411, 418, 497, 498, 520, 528, 596, 598, 599, 601
arm
  robot, 60, 61, 116
arrows, 142, 398, 564, 565, 572, 574, 610
articulation, 1, 4
artificial neural computation, 436
assembled links and joints, 611
assembly
  piston-cylinder, 10
  planetary gear train model, 550

auto-apply, 582–584
AutoCAD, 98, 164, 222, 496
automotive engine hood, 27
automotive science, 547, 550
average acceleration, 312, 389–392
average angle velocity, 261, 262, 303, 304
average angular acceleration, 316, 317
average angular velocity, 263
average velocity, 256–259, 302
average velocity components, 256
axes, 16, 46, 62, 64, 75, 168, 169, 178, 180, 182, 196, 202, 535, 606
axes frame, 168, 178, 180, 182
axial rotation, 458, 463, 464
axial rotational camshaft, 458
axis of camshaft rotation, 458
axis of motion, 483, 485
axis of rotation, 458, 516, 519, 521
axis of transmission, 483, 485

B

backlash, 530
backpropagation learning, 368
balance, 25, 29, 495, 599
balance point, 599
bar mechanism, 41, 83, 84, 97, 106, 107, 110, 113–117, 119, 131, 174, 183, 184, 186, 188, 201, 204–206, 208–210, 221, 226, 227, 239–241, 245–248, 252, 253, 277, 281–283, 285, 286, 290, 291, 295–297, 299, 302, 307, 317, 321, 346, 347, 350, 361, 368, 369, 371, 374, 389, 395, 423, 567, 595, 606, 611, 615, 616, 618, 624, 631
bar mechanisms analytical analysis, 239
base, 13, 14, 16, 17, 22–24, 70, 71, 76, 154, 161, 162, 168, 175, 177, 178, 183, 185, 213, 215, 272, 339, 483–485, 488–491, 494, 495, 498, 509, 521, 523–525, 527, 528, 531, 535–537, 539, 542, 545, 555, 556
Base and Follower, 161, 162, 178, 185, 213, 272, 339, 535
base axes, 178
base circle, 483–485, 488–490, 495, 523, 524, 527, 528, 531, 556
base circle diameter, 531, 556
base circle heading, 527
base circle radius, 495
base circle's center, 483
base diameter, 524–526, 555
base DoF (base degree of freedom), 71
base gears, 16, 17
base node, 536
base shaft, 535, 537
base spur gear, 535
belts, 161
Bernoulli, Daniel, 417
beryllium, 635
bevel gears, 36, 37, 520, 521, 536
bevel gear type, 521
bevel type, 520
bicycle, 22, 24–26, 107, 111, 112, 127
bicycle crank, 107, 111
bicycle frame, 26
Bicycle Rider Machine, 26
block diagram
  analytical velocity, 296
bodies, free, 41, 42, 56, 58, 60–62, 78–84, 131, 183, 184, 270, 338, 563, 571, 615, 616, 620
Brick, 164, 185, 189–192, 194, 196, 219, 237, 243, 276, 383, 385, 445, 597, 613, 614, 628, 631
bucket, 37, 38, 61, 62, 102, 126, 128, 129, 248, 251, 398–400, 404, 405, 407, 425, 428, 431, 439, 442, 444–447, 565, 581, 586, 592, 602–604, 608
bucket counterclockwise, 38
bucket cylinder, 37, 425, 428, 431, 439, 442, 445
bucket mechanism, 102, 248, 251, 398–400, 404, 405, 407, 442, 445, 565, 581, 586, 592, 602
bucket piston, 37, 445

C

cables, 161
CAD applications, 504
CAD assemblies, complete, 160
CAD (computer-aided design), 41, 160, 346, 352, 362, 496, 504–506, 527, 597, 601, 605
CAD files, 496, 504–506
CAD models, 504
CAD software, 41, 346, 352, 496, 505, 527, 597, 601, 605
calculation, 43, 78, 204, 212, 295, 298, 320, 332, 377, 423, 467, 542, 588, 589, 591, 597, 601, 608, 618, 620, 624
cam analysis, 453
cam and follower, 484, 499, 513
cam angle, 34
cam angular velocity, 479
cam axis, 454
cam base, 76
cam base radius, 483
cam base radius and base circle, 483
cam center, 482, 492, 502
cam configuration, 76, 484
cam contour
  convex, 496
cam curvature, 482
cam curve, 466, 483
cam cylinder, 455
cam design, 453, 494, 495
cam
  dual, 496
  exhaust, 32, 34, 455, 457, 465, 466
  flat plate, 506, 509
  intake, 455, 457, 458, 466, 467
  plate-shaped, 462
cam followers, 42, 453, 509

cam geometry, 483, 485
cam groove, 482, 509
cam joint, 74, 76, 77, 122, 161, 453, 454, 457, 459, 485, 496, 509, 568
cam link
  heart-shaped, 462
  wedge-shaped, 461
cam link and follower link, 496
cam link profile, 457
cam link shapes, 458, 461
cam link type, 461
cam machine, 466
cam mechanism
  disk, 473, 477, 511, 512
cam motion, 34, 397, 423, 513
cam profile, 453, 457, 466, 481–485, 488, 489, 491–498, 509, 512, 513
cam rotation, 484, 485, 488
cam rotational angle, 485
camshaft, radial, 457
cam shape, 458, 494
cam size, 496
cam surface, 459, 483–485, 496
cam trajectory, 469
cam types, 466, 509
Cartesian components, 320, 355, 364
casement window machine, 7, 8
center gear, 547
center of mass, 164, 180, 196, 599
center of shaft, 542
centripetal, 324, 328, 331, 337–339, 346, 392, 394
CEP (critical extreme position), 468, 469, 472–475, 480–482, 511, 512, 514
CFD (computational fluid dynamics), 418
circle, 98, 121, 221, 224, 280–282, 324, 326, 483–485, 489, 490, 521, 523, 527, 528, 530, 531, 535, 555, 556, 568, 598, 599
circle radius, 221
circuit, 159, 221, 224, 227, 229, 230, 233, 374, 375, 379, 400, 402, 404, 405, 448
circuit diagrams, 404
circular motion, 125, 324
circular pitch, 522, 523, 534, 542, 555, 556
circular tooth thickness, 530
closed chain, 56, 62, 80, 81, 83, 106, 116, 240
closed-chain mechanisms, 41, 97
closed kinematic chains, 52, 54
closed-loop, 54, 56, 128, 225, 229, 239
close mechanism, 268
closer machine, 6
closer mechanism, 5, 6, 11, 93, 132, 249, 267, 268
CNC (Computer Numerical Control), 492
collinear, 109, 110, 245, 459
collinearity, 245
collinear parallelogram double crank, 110
combustion energy, 46, 127, 128, 130
complex conjugate, 154
complex modulus, 154
components
  mechanical, 406
  mechanical systems, 160
composite materials, 597
compound gear, 515, 546, 547, 554
compound gear train, 546, 547
computational fluid dynamics, *see* CFD
computer, 43, 157, 492, 605
computer numerical control, *see* CNC
conjugate cam, 459
connected link, 269, 337, 568
connect gear, 539, 545
connecting rod, 26, 27, 32, 33, 36, 39, 40, 47, 50, 52, 57, 113, 114, 198, 223, 229, 230, 258, 270, 288, 337, 338, 355, 364, 376, 455, 575, 634
connection port, 273, 340
connections, mechanical, 20, 22, 26, 47, 50, 51, 54, 158, 161, 164, 420, 459
connection types, 161
connectivity, 98
conservation, 127
constant acceleration CPM motion, 482
constant displacement CEP motion, 473
constant displacement trajectory, 468
constant motion, 3
constant velocity CEP Motion, 473, 511
constant velocity CPM Motion, 514
constant velocity trajectory, 468
constraint force, 583, 587, 628
constraints, 131, 160, 161, 228, 258, 266, 295, 312, 367, 400, 404, 432, 468, 476, 497, 498
contact
  metal-to-metal, 417
contact surface, 484
continuity, 469, 496
control
  dampers, 31
  hydraulic elements, 417
  inaccurate motion, 54
controller, 216, 398, 400, 552
control loop
  feedback, 216
control panel, 398
control process, 400
control subsystem, 399, 400
control valves, 129, 399, 400, 402
conveyor machine, 420, 591
coordinates
  cylindrical, 63, 178
coordinate systems
  Cartesian, 62–65, 168, 169, 178, 202, 203, 328, 339, 370, 371, 394, 564, 566, 606
  cylindrical, 63
  spherical, 63
coriolis acceleration, 309, 325, 326
coriolis acceleration components, 325
corkscrew
  wing, 75, 76
corkscrew machine, 17, 76, 87
counterweight, 206, 579

couple, 127, 174, 255, 266, 330, 432
coupler, 6, 7, 11, 15, 16, 21–24, 27, 30, 33, 39, 40, 42, 97, 107, 108, 113, 118, 120, 201, 246, 247, 267, 268, 276, 285, 298, 305, 306, 332, 350, 371, 392, 395, 518, 575, 611, 628
coupler angular velocity, 395
coupler curve, 42, 118, 201, 246, 247
coupler movement, 23
coupler rod, 268, 332
CPM, constant acceleration, 482
CPM (Critical Path Motion), 468, 481, 482
crank, 8, 22–27, 41, 106–110, 114, 115, 124, 127, 128, 130, 131, 192, 198, 223, 229, 233, 234, 246, 268, 288, 300, 303, 305, 306, 317, 332, 337, 355, 364, 376, 379, 380, 391, 394, 395, 564, 579, 593, 621, 631, 632, 634
crank angle, 223, 288
crankpin, 270, 337, 355
crank radius, 223, 288
crank-rocker, 107–110, 112, 245
crankshaft, 32, 33, 46–48, 50–52, 82, 107, 113, 127, 129, 130, 223, 224, 229, 230, 288, 337, 355, 376, 395, 400, 455–457, 466, 475
crank slider mechanism, 192
critical extreme position, *see* CEP
critical path motion, *see* CPM
cross product, 326, 327, 567, 575, 610
crosssection, 276, 444
crown/pinion, 552
curvature, 482, 496
curve constraint, 497–499, 502, 505
curve crosses, 436
curves
  green, 466
cycle
  camshaft rotational, 467
  complete, 246, 456
cycloidal CEP motion, 481, 512
cycloidal CEP trajectory, 481, 482
cylinder, 10, 11, 31–34, 36–39, 46, 47, 50–52, 59, 75, 85, 91, 97, 113, 114, 120, 127, 129, 130, 161, 164, 196, 198, 199, 223–225, 230, 231, 235, 237, 256, 264, 270, 288–291, 301, 304, 306, 328, 329, 337, 338, 355, 357, 358, 455–459, 463, 516, 519, 542, 568, 572, 577, 585, 586, 597, 601, 631, 633
cylinder angular acceleration, 364
cylinder angular velocity, 291, 395
cylinder losses, 417, 420
cylinder motion, 256
cylindrical cam, 455, 464, 507, 509
cylindrical cam mechanism, 455, 507

D

damper force, 582
Darcy–Weisbach equation, 411
DC (direct current), 420
DC motor, 122, 209, 210, 216, 246, 420, 421
debugging, 620, 624
deceleration, 420, 422
dedendum, 521, 527, 528, 530, 531, 556
dedendum circle, 521, 528
deformable mechanism, 48
deformation, 46, 48–50, 596
degrees, 41, 45, 71, 72, 75–78, 83, 116, 120–122, 130, 178, 209, 210, 217, 235, 248, 249, 468, 471, 492, 496, 510, 513, 514, 519, 552
degrees of freedom, *see* DoF
Denavit, Jaques, 3
density, 383, 384, 409, 410, 412, 450, 596, 597, 600, 603, 604, 625
derivation, 107, 227, 240, 245, 324, 470, 477
derivatives, 66, 259, 274, 312, 424, 430, 436, 438, 477, 481, 482
design, 1–4, 11, 17, 18, 23, 27, 29, 31, 40–43, 48, 98, 113, 117, 118, 140, 202, 240, 247, 277, 295, 368, 383
design analysis, 277
design and analysis of machines, 609
design engineer, 18, 23, 27, 481, 482, 494, 495, 520, 530, 550
designers, 3, 26, 64, 106, 386, 417
design trajectory, 42, 397, 448
device, 2, 11, 16–18, 117, 420
diagram
  schematic, 525, 537, 538, 547
diameter, 408–410, 412, 416, 417, 419, 425, 428, 431, 439, 450–452, 521–525, 527, 531, 534, 555–557, 559, 560
diameter pitch, 522
diameter ratio, 417
diametral pitch
  standardized, 530
differential assembly elements, 553
differential case, 552
differential gear assembly, 554
differential pinions, 552
differentials, 42, 515, 554
differentiate equation, 296, 299, 369, 373, 378, 379
dimensions
  given, 248
dimensions and elements, 145
direct current, *see* DC
direction, 6, 8, 9, 29, 37, 46, 64, 65, 67, 116, 121, 122, 128, 129, 162, 172, 175, 178, 180, 202, 203, 212, 226, 229, 233, 235, 256, 257, 265, 279, 284, 286, 288, 292, 295, 296, 310, 315, 319, 320, 324–327, 329–332, 339, 346, 348, 349, 351, 353, 354, 357, 361–363, 367, 402, 407, 411, 417, 455, 457, 459, 465, 483, 485, 488, 490, 498, 521, 533, 541, 564–566, 568, 570, 572, 574, 575, 577, 578, 583, 584, 597, 598, 605, 609, 610, 613, 615, 616, 620, 622
direction of gear, 570
direction of rotation, 458
disassemble, 568
disc, 13, 461

discontinuities, 42, 162, 397, 424, 427, 431, 433, 434, 436, 437, 442, 448, 473, 477, 481
discretization, 597, 599, 601
disk, 25, 123, 124, 458, 461, 462, 473, 477, 511, 512, 516, 519
disk cam type, 466
disk plate, 458
displacement
active, 208–212, 248, 249
displacement, velocity, acceleration and jerk, *see* SVAJ
display, 149, 150, 219, 220, 232, 244, 504, 587, 628
distance
center-to-center, 524
circular, 522, 556, 566
radial, 489, 521, 528
relative, 48
distribution, 49, 108, 405, 436–438, 466, 496, 596, 597, 605
distribution parameter, 436–438
disturbances, 398
division, 154, 596, 604
DoF (Degrees of Freedom), 68, 70, 76, 210, 248, 249
domain, 62, 64, 84, 150, 159, 310
doors, 4
double acting, 416, 419, 451
double acting and single rod cylinder, 415
double-acting Piston-cylinder Subsystem, 129
double-click, 142, 147, 157, 162, 189, 190, 217, 219, 220, 272, 273, 275, 289, 293, 294, 339, 383, 433, 434, 497, 582, 584
double crank, 107, 109, 110
double rocker, 108, 110
double rod cylinder, 415, 416
downstream pressure, 418
drag, 36, 145, 150, 156, 157, 186, 192, 219, 272, 273, 275, 289, 293, 339, 340, 440, 497, 505, 584
drive, 14, 16, 17, 26, 29, 34, 120, 123–125, 216, 400, 406, 457, 459, 552
driver
tractor, 433
drive shaft, 16, 17, 552
drive wheels, 34, 120
dump, 398, 416, 425, 428, 431, 439, 577, 586, 587
dump bed, 416
dump bed truck, 416
dumpster truck, 38, 85, 89, 131, 133, 198, 199, 250, 304, 306, 307, 393–395, 577, 578, 585, 631, 633
dump truck mechanism, 586, 587
duration, 364, 431
dwell, 124, 423–429, 431, 432, 439, 451, 452, 465, 466, 473, 477, 495, 509, 511, 512
dwell period, 124
dwell segment, 495, 509
dwell trajectory, 423, 431
dynamic analysis, 1, 3, 4, 6, 17, 18, 26, 27, 41, 625
dynamic performance, 2
dynamic calculation algorithm, 618
dynamic cam-follower design, 496
dynamic model, 29, 39, 42, 43, 595, 609, 611, 615, 618, 623, 628
dynamic model solution, 595
dynamic moments, 624
dynamics, 1–4, 11, 18, 23, 31, 37, 41, 140, 158, 160, 162, 382, 411, 418, 420, 564, 611, 619, 623
dynamics courses, 382
dynamics diagram, 619, 623
dynamics of machines, 1–4, 11, 18, 31, 37, 140
dynamics studies, 3
dysprosium, 635

E

Earth, 49
Earth–sun model, 49
Earth–Sun motion, 49
eccentricity, 458, 483, 485, 509
edge
cutting, 14
edit, 155, 156, 164, 189, 275, 293
editing, 155, 156, 164, 180, 182, 187, 498, 502, 504, 506, 535, 539, 544, 583, 584, 620, 624
editing list, 164, 180, 187
editing menu, 182, 498, 502, 504, 535, 539, 544, 584
effector, 48, 54, 56, 58, 60–62, 98, 102, 103, 105, 241, 268, 332
elbows, 407, 411–413, 450
elbow shapes, 411
electrical components and circuits, 159
electrical motors, 420
electricity, 33, 97, 128, 398
electric motor system, 128
electromagnetic components, 160
electromagnetism, 128
elements
basic, 98
discrete, 597
forearm, 5, 6
functional, 127, 398
geometric, 62
major, 4
elliptical gym machine, 22, 24
end-effector, 48, 54, 98, 102, 103, 105, 241, 332
energy conversion, 128
energy
electrical, 127, 128
mechanical rotational, 2
total stored, 411
heat, 32
kinetic, 32, 128
producing thermal, 285
energy equation, 411
energy losses, 411, 417
energy method, 609
energy sources, 2, 127
energy transfer, 127–132
engine
diesel, 127, 457
four-stroke, 32, 455, 456, 465
gasoline, 33, 127, 457

inline four-cylinder, 46
steam, 33–35, 120, 121, 127
engine block, 47, 48
engine crankshaft, 107
engineering, Newton, 43, 383, 563, 574, 588, 595, 609, 610, 616, 621, 628, 631
engineering mechanics, 196
engineering note, 504, 523, 524
engineers, 2, 11, 41, 43, 48, 64, 139, 158, 239, 423, 504, 530, 550, 596
engine mechanism, 32, 33, 77, 91, 94, 224, 455, 456, 576
engine piston-cylinder mechanism, 47, 51, 53, 113, 129, 229, 232, 288, 289, 338, 376
engine shaft, 515, 516
equal length, 110
equation
analytical, 492
dynamic, 611
linear, 432, 472
nonlinear, 429
static, 622
equations of motion, 160
equilibrium, 42, 563, 575, 580, 588
error, 216, 275, 368, 377, 458, 473, 496
estimate, 43, 84, 107, 110, 131, 264, 279, 281, 282, 284, 286, 290, 295, 302, 350, 399, 405, 407, 409, 411, 421, 596
estimation, 107, 350, 447
Europe, 2
europium, 635
example
real-world, 97, 408
solved, 4, 43, 115, 498, 505
examples of cam mechanism, 42
excavator lift machine, 59, 60
excessive side force, 485
execute, 142, 150, 153
exercise, 20–22
exhaust valves, 455, 457
exit, 129, 326, 403, 415, 416
exit chamber, 129
experiments, 410, 411
experts, 295, 368
explosion, 33, 410
extension, 31, 63, 122, 129, 142, 159, 415, 416, 496, 567
external forces, 48, 563, 564, 568, 588, 615, 620
extruded solid block, 496
extrusion, 164, 191, 194, 237, 243, 276, 385, 444, 501, 502, 597, 628
extrusion cross section, 444

F

failure, 383, 422, 423, 427, 468, 528, 541
fall, 126, 423, 425, 426, 428, 431, 439, 440, 451, 452, 465, 466, 468, 469, 473, 475, 477, 480–482, 495, 509, 511, 512
fall motion, 460, 466, 477
fall trajectory, 423
faucet, 408
FBD, linkage's, 572, 574
FBD (Free Body Diagram), 41, 42, 56, 58, 60–62, 78–84, 131, 183, 184, 270, 338, 563, 571–573, 576–581, 588–593, 611, 612, 614, 615, 616, 620, 621, 633, 634
fillet, 528, 530, 554
fillet radius, 528
filter, 403
finite number of point masses, 596, 600
fit, 12, 14, 21, 22, 41, 224, 284, 287, 291, 346, 352, 362
five-bar mechanism, 84, 97, 116, 117, 239
fixed link, 50, 107
fixed rocker linkage, 232
flanged elbow, 413
flapper, 8, 9
flapping wing, 509
flat cam, 461
flat-faced follower, 465, 466, 484, 485
Fleming, 64, 65, 128, 202
Fleming, John Ambrose, 64
flex brick, 383, 385
flexible body, 383, 385
flipping, 186
flow
uniform, 408
flow control valve, 404
flow rate, 129, 404, 406, 407, 410, 417–420, 423, 439, 446, 449–451
flow types, 411
fluid, 120, 129, 158, 401–404, 406–411, 415–418
fluid density, 409, 410
fluid dynamics, 418
fluid flow, 415
fluid flow rate, 415
fluid friction, 417, 422, 458, 516, 570, 571
fluid leakage, 417
fluid power, 75
fluid speed, 408
fluid viscosity, 409, 410
flywheel, 16
fold, 6, 11, 36
folding chair machine, 7
Folding Chair Mechanism, 6, 86, 132
follower acceleration curve, 496
follower and base, 185
follower connection, 535, 537
follower
flat-face, 465, 484
knife-shaped, 494
rolling, 495
follower frames, 535
follower link, 159, 453, 454, 457–459, 465, 484–486, 496, 505, 506
follower linkages, 496
follower motion, 453–455, 460, 466, 468, 471, 482, 485, 488, 494, 495, 508–510
follower node, 536
follower SVAJ diagram, 499, 513

follower trajectory, 470, 473, 477, 511, 512
follower type, flat-faced, 459
force closure type, 459, 466
force elements, 160
force ratio, 284
forces
  applied, 383, 588, 590, 596, 599
  brake, 383
  desired, 584
  frictional, 571
  inertial, 42, 410
  transfer, 6, 16
  viscous, 410
forces and moments, 4, 42, 420, 563, 568, 569, 574, 576, 588, 595, 610, 611
forms, 1, 285, 359–361
formulas, 256, 310, 423, 453, 493, 515, 530, 531, 554, 597, 601, 607
four-bar, 41, 78, 79, 84, 97, 106, 107, 110, 111, 113–116, 118, 119, 121, 131, 174, 183, 184, 187, 197, 201, 204–206, 225, 253, 255, 277, 281–286, 290, 291, 295–297, 299, 302, 307, 309, 317, 321, 322, 346, 347, 350, 361, 368, 369, 371, 375, 389, 395, 423, 567, 595, 606, 611, 615, 616, 618, 624, 628
four-bar linkage, 106, 110, 113
four-bar linkage models, 106
four-bar mechanism
  open-chain, 204
Frame axes, 168, 178, 180, 182
frame
  Cartesian, 169
  helicopter, 37
  rotating, 271, 310, 312, 323, 325, 370, 610
Frames and Transformers, 161
free body diagram, *see* FBD
free link motion, 71
friction, 110, 285, 405, 406, 411, 412, 417, 458, 516, 570, 571
friction coefficient, 411, 571
friction forces, 570
friction losses, 422
fuel, 32, 46, 47, 97, 113, 129, 130, 455, 456
fuel consumption, 97
functionality, 1, 4, 41, 272, 339, 398, 400, 401, 420
function of Reynolds number, 412
functions, 41, 76, 139–141, 145, 147, 153, 154, 168, 196, 240, 398, 423, 429, 436, 438, 440, 467, 468, 470, 472, 477, 480, 481

G

gadolinium, 635
gain, 141, 153–157, 215, 217, 341, 359, 539, 545
gallium, 635
gas, 159, 160, 433
gate, 10, 11, 93, 235, 291, 292, 301, 328, 332, 364, 381, 414, 611, 625
gear analysis, 515
gearboxes, 550
gear connection, 537, 539, 552
gear constraint, 535, 539, 542, 545
gear contact force, 568
gear joint, 58, 74, 76, 77, 98, 161, 279, 568
gear kinematic pair, 531, 546
gear motion, 8, 16, 17, 29
gear pair, 516, 520, 531
gear ratio, 407, 449, 450, 520, 529, 532–535, 541, 546–549, 552, 554, 556, 579
gear reduction, 552
gears
  bull, 533, 534, 537, 557
  carrier, 561
  circular, 130, 521
  driver, 401, 532, 549, 583
  landing, 85, 90, 131, 134, 160, 172, 173, 199, 200, 592
  mating, 530, 568
  meshing, 520, 523, 529, 531
  partial, 8, 17, 28, 29
  planet, 547, 549, 550, 552, 561
  spider, 552
  sun, 550, 552
  super, 524
gear schematic diagram, 524
gear shaft, 516
gear speed, 547
gear teeth
  helical, 519
gear tooth, 521, 522, 528, 530, 541
geartrain, 524, 525
gear train assembly, 552
gear train, epicyclic, 547
generic AC motor torque-speed curve, 421
generic shape, 142, 504
Geneva mechanism, 123, 124, 126
geometric properties, 639
geometric shape, 585, 586
geometry, 50, 54, 160, 163, 164, 166, 190, 203, 207, 235, 339, 383, 385, 412, 459, 484, 488, 491, 492, 498, 504, 523, 524, 530, 531, 535, 539, 544, 597, 601, 606, 607
germanium, 635
graph, 150, 180, 436
graphical, 301, 352, 355, 364, 382
graphical analysis, 359, 364, 367
graphical method, 41, 201, 248, 255, 285, 288, 293, 298, 301, 302, 305, 306, 350, 357, 365, 371, 373, 381, 389, 490, 492
graphical user interface, *see* GUI
graphics, 1, 98, 164, 221
graphics and animation, 1
Grashof, 41, 97, 107, 110–113, 131, 245
Grashof case, 110, 131
Grashof condition, 107
Grashof, Franz, 3
Grashof linkage, 97
Grashof mechanism, 107, 110, 111

Grashof's Law, 107
Grashof's theorem, 41
gravity
  center of, 42, 43, 49, 66, 164, 174, 175, 185, 247, 275
grid, 13, 14
groove
  curved, 459, 462, 466, 468
ground, 41, 45, 50–52, 55, 57–60, 77, 78, 98, 107, 108, 112–114, 116, 161, 169, 171–174, 183, 185, 192, 194, 206, 212, 221, 230, 237, 243, 247, 255, 264–268, 270, 271, 277, 282, 301, 304, 305, 309, 323, 329–332, 337, 381, 536, 542, 572, 577, 579, 580, 588–593, 611, 613
ground block, 542
ground link, 41, 45, 50–52, 54, 55, 57, 59, 77, 78, 84, 98, 107–109, 116, 159, 161, 171, 174, 206, 230, 237, 243, 255, 267, 268, 282, 309, 329, 332, 536, 577, 579, 580, 611, 613
ground nodes, 57–60, 78, 98, 169, 172, 173, 183, 221, 330, 589–593
GRRR (ground–rocker–rocker–rocker) mechanism, 107
Grübler's equation, 45, 77–79, 81–84, 113, 114, 116, 211
GUI (graphical user interface), 189
gymnasium machines, 18

H
hafnium, 639
hammer strength gym machine, 20
handle
  green, 8
harmonic CEP motion, 480
harmonic motion, 480
harmonic SVAJ, 429, 452
hartenberg, Richard S, 3
heads, 348, 353, 362
heat
  dissipating, 401
  generated, 422
height, 9, 383–385
helical gear, 518–520
helical gear angle, 518–520
helical type, 519
helicopter coaxial rotors mechanism, 36
helicopters, 37
help, 1, 2, 4, 7, 18, 22, 140, 141, 145, 146, 149, 189, 204, 235, 314, 318, 417
high-order derivatives, 436, 477
high-speed SCARA parallel robot, 117
hinge joint, 71, 72, 328
history, 2, 150, 218, 274, 343, 355, 364, 423, 478, 625
holder, 33, 46–48, 51, 52, 55, 56, 113, 135, 159, 230
holder linkage, 55, 56
holder mechanism, 135
holes, 506, 597
hood mechanism, 27, 28, 88, 116–118, 133
horsehead, 106
hoses, 401, 404, 407
human anatomy, 18–20, 25, 26
humans, 2, 127, 383
humerus, 51
hydraulic actuator, 10, 36, 97, 129–131, 158, 161, 210, 248, 304, 306, 354, 397, 400, 404, 417, 420, 423, 433, 436, 564, 589, 590
hydraulic circuit, 400, 404, 405, 448
hydraulic circuit diagram, 400
hydraulic control valves, 400
hydraulic cylinder losses, 417
hydraulic direction control valve, 402
hydraulic energy, 130
hydraulic filters, 403
hydraulic flow, 404
hydraulic fluid, 120, 129, 402–404, 408, 409
hydraulic force, 577, 578, 590–592
hydraulic gate machine, 11
hydraulic gate opener, 236, 291, 328, 364
hydraulic hoses, 404
hydraulic losses, 405
hydraulic oil, 10, 402, 412, 450
hydraulic power, 399, 401, 405, 406, 415, 420
hydraulic subsystem, 38, 39, 128, 130, 400, 401, 404, 405, 407, 417, 420, 449
hydraulic system, 72, 128, 129, 211, 212, 263, 264, 394, 395, 402, 405, 423, 425, 428, 431, 439, 448, 451, 452
hydraulic valves, 418
hysteresis power losses, 422

I
idealization, 48, 596
implementation, 385–387, 585
import, 160, 163, 164, 504
incompressible flows, 129
independent variables, 141, 347, 352, 353, 362
indeterminate, 110, 575, 579
index, 64, 72, 75–77, 263, 279, 326
industries, 233
inertia, 42, 43, 49, 164, 171, 173, 312, 323, 399, 406, 410, 496, 504, 595, 604–608, 611, 612, 614, 618, 623, 625, 628, 631, 632
inertia forces, 410, 496
inertia frame, 171, 173, 312, 323
inertial frame, 171–173, 197–199, 270
inertia load, 42
inertia matrix, 164
inertia tensor, 164
infinity, 171, 172, 279, 286, 290, 414, 424, 425
influence, 309, 383, 384, 442, 495, 528, 572, 615, 620
influence of discontinuities, 442
initial condition, 188
initial position, 39, 168, 217, 291, 301, 364, 381
input
  desired, 398
input motion, 106, 116, 118, 302, 397, 405, 423, 425, 428, 431, 437, 448, 451, 452, 564
input motion trajectory, 448
instantaneous center of rotation, 279

instantaneous centers, 277, 279–284, 286, 287, 290, 291
instantaneous ground, 277
instantaneous location, 278
instantaneous velocity, 65, 257–259, 302, 389–392
instant centers, 280–282
instant ground, 277
intake, 32, 455–457, 466, 467
intake stroke, 455–457
intake valve, 455, 457
intake valve cam mechanism, 457, 458, 467
integrations, 596, 597, 601
intermittent mechanisms, 123
internal gear, 518, 525, 533, 547
internal gear coupler, 518
interpolation, 141, 147, 496
inverted slider–crank linkage, 300, 378
iron, 2, 571
ironing board mechanism, 8, 10, 86, 132
irregularities, 31, 383, 570

J

jerk, 3, 42, 43, 162, 309, 382–384, 386–388, 397, 399, 423, 425, 427–432, 442, 446, 448, 453, 467, 468, 471, 472, 481, 482, 510
jerk analysis, 309
joints, 4, 7, 8, 41, 42, 45, 47, 48, 50, 51, 55–59, 61, 62, 71, 72, 75–79, 81–84, 97–99, 101–106, 112–114, 116, 119–121, 127, 130, 139, 159–162, 171, 173–175, 183–186, 189–192, 194, 196, 209, 211, 219, 237, 243, 258, 263, 267, 268, 275, 276, 279–281, 286, 290, 295, 329, 332, 383, 397, 400, 401, 442, 444, 445, 453, 501, 502, 509, 564–566, 568, 571, 572, 574, 575, 577, 579, 581, 583, 585, 586, 596, 615, 618, 620, 623, 625, 627, 628, 630
joint types, 281, 282

K

Kempe, Alfred B., 3
Kennedy circle, 281, 282
Kennedy's theorem, 255, 277, 280–282, 286, 290, 302
kinematic and dynamic analyses, 27
kinematic and dynamic model, 29, 39, 41
kinematic chain, 51–54, 60, 61
kinematic diagram, 41, 51, 97–99, 101–105, 111, 112, 114, 117, 119–123, 131, 169, 170, 172–174, 183, 184, 192, 194, 197–199, 204, 206, 209–211, 226–230, 232, 233, 235, 236, 240, 267–270, 284, 286, 287, 289, 292, 295, 296, 298–301, 305, 306, 315, 329, 332, 333, 337, 338, 347, 348, 352, 353, 355, 361, 362, 368, 369, 373–381, 458, 577, 578, 585, 614, 616, 621
kinematic diagrams of machines, 97, 131
kinematic energy, 410
kinematic motion, 62
kinematics, 1–4, 11, 18, 23, 31, 37, 41, 45, 62, 71, 139, 140, 160–162, 164, 191, 192, 215, 217, 219, 264, 273–275, 277, 289, 293, 294, 324, 326, 340, 341, 343, 345, 355, 356, 364, 365, 383, 386, 438, 440, 453, 501, 505, 515, 538, 543, 546, 547, 578, 619, 623
kinematics analysis, 62, 71, 221, 223, 277, 294, 364, 367, 420, 571
kinematics and dynamics, 2, 3
kinematics and kinetics, 3
kinematics fundamentals, 45
kinematics of machines, 3, 4, 139
knowledge, 196, 281, 346, 383, 399, 505

L

labels, 111, 112, 440
laminar, 408–411, 413
laminar flow, 408
landing gear mechanism, 90, 134, 199, 200, 592
latch, 27
layer, 189, 408, 444
leakage, 4045, 406, 417, 418
leaks, 405, 406
left-hand side, *see* LHS
length, 21, 77, 107, 111, 183, 185, 198, 199, 206, 221, 223, 225, 226, 228, 233, 235, 240, 247, 250, 252, 288, 295, 298, 300, 305, 306, 320, 328, 337, 346, 348, 349, 351, 353–355, 358, 362, 363, 367, 368, 373, 376, 377, 379, 383, 385, 393, 395, 408, 409, 412, 416, 417, 445, 450, 516, 521, 522, 527, 537, 539, 557, 558, 578, 586, 597, 598, 605
LHS (left-hand side), 175, 385, 529
lift mechanism, 28, 38, 81, 114, 120, 126, 442, 445, 525, 529, 534, 538
Lift Table Machine, 75, 264, 589, 590
lift table mechanism, 93, 121, 136, 264, 449, 589, 632
lift vehicle mechanism, 590, 591
limitations, 42, 201, 383, 397, 418, 423, 448
limits, 397, 405, 417, 420, 422, 423, 431–433, 436, 446, 448
linear acceleration, 310, 324
linear displacement, 208–212, 248, 256, 470
linear velocity, 256, 310
line
  pitch, 524, 532
  symmetrical, 601
  tangent, 527
linkage
  mechanical, 45, 51
  planar, 77
linkage FBD, 611
link degrees of freedom, *see* link DoF
link DoF (link degrees of freedom), 70, 76
links
  binary, 98
  flexible, 383

free, 71, 266
homogenous, 597, 600, 601
non-ground, 98, 174, 611
pivoted, 107
rigid, 46, 51, 54
load
applied, 383, 399
inertial, 588
Local frame, 174, 175, 182, 183, 270, 271, 370
locus, 483, 485
lookup table, 148
loop
feedback, 420
loop vector, 226, 228, 232, 239
losses
hose, 417, 420
hysteresis, 422
leakage, 417
major, 411
mechanical, 406
minor, 411–413
rotor resistance, 422
stator resistance, 422
thermal, 406
total, 405, 413
vibrational, 406
viscous flow, 411
volumetric, 406
lubrication, 571

M

machine actuators, 97
machine animation, 171
machine assembly, 164, 182
machine design, 1, 3, 23, 41, 247, 383, 455
machinery, 54
machines
closed chain, 106
mechanical, 78
quick return, 121
machines and mechanisms, 4, 97, 196
machines fundamentals, 97
machine user, 423
magnesium, 636
magnetic ball levitation, 49
magnitude, 154, 203, 278, 295, 309, 315, 346, 348, 361, 368, 383, 384, 386–388, 423, 582
maneuverability, 37
manufacture, 2, 11, 18, 27, 418, 492, 522, 523, 530, 631
mass, 42, 43, 49, 65–68, 84, 163, 164, 595–600, 602, 604–609, 611, 613, 614, 618, 623, 625, 628, 631, 632
mass flow rate, 129
mass moments of inertia, 605
mathematical, 1, 41, 42, 48, 63, 66, 77, 139, 140, 142, 153, 154, 158, 159, 161, 196, 201, 202, 229, 240, 241, 256, 310, 397, 411, 423, 426, 448, 453, 466, 467, 470, 473, 474, 480–482, 494, 549, 561, 564, 566, 577, 596, 618, 623
mathematical models, 42, 158, 196, 201, 240, 494
mathematical techniques, 42, 397, 448
mathematics, 43, 140, 494
MathWorks, 139, 140, 159, 164, 509, 550
MATLAB®, 1, 41, 43, 139–142, 159, 203, 218, 235, 253, 258, 260, 302, 307, 311, 314, 317, 318, 332, 383–385, 389–391, 395, 442, 453, 477, 478, 492, 493, 500, 511–513, 515, 563, 595, 618, 620, 624
matrices, 140
matrix, 140, 145, 164, 326, 497, 618, 620, 622, 624, 625
measurement, 41, 162, 224, 355, 364, 399, 596
mechanics, 1, 3, 4, 160, 193, 195, 232, 238, 239, 244, 386, 501, 506, 538, 543, 627
mechanism
crank, 41, 97, 201, 248, 255, 309, 377
crank-rocker, 246
double crank, 107
double rocker, 110
flapping wing, 509
foldable stair, 34
folding, 34
meshed gears, 515, 516, 523, 524, 530–533
potato chopper, 305
quick return, 41, 97, 121, 122
ratchet, 126
scissor, 34, 36
sew, 395
slide crank, 564
Sun-planets, 16
mechanism degrees of freedom, 76
mechanism of Geneva operated roller, 125
mechanisms and machines, 42
memory intelligence, 157
menu, 151, 157, 175, 433, 434, 497, 498, 502, 535, 584
mercury, 636
mesh, 42, 76, 401, 530, 531
meshed gears, 515, 516, 523, 524, 530–533, 535, 541, 554
metal, 13, 417
methods, 42, 176, 178, 201, 221, 244, 255, 309, 468, 494, 496, 515, 596, 628
minimum, limited, 423
mission, 425, 428, 431, 437, 439, 440
model
analytical, 229, 248, 253, 295, 307, 368, 377, 395
kinematic, 42, 509, 515, 611, 618, 623
model-based Simscape™ Multibody™, 230, 237, 242
modeling
dynamic, 41, 609
physical, 159
modeling technique, 620, 624
model validation, 540, 543
module, 522, 523, 555
molybdenum, 636
moments, 3, 4, 42, 420, 563, 566, 568, 569, 574–576, 588, 595, 605, 608–611, 625, 628, 631
moments of inertia, 608, 611
Moody chart, 411, 412

motion
actuated, 417
motion actuator, 129, 162, 215–218, 250, 274, 275, 289, 293, 306, 341–343, 355, 356, 364, 386, 395, 397, 400, 445, 501, 502, 505, 539, 545, 585, 587, 625, 628
motion limits, 397, 431, 448
motion segment
polynomial CEP, 472, 482
ramp CEP, 473, 482
motion trajectory, 118, 171, 397, 420, 595
motor, 13, 16, 17, 26–28, 33, 40, 72, 97, 106, 121, 122, 127, 128, 130, 131, 158, 208, 209, 216, 227, 242, 245, 248, 261, 262, 266, 285, 302, 303, 305, 306, 316, 317, 326, 330, 346, 349, 350, 352, 354, 361, 364, 389, 391, 392, 394, 395, 397, 398, 401, 405, 407, 420–423, 448–450, 495, 502, 506, 509, 525, 531, 534, 537, 542, 555, 567, 579, 580, 588, 591, 593, 613, 618, 622, 625
motor angular velocity, 261
motor efficiency, 422
motor shaft, 567, 579
motor torque, 216, 420, 567, 579, 580, 588, 591, 593, 613, 618, 622
move, 2, 8, 10, 15, 21–23, 28, 30, 46, 50, 51, 68, 72, 78, 79, 106, 113, 114, 121, 122, 127, 129, 161, 174, 213, 256, 289, 306, 399, 400, 408, 419, 451, 455, 459, 484, 508, 510, 574, 591
movement, 9, 10, 20, 21, 30, 42, 48, 51, 68, 72, 128, 201, 277, 321, 341, 458, 459, 483, 506, 570, 628
multibody, 1–13, 16–22, 24–29, 31–34, 36–40, 139, 160–162, 171, 175, 186, 188, 191, 221, 244, 273, 276, 277, 290, 294, 339, 340, 343, 345, 356, 357, 366, 397, 448
multibody assembly, 7, 8, 10–12, 17, 18, 20–22, 33, 34, 36–40
multibody elements, 6–12, 14, 16, 17, 19, 21–24, 26–30, 32–34, 36–40
Multibody for Machine Analysis, 19
multibody kinematics, 277
multibody model, 160, 171, 183, 276, 290, 294, 345, 357, 366
multibody toolbox, 160, 161, 175
multi-domain, 159
multiplication, 596, 604
mushroom follower, 465, 484–486

N

neural networks, 436
Newton, Issac, 43, 383, 563, 574, 588, 595, 609, 616, 621, 628, 631
Newton's first law, 563, 573, 574, 588
Newton's laws, 43
Newton's second law, 383, 595, 609, 616, 621, 628, 631
nickel, 636
niobium, 636
node
free, 386, 444, 564, 587
noise, 37, 383, 406, 409, 418, 420, 422, 423, 468, 471, 481, 495, 496, 509, 520
noise generation, 383
noise losses, 406
noise reduction, 520
nonlinearity, 436
normal acceleration component, 321, 324, 329, 347, 348, 353, 361, 362, 368
normal direction, 257, 324, 332, 568, 575
number, 47, 48, 62, 76–79, 81–83, 106, 141, 153, 168, 174, 184, 189, 209, 239, 252, 278, 280, 315, 364, 373, 377, 410–413, 420, 474, 478, 488, 522, 524, 527, 529, 533, 534, 541, 542, 547, 548, 550, 554–559, 561, 575, 579, 596, 600, 603, 611, 622

O

object, 49, 215, 256, 310, 599
observation, 31, 282, 287, 291
offset, 178, 190, 458, 459, 483, 485, 494, 495, 509
omega, 273, 364, 627, 628
open chain, 55, 61, 84, 206
opener mechanism, 103, 104, 197, 235, 237, 238, 269, 291, 294, 328, 336, 350, 364, 366, 576
operation, 120, 140, 142, 153, 154, 204, 205, 401, 520, 564, 566, 596, 604
orbit, 16, 49, 547
orientation, 76, 164, 168, 169, 174, 178, 519, 585, 625
origin, 62, 63, 161, 171–173, 175, 180, 181, 183, 184, 202–204, 206, 207, 221, 224, 256, 266, 270, 276, 315, 320, 321, 323, 332, 348, 349, 353, 362, 488, 490, 575, 577, 584
O-ring, 417
oscillation, 31, 483
oscillatory motion, 113, 114, 461
osmium, 636
output, 34, 42, 48, 106, 113, 116, 120, 122, 127, 140–142, 145, 148, 150, 152–156, 162, 164, 189, 209, 210, 212, 213, 215–217, 235, 237, 240–242, 244, 247, 273, 281, 284–287, 291, 294–296, 298, 305, 339, 340, 345, 346, 352, 356, 361, 367, 368, 373, 377, 386, 397, 398, 400, 405–407, 420, 422, 433, 434, 438, 440, 442, 444, 445, 448, 541, 542, 582, 583, 611, 628
output motion, 106, 302
output power, 285
output torque, 285, 286, 305
output velocity, 442

P

paddle, 16, 17
pair, 37, 46, 51, 52, 55, 56, 62, 70–72, 75, 84, 110, 178, 180, 277–282, 374, 379, 392, 394, 415, 486, 516, 519, 520, 531, 563, 568, 588, 615
palladium, 636

paper punch machine, 506, 507
parallelogram, 109, 118, 120, 296
parallelogram mechanism, 120
parameters, 6, 18, 27, 41, 42, 68, 70, 164, 178, 189, 190, 193, 195, 207, 213, 218, 219, 228, 235, 237, 252, 253, 274–276, 284, 287, 291, 295, 307, 343, 346, 352, 361, 367, 368, 376, 383–385, 395, 433, 436, 442, 444, 453, 478, 495, 500, 502, 506, 509, 515, 523, 530, 539, 542, 544, 554, 586, 596, 603, 611, 612, 618, 623, 626, 628
parasolid, 504
Parker group, 418, 419
particle, 408, 573
parts, 6, 13, 76, 103, 398, 406, 415–417, 433, 453, 496, 597–599, 601–604, 607, 608
pendulum mechanism, 613, 614
performance, 2, 11, 21, 22, 27, 42, 159, 404, 453, 496, 516
phosphorus, 636
pictures, 1, 43
PID (proportional–integral–differential) controller, 216
Pin-in-Slot Joint, 74
pinion, 29, 30, 130, 506, 516, 517, 525, 528, 529, 531, 533, 534, 537, 539, 552, 555, 556
pinion gear, 552
pinion hub, 539
pinion shaft, 537, 539, 552
pin joints, 113
Pin-Slotted Joint, 269, 337
pipe, 8, 130, 326, 404, 407–413, 417, 450
piston, 10, 11, 31–34, 36–39, 46, 47, 50–52, 59, 72, 73, 77, 85, 97, 104, 105, 113, 114, 120, 127, 129, 130, 196, 198, 199, 223–225, 229, 230, 232, 235, 237, 249, 250, 258, 264, 270, 288–291, 301, 304, 306, 311, 328, 329, 337–339, 355–359, 367, 376, 381, 393, 395, 402, 404, 415–419, 433, 435, 445, 451, 452, 455–457, 494, 568, 572, 576, 577, 586, 592, 633, 634
piston cylinder mechanism, 290, 357
piston engine, 32, 77, 104, 105, 494, 576
pitch circle, 521, 523, 527, 528, 530, 535, 555, 568
pitch curve, 483, 484, 509
pitch diameter, 522, 524–527, 534, 555, 556
pitcher water pump, 39, 85, 89, 258, 260, 261, 311, 314, 315
Pitman arm, 107, 206, 305, 306, 394, 579
pivots, 15, 23, 32
plane, 2, 34, 62, 66, 68, 175, 203, 315, 320, 324, 327, 454, 458, 459, 485, 601, 606
planetary gear, 16, 42, 515, 518, 547–550, 552, 561
planets, 16, 547, 550
plate, 13–15, 20, 243, 302, 390, 458, 462, 484, 506, 509
plate cam, 461, 484, 506
platinum, 636
plot, 492–494, 499, 511–513, 527, 528, 620, 624
point, 34, 41, 45, 47–50, 62–65, 68, 109, 121, 148, 168, 172, 182, 202, 203, 208, 221, 241, 256, 277–279, 281, 310, 315, 320, 321, 324, 326, 330, 341, 348, 349, 351, 353, 354, 357, 359, 362, 363, 367, 408, 409, 411, 420–422, 436, 439, 440, 465, 466, 469, 470, 480–485, 489–492, 496, 497, 499, 500, 505, 516, 523, 524, 527, 568, 596, 599, 600, 604
point mass degrees of freedom, 68
point masses, 41, 49, 161, 164, 208, 596, 599, 600, 604
polar coordinates, 168, 201, 255, 257, 266, 309, 320, 370
polynomial CEP motion, 472
polynomial SVAJ motion diagram, 432
position
  angular, 215, 262, 317
  desired, 468
  relative, 164, 204, 323
position trajectory, 65, 66
potassium, 636
potato chopper machine, 15, 87, 133, 302, 304, 390, 391
potato peeler machine, 12, 14
potential energy, 28, 32
potentiometer, 216
power, 2, 33, 46, 206, 285, 398–401, 404–407, 409, 415, 417, 420, 422, 423, 427, 431, 448, 449, 495, 509, 515, 516, 520, 523, 530, 533, 541, 552
power flow, 399, 552
power losses, 404–407, 409, 415, 417, 422, 448, 533
power source, 46, 397, 398, 403, 405, 420, 422, 427, 431, 495, 509, 515, 516, 552
power supply, 400
praseodymium, 636
precision, 458, 492
preloaded structure, 78, 79
pressure
  steam, 33, 120
pressure angle, 483, 485, 486, 488, 495, 496, 509, 524, 525, 527–531, 554–556, 568
pressure differences, 409
primary gear, 36, 37
prime circle, 483, 485
prime radius, 483
principle of "D'Almbert", 609
Prism, 602, 609, 614
prismatic active joints, 159
prismatic joints, 173, 400, 401
prismatic motion, 122
process, 3, 11–13, 18, 32, 39, 42, 47, 127, 129, 130, 157, 178, 220, 239, 294, 359, 368, 398, 400, 404, 405, 422, 425, 428, 431, 439, 448, 465, 466, 468, 482, 488, 494, 495, 527
product, 156, 158, 164, 326, 327, 458, 567, 575, 606, 610
profile, 9, 34, 311, 453, 457, 466, 481, 482, 484, 485, 488, 489, 491–498, 506, 509, 512, 513, 515, 520–523, 526–528, 554
profile shape, 528
programming, 140, 141
programming of MATLAB®, 140
properties, 151, 155, 156, 164, 274, 295, 498, 502, 504, 539, 544, 596, 625
propulsion, 25

protactinium, 636
prototyping, 3
pump, 39, 85, 106, 122, 128, 130, 258, 311, 314, 398, 400–402, 404–407, 417, 420, 449, 450
pump efficiency, 406
pumpjack mechanism, 91, 106, 107, 135, 137, 206, 303, 305, 306, 391, 579, 580, 631
pump losses, 406, 407, 417, 420
pump oil, 106
pump shaft, 405
push, 8, 26, 39, 40, 120, 256, 258, 433, 459, 564

Q

quadrants, 203

R

rack, 30, 130, 131, 506, 516, 517, 528
rack cutter, 528
rack gears, 517
rack–pinion, 516, 517
radial engine, 85, 94, 173, 174
radius, 198, 199, 221, 223, 224, 250, 264, 266, 269, 283, 288, 305, 306, 321, 324, 332, 337, 341, 413, 483, 485, 488, 494, 495, 509, 528, 531, 532, 534, 535, 539, 547, 554, 556, 558–560, 575, 598, 607, 608, 611
ramp, 2, 146, 154–156, 426–429, 432, 446, 451, 452, 468, 473, 475, 481, 482, 505, 506, 511, 514, 539, 545
ramp CEP motion, 473, 475, 482, 511, 514
ramp CEP motion design, 473
ramp CEP trajectory, 473, 474
ramp motion, 473, 481, 482
ramp SVAJ, 426–429, 432, 442, 446, 451, 452, 472
ramp trajectory, 468
ratchet water lift mechanism, 126
ratchet wheel, 125
rate, 42, 65, 256, 261, 310, 315, 383, 386, 404, 406, 410, 411, 418, 431, 433–437, 439, 440, 448, 449, 451, 453
rate limiter, 433–437, 439
reference frame, 164, 167–169, 175, 180, 182, 202, 497, 504
relative acceleration, 42, 309, 323–325, 327–329, 337, 339, 345–347, 349, 354, 359, 363, 370, 374
relative motion, 51, 175, 263, 264, 271, 322, 331, 488, 516, 549, 568
relative velocities, 42, 270
relief valve, 402, 404
reservoir, 401, 402
resistance, 6, 20, 409, 422, 552, 591, 593, 596, 604
return stroke, 122, 415, 416
Reuleaux, Franz, 3
revolute joints, 189–191, 194, 237, 243, 258, 267, 268, 383, 444, 579, 628
revolute motion, 75
Reynolds number, 410–412
Reynolds, Osborne, 410
right rule, 326
rigid bodies, 3, 6–12, 14–19, 21, 23, 24, 26–30, 32–34, 36–40, 48, 51, 163, 168, 263, 271, 278–280, 297, 386, 596, 611
rigid body mechanics, 3, 4
rigid mechanism, 48
rise trajectory, 423, 431
road
  curved, 324, 552
road bumps, 31
road irregularities, 383
Roberts, Samuel, 3
robots, 160
rocker, 41, 97, 107, 110, 305
rocker-crank, 97
rods, 8, 9, 30, 32
roll, 40, 485
roller, 124, 306, 393, 465, 483–485, 494, 496, 497, 521
roller coaster, 497
rolling motion, 255, 263–265, 302
rolling velocity, 264
room, 12, 202, 398
rotate, 6–8, 11, 16–18, 22, 28, 30, 34, 37–40, 46, 68, 76, 82, 106, 107, 112–115, 121, 122, 124, 125, 127, 129, 130, 168, 209, 245, 246, 248, 256, 258, 262, 264, 266, 285, 296, 302–306, 317, 326, 330, 337, 346, 355, 359, 376, 390–392, 394, 395, 401, 420, 423, 455, 457, 458, 466, 468, 471, 488, 489, 497, 499, 508–510, 513, 525, 532–535, 541, 547, 552, 555, 556, 558–560, 566, 567, 604
rotational, 6–12, 14–17, 19–24, 26–29, 32–34, 36, 38, 40, 47, 65, 68, 70, 71, 75, 106, 107, 113, 114, 116, 118, 120, 122, 124, 126–128, 130, 160–162, 169, 171, 175, 178–180, 183, 209, 211, 213, 244, 255, 256, 261, 265, 266, 271, 274, 275, 278, 285, 289, 292–294, 302, 329, 341, 343, 356, 364, 383, 386, 389, 397, 398, 400, 405, 420, 422, 423, 426, 454, 455, 457, 459, 461–463, 466, 485, 492, 494, 505, 515–517, 521, 537, 539, 545, 547, 550, 552, 557–561, 567, 568, 585, 604, 606, 611, 613, 625, 628
rotational angle, 70, 244, 261, 492
rotational angular velocity, 532, 533, 549, 554, 615, 618, 628
rotational cam mechanism, 454, 455
rotational camshafts, 457, 458
rotational motion, 6–12, 14–17, 19, 21–24, 26–29, 459, 461–463, 466, 494, 501, 502, 505, 515–517, 539, 545, 550, 567, 568, 585, 606, 613, 625, 628
rotational ratio, 457
rotational speed, 17, 278, 405, 457, 552
rotational transformation, 183
rotational velocity, 171, 255, 256, 271, 292, 294, 302, 537, 547, 550, 557–561
rotation, clockwise, 6, 18, 22, 23, 29, 38, 39, 63, 121, 125, 203, 212, 223, 233, 245, 246, 258, 286, 288, 300, 324, 326, 346–348, 353, 362, 379, 423, 455, 457, 466, 488, 492, 527, 566

rotation motion, 130, 215, 265, 483
rotors, 37
rule, 31, 64, 65, 103, 128, 168, 202, 205, 235, 266, 280, 319, 325, 326, 546, 550, 566, 584

S
Sandor, George N., 3
saturation, 433, 434, 436, 437, 439
scale, 41, 139, 224, 284, 285, 287, 288, 291, 292, 346, 350, 352, 357, 362, 365
SCARA robot arm, 116
scissor lift table mechanism, 93, 121, 136, 264, 589
Scotch yoke mechanism, 122, 123, 128, 130
script, 41, 139, 140, 192, 218, 220, 230, 232, 237, 238, 242, 244, 253, 258, 260, 274, 275, 289, 293, 302, 307, 311, 314, 317, 318, 334, 343, 345, 355, 356, 364, 384–386, 389–392, 395, 442, 477, 478, 492, 493, 500, 511–513, 618, 620, 624, 626
s-Diagram, 491
seconds, 66, 70, 196, 262, 267, 274, 285, 298, 303, 343, 371, 391, 425, 501, 538, 543, 627
sensitivity analysis, 3, 27, 42, 43, 45, 50, 51, 63, 113, 140, 150, 158, 196, 202, 222, 240, 241, 261, 277, 284, 287, 288, 291, 295, 316, 322, 327, 339, 345, 352, 362, 368, 371, 417, 442, 448, 515, 563, 564, 566, 584, 585, 600, 609, 619, 620, 623–625
sensors, 160, 162, 164, 213, 215, 220, 272, 275, 306, 340, 341, 395, 399, 400, 539, 542, 545, 583, 584
sequence, 32, 178, 204
servomotor, 420
set, 429, 432, 437, 440, 442, 458, 459, 469, 472, 481, 516, 517, 519, 524, 547, 567, 574, 579, 583, 585, 588, 596, 597, 601, 607, 611, 615, 620, 625, 631
sewing machine, 40, 82, 83, 85, 92, 248, 252, 302, 303, 305, 306, 389–392, 507, 588, 589
sewing mechanism, 40–42, 45–51, 55–57, 59, 61, 62, 71, 72, 75–78, 81–85, 90, 97, 98, 106, 107, 109–114, 116–132, 134, 171, 182, 183, 185, 186, 188, 192, 193, 197–199, 201, 208, 210–213, 216, 221–225, 228, 230, 232, 244, 245, 248–250, 252, 255, 262, 263, 268, 270, 274, 275, 280–282, 284–288, 290, 291, 294–296, 298, 299, 301–303, 305–307, 309, 310, 315, 318, 322, 326, 329, 330, 332, 343, 345, 346, 352, 357, 361, 365, 367, 368, 373–375, 377, 381, 383, 384, 386, 389, 391, 393, 395, 397–399, 417, 419, 422, 425, 428, 431, 442, 449, 451, 453, 455, 457–460, 465–468, 471, 477, 486–488, 494, 496–499, 503, 506, 507, 509, 512, 515, 534, 539, 542, 547, 552, 556, 577, 578, 581, 585, 590, 593, 611, 620, 628, 631
shafts, 16, 37, 158, 516, 519–521, 524, 525, 532, 535, 537, 542, 546, 547, 552, 557–559
shape, 18, 29, 32, 34, 38, 48, 72, 75, 98, 108, 116, 155, 156, 164, 183, 189, 196, 233–235, 237, 243, 245, 247, 383, 400, 408, 409, 413, 429, 432, 437, 445, 458, 461, 465, 466, 472, 485, 488, 494, 496, 504, 508, 520, 521, 523, 530, 535, 571, 595–598, 601, 602, 604, 605, 607, 631, 632
shear stress, 409
signals
desired, 398
required, 361
sinusoidal, 433
SimMechanics, 160
simple gear, 515, 541–543, 546, 547
simple gear train, 541
simple harmonic, 429–432, 446, 451, 452, 480, 481
Simscape™, 41–43, 139, 158–161, 164, 165, 168, 176, 180, 187, 193, 195, 196, 213, 215, 218, 220, 227, 231, 232, 239, 240, 242, 244, 245, 250, 253, 272–274, 293, 306, 339, 340, 343, 356, 365, 395, 397, 442, 444, 453, 497, 499, 501, 504, 509, 513, 515, 535–538, 542, 543, 550–552, 557–560, 563, 582–585, 588, 590–593, 595, 618, 620, 624–628, 631
Simscape™ Multibody™, 42, 43, 201, 213, 215, 218, 220, 221, 227, 230–232, 237, 239, 240, 242, 244, 245, 248, 250, 253, 255, 272–274, 276, 290, 293–295, 302, 306, 309, 339, 340, 343–345, 356, 357, 365, 366, 383–385, 389, 395, 397, 442, 444, 447, 501, 515, 535–538, 542, 543, 550–552, 557–560, 563, 581–585, 587, 588, 590–593
simulate, 41, 139, 140, 274, 295, 343, 368, 383, 394, 395, 442, 499, 509, 513, 537, 550, 552, 557–560, 596
simulation, 1, 22, 41, 43, 141, 142, 148–150, 152, 159–162, 168, 193, 195, 227, 245, 275, 288–291, 293, 294, 345, 350, 355, 364, 365, 376, 383, 384, 386, 397, 448, 453, 499–501, 513, 515, 535, 538, 540, 543, 587, 595, 625, 627, 628
Simulink®, 41, 66, 139, 141–146, 148, 149, 151, 152, 154–160, 196, 219, 340, 341, 383, 385, 389, 397, 432–434, 437–439, 446, 497, 498, 501, 505, 506, 538, 542, 584, 628
Simulink® libraries, 505
Simulink® model, 41, 139, 142, 145, 151, 154–157, 183, 192, 193, 196, 383, 497, 498, 501, 505, 506, 538, 542, 584
Simulink® toolbox, 432
sine waves, 145
Sinusoidal wave, 436
six-bar mechanisms, 41, 97, 117, 119
sketch, 41, 98, 111, 112, 131, 132, 183, 192, 194, 206, 221, 222, 315, 329, 358, 398, 399, 424–426, 428, 431, 449, 468, 472, 524, 563, 564, 566, 567, 571–573, 580, 588, 590–593, 611
slider, 8, 41, 46, 72, 97, 103, 112–116, 124, 130, 131, 192, 201, 210–212, 218–221, 228–230,

232–235, 248, 255, 277, 281, 286–288, 290, 291, 298–300, 302, 309, 326, 352, 355, 357, 361, 363, 365
slider–crank, 41, 46, 97, 112–116, 130, 131, 193, 201, 210–212, 219, 221, 228–230, 232, 234, 235, 248, 255, 277, 281, 286–288, 290, 291, 298–300, 302, 309, 326, 352, 355, 357, 361, 363, 365, 368, 373–375, 377–380, 389, 595, 611, 620, 621, 623, 628, 631
slider–crank linkage, 300, 378
slider–crank mechanism, 97, 113, 115, 131, 201, 255, 281, 290, 298, 302, 309, 352, 368, 373, 374, 377, 389, 624
sliding surfaces, 570, 571
sliding velocity, 269, 270, 288, 292, 304–306, 328, 338, 346, 392, 393, 496
slipping, 264, 265, 279–282, 330, 337, 359, 360, 516, 552
slope, 154–156, 428, 436, 438, 469–471
slope discontinuity, 436, 438, 469, 470
slots, 15, 123, 124
software, 1, 41, 43, 98, 222, 295, 346, 352, 362, 417, 423, 467, 496
solid materials, 401
SolidWorks, 98, 222, 496, 504, 505
solution
  analytic, 242
  graphical, 301, 352, 355, 364, 382
  numerical, 240, 277, 295
Solver, 183, 186, 240, 539, 544
speed
  average, 515
spherical cam, 464
spiral cam, 462
spline, 496–498, 502, 506
spring, 10, 28, 31, 114, 125, 402, 415, 459, 460, 582
spur, 515, 516, 518–523, 527, 530, 533, 535, 546, 554, 555
spur gear, 516–526, 533, 535, 546, 554, 555
square, 98, 154–156, 417, 582, 606
stability, 31, 37
statics, 3, 4, 564
statics and dynamics, 3, 564
statics of machines, 563
stationary bike, 22, 23, 85
stationary link, 571
steam, 2, 33, 34, 120, 127
steam engine straight-line mechanism, 34
steel, 531, 556, 571, 597, 598, 602, 631, 632
steering mechanism, 29, 30
steering system, 517
step, 3, 36, 46, 102, 103, 105, 112, 180, 183, 185, 206, 213, 216, 222, 224, 225, 229, 272, 295, 340, 348, 350, 353, 368, 423–426, 451, 452, 473, 481, 493, 495, 498, 527, 548, 549, 579, 615, 620
Stephenson's mechanism, 119
stepper motor, 128, 420
streamlines, 408
stresses, 496
stroke, 32, 122, 128, 130, 131, 415–417, 420, 425, 428, 431, 433, 439, 442, 446, 451, 452, 455–457, 465, 592
structure
  pre-loaded, 78, 79, 81–84
study, 3, 4, 42, 46, 48, 106, 266, 382, 384, 405, 442, 448, 496, 505, 566, 606
subroutines, 141, 143
substitute, 41, 259, 313, 368
subsystems, 38, 128, 189, 191, 215, 217, 273–275, 340, 341, 345, 356, 365, 383, 386, 398, 442, 444, 446, 501, 502, 538, 539, 544, 545, 627
subtract, 153, 564, 566
subtraction, 204, 241, 596, 604
sum, 153–155, 225, 406, 417, 429, 432, 472, 535, 597, 606
summation, 265, 320, 339, 353, 359, 362, 363, 437, 440, 468, 521, 574, 576, 578, 580, 596, 597, 601, 603, 608, 610
sun, 16, 49, 547–550, 552, 561
surface roughness, 411, 570
surfaces
  convex, 464
  curved, 521
  flat, 62, 68, 521
suspension, 31, 32, 114, 160, 199, 247, 383
suspension mechanism, 31, 114, 115
SVAJ (displacement, velocity, acceleration and jerk), 42, 162, 171, 397, 399, 423–426, 428, 430–432, 436–442, 446, 448, 451–453, 467, 468, 470–473, 475, 477, 482, 484, 485, 488, 489, 499, 501, 502, 509–513, 564, 567, 611
SVAJ diagram, 423, 424, 428, 429, 452, 453, 468, 470, 472, 473, 475, 477, 478, 480, 488, 489, 509–512
swing valve, 414
Sylvester, James J., 3
symbol, 41, 47, 72, 75, 76, 97, 164, 206, 224, 266, 270, 279, 326, 596, 606
symmetry, 7, 20–22, 527, 601
synchronization, 420
synchronize, 496, 550
system block diagram, 127, 398, 400, 448
system diagram, 128–132, 210, 212, 237, 242, 244, 295, 296, 298–300, 368, 369, 373, 374, 377, 378, 399
system response, 368
systems
  cam-follower, 496
  conveyor, 516
  dynamic, 160
  electrical, 158
  mathematical, 139
  mechanical, 160
  model multibody, 160

T

table, 5–8, 10–13, 16, 17, 19–22, 25–29, 31, 33, 34, 36–40, 71, 75, 93, 98, 107, 110, 120, 121, 136, 140, 153, 164, 175, 176, 178, 185, 186, 189, 191, 202, 213, 244, 253, 264, 272, 275, 339, 359, 367, 384, 400, 412, 413, 440, 442, 449, 450, 458, 459, 468, 473, 480–482, 501, 524, 530, 531, 535, 538, 542, 543, 547–549, 568, 583, 585, 589, 590, 598, 603, 608, 611, 627, 632
tablet, 68
tangent, 483–485, 523, 527, 568, 570
tangential, 42, 257, 264–266, 269, 275, 309, 321, 324, 329, 331, 332, 337–339, 341, 346, 348, 349, 353, 354, 361, 362, 394, 485, 532, 535, 568, 569, 575
tangential acceleration component, 324, 346, 353, 362, 393, 394
tank, 8, 9, 400, 402, 404
techniques
  analytical, 41, 488
  graphical, 41, 229, 281, 284, 488
  model-based, 43, 229
teeth
  spiral, 520
  straight, 520
teeth contact, 524, 568
teeth profile, 520, 523
teeth specification, 530
test cases, 433, 434
theory of machines and mechanisms, 4
thermal liquid, 160
Thermodynamics, 158, 411
time
  instantaneous, 65
  periodic, 468
time delay, 436
time rate, 310, 411
time span, 258, 302, 311, 389–392, 430
time step, 618
titanium, 637
toggle position, 201, 245, 248
toilet
  flushing, 4
tools, 1, 2, 98, 150, 433, 509
tooth, 520–522, 526–528, 530
tooth length, 521
tooth number, 527
tooth profile, 528, 554
tooth surfaces, 520
tooth width, 530
torque actuators, 584, 588
torque losses, 406
torque ratio, 284–286, 302, 305, 533, 534, 554, 556
torques, 3, 4, 161, 383, 399, 420, 568, 574, 581–585, 588, 595, 610, 611, 615, 620, 625, 628, 631
torque sensors, 162, 583, 584
torso, 19–22, 24
total acceleration, 310, 328, 332, 347–350, 353, 354, 359, 362, 393, 394
total velocity, 265, 266, 304, 305, 310, 367, 541
toy mechanism, 508, 509
tractor boom mechanism, 193, 194, 251
tractor loader mechanism, 195, 196
tractor machine, 169–171
tractors, 1
tractor's kinematic diagram, 170
train, 509, 518, 537, 541, 546, 547, 549, 550, 552, 557–561
trajectory, 66, 118, 171, 215, 247, 324, 332, 397, 420, 423, 425, 426, 429–431, 440, 448, 455, 468, 469, 471, 473–475, 477, 478, 480–482, 510–512, 595, 618, 622
transformation, 63, 161, 168, 203, 213, 285, 370, 376, 539, 542, 544
transformers, 161
transition, 410, 413, 436–439, 442, 446, 480, 481
transitional motion, 8, 30, 47, 341, 517, 585, 622, 625
translational accelerations, 374, 379, 618, 620, 624
translational cam mechanisms, 455
translational coordinates, 70
translational degree of freedom, 75, 162, 415
translational motion, 65, 67, 70, 113, 122, 124, 126, 130, 213, 216, 263, 266, 274, 289, 324, 355, 356, 376, 445, 457, 459, 485, 494, 587
translational position, 213, 219, 323
translational speed, 256
translational velocity, 255, 256, 261, 270, 273, 291, 295, 301, 302, 310, 323, 331, 368, 534, 535
transmission controller, 552
transmission industry, 550
transmissions, 42, 516, 518, 550, 554
triangular geometry, 235
trigonometry, 485, 487, 488
trigonometry shapes, 485
truss, 78
tubes, 416
tunes, 400
turbulent flow, 408
twisting action, 565, 566
types, 3, 4, 41–43, 45, 71, 76, 97, 98, 101, 108, 119, 128, 131, 142, 161, 162, 168, 208, 279–282, 402, 407, 415, 418, 420, 423, 436, 448, 453, 457, 458, 465, 468, 481, 494, 509, 515, 516, 518, 520, 521, 524, 550, 554, 568, 582

U

umbrella mechanism, 11, 12
undercut, 528, 529
undercutting, 495, 528, 554
unfold, 11, 36
unit conversion, 522
United States of America, 2
unit length, 522
unit of time, 310

unit system, 522
unit vector, 203, 257, 319, 320, 325–327, 564
uranium, 637
utilities, 6

V
valve datasheet, 418
valve losses, 417, 420
valves
  directional, 404
  flush, 4, 8, 9, 85
  pressure relief, 402, 404
  reducing, 404
  safety, 402
valve sizing, 418
valve sizing coefficient, 418
vanadium, 637
variables, 76, 139–141, 226, 229, 233, 237, 244, 274, 343, 347, 349, 350, 352, 353, 355, 362, 368, 373, 377, 475, 492, 493, 611
variation, 65, 66, 69, 141, 171, 173, 247, 306, 395, 398, 418
vector analysis, 241
vector arrow, 315
vector calculus, 241
vector diagram, 206, 208, 211, 309, 315, 319, 320, 327, 328, 330, 332, 335, 336, 348–351, 353, 355, 358, 359, 362–364, 366–368, 393
vector diagram of four-bar mechanism, 206
vector diagram of slider–crank mechanism, 211
vector form, 202, 261, 564, 566
vector loop, 225, 226, 229, 232–236, 239, 241, 296, 298, 301, 369, 373, 377, 381
vector loop method, 225
vectors, 201, 204, 206, 207, 210–212, 218, 225, 226, 228, 233, 240, 248, 250, 253, 257, 266, 267, 277, 278, 281, 287, 291, 295, 307, 309, 315, 320, 326, 332, 343, 345, 346, 350, 352, 357–360, 365, 368, 564, 566, 577
vehicle bucket mechanism, 586
vehicle frame, 37, 38, 577
vehicle suspensions, 160
vehicle transmissions, 554
velocities vectors, 277
velocity
  rotating, 269
  tangential, 266
  transitional, 352, 362
velocity components, relative sliding, 346
velocity directions, 295
velocity discontinuity, 436
velocity sensor, 273, 275, 289, 294, 306, 539, 545
velocity system diagram, 298, 299, 373, 377
velocity vectors, 278, 281, 287, 291, 295, 346, 352
vertical prismatic, 501, 502
vibration, 285, 386, 409, 418, 420, 422, 423, 425, 427, 431, 433, 446, 468, 471, 481, 495, 496, 509
vibration energy, 285
vibration rate, 386
viscosity, 407, 409–411, 450
visualizations, 625
voltage, 128, 216
volume, 3, 34, 38, 120, 127, 129, 398, 406, 596, 597, 601, 603, 604, 607
volume of fluid, 406

W
walking beam, 106, 206, 305, 306, 394, 579
wall frame, 6
water pump mechanism, 124
waterwheel, 2
Watt, James, 3
Watt's mechanism, 119
wheels
  steering, 29, 30
  vehicle, 552
window lift machine, 57, 58
window mechanism, 28, 29, 75, 90
window opener mechanism, 103, 104
window track, 29
window wiper, 132
wing corkscrew machine, 75
wings flap, 509
wiper mechanism, 26, 27, 92, 136, 262, 263, 296, 317, 318, 593
worm
  non-sliding, 17, 18
worm gears, 550
worm type, 520

Y
Young's modulus, 383–385

Z
zero angular velocity and torque, 421
zero center, 255, 281, 282
zero center of velocity, 255, 281, 282
zero degrees of freedom, 78
zero initial values, 480
zero offset value, 495

MIX
Papier aus verantwortungsvollen Quellen
Paper from responsible sources
FSC® C105338

If you have any concerns about our products,
you can contact us on
ProductSafety@springernature.com

In case Publisher is established outside the EU,
the EU authorized representative is:
Springer Nature Customer Service Center GmbH
Europaplatz 3, 69115 Heidelberg, Germany

Printed by Libri Plureos GmbH
in Hamburg, Germany